# Frontiers in
# Molecular Toxicology

# Frontiers in Molecular Toxicology

EDITED BY
## Lawrence J. Marnett
Vanderbilt University

American Chemical Society, Washington, DC 1992

**Library of Congress Cataloging-in-Publication Data**

Frontiers in molecular toxicology / Lawrence J. Marnett, editor

p.     cm.

Selection of reprints from the journal Chemical research in toxicology.

Includes bibliographical references and indexes.

ISBN 0–8412–2428–5

1. Molecular toxicology.

I. Marnett, Lawrence J.

RA1220.3.F76   1992
615.9—dc20                                                92–10062
                                                              CIP

The paper used in this publication meets the minimum requirements of American National Standard for Information Sciences—Permanence of Paper for Printed Library Materials, ANSI Z39.48–1984.  ∞

# Contents

## Enzymes of Activation, Inactivation, and Repair

## Physical Methods

## Macromolecular Modification

# Preface

THE APPLICATION OF CHEMICAL CONCEPTS AND TECHNOLOGIES to the field of toxicology is one of the great scientific opportunities of our time. Elucidation of the molecular basis of toxicity, mutagenicity, carcinogenicity, and so on, requires all of the techniques of modern chemistry, including structure elucidation, chemical analysis, mechanistic and theoretical chemistry, chemical synthesis, and physical characterization. The variety of structural classes of toxic agents and the range of biological effects they induce provide a limitless array of challenging and interesting problems.

The ACS journal *Chemical Research in Toxicology* attempts to highlight the frontiers of molecular toxicology by publishing one Invited Review or Perspective in each issue. The subject matter and quality of these articles have been widely praised in the scientific community. The Invited Reviews and Perspectives from the first four volumes of *Chemical Research in Toxicology* are reproduced here. Articles are grouped into four broad categories: Toxic Agents and Their Actions; Enzymes of Activation, Inactivation, and Repair; Physical Methods; and Macromolecular Modification. Together, they convey the opportunity and excitement that exist at the interface of chemistry and toxicology. I hope that readers will find them interesting and useful in their own research.

This publication is a collaborative effort of the Journals and Books Departments of the Publications Division of the American Chemical Society. I am grateful to Robert H. Marks and M. Joan Comstock for their support and assistance in assembling and publishing this volume.

LAWRENCE J. MARNETT
Vanderbilt University
Nashville, TN 37232

# Toxic Agents
## and Their Actions

# Chapter 1

# Metabolic Activation of Valproic Acid and Drug-Mediated Hepatotoxicity. Role of the Terminal Olefin, 2-*n*-Propyl-4-pentenoic Acid

Thomas A. Baillie

*Department of Medicinal Chemistry, School of Pharmacy, BG-20, University of Washington, Seattle, Washington 98195*

Reprinted from *Chemical Research in Toxicology*, Vol. 1, No. 4, July/August, 1988

## Introduction

Valproic acid (VPA[1], 1, Figure 1), an anticonvulsant agent introduced in France in 1967 for the therapy of epilepsy and approved by the U.S. Food and Drug Administration in 1978, has now become a primary antiepileptic drug worldwide (*1*). In recent years, VPA has been shown to be effective against a broad spectrum of seizure types, and it has found use both as sole medication and as a component of polytherapy. The drug is unique within its therapeutic class, in terms of both its chemical structure and its mechanism of action, and as a consequence, it has been the focus of much basic and applied research (*2*). However, despite its numerous attributes as an antiepileptic drug, initial enthusiasm for valproate has been tempered by reports of a rare, but serious, liver injury associated with VPA therapy (*3–5*). Although sometimes reversible, this drug-mediated hepatotoxicity has led to instances of irreversible liver failure (usually characterized by hepatic steatosis with or without necrosis), and at least 80 cases of fatal VPA hepatotoxicity had been reported to the manufacturer by the end of 1982 (*6*). Since the majority of these fatalities involved young children, the American Academy of Pediatrics issued, in the same year,

a recommendation that "...physicians adopt a conservative approach and utilize VPA only in carefully selected situations" (*7*).

The biochemical mechanisms that underlie VPA-mediated liver injury are not understood, although a number of hypotheses for the toxicity have been advanced (*8*). The finding (*9*) that the incidence of VPA-related fatalities is strikingly higher in patients receiving VPA in polytherapy, rather than as sole agent, is noteworthy in that it is consistent with the view that *metabolites* of VPA (produced in increased amounts by an induced liver) may play a key role in the hepatotoxic response to this drug. In fact, this toxic metabolite theory was suggested by Gerber et al. (*3*), in one of the first published reports on fatal VPA hepatotoxicity, who drew attention to the similarity in structure between VPA and two known hepatotoxins, 4-pentenoic acid (4, Figure 1) and methylenecyclopropylacetic acid. On the basis of the fact that the latter compounds (both terminal olefins) produce a Reye-like syndrome in animals and cause hepatic steatosis as the primary tissue lesion, these authors proposed that a metabolite of VPA may well be responsible for the mitochondrial damage (*10*), impairment of fatty acid $\beta$-oxidation activity (*11*), and lipid accumulation (*4*) characteristically observed in cases of VPA hepatotoxicity. This line of reasoning was developed further by Zimmerman and Ishak (*4*), who proposed that the terminal olefin metabolite of VPA, 2-*n*-propyl-4-pentenoic acid ($\Delta^4$-VPA, 2, Figure 1), might be the responsible hepatotoxin. Interestingly, this unsaturated derivative of VPA, which was first detected as a minor metabolite in the plasma of epileptic children receiving valproate (*12*),

---

[1] Abbreviations: VPA, 2-*n*-propylpentanoic acid or valproic acid; $\Delta^2(\Delta^3$ or $\Delta^4$)-VPA, 2-*n*-propyl-2(3 or 4)-pentenoic acid; $\Delta^{2,4}$-VPA, 2-*n*-propyl-2,4-pentadienoic acid; 4,5-dihydroxy-VPA-$\gamma$-lactone, 3-*n*-propyl-5-(hydroxymethyl)tetrahydro-2-furanone; 3(4 or 5)-hydroxy-VPA, 2-*n*-propyl-3(4 or 5)-hydroxypentanoic acid; 3-oxo-VPA, 2-*n*-propyl-3-oxopentanoic acid; 3-hydroxy(3-oxo)-$\Delta^4$-VPA, 2-*n*-propyl-3-hydroxy(3-oxo)-4-pentenoic acid; AIA, 2-isopropyl-4-pentenamide or allylisopropylacetamide; CoA, coenzyme A.

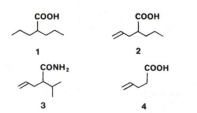

**Figure 1.** Structures of VPA (1), $\Delta^4$-VPA (2), AIA (3), and 4-pentenoic acid (4).

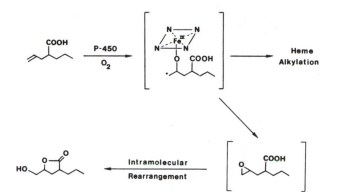

**Figure 2.** Proposed scheme for the cytochrome P-450 dependent olefin oxidation pathway of $\Delta^4$-VPA metabolism. Initial one-electron oxidation of the substrate yields a heme-bound free-radical intermediate, which partitions between heme alkylation (enzyme destruction) and epoxide formation. The unstable epoxide, in turn, undergoes intramolecular nucleophilic attack by the carboxylate group to give the final product, 4,5-dihydroxy-VPA-$\gamma$-lactone. This scheme, which is adapted from work by Ortiz de Montellano and Correia (23), is supported by the results of oxygen-18 labeling experiments (22).

appears to be present at higher levels in the serum of pediatric patients (the group most susceptible to VPA-induced liver damage) than in either youths or adults (13, 14). Moreover, when biochemical studies with $\Delta^4$-VPA were performed, it was shown that this compound was indeed cytotoxic, in a dose-dependent fashion, to rat hepatocytes in culture (15), that it inhibited $\beta$-oxidation of medium-chain fatty acids in rat liver homogenates (16), and that, following administration to rats, pronounced elevations in blood urea and SGOT levels were obtained (17). Subsequently, Kesterson, Granneman, and co-workers (18, 19) published the findings of a comprehensive study of the toxicity in rats of VPA and certain of its metabolites, in which $\Delta^4$-VPA and $\Delta^{2,4}$-VPA (the principal metabolite of $\Delta^4$-VPA) were found to be potent inducers of microvesicular steatosis and inhibitors of fatty acid $\beta$-oxidation. These authors noted that "...the liver of animals dosed with 4-en-VPA was more severely affected than that in any other group, including 4-pentenoic acid treated rats" (18). The possible clinical relevance of these findings is suggested by the paper of Kochen et al. (20), which reported grossly elevated levels of $\Delta^4$-VPA and other unsaturated metabolites of VPA in plasma and urine of a 7-year-old boy who died of VPA-mediated liver failure.

From the foregoing discussion, it is apparent that a considerable body of evidence has now accumulated which suggests that $\Delta^4$-VPA could play a key role in mediating VPA hepatotoxicity. In an attempt to examine the validity of this hypothesis, and in order to elucidate the biochemical mechanism(s) by which $\Delta^4$-VPA exerts its hepatotoxic effects, a number of studies have been carried out in our laboratories to define the metabolic origin of this compound and to elucidate its biological fate. The approach adopted in these investigations has been to examine in detail the "$\Delta^4$" pathway of VPA metabolism in vitro and in vivo in order to assess the potential for toxic metabolite generation and resulting cellular injury. The outcome of these studies, which have focused largely on the role of hepatic cytochrome P-450 and $\beta$-oxidation enzymes in the formation and metabolic activation of $\Delta^4$-VPA, may be summarized as follows.

## Metabolic Activation of $\Delta^4$-VPA

**Reactive Metabolites of $\Delta^4$-VPA Generated by Cytochrome P-450.** Since liver microsomal cytochrome P-450 enzymes are known to catalyze the oxidation of monosubstituted olefins to reactive free-radical and epoxide intermediates (21), it was of interest to examine the interaction of $\Delta^4$-VPA with this mixed-function oxidase enzyme system. Studies conducted in vitro with rat liver microsomal preparations demonstrated that both the free acid and ethyl ester forms of $\Delta^4$-VPA underwent cytochrome P-450 dependent metabolism at the olefin functionality and yielded a common $\gamma$-butyrolactone derivative, 4,5-dihydroxy-VPA-$\gamma$-lactone (Figure 2) (22). That this process involved the intermediacy of a chemically reactive species was suggested by the observation that substrate turnover was accompanied by loss of spectrophotomet-

rically detectable cytochrome P-450, in a fashion similar to that documented for a series of terminal olefin "suicide substrate" inhibitors of this enzyme, e.g., allylisopropylacetamide (AIA, 3, Figure 1) (23). Interestingly, oxygen-18 tracer experiments showed that three quite distinct ring-closure mechanisms were operative in the metabolism of $\Delta^4$-VPA, ethyl $\Delta^4$-VPA, and AIA to their respective $\gamma$-butyrolactone derivatives but indicated that autocatalytic destruction of cytochrome P-450 in each case depended solely on the efficiency of the initial olefin oxidation step (23, 24). This finding was consistent with earlier work by Oritiz de Montellano et al. (25), who showed that olefin-mediated enzyme destruction is caused by free-radical species that alkylate the prosthetic heme but that are formed *prior to* the generation of epoxide intermediates. Hence, the fate of the epoxide in such systems, whether it be hydrolysis to a vicinal diol (as with ethyl $\Delta^4$-VPA) or intramolecular capture by a proximate nucleophilic center (as with $\Delta^4$-VPA and AIA), is of little consequence with respect to the phenomenon of drug-mediated cytochrome P-450 inhibition. A scheme consistent with these observations, and accounting for the formation of enzyme-bound (heme-alkylated) and $\gamma$-lactone metabolites of $\Delta^4$-VPA, is depicted in Figure 2.

When compared to AIA, $\Delta^4$-VPA was found to be relatively weak in its ability to destroy microsomal cytochrome P-450 in vitro. However, the possibility that reactive metabolites of $\Delta^4$-VPA, generated by cytochrome P-450, might diffuse away from their site of formation in the endoplasmic reticulum and alkylate critical biomacromolecules to cause cellular damage remains to be explored.

**Reactive Metabolites of $\Delta^4$-VPA Generated by $\beta$-Oxidation.** Bioactivation of $\Delta^4$-VPA by the enzymes of fatty acid $\beta$-oxidation is also suggested on structural grounds, since 4-pentenoic acid (of which $\Delta^4$-VPA is the 2-propyl derivative) is considered to be a mechanism-based irreversible inhibitor of this enzyme complex (26). In the case of 4-pentenoic acid, $\beta$-oxidation is believed to lead to 3-oxo-4-pentenoyl-CoA, a reactive, electrophilic species that is proposed to alkylate (and thereby inactivate) 3-ketoacyl-CoA thiolase, the terminal enzyme of $\beta$-oxidation. Thus, if the structural analogy between 4-pentenoic acid and $\Delta^4$-VPA extends to their respective routes of metabolism, it would be predicted that $\Delta^4$-VPA should serve as an enzyme-activated inhibitor of fatty acid $\beta$-oxidation.

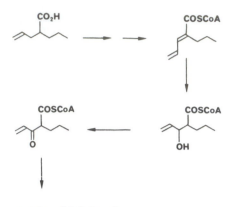

Alkylation of 3-Ketoacyl-
CoA Thiolase

**Figure 3.** Proposed metabolic activation pathway for $\Delta^4$-VPA by the fatty acid $\beta$-oxidation cycle. Following conversion of $\Delta^4$-VPA to its coenzyme A derivative, sequential steps of $\beta$-oxidation lead to $2(E)$-$\Delta^{2,4}$-VPA, 3-hydroxy-$\Delta^4$-VPA, and 3-oxo-$\Delta^4$-VPA. The latter species is believed to be a reactive, electrophilic metabolite that binds covalently to, and thereby inactivates irreversibly, 3-ketoacyl-CoA thiolase. The proposed diene and allylic alcohol intermediates have been identified as metabolites of $\Delta^4$-VPA in the isolated perfused rat liver (27) and in the rhesus monkey in vivo (29).

In order to address this intriguing mechanistic possibility, the metabolism of $\Delta^4$-VPA was examined in the isolated perfused rat liver, when products of $\beta$-oxidation were indeed observed. Not unexpectedly, metabolism occurred on both unsaturated and saturated side chains of $\Delta^4$-VPA to yield a variety of products, the most significant of which were the conjugated diene $2(E)$-$\Delta^{2,4}$-VPA and the allylic alcohol 3-hydroxy-$\Delta^4$-VPA. These metabolites correspond to the first and second intermediates, respectively, of the proposed bioactivation pathway leading to 3-oxo-$\Delta^4$-VPA, the putative toxic alkylating species (Figure 3). Although the latter compound itself was not found in samples of either perfusate or bile, the chemical properties of synthetic 3-oxo-$\Delta^4$-VPA indicated that this $\alpha\beta$-unsaturated ketone is indeed a highly reactive electrophile, which readily undergoes Michael-type addiion reactions through nucleophilic attack at the olefinic terminus (27).

In subsequent studies performed in rhesus monkeys (considered to be a good animal model for man with respect to VPA metabolism), the pharmacokinetics and metabolism of an intravenous dose of $\Delta^4$-VPA were compared to those of a similar dose of VPA (28, 29). Evidence was obtained that this nonhuman primate metabolizes both compounds via $\beta$-oxidation, although quantitative assessments of the various biotransformation products excreted into urine revealed interesting differences between the substrates. Thus, in the case of VPA, products of glucuronidation and cytochrome P-450 mediated reactions (both of which occur in the endoplasmic reticulum) accounted for ~65% of the dose that was recovered in urine over 24 h, whereas metabolites formed by mitochondrial $\beta$-oxidation were relatively minor and made up a further 6% (28). In contrast, when $\Delta^4$-VPA was given, some 34% of the dose was recovered in the form of products of microsomal enzyme activity, while a full 22% appeared to derive from mitochondrial metabolism (29). $\Delta^4$-VPA, therefore, seems to exhibit a marked preference for metabolism by $\beta$-oxidation as compared with the parent drug, a property that may contribute (through formation of 3-oxo-$\Delta^4$-VPA) to the high potency of this terminal olefin as an inhibitor of fatty acid $\beta$-oxidation in vitro (16) and

in vivo (18, 19). A scheme depicting the proposed bioactivation pathway for $\Delta^4$-VPA in mitochondria is illustrated in Figure 3.

While the above findings are strongly suggestive of a role for metabolic activation in the hepatotoxicity of $\Delta^4$-VPA, they are nevertheless indirect, and further studies are needed in order to both elucidate the mechanism(s) by which $\Delta^4$-VPA inhibits fatty acid metabolism and assess its contribution to VPA-mediated hepatic steatosis.

## Metabolic Origin of $\Delta^4$-VPA

Although the studies cited above provide considerable insight into the metabolic basis for the hepatotoxic properties of $\Delta^4$-VPA, details of the mechanism by which this olefin is generated during VPA biotransformation remained obscure until recently. Early experiments by Kochen and Scheffner (12) and by Granneman et al. (30) established that $\Delta^4$-VPA was not formed as an artifact (either in vitro or in vivo) by dehydration of 4- or 5-hydroxy-VPA, and more recent studies on the metabolism of specifically deuterium labeled analogues of VPA in the rat indicated that this terminal olefin had a biochemical origin quite different from that of its nonhepatotoxic positional isomers, $\Delta^2$- and $\Delta^3$-VPA (31).

Investigation of the products formed upon incubation of VPA with hepatic microsomes from phenobarbital-pretreated rats revealed that $\Delta^4$-VPA was generated by this in vitro system and that its formation was both oxygen- and NADPH-dependent (32). Moreover, since metyrapone, a relatively specific inhibitor of cytochrome P-450, blocked the desaturation reaction, a role for this hemoprotein in the metabolic process was indicated. Subsequent experiments with a purified and reconstituted form of cytochrome P-450 from rat liver (P-450 PB-4) served to verify the above conclusion, since VPA was metabolized by this enzyme to both 4- and 5-hydroxy-VPA and to $\Delta^4$-VPA. These results were of interest for two reasons: (i) They demonstrated that cytochrome P-450 enzymes can catalyze the oxidation of a nonactivated alkyl substituent to the corresponding olefin, a reaction not described previously. The underlying mechanism of this process is proposed to involve initial hydrogen atom abstraction by the enzyme to generate a transient free-radical intermediate, which partitions between recombination (alcohol formation) and elimination (olefin production) pathways. Recent deuterium isotope effect studies have lent support to this mechanistic interpretation and have indicated that the carbon-centered radical in question is located at the C-4 (and not at the alternative C-5) position (33). (ii) The finding that phenobarbital pretreatment of animals (32, 33) induces metabolism of VPA to $\Delta^4$-VPA clearly has important clinical implications, especially with regard to the use of VPA in polytherapy. Thus, desaturation of the parent drug to yield the hepatotoxic $\Delta^4$ metabolite, catalyzed by isoenzymes of cytochrome P-450 that have been induced specifically by coadministration of antiepileptic drugs such as phenobarbital (32, 33), phenytoin (33), or carbamazepine (33), may prove to be a key step in the sequence of events leading to VPA-mediated liver injury. Indeed, preliminary results from clinical studies designed to assess the influence of polytherapy on the "$\Delta^4$" pathway of VPA metabolism in epileptic patients have indicated that classical inducers of cytochrome P-450 (phenytoin, carbamazepine) stimulate this pathway of biotransformation, whereas an inhibitor of the enzyme (stiripentol) reduces the formation of $\Delta^4$-VPA (34, 35). In light of these observations, it appears that epileptic patients treated concomitantly with VPA and microsomal enzyme inducers may be at increased risk of liver injury and should be monitored carefully for early signs of hepatic dysfunction. Conversely, polytherapy with the newer generation of anticonvulsant drugs which act as inhibitors of cytochrome P-450 may, on theoretical grounds, prove to be a safer therapeutic strategy. Clearly, much more information is needed from controlled clinical studies before the validity of this prediction can be assessed accurately.

## Conclusions

On the basis of the findings reviewed above, it seems highly probable that the hepatotoxic effects of $\Delta^4$-VPA are a consequence of further biotransformation of this unsaturated VPA metabolite to chemically reactive intermediates that alkylate key cellular macromolecules. Enzymes of the fatty acid $\beta$-oxidation complex most likely play a pivotal role in the metabolic activation of $\Delta^4$-VPA (as they do with 4-pentenoic acid), although it is possible that products of cytochrome P-450 catalyzed olefin oxidation also contribute to the cytotoxic properties of this compound. Interestingly, both cytochrome P-450 and $\beta$-oxidation enzymes themselves appear to be targets for $\Delta^4$-VPA-mediated inhibition. However, in order to establish unequivocally that $\Delta^4$-VPA is a mechanism-based inhibitor of both cytochrome P-450 and 3-ketoacyl-CoA thiolase, it will be necessary, inter alia, to demonstrate that substrate turnover in each case is accompanied, on the one hand, by loss of catalytic activity and, on the other, by covalent modification of the enzyme through attachment of a $\Delta^4$-VPA residue at the active site. Radiolabeled $\Delta^4$-VPA has been synthesized recently for use in such work, and preliminary metabolic studies have shown that, following administration to rats by intraperitoneal injection, the compound does become covalently bound to proteins and that liver is the primary target organ for such protein alkylation (*36*). While the subcellular distribution of this binding and its role in the pathogenesis of $\Delta^4$-VPA-induced hepatotoxicity remain to be defined, the development of an in vitro model system, based on freshly isolated rat hepatocytes (*36*), should prove valuable in addressing such issues.

Finally, it should be stressed that the studies discussed in this Perspective have dealt with only one mechanism by which VPA may cause liver injury, i.e., via metabolism to, and subsequent activation of, the hepatotoxic olefin $\Delta^4$-VPA. Under normal conditions, the "$\Delta^4$" pathway represents a very minor route of VPA biotransformation, although it may assume quantitative significance in certain situations (*20*). However, other electrophilic metabolites of VPA are known to be formed in relatively large amounts, e.g., the acyl-linked glucuronide conjugate (*37*) and the coenzyme A derivative of VPA (*38, 39*), either of which might play a role in VPA-mediated liver injury. Several other viable mechanisms of VPA hepatotoxicity, which have been the subject of a recent authoritative review (*8*), include competitive inhibition of 3-ketoacyl-CoA thiolase by 3-oxo-VPA (*40*), carnitine deficiency induced by VPA administration (*41*), depletion of mitochondrial pools of free coenzyme A by VPA (*18, 38, 42*), and interference by VPA with processes of intermediacy metabolism in a liver whose function is already compromised by severe illness, inherited metabolic disorders, or exposure to multiple anticonvulsant drugs (*9*). Whatever the precise underlying factors, it is clear that a detailed knowledge of the metabolic fate of VPA, and of the properties of its hepatotoxic $\Delta^4$ metabolite, will add greatly to our understanding of VPA-mediated hepatotoxicity and may also afford information of a fundamental nature on the complex interplay between processes of foreign compound biotransformation and endogenous fatty acid metabolism (*43*).

**Acknowledgment.** I acknowledge with thanks the contributions of the following colleagues at the University of Washington who participated in the studies cited in this review: A. W. Rettenmeier, R. H. Levy, K. S. Prickett, W. P. Gordon, A. E. Rettie, D. J. Porubek, M. Boberg, S. M. Bjorge, H. Barnes, M. P. Grillo, and W. N. Howald. I also thank Ms. S. West for assistance with manuscript preparation. Financial support for my work on VPA metabolism has been provided by the National Institutes of Health (Research Grants GM 32165, NS 17111, and DK 30699), which is gratefully acknowledged.

## References

(1) Levy, R. H., Ed. (1984) "Valproate: modern perspectives". *Epilepsia (N.Y.)* **25** (*Suppl.* 1), S1–S77.

(2) Chapman, A., Keane, P. E., Meldrun, B. S., Simiand, J., and Vernieres, J. C. (1982) "Mechanism of anticonvulsant action of valproate". *Prog. Neurobiol. (Oxford)* **19**, 315–359.

(3) Gerber, N., Dickinson, R. G., Harland, R. C., Lynn, R. K., Houghton, D., Antonias, J. I., and Schimschock, J. C. (1979) "Reye-like syndrome associated with valproic acid therapy". *J. Pediatr. (St. Louis)* **95**, 142–144.

(4) Zimmerman, H. J., and Ishak, K. G. (1982) "Valproate-induced hepatic injury: analysis of 23 fatal cases". *Hepatology (Baltimore)* **2**, 591–597.

(5) Zafrani, E. S., and Berthelot, P. (1982) "Sodium valproate in the induction of unusual hepatotoxicity". *Hepatology*, **2**, 648–649.

(6) Nau, H., and Loscher, W. (1984) "Valproic acid and metabolites: pharmacological and toxicological studies". *Epilepsia (N.Y.)* **25** (**Suppl.** 1), S14–S22.

(7) Committee on Drugs (1982) "Valproic acid: benefits and risks". *Pediatrics* **70**, 316–319.

(8) Eadie, M. J., Hooper, W. D., and Dickinson, R. G. (1988) "Valproate-associated hepatotoxicity and its biochemical mechanisms". *Med. Toxicol.* **3**, 85–106.

(9) Dreifuss, F. E., Santilli, N., Langer, D. H., Sweeney, K. P., Moline, K. A., and Menander, K. B. (1987) "Valproic acid hepatic facilities: a retrospective review". *Neurology* **37**, 379–385.

(10) Jezequel, A. M., Bonazzi, P., Novelli, G., Venturini, C., and Orlandi, F. (1984) "Early structural and functional changes in liver of rats treated with a single dose of valproic acid". *Hepatology (Baltimore)* **4**, 1159–1166.

(11) Mortensen, P. B. (1980) "Inhibition of fatty acid oxidation by valproate". *Lancet*, 856–857.

(12) Kochen, W., and Scheffner, H. (1980) "On unsaturated metabolites of the valproic acid (VPA) in serum of epileptic children". In *Antiepileptic Therapy: Advances in Drug Monitoring* (Johannessen, S. I., Morselli, P. L., Pippenger, C. E., Richens, A., Schmidt, D., and Meinardi, H., Eds.) pp 111–120, Raven, New York.

(13) Abbott, F. S., Kassam, J., Acheampong, A., Ferguson, S., Panesar, S., Burton, R., Farrell, K., and Orr, J. (1986) "Capillary gas chromatography–mass spectrometry of valproic acid metabolites in serum and urine using *tert*-butyldimethylsilyl derivatives". *J. Chromatogr.* **375**, 285–298.

(14) Tatsuhara, T., Muro, H., Matsuda, Y., and Imai, Y. (1987) "Determination of valproic acid and its metabolites by gas chromatography–mass spectrometry with selected ion monitoring". *J. Chromatogr.* **399**, 183–195.

(15) Kingsley, E., Gray, P., Tolman, K. G., and Tweedale, R. (1983) "The toxicity of metabolites of sodium valproate in cultured hepatocytes". *J. Clin. Pharmacol.* **23**, 178–185.

(16) Bjorge, S. M., and Baillie, T. A. (1985) "Inhibition of medium-chain fatty acid $\beta$-oxidation in vitro by valproic acid and its unsaturated metabolite, 2-*n*-propyl-4-pentenoic acid". *Biochem. Biophys. Res. Commun.* **132**, 245–252.

(17) Schäfer, H., and Lührs, R. (1984) "Responsibility of the metabolite pattern for potential side effects in the rat being treated with valproic acid, 2-propylpenten-2-oic acid, and 2-propylpenten-4-oic acid". In *Metabolism of Antiepileptic Drugs* (Levy, R. H., Pitlick, W. H., Eichelbaum, M., and Meijer, J., Eds.) pp 73–83, Raven, New York.

(18) Kesterson, J. W., Granneman, G. R., and Machinist, J. M. (1984) "The hepatotoxicity of valproic acid and its metabolites in

rats. I. Toxicologic, biochemical and histopathologic studies". *Hepatology* **4**, 1143–1152.

(19) Granneman, G. R., Wang, S.-I., Kesterson, J. W., and Machinist, J. M. (1984) "Hepatotoxicity of VPA and its metabolites. II. Intermediary and valproic acid metabolism". *Hepatology* **4**, 1153–1158.

(20) Kochen, W., Schneider, A., and Ritz, A. (1983) "Abnormal metabolism of valproic acid in fatal hepatic failure". *Eur. J. Pediatr.* **141**, 30–35.

(21) Ortiz de Montellano, P. R. (1985) "Alkenes and Alkynes". In *Bioactivation of Foreign Compounds* (Anders, M. W., Ed.) pp 121–155, Academic, New York.

(22) Prickett, K. S., and Baillie, T. A. (1986) "Metabolism of unsaturated derivatives of valproic acid in rat liver microsomes and destruction of cytochrome P-450". *Drug. Metab. Dispos.* **14**, 221–229.

(23) Ortiz de Montellano, P. R., and Correia, M. A. (1983) "Suicidal destruction of cytochrome P-450 during oxidative drug metabolism". *Annu. Rev. Pharmacol. Toxicol.* **23**, 481–503.

(24) Prickett, K. S., and Baillie, T. A. (1984) "Evidence for the in vitro metabolism of allylisopropylacetamide to reactive intermediates. Mechanistic studies with oxygen-18". *Biomed. Mass Spectrom.* **11**, 320–331.

(25) Ortiz de Montellano, P. R., Yost, G. S., Mico, B. A., Dinizo, S. E., Correia, M. A., and Kambara, H. (1979) "Destruction of cytochrome P-450 by 2-isopropyl-4-pentenamide and methyl 2-isopropyl-4-pentenoate: mass spectrometric characterization of prosthetic heme adducts and nonparticipation of epoxide metabolites". *Arch. Biochem. Biophys.* **197**, 524–533.

(26) Schulz, H. (1983) "Metabolism of 4-pentenoic acid and inhibition of thiolase by metabolites of 4-pentenoic acid". *Biochemistry* **22**, 1827–1832.

(27) Rettenmeier, A. W., Prickett, K. S., Gordon, W. P., Bjorge, S. M., Chang, S.-L., Levy, R. H., and Baillie, T. A. (1985) "Studies on the biotransformation in the perfused rat liver of 2-*n*-propyl-4-pentenoic acid, a metabolite of the antiepileptic drug valproic acid. Evidence for the formation of chemically reactive intermediates". *Drug Metab. Dispos.* **13**, 81–96.

(28) Rettenmeier, A. W., Gordon, W. P., Prickett, K. S., Levy, R. H. Lockard, J. S., Thummel, K. E., and Baillie, T. A. (1986) "Metabolic fate of valproic acid in the rhesus monkey. Formation of a toxic metabolite, 2-*n*-propyl-4-pentenoic acid". *Drug Metab. Dispos.* **14**, 443–453.

(29) Rettenmeier, A. W., Gordon, W. P., Prickett, K. S., Levy, R. H., and Baillie, T. A. (1986) "Biotransformation and pharmacokinetics in the rhesus monkey of 2-*n*-propyl-4-pentenoic acid, a toxic metabolite of valproic acid". *Drug Metab. Dispos.* **14**, 454–464.

(30) Granneman, G. R., Wang, S. I., Machinist, J. M., and Kesterson, J. W. (1984) "Aspects of the metabolism of valproic acid". *Xenobiotica* **14**, 375–387.

(31) Rettenmeier, A. W., Gordon, W. P., Barnes, H., and Baillie, T. A. (1987) "Studies on the metabolic fate of valproic acid in the rat using stable isotope techniques". *Xenobiotica* **17**, 1147–1157.

(32) Rettie, A. E., Rettenmeier, A. W., Howald, W. N., and Baillie, T. A. (1987) "Cytochrome P-450-catalyzed formation of $\Delta^4$-VPA, a toxic metabolite of valproic acid". *Science (Washington, D.C.)* **235**, 890–893.

(33) Rettie, A. E., Boberg, M., Rettenmeier, A. W., and Baillie, T. A. (1988) "Cytochrome P-450 catalyzed desaturation of valproic acid in vitro. Species differences, induction effects and mechanistic studies". *J. Biol. Chem.* (in press).

(34) Levy, R. H., Loiseau, P. Guyot, M., Acheampong, A., Tor, J., and Rettenmeier, A. W. (1987) "Effects of stiripentol on valproate plasma level and metabolism". *Epilepsia (N.Y.)* **28**, 605.

(35) Levy, R. H., Rettenmeier, A. W., Baillie, T. A., Howald, W. N., Wilensky, A. J., Friel, P. N., and Anderson, G. (1987) "Formation of hepatotoxic metabolites of valproate in patients in carbamazepine or phenytoin". *Epilepsia (N.Y.)* **28**, 627.

(36) Porubek, D. J., Grillo, M. P., and Baillie, T. A. (1987) "Studies on the covalent binding of valproic acid (VPA) and its unsaturated metabolite, $\Delta^4$-VPA, to rat proteins". *Pharmacologist* **29**, 187.

(37) Dickinson, R. G., Hooper, W. D., and Eadie, M. J. (1984) "pH-Dependent rearrangement of the biosynthetic ester glucuronide of valproic acid to β-glucuronidase-resistant forms". *Drug Metab Dispos.* **12**, 247–252.

(38) Becker, C.-M., and Harris, R. A. (1983) "Influence of valproic acid on hepatic carbohydrate and lipid metabolism". *Arch. Biochem. Biophys.* **223**, 381–392.

(39) Brown, N. A., Farmer, P. B., and Coakley, M. (1985) "Valproic acid teratogenicity: demonstration that the biochemical mechanism differs from that of valproate hepatotoxicity". *Biochem. Soc. Trans.* **13**, 75–77.

(40) Dickinson, R. G., Bassett, M. L., Searle, J., Tyrer, J. H., and Eadie, M. J. (1985) "Valproate hepatotoxicity: a review and report of two instances in adults". *Clin. Exp. Neurol.* **21**, 79–91.

(41) Coulter, D. L. (1984) "Carnitine deficiency: a possible mechanism for valproate hepatotoxicity". *Lancet*, 689.

(42) Moore, K. H., Decker, B. P., and Schreefel, F. P. (1988) "Hepatic hydrolysis of octanoyl-CoA and valproyl-CoA in control and valproate-fed animals". *Int. J. Biochem.* **20**, 175–178.

(43) Baillie, T. A., and Rettenmeier, A. W. (1988) "Valproate: biotransformation". In *Antiepileptic Drugs* (Levy, R. H., Dreifuss, F. E., Mattson, R. H., Meldrum, B., and Penry, J. K., Eds.) 3rd ed., Raven, New York (in press).

# Chapter 2

# Photochemistry of the Psoralens

John E. Hearst

*College of Chemistry, University of California, Berkeley, California 94720*

Reprinted from *Chemical Research in Toxicology,* Vol. 2, No. 2, March/April, 1989

### Introduction

Psoralen is a three-ring heterocyclic compound with the structure

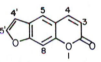

Psoralen-containing plants have been used since the beginnings of written history in the cures of various skin disorders. This is not terribly surprising, considering the large numbers of plants that contain psoralens. Identification of the active plants in these cures and recognition of the need for the use of sunlight actually occurred much later, perhaps between 300 B.C. and 600 A.D. (*1*). The physiological activity of this family of compounds was apparently described in old Arabic literature (1400 B.C.), where reference was made to the use of fruits of *Ammi majus* as a remedy for leukoderma (*2*). The purified compounds are used in the treatment of psoriasis and vitiligo (*2, 3*). When psoralens are administered, either topically to the skin or orally, they sensitize the skin to near-ultraviolet light, inducing tanning and sunburn. Such treatment results in remission of the symptoms of psoriasis.

The suggestion that psoralens photoreact with DNA and result in the cross-linkage of the two strands of the DNA helix first appeared in 1971 (*4, 5*). The synthetic organic chemistry resulting in modified psoralens has been extensive (*6*). The most commonly used psoralens are shown in Figure 1. The greater solubilities in water provided by the hydroxymethyl substitution in HMT and the aminomethyl substitution in AMT are major factors in the greater efficiency of the photochemistry using these compounds relative to the natural products. Isaacs et al. (*6*)

proved that RNA helices are also cross-linked by psoralen photochemistry.

The mechanism of the photoaddition of a psoralen to a nucleic acid helix involves several steps. The initial step is the intercalation of the psoralen into the nucleic acid helix in a dark reaction, reaction 1. While it is likely that this intercalation is somewhat sequence specific, this has not yet been demonstrated. When an intercalated psoralen absorbs a photon of wavelength between 300 and 400 nm, it is sensitized to react by cycloaddition at either the 3,4 double bond of the pyrone ring or the 4′,5′ double bond of the furan ring with the 5,6 double bond of an adjacent pyrimidine, reaction 2. The second cycloaddition, to the opposite nucleic strand from the position of the first adduct, can only form if two conditions are met. First, the monoadduct formed in reaction 2 must be a furan-side monoadduct (cycloaddition in reaction 2 must have occurred at the 4′,5′ double bond of the psoralen). Only in this case is the remaining adduct a coumarin derivative which can still absorb a photon of wavelength between 300 and 380 nm, making reaction 3 possible. Second, a pyrimidine has to be adjacent to the psoralen monoadduct on the opposite strand. Thus, for cross-link to form, the original intercalation had to occur in either a 5′ purine–pyrimidine 3′ site or in a 5′ pyrimidine–purine 3′ site in the helix. Johnson et al. (*7*) proved that a time delay of approximately 1 $\mu$s must occur between the absorption of the photon leading to the monoaddition of the psoralen to the first strand and the absorption of the second photon leading to efficient formation of cross-link to the opposite strand. This time delay has been associated with a major conformational change in the modified helix prior to cross-link formation. Figure 2 provides a schematic representation of the reaction mechanism which represents this conformational change as a kink in the helix.

2428–5/92/0007$06.00/0

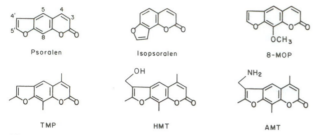

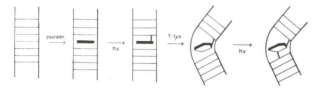

**Figure 1.** Various psoralens—psoralen, isopsoralen (angelicin), 8-methoxypsoralen (8-MOP), 4,5′,8-trimethylpsoralen (TMP) (trioxsalen), 4′-(hydroxymethyl)-4,5′,8-trimethylpsoralen (HMT), and 4′-(aminomethyl)-4,5′,8-trimethylpsoralen (AMT).

**Figure 2.** Mechanism of the photoreactions of psoralens with the DNA helix. The first step is a thermal reaction involving the intercalation of the psoralen in the helix. The second step is cyclobutane addition after the absorption of a photon. The third step is a major conformational change in the helix at the position of the psoralen monoadduct. The final step is the formation of the second cyclobutane ring upon absorption of a second photon, thus creating a covalent cross-link between the two strands of the DNA helix.

Several specificities intrinsic to the photochemistry of the psoralens with DNA and RNA have been demonstrated. The stereochemistry of the products is dictated by the DNA and RNA helices and will be discussed in detail in the next section. Piette et al. (8) have shown that in DNA 5′TpA3′ sites are far more photoreactive than the 5′ApT3′ site in model oligonucleotide DNA helices. Tessman et al. (9) have demonstrated a similar bias in reactivity in DNA by observing that the stereochemical product expected from the first of these sequences is 3–14 times more prevalent than the opposite product, depending on the detailed conditions used during the photochemical reaction and the extent of the reaction of the DNA with psoralen. These studies clearly prove that the most reactive sites in DNA are thymines which react with a psoralen on the 3′ side of the thymine. Nevertheless, if the photochemical reaction is pushed to near saturation of the DNA, addition densities higher than 1 psoralen adduct per 5 base pairs can be obtained (6). Such numbers suggest that most pyrimidines in nucleic acid helices are photoreactive, but the rates of reactivity are base and sequence dependent. It is clear that thymines and uracils are more reactive than cytosines. Another interesting specificity revealed by Kanne et al. (10) relates to the probability of furan-side monoaddition versus pyrone-side monoaddition to DNA. Kanne et al. (10) have shown that psoralen derivatives with a 4-methyl substitution show an overwhelming preference (98%) for addition to the furan double bond, while compounds with a hydrogen at the 4-carbon show nearly 20% monoaddition to the pyrone double bond.

### Structural Characterization of the Adducts

The photoadducts of HMT, TMP, and 8-MOP to DNA have been examined after isolation of the photoadducted mononucleosides by extensive enzymatic hydrolysis of photoreacted DNA. The various products were isolated on reverse-phase HPLC, a process that was greatly assisted by the availability of tritium-labeled psoralen derivatives

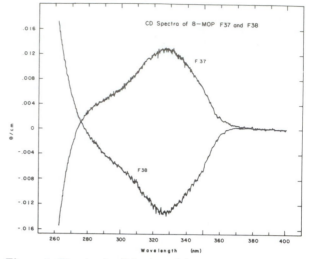

**Figure 3.** The circular dichroism of the two cys-syn furan-side monoadducts between 8-MOP and thymidine. These two hydrolysis products have opposite and equal circular dichroism spectra, as expected for a pair of enantiomers, since the deoxyribose moiety influences the overall molar ellipticity of each diastereomer to a very small degree. Which of the two possible cis-syn diastereomers forms is determined by whether the psoralen is on the 3′ or the 5′ side of the thymidine with which it reacts in the helix.

(11). In general, three nucleoside–psoralen monoaddition products have been isolated and characterized, corresponding to three deoxythymidine–psoralen adducts. In some cases two minor deoxyuridine–psoralen adducts derived from an initially formed deoxycytidine adduct by hydrolytic deamination were found. The major products are two diastereomeric thymidine adducts formed by cycloaddition between the 5,6 double bond of the thymine and the 4′,5′ (furan) double bond of the psoralen. The third thymidine adduct has been shown to be the cycloaddition product between the 5,6 double bond of the thymine and the 3,4 (pyrone) double bond of the psoralen. The pyrone adduct represented 20% of the covalently bound 8-MOP to DNA but less that 3% of the covalently bound TMP (12–14). The two prominent diastereomers have equal but opposite sign circular dichroism spectra (Figure 3), suggesting that they are identical in structure except that in one case cycloaddition has occurred on the 3′ side of the thymidine on the phosphodiester chain while in the other case cycloaddition has occurred on the 5′ side of the thymidine.

These three products are a small fraction of the cycloaddition products that are possible. Figure 4 shows some of these potential stereoisomers. The fact that only the cis-syn adducts are isolated from DNA indicates that the DNA helix itself has a structural role in determining the addition products formed. The unambiguous assignment of the configuration and stereochemistry of the addition products was first achieved by NMR. The assignment of all the protons in these nucleoside monoadducts is straightforward (12–16), establishing the chemistry as cycloaddition chemistry and positioning the reactive double bonds. The stereochemistry of the products has been established by NOE experiments. A similar analysis has been possible on the isolated thymidine–psoralen–thymidine cross-link. Figure 5 shows the NOE spectra on the TMP cross-link, establishing that both cyclobutane stereochemistries are cis-syn.

A racemic mixture of the furan-side monoadduct between 8-MOP and thymine has been isolated from DNA and the purified racemate crystallized for X-ray crystal-

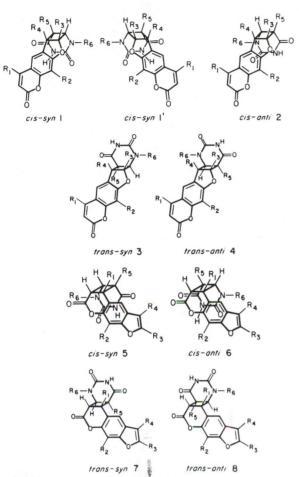

**Figure 4.** Various possible cycloaddition products between the psoralens and thymidine. 1 and 1′ are the near enantiomer pair discussed in Figure 3. Such a pair exists for each of the remaining adducts shown in this figure. 1–4 are furan-side monoadducts. 5–8 are pyrone-side monoadducts.

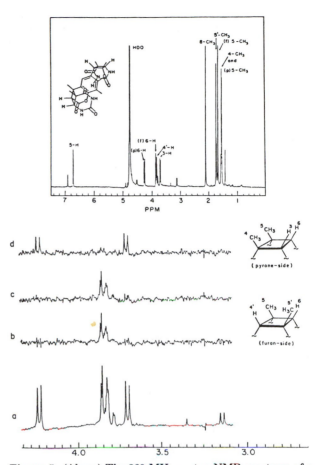

**Figure 5.** (Above) The 360-MHz proton NMR spectrum of a thymine–trioxsalen–thymine cross-link in $D_2O$. (Below) Nuclear Overhauser enhancements for the thymine–trioxsalen–thymine cross-link: (a) 3.5–4.5-ppm region of the proton NMR spectrum; (b) difference spectrum between sample irradiated in the 5′-$CH_3$ resonance (1.76 ppm) and (a), demonstrating proximity to the C6-H (3.87 ppm) and to the C4′-H (3.87 ppm) at the furan-side cyclobutane ring; (c) difference spectrum between sample irradiated in the 5-$CH_3$ (dT) resonance (1.66 ppm) and (a), demonstrating proximity to the C6-H (3.87 ppm) and to the C4′-H (3.83 ppm) at the furan-side cyclobutane ring; (d) difference spectrum between sample irradiated in the 4-$CH_3$ and 5-$CH_3$ (dT) (1.54 ppm) and (a), demonstrating proximities to C6-H (4.25 ppm) and to C3-H (3.87 ppm) at the pyrone-side cyclobutane ring.

lographic analysis. The unit cell contained six monoadducts, thymine–8-MOP, three of each enantiomer. This structure verified the assignment of the adduct as a cis-syn cyclobutane adduct. The crystal structure indicated that both the thymine ring and the psoralen ring remain planar in the photoadduct and that the three nonequivalent compounds in the unit cell showed angles of 44.1°, 50.6°, and 53.5° between the two planes, indicating considerable flexibility in the cyclobutane bridge (17, 18). On the assumption that the angles between the rings at the pyrone end of the psoralen in the DNA cross-link are similar to those determined in this crystal structure, the psoralen cross-link has been predicted to create a kink in the DNA backbone.

The two-dimensional NMR analysis has been completed on the AMT cross-link of the oligonucleotide d-GGGTACCC. Figure 6 shows the standard sequential connectivities in (A) for the unmodified oligonucleotide helix and in (B) for each of the independent four half-strands which are separated by the drug in the cross-linked oligonucleotide. By use of the methods of distance geometry, a 3D structure has been proposed and is shown as a stereopair in Figure 7. The structure includes a kink in the DNA helix backbone as well as clear asymmetry in the base-pairing stability in the region adjacent to the cross-link (19). This asymmetry is also clearly indicated by the temperature dependences of the imino resonances, which are shown in Figures 8 and 9. At the present time

it is uncertain if other structures might also be consistent with the 2D NMR data.

## Application to Structure Determination

As natural products, the psoralens have evolved with remarkable abilities to penetrate both cells and virus particles. This fact was first demonstrated with the photoreaction of trioxsalen with isolated *Drosophila melanogaster* and mouse liver interphase nuclei (20, 21) and with mouse L cells in vivo (22). The DNA isolated from these reactions was spread under denaturing conditions for electron microscopic examination. The cross-links were observed to occur in a pattern with a regular repeat of approximately 200 base pairs. This pattern was proven to be the result of a high reactivity of the chromatin DNA in the interbead or linker regions of chromatin by correlating the positions of the cross-links with the regions of micrococcal nuclease sensitivity (23).

There is a large literature reporting the photoreaction of psoralens with viruses. Much of this activity is directed at virus inactivation and vaccine production. Examples include vesicular stomatitis virus, SV-40 virus, polio 1 and

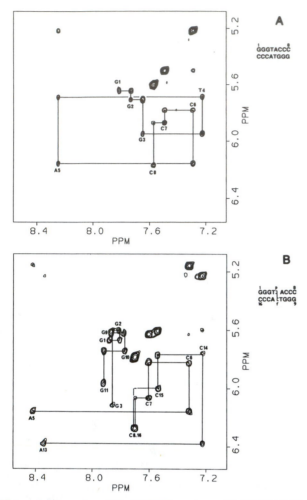

**Figure 6.** Expansions of the 2D NOE spectrum of d-GGGTACCC showing the cross-peaks of the aromatic proton to H1′ of the same nucleotide (numbered peaks) and to the H1′ of the sugar of the base on the 5′ side (connected with solid lines). (A) Unmodified octamer possessing a center of symmetry with all the protons sequentially assigned. (B) Cross-linked oligomer where the residues on either side of the drug are no longer equivalent. T4 is linked to the pyrone side ("p") and T12 is linked to the furan side ("f") of the 4′-(aminomethyl)-4,5′,8-trimethylpsoralen.

2, herpes simplex 1 and 2, influenza, vaccinia, blue tongue virus, murine sarcoma, canine hepatitis, western equine encephalitis, and the AIDS virus HIV (24). The photo-cross-linking reaction has been applied to the investigation of DNA structure in virus particles. The best example relates to the fd phage, which contains a circular single-stranded DNA genome of 6408 bases. The origin of DNA replication is known from sequencing studies to contain a complex of four hairpins which when cross-linked is readily visualized by electron microscopy. The phage particle is a long cylindrical filament with the DNA at its center. The object of the cross-linking experiment is to establish if the circular DNA is oriented in a specific way in the phage particle or if it is randomly permuted. Cross-linkage established that the origin complex is located at the end of the filamentous phage.

16S rRNA has been extensively studied by psoralen photo-cross-linkage. A cross-linkage map was first generated by using electron microscopy. These maps were generated both for the isolated 16S rRNA in vitro and for the same rRNA in the 30S subunit of the ribosome. It was concluded, to the sensitivity of this assay, that RNA has equivalent regions of secondary structure in both molecules

(25, 26). These mapping procedures are now performed for sequence resolution. These studies have provided tentative evidence for conformational switches in the rRNA cycle. Figure 10 presents a model showing the positions of the observed psoralen cross-links superimposed on the phylogenetic secondary structure map of Noller and Woese (27). In addition, sites hypothesized to be associated with the initiation of translation or mRNA binding, and with tRNA binding, are shown (28).

## Hybridization Dynamics

Site-specific placement of a furan-side psoralen mono-adduct in synthetic oligonucleotides is possible because of the greater photoreactivity of 5′TpA3′ sites than any other potential photoaddition sites. Synthetic oligonucleotides containing only one such site are added to a complementary strand of different length (to facilitate electrophoretic gel separation), and the mixture plus a psoralen is irradiated at 390 nm. The monoadducted oligonucleotide is then isolated and purified by gel electrophoresis or high-performance liquid chromatography (8, 29). These monoadducted oligonucleotides may now be used as hybridization probes which can be irreversibly bonded to their target sequence by near-ultraviolet light. Hearst (30) has presented a theory which demonstrates that the optimal condition for hybridization are at the melting temperature of the oligonucleotide from its target or at the point where half of the target is covered and half is free. This point provides maximum discrimination with respect to the partial homologies that may occur between probe and target. In addition, high probe concentration favors rapid attainment of equilibrium but also reduces discrimination. In general, it is concluded that optimal hybridization conditions occur near a probe concentration of $10^{-8}$ M, where the half-life for the hybridization reaction is 43 s. While these conditions are readily achievable, in order to separate unreacted probe from target in a hybridization reaction, a wash step is essential. For conditions favoring such rapid reaction, a rapid fixation step is essential, and it is in this context that the photo-cross-linkable probe proves useful. The hybridization of an oligonucleotide probe to a high molecular weight single-stranded target is complex for two additional reasons. First, the single-stranded target invariably contains considerable secondary structure which competes for the target sequence, and second, the complementary long strand can hybridize to the target, displacing the probe. These variables have been experimentally examined in detail (31, 32), and the phenomenon of photochemical "pumping" has been described. Figure 11 provides evidence for the effects discussed above. For the reaction of small probes with large targets, the photo-cross-linkable probes provide a unique tool for the study of the kinetics and equilibria of importance to the process. For such a study to be meaningful, the monoadduct should modify the interaction of probe with target as little as possible. Shi and Hearst (33) have demonstrated this to be the case for monoadducts of HMT in 5′TpA3′ sites. In addition, the use of tetraalkylammonium salts in the hybridization solution can be used to minimize the base compositional dependence of the thermodynamic interactions, making the hybridization dependent only on the length of the base-pairing interaction (34, 35).

## DNA–Protein Interactions

Piette and Hearst (36) have used a double-stranded circular DNA which was photoreacted with HMT as a substrate for nick-translation with *Escherichia coli* DNA polymerase I holoenzyme. The template DNA had a sin-

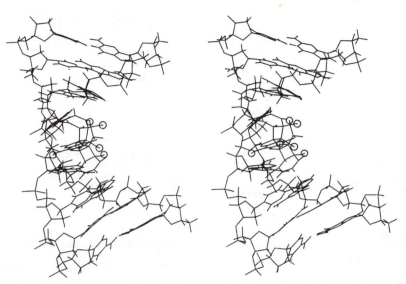

**Figure 7.** Stereo picture of the NMR-derived model for DNA octamer duplex cross-linked with (aminomethyl)trioxsalen. All of the methyl groups and the amino group of AMT are indicated by circles.

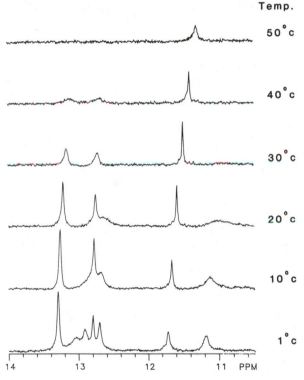

**Figure 8.** Imino proton spectra of cross-linked DNA as a function of temperature.

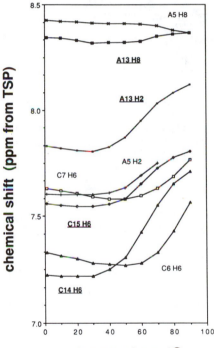

**Figure 9.** Plots of aromatic base proton chemical shifts of cross-linked oligomer in $D_2O$ as a function of temperature.

gle-stranded nick at a specific site so that the positions of pauses and stops in the nick-translation process could be mapped to sequence resolution. They showed that psoralen monoadducts on the template strand resulted in kinetic pauses in DNA nick translation but that incorporation of adenine opposite the monoadducted thymine still occurred. This suggests that a monoadducted base may still base pair to its complement strand and that such a monoadducted thymine can code for its complementary base during DNA replication with high fidelity. These data indicate that it is the psoralen cross-link and not the monoadduct that is responsible for the lethal effects of psoralen photochemistry in *E. coli*. In fact, Chanet et al. (*37*) have shown that, in yeast, psoralen monoadducts are

retained through several cell divisions with no evidence of mutagenesis, suggesting that such monoadducted DNA may properly code for its complementary strand in vivo. Physical evidence for the base-pairing capacity of the thymine monoadduct in the DNA helix has been obtained by the two-dimensional NMR study of Tomic et al. (*19*).

By site-specific placement of psoralens into oligonucleotides, it has been possible to generate specific substrates for a number of interesting enzymic processes. Included in this list have been *E. coli* RNA polymerase (*38*), T7 RNA polymerase (*39*), uvrABC excinuclease (*40*), and recA (*41*). An in vitro model for DNA cross-link repair in *E. coli* has been developed by using the last two of these examples. Figure 12 demonstrates the sequential steps in this modeled repair process (*40*). Cheng et al. (*41*) have studied the hybridization between a single-stranded mo-

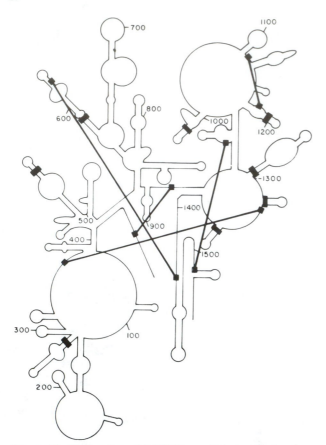

**Figure 10.** The locations of 13 HMT cross-links in 16S ribosomal RNA superimposed over the skeleton of the secondary structure model presented by Noller and Woese (27).

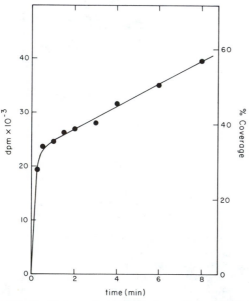

**Figure 11.** Continuous photochemical pumping of the hybridization equilibrium between a HMT-monoadducted 25-mer oligonucleotide and single-stranded M-13 mp19 target DNA. Annealing and photofixation were carried out at 45 °C in 35% formamide. Notice that 30% of the target reacts in less than a minute, but reaction of the remaining target evidently requires a slow conformational change which makes the target sequence available to the probe.

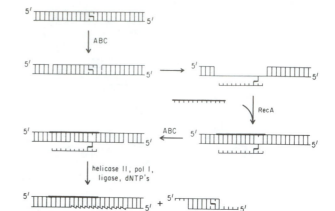

**Figure 12.** The repair pathway for DNA–psoralen cross-links suggested by in vitro experiments with site-specifically modified oligonucleotide substrates, the *E. coli* excinuclease complex uvrABC, recA, DNA polymerase, and ligase.

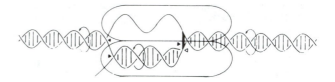

**Figure 13.** A model for the *E. coli* RNA polymerase transcription elongation complex. The horizontal arrow indicates the direction of RNA synthesis. The curved arrows indicate the unwinding and rewinding of DNA of DNA–RNA helix by the polymerase. The filled triangles denote the hypothetical unwindase and rewindase centers of the enzyme. The open triangle denotes the catalytic site of the polymerase, which is very close to the leading unwindase site. The vertical bar indicates the site of the psoralen cross-link which was used to arrest RNA polymerase elongation. The footprint of the polymerase on the DNA is indicated by the length of the two lobes of the enzyme relative to the turns of the DNA helix.

three-strand complex that occurs in this system comprised of the oligonucleotide covered with recA protein and the DNA target duplex can be photochemically fixed has been interpreted as a clear indication that base pairing is the structural interaction between the oligonucleotide and its complementary target.

A model for the ternary elongation complex between *E. coli* RNA polymerase, the DNA template, and the newly synthesized RNA molecule has been formulated from footprinting studies on such a complex arrested at a psoralen cross-link. The results of this study are summarized in Figure 13 (*38*).

### References

(1) Scheel, L. D. (1967) In *Biochemistry of Some Foodborne Microbial Toxins* (Mateles, R. I., and Wogan, G. N., Eds.) p 109, MIT Press, Cambridge, MA.

(2) Pathak, M. A., and Fitzpatrick, T. B. (1959) Relationship of Molecular Configuration to the Activity of Furocoumarins which Increase the Cutaneous Responses Following Long Wave Ultraviolet Radiation. *J. Invest. Dermatol.* **32**, 255–262.

(3) Scott, D. R., Pathak, M. A., and Moh, G. R. (1976) Molecular and Genetic Basis of Furocoumarin Reactions. *Mutat. Res.* **39**, 29–74.

(4) Cole, R. S. (1971) Psoralen Monoadducts and Interstrand Cross-links in DNA. *Biochim. Biophys. Acta* **254**, 30–39.

(5) Dall'acqua, F., Marciani, S., Ciavatta, L., and Rodighiero, G. (1971) Formation of Inter-Strand Cross-Linkings in the Photoreactions between Furocoumarins and DNA. *Z. Naturforsch.* **B26**, 561–569.

(6) Isaacs, S. T., Shen, C.-K. J., Hearst, J. E., and Rapoport, H. (1977) Synthesis and Characterization of New Psoralen Deriva-

noadducted oligonucleotide and its complementary sequence within a duplex DNA molecule, a reaction catalyzed by the recA protein of *E. coli*. The fact that the

tives with Superior Photoreactivity with DNA and RNA. *Biochemistry* **16**, 1058–1064.

(7) Johnston, B. H., Kung, A. H., Moore, C. B., and Hearst, J. E. (1981) Kinetics of Formation of Deoxyribonucleic Acid Cross-Links by 4′-(Aminomethyl)-4,5′,8-trimethylpsoralen. *J. Am. Chem. Soc.* **20**, 735–738.

(8) Gamper, H., Piette, J., and Hearst, J. E. (1984) Efficient Formation of a Crosslinkable HMT Monoadduct at the KpnI Recognition Site. *Photochem. Photobiol.* **40**, 29–34.

(9) Tessman, J. W., Isaacs, S. T., and Hearst, J. E. (1985) Photochemistry of the Furan-Side 8-Methoxypsoralen–Thymidine Monoadduct Inside the DNA Helix. Conversion to Diadduct and to Pyrone-Side Monoadduct. *Biochemistry* **24**, 1669–1676.

(10) Kanne, D., Rapoport, H., and Hearst, J. E. (1984) 8-Methoxypsoralen–Nucleic Acid Photoreaction. Effect of Methyl Substitution on Pyrone vs. Furan Photoaddition. *J. Med. Chem.* **27**, 531–534.

(11) Isaacs, S. T., Rapoport, H., and Hearst, J. E. (1982) Synthesis of Deuterium and Tritium Labeled Psoralens. *J. Labelled Compd. Radiopharm.* **19**, 345–356.

(12) Straub, K., Kanne, D., Hearst, J. E., and Rapoport, H. (1981) Isolation and Characterization of Pyrimidine–Psoralen Photoadducts from DNA. *J. Am. Chem. Soc.* **103**, 2347–2355.

(13) Kanne, D., Straub, K., Rapoport, H., and Hearst, J. E. (1982) Psoralen–Deoxyribonucleic Acid Photoreaction. Characterization of the Monoaddition Products from 8-Methoxypsoralen and 4,5′,8-Trimethylpsoralen. *Biochemistry* **21**, 861–871.

(14) Kanne, D., Straub, K., Hearst, J. E., and Rapoport, H. (1982) Isolation and Characterization of Pyrimidine–Psoralen–Pyrimidine Photoadducts from DNA. *J. Am. Chem. Soc.* **104**, 6754–6764.

(15) Cimino, G. D., Gamper, H. B., Isaacs, S. T., Hearst, J. E. (1985) Psoralens as Photoactive Probes of Nucleic Acid Structure and Function: Organic Chemistry, Photochemistry, and Biochemistry. *Annu. Rev. Biochem.* **54**, 1151–1193.

(16) Hearst, J. E., Isaacs, S. T., Kanne, D., Rapoport, H., and Straub, K. (1984) The Reaction of the Psoralens with Deoxyribonucleic Acid. *Q. Rev. Biophys.* **17**, 1–44.

(17) Peckler, S., Graves, B., Kanne, D., Rapoport, H., Hearst J. E., and Kim, S.-H. (1982) Structure of a Psoralen–Thymine Monoadduct Formed in Photoreaction with DNA. *J. Mol. Biol.* **162**, 157–172.

(18) Kim, S.-H., Peckler, S., Graves, B., Kanne, D., Rapoport, H., and Hearst, J. E. (1983) Sharp Kink of DNA at Psoralen-cross-link Site Deduced from Crystal Structure of Psoralen–Thymine Monoadduct. *Cold Spring Harbor Symp. Quant. Biol.* **47**, 361–365.

(19) Tomic, M. T., Wemmer, D. E., and Kim, S.-H., (1988) Structure of a Psoralen Cross-Linked DNA in Solution Determined by Two-Dimensional NMR. *Science* **238**, 1722–1725.

(20) Hanson, C. V., Shen, C.-K. J., and Hearst, J. E. (1976) Cross-Linking of DNA in situ as a Probe for Chromatin Structure. *Science* **193**, 62–64.

(21) Cech, T., and Pardue, M. L. (1977) *Crosslinking* of DNA with Trimethylpsoralen is a Probe for Chromatin Structure. *Cell* **11**, 631–640.

(22) Cech, T., Potter, D., and Pardue, M. L. (1978) Chromatin Structure in Living Cells. *Cold Spring Harbor Symp. Quant. Biol.* **42**, 191–198.

(23) Weisehahn, G. P., Hyde, J. E., and Hearst, J. E. (1977) The Photoaddition of Trimethylpsoralen to Drosophila melanogaster Nuclei: A Probe for Chromatin Substructre. *Biochemistry* **16**, 925–932.

(24) Hanson, C. V. (1983) Inactivation of Viruses for Use as Vaccines and Immunodiagnostic Reagents, Medical Virology II (de la Maza, L. M., and Peterson, E. M., Eds.) pp 45–79, Elsevier, New York.

(25) Wollenzien, P., Hearst, J. E., Thammana, P., and Cantor, C. R. (1979) Base-pairing between Distant Regions of the Escherichia coli 16S Ribosomal RNA in Solution. *J. Mol. Biol.* **135**, 255–269.

(26) Thammana, P., Cantor, C. R., Wollenzien, P. L., and Hearst, J. E. (1979) Crosslinking Studies on the Organization of the 16S Ribosomal RNA within the 30 S Escherichia coli Ribosomal Subunit. *J. Mol. Biol.* **135**, 271–283.

(27) Noller, H. F., and Woese, C. R. (1981) Secondary Structure of 16S Ribosomal RNA. *Science* **212**, 403–411.

(28) Thompson, J. F., and Hearst, J. E. (1983) Structure of E. Coli 16S RNA Elucidated by Psoralen Crosslinking. *Cell* **32**, 1355–1365.

(29) Shi, Y., and Hearst, J. E. (1987) Wavelength Dependence for the Photoreactions of DNA–Psoralen Monoadducts. 1. Photoreversal of Monoadducts. 2. Photo-Cross-Linking of Monoadducts. *Biochemistry* **26**, 3786–3798.

(30) Hearst, J. E. (1987) Background Hybridization Associated with the Use of Photocrosslinkable Oligonucleotide Hybridization Probes for the Identification of Unique Sequences in the Human Genome. *Photobiochem. Photobiophys., Suppl.*, 23–32.

(31) Gamper, H. B., Cimino, G. D., Isaacs, S. T., Ferguson, M., and Hearst, J. E. (1986) Reverse Southern Hybridization. *Nucleic Acids Res.* **14**, 9943–9954.

(32) Gamper, H. B., Cimino, G. D., and Hearst, J. E. (1987) Solution Hybridization of Crosslinkable DNA Oligonucleotides to M13DNA: Effect of Secondary Structure on Hybridization Kinetics and Equilibria. *J. Mol. Biol.* **197**, 349–362.

(33) Shi, Y., and Hearst, J. E. (1986) Thermostability of Double-Stranded Deoxyribonucleic Acids: The Effects of Covalent additions of a Psoralen. *Biochemistry* **25**, 5895–5902.

(34) Shapiro, J. T., Stannard, B. S., and Felsenfeld, G. (1969) The Binding of Small Cations to Deoxyribonucleic Acid. Nucleotide Specificity. *Biochemistry* **8**, 3233–3241.

(35) Melchior, W. B., Jr., and von Hippel, P. H. (1973) Alteration of the Relative Stability of dA·dT and dG·dC Base Pairs in DNA. *Proc. Natl. Acad. Sci. U.S.A.* **70**, 298–302.

(36) Piette, J. G., and Hearst, J. E. (1983) Termination Sites of the in vitro Nick-Translation Reaction on DNA that had Photoreacted with Psoralen. *Proc. Natl. Acad. Sci. U.S.A.* **80**, 5540–5544.

(37) Chanet, R., Cassier, C., Magana-Schwenke, N., and Moustacchi, E. (1984) Fate of Photoinduced 8-Methoxypsoralen Monoadducts in Yeast. Evidence for Bypass of these Lesions in the Absence of Excision. *Mutat. Res.* **112**, 201–214.

(38) Shi, Y., Gamper, H., Van Houten, B., and Hearst, J. E. (1988) Interaction of E. coli RNA Polymerase with DNA in an Elongation Complex Arrested at a Site Specific Cross-link Site. *J. Mol. Biol.* **199**, 277–293.

(39) Shi, Y., Gamper, H., and Hearst, J. E. (1987) Interaction of T7 RNA Polymerase with DNA in an Elongation Complex. *J. Biol. Chem.* **263**, 527–534.

(40) Van Houten, B., Gamper, H., Holbrook, S. R., Hearst, J. E., and Sancar, A. (1986) Action Mechanism of ABC Excision Nuclease on a DNA Substrate Containing a Psoralen Crosslink at a Defined Position. *Proc. Natl. Acad. Sci. U.S.A.* **83**, 8077–8081.

(41) Cheng, S., Van Houten, B., Gamper, H. B., Sancar, A., and Hearst, J. E. (1988) Use of Psoralen-Modified Oligonucleotides To Trap Three-Stranded RecA–DNA Complexes and Repair of These Cross-Linked Complexes by ABC Excinuclease. *J. Biol. Chem.* **263**, 15110–15117.

**Chapter 3**

# Mechanisms of 3-Methylindole Pneumotoxicity

Garold S. Yost

*Department of Pharmacology and Toxicology, 112 Skaggs Hall, University of Utah,
Salt Lake City, Utah 84112*

Reprinted from *Chemical Research in Toxicology,* Vol. 2, No. 5, September/October, 1989

## Introduction

A large number of chemicals selectively damage mammalian lung tissues without significant toxicity to other organs. The selectivity of these xenobiotic compounds for pulmonary damage is explained by a number of different mechanisms. Certainly, toxicants that are inhaled such as $O_3$, $NO_2$, $H_2CO$, and acrolein can cause direct damage to the first tissue they come in contact with and cause damage because of the route of exposure (1). This mechanism of pneumotoxicity is probably the major cause of lung injury to humans. Other compounds, such as paraquat, that are selectively accumulated in pulmonary cells are toxic primarily because they are found in highest concentrations in these cells (2).

The class of toxicants for which it is most difficult to provide rational mechanisms for organ selectivity are those that primarily damage lung tissues after systemic circulation and are not accumulated in lung cells. 3-Methylindole (3MI)[1] is a fascinating example of one of these circulating pneumotoxins. Other examples include 4-ipomeanol (3), naphthalene (4), monocrotaline (5), trichloroethylene (6), and butylated hydroxytoluene (7). Most of these compounds are activated by CYPs in either hepatic or pulmonary cells and cause selective toxicity to lung tissues. 4-Ipomeanol has been one of the most thoroughly studied systemic pneumotoxins (3), and its lung selectivity is a result of bioactivation in, and selective damage to, the nonciliated bronchiolar epithelial cells (Clara cells). A different mechanism is involved in the pneumotoxicity of monocrotaline, a dihydropyrrole, which

[1] Abbreviations: 3MI, 3-methylindole; CYP, cytochrome P-450 monooxygenases; GSH, glutathione; ABT, 1-aminobenzotriazole; αMB, [(α-methylbenzyl)amino]benzotriazole; BSO, L-buthionine (S,R)-sulfoximine; D₃-3MI, [*methyl*-²H₃]-3-methylindole.

**Scheme I. Pneumotoxicity of 3-Methylindole by Fermentation of Tryptophan**

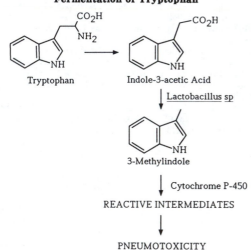

is oxidized in the liver to a toxic pyrrole, which is exported to pulmonary endothelial cells that are damaged (5, 8). Several reviews have been published about the mechanisms of bioactivation and toxicity of these and other pneumotoxins (9–11).

The etiology of the pneumotoxic process of 3MI in ruminants is an intriguing story that has been developed over many years primarily by Dr. James R. Carlson and his colleagues at Washington State University. A disease called acute bovine pulmonary edema and emphysema is produced in cattle that are exposed to drastic changes in grazing patterns (12). The disease causes significant pulmonary toxicity and results in the death of the cattle by respiratory failure. The pneumotoxicity is caused by

the ingestion of tryptophan that is metabolized by rumi-
nant bacteria to indole-3-acetic acid, which is subsequently
metabolized by an anaerobic *Lactobacillus* sp. bacteria that
produces 3MI (*13*) (Scheme I). The rate of 3MI production, and resultant bovine pneumotoxicity, is not dependent on the amount of tryptophan ingested by the
animal, but rather the fermentation conditions in the rumen that stimulate the growth of the anaerobic bacteria.
The toxicity of tryptophan and resultant 3MI formation
can be ameliorated by the administration of oral antibiotics
(*14*).

3MI is absorbed readily from the rumen and circulated
systemically, but the toxin produces damage *solely* to
pulmonary tissues (*15, 16*). Selective lung damage is
particularly surprising since the portal circulation requires
passage of 3MI through the liver prior to pulmonary exposure, and no hepatotoxicity has been reported from 3MI.
Since 3MI is not toxic per se, but must be oxidatively
transformed to reactive intermediates, the selectivity of
the toxin for pulmonary tissues is intriguing given the
paucity of pulmonary metabolism compared to hepatic
metabolism in general. Thus, the mechanisms for 3MI
pulmonary selectivity must be due to selective activation
in lung cells, poor detoxication in lung cells, superior
detoxication in hepatocytes, export of a pulmonary-se-
lective toxic intermediate from hepatocytes, or a combination of these mechanisms. The objectives of our research
have been to describe the mechanisms of pulmonary selectivity and the precise chemical transformations of 3MI
that lead to lung injury.

The pneumotoxicity of 3MI is species-selective. Ruminants are highly susceptible to 3MI; an iv dose of 30–40
mg/kg to a goat is usually fatal (*15*). Although the de-
scriptions of pneumotoxicity of 3MI have been confined
primarily to ruminants (*15, 17, 18*), rodents are also susceptible to 3MI lung damage at much higher doses (*19–21*);
the LD50 in mice (*22*) is 578 mg/kg. 3MI has considerably
lower toxicity to rabbits since the compound is not lethal
in doses less than 900 mg/kg (unpublished observations).
Exploitation of the differences in species sensitivity to 3MI
has been an important approach for mechanistic studies,
which have employed species comparisons with techniques
such as covalent binding to tissue proteins, CYP isozyme
characterization, and urinary metabolite isolation and
identification.

The exposure of humans to 3MI is significant, predominantly through intestinal production and absorption of
the chemical (*23*). Cigarette smoke also contains considerable quantities of 3MI (*24*), but the toxicity of 3MI by
the inhalation route has not been evaluated. Fecal excretion of 3MI has been linked to impaired colonic
movement (*25*), and metabolites of 3MI such as
"hydroxyskatoles" (Scheme III, structure 4) have been
identified from human urine (*26, 27*). It is likely, therefore,
that mechanistic studies of 3MI bioactivation and organ-
selective toxicity may have applications to human exposures.

### Covalent Binding: A Partial Indicator of Toxicity

The use of covalent binding of electrophilic metabolites
of toxins to proteins has been used extensively as an indicator of the relative susceptibilities of tissues and cells
to damage by certain chemicals. Protein binding has been
associated with the initial toxic events in the hepatotoxicity
of acetaminophen and other agents (*28*), but the correlation
of covalent binding with toxicity does not exist with some
compounds; an example is the acetaminophen regioisomer,
3'-hydroxyacetanilide (*29*). Covalent binding to tissue
proteins may be more appropriately viewed as an index

**Table I. Covalent Binding of 3-Methylindole**

|  | lung | liver | kidney | ref |
|---|---|---|---|---|
| in vivo[a] | | | | |
| cow | 752 | 355 | 245 | 18 |
| goat | 537 | 96 | 95 | 30 |
| mouse | 5 | 20 | 13 | 32 |
| in vitro[b] | | | | |
| cow | 0.480 | | | 18 |
| goat | 0.135 | 0.042 | | 31 |
| horse | 0.045 | 0.028 | | 31 |
| monkey | 0.030 | 0.108 | | 31 |
| mouse | 0.017 | 0.085 | | 31 |
| rat | nd[c] | 0.112 | | 31 |

[a] In units of nmol/g of tissue.  [b] In units of nmol/(mg of pro-
tein·min).  [c] nd = not detected.

of initial cellular damage than as a cause of cytotoxicity.
Regardless of the role that covalent binding plays in the
mechanisms of toxicity, the formation of covalent bonds
between activated toxins and proteins is indicative of the
presence of electrophilic intermediates. Therefore, the
covalent binding of radioactive 3MI to various tissue
proteins from susceptible and nonsusceptible species has
been used to probe the mechanisms of 3MI toxicities.

Early studies with ruminants (*18, 30*) had indicated that
the covalent binding of radioactive 3MI to pulmonary
proteins was higher than for other organs (Table I), including hepatic proteins, and higher pulmonary binding
was indicative of the organ selectivity of this toxin. The
covalent binding of radioactive 3MI to microsomal proteins
from lungs and livers of goats has been compared to that
for mice, rats, and other animals (Table I) to see if relative
covalent binding values correlated with tissue-selective
damage in nonruminant species (*31*). These studies
showed that covalent binding of 3MI to hepatic microsomes was *considerably* higher than to pulmonary microsomes from mice and rats, in spite of the fact that 3MI is
highly selective for lung damage in these species, albeit at
much higher doses than in ruminants. Radioactive 3MI
administered in vivo (ip) to mice also produced about
4-fold higher amounts of covalently bound metabolites to
liver than lung proteins (*32*). Thus, the selective covalent
binding of 3MI to pulmonary tissues can be correlated to
the high species selectively of 3MI for ruminants, but not
to the organ selectivity of the toxin in more nonsusceptible
species, mice and rats. The use of relative amounts of
covalent binding of 3MI to lung and liver proteins provides
some insights into the extreme susceptibility of ruminants
to this pneumotoxin but does not explain the fact that 3MI
is also a pneumotoxin and not a hepatotoxin in rodents.

### Glutathione Modulation of Organ Selectivity

The livers of mice produced large quantities of electrophilic metabolites, as measured by covalent binding, but
hepatotoxicity to mice and rats was not observed. The lack
of hepatotoxicity may be a result of binding to nonessential
proteins of liver tissues or may simply indicate that covalent binding is not a reliable index of tissue damage by
3MI. Since hepatic GSH levels are considerably higher
than pulmonary GSH levels, the high GSH concentrations
in hepatic tissues might play a role in the defense of the
liver, but, obviously, high hepatic GSH levels do not prevent hepatic covalent binding of 3MI. Pulmonary GSH
modulation (*30*) has demonstrated that 3MI-induced
pneumotoxicity can be increased or decreased by depletion
with diethyl maleate or supplementation with cysteine of
pulmonary GSH levels. L-Buthionine (*S,R*)-sulfoximine
(BSO) is a selective inhibitor of GSH synthesis (*33*) and
also selectively depletes hepatic GSH (*34*) without sig-

Table II. Glutathione Depletion by BSO and Covalent Binding of 3-Methylindole in Mice

| BSO pretreatment (mmol/kg) | glutathione levels[a] (% of control) | | | covalent binding[b] | | |
|---|---|---|---|---|---|---|
| | liver | lung | kidney | liver | lung | kidney |
| 0 | | | | 2.6 ± 0.6 | 0.6 ± 0.1 | 1.7 ± 0.2 |
| 1.0 | 60 ± 3 | 77 ± 15 | 105 ± 14 | 3.8 ± 1.2 | 1.3 ± 0.4 | 5.6 ± 0.5 |
| 3.0 | 17 ± 10 | 75 ± 6 | 82 ± 29 | 8.4 ± 1.4 | 1.7 ± 0.5 | 6.7 ± 2.6 |
| 6.0 | 11 ± 1 | 70 ± 9 | 64 ± 2 | 10 ± 3.7 | 1.6 ± 0.3 | 7.9 ± 0.5 |

[a] Control GSH levels were as follows: liver, 8.4 μmol/g; lung, 1.9 μmol/g; and kidney, 0.44 μmol/g. [b] In units of μg/g of tissue.

nificantly depleting pulmonary or renal GSH after ip administration to mice (32).

Recent studies (32) have utilized the selective depletion of hepatic GSH by pretreatment of mice with a low dose of BSO followed by a low dose of 3MI to evaluate the hypothesis that hepatotoxicity can be induced in mice if hepatic GSH levels are depleted. This regimen depleted hepatic GSH to 60% of controls but did not deplete renal or pulmonary GSH. Surprisingly, liver samples from these mice were not injured, as assessed by histopathological examination, but renal proximal tubules were exceptionally damaged. Increases in the BSO doses caused increased 3MI-mediated renal toxicities but never produced hepatic or pulmonary damage. Table II shows the relative depletion of GSH by BSO pretreatment and also shows the much greater increases in covalent binding of 3MI to renal proteins than to hepatic or pulmonary proteins with a low dose of BSO. Therefore, covalent binding of radioactive 3MI metabolites was again used to partially explain the organ-selective toxicity of 3MI to renal tissues after hepatic GSH depletion but could not provide a rationale for the resistance of the liver to 3MI toxicity under these conditions.

An explanation for the nephrotoxicity of 3MI subsequent to *hepatic* depletion of GSH may involve the export of a reactive intermediate of 3MI from hepatocytes that is not normally circulated if GSH levels are not depleted. Such a reactive intermediate might selectively accumulate in renal proximal cells and be toxic because it is concentrated in these cells. The nephrotoxic metabolite is probably not produced in the renal cells because 3MI is not toxic to kidneys after ip administration unless hepatic GSH is depleted. Several possible toxic intermediates and their routes of formation are discussed below. This fascinating switch in organ-selective toxicity from pulmonary to renal sites has not been observed with other pneumotoxins.

## Deuterium Isotope Effects as Mechanistic Probes

The use of deuterated analogues of toxic chemicals can be a useful method for evaluations of the importance of C–H bond breakage steps during the bioactivation process (35). Deuterium isotope effects can be observed both in vitro (36) and in vivo (37) with appropriately substituted analogues. The primary isotope effects can be masked in enzymatic determinations by equilibrium processes, and several investigators (38, 39) have utilized intramolecular isotope effects to determine the intrinsic isotope effects. Isotope effects have been utilized (39) with deuterium-labeled and tritium-labeled dihydropyridines to establish a mechanism of one-electron oxidation of the nitrogen by CYP, followed by facile loss of the 4-proton and subsequent electron transfer to produce the aromatic pyridine. The intrinsic isotope effect was shown to be small, and therefore, nitrogen rather than C–H oxidation was rate-determining.

The bioactivation of 3MI has been proposed (40) to proceed through the oxidation of the methyl group to form

Scheme II. Pathways of One-Electron Oxidations of Deuterium-Labeled 3-Methylindole

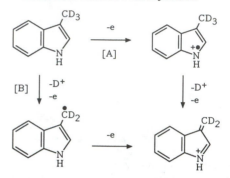

a methylene imine (7) electrophilic intermediate. The routes of formation of this intermediate could proceed through initial nitrogen or carbon oxidation steps (Scheme II). The rate of 3MI oxidation, and presumably resultant toxicity, could be decreased with methyl deuterium substituted 3MI due to a primary deuterium isotope effect. This deuterium isotope effect would be dependent on the rate of hydrogen (deuterium) abstraction in comparison to the rates of nitrogen oxidation and other contributing steps. If the rate-determining step included or followed the hydrogen abstraction step, a large isotope effect should be observed, if the effect were not masked by enzyme equilibration factors. If nitrogen oxidation was the slow step, then a small or negligible isotope effect would be observed.

$D_3$-3MI was synthesized by $LiAlD_4$ reduction of indole-3-carboxylic acid and used to evaluate possible deuterium isotope effects in vivo (22). $D_3$-3MI was found to be considerably less toxic than 3MI in mice. Evaluations of toxicity included LD50 determinations, rates of pulmonary GSH depletion, comparisons of wet lung weights, and evaluations of minimal toxic doses that caused Clara cell damage. Although these in vivo determinations did not provide good quantitative measures of a deuterium isotope effect, they were indicative of qualitative changes in toxicity that were probably caused by deuterium isotope effects. These isotope effects have been confirmed preliminarily with isolated rabbit Clara cells (41). $D_3$-3MI was not cytotoxic during a 4-h incubation at a concentration of 1.0 mM, although a 1.0 mM concentration of 3MI caused a decrease of 35% in Clara cell viability during a 4-h incubation. Work is currently in progress on the determination of the in vitro isotope effects with goat pulmonary microsomes.

The studies with $D_3$-3MI have demonstrated that the oxidation of a C–H bond on the methyl group of 3MI is a significant event in the mechanism of bioactivation to a toxic intermediate. These results indicate that, although the first oxidative step of 3MI bioactivation is probably a one-electron abstraction from the nitrogen (pathway A, Scheme II), in analogy with oxidations of other nitrogen heterocyclic compounds (39, 42), the C–H bond breakage step is a slow step and is rate-determining. Thus, deuterium isotope effects have been important in the partial elucidation of the mechanisms of 3MI metabolism and toxicity.

## CYP Suicide Substrates as Biochemical Probes

Early studies with ruminants (*43*) indicated that 3MI required bioactivation to exert its toxicity, and oxidation of the indole was proposed to be a requisite activation step. Indoles as a class of chemicals are very easily oxidized. The ease of oxidation of indoles was responsible for the inability of electrochemists to measure the oxidation potential of indole or substituted indoles because polymerization of the oxidized products occurred too rapidly for voltametric measurements. Only recently (*44*) have the oxidation potentials of substituted indoles, including 3MI, been measured by indirect techniques with the use of the single-electron oxidant radical, chlorine dioxide ($ClO_2^•$). The oxidation potential of 3MI was estimated as 1.07 V. Interestingly, this oxidation potential is less than that of the prototypic biological antioxidant, butylated hydroxytoluene ($E° = 1.21$ V) (*45*).

The oxidation of 3MI in lung cells was postulated to be mediated by CYP enzymes (*43*), but others (*46*) have recently postulated that prostaglandin H synthetase mediated cooxidation also is an important contributor to the bioactivation of 3MI to intermediates that covalently bind to goat lung microsomes. Therefore, questions about the biochemical process of activation required further investigations.

The use of mechanism-based (suicide) inhibitors of CYP enzymes offers great promise for mechanistic studies in toxicology. These inhibitors are *specific* for the CYP enzymes and can be effective irreversible inhibitors in both in vitro (*47, 48*) and in vivo (*49*) studies. 1-Aminobenzotriazole (ABT) is a potent inactivator of CYP enzymes. The inactivator operates through the formation of benzyne which cross-links two pyrrolic nitrogens on the prosthetic heme of the CYP enzymes and results in irreversible inactivation of the enzymes (*50*). An analogue of ABT, [(α-methylbenzyl)amino]benzotriazole (αMB), was synthesized to be an isozyme-selective irreversible inhibitor of CYPIIB4 [form 2 from rabbit lung]. αMB is structurally similar to benzphetamine, a selective substrate of CYPIIB4, and still retains the requisite aminobenzotriazole moiety. αMB is highly selective for inhibition of CYPIIB4 from rabbit lung (*51*).

Goat lung microsomes were utilized to evaluate the importance of CYP in the metabolic turnover of 3MI and the relationship of this turnover to covalent binding, and presumably to toxicity (*52*). ABT completely inhibited 3MI turnover at concentrations greater than 0.1 mM and inhibited approximately 50% 3MI turnover at 0.01 mM. The covalent binding of radioactive 3MI to the microsomes was inhibited to the same degree as 3MI turnover at these concentrations. Interestingly, the rate of covalent binding, 3.43 nmol/(mg·30 min), was half of the rate of 3MI turnover, 7.04 nmol/(mg·30 min) and, thus, represented a significant portion of 3MI metabolism in the goat lung microsomes. Many other metabolism enzymes were not operative in these microsomes, however, so the covalent binding/3MI turnover ratio is probably higher that one would expect in whole cells. The correlation of inhibition of 3MI turnover to covalent binding is reasonably good evidence that the metabolism and resultant toxicity of 3MI in a susceptible species and organ (goat lung) is mediated by CYP, with the assumption that covalent binding is a good index of toxicity. At the very least, the inhibition of 3MI turnover and covalent binding by CYP inhibitors demonstrates that CYP enzymes are intimately involved

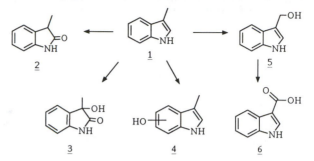

Scheme III. Major Metabolites of 3-Methylindole

in the metabolism of 3MI to reactive intermediates. The flavin monooxygenases are probably not participants in 3MI metabolism since these enzymes are not inhibited by ABT (*48*). However, the prostaglandin H synthetase mediated cooxidation of 3MI may contribute slightly to in vivo covalent binding since substrates (fatty acids) for this enzyme were not included in the goat lung microsomal studies.

The isozyme-selective inhibitor, αMB, was used to demonstrate that the goat homologue of CYPIIB4 was the primary CYP isozyme involved in 3MI turnover in goat lung microsomes (*52*). αMB was more effective than ABT at inhibiting 3MI turnover and covalent binding, and the inhibition of 3MI turnover correlated well to the inhibition of benzphetamine *N*-demethylase activity but correlated poorly to the inhibition of 7-ethoxyresorufin *O*-deethylase activity (catalyzed by CYPIA1, form 6 in rabbit lungs). CYPIA1 is an aromatic hydrocarbon hydroxylase form of CYP and is a minor isozyme in rabbit lungs. The goat homologues of rabbit CYPIA1 and CYPIIB4 have not been isolated. Thus, studies with αMB demonstrated that the metabolism of 3MI in goat lung microsomes is primarily mediated by one CYP isozyme (the goat homologue of CYPIIB4) but that at least some of the 3MI turnover is probably due to metabolism by other CYP isozymes, such as the goat homologue of CYPIA1. The organ and species selectivities of 3MI toxicities may be due primarily to the expression and activity of CYP isozymes in susceptible pulmonary cells. The use of CYP inactivators has provided additional evidence for the selective bioactivation of 3MI by CYP enzymes.

## Metabolites Point to Reactive Intermediates

The isolation and identification of the metabolites of 3MI from microsomal incubations or urinary samples could be used as a "smoking gun" to provide information about the reactive intermediates that are responsible for 3MI pneumotoxicity. Scheme III is an illustration of most of the identified metabolites of 3MI. Since large differences exist in the susceptibility of goats and mice to the pneumotoxicity of 3MI, the isolation and identification of urinary metabolites from both of these species may lead to conclusions about differences in metabolism and susceptibility. Isolations of metabolites from goat urine (*53*) showed that 3-methyloxindole (2), indole-3-carbinol (5), and indole-3-carboxylic acid (6) were major metabolites of 3MI. 3-Methyloxindole and indole-3-carbinol were administered to goats to evaluate their toxicities, and neither compound produced pneumotoxicity in this species. From these observations, it was concluded that an intermediate between 3MI and these metabolites might be responsible for the toxicity of 3MI. The oxidized methylene imine (7) was proposed as a likely candidate for such an intermediate. Proof of the presence of this intermediate was presented by the isolation and identi-

Scheme IV. Proposed Pathways of Activation of
3-Methylindole and Structures of the Glutathione Adducts
of Proposed Electrophilic Intermediates

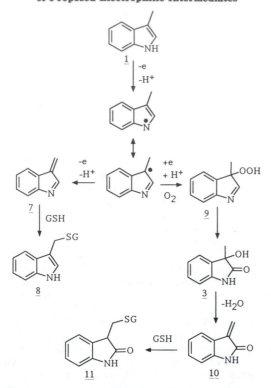

fication of the GSH adduct of 3MI (8) at the methyl group (40) from goat lung microsomal incubation that were fortified with GSH.

Comparisons of mouse urinary metabolites of 3MI with the goat urinary metabolites have been studied recently. In keeping with the large differences in susceptibilities of these two species to 3MI-mediated pneumotoxicity, significant differences in the metabolites were observed. The major metabolites from goat urine, 3-methyloxindole and indole-3-carbinol, were not found in mouse urine. Instead, a highly oxidized indole, 3-hydroxy-3-methyloxindole (3), was isolated and identified (54). This oxindole has also been isolated and identified recently from human urine (55). The compound has been produced by bacterial degradation of indoleacetic acid (56). Some evidence was presented that dehydration of 3-hydroxy-3-methyloxindole produces an electrophilic methyleneoxindole (10). Methyleneoxindole inhibits bacterial growth and readily adds GSH, presumably to the exocyclic methylene moiety, to produce the GSH adduct (11); this adduct has not been isolated, however. Thus, the isolation of 3-hydroxy-3-methyloxindole led to the proposition that methyleneoxindole might be a pneumotoxic or nephrotoxic metabolite of 3MI in mice.

The pathway for the formation of 3-hydroxy-3-methyloxindole may proceed through the formation of a free-radical intermediate that is formed by CYP and adds $O_2$ to form a hydroperoxide (9) that undergoes an oxygen rebound type of mechanism to produce the oxidized indole (Scheme IV). Support for this mechanism comes from the following items: (1) the mechanism of CYP-mediated oxidation proceeds through successive one-electron oxidative steps (42); (2) 3MI metabolism produces free radicals (57); (3) oxygen rebound mechanisms from peroxide intermediates have been postulated to be responsible for the production of metabolites of butylated hydroxytoluene (58), a known pneumotoxin; and (4) 3-hydroxy-3-

methyloxindole is isolated as an optically active compound (54), which demonstrates that the compound cannot be formed by nonenzymatic autoxidation of 3MI to a hydroperoxide.

The identification of metabolites of 3MI from a highly susceptible species, goats, that were not produced in a less sensitive species, mice, indicates that the pathways of metabolism, and perhaps bioactivation, are differentiable and may be correlated to species susceptibilities. In addition, the isolations of compounds such as 3-hydroxy-3-methyloxindole (3) and the GSH adduct (8) of the methylene imine may be good indicators of the presence of toxic intermediates such as the methyleneoxindole (10) or the methylene imine (7). Structural elucidations of these types of metabolites or trapped intermediates are indirect methods for the determinations of the reactive intermediates of 3MI. However, the structures of the metabolites provide important information about the pathways of chemical transformations in intact animals.

## Conclusions

The studies that have been presented concerning the mechanisms of 3MI-mediated pneumotoxicity have provided some answers, such as the importance of CYP metabolism and mechanisms of C–H bond breakage, but generated many more questions, such as the cause of nephrotoxicity of 3MI after GSH depletion. The use of powerful techniques such as selective hepatic GSH depletion, deuterium isotope effects, and suicide CYP inactivators have allowed us to postulate some mechanisms for the organ and species selectivity of 3MI toxicities. In addition, the pathways of bioactivation of 3MI to electrophilic reactive intermediates have been partially elucidated by the use of the aforementioned techniques. In spite of the seemingly simple chemistry of this small molecule, 3MI, the mechanisms of its activation and toxicities are complex, and the complete elucidation of these mechanisms remains a challenging task.

**Acknowledgment.** I express my gratitude to the hardworking and fun-loving graduate students Mark Nocerini, Jeannette Huijzer, and Gary Skiles; postdoctoral fellows Drs. John-Yan Jaw and Dave Kuntz; and colleagues Drs. Jim Adams, Jr., Roger Breeze, and Bill Nichols who have contributed greatly to this work. My most sincere thanks is expressed to Dr. Jim Carlson for his initial guidance and continuing help with this research. This research was supported by Grant HL13645 from the U.S. Public Health Service, National Institutes of Health. The author is a U.S. Public Health Service Research Career Development Awardee (HL02119).

## References

(1) Overton, J. A., and Miller, F. J. (1988) Absorption of inhaled reactive gases. In *Toxicology of the Lung* (Gardner, D. E., Crapo, J. D., and Massaro, E. J., Eds.) pp 477–507, Raven Press, New York.

(2) Rose, M. S., Smith, L. L., and Wyatt, I. (1976) Paraquat accumulation: tissue and species specificity. *Biochem. Pharmacol.* 25, 1763–1767.

(3) Boyd, M. R. (1976) Role of metabolic activation in the pathogenesis of chemically induced pulmonary disease: mechanism of action of the lung-toxic furan, 4-ipomeanol. *Environ. Health Perspect.* 16, 127–138.

(4) Buckpitt, A. R., and Warren, D. L. (1983) Evidence for the hepatic formation, export and covalent binding of reactive naphthalene metabolites in extrahepatic tissues in vivo. *J. Pharmacol. Exp. Ther.* 225, 8–16.

(5) Roth, R. A., and Ganey, P. E. (1988) Platelets and the puzzles of pulmonary pyrrolizidine poisoning. *Toxicol. Appl. Pharmacol.* 93, 463–471.

(6) Forkert, P. G., and Birch, D. W. (1989) Pulmonary toxicity of trichloroethylene in mice. Covalent binding and morphological manifestations. *Drug Metab. Dispos.* 17, 106–113.

(7) Witschi, H., Malkinson, A. M., and Thompson, J. A. (1989) Metabolism and pulmonary toxicity of butylated hydroxytoluene (BHT). *Pharmacol. Ther.* 42, 89–113.

(8) Bruner, L. H., Carpenter, L. J., Hamlow, P., and Roth, R. A. (1986) Effect of a mixed function oxidase inducer and inhibitor on monocrotaline pyrrole pneumotoxicity. *Toxicol. Appl. Pharmacol.* 85, 416–427.

(9) Yost, G. S., Buckpitt, A. R., Roth, R. A., and McLemore, T. L. (1989) Mechanisms of lung injury by systemically administered chemicals. *Toxicol. Appl. Pharmacol.* (in press).

(10) Boyd, M. R., Grygiel, J. J., and Minchin, R. F. (1983) Metabolic activation as a basis for organ-selective toxicity. *Clin. Exp. Pharmacol. Physiol.* 10, 87–99.

(11) Kehrer, J. P., and Kacew, S. (1985) Systematically applied chemicals that damage lung tissue. *Toxicology* 35, 251–293.

(12) Hammond, A. C., Bradley, B. J., Yokoyama, M. T., Carlson, J. R., and Dickinson, E. O. (1979) 3-Methylindole and naturally occurring acute bovine pulmonary edema and emphysema. *Am. J. Vet. Res.* 40, 1398–1401.

(13) Yokoyama, M. T., and Carlson, J. R. (1974) Dissimulation of tryptophan and related indolic compounds by ruminal microorganisms in vitro. *Appl. Microbiol.* 27, 1540–1548.

(14) Nocerini, M. R., Honeyfield, D. C., Carlson, J. R., and Breeze, R. G. (1985) Reduction of 3-methylindole production and prevention of acute bovine pulmonary edema and emphysema with lasalocid. *J. Anim. Sci.* 60, 232–238.

(15) Carlson, J. R., and Breeze, R. G. (1983) Cause and prevention of acute pulmonary edema and emphysema in cattle. In *Handbook of Natural Toxins. Plant and Fungal Toxins* (Keeler, R. F., and Tu, A. T., Eds.) Vol. 1, pp 85–115, Marcel Dekker, New York.

(16) Bradley, B. J., and Carlson, J. R. (1980) Ultrastructural pulmonary changes induced by intravenous administered 3-methylindole in goats. *Am. J. Pathol.* 99, 551–560.

(17) Bradley, B. J., Carlson, J. R., and Dickinson, E. O. (1978) 3-Methylindole-induced pulmonary edema and emphysema in sheep. *Am. J. Vet. Res.* 39, 1355–1358.

(18) Hanafy, M. S. M., and Bogan, J. A. (1980) The covalent binding of 3-methylindole metabolites to bovine tissue. *Life Sci.* 27, 1225–1231.

(19) Turk, M. A. M., Flory, W., and Henk, W. G. (1984) Dose response in 3-methylindole-induced bronchiolar epithelial necrosis in mice. *Res. Commun. Chem. Pathol. Pharmacol.* 46, 351–362.

(20) Kiorpes, A. L., Keith, I. M., and Dubielzig, R. R. (1988) Pulmonary changes in rats following administration of 3-methylindole in cremophore EL. *Histol. Histopathol.* 3, 125–132.

(21) Adams, J. D., Laegreid, W. W., Huijzer, J. C., Hayman, C., and Yost, G. S. (1988) Pathology and glutathione status in 3-methylindole-treated rodents. *Res. Commun. Chem. Pathol. Pharmacol.* 60, 323–335.

(22) Huijzer, J. C., Adams, J. D., Jr., and Yost, G. S. (1987) Decreased pneumotoxicity of deuterated 3-methylindole: Bioactivation requires methyl C–H bond breakage. *Toxicol. Appl. Pharmacol.* 90, 60–68.

(23) Fordtran, J. S., Scroggie, W. B., and Potter, D. E. (1964). Colonic absorption of tryptophan metabolites in man. *J. Lab. Clin. Med.* 64, 125–132.

(24) Hoffman, D., and Rathkamp, G. (1970) Quantitative determination of 1-alkylindoles in cigarette smoke. *Anal. Chem.* 42, 366–370.

(25) Karlin, D. A., Mastromarino, A. J., Jones, R. D., Stroehlein, J. R., and Lorentz, O. (1985) Fecal skatole and indole and breath methane and hydrogen in patients with large bowel polyps or cancer. *Cancer Res. Clin. Oncol.* 109, 135–141.

(26) Nakaro, A., and Ball, M. (1960) The appearance of a skatole derivative in the urine of schizophrenics. *J. Nerv. Ment. Dis.* 130, 317–419.

(27) Acheson, R. M., and Hands, A. R. (1961) 6-Sulphatoxyskatole in human urine. *Biochim. Biophys. Acta* 51, 579–581.

(28) Monks, T. J., and Lau, S. S. (1988) Reactive intermediates, and their toxicological significance. *Toxicology* 52, 1–53.

(29) Streeter, A. J., Bjorge, S. M., Axworthy, D. B., Nelson, S. D., and Baillie, T. A. (1984) The microsomal metabolism and site of covalent binding to protein of 3′-hydroxyacetanilide, a non-hepatotoxic positional isomer of acetaminophen. *Drug Metab. Dispos.* 12, 565–576.

(30) Nocerini, M. R., Carlson, J. R., and Breeze, R. G. (1983) Effect of glutathione status on covalent binding and pneumotoxicity of 3-methylindole in goats. *Life Sci.* 32, 449–458.

(31) Nocerini, M. R., Carlson, J. R., and Yost, G. S. (1985) Adducts of 3-methylindole and glutathione: species differences in organ-selective bioactivation. *Toxicol. Lett.* 28, 79–87.

(32) Yost, G. S., and Kuntz, D. J. (1989) Organ-selective switching of 3-methylindole toxicity by glutathione depletion. *Toxicol. Appl. Pharmacol.* (in press).

(33) Griffith, O. W., and Meister, A. (1979) Potent and specific inhibition of glutathione synthesis by buthionine sulfoximine (S-N-butyl homocysteine sulfoximine). *J. Biol. Chem.* 254, 7558–7560.

(34) Buckpitt, A. R., and Warren, D. L. (1983) Evidence for hepatic formation, export and covalent binding of reactive naphthalene metabolites in extrahepatic tissues in vivo. *J. Pharmacol. Exp. Ther.* 225, 8–16.

(35) Soderlund, E. J., Gordon, W. P., Nelson, S. D., Omichinski, J. G., and Dybing, E. (1984) Metabolism in vitro of tris(2,3-dibromopropyl)-phosphate: oxidative debromination and bis(2,3-dibromopropyl)phosphate formation as correlates of mutagenicity and covalent protein binding. *Biochem. Pharmacol.* 33, 4017–4023.

(36) Miwa, G. T., Harada, N., and Lu, A. Y. H. (1985) Kinetic isotope effects on cytochrome P-450-catalyzed oxidation reactions: full expression of the intrinsic isotope effect during the O-deethylation of 7-ethoxycoumarin by liver microsomes from 3-methylcholanthrene-induced hamsters. *Arch. Biochem. Biophys.* 239, 155–162.

(37) Yost, G. S., Horstman, M. G., El Walily, A. F., Gordon, W. P., and Nelson, S. D. (1985) Procarbazine spermatogenesis toxicity: deuterium isotope effects point to regioselective metabolism in mice. *Toxicol. Appl. Pharmacol.* 80, 316–322.

(38) Hales, D. B., Ho, B., and Thompson, J. A. (1987) Inter- and intramolecular deuterium isotope effects on the cytochrome P-450-catalyzed dehalogenation of 1,1,2,2-tetrachloroethane. *Biochem. Biophys. Res. Commun.* 149, 319–325.

(39) Guengerich, F. P., and Bocker, R. H. (1988) Cytochrome P-450-catalyzed dehydrogenation of 1,4-dihydropyridines. *J. Biol. Chem.* 263, 8168–8175.

(40) Nocerini, M. R., Yost, G. S., Carlson, J. R., Liberato, D. J., and Breeze, R. G. (1985) Structure of the glutathione adduct of activated 3-methylindole indicates that an imine methide is the electrophilic intermediate. *Drug. Metab. Dispos.* 13, 690–694.

(41) Nichols, W. K., Larson, D. N., and Yost, G. S. (1989) Bioactivation of 3-methylindole (3MI) by isolated rabbit lung cells. *Toxicologist* 9, 155.

(42) Ortiz de Montellano, P. R. (1986) Oxygen activation and transfer. In *Cytochrome P-450 Structure, Mechanism, and Biochemistry* (Ortiz de Montellano, P. R., Ed.) pp 217–271, Plenum Press, Nw York.

(43) Bray, T. M., and Carlson, J. R. (1979) Role of mixed-function oxidase in 3-methylindole-induced acute pulmonary edema in goats. *Am. J. Vet. Res.* 40, 1268–1272.

(44) Merenyi, G., Lind, J., and Shen, X. (1988) Electron transfer from indoles, phenol, and sulfite ($SO_3^{2-}$) to chlorine dioxide ($ClO_2$). *J. Phys. Chem.* 92, 134–137.

(45) Ronlan, A. (1971) Coupling of phenols via an anodically generated phenoxonium ion. *Chem. Commun.*, 1643–1645.

(46) Formosa, P. J., and Bray, T. M. (1988) Evidence for metabolism of 3-methylindole by prostaglandin H synthase and mixed-function oxidases in goat lung and liver microsomes. *Biochem. Pharmacol.* 37, 4359–4366.

(47) Ortiz de Montellano, P. R., and Reich, N. O. (1986) Inhibition of cytochrome P-450 enzymes. In *Cytochrome P-450 Structure, Mechanism, and Biochemistry* (Ortiz de Montellano, P. R., Ed.) pp 273–314, Plenum Press, New York.

(48) Mathews, J. M.; Dostal, L. A., and Bend, J. R. (1985) Inactivation of rabbit pulmonary cytochrome P-450 in microsomes and isolated perfused lungs by the suicide substrate 1-aminobenzotriazole. *J. Pharmacol. Exp. Ther.* 235, 186–190.

(49) Mico, B. A., Federowicz, D. A., Ripple, M. G., and Kerns, W. (1988) In vivo inhibition of oxidative drug metabolism by, and acute toxicity of, 1-aminobenzotriazole (ABT). *Biochem. Pharmacol.* 37, 2515–2519.

(50) Ortiz de Montellano, P. R., Mathews, J. M., and Langry, K. C. (1984) Autocatalytic inactivation of cytochrome P-450 and chloroperoxidase by 1-aminobenzotriazole and other aryne precursors. *Tetrahedron* 40, 511–519.

(51) Mathews, J. M., and Bend, J. R. (1986). N-Alkylaminobenzotriazoles as isozyme-selective suicide inhibitors of rabbit pulmonary microsomal cytochrome P-450. *Mol. Pharmacol.* 30, 25–32.

(52) Huijzer, J. C., Adams, J. D., Jr., Jaw, J.-Y., and Yost, G. S. (1989) Inhibition of 3-methylindole bioactivation by the cytochrome P-450 suicide substrates 1-aminobenzotriazole and $\alpha$-methylbenzylaminobenzotriazole. *Drug Metab. Dispos.* **17**, 37–42.

(53) Hammond, A. C., Carlson, J. R., and Willett, J. D. (1979) The metabolism and disposition of 3-methylindole in goats. *Life Sci.* **25**, 1301–1306.

(54) Skiles, G. L., Adams, J. D., Jr., and Yost, G. S. (1989) Isolation and identification of 3-hydroxy-3-methyloxindole, the major murine metabolite of 3-methylindole. *Chem. Res. Toxicol.* **2**, 254–259.

(55) Albrecht, C. F., Chorn, D. J., and Wessels, P. L. (1989) Detection of 3-hydroxy-3-methyloxindole in human urine. *Life Sci.* (in press).

(56) Still, C. C., Fukuyama, T. T., and Moyed, H. S. (1965) Inhibitory oxidation products of indole-3-acetic acid. *J. Biol. Chem.* **240**, 2612–2618.

(57) Bray, T. M., and Kubow, S. (1985) Involvement of free radicals in the mechanism of 3-methylindole-induced pulmonary toxicity: an example of metabolic activation in chemically induced long disease. *Environ. Health Perspect* **64**, 61–67.

(58) Thompson, J. A., and Wand, M. D. (1985) Interaction of cytochrome P-450 with a hydroperoxide derived from butylated hydroxytoluene. *J. Biol. Chem.* **260**, 10637–10644.

# Chapter 4

# Redox Chemistry of Anthracycline Antitumor Drugs and Use of Captodative Radicals as Tools for Its Elucidation and Control

Giorgio Gaudiano and Tad H. Koch*

*Department of Chemistry and Biochemistry, The University of Colorado, Boulder, Colorado 80309-0215*

Reprinted from *Chemical Research in Toxicology*, Vol. 4, No. 1, January/February, 1991

## Introduction

The anthracycline antitumor drugs, most notably daunorubicin (daunomycin, **1a**) and doxorubicin (adriamycin, **1b**), likely derive a portion of their biological activity through bioreductive activation (*1, 2*). Several enzymes are known to reduce the quinone functionality to semiquinone (**2**) and hydroquinone (**3**) redox states (*3, 4*). In aerobic medium both of these states react rapidly with molecular oxygen to produce superoxide with regeneration of the quinone state (Scheme I) (*5–7*). Subsequent reaction of superoxide in a Haber–Weiss cycle with Fe(III)/Fe(II) catalysis yields sequentially hydrogen peroxide and the cytotoxic hydroxyl radical. The iron catalysis is possibly facilitated through strong complexation of the anthracycline with iron (*8*). Hydroxyl radical may also result simply from reaction of semiquinone with hydrogen peroxide without iron catalysis (*9*). The overall process has come to be known as redox recycling, and the materials produced, as reactive oxygen species. The catalytic formation of reactive oxygen species likely contributes to the cytotoxicity of these anthracyclines.

Two forms of toxicity are of primary concern to the oncologist, tumor cytotoxicity and cardiotoxicity. Chronic cardiotoxicity ultimately limits the chemotherapy and results in part from oxidation of cardiac cell membrane lipids, among other critical molecules, by the reactive oxygen species (*10*). Heart cells are particularly vulnerable becuase they are poorly equipped to defend against reactive oxy radicals. Whether reactive oxygen species also contribute to tumor cytotoxicity is presently being debated (*11, 12*). Development of methodology for separation of cardiotoxicity from tumor toxicity remains an important goal. One approach in clinical development is the addition

2428–5/92/0021$06.00/0
© 1991 American Chemical Society

**Scheme I. Catalytic Production of Superoxide**

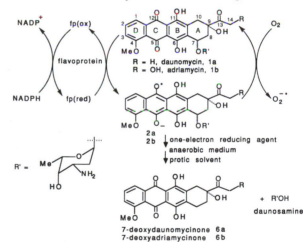

of a cell membrane transportable iron chelating agent, *d*-1,2-bis(3,5-dioxopiperazin-1-yl)propane (ICRF-187), which slows the Haber–Weiss cycle, possibly by competing with the anthracycline for iron cation (*13, 14*).

In an anaerobic medium a reduced state of the anthracyclines undergoes glycosidic cleavage. If the reduced state is the semiquinone, the aglycon transient formed is the 7-deoxy semiquinone methide (**4**), and if the reduced state is the hydroquinone, the aglycon transient is the 7-deoxy quinone methide (**5**). Semiquinone methide and quinone methide have both been discussed as reactive intermediates in coupling the aglycon to critical biological molecules (Scheme II); interesting possibilities are DNA, DNA-to-

**Scheme II. Proposed Radical and Nucleophilic Coupling to Macromolecules**

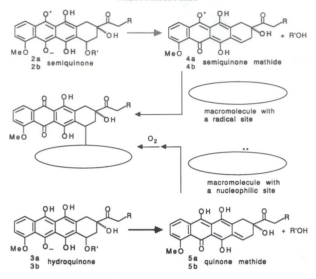

poisomerase complex, and proteins in cell membranes. Topoisomerase enzymes unwind DNA during replication, and inhibition, resulting in DNA strand breaks, is thought to be an important mechanism for anthracycline cytotoxicity (15). Inhibition of topoisomerase, however, may not necessarily involve redox chemistry and/or covalent bond formation. Cell membranes are implicated from the cytotoxicity of anthracyclines which are immobilized with respect to cell membrane transport (16). The redox state at which glycosidic cleavage actually occurs remains under continuous debate. The nature of the reactive state dictates subsequent chemical possibilities. Semiquinone methide, unless it is reduced to quinone methide, might logically couple with a radical site in a biological molecule. Quinone methide might couple with a nucleophilic or electrophilic site (vide infra). In competition with covalent bond formation is reaction with the ever-present electrophile, the proton. Protonation gives 7-deoxydaunomycinone (6a) which as an external drug is nontoxic and inactive (17). The possibility of covalent bond formation between the aglycon and, for example, DNA is an attractive proposal for tumor cell cytotoxicity. Because the aglycon is also capable of redox recycling, covalent bond formation would locate a catalyst for the production of hydroxyl radical adjacent to the DNA. Since hydroxyl radical does not diffuse far from its origin before it reacts, such a location is ideal for oxidative cleavage of DNA. Of course, covalently bound aglycon must be reducible by the available enzymes.

Over the past decade, we have been exploring the redox chemistry of anthracyclines using primarily a chemical one-electron reducing agent developed in our laboratories. The reagent is the captodative free radical, 3,5,5-trimethyl-2-oxomorpholin-3-yl (TM-3), formed in solution through spontaneous bond homolysis of its dimer, bi-(3,5,5-trimethyl-2-oxomorpholin-3-yl) (TM-3 dimer) (18–20). A more water-soluble version, 3,5-dimethyl-5-(hydroxymethyl)-2-oxomorpholin-3-yl (DHM-3), from bond homolysis of bi[3,5-dimethyl-5-(hydroxymethyl)-2-oxomorpholin-3-yl] (DHM-3 dimer), has also been developed (21, 22). These reducing agents are of particular interest because the redox chemistry with the anthracyclines is often quantitative, and the reagents can possibly be employed in vivo as well as in vitro. Both TM-3 dimer and DHM-3 dimer show low mouse toxicity.

Goals of our investigations continue to be the elucidation of reaction pathways which might be possible in vivo, especially ones that might provide differential cytotoxicity to tumor cells, and the development of reagents and protocols that might be useful for modulating the redox chemistry of bioreductively activated antitumor drugs to therapeutic advantage. This paper does not attempt to review the subject of anthracycline chemistry, biochemistry, and/or medicinal chemistry but to describe a useful one-electron reducing agent as a tool for elucidating possible in vivo redox processes. Excellent comprehensive reviews on various aspects of the anthracyclines are available (23–28).

## Captodative Radicals as One-Electron Reducing Agents

Captodative radicals are radicals that have electronic stabilization through the synergistic interaction of an electron-donating and an electron-withdrawing substituent (29, 30). Such radicals have also been called merostabilized (31) and push–pull stabilized (32). One of the best combinations of electron-donating and electron-withdrawing substituents is the amino and carboxyl functional groups. We have been particularly intrigued with the stability and properties of 2-oxomorpholin-3-yl radicals, which contain these two functional groups. Such radicals are best illustrated by 3,5,5-trimethyl-2-oxomorpholin-3-yl (TM-3), which also bears sterically stabilizing methyl substituents. Captodative resonance can be illustrated in valence bond terms by the zwitterionic resonance structure 7, among other structures.

2-Oxomorpholin-3-yl radical dimers are prepared in synthetically useful quantities by photoreduction of the corresponding 5,6-dihydro-1,4-oxazin-2-ones (33). For example, irradiation of the n–π* band of 5,6-dihydro-3,5,5-trimethyl-1,4-oxazin-2-one (8a) in 2-propanol solvent yields the TM-3 radical, which dimerizes to yield a mixture of the meso and dl isomers of TM-3 dimer. When the irradiation is performed at reduced temperature, the dimers precipitate and are collected in 50–70% yield, almost analytically pure, by filtration. The meso and dl isomers are readily separated by low-temperature flash or suction chromatography (34). Pure stereoisomers are useful for experiments in which kinetics are measured (vide supra).

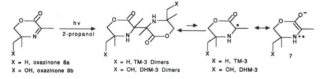

X = H, oxazinone 8a          X = H, TM-3 Dimers          X = H, TM-3
X = OH, oxazinone 8b         X = OH, DHM-3 Dimers        X = OH, DHM-3

TM-3 forms spontaneously in solution upon dissolving its dimers, meso- and dl-bi(3,5,5-trimethyl-2-oxomorpholin-3-yl), in solvents ranging from benzene to 10% methanol in water and is readily characterized from its relatively simple EPR spectrum. Three observations are consistent with the captodative nature of the electronic stabilization. The methyl EPR hyperfine coupling constant, which is a measure of the spin density localized at the 3-position, is solvent dependent, ranging from 13.7 G in ethanol to 18 G in benzene, and is linearly related to the free energy of bond homolysis (35). The equilibrium constant for bond homolysis is solvent dependent, ranging from $6 \times 10^{-16}$ M in benzene to $3 \times 10^{-11}$ M in ethanol at 25 °C, and the free energy is linearly related to the solvent Z parameter. Inductively electron-withdrawing substituents located at the 3-position facilitate bond homolysis, and inductively electron-withdrawing substituents located at the 5-position suppress bond homolysis (*36*). The solvent and substituent effects are consistent with the dipolar nature of the radical predicted by the captodative interaction. Polar solvents, especially hydrogen-bonding solvents, stabilize the radical more than the dimer, resulting in a larger equilibrium constant for bond homolysis. The solvent stabilization results in a larger contribution from the dipolar resonance structure 7, delocalizing the odd electron and lowering the hyperfine coupling constant. The favorable radical solvation energy more than compensates for the resulting charge separation.

A continuing question is what is the resonance energy for the captodative effect. The estimation is clouded by uncertainty with regard to models, to additivity of substituent effects on radicals with no captodative stabilization, to the effect of substituents on radical precursors, and to medium effects. Recent experimental (*37*) and theoretical (*38–40*) evaluations suggest that captodative radical stabilization is not general; however, it does appear to be in the range of 4–5 kcal/mol with carbonyl and amino substituents and will be affected by medium, especially protic solvent. Protic solvent probably magnifies the effect by an additional 3–4 kcal/mol in TM-3 as indicated by the effect of solvent on the bond dissociation energy.

The chemical property of the oxomorpholinyl radicals which is most interesting with respect to the anthracyclines is their ability to serve as selective one-electron reducing agents. Reductions which TM-3 will perform quantitatively include isatin (9) to isatide (10), methyl viologen (11) to its radical cation (12), diphenylpicrylhydrazyl (13) to diphenylpicrylhydrazine (14), Ag$^+$ to Ag$^0$, Fe$^{3+}$ to Fe$^{2+}$, benzil (15) to benzoin (16), di-*tert*-butyl nitroxide (17) to di-*tert*-butylhydroxylamine (18), molecular oxygen to hydrogen peroxide, 2-benzoyl-4,4-dimethyl-2-oxazoline (19) to 2-(hydroxyphenylmethyl)-4,4-dimethyl-2-oxazoline (20), and anthracyclines to their hydroquinones (Schemes III, IV, and VI) (*20, 34*). Disproportionation of TM-3 to form oxazinone 8a and 3,5,5-trimethyl-2-oxomorpholine (29), which is also a redox process, occurs very slowly at ambient temperature. Electrochemical reduction of the oxazinone 8a is irreversible and does not yield a thermodynamic reduction potential. Bracketing experiments place the $E^{\circ\prime}$ for the reduction of oxazinone 8a to TM-3 dimer in the range of −0.5 to −0.6 V vs the standard hydrogen electrode. Although oxomorpholinyl radicals react as reducing agents, this is not a general property of captodative radicals. In simplistic terms, oxomorpholinyl radicals bear the odd electron in a $\pi^*$ orbital, and consequently, transfer of the electron to a low-energy vacant orbital of an acceptor molecule is exothermic.

For many of the reactions, reduction is faster than recombination of TM-3 radicals, and consequently, the rate of reduction is the rate of bond homolysis. The rate of bond homolysis is a function of the stereoisomer, the solvent, and the pH. As reported in Table I, the rate constant for bond homolysis of the *dl* isomer varies from $3.4 \times 10^{-7}$ s$^{-1}$ in benzene to $3.4 \times 10^{-3}$ in methanol at 25 °C. Consequently, the half-life varies from 570 h in benzene to 200 s in methanol; in 10% methanol/90% water the half-life is approximately 50 s. Bond homolysis is inhibited at pH much less than 5 because at least one

## Scheme III. Reduction Reactions of TM-3 Dimer

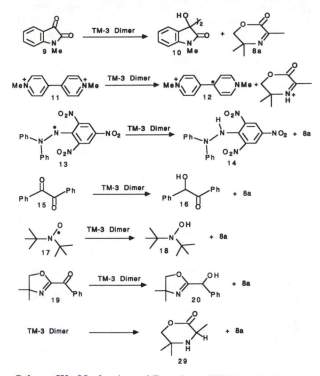

## Scheme IV. Mechanism of Reaction of TM-3 with Oxygen

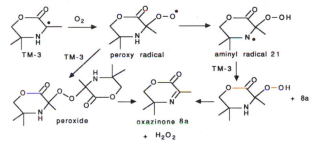

### Table I. Rate Constants for Bond Cleavage of *meso*- and *dl*-TM-3 Dimers

| stereoiso-mer | solvent | temp, °C | rate constant, s$^{-1}$ | half-life | ref |
|---|---|---|---|---|---|
| *dl* | CH$_3$OH | 25 | $3.4 \times 10^{-3}$ | 200 s | 50 |
| *dl* | CH$_3$OH | 20 | $1.9 \times 10^{-3}$ | 360 s | 50 |
| *dl* | CH$_3$OH | 15 | $9.0 \times 10^{-4}$ | 770 s | 50 |
| *dl* | CH$_3$OH | 5 | $2.6 \times 10^{-4}$ | 2700 s | 50 |
| *dl* | EtOH | 25 | $1.2 \times 10^{-3}$ | 580 s | 35 |
| *dl* | glyme | 25 | $7.8 \times 10^{-6}$ | 25 h | 35 |
| *dl* | CHCl$_3$ | 30 | $3.7 \times 10^{-6}$ | 52 h | 34 |
| *dl* | PhH | 71 | $2.3 \times 10^{-4}$ | 50 min | 35 |
| *dl* | PhH | 25 | $3.4 \times 10^{-7}$ | 570 h | 35 |
| meso | CH$_3$OH | 25 | $2.2 \times 10^{-3}$ | 315 s | unpublished |
| meso | CHCl$_3$ | 30 | $1.0 \times 10^{-5}$ | 19 h | 34 |

of the amine functions of the dimer is protonated and protonation eliminates the dative group.

The mechanism by which the oxomorpholinyl radicals serve as reducing agents is not completely established. Isotope effects on the radical and substituent effects on the substrate point to single electron transfer (*20*). Certainly for reduction of metal ions, electron transfer appears reasonable. Mechanistic studies of reduction of molecular oxygen also suggest a covalent mechanism with two pathways as shown in Scheme IV (*41*). Both pathways start with covalent bond formation between molecular oxygen and the 3-position of TM-3. One pathway involves subsequent intramolecular transfer of a hydrogen from the nitrogen to the peroxy radical with formation of the transient aminyl radical 21, characterized by a strong three-line EPR signal. The other pathway involves combination of the peroxy radical with a second TM-3 radical. In some cases, although not with TM-3, the intermediate peroxides have been isolated and characterized spectroscopically. The overall process is quantitative. TM-3 does not react rapidly, if at all, with hydrogen peroxide. The exact pathway by which TM-3 transfers electrons to daunomycin is not known; however, characteristic semiquinone and hydroquinone EPR and visible spectra, respectively, have been observed for these reduced states during the reduction, suggesting single electron transfer (42).

No evidence supports direct hydrogen atom transfer from TM-3 to a reducible substrate. A possible rationale is that direct hydrogen atom transfer correlates TM-3 with the $n,\pi^*$ excited state of oxazinone 8a. Correlation with an excited state is likely because the nitrogen of TM-3 is probably $sp^2$ hybridized to accommodate the resonance interaction, and consequently, the N–H bond lies in the $\sigma$-plane (Scheme V). Such an intended correlation would raise the activation energy for direct hydrogen atom transfer above that for electron transfer or covalent bond formation (43). For these latter mechanisms correlation is with a ground state.

### Scheme V. Rationale for Electron-Transfer Mechanism

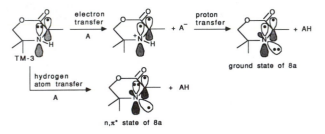

Because of our interest in conducting anthracycline redox chemistry in vivo with the oxomorpholinyl radicals (vide infra), a material more water soluble than TM-3 dimer was needed. After several possibilities were explored, the addition of a hydroxyl group to one of the methyls in the 5-position to achieve higher water solubility was pursued. The synthesis of bi[3,5-dimethyl-5-(hydroxymethyl)-2-oxomorpholin-3-yl] (DHM-3 dimer) was analogous to the synthesis of TM-3 dimer; however, the additional chiral centers resulted in the formation of six diastereoisomeric radical dimers, with one diastereoisomer predominating (21). Because separation of the stereoisomers is difficult and bond homolysis gives a single racemic radical from all six dimers, reductions are commonly performed with the mixture. The major diastereoisomer was isolated and the rate of bond homolysis measured. The rate constant for bond homolysis in methanol at 25 °C is $1.2 \times 10^{-3}$ s$^{-1}$ (half-life 580 s), and in water at 25 °C, $6.5 \times 10^{-3}$ s$^{-1}$ (half-life 110 s). The smaller rate constant relative to that for TM-3 dimer is consistent with the general effect of an inductively electron-withdrawing substituent at the 5-position (vide supra) (36). Although DHM-3 radical still reduces molecular oxygen to hydrogen peroxide, the reduction appears to be slower than the rate of bond homolysis for as yet unexplained reasons.

## Reduction of Daunomycin: Formation of a Quinone Methide versus a Semiquinone Methide

Reasonable circumstantial evidence exists for in vivo covalent bond formation between daunomycin and DNA (44, 45). The results of in vitro experiments also point to covalent binding upon reductive activation (46, 47). The mechanism for the bonding has been intensely investigated, and at present no clear mechanism has been established. As indicated earlier, both semiquinone methide (4) and quinone methide (5) transients have been proposed as reactive intermediates, the semiquinone methide resulting from glycosidic cleavage of the semiquinone anion and the quinone methide resulting from glycosidic cleavage of the hydroquinone anion. Reduction of the anthracyclines especially in methanol solvent with TM-3 dimer has proven to be an excellent method for generating the redox transients and probing their chemical reactivity.

Upon mixing 10 mol equiv of dl-TM-3 dimer with $1 \times 10^{-4}$ M daunomycin in freeze–pump–thaw degassed Tris-buffered methanol (apparent pH of 8) at 25 °C and scanning the visible absorption spectrum, destruction of daunomycin ($\lambda_{max}$ 480 nm) with initial formation of the hydroquinone anion 3a ($\lambda_{max}$ 420 nm) is apparent (Figure 1). The semiquinone anion 2a, although formed first, does not exist in sufficient concentration for visible spectroscopic observation in protic solvent; absorption at 382 and 510 nm has been reported for the semiquinone anion of daunomycinone in dimethylformamide solvent (48) and at 480 and 700 nm for the semiquinone of adriamycin in water (49). The semiquinone anion is readily observed by EPR spectroscopic monitoring of reaction of 1 mol equiv of TM-3 dimer with daunomcyin in methanol at 0 °C and characterized from the hyperfine coupling (50). A similar EPR spectrum has been observed upon chemical or enzymatic reduction of daunomycin in DMSO–water solvent (51). Semiquinone is never present in high concentration either because it is rapidly reduced by TM-3 radical or because it rapidly disproportionates to hydroquinone and quinone (49, 52). Soon after visible absorption by hydroquinone 3a appears (Figure 1a), it disappears with absorption by quinone methide 5a at 380 and 610 nm. The new absorption, which is substantially intense, was assigned to the quinone methide rather than the semiquinone methide because its intensity does not correlate with any new EPR signal, and the chemical reactivity is consistent with the structure. Although the structure for the quinone methide is commonly shown as a C-ring quinone methide (Scheme VI), it has never been fully characterized and a B-ring quinone methide is also possible (53). This experiment provided the first glimpse of the quinone methide predicted by the theory of bioreductive activation (1, 2). Further monitoring of the visible absorption (Figure 1b) showed the disappearance of the quinone methide bands with reappearance of a band at $\lambda_{max}$ 480 nm for formation of 7-deoxydaunomycinone (6a). The 7-deoxyaglycon results from solvent protonation at the 7-position; the process has been characterized by deuterium incorporation from deuterated protic solvent (54). Consequently, the quinone methide is simply a higher energy tautomer of the 7-deoxyaglycon. With excess reducing agent 7-deoxydaunomycinone is reduced to its hydroquinone, which also absorbs at 420 nm as shown in Figure 1c. Spectroscopic monitoring of a reaction mixture equimolar in daunomycin and dl-TM-3 dimer at $2 \times 10^{-4}$ M showed no buildup of a band for hydroquinone and clean isosbestic points for formation and destruction of quinone methide. With these conditions the kinetics of the reaction could be modeled with the consecutive first-

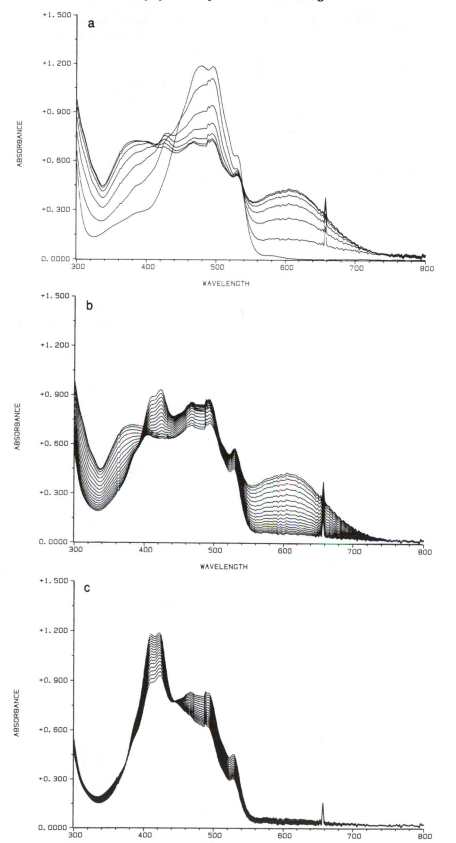

**Figure 1.** UV–vis spectral changes during reaction of a methanolic solution, $1.0 \times 10^{-4}$ M in daunomycin and $1.0 \times 10^{-3}$ M in TM-3 dimer, buffered to an apparent pH of 8 and thermostated at 25 °C.  Panel a shows scans every 10 s for the first 60 s; absorption at 480 nm is by daunomycin (**1a**), at 420 nm by daunomycin hydroquinone (**3a**), and at 380 and 610 nm by 7-deoxydaunomycinone quinone methide (**5a**).  Panel b shows scans from 70 to 240 s; absorption at 420 nm is by the hydroquinone of 7-deoxydaunomycinone, at 480 nm by 7-deoxydaunomycinone (**6a**), and at 380 and 610 nm by **5a**.  Panel c shows scans from 240 to 400 s with the bands assigned as in panel b.

**Scheme VI. Reductive Activation of Daunomycin with TM-3 Dimer**

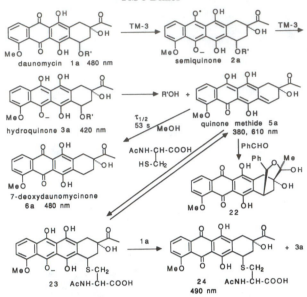

order rate law, with the two slow steps being bond homolysis of *dl*-TM-3 dimer and solvent protonation of quinone methide. The protonation occurred with a rate constant of $1.3 \times 10^{-2}$ s$^{-1}$, giving a half-life for the quinone methide of 53 s at 25 °C. The rate constant as a function of temperature gave an enthalpy of activation for protonation of 18 kcal/mol and an entropy of activation of −6 cal/(deg·mol) (50). Reduction in methanol-*d* solvent showed a substantial primary deuterium kinetic isotope effect of 9 with a half-life for the quinone methide of 490 s. Similar chemistry has been observed in aqueous medium; however, 7-deoxydaunomycinone is not water soluble, and precipitation inhibits the complete spectroscopic monitoring of the reaction (54). At 25 °C in aqueous medium the half-life of the quinone methide is 15 s. Reductions of adriamycin (**1b**) and 4-demethoxydaunomycin (idarubicin) occur similarly with small differences in the rate constants for protonation as shown in Table II (55).

**Table II. Rate Constants for Isomerization of Quinone Methides to 7-Deoxyaglycons**

| anthracycline | solvent | rate constant,$^a$ s$^{-1}$ | half-life, s | ref |
|---|---|---|---|---|
| 5-iminodaunomycin (38) | MeOH | 0.022 | 32 | 84 |
| daunomycin (**1a**) | H$_2$O | 0.046 | 15 | 42, 50 |
| daunomycin (**1a**) | MeOH | 0.013 | 53 | 50 |
| daunomycin (**1a**) | MeOD | 0.00142 | 490 | 50 |
| adriamycin (**1b**) | H$_2$O | 0.030 | 23 | 54 |
| adraimycin (**1b**) | MeOH | 0.011 | 63 | 55 |
| 4-demethoxydaunomycin | MeOH | 0.0086 | 86 | 55 |
| aclacinomycin A (25) | MeOH | 0.00089 | 780 | 66 |
| 11-deoxydaunomycin (11) | MeOH | 0.000095 | 7300 | 55 |

$^a$ The pH in water and the apparent pH in MeOH was 8, and the reaction temperature was 25 °C. Structures not given elsewhere are as follows:

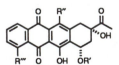

4-demethoxydaunomycin: R″ = OH; R‴ = H
11-deoxydaunomycin (11): R″ = H; R‴ = OMe
R′ as defined in Scheme I for both anthracyclines

Although the visible and EPR spectroscopic data are most consistent with glycosidic cleavage occurring at the hydroquinone state, with enzymatic, one-electron reduction in vivo at very low concentration, glycosidic cleavage could occur at the semiquinone state to give the semiquinone methide. Proving or disproving such a possibility remains a challenge. Results from recent experiments suggest an additional pathway to semiquinone methide (vide infra).

The structure of the quinone methide suggests a balance between reactivity as a nucleophile and as an electrophile. On the bottom side of the structure, the functionality is an unsaturated ketone, a Michael acceptor, and on the top side it is a nucleophilic and basic enol. Facile protonation to form the 7-deoxyaglycon confirms the base character; aldol reaction with the electrophiles benzaldehyde (56) and *p*-carboxybenzaldehyde (54) supports the nucleophilic character. The aldol product **22** from reaction with the *re* face predominates; the diastereomer of **22** has never been characterized. Reaction at the *re* face is possibly directed thermodynamically by the acetyl group at the 13-position through subsequent hemiacetal formation. To date, no biological electrophiles including the disulfide functional group have had sufficient reactivity to trap the quinone methide. The quinone methide also reacts with nucleophiles, most notably mercaptans such as *N*-acetyl-cysteine (57). The Michael addition at the 7-position yields the adduct in its hydroquinone state (**23**). Because the nucleophile is also a good leaving group, re-formation of the quinone methide also occurs. The adduct can be stabilized through subsequent oxidation to form the adduct in its quinone state (**24**). For in vitro experiments the oxidizing agent is commonly the starting quinone (54, 57). Consequently, the overall process is a chain reaction. In vivo, molecular oxygen might serve as the oxidizing agent. The concentration of molecular oxygen as a function of time is crucial because molecular oxygen also inhibits formation of the quinone methide. No nucleotides or nucleic acid bases have been sufficiently reactive to trap the quinone methide in vitro.

## Reduction of 11-Deoxyanthracyclines

Three anthracyclines missing the hydroxyl group in the 11-position are of interest with regard to understanding formation and reactivity of the anthracycline quinone methide. These are 11-deoxydaunomycin (**24**), aclacinomycin A (**25**), and 7-*con-O*-methylnogarol (menogaril, **26**) (58, 59). Aclacinomycin A and menogaril (60) are in various stages of clinical investigation. An important initial observation was formation of the 7,7′-aglycon dimer, bi-(7-deoxyaklavinon-7-yl) (**28**), as well as 7-deoxyaklavinone (**27**), upon anaerobic reduction of aclacinomycin A (Scheme VII) (61, 62). Aglycon dimer formation has been interpreted as evidence for the intermediacy of semiquinone methide from glycosidic cleavage at the semiquinone state (63, 64). Subsequently, aglycon dimers were characterized from reduction of 11-deoxydaunomycin (55) and menogaril (65). EPR and visible spectroscopic monitoring of the reduction of aclacinomycin A by TM-3 dimer in methanol as described for the reduction of daunomycin again revealed the formation of semiquinone and quinone methide ($\lambda_{max}$ 350, 550 nm) transients, respectively, but not semiquinone methide (66). The rate of disappearance of the quinone methide followed a combined first- and second-order rate law consistent with pseudo-first-order protonation to form **27** and second-order dimerization to form **28**. Actually the aglycon dimer is formed in a half quinone–half hydroquinone state from one quinone methide serving as a nucleophile and the other as an electrophile.

**Scheme VII. Reduction of Aclacinomycin A and Formation of Aglycon Dimer**

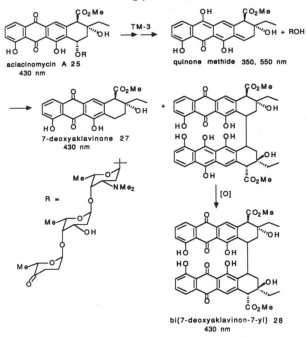

aclacinomycin A **25**
430 nm

quinone methide 350, 550 nm + ROH

7-deoxyaklavinone **27**
430 nm

bi(7-deoxyaklavinon-7-yl) **28**
430 nm

Upon admission of molecular oxygen the mixed-state dimer is rapidly oxidized to the bis-quinone **28**. These mechanistic results now suggest that aglycon dimer formation does not establish the intermediacy of semiquinone methide.

The behavior of menogaril (**26**) upon reduction differs in some respects. Initial inspection of the structure revealed the possibility that the quinone methide functionality could be formed on both sides of the molecule by elimination of the methoxy group at the 7-position and by breaking the C5'–O bond and that menogaril might function as a bioreductively activated cross-linking agent. However, no evidence for reductive cleavage of the C5'–O bond has been found. Menogaril semiquinone formed from reaction with TM-3 in methanol is never present in sufficient concentration for complete EPR characterization as has been accomplished for daunomycin and aclacinomycin semiquinones. A signal was observed, but only the benzylic hyperfine splittings were resolved. The primary difference in menogaril redox chemistry is that the hydroquinone ($\lambda_{max}$ 420 nm) and quinone methide ($\lambda_{max}$ 375, 604 nm) transients are much longer lived (*67*). In fact, the quinone methide formed in methanol solvent at an apparent pH of 8 is not protonated by the solvent to give 7-deoxynogarol (**30**) but only dimerizes to form bi(7-deoxynogarol-7-yl) (**31**) in the mixed redox state (Scheme VIII). Surprisingly, 7-deoxynogarol is formed as the major reduction product in vitro when an excess of reducing agent is employed. Under this condition, some materials are always present in hydroquinone states, and these catalyze the formation of 7-deoxynogarol from the quinone methide. The kinetics of decay of the quinone methide in the presence of excess reducing agent can actually be fit with an autocatalytic rate law with the product, 7-deoxynogarol (in its hydroquinone state), catalyzing formation of additional 7-deoxynogarol. The autocatalytic effect is probably only relevant to in vitro reductions. The second-order rate constants for dimerization of the quinone methides from reductive cleavage of aclacinomycin A, menogaril, and 11-deoxydaunomycin in methanol at 25 °C are 23, 11, and 1 $M^{-1}$ $s^{-1}$, respectively.

The presence of water greatly diminishes the observability of the quinone methide from reduction of menogaril. The predominant effect is upon the rate of formation from the hydroquinone. With water present or as the solvent, the rate of formation of the quinone methide from hydroquinone becomes much slower than the rates of dimerization and tautomerization. Aglycon dimer formation is still a major pathway in water with substoichiometric amounts of reducing agent, suggesting that the quinone methide, although not observable, is sufficiently long lived for bimolecular reactivity.

Aclacinomycin A and 11-deoxydaunomycin were the first anthracyclines with which nucleophilic addition of mercaptans to the quinone methides was observed (*68*), and with these quinone methides the yields of adducts were substantial. Trapping experiments with less nucleophilic materials such as nucleosides remain unsuccessful. The quinone methide from reduction of menogaril, because it is the longest lived, has recently been successfully trapped by the nitrogen nucleophiles imidazole (*69*) and 2'-deoxyguanosine (*70*) by using the chain process described earlier. With a large excess of 2'-deoxyguanosine present in aqueous medium, reduction with 20 mol % DHM-3 dimer yielded 17% of the adduct (**32**) along with 30% 7-deoxynogarol (**30**) and 23% bi(7-deoxynogarol-7-yl) (**31**) (Scheme IX). This successful trapping provided the first model reaction for possible reductive linkage of an anthracycline aglycon to DNA.

## Formation of Cyclomers from Reduction of Daunomycin Derivatives

As established by the results described above, the quinone methides from reductive cleavage of daunomycin and adriamycin are too reactive with protons from solvent to bind competitively with nitrogen and oxygen nucleophilic sites in small molecule models for biological macromolecules. In vivo, covalent binding by nucleophilic addition might be competitive if the quinone methide were generated in association with the macromolecule. This possibility has been tested by locating a nucleophile with propitious geometry within the daunomycin molecule through derivatization of the carbonyl at the 13-position (*71*). Oxygen and nitrogen nucleophiles were introduced through preparation of oxime (**33a**) and semicarbazone (**33b**) derivatives, respectively.

Anaerobic reduction of the oxime derivative **33a** in water at pH 9.0 with an excess of DHM-3 dimer yielded 10% 7-deoxydaunomycinone oxime (**34a**) and 90% 7,13-dideoxy-7,13-(epoxynitrilo)daunomycinone (cyclooxime, **35a**) upon aerobic workup. The ratio of **35a** to **34a** was pH dependent, ranging from 9/1 in methanol at apparent pH of 9.3 to 1/10 at apparent pH of 7.0. The pH dependency is consistent with the effect of pH on the availability of protons for formation of **34a** and the nucleophilicity of the oxime for formation of **35a**, both from the quinone methide **36a** (Scheme X).

Similar reduction of the semicarbazone derivative **33b** at pH 11 in aqueous medium containing 10% methanol with TM-3 dimer followed by aerobic workup yielded 20% 7-deoxydaunomycinone semicarbazone (**34b**), 50% 7,13-dideoxy-7,13-(iminocarbonyliminonitrilo)daunomycinone (cyclosemicarbazone, **35b**), and an unidentified product. Again the product ratio (**35b** to **34b**) was pH dependent, ranging from 3/1 at pH 11 to 1/5 at pH 8.2. The semicarbazone derivative **33b** bears two nucleophilic nitrogens and could have formed both three- and five-atom bridges to the 7-position of the quinone methide; the major pathway was to form the five-atom bridge.

**Scheme VIII. Reductive Activation of Menogaril with TM-3 Dimer**

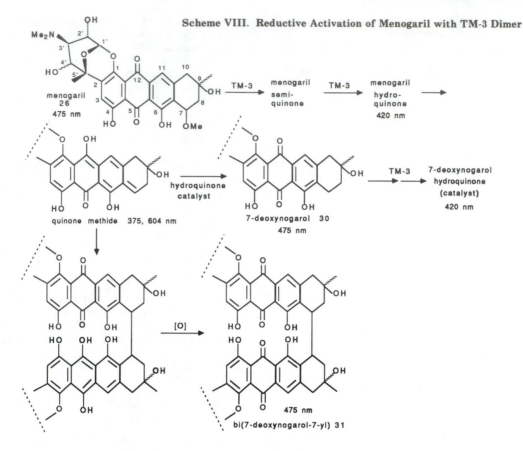

**Scheme IX. Reaction of 7-Deoxynogarol Quinone Methide with Guanosine**

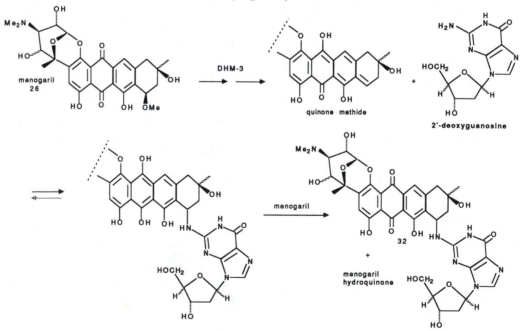

The possibility of equilibrium between the cyclomers in their hydroquinone states and the quinone methides **36a** and **36b** is of interest because of the biological importance of creating irreversible linkages. Reduction of cyclooxime **35a** in methanol at apparent pH of 8.2 at ambient temperature with TM-3 dimer yielded 60% of 7-deoxydaunomycinone oxime (**34a**) after 12 min and 96% of **34a** after 90 min. Similar reduction of cyclosemicarbazone (**35b**)

yielded only recovered **35b**. The irreversibility of cyclosemicarbazone formation may result from the higher stability of the 5-atom bridge and the poorer leaving ability of the semicarbazone substituent. Reductive cleavage has been discussed earlier in terms of stereoelectronic effects on the leaving group (72).

The electronic nature of the amino group of the semicarbazone is similar to the electronic nature of the 2-amino

## Scheme X. Intramolecular Trapping of the Quinone Methide

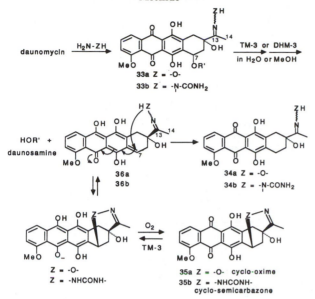

33a Z = -O-
33b Z = -N-CONH₂

36a
36b

34a Z = -O-
34b Z = -N-CONH₂

Z = -O-
Z = -NHCONH-

35a Z = -O- cyclo-oxime
35b Z = -NHCONH-
   cyclo-semicarbazone

group of guanosine. Consequently, the irreversible formation of the cyclosemicarbazone suggests the possibility of covalent bond formation between nucleic acids and daunomycin aglycon if the reactants are propitiously located at the time of formation of the quinone methide. The proper location might logically be daunomycin intercalated in DNA. Both daunomycin and adriamycin show high binding constants for intercalation (73). However, enzymatic reduction is inhibited by intercalation (74). Furthermore, even reduction of daunomycin intercalated in DNA with the small molecule reducing agent, carbon dioxide radical anion, does not lead to covalent bond formation but only to production of intercalated 7-deoxydaunomycinone (75). Reduction with carbon dioxide radical anion was performed anaerobically, and the lack of covalent binding may have resulted from the lack of an oxidizing agent such as molecular oxygen to prevent cleavage of the covalent bond at the hydroquinone state. Alternatively, the geometry of intercalated anthracycline may not be favorable for covalent bond formation.

## Formation of Leucodaunomycin, a Semistable Form of Daunomycin Hydroquinone

A problem associated with the proposal of the quinone methide as a reactive intermediate in covalent binding to

DNA is that the reductive activation may occur in a region of the cell remote from the DNA. The quinone methide may then be protonated by the medium to form the 7-deoxyaglycon before it can diffuse to the DNA. This would be particularly problematical with daunomycin and adriamycin, which form short-lived quinone methides. The chemistry that occurs upon reduction of daunomycin in aprotic medium and at lower pH suggests a solution.

Reduction of daunomycin with sodium dithionite in water at pH 3.5 gives, in addition to 7-deoxydaunomycinone, a mixture of four diastereoisomers which are tautomers of daunomycin hydroquinone (Scheme XI) (76). A similar result occurs when daunomycin is reduced with TM-3 dimer in acetonitrile solvent. These stereoisomers now bear the common name leucodaunomycin (37) because they are less colored than daunomycin. The process of tautomerization of the hydroquinone state was first observed in the reduction of 7-deoxydaunomycin to its hydroquinone (77). Leucodaunomycin is characterized by twin visible absorption bands at 420 and 440 nm. The leucodaunomycin isomers can be isolated and separated chromatographically. When they are redissolved in anaerobic pH 7 water, they form quinone methide as shown by the isolation of its tautomer 6a and by trapping with N-acetylcysteine to form 7-(N-acetyl-L-cystein-S-yl)-7-deoxydaunomycinone hydroquinone. Oxidation of the hydroquinone by daunomycin or by molecular oxygen yields 7-(N-acetyl-L-cystein-S-yl)-7-deoxydaunomycinone (24). The rates at which two of the leuco isomers yield 7-deoxydaunomycinone have been measured at 25 °C at pH 7.4. One of the isomers has a half-life of 530 s, which is 24 times the half-life of the quinone methide 5a. Possibly, one of the other isomers has a half-life even longer. Tautomerization of daunomycin hydroquinone to leucodaunomycin in competition with glycosidic cleavage at low pH or in an aprotic environment is consistent with the glycosidic cleavage occurring at a zwitterionic state and tautomerization occurring at a cationic state. The cationic center is the protonated amino group of the daunosamine.

A scenario that might explain covalent binding of daunomycin or adriamycin to DNA is as follows. The anthracycline is reduced enzymatically in one part of the cell where the appropriate enzymes reside. The environment, either acidic or hydrophobic, facilitates tautomerization to a relatively stable isomer of leucodaunomycin. Some preliminary results suggest that the pH necessary to achieve a leuco isomer may be as high as 6. The leucodaunomycin then diffuses to the DNA or other critical biological molecule and subsequently tautomerizes, with formation of the reactive quinone methide 5a. Covalent binding then proceeds. Interestingly, the environment of

## Scheme XI. Formation and Reaction of Leucodaunomycin

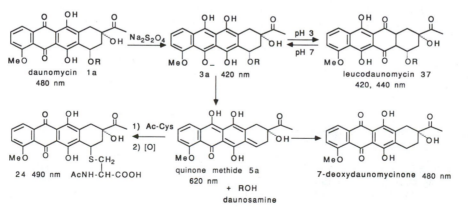

daunomycin 1a
480 nm

3a  420 nm

leucodaunomycin 37
420, 440 nm

24  490 nm
AcNH-CH-COOH

quinone methide 5a
620 nm
+ ROH
daunosamine

7-deoxydaunomycinone  480 nm

the tumor cell is in general more acidic than that of the normal cell (78). Possibly, the difference in pH of the two types of cells provides differential cytotoxicity.

## Reduction of 5-Iminodaunomycin

Because chronic cardiotoxicity is the predominant total dose limiting side effect of chemotherapy with adriamycin, significant effort has been devoted to elucidating the origin of the chronic cardiotoxicity. The prevailing theory implicates redox cycling and the production of reactive oxygen species. An important observation consistent with this theory is that 5-iminodaunomycin (38) is much less cardiotoxic than daunomycin and is not an efficient catalyst for the in vivo or in vitro production of reactive oxygen species (79–81). Inefficient catalysis of oxygen reduction is proposed to result from the more negative reduction potential and/or some difficulty in reoxidizing reduced 38. 5-Iminodaunomycin is a little more difficult to reduce than daunomycin; the half-wave potentials versus SCE are −0.67 and −0.64 V, respectively (82). However, 38 is reduced in vitro by NADPH cytochrome P-450 prepared from rat liver microsomes (83). Reduction with sodium dithionite in methanol solvent yields a quinone methide transient which has a half-life with respect to protonation to form 5-imino-7-deoxydaunomycinone (39) of 32 s, similar to the half-life of 5a (84). The reduction is suppressed by molecular oxygen analogously to the suppression of the reduction of daunomycin. The primary difference in the chemical reduction of 38 resides in the next phase, the reduction of 5-imino-7-deoxydaunomycinone. Reduction of 39 with dithionite leads to rapid deamination to form 2-acetyl-2,11-dihydroxy-7-methoxy-1,2,3,4-tetrahydro-5,12-naphthacenedione (40) (Scheme XII) (84, 85). Reductive deamination occurs with a half-life of only 17 s at 25 °C in methanol. Although naphthacenedione 40 is a quinone, a B-ring quinone, it is a poor catalyst for the production of reactive oxygen species. Reduction of 40 rapidly yields 8-acetyl-1-methoxy-7,9,10,12-tetrahydro-6,8,11-trihydroxy-5(8H)-naphthacenone (41), which reoxidizes with a half-life of 35 min. In contrast, the hydroquinone of 7-deoxydaunomycinone is relatively stable and is rapidly oxidized by molecular oxygen. Consequently, a rationale for the poor catalytic ability of 5-iminodaunomycin to produce reactive oxygen species is the facile reductive deamination in its 7-deoxy hydroquinone state.

Although production of reactive oxygen species has been implicated in the cardiotoxicity of adriamycin, it has also been implicated in at least part of the cytotoxicity to tumor cells. Correspondingly, 5-iminodaunomycin is less cardiotoxic and less toxic to some tumor cells. These observations together with the studies of chemical reduction suggest that part of the tumor cell toxicity of adriamycin and daunomycin resides in the ability of their 7-deoxyaglycons to serve as catalysts for the intracellular production of reactive oxygen species. As extracellular materials, the 7-deoxyaglycons are inactive, possibly because they are not easily transported. Another possibility that remains to be investigated is that 5-iminodaunomycin is unable to form a leuco isomer upon reduction at lower pH because of competitive deamination.

## Formation of a Semiquinone Methide from Daunomycin

As mentioned above, the earliest proposed mechanism for covalently linking the aglycon of daunomycin and adriamycin to DNA invoked the respective semiquinone

### Scheme XII. Reductive Activation and Deactivation of 5-Iminodaunomycin

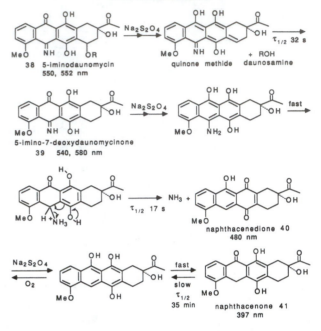

methides 4a and 4b. The in vitro studies reviewed here and those of others provide little or no evidence for formation of the semiquinone methide from reduction of the anthracyclines in one- or two-electron steps. Even the formation of aglycon dimers from reduction of the 11-deoxyanthracyclines, often thought to implicate semiquinone methide, occurs from reaction of the quinone methide. Semiquinone methide is theoretically an attractive transient for the coupling to macromolecules. It should be a stabilized radical which would efficiently combine irreversibly with radical sites in macromolecules rather than abstract hydrogen atoms. The radical sites might logically be created by H-atom abstraction by reactive oxygen species formed through anthracycline catalysis.

Investigation of the reaction of the quinone methide 5a with molecular oxygen has revealed a logical pathway to semiquinone methide (86). Significant concentrations of 5a are produced upon anaerobic reduction of daunomycin in methanol-d solvent with sodium dithionite. Admission of molecular oxygen results in almost instantaneous consumption of the quinone methide, as indicated by the color change from dark gray to orange. The products formed are 7-deoxydaunomycinone (6a), daunomycinone (42), 7-epidaunomycinone (43), 7-deoxy-7-ketodaunomycinone (44), 7-deoxy-7,13-epidioxydaunomycinol (45), and two diastereomers of bi(7-deoxydaunomycinon-7-yl) (46) (Scheme XIII). The major products are 6a, 45, and 46. The ratio of 46 to 42 + 43 + 44 + 45 is dependent upon the partial pressure of oxygen, increasing with decreasing partial pressure. The yield of daunomycinone plus epidaunomycinone significantly exceeds the yield of 7-deoxy-7-ketodaunomycinone, and the same products are formed when the quinone methide is generated in water. When quinone methide is generated by raising the pH of an aqueous solution of leucodaunomycin in the presence of air, the same products are formed except 46 (87).

7-Deoxydaunomycinone most likely results from solvent protonation of the quinone methide. Pathways that rationalize formation of the other products are shown in Scheme XIV. Quinone methide reacts by combination

**Scheme XIII. Reaction of Quinone Methide with Molecular Oxygen**

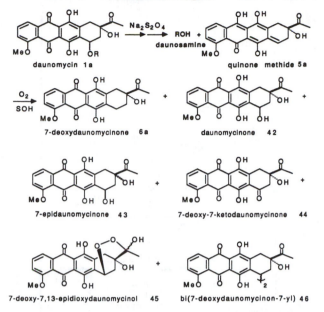

with molecular oxygen to form the semiquinone peroxy radical **47** and/or by electron transfer to molecular oxygen to form superoxide and the semiquinone methide **4a**. Reaction of **47** with another molecule of quinone methide gives a bis-semiquinone peroxide, **48**, which is at the proper redox state to fragment to daunomycinone and epidaunomycinone. Stepwise combination of two semiquinone methides with two molecules of oxygen yields the bis-quinone tetraoxide **49**, which would likely fragment to ketodaunomycinone and a mixture of daunomycinone and epidaunomycinone plus singlet or triplet molecular oxygen. Tetraoxides have been implicated as intermediates in hydrocarbon oxidations (*88*). Combination of semiquinone methide with superoxide produces diastereomeric quinone hydroperoxides (**50**), one with the proper stereochemistry to cyclize to epidioxydaunomycinol and the other which must fragment to ketodaunomycinone. Combination of two semiquinone methides is now the logical pathway to bi(7-deoxydaunomycinon-7-yl) (**46**) because **46** is not formed from quinone methide in the absence of oxygen. Variation in the relative yield of aglycon dimer as a

function of oxygen partial pressure and the quinone methide concentration results from the competition between semiquinone methide combining with another semiquinone methide or with molecular oxygen. Aglycon dimer is not observed when quinone methide is produced from leucodaunomycin because the concentration of the quinone methide and, correspondingly, semiquinone methide is never high.

With these observations, semiquinone methide reemerges as a contender for the transient involved in covalent binding to biological macromolecules. The scenario begins with radical sites in the macromolecules being created by reactive oxygen species formed through anthracycline catalysis. Upon reduction of the oxygen concentration, competitive reductive glycosidic cleavage occurs to form quinone methide. Quinone methide reacts with the available oxygen to form semiquinone methide, which then combines with radical sites in the macromolecule. The bond formed is likely a carbon–carbon bond which will not cleave upon subsequent reduction of the attached aglycon. Further catalytic production of reactive oxygen species by the attached aglycon will cleave the macromolecule.

Unless the redox enzymes are in the vicinity of the target macromolecule, the probability of the above scenario leading to efficient covalent bond formation is low. The problem with radicals, probably including semiquinone methide, is that they do not diffuse very far. TM-3 and DHM-3 are exceptions. Of course, leucodaunomycin could serve as a protected source of semiquinone methide and be formed in a region of the cell remote from the macromolecular target. Also radical sites in the target could be formed by production of superoxide in one region of the cell, diffusion of superoxide to the target, and Haber–Weiss chemistry in the vicinity of the target. A cartoon illustrating this possibility is shown in Scheme XV. Another possibility which remains to be explored is that the actual cytotoxic species is the quinone hydroperoxide **50**, epidioxydaunomycinol **45**, or some other anthracycline peroxide. These species bear functionality which can both create a radical site and bind to it. For example, Fe(II)-catalyzed decomposition of **50** might yield a hydroxyl radical for H-atom abstraction and 7-deoxydaunomycinon-7-yloxy radical for combination with the radical site in the macromolecule.

As mentioned earlier, cytotoxicity may require more than covalent binding of the aglycon to a macromolecule. A possibility is subsequent cutting of the macromolecule.

**Scheme XIV. Proposed Mechanism for Reaction of Quinone Methide with Molecular Oxygen**

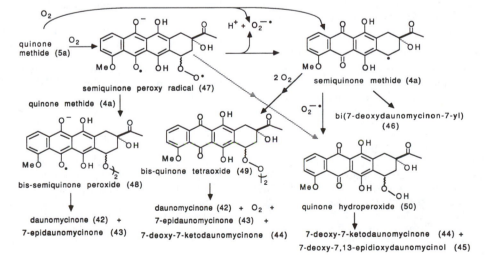

**Scheme XV. Logistics of Covalent Bond Formation to Macromolecule and Subsequent Cutting**

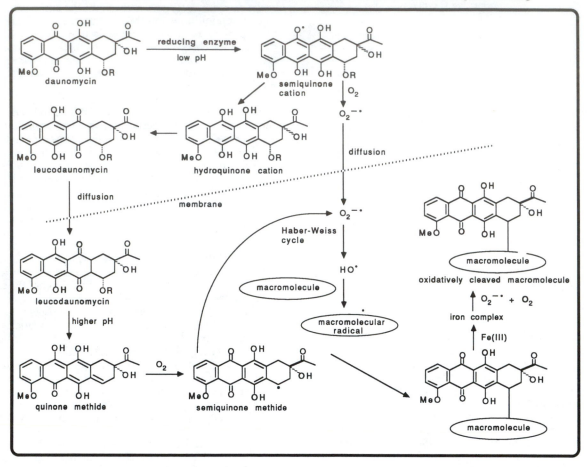

If reducing enzymes are not available at the site of the macromolecule, aglycon-catalyzed redox recycling will not occur. However, the aglycon is known to be a strong chelating agent for iron, competitive with iron storage proteins (89). Consequently, the purpose of covalent binding may only be to locate an iron cation adjacent to the macromolecule. The hydrogen peroxide which diffuses to the macromolecule will then be efficiently converted to hydroxyl radical for subsequent cutting of the macromolecule (90). The hydrogen peroxide would be produced by other anthracyclines in the reducing region of the cell. This scenario also relates to an aspect of the chemistry of 5-iminodaunomycin and its lesser cytotoxicity. Although 5-iminodaunomycin forms an Fe(III) complex, the complex is less efficient at producing hydroxyl radical from hydrogen peroxide (91).

### Utilization of Captodative Radicals in Vivo

As already mentioned several times, a problem with bioreductive activation as a mechanism for cytotoxicity is that reducing enzymes are bulky and are probably not located sufficiently near target molecules for efficient reaction of the transients with the targets. This raises the possibility of in vivo activation with a chemical reducing agent such as DHM-3 dimer. DHM-3 dimer is relatively nontoxic in mice when administered both intraperitoneally and intravenously as single bolus injections (92). The first application explored was the use of TM-3 and DHM-3 dimers as antidotes for the anthracyclines because they reduce the anthracyclines to the inactive and nontoxic 7-deoxyaglycons (17). The goal was to effect extracellular reductive cleavage because intracellularly, the aglycons

may still be cytotoxic. The initial application was with regard to the medical problem of extravasation necrosis (93). If intravenous administration results in some leakage around the catheter, significant necrosis occurs. The model employed was the swine model because of the similarity with human skin. When DHM-3 dimer was introduced 15 min after a subcutaneous injection of adriamycin, the total toxicity was reduced by 95%. Subsequent to administration of DHM-3 dimer, precipitation of 7-deoxyadriamycinone was apparent. DHM-3 dimer was also effective at reducing the skin toxicity of daunomycin, 4-demethoxydaunomycin, 4'-epidoxorubicin, 5-iminodaunomycin, aclacinomycin A, menogaril, and mitomycin C (94). Mitomycin C is another quinone type antitumor drug which is bioreductively activated; with this drug, covalent binding to DNA has been unambiguously established (95). If the time before administration of the antidote exceeded an hour, the antidote was ineffective. No other antidote to quinone antitumor drug extravasation necrosis has proven to be as effective as DHM-3 dimer in animal studies.

Next, the concept of adriamycin chemotherapy modulated with DHM-3 dimer was explored (96). Mice were given L1210 tumor intraperitoneally on day 0, 25 mg/kg of body weight adriamycin on day 1, and 50 mg/kg DHM-3 dimer 15 min later, both intraperitoneally. The dose of adriamycin was a lethal intraperitoneal dose in the absence of DHM-3 dimer. This protocol resulted in a 5 times improvement in therapeutic index with greater than 70% long-term survivors versus the optimum sublethal dose of adriamycin without modulation with DHM-3 dimer. In general, sublethal doses of adriamycin with DHM-3 modulation were ineffective. Injection of the DHM-3 dimer more than 30 min after the adriamycin was also ineffective. A control experiment showed that the DHM-3 dimer itself is not an active antitumor drug. Heart, lung, liver, and kidney tissues of animals receiving adriamycin plus DHM-3 dimer showed lower levels of adriamycin and higher levels of adriamycin aglycons relative to animals receiving only adriamycin. Clearly, the result of this experiment poses more questions than it answers. Of particular relevance to this discussion, does the improved therapeutic index involve reductive activation of adriamycin by DHM-3? Does DHM-3 dimer even pass through cell membranes? An explanation of the improved therapeutic index may simply be that the modulated therapy achieved higher intratumor levels of the adriamycin because the tumor was localized in the peritoneal cavity and that the excess toxic adriamycin never left the peritoneal cavity because it was reductively cleaved to its 7-deoxyaglycon by DHM-3. Even if this is the explanation, techniques for achieving higher intratumor levels of anthracyclines without increasing toxic side effects may be of significant value in chemotherapy.

Further progress with in vivo experiments depends upon developing drugs and methodology for intravenous administration of modulating and/or antidote materials. A major problem will likely be the deactivation of drugs such as DHM-3 by molecular oxygen and/or hemoglobin-bound molecular oxygen (vide supra). Thermodynamically, reduction of molecular oxygen is easier than reduction of the anthracyclines. Consequently, the problem must be solved kinetically.

## Summary

The oxomorpholinyl radicals are unique materials in organic and medicinal chemistry. Their closest parallel lies in inorganic chemistry with dithionite, which exists in equilibrium with sulfur dioxide radical anion, also a one-electron reducing agent (97). However, dithionite is a more powerful reducing agent and is probably more toxic. The rate of release of the oxomorpholinyl radical from its dimer is medium and is structure dependent, which provides for some level of control. The oxomorpholinyl radicals TM-3 and DHM-3 are selective one-electron reducing agents for the anthracyclines, generating sequentially semiquinone and hydroquinone redox states. Formation of the reduced states of the anthracyclines is probably relevant to their cytotoxic activity. Semiquinones and hydroquinones react rapidly with molecular oxygen to yield superoxide. Hydroquinone redox states with anaerobic conditions in protic media at pH 7–8 undergo glycosidic cleavage to form quinone methides; in aprotic media or at pH less than 4, they tautomerize to leuco forms. Quinone methides react with protons from solvent to form 7-deoxyaglycons, with some nucleophiles to form adducts, and with molecular oxygen to form semiquinone methide. The reactivity of the quinone methide is a function of substitution; nucleophilic addition is facilitated by the absence of a hydroxyl group at the 11-position and by proper location of the nucleophile. Quinone methides and semiquinone methides are both viable transients for covalently linking anthracycline aglycons to biological macromolecules. DHM-3 dimer is of possible pharmaceutical value for the detoxification of quinone antitumor drugs and for the improvement of chemotherapy through modulating the redox chemistry of the quinone antitumor drugs.

**Acknowledgment**. We gratefully acknowledge financial support from the USPHS in the form of Grant CA 24665, from the NSF in the form of Grant CHE-8903637, and from the University of Colorado Council on Research and Creative Work in the form of a Faculty Fellowship. We thank Barbara Schweitzer for preparing Figure 1.

## References

(1) Moore, H. W. (1977) Bioactivation as a model for drug design bioreductive alkylation. *Science* **197**, 527–532.
(2) Moore, H. W., and Czerniak, R. (1981) Naturally occurring quinones as potential bioreductive alkylating agents. *Med. Res. Rev.* **1**, 249–280.
(3) Bachur, N. R., Gee, M. V., and Friedman, R. D. (1982) Nuclear catalyzed antibiotic free radical formation. *Cancer Res.* **42**, 1078–1081.
(4) Fisher, J., Ramakrishnan, K., and Becvar, J. E. (1983) Direct enzyme-catalyzed reduction of anthracyclines by reduced nicotinamide adenine dinucleotide. *Biochemistry* **22**, 1347–1355.
(5) Bachur, N. R., Gordon, S. L., and Gee, M. V. (1977) Anthracycline antibiotic augmentation of microsomal electron transport and free radical formation. *Mol. Pharmacol.* **13**, 901–910.
(6) Bachur, N. R., Gordon, S. L., and Gee, M. V. (1978) A general mechanism for microsomal activation of quinone anticancer agents to free radicals. *Cancer Res.* **38**, 1745–1750.
(7) Lown, J. W., Chen, H.-H., Plambeck, J. A., and Acton, E. M. (1982) Further studies on the generation of reactive oxygen species from activated anthracylines and the relationship to cytotoxic action and cardiotoxic effects. *Biochem. Pharmacol.* **31**, 575–581.
(8) Gianni, L., Corden, B. J., and Myers, C. E. (1983) The biochemical basis of anthracycline toxicity and antitumor activity. *Rev. Biochem. Toxicol.* **5**, 1–82.
(9) Lown, J. W., and Chen, H.-H. (1981) Evidence for the generation of free hydroxyl radicals from certain quinone antitumor antibiotics upon reductive activation in solution. *Can. J. Chem.* **59**, 390–395.
(10) Doroshow, J. H. (1983) Effect of anthracyline antibiotics on oxygen radical formation in rat heart. *Cancer Res.* **43**, 460–472.
(11) Doroshow, J. H. (1986) Role of hydrogen peroxide and hydroxyl radical formation in the killing of Ehrlich tumor cells by anticancer quinones. *Proc. Natl. Acad. Sci. U.S.A.* **83**, 4514–4518.
(12) Alegria, A. E., Samuni, A., Mitchell, J. B., Riesz, P., and Russo, A. (1989) Free radicals induced by adriamycin-sensitive and adriamycin-resistant cells: a spin trapping study. *Biochemistry* **28**, 8653–8658.
(13) Speyer, J. L., Green, M. D., Kramer, E., Rey, M., Sanger, J., Ward, C., Dubin, N., Ferrans, V., Stecy, P., Zeleniuch-Jacquotte, A., Wernz, J., Feit, F., Slater, W., Blum, R., and Muggia, F. M. (1988) Protective effect of the bispiperazinedione ICRF-187 against doxorubicin-induced cardiac toxicity in women with advanced breast cancer. *N. Engl. J. Med.* **319**, 745–752.
(14) Vile, G. F., and Winterbourn, C. C. (1990) *dl*-N,N'-Dicarboxamidomethyl-N,N'-dicarboxymethyl-1,2-diaminopropane (ICRF-198) and *d*-bis(3,5-dioxopiperazine-1-yl)propane (ICRF-187) inhibition of Fe$^{3+}$ reduction, lipid peroxidation, and CaATPase inactivation in heart microsomes exposed to adriamycin. *Cancer Res.* **50**, 2307–2310.
(15) Liu, L. F. (1989) DNA topoisomerase poisons as antitumor drugs. *Annu. Rev. Biochem.* **58**, 351–375.
(16) Tritton, T. R., and Yee, G. (1982) The anticancer agent adriamycin can be actively cytotoxic without entering cells. *Science* **217**, 248–250.
(17) Banks, A. R., Jones, T., Koch, T. H., Friedman, R. D., and Bachur, N. R. (1983) Prevention of adriamycin toxicity. *Cancer Chemother. Pharmacol.* **11**, 91–93.
(18) Koch, T. H., Olesen, J. A., and DeNiro, J. (1975) An unusually weak carbon–carbon single bond. *J. Am. Chem. Soc.* **97**, 7285–7288.
(19) Bennett, R. W., Wharry, D. L., and Koch, T. H. (1980) Formation kinetics of an amino carboxy type merostabilized free radical. *J. Am. Chem. Soc.* **102**, 2345–2349.
(20) Burns, J. M., Wharry, D. L., and Koch, T. H. (1981) Electron-transfer chemistry of the merostabilized 3,5,5-trimethyl-2-morpholinon-3-yl radical. *J. Am. Chem. Soc.* **103**, 849–856.
(21) Gaudiano, G., and Koch, T. H. (1987) Bi[3,5-dimethyl-5-(hydroxymethyl)-2-oxomorpholin-3-yl](DHM-3 dimer). A water-soluble, one-electron reducing agent. *J. Org. Chem.* **52**, 3073–3081.
(22) Averbuch, S. D., Bachur, N. R., Gaudiano, G., and Koch, T. H. (Dec 30, 1986) An Oxomorpholinyl Dimer and Rescue of Anthracycline Damage. U.S. Patent Number 4,632,922.

(23) Arcamone, F. (1981) *Doxorubicin Anticancer Antibiotics*, Academic Press, New York.

(24) Abdella, B. R. J., and Fisher, J. (1985) A chemical perspective on the anthracycline antitumor antibiotics. *Environ. Health Perspect.* **64**, 3–18.

(25) Powis, G. (1987) Metabolism and reactions of quinoid anticancer agents. *Pharmacol. Ther.* **35**, 57–162.

(26) Fisher, J. F., and Aristoff, P. A. (1988) The chemistry of DNA modification by antitumor antibiotics. *Prog. Drug Res.* **32**, 411–498.

(27) Lown, J. W., Ed. (1988) *Anthracyclines and Anthracenedione-Based Anti-Cancer Agents*, Elsevier, Amsterdam.

(28) Powis, G. (1989) Free radical formation by antitumor quinones. *Free Radicals Biol. Med.* **6**, 63–101.

(29) Viehe, H. G., Merényi, R., Stella, L., and Janousek, Z. (1979) Captodative substituent effects in syntheses with radicals and radicophiles. *Angew. Chem., Int. Ed. Engl.* **18**, 917–932.

(30) Viehe, H. G., Janousek, Z., Merényi, R., and Stella, L. (1985) The captodative effect. *Acc. Chem. Res.* **12**, 148–154.

(31) Baldock, R. W., Hudson, P., and Katritzky, A. R. (1974) Stable free radicals. Part I. New principle governing the stability of organic free radicals. *J. Chem. Soc., Perkin Trans. 1*, 1422–1427.

(32) Balaban, A. T., Caproiu, M. T., Negoita, N., and Baican, R. (1977) Factors affecting stability and equilibrium of free radicals—IX. Non-equivalence of aryl groups in 1,1-diaryl-2-benzenesulfonylhydrazyls and related compounds. *Tetrahedron* **33**, 2249–2253.

(33) Koch, T. H., Olesen, J. A., and DeNiro, J. (1975) Photochemical reactivity of imino lactones. Photoreduction and photoelimination. *J. Org. Chem.* **40**, 14–19.

(34) Bennett, R. W., Wharry, D. L., and Koch, T. H. (1980) Formation kinetics of an amino carboxy type merostabilized free radical. *J. Am. Chem. Soc.* **102**, 2345–2349.

(35) Olson, J. B., and Koch, T. H. (1986) Kinetic and thermodynamic parameters for the formation of 3,5,5-trimethyl-2-oxomorpholin-3-yl (TM-3). A negative activation energy for radical combination. *J. Am. Chem. Soc.* **108**, 756–761.

(36) Himmelsbach, R. J., Barone, A. D., Kleyer, D. L., and Koch, T. H. (1983) Substituent effects on the formation of aminocarboxy-type captodative free radicals. *J. Org. Chem.* **26**, 2989–2994.

(37) Bordwell, F. G., and Lynch, T.-Y. (1989) Radical stabilization energies and synergistic (captodative) effects. *J. Am. Chem. Soc.* **111**, 7558–7562.

(38) Leroy, G., Peeters, D., Sana, M., and Wilante, C. (1986) A theoretical approach to substituent effects in radical chemistry. In *Substituent Effects in Radical Chemistry* (Viehe, H. G., Janousek, Z., and Merényi, R., Eds.) pp 1–48, Reidel, Boston.

(39) Katritzky, A. R., Zerner, M. C., and Karelson, M. M. (1986) A quantitative assessment of the merostabilization energy of carbon-centered radicals. *J. Am. Chem. Soc.* **108**, 7213–7214.

(40) Pasto, D. J. (1988) Radical stabilization energies of disubstituted methyl radicals. A detailed analysis of the captodative effect. *J. Am. Chem. Soc.* **110**, 8164–8175.

(41) Gaudiano, G., and Koch, T. H. (1986) Oxidation of 3,5,5-trimethyl-2-oxomorpholin-3-yl (TM-3) with molecular oxygen. Generation of a persistent aminyl radical. *J. Am. Chem. Soc.* **108**, 5014–5015.

(42) Kleyer, D. L., and Koch, T. H. (1983) Spectroscopic observation of the tautomer of 7-deoxydaunomycinone from elimination of daunosamine from daunomycin hydroquinone. *J. Am. Chem. Soc.* **105**, 2504–2505.

(43) Dauben, W. G., Salem, L., and Turro, N. J. (1975) A classification of photochemical reactions. *Acc. Chem. Res.* **8**, 41–54.

(44) Kacinski, B. M., and Rupp, W. D. (1984) Interaction of the UVRABC endonuclease *in vivo* and *in vitro* with DNA damage produced by antineoplastic anthracyclines. *Cancer Res.* **44**, 3489–3492.

(45) Lambert, B., Laugaa, P., Roques, B., and LePecq, J. (1986) Cytotoxicity and SOS-inducing ability of ethidium and photoactivable analogs on *E. coli* ethidium-bromide-sensitive (Ebs) strains. *Mutat. Res.* **166**, 243–254.

(46) Sinha, B. K., and Gregory, J. L. (1981) Role of one-electron and two-electron reduction products of adriamycin and daunomycin in DNA binding. *Biochem. Pharmacol.* **30**, 2622–2629.

(47) Sinha, B. K., Trush, M. A., Kennedy, K. A., and Mimnaugh, E. G. (1984) Enzymatic activation and binding of adriamycin to nuclear DNA. *Cancer Res.* **44**, 2892–2896.

(48) Anne, A., and Moiroux, J. (1985) One-electron and two-electron reductions of daunomycin. *Nouv. J. Chim.* **9**, 83–89.

(49) Land, E. J., Mukherjee, T., Swallow, A. J., and Bruce, J. M. (1985) Possible intermediates in the action of adriamycin. A pulse radiolysis study. *Br. J. Cancer* **51**, 515–523.

(50) Kleyer, D. L., and Koch, T. H. (1984) Mechanistic investigation of reduction of daunomycin and 7-deoxydaunomycinone with bi(3,5,5-trimethyl-2-oxomorpholin-3-yl). *J. Am. Chem. Soc.* **106**, 2380–2387.

(51) Schreiber, J., Mottley, C., Sinha, B. K., Kalyanaraman, B., and Mason, R. P. (1987) One-electron reduction of daunomycin, daunomycinone, and 7-deoxydaunomycinone by the xanthine/xanthine oxidase system: detection of semiquinone free radicals by electron spin resonance. *J. Am. Chem. Soc.* **109**, 348–351.

(52) Svingen, B. A., and Powis, G. (1981) Pulse radiolysis studies of antitumor quinones: radical lifetimes, reactivity with oxygen, and one-electron reduction potentials. *Arch. Biochim. Biophys.* **209**, 119–126.

(53) Houée-Levin, C., Gardés-Albert, M., and Ferradine, C. (1986) Pulse radiolysis study of daunorubicin redox reactions: redox cycles or glycosidic cleavage? *J. Free Radicals Biol. Med.* **2**, 89–97.

(54) Fisher, J., Abdella, B. R. J., and McLane, K. E. (1985) Anthracycline antibiotic reduction by spinach ferredoxin-NADP$^+$ reductase and ferredoxin. *Biochemistry* **24**, 3562–3571.

(55) Boldt, M., Gaudiano, G., and Koch, T. H. (1987) Substituent effects on the redox chemistry of anthracycline antitumor drugs. *J. Org. Chem.* **52**, 2146–2153.

(56) Kleyer, D. L., and Koch, T. H. (1983) Electrophilic trapping of the tautomer of 7-deoxydaunomycinone. A possible mechanism for covalent binding of daunomycin to DNA. *J. Am. Chem. Soc.* **105**, 5154–5155.

(57) Ramakrishnan, K., and Fisher, J. (1986) 7-Deoxydaunomycinone quinone methide reactivity with thiol nucleophiles. *J. Med. Chem.* **29**, 1215–1221.

(58) Wiley, P. F., Kelly, R. B., Caron, E. L., Wiley, V. H., Johnson, J. H., Mackeller, F. A., and Mizsak, S. A. (1977) Structure of nogalamycin. *J. Am. Chem. Soc.* **99**, 542–549.

(59) Arora, S. K. (1983) Molecular structure, absolute stereochemistry, and interaction of nogalamycin, a DNA-binding anthracycline antitumor antibiotic. *J. Am. Chem. Soc.* **105**, 1328–1332.

(60) Dodion, P., Sessa, C., Joss, R., Crespeigne, N., Willems, Y., Kitt, M., Abrams, J., Finet, C., Brewer, J. E., Adams, W. J., Earhart, R. H., Roxencweig, M., Kenis, Y., and Cavalli, F. (1986) Phase I study of intravenous menogaril administered intermittently. *J. Clin. Oncol.* **4**, 767–774.

(61) Oki, T., Komiyama, T., Tone, H., Inui, T., Takeuchi, T., and Umezawa, H. (1977) Reductive cleavage of anthracycline glycosides by microsomal NADPH–cytochrome C reductase. *J. Antibiot.* **30**, 613–615.

(62) Komiyama, T., Oki, T., Inui, T., Takeuchi, T., and Umezawa, H. (1979) Metabolism of aclacinomycin A. Part II. Reduction of anthracyline glycoside by NADPH–cytochrome P-450 reductase. *Gann* **70**, 403–410.

(63) Komiyama, T., Oki, T., and Inui, T. (1979) A proposed reaction mechanism for the enzymatic reductive cleavage of glycosidic bond in anthracycline antibiotics. *J. Antibiot.* **32**, 1219–1222.

(64) Berg, H., Horn, G., Jacob, H.-E., Fiedler, U., Luthardt, U., and Tresselt, D. (1986) Polarographic modelling of metabolic processes of cancerostatic anthracyclines. *Bioelectrochem. Bioenerg.* **16**, 135–148.

(65) Boldt, M., Gaudiano, G., Haddadin, M. J., and Koch, T. H. (1988) Formation and autocatalytic destruction of the quinone methide from reductive cleavage of menogaril. *J. Am. Chem. Soc.* **110**, 3330–3332.

(66) Kleyer, D. L., Gaudiano, G., and Koch, T. H. (1984) Spectroscopic and kinetic evidence for the tautomer of 7-deoxyaklavinone as an intermediate in the reductive coupling of aclacinomycin A. *J. Am. Chem. Soc.* **106**, 1105–1109.

(67) Boldt, M., Gaudiano, G., Haddadin, M. J., and Koch, T. H. (1989) Formation and reaction of the quinone methide from reductive cleavage of the antitumor drug menogaril. *J. Am. Chem. Soc.* **111**, 2283–2292.

(68) Ramakrishnan, K., and Fisher, J. (1983) Nucleophilic trapping of 7,11-dideoxyanthracyclinone quinone methides. *J. Am. Chem. Soc.* **105**, 7187–7188.

(69) Gaudiano, G., Egholm, M., Haddadin, M. J., and Koch, T. H. (1989) Trapping of the quinone methide from reductive cleavage of menogaril with the nitrogen nucleophile imidazole. *J. Org. Chem.* **54**, 5090–5093.

(70) Egholm, M., and Koch, T. H. (1989) Coupling of the anthracycline antitumor drug menogaril to 2'-deoxyguanosine through reductive activation. *J. Am. Chem. Soc.* 111, 8291–8293.

(71) Gaudiano, G., Frigerio, M., Bravo, P., and Koch, T. H. (1990) Intramolecular Trapping of the Quinone Methide from Reductive Cleavage of Daunomycin with Oxygen and Nitrogen Nucleophiles. *J. Am. Chem. Soc.* 112, 6704–6709.

(72) Malatesta, V., Penco, S., Sacchi, N., Valentini, L., Vigevani, A., and Arcamone, F. (1984) Electrochemical deglycosidation of anthracyclines: stereoelectronic requirements. *Can. J. Chem.* 62, 2845–2850.

(73) Byrn, S. R., and Dolch, G. D. (1978) Analysis of binding of daunorubicin and doxorubicin to DNA using computerized curve-fitting procedures. *J. Pharm. Sci.* 67, 688–693.

(74) Youngman, R. J., and Elstner, E. F. (1984) On the interaction of adriamycin with DNA. Investigation of spectral changes. *Arch. Biochem. Biophys.* 231, 424–429.

(75) Rouscilles, A., Houée-Levin, C., Gardés-Albert, M., and Ferradine, C. (1989) γ-Radiolysis study of the reduction by COO⁻ free radicals of daunorubicin intercalated in DNA. *Free Radicals Biol. Med.* 6, 37–43.

(76) Bird, D. M., Boldt, M., and Koch, T. H. (1989) Leucodaunomycin, a tautomer of daunomycin hydroquinone. *J. Am. Chem. Soc.* 111, 1148–1150.

(77) Brand, D. J., and Fisher, J. (1986) Tautomeric instability of 10-deoxydaunomycinone hydroquinone. *J. Am. Chem. Soc.* 108, 3088–3096.

(78) Vaupel, P., Kallinowski, F., and Okunieff, P. (1989) Blood flow, oxygen and nutrient supply, and metabolic microenvironment of human tumors: a review. *Cancer Res.* 49, 6449–6465.

(79) Lown, J. W., Chen, H.-H., Plambeck, J. A., and Acton, E. M. (1979) Diminished superoxide anion generation by reduced 5-iminodaunorubicin relative to daunorubicin and the relationship to cardiotoxicity of the anthracycline antitumor agents. *Biochem. Pharmacol.* 28, 2563–2568.

(80) Davies, K. J. A., Doroshow, J. H., and Hochstein, P. (1983) Mitochondrial NADH dehydrogenase-catalyzed oxygen radical production by adriamycin, and the relative inactivity of 5-iminodaunorubicin. *FEBS Lett.* 153, 227–230.

(81) Jensen, R. A., Acton, E. M., and Peters, J. H. (1984) Electrocardiographic and transmembrane potential effects of 5-iminodaunorubicin in the rat. *Cancer Res.* 44, 4030–4039.

(82) Lown, J. W., Chen, H.-H., Plambeck, J. A., and Acton, E. M. (1982) Further studies on the generation of reactive oxygen species from activated anthracyclines and the relationship to cytotoxic action and cardiotoxic effects. *Biochem. Pharmacol.* 31, 575–581.

(83) Bachur, N. R., Gordon, S. L., Gee, M. V., and Kon, H. (1979) NADPH cytochrome P-450 reductase activation of quinone anticancer agents to free radicals. *Proc. Natl. Acad. Sci. U.S.A.* 76, 954–957.

(84) Bird, D. M., Boldt, M., and Koch, T. H. (1987) A kinetic rationale for the inefficiency of 5-iminodaunomycin as a redox catalyst. *J. Am. Chem. Soc.* 109, 4046–4053.

(85) Ihn, W., Tresselt, D., Horn, G., and Berg, H. (1984) Electrochemical reduction of 5-iminodaunorubicin. Isolation and structures of the reduction products. *Stud. Biophys.* 104, 101–102.

(86) Gaudiano, G., and Koch, T. H. (1990) Reaction of the quinone methide from reductive glycosidic cleavage of daunomycin with molecular oxygen. Evidence for semiquinone methide formation. *J. Am. Chem. Soc.* 112, 9423–9425.

(87) Bird, D. M., Gaudiano, G., and Koch, T. H. (1991) Leucodaunomycins, new intermediates in the redox chemistry of daunomycin. *J. Am. Chem. Soc.* 113, 308–315.

(88) Niu, Q., and Mendenhall, G. D. (1990) Structural effects on the yields of singlet molecular oxygen ($^1\Delta_g O_2$) from alkylperoxyl radical recombination. *J. Am. Chem. Soc.* 112, 1656–1657.

(89) Thomas, C. E., and Aust, S. D. (1986) Release of iron from ferritin by cardiotoxic anthracycline antibiotics. *Arch. Biochem. Biophys.* 248, 684–689.

(90) Muindi, J., Sinha, B. K., Gianni, L., and Myers, C. E. (1985) Thiol-dependent DNA damage produced by anthracycline–iron complexes. The structure-activity relationships and molecular mechanisms. *Mol. Pharmacol.* 27, 356–365.

(91) Myers, C. E., Muindi, J. R. F., Zweier, J., and Sinha, B. (1987) 5-Iminodaunomycin. *J. Biol. Chem.* 262, 11571–11577.

(92) S. D. Averbuch, unpublished data.

(93) Averbuch, S. D., Gaudiano, G., Koch, T. H., and Bachur, N. R. (1986) Doxorubicin-induced skin necrosis in the swine model. Protection with a novel radical dimer. *J. Clin. Oncol.* 4, 88–94.

(94) Averbuch, S. D., Boldt, M., Gaudiano, G., Stern, J. B., Koch, T. H., and Bachur, N. R. (1988) Experimental chemotherapy-induced skin necrosis in swine. Mechanistic studies of anthracycline antibiotic toxicity and protection with a radical dimer compound. *J. Clin. Invest.* 81, 142–148.

(95) Tomasz, M., Lipman, R., McGuinness, B. F., and Nakanishi, K. (1988) Isolation and characterization of a major adduct between mitomycin C and DNA. *J. Am. Chem. Soc.* 110, 5892–5896.

(96) Averbuch, S. D., Gaudiano, G., Koch, T. H., and Bachur, N. R. (1985) Radical dimer rescue of toxicity and improved therapeutic index of adriamycin in tumor-bearing mice. *Cancer Res.* 45, 6200–6204.

(97) Tsukahara, K., and Wilkins, R. G. (1985) Kinetics of reduction of eight viologens by dithionite ion. *J. Am. Chem. Soc.* 107, 2632–2635.

## Chapter 5

# The Fecapentaenes, Potent Mutagens from Human Feces

David G. I. Kingston,*,† Roger L. Van Tassell,‡ and Tracy D. Wilkins*,‡

*Departments of Chemistry and Anaerobic Microbiology, Virginia Polytechnic Institute and State University, Blacksburg, Virginia 24061*

Reprinted from *Chemical Research in Toxicology,* Vol. 3, No. 5, September/October, 1990

## Introduction

The importance of carcinogens in the etiology of cancer has been known since studies in the 1700s linked nasal cancer with the immoderate use of snuff (1) and scrotal cancer with the chronic exposure of chimney sweeps to soot (2). The actual mechanisms by which carcinogens accomplish their nefarious work are not always known, however, and recent models for carcinogenesis reveal that the two-stage "initiation–promotion" approach cannot satisfy the scientific observations of the 1980s and 1990s. It is no longer that simple. In the days before the advent of molecular biology, environmental carcinogens and mutagens were believed to be the primary factors in the formation of human tumors. With the emergence of gene cloning and amplification techniques, however, several other factors have risen to the forefront of medical concern. Endogenous genotoxic processes, desmutagens, oncogenes and antioncogenes, and karyotype instability, to name a few, are now considered key factors in a multistage process involving not only initiation and promotion but "demotion" and "progression" as well (3–6). Nevertheless, as researchers strive to unravel this complex scenario, the minimization of carcinogens and mutagens in the environment will remain one of the primary focal points for prevention of cancers in humans.

Three major areas of the human body are exposed to the external environment; lungs, skin, and intestines. Because of their surface areas, in all three cases neoplasia of these organ sites are caused by long-term exposure to weak carcinogens and tumor promoters present in the environment. Lung cancer is caused most often by continued exposure to the carcinogens and tumor promoters present in tobacco smoke. Skin cancer is caused, only after many years, by exposure to another very weak carcinogen, sunlight. Colorectal cancer is caused by, or at least is believed to involve, a variety of environmental genotoxins and promoters which originate from the diet (7, 8). However, in the case of colorectal cancer, we believe that tumors also are caused by a variety of genotoxic and promoting agents produced anaerobically in vivo by the colonic bacterial microflora.

Second to the lungs, the small intestines and colon have the largest surface area of the body which is exposed to the environment. One of the major environmental differences between the small intestine and colon is the presence of at least 100–1000-fold more bacteria per gram of intestinal contents in the colon (9). The growth of these bacteria reduce the colonic environment to one totally lacking oxygen. Consequently, although human beings are aerobes, our colons contain many of the strictest anaerobes on earth. At the boundary between the aerobic and anaerobic environment is the colonic epithelium which is coated by a protective mucinous layer and is the primary substrate for colonic bacterial growth (10). On the lumen side the oxidation reduction potential is about −400 meV whereas, only a millimeter away, the mammalian cells are aerobic with a positive oxidation reduction potential. In addition to this redox gradient, these epithelial cells are continually exposed to the myriad of metabolites produced in this anaerobic environment by the hundreds of species of bacteria found in the human colon (11). As these metabolites interact with the redox gradient the potential is great for the formation of reactive intermediates, many of which may act as ultimate carcinogens or promoting agents.

Over 12 years ago we started our research to learn more about this interaction of the human "aerobe" and the bacterial anaerobes which we believed played a role in

---

† Department of Chemistry.
‡ Department of Anaerobic Microbiology.

2428–5/92/0036$06.00/0

human colorectal cancer. Our initial hypothesis was that products of colonic bacteria might cause mutations leading eventually to this disease. Dr. Robert Bruce in Toronto had just reported that ether extracts of human feces were mutagenic in the Ames test (*12*). We believed that if this mutagenicity was involved in colon cancer, then the amount of fecal mutagenicity within individuals of a population should correlate with the known risk level of that population for colon cancer. Our study of South African blacks and whites was the first of several studies which showed that this indeed was the case (*13*). This was also the beginning of an exciting scientific inquiry to determine what compounds were responsible for genotoxicity within the human colon. Our work and that of others in this area has been reviewed on several occasions (*14–19*); this review however attempts to give a complete summary of all previous work.

## Isolation and Structural Determination of the Mutagens

Bruce's original finding of mutagenicity in the ether extracts of human feces was soon followed by the important observation that the major mutagenicity was associated with a substance or substances having an intense UV absorption spectrum (*20, 21*). As noted above, the occurrence of mutagenic ether extracts of feces correlates with populations at risk for colon cancer, but approximately 40% of the Caucasian population has been shown to yield extracts mutagenic in the *Salmonella* test strain TA-100 without metabolic activation (*12, 22*). We were fortunate in identifying a donor who gave consistently high levels of mutagens in her feces, and almost all of our work has been carried out with feces from this donor.

Even with this excellent donor the levels of mutagens isolable were initially too small for chemical characterization, but the discovery that fecal mutagenicity could be increased dramatically by adding bile and incubating the feces provided the breakthrough which allowed both our group and the Toronto group to obtain adequate amounts of the mutagenic fraction (*23*). The isolation of mutagens was still a challenging task, since they proved to be unstable to both air and acid, but the discovery of the stabilizing effect of butylated hydroxytoluene (BHT) on mutagen-containing solutions enabled us to circumvent the oxidation problem (*22*). A combination of chromatographic techniques then yielded a mutagen fraction which was originally thought to be homogeneous (*24*), but which was later shown to consist of a mixture of stereoisomers (*25–27*). Similar studies by the Toronto group, using pooled feces from several donors, yielded a similar mixture of mutagens and also a second mutagen which was shown to be a homologue of those in the first group (*28, 29*).

The structure elucidation of the mutagens relied heavily on spectroscopic techniques. The UV spectrum of the chromophore was recognized as that of a pentaene (*20, 21*), and the crucial structural studies in our group made use of the fact that hydrogenation of this pentaene yielded a compound which could be derivatized in various ways and analyzed by GC–MS. The dimethyl ether of the hydrogenated mutagen, for example, showed major peaks in its mass spectrum consistent with its assignment as the dimethyl ether of 3-(dodecyloxy)-1,2-propanediol (Figure 1), and this assignment was confirmed by synthesis (*24, 25*).

The structure of the mutagen itself was deduced from its [1]H NMR spectrum, which showed peaks characteristic of a vinylic ethyl group, and by ozonolysis studies, which yielded propanal [isolated as its (2,4-dinitrophenyl)-hydrazone]. The stereochemistry at the 2-position of the

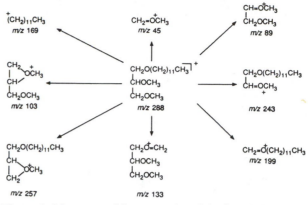

**Figure 1.** Mass spectral fragmentation of the dimethyl ether of the hydrogenated mutagen.

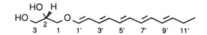

**Figure 2.** The structure of the major component of fecapentaene 12.

glycerol unit was determined by HPLC comparison of the (+)-α-methoxy-α-(trifluoromethyl)-α-phenylacetyl diester (MTPA diester) of hydrogenated mutagen with the MTPA diesters of synthetic racemic 3-(dodecyloxy)-1,2-propanediol and of synthetic (*S*)-3-(dodecyloxy)-1,2-propanediol. The MTPA diester of the hydrogenated natural product cochromatographed with that of the synthetic *S* product, establishing the stereochemistry of the natural product as *S* (Figure 2).

Parallel studies by the Toronto group, carried out primarily by [1]H NMR spectroscopy, led to the establishment of the same structure for one component of their mixture (except that the stereochemistry at C-2 was not determined) and to determination of the structure of a new mutagen with a tetradecapentaene side chain instead of the dodecapentaene side chain of the first compound. This group coined the names fecapentaene 12 and fecapentaene 14 to distinguish the two types of fecal mutagen, and these names have been adopted by all workers in this area (*28, 29*).

The question of the stereochemistry of the double bonds in the fecapentaenes has been addressed both by [1]H NMR and by comparison of the natural product with synthetic materials. The [1]H NMR signal of the C-1' proton of fecapentaene 12 appears as a doublet at 6.56 ppm with a coupling constant of 12 Hz, indicative of an *E* configuration for the first double bond (*25*). A careful analysis of this spectrum, together with the spectra of synthetic 1-*E* and 1-*Z* isomers of fecapentaene-12, indicated that the natural product contains a small amount (around 10%) of a 1-*Z* isomer (*27*). Further comparisons with synthetic materials (*27, 30*) indicated that a major component of fecapentaene 12 has the all-*E* configuration, but that 5-*Z*, 1-*Z*, and probably 3-*Z* components are also present.

In addition to the fecapentaenes, the presence of a fecahexaene and a fecatetraene in feces has been suggested, based on the UV spectrum of HPLC peaks (*31*). The compounds involved have not yet been isolated and characterized, however.

## Synthesis of Fecapentaenes and Their Analogues

The first synthesis of racemic fecapentaene 12 was achieved in our group by Gunatilaka et al. using Wittig chemistry to create the 5',6' double bond in the product

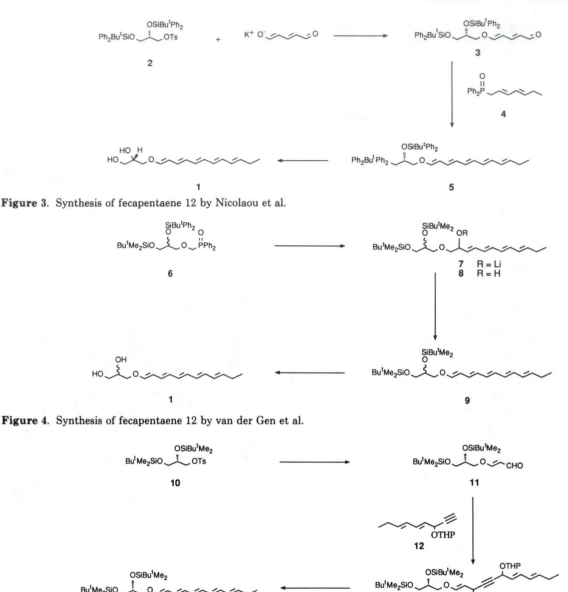

**Figure 3**. Synthesis of fecapentaene 12 by Nicolaou et al.

**Figure 4**. Synthesis of fecapentaene 12 by van der Gen et al.

**Figure 5**. Synthesis of fecapentaene 12 by Pfaendler et al.

*(32)*. Although the overall yield in this synthesis was rather low, it did serve to provide independent proof of the structure of fecapentaene 12.

An independent synthesis by Nicolaou et al. *(33)* proceeded by a similar but somewhat better pathway (Figure 3). In this pathway, the protected dienal 3 was formed efficiently by displacement of the tosylate group in the chiral precursor 2 with potassium glutaconate. Horner–Wittig reaction of the aldehyde 3 with the anion of the phosphine oxide 4 yielded the pentaene 5, which was deprotected to form 1, identical with the natural product. A similar synthetic pathway has been used to prepare [6-²H]- and [6-³H]fecapentaene 12 *(34)*, and this pathway was also used in the synthesis of gram quantities of material for biological testing *(35)*.

A third synthesis of fecapentaene 12 was developed by van der Gen and his collaborators, using the Horner–Wittig reaction to prepare the 1′,2′ double bond (Figure 4) *(36, 37)*. In this synthesis the protected phosphine oxide 6 is prepared by reaction of (hydroxymethyl)diphenyl-phosphine oxide and allyl bromide, followed by hydroxy-

lation and protection of the hydroxyl groups. The lithium salt of phosphine oxide 6 is then reacted with *all-trans*-2,4,6,8-undecatetraenal to give the adducts 7, which were heated with potassium *tert*-butoxide to yield the protected fecapentaenes 9. Deprotection with tetrabutylammonium fluoride then completed the synthesis, yielding a mixture of *E* and *Z* isomers at the 1′-position. An improved version of this synthesis starts with chiral phosphine oxide 6 and involves quenching the lithium salts 7 with ammonium chloride to give the diastereomeric alcohols 8. Separation of these alcohols was achieved by flash chromatography, and elimination followed by deprotection of the separated alcohols yielded (1′*E*)-(*S*)- and (1′*Z*)-(*S*)-fecapentaene in good overall yields *(38)*.

The fourth synthesis of fecapentaene 12 avoids the use of the Wittig or Horner–Wittig reaction and instead uses a modified Whiting reaction as the key step (Figure 5) *(39, 40)*. The propenal 11 was prepared by reaction of the racemic tosylate 10 with the tetrabutylammonium salt of malondialdehyde. Reaction of 11 with the lithium salt of the alkyne 12 yielded the key adduct 13, which was re-

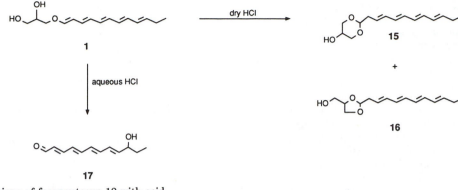

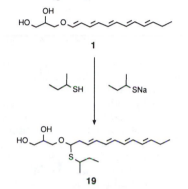

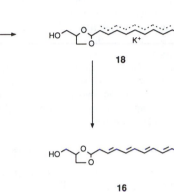

**Figure 6.** Reactions of fecapentaene 12 with acid.

**Figure 7.** Reactions of fecapentaene 12 with base.

duced with lithium aluminum hydride to the *all-trans*-fecapentaene 12 derivative 14 in excellent yield. Deprotection of 14 gave crystalline racemic fecapentaene 12 (1). The use of chiral precursors to 10 led to the synthesis of both natural (S)-(+)-fecapentaene 12 and its enantiomer, and the use of a homologue of alkyne 12 led to the synthesis of racemic and (S)-(+)-fecapentaene 14 (40).

The synthesis of various analogues of the fecapentaenes has been accomplished by modifications of the methods used for the fecapentaenes themselves. Thus van der Gen prepared the triene and tetraene analogues as well as fecapentaene 12 itself (37), and these analogues were also prepared by Govindan et al. (41). Simple enol ether analogues in which the glyceryl moiety is replaced by a methyl group have also been prepared, using a modification of the Peterson reaction as the key step (42).

### Chemistry of the Fecapentaenes

The fecapentaenes, as both pentaenes and enol ethers, are very unstable compounds, decomposing in the presence of acids, air, and light. Their acid sensitivity was evident even before their structures were known, as our early attempts to obtain ¹H NMR spectra in CDCl₃ were frustrated by the decomposition brought about by the traces of acid often present in this solvent. This decomposition in acid was made use of by the Toronto group, who reduced the hydrolysis products to saturated alcohols for characterization (29). Interestingly, the lability of unsaturated enol ethers with acids does not increase with increasing numbers of double bounds. Thus fecapentaene 12 (1) reacted only sluggishly with dry hydrochloric acid in DMSO at room temperature, yielding the cyclic acetals 15 and 16 only incompletely after 150 h (Figure 6). In contrast, the simple monoenol ether analogue of 1 was completely converted into the saturated analogues of 15 and 16 within 30 min (43). Under aqueous acidic conditions (aqueous HCl in tetrahydrofuran) the expected hydrolysis of the enol ether to a simple tetraenal does not take place;

instead, an unusual hydrolysis occurs to give the 10-hydroxyaldehyde 17 as the sole product (Figure 6). In aqueous methanolic HCl the corresponding methoxyaldehyde is formed, again as the sole product (44).

In the presence of base fecapentaene 12 also displays an unusual reactivity. Treatment with potassium *tert*-butoxide in DMSO solution yields the dioxolane 16 (as a mixture of cis and trans isomers) as the only isolated product. Reaction is complete within 5 min at room temperature and presumably proceeds through the anion 18 (Figure 7) (43). Reaction of fecapentaene 12 or its acetonide derivative with external nucleophiles gave products indicative of nucleophilic addition to the pentaene system; thus reaction of fecapentaene 12 with sodium 2-butanethiolate in the presence of excess 2-butanethiol in DMSO gave the adduct 19 in reasonable yield (43).

The sensitivity of the fecapentaenes to air appears to vary considerably with the experimental conditions. Thus a solution of *all-E*-fecapentaene 12 in dimethyl sulfoxide (DMSO) is stable without agitation for about 3 h, but a DMSO/pH 7.2 buffer solution shows noticeable decomposition after 30 min (17). A crystalline sample of fecapentaene 12 decomposes completely on standing in air for 30 min (39). The fecapentaenes are also unstable in ultraviolet light. Thus fecapentaene 12 decomposes almost completely on irradiation for as little as 30 s; the decomposition products have not been characterized (45).

Because of the lability of the fecapentaenes, careful consideration must be given to their analytical chemistry and to quality control considerations in their bioassay. Analysis of the fecapentaenes is most often carried out by HPLC, since they are strongly UV-absorbing and can thus be detected fairly readily. Several systems for HPLC analysis have been reported; a typical one employs a reversed-phase column with a solvent system employing acetonitrile, water, methanol, and THF (26). A combined HPLC–isotope dilution method for analysis of fecapentaenes in human feces has also been described (31), and a fast and easy method for fecal extraction and subsequent HPLC analysis has been developed by using triethyl-amine-stabilized eluents and stock solutions (46). A recent paper proposed an alternate method to stabilize the fecapentaenes, namely, their chemical conversion by methanolic HCl followed by sodium borohydride to a relatively stable 10-methoxytetraenol (47).

Quality control considerations in biological work with the fecapentaenes have been addressed in a recent publication. In this study the decomposition of fecapentaene 12 was investigated under various conditions, and it was concluded that unchecked decomposition can profoundly affect biological test results. The authors commented that "we found fecapentaene 12 to be more unstable and difficult to handle than any other substance we have as yet attempted to test for carcinogenic activity" (45). A particular concern was that decomposition of fecapentaene 12 occurred under very variable conditions, and the authors recommend storage in amber vessels, avoidance of strong light, nucleophiles, and acidic conditions, the use of minimum quantities, and maximum protection from air, as standard methods for handling fecapentaene 12. The use of protective agents such as vitamin E or butylated hydroxytoluene was also recommended.

## Biological Activity of the Fecapentaenes

The fecapentaenes are directly active in most short-term systems in which they have been tested. They are potent mutagens in the Ames test on both TA98 and TA100 (41, 48, 49), and in mammalian cells they induce a broad spectrum of genotoxic effects. These include unscheduled DNA synthesis and cellular transformations (48), mutations and sister chromosome exchanges (50, 51), single-stranded breaks in DNA (50, 52), and nuclear aberrations and mitotic figures (53, 54). In addition, the fecapentaenes are the most prevalent mutagens found in the colon (55), accounting for a majority of the direct-acting fecal mutagenicity (56).

There have been three studies to determine the carcinogenic potential of the fecapentaenes in experimental animals. The first study completed at NCI (57) found no evidence that fecapentaenes could induce tumors in the colons or in other sites of rats or mice. However, as has been recently described in detail by Streeter and co-workers (45) and has been observed over the years in our laboratory, the fecapentaenes are notoriously unstable compounds which require special precautions when conducting carcinogenicity tests or other biological assays. Even when the utmost care is taken to control for decomposition of the fecapentaenes before and after administration to animals, there is currently no way to determine if the fecapentaenes reach their cellular "targets" in a biologically active form. Nevertheless, a recent collaborative effort between our research group and those of Dr. Van der Gen and of Dr. John Weisburger of the American Health Foundation showed that fecapentaene 12 does have weak carcinogenic activity in newborn mice (58). Low but significant numbers of lung, liver, and stomach tumors and subcutaneous sarcomas were observed when newborn mice were injected with fecapentaene 12 intraperitoneally. Finally, a recent abstract by Shamsuddin and Ullah (59) reports the formation of tumors in 30% of male F-344 rats treated with fecapentaene 12 intrarectally, although no large intestinal tumors were observed. These observations indicate that the fecapentaenes are almost certainly carcinogens. The inability of some conventional bioassays to demonstrate the carcinogenic potential of the fecapentaenes may be due to their extreme lability, as noted above,

| 20 | n = 0 | | 25 | n = 1 |
| 21 | n = 1 | | 26 | n = 2 |
| 22 | n = 2 | | 27 | n = 3 |
| 23 | n = 3 | | 28 | n = 4 |
| 24 | n = 4 | | 29 | n = 5 |

**Figure 8.** Enol ethers and aldehydes assayed for mutagenicity.

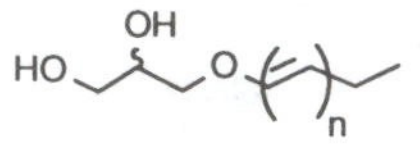

| 30 | n = 1 |
| 31 | n = 2 |
| 32 | n = 3 |
| 33 | n = 4 |
| 34 | n = 5 |

**Figure 9.** Fecapentaene analogues assayed for mutagenicity.

but it may also be the result of the involvement of protective mechanisms in adult animals which are not present in infants. If we may extrapolate this to humans—an admittedly risky undertaking—we can thus theorize that conditions which decrease natural cellular defenses and promote cell proliferation in the colon may play a role in the in vivo expression of fecapentaene carcinogenicity. Such conditions could include general disease states such as polyposis, bacterially induced colitis, chemically induced GI disorders, and inflammatory bowel diseases or natural conditions such as aging or immunodeficiencies.

## Structure–Activity Relationships and Mechanism of Action of the Fecapentaenes

As noted earlier, the fecapentaenes are potent genotoxins, showing direct-acting mutagenic activity against various *Salmonella typhimurium* tester strains. Several studies have been carried out to determine structural effects on the mutagenicity of the fecapentaenes and their analogues, and some tentative conclusions about their mechanism of action can be made.

A series of methyl enol ethers **20–24** and the unsaturated aldehydes **25–29** (Figure 8) were prepared by the Toronto group in an initial study of structure–activity relationships (42). Mutagenicity of these compounds was determined against *S. typhimurium* TA100, and the pentaene **24** showed slightly less activity than that of fecapentaene 14. The tetraenyl ether **23** showed about 100-fold less activity than **24**, and the remaining ethers **20–22** were inactive or nearly so. The aldehydes **29** and **28** also showed some activity, but less than the corresponding ethers, and it was implied that hydrolysis to the aldehyde was not a factor in the mutagenicity of the fecapentaenes.

A second study from our group reported the synthesis and biological activity of the glyceryl enol ethers **30–34** (Figure 9), analogous to fecapentaene 12 (41). The synthetic pentaenyl ether **34**, prepared as a mixture of stereoisomers at the 2-position and at the 5′-position, showed an essentially identical mutagenicity to natural fecapentaene 12 in both *S. typhimurium* TA98 and TA100, and it also showed activity in TA104 comparable to its activity in TA100. The tetraene **33** was about 30-fold less active than **34** in TA98 and about 20-fold less active than **34** in TA100 and TA104. The tetraenal **28**, corresponding to the presumed hydrolysis product of **34**, was also tested and found to be 200-fold less active than **34** in TA98 and about 800-fold less active in TA100 and TA104.

A study by the SRI group determined the mutagenicity of *all-trans*-fecapentaene 12, its "cis isomer", fecapentaene

14, and the methoxy analogues of fecapentaene 12 and 14 (*49*). In this study fecapentaene 12, "*cis*-fecapentaene 12", and fecapentaene 14 were found to have essentially the same mutagenicity against TA100. Surprisingly, however, the activities of the methoxy analogues were found to be significantly higher than those of the fecapentaenes, in contrast to the findings of the Toronto group (*42*). The reason for this discrepancy has not been explained, but it may relate to the instability of the pentaenes or to the different conditions of the assay. It may be noted, for example, that the SRI group dissolved their test compounds in DMSO, and DMSO has been shown to bring about changes in some test substances under the conditions of the Ames assay (*60*). The mutagenicity of *all-trans*-fecapentaene 12 has also been determined in an independent study, and it was shown to have comparable mutagenicity to natural fecapentaene 12 in TA100 (*61*).

The mutagenicity of synthetic fecahexaene 14, fecapentaene 12, and fecatetraene 10 have recently been determined. Surprisingly, fecahexaene 14 turned out to be less mutagenic than fecapentaene 12, although the instability of this compound makes this conclusion somewhat tentative. The sulfur analogue of fecapentaene 12 was nonmutagenic, and the aldehyde 17 was only very slightly mutagenic (*62*).

These results thus indicate that mutagenicity in the fecapentaene series is critically dependent on the number of conjugated double bonds in the enol ether system. Other factors such as the stereochemistry of the double bounds in the alkenyl chain and the length of the hydrocarbon "tail" have only secondary effects on mutagenicity. The nature of the ether group is also not critical to mutagenicity, and the substitution of the glyceryl unit by a methoxy group appears actually to increase mutagenicity. However, the oxygen of the enol ether linkage is crucial, with activity nonexistent in the sulfur analogue.

The aldehyde 28 corresponds to the presumed conjugated acid hydrolysis product of the fecapentaenes. Although it is significantly less mutagenic than the fecapentaenes, it was considered possible that the enol ether linkage simply served as a "carrier group" to enable the reactive aldehyde to pass into the cell. Under this hypothesis, the enol ether would hydrolyze in the cell to a homologue of 28 or to its 10-hydroxy derivative 17, and this compound would be the actual genotoxic agent. If this scenario were correct, aldehyde 28 might be expected to be highly reactive toward DNA and similar compounds. However, in model studies in our laboratory we found that aldehyde 28 is actually significantly *less* reactive than simple unsaturated aldehydes. Thus although both acrolein (*63, 64*) and crotonaldehyde (*65*) form adducts with nucleosides, the aldehyde 28 did not form any adducts even under prolonged incubation (*66*). It thus seems clear that the fecapentaenes do not act as mutagens simply by hydrolyzing to the corresponding aldehyde.

The initial interpretation of the mutagenicity of the fecapentaenes was that they first underwent protonation or reaction with an electrophilic moiety to give a delocalized carbocation, which would then serve as an electrophilic species for subsequent reaction with DNA or other nucleophiles (*42*), and this hypothesis has been the subject of some theoretical calculations (*67*). However, this interpretation has been called into question by the finding that the mutagenicity of the fecapentaenes does not parallel their acid lability. As mentioned earlier, increasing the number of double bonds in the fecaenes (at least up to five double bonds) makes them more mutagenic and more reactive toward nucleophiles, but less reactive toward

acidic reagents (*43*). The reactivity of fecapentaene 12 toward nucleophiles suggests that a part of the explanation for its biological activity may lie in its direct reaction with DNA or other nucleophilic species. Indeed, fecapentaene 12 has been shown to react directly with glutathione, causing cellular depletion of this protective agent (*68*), but the structure of the adduct formed has not been determined. Fecapentaene 12 has also been shown to form strand breaks and cross-links in plasmid DNA, but again the structural identification of the presumptive DNA adduct(s) has not yet been achieved (*66*). However, the mutagenicity of fecapentaene 12 toward the frame-shift mutant TA98 (*41, 48*) suggests that a direct interaction of the mutagen with DNA at a specific site is indeed occurring.

The actual mechanism of action of fecapentaene 12 is undoubtedly more complicated than the simple explanation implied above, however. Thus the mutagenicity of fecapentaene 12 changes when it is assayed under anaerobic conditions. Under aerobic conditions fecapentaene 12 is less mutagenic to the frame-shift conditions mutant TA98 than it is under anaerobic conditions. However, with the base-substitution mutant TA100 fecapentaene 12 is mutagenic *only* under aerobic conditions. These results thus suggest that fecapentaene 12 may have two mechanisms of action, with activity as a base-substitution mutagen being oxygen-dependent (*69*). This conclusion correlates nicely with the recent finding that fecapentaene 12 and fecapentaene 14 hydroxylate the C-8 position of guanine residues in DNA in vitro, presumably through the formation of oxygen radicals of an as yet undetermined nature (*70*). It is thus likely that mutagenicity to TA100 comes about through a base-pair substitution involving an oxygen-mediated formation of damaged purines.

## Origin of the Fecapentaenes

During the early work on the fecapentaenes many researchers did not believe that such compounds could be formed in the extremely anaerobic environment of the colonic lumen. However, at the time we could show that fecapentaene levels dramatically increased when feces were incubated at 37 °C. In addition, we showed that this increase did not occur if feces were supplemented with antibiotics, heat sterilized, or $\gamma$-irradiated prior to incubation (*23*). Yet when sterilized feces were inoculated with fresh feces a dramatic increase in fecapentaene levels was once again observed. Later we were able to show that incubation of fecal extracts with pure cultures of certain strains of colonic *Bacteroides* resulted in "in vitro production" of fecapentaenes (*71*). The membrane fraction of lysed cells of the *Bacteroides* did the reaction even faster, and bile increased the rate another 25–50-fold (*72*). We could not get these *Bacteroides* to produce fecapentaenes in the absence of fecal extracts, and these fecal extracts varied greatly in their ability to support this reaction. To make a long microbiological story short, we determined that there were three major requirements for production of fecapentaenes in vitro and in vivo. The first is the enzymatic machinery which is provided in the form of bacterial membranes. The second is an incubation "environment" in which all the reactants are soluble and can react with the appropriate associations; this is provided by the presence of the bile salts. The third and key requirement is a precursor or precursors which are present in feces and colonic contents of certain individuals and are processed by the *Bacteroides* membranes into the fecapentaenes (*15*). The ultimate origin of these precursors still remains unknown, but as will be discussed later, they are probably products of host (mammalian) metabolism.

More recently, epidemiological studies have shown that the fecapentaenes and their precursors are excreted in the feces of most (>75%) white North Americans (56, 73). Although the relative amounts of endogenous fecapentaene 12 and fecapentaene 14 vary considerably among individuals, their ratios within any one individual can remain constant for years (14). In a collaborative effort coordinated by Dr. Mark Schiffman of NCI we showed that both the fecapentaenes and their precursors are excreted in significantly lower concentrations in individuals who have colorectal cancer than in normal individuals (14). Although the diagnostic workup or the diseased state itself might be expected to alter the fecal concentrations of these compounds in vivo, control experiments examining the effects of colorectal bleeding, cleansing regimens, and colonoscopy revealed no insight into why fecapentaene levels were lower in colorectal cancer cases. Although these lower levels suggest that the fecapentaenes may be somehow associated with protective mechanisms—as will also be discussed later—the overall implications of these recent observations are unclear. What is clear, however, is that excretion of the fecapentaenes and their precursors are not good markers for risk for colorectal cancer. Indeed, the best correlation between colon cancer and mutagenicity was observed with non-fecapentaene TA98 mutagenicity (74). Thus if any association exists between the occurrence of these genotoxic lipids and colorectal cancer it most likely occurs early in the natural history of the disease when the tumors are initiated, and not years later at the time of diagnosis.

## Isolation and Structure Determination of Precursors of the Fecapentaenes

Purification and structural elucidation of the precursors of fecapentaenes from feces have been hindered by the same obstacles encountered during purification and structural elucidation of the fecapentaenes themselves: their instability to oxygen and mild acids, their low concentrations in feces, and their molecular heterogeneity. As described in a preliminary report (75), we observed that the precursors had the same UV absorbance characteristics as the fecapentaenes, contain alkyl or ester groups—as opposed to hydroxyl groups—covalently linked to the sn-2 and sn-3 positions of the glycerol backbone, and behave like phospholipids, with both polar and nonpolar character. In a recent report (76) we described the methods for isolation and purification of a major form of precursor from feces using a series of extractions with organic solvents, precipitation with acetone, and fractionation by silica and amine HPLC. In addition, we characterized the purified precursor as to its chromatographic properties by comparison to fecapentaene 12 and a synthetic model ether phospholipid. Although these results supported our beliefs that the precursors were phospholipids, we were not certain until we examined the effects of various lipolytic enzymes on the purified compounds.

The purified precursor was converted to fecapentaene 12 when incubated with whole cells or purified membranes of certain Bacteroides species. Using micellar enzymological techniques, we used commercial lipolytic enzymes to model probable conditions within the human colon (76). Using a combination of standard techniques for producing liposomes and the anaerobic techniques required for handling oxygen-labile polyenyl compounds, we incorporated purified precursor into artificial membranes. When these "precursor" liposomes or micelles were incubated with combinations of commercial enzymes, hydrolysis to

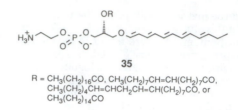

**35**

R = CH$_3$(CH$_2$)$_{16}$CO, CH$_3$(CH$_2$)$_7$CH=CH(CH$_2$)$_7$CO, CH$_3$(CH$_2$)$_4$CH=CHCH$_2$CH=CH(CH$_2$)$_7$CO, or CH$_3$(CH$_2$)$_{14}$CO

**Figure 10.** The structure of one component of plasmalopentaenes 12.

fecapentaene occurred only with a mixture of lipase and phospholipase C. None of the individual classes of phospholipases A, B, C, or D, sphingomyelinase, or any of several lipases alone converted the purified precursor to fecapentaenes. In all cases, the form of fecapentaene produced by bacterial membranes or commercial enzymes from the purified precursor was fecapentaene 12; this is reasonable, since our donor produces very little fecapentaene 14, as previously noted. On the basis of these and other observations we believe that the Bacteroides strains have a lipase and phospholipase C intrinsically complexed within their membranes—whose normal role is most likely turnover of bacterial membrane phospholipids—but which fortuitously hydrolyze precursors to the fecapentaenes within the colonic milieu.

Final structure elucidation of the precursors was accomplished by chemical methods (77). Deacylation of the precursors followed by methylation of the resulting fatty acids with diazomethane yielded a mixture of esters which was analyzed by capillary GC and GC–MS. These methods indicated the presence of two major and two minor esters, with the major esters identified as methyl hexadecanoate and methyl octadecanoate, and the minor esters as methyl 9-octadecenoate and methyl 9,12-octadecadienoate. The group at the sn-3 position of glycerol was identified as a phosphoethanolamine moiety by hydrolysis of the precursors to 3-glycerophosphoethanolamine. The precursors thus have the general structure **35** (Figure 10); because of their similarity to the plasmalogens we have named them plasmalopentaenes. Presumably, analogous precursors to fecapentaene 14 also exist, but we have not detected them in the feces of our donor.

## Biological Activity of the Plasmalopentaenes

The purified plasmalopentaenes 12 were tested in the two bacterial assays used in our laboratory, (1) our standard Salmonella assay on TA98 or TA100 with and without microsomal activation and (2) the SOS Chromatest, a "chromogenic" genotoxicity assay designed to detect an induction of DNA repair in Escherichia coli. Whereas fecapentaene 12 is a potent dose–response mutagen in the Salmonella assay at 1–5 µg/plate and in the SOS Chromatest at 10–100 ng/well, the plasmalopentaenes 12 were not mutagenic under any test conditions at concentrations up to 100 µg/plate (Van Tassell and Wilkins).[1] When synthetic plasmalopentaene becomes available, we will be able to more fully examine its genotoxic potential.

## Origin of the Plasmalopentaenes

For years we have considered the three probable origins of the plasmalopentaenes to be the diet, the colonic microflora, or the host itself. Using autopsy samples from humans and pigs—the only other animal shown to excrete fecapentaenes (78)—we showed that plasmalopentaenes,

---

[1] R. L. Van Tassell and T. D. Wilkins, unpublished results.

as well as the fecapentaenes, were distributed in uniform concentrations throughout the colons of both species (*79*). We found no plasmalopentaene from the sterilized conventional pig feed. Thus we concluded that plasmalopentaenes probably do not originate in the diet. If they entered the colon as components of digested food, a dramatic decrease in their levels—with a corresponding increase in fecapentaene levels—from proximal to distal regions would be expected. This was not observed.

During our early work, we felt that the bacterial microflora itself might produce the plasmalopentaenes. Ether lipids are common membrane components of anaerobic bacteria and should be present in relatively high amounts in the colon contents. However, after years of trying to isolate a precursor-producing bacterium, we recently obtained results which appear to rule out this possibility as well. Using standard HPLC screening techniques, we analyzed the feces of 12 neonatal germ-free pigs. In 11 of the 12 pigs we detected the plasmalopentaenes; expectedly we did not detect any fecapentaenes since there were no bacterial enzymes to carry out the requisite hydrolysis (*80*). Since the plasmalopentaenes were present in the pig feces in the absence of a bacterial flora and neither fecapentaenes nor plasmalopentaenes were detected in the diets, we now believe that the precursors are likely to be endogenous compounds made by the host.

Using classical membrane extraction methods, we have screened human and porcine tissues for the presence of plasmalopentaenes. Segments of human colons were obtained through the National Disease Research Interchange (Philadelphia, PA). Porcine tissues, which included small intestines, colons, livers, heart, kidney, gall bladder, and bile, were obtained through the Virginia–Maryland Regional College of Veterinary Medicine associated with our university, Virginia Polytechnic Institute and State University. We extracted all tissues using the methods we used routinely for extracting the precursors from feces (*80*). So far, however, we have not yet isolated any precursors from any tissue extracts or other biological samples. Thus we are now focusing on what we feel may be some of the reasons for our inability to isolate these plasmalogens. Normal plasmalogens are notoriously difficult to isolate because of the instability of their enol ether linkages. In the case of the plasmalopentaenes, the conjugation in the ether-linked groups would promote further instability which would likely make this class of plasmalogens even more difficult to isolate. Within the highly buffered and reduced environment found within feces and colon contents the plasmalopentaenes are presumably sufficiently stabilized to allow them to accumulate in their native forms long enough to survive extraction. However, in tissues, where the redox potentials are much greater and where enzymes specialize in rapid phospholipid turnover, the plasmalopentaenes probably have a very transient lifetime and may thus not accumulate to any appreciable extent. We are thus currently modifying our extraction techniques and designing more sensitive detection methods based on spectrophotometric and monoclonal technology to be used directly on tissues and tissue explants.

In conclusion, we now believe that these novel ether lipids may be produced by mammalian cells, particularly the colonic epithelial cells, as protective agents against the continual oxidation–reduction battle that is occurring at this border between oxidized and anaerobic environments. The pentaenyl portion of the plasmalopentaenes would make a very good sink to rapidly scavenge toxic radicals formed at this interface. Plasmalogens with a single double bond have been shown to protect cells from the effects of oxidative radicals, so our idea is not completely without

support (*81, 82*). Consequently, the plasmalopentaenes may represent a first line of cellular defense in colonic cells, and they may thus be involved more directly in the etiology of colorectal cancer than the fecapentaenes themselves.

**Acknowledgment.** We could not have carried out the work described in this review without the dedicated help of many excellent co-workers, whose names are included in the references to the work from our groups. We also gratefully acknowledge the continuous support of this work by the National Cancer Institute (Grant CA23857) since 1978.

## References

(1) Hill, J. (1761) *Cautions Against the Immoderate Use of Snuff*, Baldwin and Jackson, London; cited in Redmond, E. (1970) Tobacco and Cancer: The First Clinical Report. *N. Engl. J. Med.* **282**, 18–23.

(2) Pott, P. (1775) *Chirurigical Observations*, p 63, Hawes, Clarke, and Collins, London.

(3) Roe, F. J. C. (1989) Non-genotoxic carcinogenesis: implications for testing and extrapolation to man. *Mutagenesis* **4**, 407–411.

(4) Davis, D. L. (1989) Natural anticarcinogens, carcinogens, and changing patterns in cancer: some speculations. *Environ. Res.* **50**, 322–340.

(5) Pirot, M. (1989) Progression: the terminal stage in carcinogenesis. *Jpn. J. Cancer Res.* **80**, 599–607.

(6) Astrin, S., and Costanzi, C. (1989) The molecular basis for colon cancer. *Semin. Oncol.* **16**, 138–147.

(7) Reddy, B. S. (1989) Overview of diet and colon cancer. *Prog. Clin. Biol. Res.* **279**, 111–121.

(8) Weisburger, J. H. (1989) Dietary prevention of colorectal cancer. In *Colorectal Cancer: from pathogenesis to prevention* (Seitz, H. K., Simanowshi, U. A., and Wright, N. A., Eds.) pp 261–373, Springer-Verlag, New York.

(9) Moore, W. E. C., Holdeman, L. V., and Cato, E. P. (1978) Some current concepts in intestinal bacteriology. *Am. J. Clin. Nutr.* **31**, 533–542.

(10) McCarthy, R. E., and Salyers, A. A. (1988) The effect of dietary fiber utilization on the colonic microflora. In *Role of the Gut Flora in Toxicity and Cancer* (Rowland, I. R., Ed.) pp 298–313, Academic Press, New York.

(11) Hill, M. J. (1989) Gut flora and cancer in humans and laboratory animals. In *Role of the Gut Flora in Toxicity and Cancer* (Rowland, I. R., Ed.) pp 461–502, Academic Press, New York.

(12) Bruce, W. R., Varghese, A. J., Furrer, R., and Land, P. C. (1977) A mutagen in the feces of normal humans. In *Origins of Human Cancer* (Hiatt, H. H., Watson, J. D., and Winsten, J. A., Eds.) pp 1641–1646, Cold Spring Harbor, New York.

(13) Ehrich, M. F., Aswell, J. E., Van Tassell, R. L., Wilkins, T. D., Walker, A. R. P., and Richardson, N. J. (1979) Mutagens in the feces of 3 South African populations at different risk levels for colon cancer. *Mutat. Res.* **64**, 231–240.

(14) Wilkins, T. D., and Van Tassell, R. L. (1983) Production of Intestinal Mutagens. In *Human Intestinal Microflora in Health and Disease* (Hentges, D. J., Ed.) pp 265–288, Academic Press, New York.

(15) Van Tassell, R. L., Schram, R. M., and Wilkins, T. D. (1986) Microbial biosynthesis of fecapentaenes. In *Genetic Toxicology of the Diet* (Knudsen, I., Ed.) pp 199–211, Alan R. Liss, New York.

(16) Krepinsky, J. J., Bruce, W. R., Gupta, I., Suzuki, K., Rafter, J. J., and Child, P. (1986) Diet-related factors in the origin of cancer of the large bowel—a molecular scientist's point of view. In *Genetic Toxicology of the Diet* (Knudsen, I., Ed.) pp 183–197, Alan R. Liss, New York.

(17) Krepinsky, J. J. (1988) Formation and biological effect of fecapentaenes. *Prog. Biochem. Pharmacol.* **22**, 35–47.

(18) Vennitt, S. (1988) Mutagens in human feces and cancer of the large bowel. In *Role of the Gut Flora in Toxicity and Cancer* (Rowland, I. R., Ed.) pp 399–460, Academic Press, New York.

(19) Van Tassell, R. L., Kingston, D. G. I., and Wilkins, T. D. (1990) Dietary genotoxins and the human microflora. In *Mutation and the Environment, Part E* (Mendelsohn, M. L., and Albertini, R. J., Eds.) pp 149–158, Wiley-Liss, New York.

(20) Bruce, W. R., Varghese, A. J., Land, P. C., and Krepinsky, J. J. (1981) *Banbury Rep.* **7**, 227–238.

(21) Kingston, D. G. I., Wilkins, T. D., Van Tassell, R. L.,

MacFarlane, R. D., and McNeal, C. (1981) *Banbury Rep.* **7**, 215–226.

(22) Wilkins, T. D., Lederman, M., Van Tassell, R. L., Kingston, D. G. I., and Henion, J. D. (1980) *Am. J. Clin. Nutr.* **33** (Suppl. 11), 2513–2520.

(23) Lederman, M., Van Tassell, R. L., West, S. E. H., Ehrich, M. F., and Wilkins, T. D. (1980). *In vitro* production of human fecal mutagen. *Mutat. Res.* **79**, 115–124.

(24) Hirai, N., Kingston, D. G. I., Van Tassell, R. L., and Wilkins, T. D (1982) Structure elucidation of a potent mutagen from human feces. *J. Am. Chem. Soc.* **104**, 6149–6150.

(25) Hirai, N., Kingston, D. G. I., Van Tassell, R. L., and Wilkins, T. D. (1985) Isolation and structure elucidation of fecapentaenes-12, potent mutagens from human feces. *J. Nat. Prod.* **48**, 622–630.

(26) Baptista, J., Bruce, W. R., Gupta, I., Krepinsky, J. J., Van Tassell, R. L., and Wilkins, T. D. (1984) On distribution of different fecapentaenes, the fecal mutagens, in the human population. *Cancer Lett.* **22**, 299–303.

(27) Kingston, D. G. I., Piccariello, T., Duh. C.-Y., Govindan, S. V., Wilkins, T. D., Van Tassell, R. L., Van der Gen, A., de Wit, P. P., and Van der Steeg, M. (1988) On the stereochemistry of fecapentaene-12. *J. Nat. Prod.* **51**, 176–179.

(28) Bruce, W. R., Baptista, J., Che, T. C., Furrer, R., Gingerich, J. S., Gupta, I., Krepinsky, J. J., Yates, P., and Grey, A. A. (1982) General structure of "fecapentaenes"—the mutagenic substances in human feces. *Naturwiss.* **69**, 557–558.

(29) Gupta, I., Baptista, J., Bruce, W. R., Che, C. T., Furrer, R., Gingerich, J. S., Grey, A. A., Marai, L., Yates, P., and Krepinsky, J. J. (1983) Structures of fecapentaenes, the mutagens of bacterial origin isolated from human faces. *Biochemistry* **22**, 241–245.

(30) Baptista, J., Krepinsky, J. J., and Pfaendler, H. R. (1987) Natural fecapentaene-14 and one fecapentaene-12 component are all-trans stereoisomers. *Angew. Chem., Int. Ed. Engl.* **26**, 1186–1187.

(31) Peters, J. H., Nolen, H. W., III, Gordon, G. R., Bradford, W. W., III, Bupp, J. E., and Reist, E. J. (1989) Combined chromatographic–isotopic dilution analysis of fecapentaenes in human feces. *J. Chromatogr.* **488**, 301–313.

(32) Gunatilika, A. A. L., Hirai, N., and Kingston, D. G. I. (1983) Synthesis of racemic facepentaene-12, a potent mutagen from human feces, and its regioisomer. *Tetrahedron Lett.* **24**, 5457–5460.

(33) Nicolaou, K. C., Zipkin, R., and Tanner, D. (1984) Total synthesis of the potent mutagen (S)-3-(dodeca-1,3,5,7,9-pentaenyloxy)propane-1,2-diol. *J. Chem. Soc., Chem. Commun.*, 349–350.

(34) Kassaee, M. Z., and Kingston, D. G. I. (1987) Synthesis of [6-³H]fecapentaene-12. *J. Labelled Compd. Radiopharm.* **24**, 1071–1076.

(35) Bradford, W. W., and Reist, E. J. (1985) Synthesis of fecapentaene-12, a mutagen isolated from human feces. *Abstracts of the 189th ACS National Meeting*, Miami Beach, FL. Abstract MEDI 058.

(36) Van Schaik, T. A. M., and van der Gen, A. (1983) Synthesis of 3-(1-alkenyloxy)-1,2-propanediols, enol ethers of glycerol, by the Horner–Wittig reaction. *Recl. Trav. Chim. Pays-Bas* **102**, 465–466.

(37) de Wit, P. P., van Schaik, T. A. M., and van der Gen, A. (1984) A convenient synthesis of fecapentaene-12 by the Horner–Wittig reaction. *Recl. Trav. Chim. Pays-Bas* **103**, 369–370.

(38) de Wit, P. P., van der Steeg, M., and van der Gen, A. (1985) Synthesis of the E- and Z-isomers of the potent mutagen (S)-fecapentaene-12 by the Horner–Wittig method. *Recl. Trav. Chim. Pays-Bas* **104**, 307–308.

(39) Pfaendler, H. R., Maier, F. K., and Klar, S. (1986) Synthesis of crystalline (±)-fecapentaene. *J. Am. Chem. Soc.* **108**, 1138–1139.

(40) Pfaendler, H. R., Maier, F. K., Klar, S., and Göggelmann, W. (1988) Racemic and enantiomeric *all-trans*-fecapentaene-12 and -14. *Liebigs Ann. Chem.*, 449–454.

(41) Govindan, S. V., Kingston, D. G. I., Gunatilaka, A. A., Van Tassell, R. L., Wilkins, T. D., de Wit, P. P., Van der Steeg, M., and Van der Gen, A. (1987) Synthesis and biological activity of analogs of fecapentaene-12. *J. Nat. Prod.* **50**, 75–83.

(42) Gupta, I., Suzuki, K., Bruce, W. R., Krepinsky, J. J., and Yates, P. (1984) A model study of fecapentaenes: mutagens of bacterial origin with alkylating properties. *Science* **225**, 521–523.

(43) de Wit, P. P., van der Steeg, M., and van der Gen, A. (1986) Remarkable electrophilic properties of the pentaenol ether system of fecapentaene-12. *Tetrahedron Lett.* **27**, 6263–6266.

(44) Vertegaal, L. B. J., van der Steeg, M., and van der Gen, A. (1989) Anomalous solvolysis of a polyenol ether of glycerol. *Tetrahedron Lett.* **30**, 5639–5642.

(45) Streeter, A. J., Donovan, P. J., Anjo, T., Ohannesian, L., Sheffels, P. R., Wu, P. P., Keefer, L. K., Andrews, A. W., Bradford, W. W., III, Reist, E. J., and Rice, J. M. (1989) Decomposition and quality control considerations in biological work with fecapentaene preparations. *Chem. Res. Toxicol.* **2**, 162–168.

(46) Kleinjans, J. C. S., Pluijmen, M. H. M., Hageman, G. J., and Verhagen, H. (1989) Stabilization and quantitative analysis of fecapentaenes in human feces, using synthetic fecapentaene-12. *Cancer Lett.* **44**, 33–37.

(47) Kivits, G. A. A., de Boer, B. C. J., Nugteren, D. H., van der Steeg, M., Vertegaal, L. B. J., and van der Gen, A. (1990) Quantitative HPLC analysis of the level of fecapentaenes and their precursors in human feces by a chemical conversion method. *J. Nat. Prod.* **53**, 42–49.

(48) Curren, R. D., Putman, D. L., Yang, L. L., Haworth, S. R., Lawlor, T. E., Plummer, S. M., and Harris, C. C. (1987) Genotoxicity of fecapentaene-12 in bacterial and mammalian cell assay systems. *Carcinogenesis* **8**, 349–352.

(49) Peters, J. H., Riccio, E. S., Stewart, K. R., and Reist, E. J. (1988) Mutagenic activities of fecapentaene derivatives in the Ames/Salmonella test system. *Cancer Lett.* **39**, 287–296.

(50) Plummer, S. M., Grafstrom, R. C., Yang, L. L., Curren, R. D., Linnainmaa, K., and Harris, C. C. (1986) Fecapentaene-12 causes DNA damage and mutations in human cells. *Carcinogenesis* **7**, 1607–1609.

(51) Bauchinger, M., Bartha, R., Schmid, E., and Pfaendler, H. R. (1988) Fecapentaene causes sister-chromatid exchanges in human lymphocytes. *Mutat. Res.* **209**, 29–31.

(52) Hinzman, M. J., Novotney, A. Ullah, A., and Shamsuddin, A. M. (1987) Fecal mutagen fecapentaene-12 damages mammalian colon epithelial DNA. *Carcinogenesis* **8**, 1475–1479.

(53) Vaughan, D. J., Furrer, R., Baptista, J., and Krepinsky, J. J. (1987) The effect of fecapentaenes on nuclear aberrations in murine colonic epithelial cells. *Cancer Lett.* **37**, 199–203.

(54) Schmid, E., Bauchinger, M., Braselmann, H., Pfaendler, H. R., and Göggelmann, W. (1987) Dose–response relationship for chromosome aberrations induced by fecapentaene-12 in human lymphocytes. *Mutat. Res.* **191**, 5–7.

(55) Dion, P. and Bruce, W. R. (1983) Mutagenicity of different fractions of extracts of human feces. *Mutat. Res.* **119**, 151–160.

(56) Schiffman, M. H., Van Tassell, R. L., Andrews, A. W., Wacholder, S., Daniels, J., Robinson, A., Smith, L., Nair, P. P., and Wilkins, T. D. (1989) Fecapentaene concentration and mutagenicity in 718 North American stool samples. *Mutat. Res.* **222**, 351–357.

(57) Ward, J. M., Anjo, T., Ohannesian, L., Keffer, L. K., Devor, D. E., Donovan, P. J., Smith, G. T., Henneman, J. R., Streeter, A. J., Konishi, N., Rehm, S., Reist, E. J., Bradford, W. W., III, and Rice, J. M. (1988) Inactivity of fecapentaene-12 as a rodent carcinogen or tumor intiator. *Cancer Lett.* **42**, 49–59.

(58) Weisburger, J. H., Jones, R. C., Wang, C.-X., Backlund, J.-Y. C., Williams, G. M., Kingston, D. G. I., Van Tassell, R. L., Keyes, R. F., Wilkins, T. D., de Wit, P. P., van der Steeg, M., and van der Gen, A. (1990) Carcinogenicity tests of fecapentaene-12 in mice and rats. *Cancer Lett.* **49**, 89–98.

(59) Shamsuddin, A. M., and Ullah, A. (1990) Carcinogenicity studies of fecapentaene-12 in F-344 rats. *Proc. Am. Assoc. Cancer Res.* **31**, 87 (Abstract 516).

(60) Vaughan, D. J., Baptista, J. A., Perdomo, G. R., and Krepinsky, J. J. (1989) The involvement of dimethyl sulfoxide in a bacteriotoxic response of the Ames tester strains TA98 and TA100. *Mutat. Res.* **226**, 39–42.

(61) Göggelmann, W., Maier, F. K., and Pfaendler, H. R. (1986) Mutagenicity of synthetic racemic fecapentaene-12. *Mutat. Res.* **174**, 165–167.

(62) Voogd, C. E., Vertegaal, L. B. J., van der Steeg, M., van der Gen, A., and Mohn, G. R. (1990) Structure, chemical reactivity, and in vitro mutagenic activity in a series of fecapentaene analogs. *Mutat. Res.* **243**, 195–199.

(63) Sodium, R. S., and Shapiro, R. (1988) Reacting acrolein with cytosine and adenine derivatives. *Bioorg. Chem.* **16**, 272–282.

(64) Galliani, G., and Pantarotto, C. (1983) The reaction of guanosine and 2'-deoxyguanosine with acrolein. *Tetrahedron Lett.* **24**, 4491–4492.

(65) Chung, F.-L., and Hecht, S. S. (1983) Formation of cyclic 1,N²-adducts by reaction of deoxyguanosine with α-acetoxy-N-nitrosopyrrolidine, 4-(carbethoxynitrosamino)butanal, or croton-

aldehyde. *Cancer Res.* **43**, 1230–1235.

(66) Piccariello, T. (1989) Studies relating to fecapentaene-12. Ph.D. Dissertation, Virginia Polytechnic Institute and State University, Blacksburg, VA; pp 131–147.

(67) Marcoccia, J. F., Csizmadia, I. G., Yates, P., and Krepinsky, J. J. (1988) An ab initio study of model compounds of fecapentaenes. *J. Mol. Struct.* (*THEOCHEM*) **167**, 359–394.

(68) Dypbukt, J. M., Edman, C. C., Sundquist, K., Kakefuda, T., Plummer, S. M., Harris, C. C., and Grafström, R. C. (1989) Reactivity of fecapentaene-12 toward thiols, DNA, and these constituents in human fibroblasts. *Cancer Res.* **49**, 6058–6063.

(69) Venitt, S., and Bosworth, D. (1988) The bacterial mutagenicity of synthetic all-trans fecapentaene-12 changes when assayed under anaerobic conditions. *Mutagenesis* **3**, 169–173.

(70) Shioya, M., Wakabayashi, K., Yamashita, K., Nagao, M., and Sugimura, T. (1989) Formation of 8-hydroxydeoxyguanosine in DNA treated with fecapentaene-12 and -14. *Mutat. Res.* **225**, 91–94.

(71) Van Tassell, R. L., MacDonald, D. K., and Wilkins, T. D. (1982) Production of a fecal mutagen by *Bacteriodes* spp. *Infect. Immun.* **37**, 975–980.

(72) Van Tassell, R. L., MacDonald, D. K., and Wilkins, T. D. (1982) Stimulation of mutagen production in human feces by bile and bile acids. *Mutat. Res.* **103**, 233–239.

(73) Schiffman, M. H., Van Tassell, R. L., Robinson, A., Smith, L., Daniels, J., Hoover, R. N., Weil, R., Rosenthal, J., Nair, P. P., Schwartz, S., Pettigrew, H., Curiale, S., Batist, G., Block, G., and Wilkins, T. D. (1988) Case-control study of colorectal cancer and fecapentaene excretion. *Cancer Res.* **49**, 1322–1326.

(74) Schiffman, M. H., Andrews, A. W., Van Tassell, R. L., Smith, L., Daniel, J., Robinson, A., Hoover, R. N., Rosenthal, J., Weil, R., Nair, P. P., Schwartz, S., Pettigrew, H., Batist, G., Shaw, R., and Wilkins, T. D. (1989) Case-control study of colorectal cancer and fecal mutagenicity. *Cancer Res.* **49**, 3420–3424.

(75) Van Tassell, R. L., and Wilkins, T. D. (1986) Precursors of fecapentaenes—a preliminary report. *Ann. Inst. Super. Sanita* **22**, 933–980.

(76) Van Tassell, R. L., Piccariello, T., Kingston, D. G. I., and Wilkins, T. D. (1989) The precursors of fecapentaenes: purification and properties of a novel plasmalogen. *Lipids* **24**, 454–459.

(77) Kingston, D. G. I., Duh, C.-Y., Piccariello, T., Keyes, R. F., Van Tassell, R. L., and Wilkins, T. D. (1989) Isolation and structure elucidation of plasmalopentaene-12, the biological precursor of fecapentaene-12. *Tetrahedron Lett.* **30**, 6665–6668.

(78) Pertel, R. (1985) Intestinal-microflora/host-diet interactions in the production of mutagens in the mini-pig. In *Abstracts: Fourth International Conference on Environmental Mutagens*, p 331, Stockholm, June 24–28, 1985.

(79) Schiffman, M. H., Bitterman, P., Viciana, A. L., Schairer, C., Russell, L., Van Tassell, R. L., and Wilkins., T. D. (1988) Fecapentaenes and their precursors throughout the bowel—results of an autopsy study. *Mutat. Res.* **208**, 9–15.

(80) Van Tassell, R. L., and Wilkins, T. D. (1989) Precursors of fecapentaenes in germ-free pigs. Annual Meeting of the American Society for Microbiology, New Orleans, LA, May 14–19, 1989.

(81) Zoeller, R. A., Morand, O. H., and Raetz, C. R. H. (1989) A possible role for plasmalogens in protecting animal cells against photosensitized killing. *J. Biol. Chem.* **263**, 11590–11596.

(82) Morand, O. H., Zoeller, R. A., and Raetz, C. R. H. (1989) Disapperance of plasmalogens from membranes and animal cells subjected to photosensitized oxidation. *J. Biol. Chem.* **263**, 11597–11606.

# Chapter 6

# Mammalian Synthesis of Nitrite, Nitrate, Nitric Oxide, and N-Nitrosating Agents

Michael A. Marletta

*College of Pharmacy, The University of Michigan, Ann Arbor, Michigan 48109-1065*

Reprinted from *Chemical Research in Toxicology,* Vol. 1, No. 5, September/October, 1988

## Introduction

In recent years, a great deal of interest has been focused on exposure to nitrite ($NO_2^-$) and nitrate ($NO_3^-$), primarily because of the mounting evidence that $N$-nitrosamines can cause human cancer (*1–3*). Nitrite and $NO_3^-$ can participate in the endogenous formation of $N$-nitrosamines and, therefore, may play a critical role in the induction of cancer by this class of compounds. Exposure to $NO_2^-$ and $NO_3^-$ has mainly been considered to be environmental, namely, $NO_3^-$ in various foods and $NO_2^-$ addition to foods as a preservative, such as in cured meat products. As will be discussed below, a number of experiments showed that mammals synthesize $NO_3^-$ as judged by the amount excreted in the urine compared to the amount ingested. The main reasons for the interest in this process in my laboratory have been 3-fold: (i) mammalian synthesis of $NO_3^-$ may play some role in the etiology of $NO_3^-$ carcinogenesis; (ii) the biochemistry of this pathway is a novel mammalian reaction with no known biological precedent and is, therefore, of some general interest; and (iii) during the course of our studies it has become clear that the pathway plays an important role in both immune system status and smooth muscle relaxation, hence vasodilation, and therefore, the rational control of this pathway will be of some therapeutic value.

A number of excellent and thorough reviews have appeared on the occurrence and chemistry of $N$-nitrosamines and related compounds since the initial report by Magee and Barnes (*4*) on the carcinogenic activity of $N$-nitroso-dimethylamine (NDMA).[1] The research effort has been

extensive, and only the major findings will be summarized here. This class of compounds has been described as versatile carcinogens, in part, because some are direct acting, such as the nitrosamides, whereas the indirect acting dialkylnitrosamines require metabolic activation (*5*). In addition, they show a high degree of organ-specific effects (*5*), and many are acutely toxic, often leading to extensive liver damage (*5*). A number of DNA adducts have been isolated, with the most extensive studies carried out on the methylating agent derived from the metabolic activation NDMA (*6*). Most relevant to the topic here is the endogenous formation of nitrosamines. A great deal is known about the formation and exposure to preformed nitrosamines (*5*), but comparatively little about the endogenous formation of nitrosamines. In 1962 Druckrey and Preussmann (*7*) proposed that the stomach could provide the acidic environment required for the formation of $N$-nitrosamines from $NO_2^-$ and secondary amines. Studies that followed showed that feeding of amines and $NO_2^-$ led to tumor formation (*8*). Mirvish studied the kinetics of this acid-catalyzed reaction with a variety of amines (*9*). The nitrosating agents under these conditions are $N_2O_4$ and $N_2O_3$. Therefore, in terms of endogenous formation of $N$-nitrosamines, it is generally considered that $NO_3^-$ reduction to $NO_2^-$ or direct exposure to $NO_2^-$ can potentially lead to in vivo nitrosation. The reaction can also be supported by nitrogen oxide gases ($NO_x$).

Endogenous $NO_2^-$ and/or $NO_3^-$ formation has been, until recently, an unanswered question and represents the major focus of this review. Metabolic balance studies carried out about 70 years ago suggested that humans were capable of the synthesis of $NO_3^-$ (*10*). Intact animal experiments can only determine $NO_3^-$ levels, and typically urinary $NO_3^-$, because in the presence of oxyhemoglobin any $NO_2^-$ formed will be oxidized to $NO_3^-$ (*11*). The key

---

[1] Abbreviations: LPS, *Escherichia coli* lipopolysaccharide; EDRF, endothelium-derived relaxing factor; IFN-$\gamma$, interferon-$\gamma$; NDMA, *N*-nitrosodimethylamine; NNM, *N*-nitrosomorpholine; BCG, *Bacillus Calmette-Guerin*.

2428–5/92/0046$06.00/0

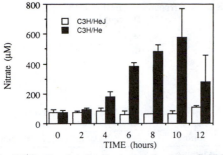

**Figure 1.** LPS-stimulated nitrate synthesis in C3H/He and C3H/HeJ mice. Nitrate levels in blood were measured after an intraperitoneal injection of LPS (10 μg) was administered at time = 0. Each point represents the value per individual mouse ± SEM averaged for three mice (16).

experiments, showing $NO_3^-$ synthesis was indeed a mammalian process, were primarily carried out by Tannenbaum and co-workers (12, 13). In general, they found that humans and rats, when fed a low-$NO_3^-$ diet (about 180 μmol/day for humans), excreted levels well above the ingested amount (12). Furthermore, when the same type of experiments were carried out comparing conventional rats to germ-free rats, the same results were observed; namely, excretion was greater than ingestion (13). This last finding is important in terms of ruling out gut microflora participation in the reaction. During the course of the human experiments, one of the subjects became ill and showed great enhancement in $NO_3^-$ excretion. This led to a number of experiments in rats that ultimately showed that the urinary $NO_3^-$ levels could be elevated about 10-fold when fever induced by an intraperitoneal injection of LPS was administered (14). These results suggested that this elevated synthesis might be related to the immunostimulation known to be brought about by LPS (15). At this time my laboratory entered the area and began a series of experiments that were aimed at determining the origin of this unusual pathway in mammals.

## Macrophage Synthesis of Nitrite and Nitrate

**Initial Studies with Animal Models and Isolated Cells.** A number of approaches to the problem of mammalian synthesis of $NO_2^-$ and $NO_3^-$ were suggested by the results summarized above. We considered the LPS induction of $NO_3^-$ synthesis an important clue to the problem and devised a series of experiments based on mouse immunological models (16). The strains chosen had been characterized as having specific immunocellular defects that could be related back to a particular cell type (17). The three key experiments involved the C3H/He and C3H/HeJ strains, CBA/N and CBA/J strains, and *nu/nu* and *nu/+* strains. The C3H/HeJ strain does not respond to LPS because of a mutation affecting the macrophages of that animal (17). The CBA/N is a B-cell-defective mutant (18), and the *nu/nu* (nude mouse) has immature and nonfunctional T-cells (19). The LPS mouse model then, in principle, could provide two very valuable pieces of information concerning the following questions: (1) was the immune system involved in the LPS-induced stimulation of urinary $NO_3^-$ excretion, and (2) if so what cell or cell types were required for the response? The results were very clear. There was indeed involvement of the immune system in the LPS-induced response, and the cell type responsible for the synthesis of $NO_3^-$ was the macrophage (16). The parent and mutant CBA and *nu* strains synthesized $NO_3^-$ in response to LPS equally well, indicating that functional T- or B-cells were not required for the synthesis (16). However, LPS-induced $NO_3^-$ synthesis was

**Table I. Nitrite and Nitrate Synthesis by Macrophage Cultures[a]**

| cells + treatment | final concentration | | $NO_2^- + NO_3^-$ (nmol/$10^6$ cells) |
|---|---|---|---|
| | $NO_2^-/NO_3^-$ (μM) | $NO_2^-$ (μM) | |
| C3H/He | 17.4 ± 0.4 | 8.3 ± 1.9 | 0.2 |
| C3H/He + LPS | 152.7 ± 1.8 | 91.3 ± 4.9 | 30.0 |
| C3H/HeJ | 15.6 ± 1.2 | | |
| C3H/HeJ + LPS | 15.6 ± 1.1 | | |
| C3H/He + AB/C[b] | 17.8 ± 0.7 | 6.8 ± 0.4 | |
| C3H/He + AB/C + LPS | 96.3 ± 7.0 | 47.9 ± 2.8 | 17.6 |
| C3H/He + Sp cells[c] | 19.6 ± 2.8 | 7.3 ± 0.7 | |
| C3H/He + Sp cells[c] + LPS | 271.2 ± 4.9 | 111.1 ± 1.8 | 56.2 |
| Sp cells[d] | 16.1 ± 0.7 | 8.1 ± 0.4 | |
| Sp cells[d] + LPS | 15.6 ± 0.5 | 7.0 ± 0.3 | |
| cell culture medium + LPS | 16.5 ± 0.4 | | |

[a] Thiogycolate-elicited C3H/He or C3H/HeJ macrophages were cultured with or without LPS (10 μg/mL) for 24 h. Nitrite and nitrate concentration values were the levels present in the supernatant after a 24-h incubation using equal numbers of macrophages and equal volumes. Each value is the mean ± SEM of three cultures. The nmol/$10^6$ cell values were derived by dividing the nitrite/nitrate present in each 24-h supernatant (minus the background nitrite/nitrate in the medium) by the number of cells per plate. [b] Macrophage cultures were depleted of T-cells by antibody and complement lysis. [c] C3H/He spleen cells ($5.5 \times 10^6$) were added to each macrophage culture at the start of the experiment. [d] C3H/He spleen cells ($1.2 \times 10^7$) were cultured with and without LPS (10 μg/mL) for 24 h.

absolutely dependent on macrophages as evidenced by our results with the C3H strains, shown in Figure 1 (16). The results shown were obtained by measuring $NO_3^-$ levels in blood, but the same pattern was observed when urinary levels were measured (16).

To confirm that the cell type responsible for the immunostimulated synthesis was the macrophage, thioglycolate-elicited macrophages were isolated from the peritoneal cavity of C3H/He and C3H/HeJ mice and treated with LPS in culture. The results again were very clear. Macrophages from the LPS-responsive mice (C3H/He) synthesized $NO_3^-$ when treated with LPS, and macrophages from the nonresponsive C3H/HeJ strain were negative (Table I) (16). The experiments with cells in culture are, of course, carried out in the absence of hemoglobin; therefore, $NO_2^-$ levels could now be determined as well. The first experiments measured the levels 24 h after LPS treatment. Of the total, 60% was $NO_2^-$, and a 3:2 ratio of $NO_2^-$ to $NO_3^-$ was observed at all time points. These results were somewhat surprising. We had initially expected that the precursor, unknown to us at that time, would first yield $NO_2^-$ and then this $NO_2^-$ would go on to be oxidized to $NO_3^-$. These findings, however, suggested that the $NO_2^-$ and $NO_3^-$ were derived independently and partitioned from a common intermediate. With cells in culture, a lag phase (discussed below) is first observed, followed by $NO_2^-/NO_3^-$ synthesis that begins and is linear for up to 48 h (20). As mentioned, the ratio of $NO_2^-$ to $NO_3^-$ is 3:2 at all time points.

While our experiments showed that T-cells themselves were not competent to synthesize $NO_2^-$ and $NO_3^-$, T-cells are well known to augment a number of the responses of activated macrophages (21). Therefore, we carried out some experiments to see if T-cells influenced $NO_2^-/NO_3^-$ synthesis by macrophages. Once again we isolated thioglycolate-elicited macrophages and compared the $NO_2^-/NO_3^-$ levels in macrophages alone versus macrophages cocultured with a T-cell-enriched fraction. As is shown in Table I, the macrophages cocultured with T-cells synthesized significantly more $NO_2^-/NO_3^-$ (16). The results in Table I again clearly show, this time in a cell culture experiment, that the T-cell preparation itself does not synthesize any $NO_2^-/NO_3^-$ in response to LPS.

Our initial experiments utilizing the C3H/He and C3H/HeJ intact animal model also allowed us to probe the involvement of a T-cell-dependent mechanism. The potent immunostimulant BCG is a very strong activator of macrophages and acts predominantly through a T-cell-dependent activation mechanism (22). It also has been reported that BCG treatment renders C3H/HeJ mice responsive to LPS (17). The experiment we carried out had two parts. The first compared the effect of BCG treatment on $NO_2^-/NO_3^-$ synthesis in C3H/He and C3H/HeJ mice, and second, we measured the effect of subsequent LPS treatment on $NO_2^-/NO_3^-$ synthesis in these same mice. The effect of BCG treatment on both strains of mice was dramatic. The background level of $NO_3^-$ excretion in these mice was 2 $\mu$mol/day and at the peak of the BCG infection reached up to 90–100 $\mu$mol/day (16). Additionally, the C3H/HeJ mice that normally do not respond to LPS were now rendered LPS responsive (16). Furthermore, C3H/HeJ macrophages isolated from the BCG-treated mice were responsive to LPS in culture (20).

The BCG and T-cell experiments both suggested the involvement of T-cell-derived lymphokines in this immunostimulated synthesis of $NO_2^-$ and $NO_3^-$. To directly test this idea, we isolated a crude lymphokine preparation from mice treated with concanavalin A, followed by isolation of a spleen cell supernatant. This supernatant, prepared from a fraction that is mainly T-cells in the spleen, contains a number of endogenous lymphokines. This lymphokine mixture showed a concentration-dependent stimulation of $NO_2^-/NO_3^-$ synthesis in macrophage cell cultures (20). The predominant lymphokine in this spleen cell preparation has been shown to be interferon-$\gamma$ (23–25). Therefore, we tested pure recombinant murine interferon-$\gamma$ and found that it acts directly to stimulate $NO_2^-/NO_3^-$ synthesis for cells in culture, in a concentration-dependent manner (20). Both C3H/He and C3H/HeJ cells responded to IFN-$\gamma$ alone and in a synergistic way to the dual stimuli of LPS and IFN-$\gamma$ (20). There is always some loss of viability with LPS treatment, but very little with IFN-$\gamma$; consequently, the costimulation with LPS and IFN-$\gamma$ allows for maximum stimulation while limiting the loss in viability (20).

**Studies with Cell Lines.** We have also examined a number of established murine macrophage cell lines for the ability to synthesize $NO_2^-/NO_3^-$ in response to various stimuli (26). Not all the cell lines responded to LPS or IFN-$\gamma$ alone, but they all synthesized $NO_2^-/NO_3^-$ when treated with the two together. This finding that established cell lines respond differently than primary cultures to various stimuli is not unusual. The RAW 264.7 line very closely resembled freshly isolated, LPS-responsive C3H/He macrophages; consequently, this cell line has been used extensively in our biochemical studies. However, the J774A.1 response is quite similar to that of the freshly isolated cells, and we have made use of this line as well (26).

## Macrophage Synthesis of *N*-Nitrosamines

**_N_-Nitrosamine Synthesis by Peritoneal Cells and Cell Lines.** The results showing that about 60% of the total $NO_2^-/NO_3^-$ is $NO_2^-$ is important with regard to considerations of the potential contribution of macrophages toward the endogenous formation of *N*-nitrosamines. These findings suggest that macrophages could contribute in the following ways: (1) The $NO_2^-$ portion would become oxidized by oxyhemoglobin to $NO_3^-$, and

**Scheme I. *N*-Nitrosomorpholine (NNM) by Activated J774 Macrophages**

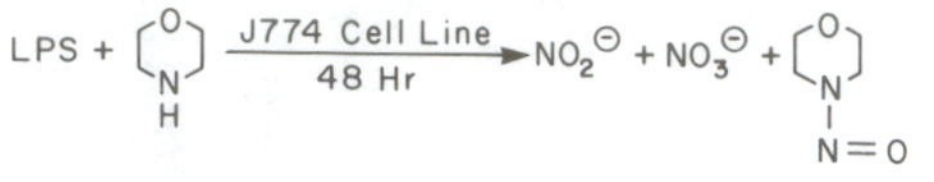

**Table II. N-Nitrosation of Amines by Activated RAW 264.7 Macrophages[a]**

| amine | $NO_2^-$ ($\mu$M) | *N*-nitrosamine (nM) |
|---|---|---|
| diethylamine | 57 | 4 |
| dibenzylamine | 77 | 23 |
| methylbenzylamine | 60 | 255 |
| morpholine | 60 | 1680 |

[a] Incubations were carried out for 72 h and contained the specific amine (5 mM), LPS (10 $\mu$g/mL), and interferon-$\gamma$ (500 units/mL).

then this $NO_3^-$ would contribute to the total body burden of $NO_3^-$. Upon partial reduction of this $NO_3^-$ back to $NO_2^-$ by bacteria, *N*-nitrosation would occur catalyzed by acidic conditions, such as in the stomach. (2) The $NO_2^-$ synthesized could directly participate in nitrosations in the macrophage itself, especially considering the relatively acidic conditions of the lysosomes (pH = 4.5) (27). (3) The nitrosation is independent of $NO_2^-$, but the nitrosating agent could be derived from an intermediate between the precursor and the products, $NO_2^-$ and $NO_3^-$.

The first series of experiments were designed to answer the question whether immunostimulated macrophages were capable of generating *N*-nitrosamines. These initial experiments were carried out with the J774A.1 cell line that was cultured normally except that a secondary amine, morpholine, was added to the cell culture medium. The results of these experiments were particularly interesting (28). Macrophages, activated with LPS, synthesized *N*-nitrosomorpholine (NNM) (Scheme I). The levels produced depended on the conditions of the experiment and ranged from 100- to 1000-fold lower than the amount of $NO_2^-/NO_3^-$ synthesized. A number of amines were tested as substrates for nitrosation with activated RAW 264.7 cells, and the results of these experiments are shown in Table II. Also included in this table are the corresponding amounts of $NO_2^-$ synthesized. The yield of N-nitrosated products paralleled the chemical reactivity of these compounds toward chemical nitrosation, with morpholine > methylbenzylamine > dibutylamine > diethylamine. The yield of $NO_2^-$ was not influenced by the amine.

To probe mechanism 2 mentioned above, $NO_2^-$ was directly added to macrophage cultures containing the morpholine. Somewhat surprisingly, nitrosation was independent of the added $NO_2^-$, even when the added $NO_2^-$ (150 $\mu$M) was about three times the concentration that would have accumulated at the end of a 24-h incubation (28). This result suggests that mechanism 2, the reaction of $NO_2^-$ synthesized by the macrophage with amines to generate *N*-nitrosamines, is not occurring. We cannot rule out the possibility that the added $NO_2^-$ does not enter the cell; however, on the basis of other experiments this explanation for these observations seems unlikely. Another important finding is that only activated macrophages can carry out this N-nitrosation reaction. Morpholine or other amines added to unstimulated cells did not synthesize any nitrosamine (28). This requirement for activated cells was observed in three different cell lines, J774A.1, PU5-1.8, and WEHI-3 (28). These results suggest that the nitrosation reaction is a property specific to the activated cell and the nitrosation itself results from the reaction of the secondary

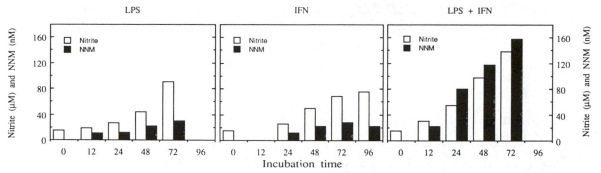

**Figure 2.** Nitrite and *N*-nitrosomorpholine by C3H/He macrophages. Each of the three culture plates contained freshly isolated macrophages [(1–2) × 10⁶ cells/mL]. The left panel shows cells treated with LPS (1 μg/mL), the center panel with IFN-γ (500 units/mL), and the right panel with both LPS and IFN-γ. The cells were treated at time = 0, and morpholine (2.5 mM) was also added at that time (28).

amine with an intermediate generated from the precursor and between the final products, $NO_2^-$ and $NO_3^-$. While all the studies above have been carried out with LPS-stimulated cells, the typical response with LPS and IFN-γ costimulation was observed. Namely, with freshly isolated C3H/He cells, LPS and IFN-γ alone led to $NO_2^-$ synthesis and NNM synthesis, and the dual treatment led to an enhanced synthesis of both compounds (Figure 2) (28). These overall results regarding the catalysis of N-nitrosations by activated macrophages and the potential role of this process in chemically induced carcinogenesis will be discussed below.

### Biochemistry of the Pathway

**General Characteristics.** At this point we are making considerable headway on the enzymology of this unusual biochemical pathway. Initially, a number of general characteristics of the reaction were studied. When cells in culture are treated with LPS and/or IFN-γ, there is a lag phase before synthesis of $NO_2^-$ and $NO_3^-$ begins. This lag phase is dependent on the concentration of the stimulant and on whether LPS or IFN-γ were present alone or together but typically ranged from 8 to 12 h (20). Once the synthesis of $NO_2^-/NO_3^-$ begins, the formation continues linearly for about 48 h (20). The observed time lag suggested that protein synthesis might be required before the synthesis of $NO_2^-/NO_3^-$ could take place. This turned out to be the case, as experiments with the protein synthesis inhibitor cycloheximide demonstrated. Cycloheximide inhibited $NO_2^-/NO_3^-$ synthesis in a concentration-dependent manner, with the maximal inhibition at 0.5 μg/mL. Although cycloheximide was toxic at this concentration, loss in viability cannot explain the nearly 100% inhibition observed. Cycloheximide was effective only when present at the outset of the experiment. If it was added after the lag phase was over, then it was not inhibitory (29).

As mentioned above, after the lag phase, synthesis of $NO_2^-$ and $NO_3^-$ continues linearly for 48 h with cells in culture. A chemically reasonable expectation was that the $NO_3^-$ would be derived by oxidation of $NO_2^-$ which would come from the precursor, whatever that might be. The results proved otherwise. After the lag, synthesis begins and the ratio of $NO_2^-$ to $NO_3^-$ is constant (3:2) at all times observed, including as early as 12 h after stimulation (20, 26). Furthermore, $NO_2^-$ or $NO_3^-$ added to the cells is not interconverted over the time course of the experiment (20), and if either anion is added to cultures at the time of stimulation, when analyzed at the end of the experiment, the expected ratio of $NO_2^-$ to $NO_3^-$ was observed plus the added anion (20). Therefore, under the cell culture con-

ditions $NO_2^-$ and $NO_3^-$ are not interconverted and are apparently not derived from one another. The most likely interpretation, and one that is further supported below, is that the $NO_2^-$ and $NO_3^-$ are partitioning from a common intermediate.

The macrophage is a cell vital to mammalian defense against invading organisms. As such, a large number of complex biochemical changes accompany the activation of the resting macrophage to one that is activated and then competent to kill bacteria, fungi, and other invading cells such as tumor cells. One mechanism of killing that becomes greatly enhanced once the macrophage is activated in the so-called respiratory burst (30). This pathway involves the formation of superoxide ($O_2^-$) with subsequent formation of hydrogen peroxide (HOOH), hypochlorous acid (HOCl), and, in the presence of divalent metals, hydroxyl radical (·OH). Because these compounds can either promote or themselves directly carry out oxidation chemistry, we could not rule out that the $NO_2^-/NO_3^-$ were synthesized via a somewhat random solution oxidation of amine precursors by the components of the respiratory burst. We tested this hypothesis with a number of experiments. The most direct study used a parent macrophage cell line, the J774.16 line that is competent for the respiratory burst, and a mutant of that line, the J774 C3C line that lacks the respiratory burst (31, 32). Both cell lines, in response to LPS and IFN-γ, synthesized equivalent amounts of $NO_2^-$ and $NO_3^-$ in the expected ratio of 3:2 (33). The following experiments were a series of indirect studies, but all proved negative. With the same two cell lines and primary cultures of C3H/He macrophages, superoxide dismutase (300 units/mL), catalase (2000 units/mL), or mannitol (100 mM), when added to the cell culture medium, had no effect on the total $NO_2^-/NO_3^-$ synthesized or on the ratio of $NO_2^-$ to $NO_3^-$. Phorbol myristate acetate is a compound that activates the respiratory burst (30). Treatment of the J774.16 and J774 C3C lines or C3H/He macrophages with phorbol myristate acetate either alone or in cells actively synthesizing $NO_2^-$ and $NO_3^-$ did not produce or enhance $NO_2^-/NO_3^-$ synthesis (33). Generating $O_2^-$/HOOH in the culture medium with xanthine oxidase or the addition of allopurinol to the medium did not affect $NO_2^-/NO_3^-$ synthesis (33). Allopurinol was added to inhibit intracellular formation of HOOH and $O_2^-$ from xanthine oxidase. All of the above experiments are consistent with the lack of involvement of the respiratory burst in macrophage synthesis of $NO_2^-$ and $NO_3^-$.

**Precursor/Product Studies.** Early experiments with cells in culture suggested that free amino acids might be the precursor(s) to $NO_2^-$ and $NO_3^-$. On the basis of those

**Table III. Biochemical Precursor to Nitrite and Nitrate[a]**

| medium | $NO_2^- + NO_3^-$ (nmol) | % |
|---|---|---|
| SMEM | 175 | 100 |
| MEM | 54 | 31 |
| MEM + arginine (2 mM) | 212 | 121 |
| MEM + glutamine (4 mM) | 40 | 23 |
| MEM + ornithine (2 mM) | 44 | 27 |
| MEM + citrulline (2 mM) | 69 | 39 |

[a] SMEM is powdered Eagle's minimal essential medium without phenol red, supplemented with sodium bicarbonate (2.0 g/L), sodium pyruvate (110 mg/L), glucose (3.5 g/L), L-glutamine (584 mg/L), penicillin (50 units/mL), streptomycin (50 µg/mL), Hepes (15 mM), and 10% calf serum (final pH 7.3–7.4). MEM is the same media except it was free of all amino acids and was vitamin and L-glutamine free. Activated RAW 264.7 cells (1 × 10^6 cells/mL) were incubated under the conditions above, and cell culture supernatants were analyzed as described (33). Other nonprecursors of note include ammonia, urea, hydroxylamine, acetohydroxamate, and all other common amino acids.

preliminary results we carried out experiments with RAW 264.7 cells. The cells were activated with LPS and IFN-γ as usual, and then the medium was replaced with a medium that was able to keep the cells viable during the course of the experiment but did not contain any amino acids (33). As shown in Table III, when all the amino acids are removed from the medium, the level of $NO_2^-/NO_3^-$ synthesized drops to about 30% of the value in the complete medium. Therefore, this experiment was repeated where the replacement medium contained only specific single amino acids. The first experiments focused on the nitrogenous amino acids. In fact, the first studies were with L-glutamine as it was hypothesized initially that ammonia was a likely precursor to $NO_2^-/NO_3^-$. L-Glutamine did not produce any $NO_2^-/NO_3^-$ over the 30% residual synthesis. We refer to the 30% residual synthesis in the absence of free amino acids as background synthesis. In fact, the only amino acid that restored the $NO_2^-/NO_3^-$ levels to their former levels was L-arginine. As can be seen in Table III, when L-arginine is the only amino acid added back at 4 mM, the levels of $NO_2^-/NO_3^-$ synthesized were 120% that of normal media. L-Arginine is normally present in cell culture media at 0.6 mM, so at 4 mM we have assumed that the increased amount of $NO_2^-$ and $NO_3^-$ over 100% is due to saturation of the enzyme(s) involved. Supporting this is the finding that if arginine is added back to a final concentration of 0.6 mM, then the levels synthesized are about 100%. Nonprecursors of note include ornithine, urea, ammonia, and guanidine (33). Activated macrophages express and excrete the enzyme arginase which will produce ornithine and urea from arginine (34); therefore, finding these two compounds as nonprecursors was particularly important.

The experiments described above show that L-arginine is the major, and perhaps the only, precursor to $NO_2^-$ and $NO_3^-$. Hibbs and co-workers have reported results very similar to our own (35). Experiments with ^15N-labeled arginine showed that L-arginine is the only precursor, and consequently, this represents a very unusual pathway of amino acid metabolism. L-Arginine has four nitrogen atoms, any one of which is a potential candidate for oxidation. The evidence that ammonia is not a precursor, plus other evidence below, rules out the pathway of arginine hydrolysis first, followed by oxidation of the ammonia generated. The labeling experiments were carried out with [*guanidino*-^15N_2]-L-arginine (see Scheme II). This molecule has the two chemically equivalent nitrogens of the guanidino end of arginine labeled, enriched 95%. Experiments were carried out with substitution of this labeled L-arginine for the unlabeled arginine. The analysis was

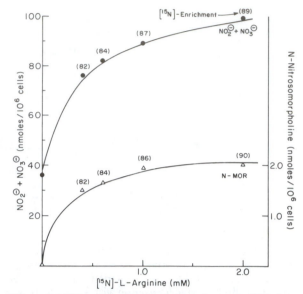

**Figure 3.** ^15N enrichment of nitrite, nitrate, and N-nitrosomorpholine. Activated RAW 264.7 macrophages (1.2 × 10^6 cells/mL) were incubated with various concentrations of [*guanidino*-^15N_2]-L-arginine and morpholine (5 mM) for 48 h. Cell culture supernatants were analyzed as described (33).

**Scheme II. Origin of the Nitrogen Atom in the Nitrite and Nitrate Derived from [*guanidino*-^15N_2]-L-Arginine[a]**

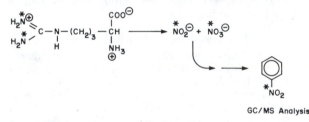

[a] The asterisk indicates ^15N-labeled nitrogens.

by GC/MS after conversion of the $NO_2^-/NO_3^-$ to $NO_3^-$, followed by nitration of benzene to yield nitrobenzene and determination of the ^15N enrichment of the nitrobenzene. Results are shown in Figure 3. At the highest arginine concentration tested (2 mM), the $NO_2^-$ and $NO_3^-$ enrichment was 89% (33). Therefore, very little dilution of the added labeled arginine occurred, leading to the conclusion that L-arginine is the only precursor and that one or both of the two chemically equivalent guanidino nitrogens are oxidized to the product anions. The slight dilution of the label is probably due to endogenous L-arginine in the cell, as the figure shows that as the concentration of [^15N]arginine is lowered, the final enrichment of the $NO_2^-$ and $NO_3^-$ drops as well, consistent with some competing endogenous arginine.

The same type of experiments were carried out with respect to the nitrosation reaction (33). These results are also shown in Figure 3. The ^15N enrichment in the product nitrosamine NNM was 90%, showing that the nitrosyl group transferred is from those same chemically equivalent nitrogens (Scheme III). Analogous to the results with $NO_2^-/NO_3^-$, as the concentration of labeled L-arginine is lowered, the enrichment in NNM falls as well. As was stated above, we determined that nitrosation was independent of $NO_2^-$. The labeled arginine results, then, are consistent with the hypothesis that secondary amines are reacting with some intermediate or intermediates between the precursor, L-arginine, and the product anions, $NO_2^-$ and $NO_3^-$.

Scheme III. Origin of the Nitrosyl Group in Macrophage-Catalyzed N-Nitrosations[a]

**Scheme III. Origin of the Nitrosyl Group in Macrophage-Catalyzed N-Nitrosations[a]**

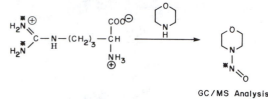

GC/MS Analysis

[a] The asterisk indicates $^{15}$N-labeled nitrogens.

The next question we chose to answer is what happened to the rest of the amino acid. This would, of course, tell us if only one or both of the labeled nitrogens eventually lead to $NO_2^-$ and $NO_3^-$. These experiments were carried out with [U-$^{14}$C]-L-arginine (33). Activated RAW 264.7 cells were incubated with this labeled arginine, and after 48 h, the cell culture medium was analyzed by TLC (33). Only one new spot containing radioactivity was detected on the plate, and it comigrated with L-citrulline. In another reaction the cell culture supernatant was purified by ion-exchange chromatography, and the $R_f$ containing the radioactivity was the same as authentic L-citrulline in two different solvent systems (33). When the amount of L-citrulline was compared to the amount of $NO_2^-$/$NO_3^-$ synthesized, we found that more citrulline was present at every time point (33). Activated macrophages secrete the enzyme arginase which would generate L-ornithine and urea. In a separate experiment with labeled ornithine we found that it is carbamoylated by macrophages leading to L-citrulline, so it appears that the additional L-citrulline synthesized in excess of $NO_2^-$/$NO_3^-$ results from this second pathway. Therefore, we can conclude that only one of the two chemically equivalent guanidino nitrogens of L-arginine is oxidized, leading to the products $NO_2^-$, $NO_3^-$, and L-citrulline. Oxidation of a guanidino nitrogen is an unprecedented reaction in biochemical systems and would require a very potent oxidizing agent.

A cell-free system containing the enzymatic activity would be useful for a number of experiments, ultimately including the purification of the enzyme(s) involved in this pathway. Various techniques for breaking cells that had been previously activated were studied, and the most efficient in terms of activity recovery turned out to be sonication. Differential centrifugation of this cellular lysate led to activity recovery exclusively in the 100000$g$ supernatant (46). Considering that the precursor to $NO_2^-$ and $NO_3^-$ is L-arginine, it is quite reasonable that the activity would be found in the cytosol of the cell. It was expected that reducing equivalents would be necessary to support the cell-free reaction. NADPH, NADH, and L-ascorbic acid were added to the cell-free system, and only NADPH was able to support the reaction. In addition, we found that the addition of $Mg^{2+}$ enhanced the activity. For most of the ongoing biochemical studies we are using this cell-free system (46).

**Intermediates in the Reaction.** As the facts regarding the pathway unfold, the most reasonable chemical scheme would seem to involve N-hydroxylation of L-arginine as the first step in the pathway. The product of this hydroxylation of arginine would lead to $N^G$-hydroxy-L-arginine. This compound has been recently synthesized at the University of Michigan.[2] They have also synthesized the N-hydroxy analogue of L-homoarginine.[2] We have previously found that homo-L-arginine is about 80% effective

as L-arginine in $NO_2^-$/$NO_3^-$ synthesis (33). Preliminary experiments with both $N^G$-hydroxy-L-arginine and $N^G$-hydroxy-L-homoarginine showed them to be precursors to $NO_2^-$ and $NO_3^-$.[2] Rigorous proof that this is an intermediate along the pathway is under way, but taken together with all our results it is reasonable that the first step involves this very unusual N-hydroxylation of L-arginine, apparently by a specific monooxygenase-like enzyme.

We have recently directed our attention toward nitric oxide as an intermediate. There are a number of reasons to consider nitric oxide as an intermediate that will be discussed below. Using the cell free activity, we have recently shown the formation of nitric oxide (46). Nitric oxide formation is dependent on NADPH, on supernatant from activated cells, and on L-arginine. Nitric oxide is very reactive toward molecular oxygen, and as discussed below, this is the first step in the solution decomposition of N=O which ultimately yields $NO_2^-$ and $NO_3^-$. Therefore, in a closed system, but in the presence of oxygen, because it is required for $NO_2^-$/$NO_3^-$ synthesis, we measured the rate of N=O formation and found it reaches a steady-state level that reflects its formation and decomposition (46). In a companion experiment we looked at the ratio of $NO_2^-$ to $NO_3^-$ and found it to be vastly different (1:2.2) than that seen in experiments with intact cells (3:2). This change in the ratio is discussed below. We also carried out experiments showing that the N=O generated from L-arginine in this cell-free system is $^{15}$N-labeled when [*guanidino*-$^{15}N_2$]-L-arginine is used (46).

## Discussion

What we have described above is a unique pathway of oxidation of the amino acid L-arginine expressed in macrophages treated with immunostimulants such as LPS and IFN-γ. Strong immunostimulants, such as BCG, provide an indication of the quantitative importance of the pathway (16). When C3H/He or C3H/HeJ mice are treated with BCG, urinary $NO_3^-$ excretion increases almost 50-fold from 2 μmol/24 h to 90–100 μmol/24 h (16). With cells in culture, the pathway is only expressed when the cells are activated. Toxicological concern arises because we have shown that if secondary amines are present in the cell culture medium, the formation of carcinogenic N-nitrosamines occurs (28). The potential for this pathway to contribute to the endogenous formation of nitrosamines is unknown. The biological role of this pathway is an important and, at this time, unanswered question. Macrophages activated with LPS and/or IFN-γ undergo complex biochemical changes. After activation the cells are immunologically competent to kill bacteria, fungi, and tumor cells (35). This cytocidal activity or, in some cases, cytostatic activity has been recently shown by Hibbs and colleagues to require L-arginine (36). These findings suggest a role for the L-arginine to $NO_2^-$/$NO_3^-$ pathway in these cell-killing activities of activated macrophages. In fact, the cytostatic activity of $NO_2^-$ on a number of microorganisms has been known and taken advantage of for some time. Experiments with $NO_2^-$ have shown this is not the explanation for this pathway. On the other hand, we had hypothesized that perhaps macrophages were synthesizing a metabolite of L-arginine that had cell-killing or growth-inhibitory properties (33). It was reported that the arginine requirement for growth inhibition of tumor cells was in part due to the inhibition of DNA synthesis (36). An attractive candidate for this DNA synthesis inhibition activity is $N^G$-hydroxy-L-arginine because the guanidino moiety is now N-hydroxylated and is, at that end of the molecule, an N-hydroxyguanidine. N-

[2] P. Nanjappan, P. S. Yoon, M. A. Marletta, and R. W. Woodard, unpublished results.

**Scheme IV. Pathway of L-Arginine to Nitrite, Nitrate, and L-Citrulline**

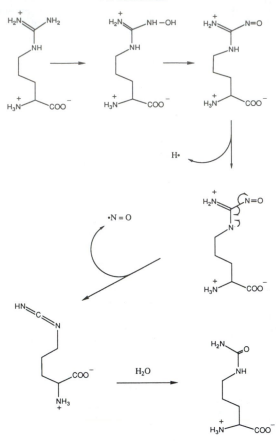

**Scheme V. Solution Decomposition Reactions of Nitric Oxide**

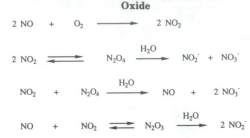

Hydroxyguanidine is as potent a DNA synthesis inhibitor as the well-known and clinically useful compound *N*-hydroxyurea. So we speculated that perhaps this pathway represented nature's equivalent to *N*-hydroxyurea (*33*). A large number of *N*-hydroxyguanidines have been synthesized and have a relatively broad range of antitumor and antiviral activity (*37*).

Very recently we have identified N=O as an intermediate in the reaction pathway (*46*). Scheme IV details the reaction sequence as it has now been pieced together. This will be discussed below. The results involving N=O are particularly important in that N=O has recently been identified as a chemical signaling agent released by endothelial cells (*39*). The N=O released, called EDRF (endothelium-derived relaxing factor) prior to its identification, is responsible for vascular smooth muscle relaxation or, in other words, vasodilation (*39*). The proposed mechanism of action of EDRF is quite interesting. The actual relaxation of smooth muscle is brought about by cGMP (*39*). EDRF leads to the increase in cGMP by activating the enzyme guanylate cyclase presumably by binding to a nonactive site heme (*40, 41*). Vasodilatory agents, such as nitroglycerin and amyl nitrite, act to mimic the biological formation of N=O by producing N=O through interaction with endogenous thiols. Moncada and co-workers identified N=O as EDRF (*39*), and at the time our work on N=O as an intermediate in the macrophage pathway was completed, they showed that the N=O from endothelial cells was derived from L-arginine by carrying out experiments identical with our own with [*guanidino*-$^{15}$N$_2$]-L-arginine (*42*). These findings are significant in that it has now been shown that two different mammalian cell types, which carry out quite different biological functions,

express this same unusual pathway involving the oxidation of L-arginine to N=O and ultimately to NO$_2^-$ and NO$_3^-$. We do not know what effect N=O has on a resting macrophage, but our present speculation is that it acts as an intracellular signal in the macrophage leading to the activated cell that is now competent for its vital cell-killing activities. The role then for L-arginine in this activation process is to serve as a substrate leading to N=O formation. Nitrite, NO$_3^-$, and citrulline represent the true end products of the reaction. We do not yet know if N=O leads to cGMP increases in the macrophage or if N=O can directly activate a resting macrophage, but these experiments are under way.

Scheme IV is consistent with all the evidence to this point. Only L-arginine and close structural homologues can serve as substrates (*33*). L-Homoarginine is one of the best analogues, being about 80% as efficient as L-arginine (*33*). The enzyme(s) is (are) specific for the L (*S*) stereochemistry at the $\alpha$-carbon. Both in cells and in the cell-free system, D-arginine is not a substrate and in addition it is not an inhibitor (*33*). Our results show that ammonia is not a precursor, which rules out hydrolysis of L-arginine followed by oxidation of ammonia. The experiments with $^{15}$N-labeled arginine show which nitrogen is oxidized. This leads to the conclusion that the first intermediate must be an N$^G$-hydroxylated arginine as shown in the scheme. Therefore, we synthesized this compound (and the homo analogue) and have indeed shown that they are precursors to NO$_2^-$ and NO$_3^-$.[2] The next two steps are chemically reasonable and would account for the products observed, although they remain speculative at this point. The first is a 2-e$^-$ oxidation to a nitrosoamidine intermediate followed by a second oxidation (loss of a hydrogen atom). The radical species produced should fragment as shown, leading to the direct formation of N=O. The amino acid moiety remaining is a carbodiimide, and a likely reaction, with chemical precedent, would be nucleophilic attack by water to form L-citrulline.

Scheme V shows the solution decomposition reactions of N=O. These reactions are also consistent with our results. The reactions, as indicated, allow for the independent formation of NO$_2^-$ and NO$_3^-$. Furthermore, depending on the absolute concentrations of the intermediates and the rate constants of each of the reactions, it is expected that the ratio of NO$_2^-$ to NO$_3^-$ would vary. That is consistent with our observations. The specific activity of the cell-free system is greater than that of intact cells, and our ratio, as mentioned above, changes from 3:2 for intact cells to 1:2.2 for the cell-free system. The solution decomposition also accounts for the observed N-nitrosation reactions. Two of the products are excellent N-nitrosating agents, N$_2$O$_4$ and N$_2$O$_3$. Both would transfer a labeled nitrosyl group as our labeling studies have shown.

So what role does this pathway play in the endogenous formation of *N*-nitrosamines? That very interesting and important question remains unanswered, but we can speculate on the importance. First, with respect to the pathway itself, there are at least two different mammalian cell types that synthesize N=O from L-arginine. The function in endothelial cells is to signal the formation of cGMP leading to smooth muscle relaxation. We speculate that N=O is serving as an intracellular signal in macrophages leading to the activation of these cells to the bactericidal/tumoricidal state. The regulation of the pathway in the two cell types appears to be very different. In endothelial cells stimulation with bradykinin or acetylcholine leads to a rapid burst of a small amount of N=O (*39*). Calcium ionophores can lead to a more extended synthesis, but still in relatively small amounts (*39, 43*). The enzyme, though, is apparently present and is either rapidly activated or perhaps stimulated by release of sequestered substrate. In the macrophage, however, the activity is not expressed until protein synthesis has occurred, which typically is about 8–12 h. Once the synthesis begins, it continues linearly over a much longer period of time.

The enzymology of the pathway is proving to be somewhat difficult. If indeed the biological role for this pathway is the formation of N=O, then it seems most reasonable that the enzyme is either a multifunctional protein or an oligomeric structure. The exact details of providing the reducing equivalents to the enzyme(s) also are not known, but it is possible that a reductase may also be involved. The first step, hydroxylation of the guanidine moiety of L-arginine, is chemically difficult and most likely will involve a metalloprotein of some sort. We expect to have answers regarding the enzymology of the process very soon.

While we have yet to show $NO_2^-/NO_3^-$ formation by human-derived macrophages in culture, there is evidence that immunostimulation leads to an increase in urinary $NO_3^-$ excretion (*43*). The action of the vasodilatory drugs in humans suggests that the endothelial cell pathway is functioning in humans. Endothelial cells studied to this point include those isolated from rabbit, bovine, and porcine aorta. The experiments with animals and humans maintained on a low-$NO_3^-$ diet show very clearly that they excrete more than they ingest (*12, 13*). This basal level may very well be due to the N=O produced by endothelial cells to regulate blood vessel homeostasis. When rats or mice are immunostimulated, an increase in urinary $NO_3^-$ is observed above this basal level, and with strong immunostimulants the increase can be very large. On the basis of all that has been summarized in this account it is clear that N=O is a mammalian metabolite. The solution decomposition of N=O will involve the formation of N-nitrosating agents. Indeed, we have demonstrated the formation of N-nitrosamines by activated macrophages. As stated above, the role this process plays in the endogenous formation of N-nitrosamines is unknown. The two critical points in this consideration are the direct participation of the $N_2O_4$ or $N_2O_3$ derived from the solution decomposition of N=O or the indirect participation by generally increasing the total body burden of $NO_3^-$. The role of this pathway in chronic inflammation and carcinogenesis is also unknown, but some clinical evidence suggests a possible link (*44*). The pathway has apparently evolved to control at least one and possibly two very important aspects of biological function, namely, immune system status and smooth muscle relaxation. Understanding the enzymology of this process should point the way toward the rational design of drugs to control these very important functions.

**Acknowledgment.** I am grateful to the many people who have contributed to this work over the last few years. I especially thank Dr. Dennis Stuehr, who as a graduate student initiated these studies in my laboratory, and Dr. Radha Iyengar, whose results extended the initial findings into some very interesting directions. My association with Prof. Steven R. Tannenbaum of MIT deserves mention on two accounts. It was he who brought the problem of mammalian synthesis of nitrate to my attention, and he has been a source of valuable discussions ever since. I also thank Dr. John S. (Pete) Wishnok for productive discussions and now for his role as a valued collaborator. My new colleagues at the University of Michigan have been very helpful, and the most notable include Profs. Ron Woodard, Jim Coward, and Jules Shafer. I also gratefully acknowledge financial support from the National Cancer Institute (CA26731) and the College of Pharmacy of the University of Michigan.

**Registry No.** $NO_2^-$, 14797-65-0; $NO_3^-$, 14797-55-8; NO, 10102-43-9; arginine, 74-79-3.

## References

(1) Magee, P. N., Ed. (1982) *Nitrosamines and Human Cancer*, Banbury Report 12, Cold Spring Harbor Press, Cold Spring Harbor, NY.

(2) Craddock, V. (1983) "Nitrosamines and human cancer: proof of an association?". *Nature* **306**, 638.

(3) Bartsch, H., and Montesano, R. (1984) "Relevance of nitrosamines to human cancer". *Carcinogenesis* **5**, 1381–1393.

(4) Magee, P. N., and Barnes, J. N. (1956) "The production of malignant primary hepatic tumors in rat by feeding dimethylnitrosamine". *Br. J. Cancer* **10**, 451–458.

(5) Preussmann, R., and Stewart, B. W. (1984) "*N*-Nitroso carcinogens". In *Chemical Carcinogens* (Searle, C. E., Ed.) 2nd ed., Vol. 2, pp 643–828, American Chemical Society, Washington, DC.

(6) Magee, P. N., and Barnes, J. M. (1967) "Carcinogenic nitroso compounds". *Adv. Cancer Res.* **10**, 163–246.

(7) Druckrey, H., and Preussmann, R. (1962) "Zur entstehung carcinogener nitrosamine am biespiel des tabakrauchs". *Naturwissenschaften* **49**, 498–499.

(8) Sander, J., Schweinsberg, F., Ladenstein, M., Benzing, H., and Wahl, S. H. (1973) "Messung der renalen nitrosaminausscheidung am hund zum nachweis einer nitrosaminbildung in vivo". *Hoppe-Seyler's Z. Physiol. Chem.* **354**, 384–390.

(9) Mirvish, S. S. (1975) "Formation of *N*-nitroso compounds: chemistry, kinetics and in vivo occurrence". *Tox. Appl. Pharmacol.* **31**, 325–351.

(10) Mitchell, H. H., Shonle, H. A., and Grindley, H. S. (1916) "The origin of the nitrates in the urine". *J. Biol. Chem.* **24**, 461–490.

(11) Kosaka, H., Imaizumi, K., Imai, K., and Tyuma, I. (1979) "Stoichiometry of the reaction of oxyhemoglobin with nitrite". *Biochim. Biophys. Acta* **581**, 184–188.

(12) Green, L. C., Ruiz de Luzuriaga, K., Wagner, D. A., Rand, W., Istfan, N., Young, V. R., and Tannenbaum, S. R. (1981) "Nitrate biosynthesis in man". *Proc. Natl. Acad. Sci. U.S.A.* **78**, 7764–7768.

(13) Green, L. C., Tannenbaum, S. R., and Goldman, P. (1981) "Nitrate synthesis in the germfree and conventional rat". *Science (Washington, D.C.)* **212**, 56–58.

(14) Wagner, D. A., Young, V. R., and Tannenbaum, S. R. (1983) "Mammalian nitrate biosynthesis: incorporation of $^{15}NH_3$ into nitrate is enhanced by endotoxin treatment". *Proc. Natl. Acad. Sci. U.S.A.* **80**, 4518–4531.

(15) Berry, L. J. (1977) "Bacterial toxins". *Crit. Rev. Toxicol.* **5**, 239–318.

(16) Stuehr, D. J., and Marletta, M. A. (1985) "Mammalian nitrate biosynthesis: mouse macrophages produce nitrite and nitrate in response to *Escherichia coli* lipopolysaccharide". *Proc. Natl. Acad. Sci. U.S.A.* **82**, 7738–7742.

(17) Vogel, S. N., Weinblatt, A. C., and Rosenstreich, D. L. (1981) "Inherent macrophage defects in mice". In *Immunologic Defects in Laboratory Animals* (Gershwin, M. E., and Merchant, B., Eds.) Vol. 1, pp 327–357, Plenum, New York.

(18) Rosenstreich, D. L., Vogel, S. N., Jaques, A., Wahl, L. M., Scher, I., and Mergenhagen, S. E. (1978) "Differential endotoxin sensitivity of lymphocytes and macrophages from mice with an X-linked defect in B-cell maturation". *J. Immunol.* **121**, 685–690.

(19) Kindred, B. (1981) "Deficient and sufficient immune systems in the nude mouse". In *Immunologic Defects in Laboratory Animals* (Gershwin, M. E., and Merchant, B., Eds.) Vol. 1, pp 215–265, Plenum, New York.

(20) Stuehr, D. J., and Marletta, M. A. (1987) "Induction of nitrite/nitrate synthesis in murine macrophages by BCG infection, lymphokines, or interferon-γ". *J. Immunol.* **139**, 518–525.

(21) Vogel, S. N., Weedon, L. I., Wahl, L. M., and Rosenstreich, D. L. (1982) "BCG-induced enhancement of endotoxin sensitivity in C3H/HeJ mice. II. T cell modulation of macrophage sensitivity to LPS in vitro". *Immunobiol.* **160**, 479–493.

(22) North, R. J. (1974) "T-Cell dependence on macrophage activation and immobilization during infection with *Mycobacterium tuberculosis*". *Infect. Immun.* **10**, 66–71.

(23) Schultz, R. M., and Kleinschmidt, W. J. (1983) "Functional identity between murine γ interferon and macrophage activating factor". *Nature (London)* **305**, 239–240.

(24) Nathan, C. F., Murray, H. W., Wiebe, M. E., and Rubin, B. Y. (1983) "Identification of interferon-γ as the lymphokine that activates human macrophage oxidative metabolism and antimicrobial activity". *J. Exp. Med.* **158**, 670–689.

(25) Schreiber, R. D., Pace, J. L., Russell, S. W., Altman, A., and Katz, D. H. (1983) "Macrophage-activating factor produced by a T cell hybridoma: physiochemical and biosynthetic resemblance to interferon-γ". *J. Immunol.* **131**, 826–832.

(26) Stuehr, D. J., and Marletta, M. A. (1987) "Synthesis of nitrite and nitrate in murine macrophage cell lines". *Cancer Res.* **47**, 5590–5594.

(27) Ohkuma, S., and Poole, B. (1978) "Fluorescence probe measurement of the intralysosomal pH in living cells and the perturbation of pH by various agents". *Proc. Natl. Acad. Sci. U.S.A.* **75**, 3327–3331.

(28) Miwa, M., Stuehr, D. J., Marletta, M. A., Wishnok, J. S., and Tannenbaum, S. R. (1987) "Nitrosation of amines by stimulated macrophages". *Carcinogenesis* **8**, 955–958.

(29) Stuehr, D. J., and Marletta, M. A. (1987) "Further studies on murine macrophage synthesis of nitrite and nitrate". In *Relevance of N-Nitroso Compounds to Human Cancer: Exposures and Mechanisms* (Bartsch, H., O'Neill, I. K., and Schulte-Hermann, R., Eds.) pp 335–339, International Agency for Research on Cancer, IARC Scientific Publication 84, Lyon, France.

(30) Johnston, R. B., Godzick, C. A., and Cohn, Z. A. (1978) "Increasing superoxide anion production by immunologically activated and chemically elicited macrophages". *J. Immunol.* **121**, 809–816.

(31) Damiani, G., Kiyotaki, C., Soeller, W., Sasada, M., Peisach, J., and Bloom, B. R. (1980) "Macrophage variants in oxygen metabolism". *J. Exp. Med.* **152**, 808–822.

(32) Kiyotaki, C., Peisach, J., and Bloom, B. R. (1984) "Oxygen metabolism in cloned macrophage cell lines: glucose dependence of superoxide production, metabolic and spectral analysis". *J. Immunol.* **132**, 857–866.

(33) Iyengar, R., Stuehr, D. J., and Marletta, M. A. (1987) "Macrophage synthesis of nitrite, nitrate, and N-nitrosamines: precursors and role of the respiratory burst". *Proc. Natl. Acad. Sci. U.S.A.* **84**, 6369–6373.

(34) Currie, G. A. (1978) "Activated macrophages kill tumor cells by releasing arginase". *Nature (London)* **273**, 758–759.

(35) Hibbs, J. B., Jr., Taintor, R. R., and Vavrin, Z. (1987) "Macrophage cytotoxicity: role for L-arginine deiminase and imino nitrogen oxidation to nitrite". *Science (Washington, D.C.)* **235**, 473–476.

(36) Adams, D. O., and Hamilton, T. A. (1984) "The cell biology of macrophage activation". *Annu. Rev. Immunol.* **2**, 283–318.

(37) Hibbs, J. B., Jr., Vavrin, Z., and Taintor, R. R. (1987) "L-Arginine is required for expression of the activated macrophage effector mechanism causing selective metabolic inhibition in target cells". *J. Immunol.* **138**, 550–565.

(38) Tai, A. W., Lien, E. J., Lai, M. M. C., and Khwaga, T. A. (1984) "Novel N-hydroxyguanidine derivatives as anticancer and antiviral agents". *J. Med. Chem.* **27**, 236–238.

(39) Palmer, R. M. J., Ferrige, A. G., and Moncada, S. (1987) "Nitric oxide release accounts for the biological activity of endothelium-derived relaxing factor". *Nature (London)* **327**, 524–526.

(40) Ignarro, L. J., and Kadowitz, P. J. (1985) "The pharmacological and physiological role of cyclic GMP in vascular smooth muscle relaxation". *Annu. Rev. Pharmacol. Toxicol.* **25**, 171–191.

(41) Craven, P. A., and DeRubertis, F. R. (1978) "Restoration of the responsiveness of purified guanylate cyclase to nitrosoguanidine, nitric oxide, and related activators by heme and hemeproteins". *J. Biol. Chem.* **253**, 8433–8443.

(42) Craven, P. A., and DeRubertis, F. R. (1983) "Requirement for heme in the activation of purified guanylate cyclase by nitric oxide". *Biochim. Biophys. Acta* **745**, 310–321.

(43) Palmer, R. M. J., Ashton, D. S., and Moncada, S. (1988) "Vascular endothelial cells synthesize nitric oxide from L-arginine". *Nature (London)* **333**, 664–666.

(44) Wagner, D. A., and Tannenbaum, S. R. (1982) "Enhancement of nitrate biosynthesis by *Escherichia coli* lipopolysaccharide". In *Nitrosamines and Human Cancer* (Magee, P., Ed.) Banbury Report 12, pp 437–443, Cold Spring Harbor Press, Cold Spring Harbor, NY.

(45) Roediger, W. E. W., Lawson, M. J., Nance, S. H., and Radcliffe, B. C. (1986) "Detectable colonic nitrite levels in inflammatory bowel disease—mucosal or bacterial malfunction?". *Digestion* **35**, 199–204.

(46) Marletta, M. A., Yoon, P. S., Iyengar, R., Leaf, C. D., and Wishnok, J. S. (1988) "Macrophage oxidation of L-arginine to nitrite and nitrate: nitric oxide is an intermediate". *Biochemistry* (in press).

## Chapter 7

# Mechanisms of Immunotoxicity to Isocyanates

Meryl H. Karol* and Ruzhi Jin

*Department of Environmental and Occupational Health, Graduate School of Public Health,
University of Pittsburgh, Pittsburgh, Pennsylvania 15261*

Reprinted from *Chemical Research in Toxicology,* Vol. 4, No. 5, September/October, 1991

### Introduction

Among the numerous chemicals associated with respiratory hypersensitivity, greatest attention has been given to isocyanates, and particularly to toluene diisocyanate (TDI).[1] The latter is considered to be the principal cause of occupational asthma in the Western world (*1*). Despite recognition of the problem, and an extensive number of investigations, the pathogenesis of this disorder remains uncertain.

Pulmonary hypersensitivity can be manifest either as isolated bronchospastic responses or as asthmatic responses to low, nonirritating concentrations of an inhaled agent. The isocyanates most frequently involved in such responses are TDI, diphenylmethane 4,4'-diisocyanate (MDI), and hexamethylene diisocyanate (HDI). Clinical and epidemiological studies have produced a strong suggestion of involvement of immunologic mechanisms in the sensitization response. The spectrum of symptoms are similar to those observed with other inhaled allergens and include the following: immediate- and late-onset reductions in pulmonary function and development of airway hyperreactivity. Additionally, immunoglobulin E (IgE) class antibodies have been detected by several investigators in the serum of individuals with isocyanate lung sensitivity (*1–3*). Other findings suggestive of immunologic involvement are the existence of a latent period between first exposure and occurrence of sensitivity, and the comparatively small incidence of cases in view of the tens of thousands of persons occupationally exposed.

Differences exist between factors associated with classical immunologic lung sensitivity and those associated with sensitivity due to isocyanates. Included in the latter category are the following: dissociation of atopy and sensitivity to isocyanates and the frequent failure to detect IgE antibody in individuals diagnosed with isocyanate sensitivity.

This review focuses on mechanisms of isocyanate sensitivity. It evaluates information obtained from animal models as well as clinical studies and highlights toxicologic concepts derived from investigations of isocyanate sensitization. Lastly, it explores possible reasons for the current controversy surrounding the pathogenesis of isocyanate sensitization.

### The Animal Model

In 1980, an animal model of TDI sensitivity was first reported (*4*). Principal features of the model included the following: use of the inhalation route to achieve both sensitization and elicitation of pulmonary responses; use of a small reactive chemical (hapten) to produce sensitization; and avoidance of immunologic adjuvants to achieve sensitization. Continued development (*5, 6*) of the guinea pig model indicated the following: (1) antibodies to TDI were always produced in sensitized animals; (2) the magnitude of the antibody response was dependent upon the airborne concentration of TDI, and a threshold concentration of TDI was identified for this response; and (3) a threshold concentration of isocyanate was also required for production of pulmonary sensitization to TDI. The threshold concentration to induce antibodies was the same

---

[1] Abbreviations: TDI, toluene diisocyanate; MDI, diphenylmethane 4,4'-diisocyanate; HDI, hexamethylene diisocyanate; IgE, immunoglobulin E; GSA, guinea pig serum albumin; RAST, radioallergosorbent test; BPC, bronchial provocation challenge: ELISA, enzyme-linked immunosorbent assay.

2428–5/92/0055$06.00/0

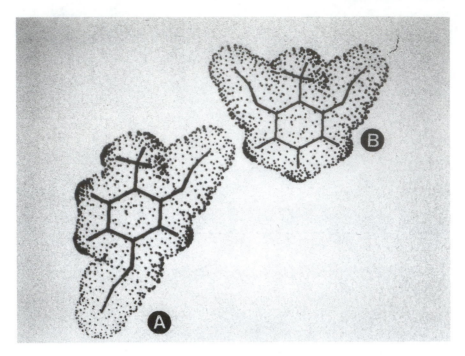

**Figure 1.** Structures of the 2,4 (A) and 2,6 (B) isomers of toluene diisocyanate.

**Table I. Antibodies to 2,4 and 2,6 TDI following Isocyanate Exposure of Guinea Pigs**

| | immunization | | | ELISA titer[a] | | | |
|---|---|---|---|---|---|---|---|
| animal no. | chemical | exposure route | dose, mg | 2,4 TDI[b] | 2,6 TDI | TD 80 | GSA |
| 20 | 2,4 TDI | id | 122 | 10240 | 2560 | 2560 | <40 |
| 21 | 2,4 TDI | id | 122 | 2560 | 1280 | 1280 | <40 |
| 30 | 2,6 TDI | id | 122 | 5120 | 10240 | 2560 | <40 |
| 31 | 2,6 TDI | id | 122 | 5120 | 10240 | 2560 | <40 |
| 40 | 80:20 TDI | id | 122 | 5120 | 5120 | 20480 | 320 |
| 41 | 80:20 TDI | id | 122 | 5120 | 2560 | 40960 | 640 |
| 50 | 80:20 TDI | inh[d] | 128[c] | 10240 | 5120 | 20480 | <40 |
| 51 | 80:20 TDI | inh[d] | 128[c] | 1280 | 640 | 2560 | <40 |

[a] Microtiter plates were coated by addition of 50 $\mu$L of antigen conjugate at 5 $\mu$g/mL (9). Bound antibody was detected by using alkaline phosphatase labeled rabbit anti-guinea pig IgG (H + L). Titer is the highest serum dilution yielding a significant absorbance value (9). [b] The moles of TDI hapten per mole of GSA of each conjugate was determined spectrophotometrically (10). In all cases the hapten:protein density was between 35 and 40. [c] Dose was calculated on the basis of a breathing frequency of 100/min and tidal volume of 2 mL during the first hour, and half these values during hours 2 and 3 when respiratory irritation was apparent (4). Retention of TDI vapor was assumed to be 10%. [d] By inhalation.

as that required for production of pulmonary sensitization (5).

The production of antibodies in the animal model in association with development of pulmonary sensitivity suggested a causal relationship. Accordingly, renewed attention was focused on possible inadequacies of current methodologies for identifying specific antibodies in sera from individuals displaying symptoms of isocyanate pulmonary sensitivity.

## Immunogenicity of 2,4 and 2,6 TDI Isomers

Commercial TDI is formulated from 2,4 and 2,6 isomers (see Figure 1) most frequently in a ratio of 4:1. The reactivities of the isomers differ, with the 2,4 isomer being more readily hydrolyzed (7). The consequence of this differential reactivity is enrichment of the 2,6 isomer relative to the 2,4 in the industrial atmosphere (8), particularly at the end of a process line.

On the basis of the finding in the animal studies that a direct relationship existed between the concentration of TDI at exposure and the amount of anti-TDI antibody produced, it was postulated that the frequent inability to detect antibodies to TDI in clinical cases may have resulted from the routine use of diagnostic antigens containing

predominantly 2,4 TDI whereas individuals may have been exposed to atmospheres in which 2,6 TDI was the predominant isomer. Since no information existed on the immunogenic properties of the individual isomers, we employed the animal model to examine the immunogenicity and cross-reactivity of antibody populations produced to the individual TDI isomers. These studies resulted in the development of protein conjugates of 2,4 and 2,6 TDI which were later applied to the investigation of antibodies in clinical samples.

Antibodies were produced in different sets of guinea pigs by intradermal injection with 100 $\mu$L of the separate TDI isomers or with the 80:20 isomer mixture (Table I). Other groups received inhalation exposure to the 80:20 isomer mixture. Inhalation exposure conditions were 3 h each day on 5 consecutive days to 2.0 ± 0.6 ppm (5). Blood was taken from all animals 20 days later to assess antibody production.

Both isomers were found to be immunogenic. Antibody titers were measured by ELISA using isocyanate–guinea pig serum albumin (GSA) conjugates as immobilized antigens and alkaline phosphatase labeled anti-guinea pig IgG to detect bound antibody. Data are shown in Table I. In all cases, the antibody titer to the homologous (immun-

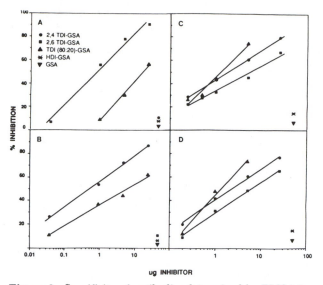

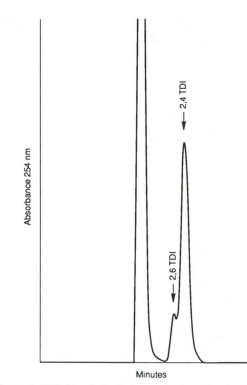

**Figure 2.** Specificity of antibodies determined by ELISA inhibition. Test conjugates (25 µL) (see symbols) were added to ELISA plates coated with (A) 2,6 TDI-GSA, (B) 2,4 TDI-GSA, or (C, D) TD 80–GSA, each at 5 µg/mL. The homologous serum was added, and assays were performed following ELISA methodology (9). Serum A, anti 2,6 TDI, animal 30; serum B, anti 2,4 TDI, animal 20; serum C, anti TD 80 (id), animal 41; serum D, anti TD 80 (inhalation), animal 50.

izing) isomer was in excess of 1/2000. In addition, titers induced by the 2,4 isomer (animals 20 and 21) were comparable to those induced by 2,6 TDI (animals 30 and 31). Little, if any, antibody was found to react with the GSA portion of the conjugate.

Antibodies to each isomer showed cross-reaction with the respective opposite isomer. In all cases, titers with the heterologous isomer were within a 4-fold dilution of that obtained with the immunizing isomer. These cross-reactions could not be attributed to interaction with GSA.

Specificity of antibody was investigated by using ELISA inhibition methodology. Antibodies produced by exposure to the 2,4 isomer were inhibited from reaction with the homologous 2,4 isomer conjugate by addition of 1 µg of 2,4 TDI–GSA (Figure 2, panel B). By comparison, 50 µg of the 2,6 TDI conjugate failed to effect any inhibition. Corresponding results were obtained with the 2,6 isomer homologous system (panel A). In the latter case, 50 µg of 2,4 TDI–GSA failed to affect binding of the 2,6 conjugate with antibody, whereas 0.87 µg of 2,6 TDI–GSA gave 50% inhibition of binding. These results demonstrated isomer specificity of antibody raised to 2,4 or 2,6 TDI.

Since the commercially employed TDI is most frequently a mixture of 2,4 and 2,6 TDI in a 4:1 proportion (i.e., TD 80), particular interest was focused on the specificity of antibody raised to the isomer mixture. Sera were obtained from animals that had received exposure to TD 80 either by intradermal injection (animals 40 and 41) or by inhalation of TDI vapors (animals 50 and 51). The 4:1 isomer composition within the inhalation chamber was confirmed by HPLC analysis of samples taken during exposure of animals (see Figure 3).

In all cases, a conjugate prepared from the 80:20 TDI mixture better detected antibody induced by TD 80 than did conjugates prepared from individual isomers (Table I). Titers assessed by using the 2,4 conjugate were greater than those obtained through use of the 2,6 preparation. Inhibition assays (Figure 2, panels C and D) revealed antibodies to be approximately 4 times more reactive with the 2,4 isomer (circles) than with the 2,6 species (squares).

**Figure 3.** HPLC analysis of air samples from the chamber during exposure of guinea pigs to TD 80. Samples were analyzed on a Lichrosorb Rp-18 column, 10-µm, particles, with acetonitrile/water for elution. The first peak indicates excess (*p*-nitrobenzyl)-*N*-*n*-propylamine hydrochloride, followed by elution of 2,6 and 2,4 TDI ureas (11).

**Table II. Isomer Specificity of Antibodies Produced in Guinea Pigs to the 80:20 Mixture of 2,4 and 2,6 TDI**

| | | | µg of TDI conjugate required for 50% inhibn[a] | | | |
|---|---|---|---|---|---|---|
| animal no. | exposure route | TD 80 dose, mg | TD 80–GSA | 2,4 TDI–GSA | 2,6 TDI–GSA | GSA |
| 40 | id | 122 | 1.7 | 1.8 | 6.4 | >50 |
| 41 | id | 122 | 1.2 | 1.8 | 5.4 | >50 |
| 50 | inh[b] | 128 | 1.3 | 2.3 | 6.0 | >50 |
| 51 | inh[a] | 128 | 0.4 | 0.4 | 2.9 | >50 |

[a] ELISA plates were coated with 5 µg/mL TD 80–GSA. Assays were performed at a 1:200 dilution of serums 40 and 41, and at a 1:40 dilution of serums 50 and 51. [b] By inhalation.

This conclusion is based on the finding that 4 times more 2,6 conjugate was required to inhibit binding of antibody to the TD 80 conjugate when compared with the required amount of 2,4 TDI (see Table II). Indeed, for inhibition of antibody binding, the 2,4 TDI conjugate was almost as effective as the TD 80 conjugate (prepared from the 80:20 isocyanate mixture). The route of exposure did not appear to influence the specificity of induced antibodies.

Taken together, the results suggested that each isomer was equally immunogenic in this animal system. They further indicated that the specificity of the antibody population was dependent upon, and thereby reflected, the isomer composition at exposure. If these findings are verified in an industrial situation, the antibody response can be proposed as a specific and quantitative *biological marker* of exposure.

## Specificity of Human Anti-TDI Antibodies

To confirm diagnoses, serologic tests for IgE antibody have traditionally been performed on serum from individuals with suspected TDI hypersensitivity. Routinely,

Table III.  Use of 2,4 and 2,6 TDI Conjugates in RAST for IgE Antibody

| patient | RAST disk | | |
| --- | --- | --- | --- |
| | 2,4 TDI–HSA[a] | 2,6 TDI–HSA | TD 80–HSA |
| A | 10.4[b] | 6.8 | 10.2 |
| B | 16.3 | 10.8 | 19.0 |
| C | 11.3 | 19.5 | 15.5 |

[a] The hapten:protein molar composition of each conjugate was determined by UV absorbance spectrometry to be between 35 and 40.  [b] Percent of [125]I-rabbit anti-human IgE which is bound to the antigen-coated disk.  Normal values are <5.0%.

Table IV.  RAST Inhibition Assay To Assess IgE Specificity

| patient | μg of conjugate for 50% inhibn of binding to TD 80 disk | | | |
| --- | --- | --- | --- | --- |
| | TD 80[a] | 2,4 TDI | 2,6 TDI | HSA |
| A | 200 | 130 | 364 | >500 |
| B | 44 | 45 | 131 | >500 |
| C | 62[b] | 193[b] | 56[b] | >500[b] |

[a] Each conjugate was prepared by reaction of TDI with HSA. The hapten density of each conjugate averaged between 35 and 40 mol/mol of protein.  [b] Micrograms for 30% inhibition.

a radioallergosorbent test (RAST) for IgE antibodies is employed utilizing a conjugate prepared from reaction of TD 80 with human serum albumin.  We questioned whether detection of such an antibody would be enhanced by selective use of conjugates of individual TDI isomers. To test this hypothesis, RAST assays were performed on sera from three individuals with clinical indication of TDI sensitivity.

Each of the sera showed elevated binding with the three TDI conjugate coated RAST disks (Table III).  Sera from patients A and B were considerably more reactive with 2,4 TDI and TD 80 coated disks than with disks coated with 2,6 TDI.  The opposite was found for serum from patient C.

To further probe the isomer specificity of the IgE antibodies, RAST inhibition assays were performed by using a procedure analogous to ELISA inhibition.  As seen in Table IV, antibodies from patients A and B were more reactive with 2,4 TDI than with the 2,6 isomer since less 2,4 TDI antigen was necessary to prevent binding of antibody to TD 80 coated disks.  The opposite result was obtained with serum from patient C.  In the latter instance, the 2,6 TDI conjugate competed more favorably than did the 2,4 TDI antigen for antibody.  In each case, binding was not attributed to HSA since the latter showed no ability to inhibit antibody binding (Table IV, last column).

These results indicated differences among patients in the specificity of their IgE antibodies toward the TDI isomers.  The reactions of other classes of antibody were not studied.  Results from the animal model had indicated that both antibody titer and specificity reflected the quantities of TDI isomers at exposure.  It would appear reasonable to propose that analogous assessment of *antibody in clinical samples be used to estimate human exposure conditions*.  In the above example, it could be inferred that individuals A and B were exposed to TDI atmospheres in which 2,4 TDI was predominant, whereas individual C would have been exposed to atmospheres in which 2,6 TDI prevailed.  As stated previously, such an atmosphere would exist toward the end of a processing operation.  The results imply a need to broaden RAST testing by selection of assays that permit detection of antibody specific for the 2,4 or 2,6 TDI isomers.  Current procedures use only TD 80 preparations and as such may be inadequate to detect antibodies with 2,6 TDI specificity.

## Relationship of IgE Antibody to Clinical Sensitivity

The clinical literature provides some evidence for a relationship between the amount of isocyanate-specific IgE in serum and observed isocyanate sensitivity (12, 13).  We described an individual with respiratory symptoms after 1.5 years of employment in a TDI manufacturing company (12).  Bronchial provocation challenge (BPC) (13) identified TDI as the occupational chemical causing the distress.  Eleven months following the individual's removal from further TDI exposure, BPC revealed loss of the pulmonary sensitivity (13).

The antibody response of the patient to TDI was monitored during and following the responsive period (3, 12). Whereas the patient had elevated RAST values at the time of clinical sensitivity to TDI, values decreased by 50% during the 11 months away from exposure.  Following an additional 12 months, during which time complete avoidance of TDI was maintained, RAST was negative.

A second study provided further evidence for a relationship between isocyanate pulmonary sensitivity and IgE antibodies.  This study described a worker who became asthmatic after cutting a polyurethane plate containing MDI (14).  MDI RAST was highly elevated during the period of symptomatology.  After 1 year without occupational exposure, the titer had decreased by 50%.  Bronchial provocation challenge was negative at that time.

The above results, together with those obtained from the animal model, demonstrate a relationship between isocyanate exposure, pulmonary sensitivity, and the presence of isocyanate-specific IgE antibody.  However, proof of a causal relationship between the latter two factors must await further investigation.

## Diagnosis of Isocyanate Sensitivity and Asthma

Numerous investigators have reported limited success detecting antibodies in clinically diagnosed cases of isocyanate sensitivity (15).  As discussed above, some of the difficulty may have been attributable to the exclusive use of TD 80 conjugates in serologic assays, rather than use of 2,6 conjugates in selected cases.  However, another cause may be inappropriate *diagnosis* of isocyanate sensitivity.

Occupational asthma is defined as reversible airway obstruction encountered in the workplace and induced by inhaled dust, vapor, fumes, or gases (16).  Diagnosis of isocyanate sensitivity is currently performed by a number of procedures (15).  Since a physical exam may not indicate symptoms, considerable emphasis is placed on a personal history, frequently obtained by questionnaire.  Bronchial provocation challenge with the putative agent is seldom used to confirm the diagnosis.

The questionnaire elicits information related to the nature of the symptoms (dyspnea, wheezing, cough, and chest tightness) and their time of occurrence (worse at work or after a shift, improved over weekends and holidays).  The inadequacy of this diagnostic procedure was demonstrated in a study which compared diagnosis made on the basis of the questionnaire with that based on the response to a specific inhalation challenge (17).  Agents used for inhalation challenge included flour and grain dusts and red cedar.  Of the 75 subjects considered sensitive on the basis of the questionnaire, less than half (46%) responded to challenge.  It was concluded that a questionnaire suggesting occupational asthma, even though administered by physicians with experience in occupational asthma, was unsatisfactory as a diagnostic tool.

**Table V. Characteristics of Pulmonary Responses to Bronchial Provocation Challenge with Allergens**[a]

| type of response | features |
|---|---|
| immediate | onset of reaction within minutes after exposure; maximum decrease in $FEV_1$ within minutes; progressive recovery in 1st or 2nd hour |
| late onset | onset at 3–4 h; maximum reaction around 8 h; lasts ≥3 h |
| dual | early and late reactions with almost complete recovery after the early |

[a] Adapted from refs 16 and 18.

Unfortunately, questionnaires have frequently been used as the *sole* diagnostic criterion for isocyanate asthma. Investigation of the mechanism of isocyanate sensitivity is possible only if there is certainty in the diagnosis of the disorder. However, it is readily agreed that diagnosis of this disorder is difficult (*15–18*). Faulty diagnosis must be considered as one possible explanation for the absence of immunologic findings in some individuals diagnosed with isocyanate asthma.

## Bronchial Provocation Challenge

Bronchial provocation challenge (BPC) has been considered the most relevant method of diagnosis and evaluation of occupational asthma. Although concern has been expressed regarding ethical considerations in attempting to provoke clinical manifestations of asthma, the practice is considered a "gold standard" in diagnosing sensitization to airborne allergens (*18*).

Patterns of response to BPC with allergens have been described (see Table V) (*15–20*). Dependent upon their time of occurrence, reactions have been classified as immediate onset, late, or dual. The late response has several varieties based upon the time of maximal response and the duration of response (*19, 20*). The mechanism underlying the immediate response has been attributed to IgE antibody. The mechanism of the late-onset response is less clear and has been attributed to both IgE (*20*) and non-IgE antibody (*1*). Whereas high molecular weight allergens typically elicit immediate, or dual, responses, low molecular weight agents produce mainly late-onset responses (*15*).

Responses to BPC with isocyanates have been studied by several investigators. By use of a procedure in which subjects paint with a varnish plus TDI, both immediate and late-onset reactions were produced (*16*). Responsiveness to TDI was also studied by BPC with TDI vapors (*19, 21, 22*). Late-onset responses were frequently produced and in many instances were accompanied by increased airway responsiveness (*19*). The preponderance of late-onset reactions seen with isocyanates is typical of the reactions seen with other low molecular weight chemicals (*15*).

On the basis of results of BPC, one would expect similar mechanisms of pathogenesis of isocyanate sensitivity and sensitivity to other haptenic allergens. However, atypical reactions to BPC have been noted more frequently after exposure to isocyanates (TDI, MDI, and HDI) than following challenge with either high molecular weight allergens (*18*) or western red cedar (*20*). Such patterns were characterized by the following: (a) progressive falls in $FEV_1$ with maximum decreases 5–6 h after exposure; (b) immediate and maximum fall in $FEV_1$ in the first minutes after exposure without significant recovery thereafter (for up to 8 h); and (c) slow recovery (over several hours) after an immediate response. The reasons for atypical reactions to BPC with isocyanates are uncertain. As discussed be-

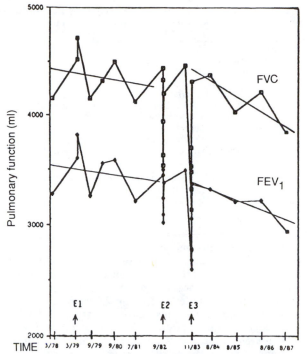

**Figure 4.** Pulmonary function measurements of a TDI worker during a 9-year study. Three accidental exposures (E1–E3) were reported during the period of study. Regression lines obtained by least-squares analysis of individual measurements before E2 and following E3.

low, it is possible that they represent toxic chemical reactions of isocyanates with functional biological molecules.

## Bronchoconstrictive Responses to Isocyanates

Isocyanates cause airway constriction in naive individuals (*23*) and animals (*3*), with the degree of response dependent upon both the concentration and duration of isocyanate exposure. The OSHA standard of 5 ppb for TDI is well below a concentration that will induce smooth muscle contraction (*24*). However, processing malfunctions, equipment breakdown, or faulty ventilation can contribute to increased airborne isocyanate concentrations and the likelihood of pulmonary reactions in the workplace. If the exposure is not recognized, such airway-constrictive responses may mistakenly be identified as sensitivity.

A longitudinal study examined airway-constrictive episodes by following the forced expiratory volume in one second ($FEV_1$) and forced vital capacity (FVC) in isocyanate workers over a 10-y period (*25*). Acute changes in lung function were documented at times of increased isocyanate exposure, i.e., following spills or equipment breakdown (Figure 4). In the case cited, decreased $FEV_1$ did not represent sensitization to TDI but might have been diagnosed as such by a questionnaire. In this case, bronchoconstriction in the workplace was the result of TDI "irritation".

## Measurement of Airborne Isocyanate Concentrations

A third problem with BPC as a diagnostic procedure is the difficulty determining airborne isocyanate concentrations. Ideally, this measurement is derived by placing a sampling device in the individual's breathing zone. However, most clinical studies use an area monitor to determine atmospheric concentrations. The device frequently em-

Table VI. Comparison of Characteristics of Late-Onset
Pulmonary Responses to TDI and to Environmental
Allergens[a]

similarities
    time of occurrence
    blocked by corticosteroids
    blocked by cromolyn sodium
    atopic background unrelated
    sensitization associated with large exposures to agents
    increase in circulating eosinophils after BPC
    increase in CD8/CD4 cells
    airway inflammation during late-onset reaction
    airway hyperreactivity following late-onset reaction

differences
    detection of specific IgE antibodies in only a small percentage
    of TDI-sensitized individuals
    occasional unusual patterns of late-onset reactions

[a] Adapted from ref 22.

ploys a tape impregnated with an isocyanate trapping reagent. The airborne isocyanate concentration is determined by readout of the color intensity of the continuous recording. This procedure poses several problems. It responds differently to the 2,4 and 2,6 isomers (26). Additionally, it integrates concentrations over a period of time and may miss peak exposures. Most importantly, because it is an area monitor, it often does not detect the concentrations to which the worker has been exposed.

## Conclusions

Events triggered in sensitized individuals by TDI exposure include functional changes in the lung as well as in peripheral cell populations (22). Many of these changes are consistent with those detected in sensitized subjects upon exposure to recognized aeroallergens. Comparison of late-onset responses to TDI with those caused by other recognized allergens indicates numerous similarities (Table VI). The mechanism of response to most aeroallergens is considered to be immunologic (16, 18). IgE antibodies have been detected to TDI.

Animal model studies have indicated that inhalation of TDI results in a concentration-dependent production of hypersensitivity antibodies (5), as occurs with other inhaled allergens. In view of the remarkable number of similarities, and few differences, between TDI sensitivity and sensitivity to other chemical allergens, it is reasonable to conclude that the pathogenesis of TDI sensitivity is frequently immunologic and to propose possible reasons for the failure to identify IgE antibodies in a larger proportion of responsive individuals. The possible reasons fall into two categories: IgE assessment methodology and the nature of bronchospastic reactions to isocyanates.

Both through use of animal models and from clinical samples, progress has been made in defining antigen conjugates which are effective in detecting isocyanate antibodies. The studies described here indicate that attention must be given to the isomeric composition of isocyanate conjugate antigens.

Animal models and industrial exposure of workers have also indicated that TDI can produce smooth muscle contraction, bronchospasm, and airway hyperreactivity. In many cases, these effects have been shown to be dose (concentration) dependent (3–6, 23, 27). In view of the difficulties in both generating stable isocyanate atmospheres and measuring isocyanate concentrations in the atmosphere, the possibility must be entertained that many of the symptoms noted in the industrial setting, as well as those evoked upon bronchial provocation challenge, may be caused by exposure to elevated concentrations of iso-

cyanates and may be irritant in nature. This uncertainty is heightened by the absence of information on the relative irritant properties of the 2,4 and 2,6 isomers.

Recent studies on methyl isocyanate have indicated in vivo reaction of the isocyanate to form mercapturic acid conjugates (28). Of particular interest is the recognized carbamoylating ability of such conjugates. The extensive amount of glutathione in the epithelial lining fluid of the lung may function to capture inhaled isocyanates and transport them to sites where they may react with other nucleophilic proteins and cause toxic manifestations, such as smooth muscle contraction (27). Progress in this area will be made by further assessment of the dose-dependent toxicity of individual isocyanate isomers.

The mechanism of TDI sensitivity, as well as sensitivity to other isocyanates, is still unclear. The evidence in favor of an immunologic etiology far outweighs that against it. It is anticipated that increased awareness of the chemical and biological reactivities of the isocyanate isomers, further use of animal models, and continued refinement of clinical methodology will result in elucidation of the etiology of isocyanate sensitivity.

**Acknowledgment.** We thank the many students, colleagues, and collaborators who have contributed to these studies. The computer-generated structures were kindly provided by Dr. Charles L. Brooks, Department of Chemistry, Carnegie Mellon University, Pittsburgh, PA. Support from Mobay Corp. and NIEHS ES01532 is gratefully acknowledged.

## References

(1) Cartier, A., Grammer, L., Malo, J.-L., Lagier, F., Ghezzo, H., Harris, K., and Patterson, R. (1989) Specific serum antibodies against isocyanates: Association with occupational asthma. *J. Allergy Clin. Immunol.* 84, 507–514.
(2) Karol, M. H., And Alarie, Y. C. (1980) Antigens which detect IgE antibodies in workers sensitive to toluene diisocyanate. *Clin. Allergy* 10, 101–109.
(3) Karol M. H. (1986) Respiratory effects of inhaled isocyanates. *CRC Crit. Rev. Toxicol.* 16, 349–379.
(4) Karol, M. H., Dixon, C., Brady, M., and Alarie, Y. (1980) Immunologic sensitization and pulmonary hypersensitivity by repeated inhalation of aromatic isocyanates. *Toxicol. Appl. Pharmacol.* 53, 260–270.
(5) Karol, M. H. (1983) Concentration-dependent immunologic response to toluene diisocyanate (TDI) following inhalation exposure. *Toxicol. Appl. Pharmacol.* 68, 229–241.
(6) Karol, M. H. (1988) The development of an animal model for TDI asthma. *Bull. Eur. Physiopathol. Respir.* 23, 571–576.
(7) Brown, W. E., Green, A. H., Cedel, T. E., and Cairns, J. (1987) Biochemistry of protein–isocyanate interaction: A comparison of the effects of aryl vs. alkyl isocyanates. *Environ. Health Perspect.* 72, 5–11.
(8) Rando, R. J., Abdel-Kader, H. M., and Hammad, Y. Y. (1984) Isomeric composition of airborne TDI in the polyurethane foam industry. *Am. Ind. Hyg. Assoc. J.* 45, 199–203.
(9) Jin, R., and Karol, M. H. (1988) Diisocyanate antigens that detect specific antibodies in exposed workers and guinea pigs. *Chem. Res. Toxicol.* 1, 288–293.
(10) Karol, M. H., Jin, R., and Rubanoff, B. (1989) Clinical and experimental evaluation of isocyanate lung injury. *Comments Toxicol.* 3, 117–130.
(11) Dunlap, K. L., Sandridge, R. L., and Keller, J. (1976) Determination of isocyanates in working atmospheres by high speed liquid chromatography. *Anal. Chem.* 48, 497–499.
(12) Karol, M. H. (1981) Survey of industrial workers for antibodies to toluene diisocyanate. *J. Occup. Med.* 23, 741–747.
(13) Butcher, B. T., O'Neil, C. E., Reed, M. A., Salvaggio, J. E., and Weill, H. (1982) Development and loss of toluene diisocyanate reactivity: Immunologic, pharmacologic and provocative challenge studies. *J. Allergy Clin. Immunol.* 70, 231–235.
(14) Chang, K. C., and Karol, M. H. (1984) Diphenylmethane diisocyanate (MDI)-induced asthma: Evaluation of the IgE response and application of an animal model of isocyanate sensi-

tivity. *Clin. Allergy* 14, 329–339.

(15) Butcher, B. T., Bernstein, I. L., and Schwartz, H. J. (1989) Guidelines for the clinical evaluation of occupational asthma due to small molecular weight chemicals. *J. Allergy Clin. Immunol.* 84, 834–838.

(16) Pepys, J., and Hutchcroft, B. J. (1975) Bronchial provocation tests in etiologic diagnosis and analysis of asthma. *Am. Rev. Respir. Dis.* 112, 829–859.

(17) Malo, J.-L., Ghezzo, H., L'Archeveque, J., Lagier, F., Perrin, B., and Cartier, A. (1991) Is the clinical history a satisfactory means of diagnosing occupational asthma? *Am. Rev. Respir. Dis.* 143, 528–532.

(18) Cartier, A., Bernstein, I. L., Burge, P. S., Cohn, J. R., Fabbri, L. M., Hargreave, F. E., Malo, J.-L., McKay, R. T., and Salvaggio, J. E. (1989) Guidelines for broncho-provocation on the investigation of occupational asthma. *J. Allergy Clin. Immunol.* 84, 823–829.

(19) Mapp, C., Palato, R., Maestrelli, P., Hendrick, D. J., and Fabbri, L. M. (1985) Time-course of the increase in airway responsiveness associated with late asthmatic reactions to toluene diisocyanate in sensitized subjects. *J. Allergy Clin. Immunol.* 75, 568–572.

(20) Perrin, B., Cartier, A., Ghezzo, H., Grammer, L., Harris, K., Chan, H., Chan-Yeung, M., and Mato, J. (1991) Reassessment of the temporal patterns of bronchial obstruction after exposure to occupational sensitizing agents. *J. Allergy Clin. Immunol.* 87, 630–639.

(21) Fabbri, L. M., Boschetto, P., Zocca, E., Milani, G., Pivirotto, F., Plebani, M., Burlina, A., Licata, B., and Mapp, C. E. (1987) Bronchoalveolar neutrophilia during late asthmatic reactions induced by toluene diisocyanate. *Am. Rev. Respir. Dis.* 136, 36–42.

(22) Finotto, S., Fabbri, L. M., Rado, V., Mapp, C. E., and Maestrelli, P. (1991) Increase in numbers of CD8 positive lymphocytes and eosinophiles in peripheral blood of subjects with late asthmatic reactions induced by toluene diisocyanate. *Br. J. Ind. Med.* 48, 116–121.

(23) Patterson, R., Hargreave, F. E., Grammer, L. C., Harris, K. E., and Dolovich, J. (1987) Toluene diisocyanate respiratory reactions. *Int. Arch. Allergy Appl. Immunol.* 84, 93–100.

(24) McKay, R. T., and Brooks, S. M. (1983) Effect of toluene diisocyanate on beta adrenergic receptor function. *Am. Rev. Respir. Dis.* 128, 50–53.

(25) Roberts, D. (1989) An epidemiologic investigation of $FEV_1$ and serology among employees of a TDI manufacturing plant. MPH Thesis, University of Pittsburgh.

(26) Rando, R. Y., Hammad, Y. Y., and Chang, S. (1989) A diffusive sampler for personal monitoring of toluene diisocyanate (TDI) exposure; Part 1: design of the dosimeter. *Am. Ind. Hyg. Assoc. J.* 50, 1–7.

(27) Mapp, C. E., Graf, P. D., Boniotti, A., and Nadel, J. A. (1991) Toluene diisocyanate contracts guinea pig bronchial smooth muscle by activating capsaicin-sensitive sensory nerves. *J. Pharmacol. Exp. Ther.* 256, 1082–1085.

(28) Slatter, J. G., Rashed, M. S., Pearson, P. G., Han, D.-H., and Baillie, T. A. (1991) Biotransformation of methyl isocyanate in the rat. Evidence for glutathione conjugation as a major pathway of metabolism and implications for isocyanate-mediated toxicities. *Chem. Res. Toxicol.* 4, 157–161.

## Chapter 8

# Mechanism of Drug-Induced Lupus

Jack P. Uetrecht

*Faculties of Pharmacy and Medicine, University of Toronto and Sunnybrook Medical Centre, Toronto, Ontario, M5S 1A1 Canada*

Reprinted from *Chemical Research in Toxicology*, Vol. 1, No. 3, May/June, 1988

### Introduction

Systemic lupus erythematosus, SLE, or simply lupus, is an autoimmune disease in which antibodies are produced to several different endogenous antigens (i.e., patients produce antibodies against their own tissue). The antibodies most commonly used diagnostically bind to nuclear antigens and are therefore known as antinuclear antibodies (ANA).[1] Although ANA are virtually always present in SLE, they can also be present in other diseases and even in "normal" people. SLE is usually a disease of young women. Common manifestations include fever, anemia, joint pains, and a characteristic skin rash. Death can result from serious involvement of the brain or kidneys. For a detailed description of lupus, the recent book by Wallace and Dubois provides an excellent reference (1).

The cause of most lupus is presently unknown, and it is therefore called idiopathic lupus. A similar disease is also associated with the use of certain drugs and is known as drug-induced lupus (2-4). The manifestations of drug-induced lupus are similar to those of idiopathic lupus although drug-induced lupus tends to be milder. The major diagnostic feature which differentiates drug-induced lupus is its association with exposure to a specific drug and the rapid resolution of symptoms when that drug is discontinued. It is not known to what degree idiopathic and drug-induced lupus share a common mechanism of pathogenesis. It has been suggested that idiopathic lupus is due to environmental chemicals similar to drug-induced lupus, but this is controversial (5).

**Drugs Associated with Lupus.** The major drugs associated with the induction of lupus are procainamide and hydralazine. Procainamide is used to treat cardiac arrhythmias. The incidence of ANA in patients treated chronically with procainamide has been estimated at from 50% to greater than 90% (6, 7). However, only 20–30% of patients develop symptoms of lupus, and the presence of ANA alone is not an indication for discontinuation of the drug.

Hydralazine is used to treat hypertension. The incidence of hydralazine-induced lupus is somewhat less. Earlier reports placed the incidence of ANA at about 50%, with the incidence of clinical lupus at 12% (8); however, it was recognized that the incidence was related to dose, and, with the introduction of other effective antihypertensive agents, it has been possible to reduce the daily dose to less than 200 mg. With this lower dose the incidence of hydralazine-induced lupus has dropped to about 3% (9).

Many other drugs have been associated with the induction of lupus. These drugs include isoniazid, phenytoin, practolol, nomifensine, sulfonamides, carbamazepine, propylthiouracil, penicillamine, and p-aminosalicylic acid. Some of these drugs are associated with characteristic types of autoimmunity in addition to the more common manifestations of lupus. For example, practolol is associated with an oculocutaneous syndrome (10), and penicillamine is associated with the myasthenia gravis syndrome in which there are antibodies to the acetylcholine receptor on the motor end-plate (11). This antibody leads to muscle weakness.

### Pathogenesis of Idiopathic Lupus

To gain an understanding of the mechanism of drug-induced lupus, it is useful to review what is known about the mechanism of idiopathic lupus and then compare the two syndromes in search of common features. Lupus involves a defect in the regulation of the immune system (1). Regulation of the immune system depends upon a delicate

---

[1] Abbreviations: ANA, antinuclear antibodies; MHC, major histocompatibility complex; HLA, human lymphocyte antigen; MPO, myeloperoxidase.

balance between a large number of different types of cells (including macrophages and several types of lymphocytes) as well as many factors which are elaborated by cells (including immunoglobulin, interleukins, interferon, and other lymphokines). A common trap in the study of lupus has been to assume that an observed defect is the primary defect which causes the syndrome rather than the result of another defect. This is due, in part, to the complexity of the immune system which contains many feedback loops. With this in mind, the following are some of the major defects that have been found in idiopathic lupus.

**1. Polyclonal Activation of B-Lymphocytes (*12*).** B-Lymphocytes are responsible for the synthesis of antibodies. The production of antibodies that react with a person's own tissue is also seen in normal people, but the number of different autoantibodies and the quantity of production are low. In idiopathic lupus the number of B-cells which spontaneously produce immunogobulin is increased, but their ability to respond to stimuli appears to be decreased. T-lymphocytes exert a major influence on the activation of B-lymphocytes and the production of antibodies.

**2. Impaired T-Lymphocytes Function (*13*).** There are several types of T-cells. The helper T-cell is involved in the activation of B-cells, and the suppressor T-cell is involved in limiting the activation of B-cells. In principle, the extent of B-cell activation should be controlled by the balance between helper and suppressor T-cells. In general, the total number of T-cells is decreased in lupus. Although a decrease in the ratio of suppressor to helper T-cells has been reported, this has not been found by all investigators. Antibodies have been found in the serum of many lupus patients which inactivate suppressor T-cells, but this has also not been found by all investigators.

**3. Depressed Production of Interleukin-2 (*14*).** Interleukin-2 is an important lymphokine required by B-cells for the production of immunoglobulin. It is decreased in both animal models of lupus and human idiopathic lupus. This is curious because one would expect decreased B-cell activation in the face of decreased interleukin-2 levels.

It is also useful to review the factors known to be associated with the development of idiopathic lupus. They include the following.

**1. Genetic Factors.** Genetic factors are important in the development of lupus. It has been estimated that first degree relatives of patients with lupus have a 100- to 200-fold greater risk of developing lupus than the general population (*15*). Asymptomatic relatives often have serologic abnormalities such as ANA. Although twin studies demonstrate the importance of genetic factors in lupus, they also suggest other factors are involved because the concordance is not 100% (i.e., if one individual of a set of identical twins develops lupus, the other individual in that set also usually develops lupus but not always). One such study found a concordance in monozygotic twins of 69% for clinical lupus (*16*). Animal models demonstrate a very strong role for genetic factors in that virtually all individuals of some strains of mice develop lupus (*17*).

Cells use specific glycoproteins, which are an integral part of the cell membrane, to differentiate "self" from "nonself". The structure of these proteins is inherited and a specific combination of many of these proteins characterizes an individual. Collectively, the proteins are known as the major histocompatibility complex (MHC), and in humans they are called the human leukocyte antigens (HLA). It appears as if HLA types DR2 and DR3 are associated with an increased risk of lupus (*18*), but not all investigators have found this association (*19*). The MHC is composed of three classes. The class I MHC proteins occur on virtually all nucleated cells. The class II MHC proteins are not found on all cells but are variably expressed on macrophages and some other cells. The class II MHC proteins are involved in the recognition of processed antigen by T-cells and will be discussed later while the class III MHC are less important for this discussion.

**2. Infectious Agents.** It has been suspected for a long time that idiopathic lupus is due to a virus. There are several similarities between lupus and AIDS. However, all attempts to isolate a virus which can cause lupus have failed (*20*). Despite this failure, there remains a strong suspicion that lupus is caused by a virus.

It is also possible that bacteria are an important etiologic agents for lupus. The finding that anti-DNA antibodies also bind to certain bacterial antigens suggests that lupus may be the outcome of the body's reaction to certain bacterial infections (*21*). It also suggests that DNA may not be the antigen which induces the formation of anti-DNA antibodies. In general, it appears that the portion of a protein that is recognized by an antibody (known as the epitope) can be as small as a few amino acids; therefore, it is not surprising that many seemingly unrelated proteins can share a common epitope.

**3. Hormonal Factors.** Lupus is 10 times more common in women than in men (*22*). The major basis for this difference is sex hormones. Estrogens augment and androgens suppress immune reactivity. This is well demonstrated in murine models of lupus where female mice develop glomerulonephritis several months earlier than males. Castrated male mice develop lupus at an age comparable to the female mice, and androgens suppress the disease and allow female mice to live to an age comparable to male mice. Other autoimmune diseases, such as myasthenia gravis, autoimmune thyroid disease, and rheumatoid arthritis, are also more common in females.

**4. Environmental Agents.** Reidenberg et al. reported a case in which a laboratory technician developed lupus after chronic exposure to hydrazine (*23*). The symptoms cleared when exposure to hydrazine was discontinued and mitogen stimulated IgG synthesis by the patient's cultured lymphocytes was inhibited by hydrazine. (A mitogen is an agent which can stimulate a specific type of lymphocyte to proliferate.) A genetic factor also appeared to be involved because two asymptomatic family members were positive for ANA.

Alfalfa sprouts produce a lupus-like syndrome in monkeys (*24*). The toxic principle of the alfalfa is L-canavanine which is a toxic analogue of arginine found in most legumes. It is unknown whether alfalfa sprouts can also induce autoimmunity in man. However, it is unlikely that most idiopathic lupus is due to such toxic agents.

In summary, the etiology of idiopathic lupus is unknown. Genetic factors appear to be very important, but some other stimulus, such as an infectious agent or environmental agent, may also be required. It may be that idiopathic lupus is simply a clinical syndrome that has several different causes. Whatever the stimulus leading to the production of autoantibodies, it appears that the final common pathway leading to most of the tissue damage is the formation of immune complexes (the combination of antigen and antibody) and deposition of these complexes in various tissues (*1*). These complexes lead to complement activation and tissue damage. DNA is the major antigen detected in the complexes observed in patients with idiopathic lupus. Other autoantibodies can be induced which bind to cell membranes leading to direct cell damage; for example, some lupus patients have anti-red cell antibodies which lead to hemolytic anemia.

## Comparison of Idiopathic and Drug-Induced Lupus

The diagnosis of lupus is based on the presence of a combination of clinical and laboratory findings, none of which is sufficient, by itself, to make the diagnosis. A set of criteria have been developed to determine the diagnosis (25). The degree of certainty in the diagnosis is not absolute and is dependent upon the number of criteria which are met.

The clinical syndromes of drug-induced and idiopathic lupus are similar, although drug-induced lupus tends to be less severe (2–4). In particular, the life-threatening involvement of the brain and kidneys is less common in drug-induced lupus. The large degree of overlap between the syndromes prevents their differentiation on the basis of clinical symptoms alone. One possible reason for the milder course of drug-induced lupus is that the offending drug is usually discontinued before more serious manifestations of the disease develop, but this explanation is probably an oversimplification. The large female to male ratio observed in idiopathic lupus is not seen in drug-induced lupus. Part of the difference may be due to the larger number of males which are exposed to the offending drugs since both procainamide and hydralazine are used for cardiovascular problems which are more common in males. Despite this source of bias, sex does not appear to be as important a factor in drug-induced lupus. The median age of patients with drug-induced lupus is higher than that for idiopathic lupus, again probably reflecting the age of the patients usually exposed to the drugs which are most commonly associated with the drug-induced syndrome. In contrast to idiopathic lupus, which is associated with the HLA types DR-2 and DR-3, hydralazine-induced lupus is reported to be associated with HLA type DR-4 (26), although this has been disputed (27).

There is no laboratory test which will differentiate drug-induced from idiopathic lupus; however, there are differences seen in the laboratory tests which are used to diagnose lupus. Both are associated with ANA, but most of the ANA in drug-induced lupus bind to histone protein while the ANA in idiopathic lupus are much more heterogeneous (28–30). Antiguanosine antibodies, which bind to single-stranded DNA, have also been reported in procainamide-induced lupus and were reported to correlate with disease activity (31). Although some ANA in drug-induced lupus bind to single-stranded DNA, they are said never to bind to double-stranded DNA (32). There have been reports of anti-double-stranded DNA, but it can always be argued that the DNA was contaminated or that it was really idiopathic lupus rather than drug-induced lupus. In contrast, antibodies to double-stranded DNA are common in idiopathic lupus and correlate well with disease activity.

In summary, there are general differences in the clinical and laboratory findings between drug-induced and idiopathic lupus; however, at present, there are none which allow absolute differentiation, and there is a great deal of overlap between the two syndromes.

## Possible Mechanisms of Drug-Induced Lupus

There has been much speculation concerning the mechanism of drug-induced lupus, but the evidence for any one theory is not conclusive.

**Lupus Diathesis.** Alarcon-Segovia proposed that a drug could cause the expression of idiopathic lupus in a person who already had a "lupus diathesis" (33). He defined a lupus diathesis as having a past history or a family history of one of many clinical or laboratory abnormalities such as arthritis, myalgia, drug reaction, pleuritic pain, epilepsy, or unexplained leukpenia. Other workers have not found such a relationship, and it seems unlikely that such a relationship exists because hydralazine can be used in patients with idiopathic lupus without exacerbation of their disease (34). In fact, just the opposite seems to be true; there is no evidence that patients with drug-induced lupus have any higher incidence of other autoimmune diseases, or that patients with lupus are any more prone to have side effects from drugs that are associated with drug-induced lupus. Likewise, the drugs associated with lupus in man do not accelerate the development of lupus in animal models of lupus. Furthermore, it is unlikely that 20–30% of the population has a lupus diathesis, and yet that is the incidence of procainamide-induced lupus observed with chronic procainamide therapy.

Although it does not appear as if the genetic factors which lead to an increased risk of idiopathic lupus also predispose to drug-induced lupus, it is likely that genetic factors are important in predisposition to drug-induced lupus. As mentioned earlier, idiopathic lupus is associated with HLA DR-2 and DR-3 while hydralazine-induced lupus is associated with DR-4. It is clear that genetic factors that control drug metabolism influence an individual's risk of developing some types of drug-induced lupus. The rate of drug acetylation is genetically determined and rapid acetylators have a much lower incidence of hydralazine-induced lupus (3). The probable explanation for this observation is that acetylation of hydralazine leads to an inactive product. Acetylator phenotype also has an effect on the risk of developing procainamide-induced lupus, but the effect is less dramatic because the major route of elimination of procainamide is excretion of unchanged drug in the urine. It has been suggested that acetylator phenotype also influences the risk of idiopathic lupus (5), but most investigators believe that idiopathic lupus is independent of acetylator phenotype (35).

Taken to the extreme, genetic factors must be important in determining the risk of drug-induced lupus because the drugs which cause lupus in man do not, in general, induce lupus in most animals. In the one known example of a drug which causes lupus in man and in an animal (i.e., propylthiouracil in the cat), the incidence in cats is much higher than it is in man (36).

**Interaction of Drugs with Nuclear Antigens.** As mentioned earlier, most of the antinuclear antibodies in both procainamide- and hydralazine-induced lupus bind to histone protein. Hydralazine has been reorted to complex with soluble nucleoprotein (a DNA–histone complex) and change its pysical properties (37). Buxman also found that hydralazine and isoniazid reacted with protein (including histone protein) in vitro in the presence of transglutaminase (38). Thus, a possible mechanism of drug-induced lupus is that hydralazine could act as a hapten and react with histone protein, leading to the formation of antihistone antibodies. On the other hand, the conditions of these experiments are unlikely to occur in vivo. Procainamide is not very reactive, but we have demonstrated that it is metabolized by hepatic microsomes to reactive hydroxylamine and nitroso metabolites and the nitroso metabolite reacts with histone protein in vitro (39, 40).

Schoen and Trentham suggested that drugs could act as adjuvants to perturb immune regulation and induce autoreactive lymphocytes (41). One argument used to support this suggestion is that interaction of drug with nuclear antigen should only induce antinulcear antibodies and drug-induced lupus can be associated with other autoantibodies such as circulating anticoagulants, cryoglobulinemia, and lymphocytotoxic antibodies. On the contrary, if the ANA are due to reactive metabolites of the drugs acting as haptens by binding to nuclear protein, it would be expected that they would also bind to other macromolecules and result in other autoantibodies. Furthermore, as mentioned earlier, seemingly unrelated proteins can share a common epitope; thus, the antigen which induces antihistone antibodies in drug-induced lupus may not have any direct relationship to histone protein. Consistent with this possibility is the observation that most ANA that bind to histone protein in idiopathic lupus also bind to a cell membrane antigen on lymphocytes and granulocytes (42, 43). This observation suggests new hypotheses involving leukocytes for the induction of antihistone antibodies.

Attempts to induce antinuclear antibodies by immunizing animals with procainamide bound to protein were unsuccessful (44). Injection of hydralazine–albumin conjugates into rabbits resulted in induction of antibodies to nuclear antigens, but there was no evidence of pathological changes characteristic of lupus (45).

If the mechanism of drug-induced lupus involves reactive metabolites acting as haptens, one might expect to find antibodies which bind to drug or metabolite bound to protein in the serum of patients with drug-induced lupus. Most attempts to find antibodies that bind to drug have been negative (46). This should not be surprising because the structure of the metabolite–protein adduct is likely to be very different than the parent drug (47). Only if the hapten concentration is very high it is likely that antibodies will be produced to the parent drug. However, one would expect to find antibodies which bind to metabolite–protein adducts. Very few studies have been reported in which these antibodies have been sought, presumably because of the difficulty in obtaining the correct reactive metabolite bound to the correct protein or other macromolecule. However, in the case of practolol, which causes a unique form of drug-induced lupus, such antibodies were found (48, 49).

**Interference with Regulation of the Immune System.** Another reasonable hypothesis for the mechanism of drug-induced lupus, as mentioned in the previous section, is that the drug or its metabolites could interfere with regulation of the immune system. Although the control of immunoglobulin by B-cells is complex, a major factor is the balance between the actions of helper and suppressor T-cells. Ochi et al. reported that procainamide selectively inhibited suppressor T-cell activity in vitro (50); yet, Miller and Salem found suppressor T-cell activity of patients who were on chronic procainamide therapy to be normal (in contrast to their patients with idiopathic lupus who had reduced suppressor T-cell activity) (51). Shamess et al. also found normal suppressor T-cell activity in patients taking procainamide who had ANA (52). Yu and Ziff found that cells from patients treated with procainamide had a normal ratio of helper to suppressor cells but diminished B-cell response, even in the presence of T-cells from normal subjects (53). Drug was not present during these incubations; therefore, the observations represent an irreversible effect of the drug or its metabolites on the cells. As with idiopathic lupus, lymphocytotoxic antibodies have been found in procainamide- and hydalazine-induced lupus, but it is not clear what role they play in the pathogenesis of drug-induced lupus (54, 55). In short, although the drugs which are associated with lupus are reported to have an effect on cells involved in mediating immune function, the nature of the effect is controversial and its signifiance is unknown.

We found that the hydroxylamine metabolite of procainamide had toxic effects (as determined by exclusion of dye) on lymphocytes at micromolar concentrations while the parent drug had little effect even at a concentration of 5 mM (56). In addition to the effects on cell viability, the hydroxylamine of procainamide also induced DNA strand breaks and decreased the response of lymphocytes to the mitogen phytohemagglutinin. Although it was the hydroxylamine which was incubated with the cells, it appears that the actual toxic species is the nitroso derivative which is readily formed in the incubation.

**Inhibition of the Complement System.** It is known that individuals with genetic deficiencies in the complement system, especially C4 and C2, have a high incidence of lupus and other autoimmune diseases (1). This may be due to the role of the complement system in clearing immune complexes. Sim et al. observed that hydralazine and isoniazid inhibit C4 binding (57, 58). This appears to be due to a reaction of these hydrazine derivatives (which are good nucleophiles) with a thio ester at the active site of C4. Procainamide was not active. The hydroxylamine of procainamide would be expected to be more reactive than the parent drug, but we were unable to detect inhibition of complement mediated hemolysis by the hydroxylamine (unpublished observation). Sim et al. were able to detect inhibition of complement activation by the hydroxylamine of procainamide using a more sensitive method which measures covalent binding of activated C3 and C4 (59). The concentration required was about 1 mM; however, we have been unable to detect any hydroxylamine in the blood of patients taking procainamide (detection limit well under 1 $\mu$M). It is possible that locally high concentrations of the hydroxylamine could exist in the vicinity of leukocytes since it is formed by leukcotyes as described in the next section. This could conceivably affect the control of complement activation.

## Metabolism of Procainamide and Other Drugs by Monocytes

Our observation that procainamide is metabolized by hepatic microsomes to chemically reactive metabolites which covalently bind to histone protein (the major antigen to which ANA in drug-induced lupus bind) and which are toxic to lymphocytes, provide an attractive hypothesis for the initial step in procainamide-induced lupus. Furthermore, several of the other drugs which have been implicated in drug-induced lupus are also aromatic amines or hydrazines which might be metabolized to similar reactive metabolites (3). A major problem with this hypothesis was that these metabolites were further metabolized in the liver and little if any escaped to the circulation (39). Early attempts to demonstrate metabolism of procainamide by mononuclear leukocytes (a mixture of lymphocytes and monocytes) met with failure. However, these experiments were based on an assumption that the metabolism would be due to cytochromes P450. Neutrophils and monocytes also contain myeloperoxidase and a superoxide generating system. When these cells are activated, such as during the phagocytosis of bacteria, they release myeloperoxidase and generate superoxide (60). The superoxide is converted to hydroxy peroxide, which combines with the myeloperoxidase to form a strong oxidant. In the presence of chloride ion, a strong chlorinating agent, with properties similar to hypochlorite, is also formed (61).

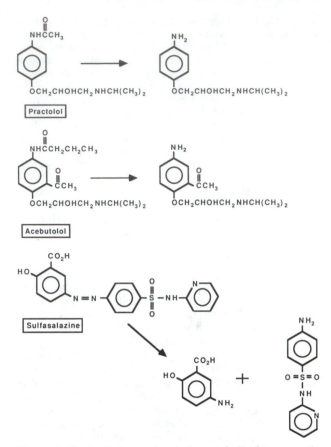

Figure 1. Oxidation of procainamide by activated leukocytes.

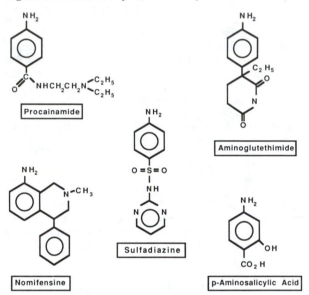

Figure 2. Aromatic amines that are associated with drug-induced lupus.

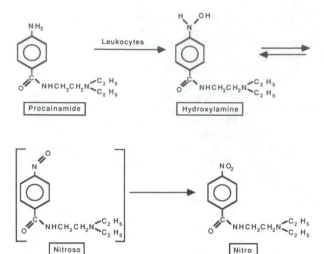

Figure 3. Drugs that are extensively metabolized to aromatic amines and are associated with drug-induced lupus.

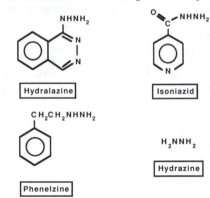

Figure 4. Hydrazines that are associated with drug-induced lupus.

When the above experiment was repeated with the addition of opsonized zymosan or phorbol ester to activate the cells, monocytes and neutrophils oxidized procainamide to the hydroxylamine metabolite (62). Neutrophils further oxidized the hydroxylamine to the nitro derivative, the nitroso derivative being the presumed intermediate as shown in Figure 1. Such metabolism provides the reactive metabolites with direct access to the immune system. Because we have found these metabolites to be toxic to lymphocytes, this provides a possible mechanism to explain the effects of procainamide on lymphocytes observed with chronic procainamide therapy. We have proposed that metabolism of procainamide to reactive metabolites by monocytes or neutrophils is the initial step in the mechanism by which it induces lupus.

The other aromatic amines that we have studied, dapsone and sulfadiazine, were also metabolized to hydroxylamines by activated leukocytes or by a combination of purified myeloperoxidase and hydrogen peroxide (63, 64). Several other aromatic amines have been implicated in drug-induced lupus including aminosalicylic acid, nomifensine, and aminoglutethimide (65–67). Their structures are shown in Figure 2. In addition, sulfasalazine, practolol, and acebutolol are extensively metabolized to aromatic amines and also induce lupus (68–74). The structures of these drugs are shown in Figure 3.

The second class of drugs which are associated with lupus are the hydrazines such as hydralazine, isoniazid, phenelzine, and hydrazine (8, 9, 75, 76, 23). As mentioned earlier, these compounds are good nucleophiles and are reactive without further metabolism. However, hydrazines are easily oxidized and would be expected to be oxidized by the combination of myeloperoxidase and hydrogen peroxide. A possible association was found between the formation of the metabolite phthalazinone and the risk of developing lupus (77). Although phthalazinone is not very reactive, it is presumably the stable product of a highly reactive intermediate. We have observed that hydralazine is metabolized to phthalazinone by activated leukocytes or by myeloperoxidase/hydrogen peroxide (unpublished observation). The structures of the hydrazines which have

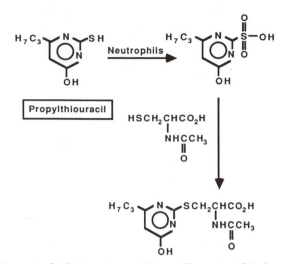

Figure 5. Oxidation of propylthiouracil by activated leukocytes and reaction of the sulfonic acid product with *N*-acetylcysteine.

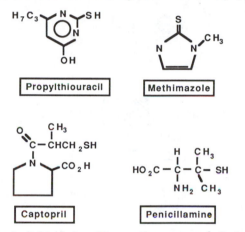

Figure 6. Sulfhydryl or thiono sulfur compounds that are associated with drug-induced lupus.

been associated with drug-induced lupus are shown in Figure 4.

Another class of drugs that has been implicated in causing drug-induced lupus contains sulfur as a sulfhydryl or thiono group. Examples are propylthiouracil, methimazole, penicillamine, and captopril (78–85). As mentioned earlier, the one good animal model of drug-induced lupus is propylthiouracil in the cat. Such a functional group would also be expected to be easily oxidized by hydrogen peroxide and a peroxidase. We investigated propylthiouracil and found that it was metabolized by either activated neutrophils or myeloperoxidase/hydrogen peroxide to several metabolites (86). Of the reactive metabolites formed, the most stable is a sulfonic acid. The sulfonic acid is chemically reactive and reacts with glutathione and other sulfhydryl containing amino acids (Figure 5). The structures of sulfhydryl or thiono containing drugs which have been associated with drug-induced lupus are shown in Figure 6.

The last class of drug associated with lupus which we have studied is the hydantoin anticonvulsants and related drugs. This class includes phenytoin, mephenytoin, ethosuximide, and trimethadione (87–89). Phenytoin is an important anticonvulsant but is also associated with a large number of serious side effects including drug-induced lupus. It has been proposed that this toxicity is due to a reactive arene oxide metabolite (90). This metabolite has never been observed, but the observation of phenol and

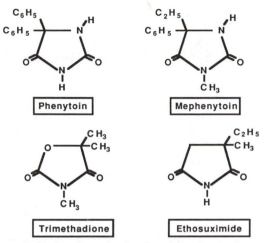

Figure 7. Hydantoin and related anticonvulsants that are associated with drug-induced lupus.

dihydrodiol metabolites suggests that it is formed. One problem with this hypothesis is that there are other drugs, such as trimethadione and ethosuximide, with similar heterocyclic rings and very similar patterns of toxicity that do not contain an aromatic ring, and therefore, could not form an arene oxide. Mephenytoin is also very similar to phenytoin, both in structure and toxicity. Also, like phenytoin, it is metabolized to a phenol, presumably through an arene oxide intermediate. Unlike phenytoin, mephenytoin is chiral and only the *S* enantiomer is metabolized to a phenol. In contrast to expectations, it appears as if it is the *R* enantiomer which is more toxic (91). With this evidence against the arene oxide mechanism of toxicity, we studied the oxidation of phenytoin by the myeloperoxidase/hydrogen peroxide/chloride system. We found that it was chlorinated to *N*,*N*'-dichlorophenytoin (92). Although we could not detect this as a metabolite when phenytoin was incubated with activated leukocytes, synthetic *N*,*N*'-dichlorophenytoin disappeared within seconds when incubated with leukocytes. When radio-labeled phenytoin was incubated with activated neutrophils, a metabolite was formed which covalently bound to neutrophils. We believe that this metabolite was *N*,*N*'-dichlorophenytoin because activation of the cells was necessary for covalent binding and *N*,*N*'-dichlorophenytoin was the only metabolite which we detected in the myeloperoxidase/hydrogen peroxide/chloride system. The structures of these anticonvulsants which are associated with drug-induced lupus are shown in Figure 7.

In summary, we have studied individual drugs from several chemical classes which are associated with drug-induced lupus. We found that they were metabolized to chemically reactive metabolites by either myeloperoxidase/hydrogen peroxide/chloride or activated leukocytes.

## Implications of Drug Metabolism by Leukocytes

A major function of the monocyte and other macrophages is to process antigen and present it, along with the class II major histcompatibility antigen (MHC), to T-cells to initiate an immunological response (93). Although an antigen can initiate an immunologic response without the presence of macrophages, the process is much less efficient. The processing of a protein antigen appears to involve the following steps: the protein is hydrolyzed inside the macrophage to peptides, and those peptides which bind to the class II MHC are protected from further hydrolysis. The peptide–MHC complex is then presented on the

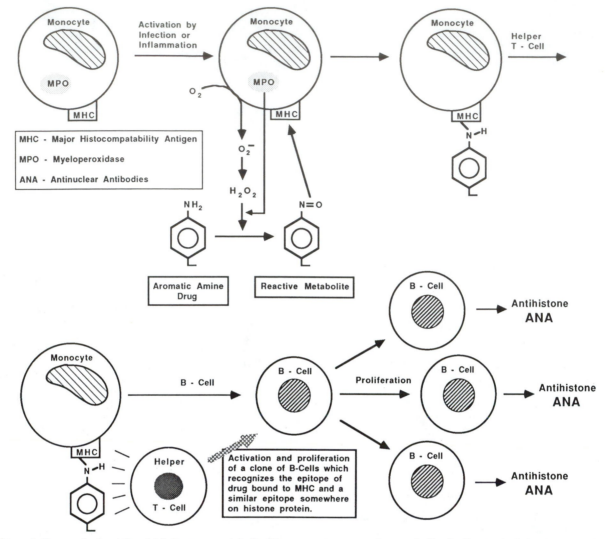

**Figure 8.** Proposed scheme by which drugs are metabolized by monocytes to reactive metabolites leading to the induction of antihistone antibodies.

surface of the cell where it is detected by T-cells. The T-cells are, in turn, stimulated to initiate an immunologic reaction against the protein. Because of the processing, the antibodies formed can have different epitopes than the original protein.

When monocytes are activated, the superoxide is formed on the outside of the cell; or if the cell is in the process of phagocytosis, the superoxide is released into the phagosome which is formed by the invagination of the cell membrane (94). Likewise, myeloperoxidase is released either outside the cell or into a phagosome. Any reactive metabolite formed during this process is in contact with the outer surface of the cell membrane. Therefore, it is possible that the reactive metabolite could react directly with MHC or some other membrane protein on the monocyte and bypass the processing steps. It has been proposed that direct binding of a molecule to MHC would be a strong stimulus for the production of autoantibodies (89, 95). If the antibody induced also recognized an epitope on histone protein or some other protein, an autoimmune syndrome could be induced. In the case of procainamide, the reason for the cross reactivity to histone protein could involve the tertiary amine at the other end of the molecule which has a positive charge and could mimic a portion of histone protein (histone protein is characterized by a large number of basic amino acids). This could occur without

causing a general increase in B-cell activation as is generally observed in idiopathic lupus. In as much as many of the ANA also bind to an antigen present on the cell membrane of leukocytes (42, 43), this could be the initial step in the induction of ANA. This hypothesis is also consistent with the observation of Gorsulowsky et al. that clinical signs of lupus correlated with ANA which bound only to leukocytes (96). Alternatively, binding of a reactive metabolite to the cell membrane of leukocytes could interfere with the control function of these cells. This hypothesis is presented graphically in Figure 8.

## Concluding Comments

I have tried to present a brief description of drug-induced lupus and its relationship to idiopathic lupus, as well as some of the proposed mechanisms of drug-induced lupus. In addition, I have presented my own hypothesis involving the metabolism of drugs to reactive metabolites by monocytes and neutrophils. This hypothesis is attractive, at least to me, for the following reasons.

1. It would provide an explanation for the structural features common to most of the drugs which are associated with drug-induced lupus. It seems unlikely that drug-induced lupus is due to the binding of some drugs to some classical receptor; the number and variety of drugs is too great. The drugs and other chemicals which are associated

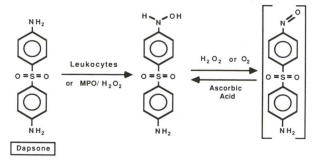

**Figure 9.** Oxidation of dapsone by activated leukocytes.

with immune-mediated toxicity are, in general, chemically reactive and bind to protein. A covalent bond between drug and protein is usually required and interactions such as hydrophobic binding do not result in an immune response. This was first discovered during the classical immunochemical studies of Landsteiner and others and is discussed at length in a recent review by Park et al. (95). Hydrazines and sulfhydryl-containing drugs are good nucleophiles and can react with macromolecules without further activation. Other drugs, such as aromatic amines, are not reactive but are easily oxidized to reactive metabolites. Although such activation can occur readily in the liver, the reactive metabolites formed do not escape the liver in significant concentrations. Thus, formation of reactive metabolites by monocytes and neutophils, which are critical to control of the immune sytem, is a much more plausible mechanism. Almost any drug can be metabolized by cytochromes P450 in the liver, but the one structural feature which most drugs associated with lupus share is the presence of a nitrogen or sulfur heteroatom in a functional group such as an aromatic amine, hydrazine, sulfhydryl, or thiono sulfur. Such groups are readily oxidized by myeloperoxidase. The β-blocking drugs provide a good illustration of this point. Although several β-blockers have been reported to be associated with drug-induced lupus, when this was studied carefully, it was found that only practolol and acebutolol caused a significant increase in the incidence of ANA (74). These are also the only β-blockers which are hydrolyzed to simple aromatic amines. This study also illustrates that a case report of drug-induced lupus cannot be assumed to be proof that a given drug causes lupus, a factor which makes the search for common structural features in drugs that cause lupus more difficult. Other drugs not falling into one of these classes, but which have been associated with drug-induced lupus, may also be metabolized by myeloperoxidase; however, at the present time, this remains to be demonstrated. One drug which will be difficult to fit into this scheme, if its association with drug-induced lupus is confirmed, is lithium since it cannot be metabolized. At present lithium's association with lupus is not convincing (97).

The converse is also true; i.e., virtually all drugs containing such easily oxidized groups have been associated with the induction of lupus. One exception that I have some difficulty with is dapsone. Dapsone is an aromatic amine and, as mentioned earlier, it is readily oxidized by activated leukocytes (Figure 9). It is also used for chronic therapy (this is important because it usually requires a month of therapy before lupus is induced). Yet, with the exception of one case report that is not very convincing (98), dapsone has not been associated with the induction of lupus.

2. The same drugs which are associated with drug-induced lupus are usually associated with other therapeutic and toxic effects which can be explained logically by the formation of reactive metabolites by myeloperoxidase and thyroid peroxidase. One such toxic effect is agranulocytosis. All of the drugs shown in the previous figures are associated with drug-induced agranulocytosis (i.e., a severe depression in neutrophil counts to the extent that the risk of a severe infection is high) (99, 100). In some cases this is due to a toxic effect on the stem cells in the bone marow, and in others it is due to the production of antibodies which destroy circulating neutrophils (101). In either case, the formation of a reactive metabolite by neutrophils could initiate such processes. Consistent with this hypothesis, the hydroxylamine of dapsone that we demonstrated is formed by activated neutrophils has also been shown by others to be toxic to bone marrow cells (102).

These same drugs are also associated with a generalized type of idiosyncratic drug reaction which consists of fever, skin rash, lymphadenopathy, and various other organ involvement such as liver, kidney, and heart (88, 103–106). These reactions also appear to be mediated by the immune system (95). In the case of sulfonamides, we have evidence that the idiosyncratic reactions are due to hydroxylamine and nitroso metabolites (103).

These drugs are probably also metabolized by thyroid peroxidase. Thyroid peroxidase is similar to myeloperoxidase and is responsible for the synthesis of thyroxine in the thyroid gland. Oxidation of propylthiouracil and methimazole by thyroid peroxidase is probably responsible for their therapeutic effects of in treating hyperthyroidism, i.e., these two drugs are metabolized by thyroid peroxidase to reactive metabolites which inhibit further formation of thyroxine (107). Several of the other drugs which induce lupus can also depress thyroid function and dapsone causes thyroid cancer in animals (108–110). The specificity of propylthiouracil and methimazole for inhibition of thyroid peroxidase presumably reflects their greater degree of active uptake and higher concentration in the thyroid gland. Yet, they are metabolized by activated leukocytes, and the most common serious side effect of propylthiouracil and methimazole is agranulocytosis.

Inhibition of myeloperoxidase can also lead to therapeutic effects and this appears to be the basis of the beneficial effect of dapsone in treating dermatitis herpetiformis (111) and may contribute to the effects of drugs such as sulfasalazine and aminosalicylic acid on ulcerative colitis.

The similarity in the toxic and some of the therapeutic effects of these drugs could be a coincidence, but this seems improbable.

Obviously, even if it can be demonstrated that drug-induced lupus involves the formation of reactive metabolites by monocytes and/or neutrophils, this does not provide a complete picture of the mechanism. This would only represent the initial step and much of the rest of the mechanism would remain a "black box" at the present time. On the other hand, it would provide a starting point for future investigations and would explain an important risk factor for drug-induced lupus. As stated earlier, metabolism mediated by leukocyte-derived myeloperoxidase does not occur unless the cells are activated. Consequently, one risk factor would be the presence of an infection or other inflammatory condition which activates these cells. This would occur in everyone at some point but to different degrees. It must be remembered that a small difference in one factor can determine if a person will develop drug-induced lupus. For example, the slow acetylator phenotype is a strong risk factor in hydralazine-induced lupus, despite the fact that rapid acetylators are also exposed to significant concentrations of unmetabolized hydralazine. This hypothesis could also be used to help predict which drugs are likely to cause lupus and other idiosyncratic drug reactions such as agranulocytosis.

**Acknowledgment.** This work was supported by a grant from the Medical Research Council of Canada (MA-9336).

## References

(1) Wallace, D. J., and Dubois, E. L., Eds. (1987) *Dubois' Lupus Erythematosus*, Lea & Febiger, Philadelphia.

(2) Lee, S. L., and Chase, P. H. (1975) "Drug-induced systemic lupus erythematosus: a critical review". *Seminars Arthritis Rheum. 5*, 83–103.

(3) Uetrecht, J. P., and Woosley, R. L. (1981) "Acetylator phenotype and lupus erythematosus". *Clin. Pharmacokin. 6*, 118–134.

(4) Cush, J. J., and Goldings, E. A. (1985) "Southwestern internal medicine conference: drug-induced lupus: clinical spectrum and pathogenesis". *Am. J. Med. Sci. 290*, 36–45.

(5) Reidenberg, M. M., and Martin, J. H. (1975) "The acetylator phenotype of patients with systemic lupus erythematosus". *Drug Metab. Dispos. 2*, 71–73.

(6) Blomgren, S. E., Condemi, J. J., Bignall, M. C., and Vaughan, J. H. (1969) "Antinuclear antibody induced by procainamide". *New Engl. J. Med. 281*, 64–66.

(7) Woosley, R. L., Drayer, D. E., Reidenberg, M. M., Nies, A. S., Carr, K., and Oates, J. A. (1978) "Effect of acetylator phenotype on the rate at which procainamide induces antinuclear antibodies and the lupus syndrome". *New Engl. J. Med. 298*, 1157–1159.

(8) Perry, H. M., Jr. (1973) "Late toxicity to hydralazine resembling systemic lupus erytematosus or rheumatoid arthritis". *Am. J. Med. 54*, 58–72.

(9) Bing, R. F., Russell, G. I., Thurston, H., and Swales, J. D. (1980) "Hydralazine in hypertension: is there a safe dose?" *Br. Med. J. 281*, 353–354.

(10) Wright, P. (1975) "Untoward effects associated with practolol administration: oculomucocutaneous syndrome". *Br. Med. J. 1*, 595–598.

(11) Steen, V. D., Blair, S., and Medsger, T. A., Jr. (1986) "The toxicity of D-penicillamine in systemic sclerosis". *Ann. Intern. Med. 104*, 699–705.

(12) Steinberg, A. D. (1984) "Modern concepts of systemic lupus erythematosus". *Prog. Clin. Rheum. 1*, 1–31.

(13) Tsokos, G. C., and Balow, J. E. (1984) "Cellular immune responses in systemic lupus erythematosus". *Prog. Allergy 35*, 93–161.

(14) Miyasaka, N., Nakamura, T., Russell, I. J., and Talal, N. (1984) "Interleukin-2 deficiencies in rheumatoid arthritis and systemic lupus erythematosus". *Clin. Immunol. Immunopathol. 31*, 109–117.

(15) Christian, C. L. (1978) "Clues from genetic and epidemiologic studies". *Arthritis Rheum. 21*, S130–S133.

(16) Block, S. R., Winfield, J. B., Lockshin, M. D., D'Angelo, W. A., and Christian, C. L. (1975) "Studies of twins with systemic lupus erythematosus. A review of the literature and presentation of 12 additional sets". *Am. J. Med. 59*, 533–552.

(17) Dixon, F. J. (1985) "Murine lupus, a model for human autoimmunity". *Arthritis Rheum. 28*, 1081–1088.

(18) Gibofsky, A. M., Winchester, R. J., Patarroyo, M., Fotino, M., and Kunkel, H. G. (1978) "Disease association of the Ia-like human alloantigens: containing patterns in rheumatoid arthritis and systemic lupus erythematosus". *J. Exp. Med. 148*, 1728–1732.

(19) Bell, D. A., and Maddison, R. J. (1980) "Serologic subsets in systemic lupus erythematosus: an examination of autoantibodies in relationship to clinical features of disease and HLA antigens". *Arthritis Rheum. 23*, 1268–1273.

(20) Christian, C. L. (1982) "Role of viruses in etiology of systemic lupus erythematosus". *Am. J. Kidney Dis. 2*, 114–118.

(21) Schwartz, R. S., and Stollar, B. D. (1985) "Origins of anti-DNA autoantibodies". *J. Clin. Invest. 75*, 321–327.

(22) Talal, N., Dauphinee, M., Ahmed, S. A., and Christadoss, P. (1983) "Sex factors in immunity and autoimmunity". In *Progress in Immunology V* (Yamamura, Y., and Tada, T., Eds.) pp 1589–1600, Academic, New York.

(23) Reidenberg, M. M., Durant, P. J., Harris, R. A., de Boccardo, G., Lahita, R., and Stenzel, K. H. (1983) "Lupus erythematosus-like disease due to hydrazine". *Am. J. Med. 758* 365–370.

(24) Malinow, M. R., Bardana, E. J., Jr., Pirofsky, B., Craig, S., and McLaughlin, P. (1982) "Systemic lupus erythematosus-like syndrome in monkeys fed alfalfa sprouts: role of a nonprotein amino acid". *Science (Washington, D.C.) 216*, 415–417.

(25) Tan, E. M., Cohen, a. S., Fries, J. F., Masi, A. T., McShane, D. J., Rothfield, N. F., Schaller, J. G., Talal, N., and Winchester, R. J. (1982) "The 1982 revised criteria for the classification of systemic lupus erythematosus". *Arthritis Rheum. 25*, 1271–1277.

(26) Batchelor, J. R., Welsh, K. I., Tinoco, R. M., Dollery, C. T., Hughes, G. R. V., Bernstein, R., Ryan, P., Naish, P. F., Aber, G. M., Bing, R. F., and Russell, G. I. (1980) "Hydralazine-induced systemic lupus erythematosus: influence of HLA-DR and sex on susceptibility". *Lancet 1*, 1107–1109.

(27) Brand, C., Davidson, A. Littlejohn, G., and Ryan, P. (1984) "Hydralazine-induced lupus: no association with HLA-DR4". *Lancet 1*, 462.

(28) Fritzler M. J., and Tan, E. M. (1978) "Antibodies to histones in drug-induced lupus and idiopathic lupus erythematosus". *J. Clin. Invest. 62*, 560–567.

(29) Portanova, J. P., Rubin, R. L., Joslin, F. G., Agnello, V. D., and Tan, E. M. (1982) "Reactivity of anti-histone antibodies induced by procainamide and hydralazine". *Clin. Immunol. Immunopathol. 25*, 67–79.

(30) Rubin, R. L., Reimer, G., McNally, E. M., Nusinow, S. R., Searles, R. P., and Tan, E. M. (1986) "Procainamide elicits a selective autoantibody immune response". *Clin. Exp. Immunol. 63*, 58–67.

(31) Weisbart, R. H., Yee, W. S., Colburn, K. K., Whang, S. H., Heng, M. K., and Boucek, R. J. (1986) "Antiguanosine antibodies: a new marker for procainamide-induced systemic lupus erythematosus". *Ann. Intern. Med. 104*. 310–313.

(32) Deng, J. S., Rubin, R. L., Lipscomb, M. F., Sontheimer, R. D., and Gilliam, J. N. (1984) "Reappraisal of the specificity of the *Crithidia luciliae* assay for nDNA antibodies: evidence for histone antibody kinetoplast binding". *Am. J. Clin. Pathol. 82*, 448–452.

(33) Alarcon-Segovia, D., Worthington, J. W., Ward, L. E., and Wakim, K. G. (1965) "Lupus diathesis and the hydralazine syndrome". *New Eng. J. Med. 272*, 462–466.

(34) Reza, M. J., Dornfeld, L., and Goldberg, L. S. (1975) "Hydralazine therapy in hypertensive patients with idiopathic systemic lupus erythematosus". *Arthritis Rheum. 18*, 335–338.

(35) Morris, R. J., Freed, C. R., and Kohler, P. F. (1979) "Drug acetylation phenotype unrelated to development of sponaneous systemic lupus erythematosus". *Arthritis Rheum. 22*, 777–780.

(36) Aucoin, D. P., Peterson, M. E., Hurvitz, A. I., Drayer, D. E., Lahita, R. G., Quimby, F. W., and Reidenberg, M. M. (1985) "Propylthiouracil-induced immune-mediated disease in the cat". *J. Pharmacol. Exp. Ther. 234*, 13–18.

(37) Tan, E. M. (1974) "Drug-induced autoimmune disease". *Fed. Proc. 33*, 1894–1897.

(38) Buxman, M. M. (1979) "The role of enzymatic coupling of drugs to proteins in induction of drug specific antibodies". *J. Invest. Dermatol. 73*, 256–258.

(39) Uetrecht, J. P., Sweetman, B. J., Woosley, R. L., and Oates, J. A. (1984) "Metabolism of procainamide to a hydroxylamine by rat and human hepatic microsomes". *Drug Metab. Dispos. 12*, 77–81.

(40) Uetrecht, J. P. (1985) "Reactivity and possible significance of hydroxylamine and nitroso metabolites of procainamide". *J. Pharmacol. Exp. Ther. 232*, 420–425.

(41) Schoen, R. T., and Trentham, D. E. (1981) "Drug-induced lupus: an adjuvant disease?" *Am. J. Med. 71*, 5–8.

(42) Searles, R. P., Messner, R. P., and Bankhurst, A. D. (1979) "Cross-reactivity of antilymphocyte and antinuclear antibodies in systemic lupus erythematosus". *Clin. Immunol. Immunopathol. 14*, 292–299.

(43) Rekvig, O. P., and Hannestad, K. (1980) "Human antibodies that react with both cell nuclei and plasma membranes display specificity for the octamer of histones H2A, H2B, H3, and H4 in high salt". *J. Exp. Med. 152*, 1720–1733.

(44) Gold, E. F., Ben-Fraim, S., Faivisewitz, A., Steiner, Z., and Klajman, A. (1977) "Experimental studies on tte mechanism of induction of anti-nuclear antibodies by procainamide". *Clin. Immunol. Immunopathol. 7*, 176–186.

(45) Yamauchi, Y., Litwin, A., Adams, L., Zimmer, H., and Hess, E. V. (1975) "Induction of antibodies to nuclear antigens in rabbits by immunization with hydralazine-human serum albumin conjugates". *J. Clin. Invest. 56*, 958–969.

(46) Litwin, A., Adams, L. E., Zimmer, H., Foad, B., Loggie, J. H. M., and Hess, E. V. (1981) "Prospective study of immunologic effects of hydralazine in hypertensive patients". *Clin. Pharmacol. Ther. 29*, 447–456.

(47) Parker, C. W. (1982) "Allergic reactions in man". *Pharmacol. Rev. 34*, 85–104.

(48) Amos, H. E., Lake, B. G., and Atkinson, H. A. C. (1977) "Allergic drug reactions: an in vitro model using a mixed function oxidase complex to demonstrate antibodies with specificity for a practolol metabolite". *Clin. Allergy 7*, 423–428.

(49) Amos, H. E., Lake, B. G., and Artis, J. (1978) "Possible role of antibody specific for a practolol metabolite in the pathogenesis of oculomucocutaneous syndrome". *Br. Med. J. 1*, 402–407.

(50) Ochi, T., Goldings, E. A., Lipsky, P. E., and Ziff, M. (1983) "Immunomodulatory effect of procainamide in man: inhibition of human suppressor T-cell activity in vitro". *J. Clin. Invest. 71*, 36–45.

(51) Miller, K. B., and Salem, D. (1982) "Immune regulatory abnormalities produced by procainamide". *Am. J. Med. 73*, 487–492.

(52) Shamess, C. J., Klein, S., and Keystone E. C. (1982) "Antigen-specific suppressor cell activity in procainamide therapy". *Arthritis Rheum. 25*, 238–239.

(53) Yu, C., and Ziff, M. (1985) "Effects of long-term procainamide therapy on immunogloblin synthesis". *Arthritis Rheum. 28*, 276–284.

(54) Bluestein, H. G., Weisman, M. H., Zvaifler, M. J., and Shapiro, R. F. (1979) "Lymphocyte alteration by procainamide: relation to drug-induced lupus erythematosus syndrome". *Lancet 2*, 816–819.

(55) Ryan, P. F. J., Hughes, G. R. V., Bernstein, R., Mansilla, R., and Dollery, C. T. (1979) "Lymphocytotoxic antibodies in hydralazine-induced lupus erythematosus". *Lancet 2*, 1248–1249.

(56) Rubin, R. L., Uetrecht, J. P., and Jones, J. E. (1987) "Cytotoxicity of oxidative metabolites of procainamide". *J. Pharmacol. Exp. Ther. 242*, 833–841.

(57) Sim, E., Gill, E. W., and Sim, R. B. (1984) "Drugs that induce systemic lupus erythematosus inhibit complement component C4". *Lancet 2*, 422–424.

(58) Sim, E., and Law, S. A. (1985) "Hydralazine binds covalently to complement component C4". *Fed. Eur. Biochem. Soc. 184*, 323–327.

(59) Sim, E., Stanley, L., Gill, E. W., Jones, A. (1988) "Metabolites of procainamide and practolol inhibit complement components C3 and C4". *Biochem. J.* (in press).

(60) Forman, H. J., and Thomas, M. J. (1986) "Oxidant production and bactericidal activity of phagocytes". *Annu. Rev. Physiol. 48*, 669–680.

(61) Winterbourn, C. (1985) "Comparative reactivities of various biological compounds with myeloperoxidase–hydrogen peroxide–chloride, and similarity of the oxidant to hypochlorite". *Biochim. Biophys. Acta 840*, 204–210.

(62) Uetrecht, J., Zahid, N., and Rubin, R. (1988) "Metabolism of procainamide to a hydroxylamine by human neutrophils and mononuclear leukocytes". *Chem. Res. Toxicol. 1*, 74–78.

(63) Uetrecht, J., Shear, N., and Biggar, W. (1986) "Dapsone is metabolized by human neutrophils to a hydroxylamine". *Pharmacologist 28*, 239.

(64) Uetrecht, J., Zahid, N., Shear, N. H., and Biggar, W. D. (1988) "Metabolism of dapsone to a hydroxylamine by human neutrophils and mononuclear cells". *J. Pharmacol. Exp. Ther. 245*, 274–279.

(65) Davies, D. M. (1985) *Textbook of Adverse Reactions*, p 4658 Oxford University Press, Oxford.

(66) Garcia-Morteo, O., and Maldonado-Cocco, J. A. (1983) "Lupus-like syndrome during treatment with nomifensine", *Arthritis Rheum. 26*, 936.

(67) McCracken, M., Benson, E. A., and Hickling, P. (1980) "Systemic lupus erythematosus induced by aminoglutethimide". *Br. Med. J. 281*, 1254.

(68) Griffiths, I. D., and Kane, S. P. (1977) "Sulfasalazine-induced lupus syndrome in ulcerative colitis". *Br. Med. J. 2*, 1188–1189.

(69) Vanheule, B. A., and Carswell, F. (1983) "Sulphasalazine-induced systemic lupus erythematosus in a child". *Eur. J. Pediatr. 140*, 66–68.

(70) Raftery, E. B., and Denman, A. M. (1973) "Systemic lupus erythematosus syndrome induced by practolol". *Br. Med. J. 2*, 452–455.

(71) Milner, G. R., Holt, P. J. L., Bottomley, J., and Maciver, J. E. (1977) "Practolol therapy associated with a systemic lupus erythematosus-like syndrome and an inhibitor to factor XIII". *J. Clin. Pathol. 30*, 770–773.

(72) Cody, R. J., Calabrese, L. H., Clough, J. D., Tarazi, R. C., and Bravo, E. L. (1979) "Development of antinuclear antibodies during acebutolol therapy". *Clin. Pharmacol. Ther. 25*, 800–805.

(73) Booth, R. J., Bullock, J. Y., and Wilson, J. D. (1980) "Antinuclear antibodies in patients on acebutolol". *Br. J. Clin. Pharmacol. 9*, 515–517.

(74) Booth, R. J., Wilson, J. D., and Bullock, J. Y. (1982) "β-Adrenergic-receptor blockers and antinuclear antibodies in hypertension". *Clin. Pharmacol. Ther. 31*, 555–558.

(75) Rothfield, N. F., Bierer, W. F., and Garfield, J. W. (1978) "Isoniazid induction of antinucler antibodies". *Ann. Intern. Med. 88*, 650–652.

(76) Swartz, C. (1978) "Lupus-like reaction to phenelzine". *J. Am. Med. Assoc. 239*, 2693.

(77) Timbrell, J. A., Facchini, V., Harland, S. J., and Mansilla-Tinoco, R. (1984) "Hydralazine-induced lupus: is there a toxic metabolic pathway?" *Eur. J. Clin. Pharmacol. 27*, 555–559.

(78) Amrhein, J. A., Kenny, F. M., and Ross, D. (1970) "Granulocytopenia, lupus-like syndrome and other complications of propylthiouracil therapy". *J. Pediatr. 76*, 54–63.

(79) Oh, B. K., Overveld, G. P., and Macfarlane, J. D. (1983e "Polyarteritis induced by propylthiouracil. Case report". *Br. J. Rheumatol. 22*, 106–108.

(80) Librick, L., Sussman, L., Bejar, R., and Clayton, G. W. (1970) "Thyrotoxicosis and collagen-like disease in three sisters of American-Indian extraction". *J. Pediatr. 76*, 64–68.

(81) Harkcom, T. M., Conn, D. L., and Holley, K. E. (1978) "D-Penicillamine and lupus erythematosus-like syndrome". *Ann. Intern. Med. 89*, 1012.

(82) Walshe, J. M. (1981) "Penicillamine and the SLE syndrome". *J. Rheumatol., Suppl. 8*, 155–160.

(83) Chalmers, A., Thompson, D., Stein, H. E., Reid, G., and Patterson, A. C. (1982) "Systemic lupus erythematosus during penicillamine therapy for rheumatoid arthritis". *Ann. Intern. Med. 97*, 659–663.

(84) Reidenberg, M. M., Case, D. B., Drayer, D. E., Reis, S., and Lorenzo, B. (1984) "Development of antinuclear antibody in patients treated with high doses of captopril". *Arthritis Rheum. 27*, 579–581.

(85) Patri, P., Nigro, A., and Rebora, A. (1985) "Lupus erythematosus-like eruption from captopril". *Acta Derm. Venereol. 65*, 447–448.

(86) Waldhauser, L., and Uetrecht, J. (1988) "Propylthiouracil is metabolized by activated neutrophils-implications for agranulocytosis". *FASEB J. 2*, A1134.

(87) Beernink, D. H., and Miller, J. J., III. (1973) "Anticonvulsant-induced antinuclear antibodies and lupus-like disease in children". *Pediatr. Pharmacol. Ther. 82*, 113–117.

(88) Singsen, B. H., Fishman, L., and Hanson, V. (1976) "Antinuclear antibodies and lupus-like syndromes in children receiving anticonvulsants". *Pediatrics 57*, 529–534.

(89) Gleichman, H. (1982) "Systemic lupus erythematosus triggered by diphenylhydantoin". *Arthritis Rheum. 25*, 1387–1388.

(90) Martz, C., Failinger, C., and Blake, D. A. (1977) "Phenytoin teratogenesis: correlation between embryopathic effect and covalent binding of putative arene oxide metabolite in gestational tissue". *J. Pharmacol. Exp. Ther. 203*, 231–239.

(91) Wells, P. G., Jupfer, A., Lawson, J. A., and Harbison, R. D. (1982) "Relation of in vivo drug metabolism to stereoselective fetal hydantoin toxicology in mouse: evaluation of mephenytoin and its metabolite, nirvanol". *J. Pharmacol. Exp. Ther. 221*, 228–234.

(92) Uetrecht, J., and Zahid, N. (1988) "N-Chlorination of phenytoin by myeloperoxidase to a reactive metabolite". *Chem. Res. Toxicol. 1*, 148–151.

(93) Unanue, E. R., and Allen, P. M. (1987) "The basis for the immunoregulatory role of macrophages and other accessory cells". *Science (Washington, D.C.) 236*, 551–557.

(94) Badwey, J. A., and Karnovsky, M. L (1980) "Active oxygen species and the functions of phagocytic leukocytes". *Annu. Rev. Biochem. 49*, 695–726.

(95) Park, B. K., Coleman, J. W., and Kitteringham, N.R. (1987) "Drug disposition and drug hypersensitivity". *Biochem. Pharmacol. 36*, 581–590.

(96) Gorsulowsky, D. C., Bank, P. W., Goldberg, A. D., Lee, T. G., Heinzerling, R. H., and Burnham, T. K. (1985) "Antinuclear antibodies as indicators for the procainamide-induced systemic lupus erythematosus-like syndrome and its clinical presentations". *J. Am. Acad. Dermatol. 12*, 245–253.

(97) Presley, A. P., Kahn, A., and Williamson, N. (1976) "Antinuclear antibodies in patients on lithium carbonate". *Br. Med. J. 2*, 280–281.

(98) Vandersteen, P. R., and Jordon, R. E. (1974) "Dermatitis herpetiformis with discoid lupus erythematosus: occurrence of sul-

fone-induced discoid lupus erythematosus". *Arch. Dermatol. 110*, 95–98.

(99) Ognibene, A. J. (1970) "Agranulocytosis due to dapsone". *Ann. Intern. Med. 72*, 521–524.

(100) Davies, D. Mm. (1985) *Textbook of Adverse Reactions*, + 580, Oxford University Press, Oxford.

(101) Pisciotta, V. (1978) "Drug-induced agranulocytosis". *Drugs 15*, 132–143.

(102) Weetman, R. M., Boxer, L. A., Brown, M. P., Mantich, N. M., and Baehner, R. L. (1980) "In vitro inhibition of granulopoiesis by 4-amino-4'-hydroxylaminodiphenyl sulfone". *Br. J. Haematol. 45*, 361–370.

(103) Reider, M. J., Uetrecht, J., Shear, N. H., and Spielberg, S. P. (1988) "Synthesis and in vitro toxicity of hydroxylamine metabolites of sulfonamides". *J. Pharmacol. Exp. Ther. 244*, 724–728.

(104) Kromann, N., Vilhelmsen, R., and Stahl, D. (1982) "The dapsone syndrome". *Arch. Dermatol. 118*, 531.

(105) Spielberg, S. P., Gordon, G. B., Blake, D. A., Goldstein, D. A., and Herlong, H. F. (1981) "Predisposition to phenytoin hepatotoxicity assessed in vitro". *New Engl. J. Med. 305*, 722–7279

(106) Mihas, A. A., Holley, P., Koff, R. S., and Hirschowitz, B. I. (1976) "Fulminant hepatitis and lymphocyte sensitization due to propylthiouracil". *Gastroenterology 70*, 770–774.

(107) Engler, H., Taurog, A., and Nakashima, T. (1982) "Mechanism of inactivation of thyroid peroxidase by thioureylene drugs". *Biochem. Pharmacol. 31*, 3801–3806.

(108) Hughes, S. W. M., and Burley, D. M. (1970) "Aminoglutethimide: side-effect turned to therapeutic advantage". *Postgrad. Med. J. 46*, 409–416.

(109) Swarm, R. L., Roberts, G. K. S., Levy, a. C., and Hines, L. R. (1973) "Observations on the thyroid gland in rats following the administration of sulfamethoxazole and trimethoprim". *Toxicol. Appl. Pharmacol. 24*, 351–363.

(110) Griciute, L., and Tomatis, L. (1980) "Carcinogenicity of dapsone in mice and rats". *Int. J. Cancer 25*, 123–129.

(111) Stendahl, O., Molin, L., and Dahlgren, C. (1978) "The inibition of polymorphonuclear leukocyte cytotoxicity by dapsone". *J. Clin. Invest. 62*, 214–220.

# Chapter 9

# Molecular and Biochemical Mechanisms of Chemically Induced Nephrotoxicity: A Review

Jan N. M. Commandeur and Nico P. E. Vermeulen*

*Department of Pharmacochemistry (Molecular Toxicology), Free University, De Boelelaan 1083, 1081 HV Amsterdam, The Netherlands*

Reprinted from *Chemical Research in Toxicology,* Vol. 3, No. 3, May/June, 1990

## Contents

## I. Introduction

The mammalian kidney is an organ with several physiologially, pharmacologically, and toxicologically important functions (*1*). The primary functions of the kidney are (a) volume regulation, (b) regulation of acid–base balance, (c) regulation of electrolyte balance, (d) excretion of waste products, (e) endocrine functions, including elaboration of renin, erythropoietin, 1,25-dihydroxyvitamin $D_3$, and (f) synthesis of vasoactive prostaglandins and kinins. Toxicity to the kidney (nephrotoxicity) may impair some or all of these functions, thus having severe or even fatal consequences to the organism. Derangement of volume regulation can lead either to dehydration due to

---

* To whom correspondence should be addressed.

2428–5/92/0073$07.00/0

Table I. Mechanisms for Initiation of Toxicity

| initiators | types of reaction or interaction | targets |
|---|---|---|
| electrophiles | covalent binding | proteins and nonprotein thiols, nucleic acids, (phospho)lipids |
| radicals | covalent binding | proteins and nonprotein thiols, nucleic acids |
| | hydrogen abstraction | protein and nonprotein thiols, nucleic acids, (phospho)lipids |
| | reduction | molecular oxygen |
| reactive oxygen species | | |
| $O_2^{\bullet-}$ | reduction | metal ions [Fe(III) → Fe(II)] |
| $H_2O_2$ | oxidation | proteins and nonprotein thiols |
| OH$^{\bullet}$ | hydrogen abstraction | protein and nonprotein thiols, nucleic acids, (phospho)lipids |
| heavy metals | complexation | proteins and nonprotein thiols, nucleic acids |
| polycations | charge–charge interaction | phospholipids |
| light hydrocarbons | high-affinity binding | $\alpha_{2\mu}$-globulin |
| ischemia | lack of oxygen | mitochondrial respiration |

massive urine production or to retention of water and salt: the resulting hypertension may lead to congestive heart failure and pulmonary congestion. Failure of regulation of the acid–base balance may lead to progressive metabolic acidosis. Derangement of the electrolyte balance can lead to hyperkalemia and hypocalcemia, both of which may cause dangerous cardiac arrhythmias and altered myocardial contractility. Retention of waste products is reflected in elevated blood urea nitrogen and creatinine levels. Liberation of ammonia from urea may be responsible for inflammatory lesions of the gastrointestinal mucosa. Impairment of synthesis of 1,25-dihydroxyvitamin $D_3$ may lead to disturbance of calcium and phosphate homeostasis, resulting in bone diseases (1).

Over the last decade, the knowledge of the nephrotoxic potential of many drugs and xenobiotics has increased rapidly. The vulnerability of the kidney to xenobiotics appears to be inherent in the functional organization of the kidney. A high blood flow (25% of the cardiac output) and the ability to concentrate xenobiotics play important roles in the onset of nephrotoxicity. The amount and concentration of potentially toxic agents in the kidney, therefore, can become much higher than in other tissues.

Over the last decades, it also has become clear that many parent compounds are not toxic themselves but instead are converted to chemically reactive intermediates by drug metabolizing enzymes. These toxic metabolites may be inactivated by protective mechanisms or, at sufficiently high levels, they may react with critical cellular targets which ultimately may lead to toxicity. The organ selectivity of a toxicant may thus depend either on distribution of the parent compound and/or on tissue-specific balances of toxication and detoxication (2). For nephrotoxins, metabolism in nonrenal tissues may even play a crucial role in bioactivation mechanism by forming proximate toxic metabolites or stable reactive intermediates.

In this review, the molecular aspects determining the vulnerability of the kidney toward nephrotoxicants will be discussed. An attempt is made to classify known nephrotoxicants according to a limited number of more general mechanisms.

## II. Biochemical Mechanisms for Chemically Induced Nephrotoxicity

On the basis of the different molecular and biochemical events, the process leading to nephrotoxicity might be divided into three major stages: (1) a first or "initiation" phase involving the primary interaction of a toxicant with cellular biomacromolecules; (2) a second or "propagation" phase involving reversible disturbances of biochemical processes due to the toxicant–biomacromolecule interactions; and (3) a "termination" phase involving irreversible biochemical processes leading finally to cell death. In this section the effects of known nephrotoxicants on different biochemical processes will be reviewed.

*II.1. General Mechanisms of Initiation and Propagation of Nephrotoxicity.* Different types of toxicants that might be involved in the initiation of the process leading to nephrotoxicity are shown in Table I. These toxicants may have irreversible or reversible interactions with critical biological macromolecules, resulting in different types of propagation of nephrotoxicity.

*II.1.a. Irreversible Molecular Interactions. Electrophiles* may bind covalently to nucleophilic sites abundantly present in biological macromolecules, such as proteins, lipids, RNA, and DNA, and as a result they may cause functional inactivation of these macromolecules. The nature and localization of the intracellular targets and the biochemical consequences of their inactivation will be discussed in the next section. The selectivity of **covalent binding to biomacromolecules** will depend on the chemical "hardness" and "softness" of both the electrophile and the nucleophilic sites in macromolecules (3). Hard electrophiles react selectively with hard nucleophiles. Electrophiles in increasing order of hardness are as follows: aldehydes, polarized double bonds < epoxides, strained-ring lactones, alkyl sulfates, alkyl halides < aryl carbonium ions < benzylic carbonium ions, nitrenium ions < alkyl carbonium ions. Nucleophilic sites in order of increasing hardness are as follows: thiol groups (cysteinyl residues in proteins and glutathione) < sulfur atoms of protein methionyl residues < primary and secondary amino groups in protein < amino groups of purine bases in DNA and RNA < oxygen atoms of purines and pyrimidines in DNA and RNA < phosphate oxygen of RNA and DNA.

To protect against covalent binding, cells possess high concentrations of nonprotein thiols, mainly glutathione (GSH, 5–10 mM). However, after excessive covalent binding, nonprotein thiols can become depleted and critical protein thiols might be attacked subsequently. Hard electrophiles can also be inactivated by the soft nucleophile GSH when the conjugation reaction is catalyzed by the glutathione S-transferases; however, depletion of the cofactor GSH will lead to destruction of vital macromolecules.

Depending on their chemical properties, *free radicals* may also bind covalently to cellular macromolecules, abstract hydrogen atoms from macromolecules, and/or donate their unpaired electron to molecular oxygen. Abstraction of hydrogen atoms of lipids and subsequent reaction of the lipid radicals with molecular oxygen may initiate the process of **lipid peroxidation**, which may have strong effects on fluidity, permeability, and/or integrity of membranes, and as a result on various membrane-associated processes (4). Hydrogen abstraction by free radicals of thiol groups of macromolecules may also lead to thiol oxidation or the formation of mixed disulfides. The sulfhydryl/disulfide status is normally maintained by the GSH/GSH-reductase/NADPH complex. Depletion

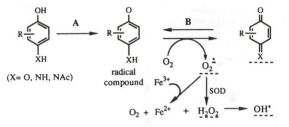

**Figure 1.** Formation of reactive species (indicated by dashed lines) as a result of redox cycling initiated by one-electron oxidation (A) or one-electron reduction (B) of suitable substrates. The hydroxyl radical formed by the Fenton reaction might abstract hydrogen atoms from lipids or proteins, initiating lipid peroxidation and oxidative stress.  Not fully substituted quinones also may initiate toxicity by covalent binding to macromolecules. SOD, superoxide dismutase.

of reducing equivalents (NADPH, GSH) may shift the redox state of cellular constituents, such as proteins, pyridine nucleotides, and flavins, to the oxidized state. This so-called **oxidative stress** can lead to decreases or increases of several activities, because the activity of a number of proteins depends on their sulfhydryl/disulfide status (5–7).

Free radicals formed by enzymic one-electron reduction or oxidation of various compounds [nitro compounds, (hydro)quinones, pyridinium cation compounds] under certain circumstances can donate their free electron to molecular oxygen under the formation of a superoxide anion radical ($O_2^{\bullet -}$). The reduced compound returns to its original state and may again be reduced (Figure 1); this process is called **redox cycling** (8). The superoxide anion radical is not able to abstract hydrogen atoms from macromolecules, but it can reduce ferric ($Fe^{3+}$) to ferrous ions ($Fe^{2+}$). The superoxide anion radical, however, is generally converted by superoxide dismutase (SOD) to $H_2O_2$. The ferrous ion can catalyze the Fenton reaction, involving conversion of $H_2O_2$ to a hydroxyl radical and a hydroxide anion. The hydroxyl radical is believed to be responsible for most of the serious damage caused by redox cycling, e.g., peroxidation of membranal lipids and protein and DNA damage (8).

Formation of superoxide anion radicals may also result from the xanthine/xanthine oxidase system which can be operative under certain pathological conditions. Formation of $H_2O_2$ and/or superoxide anion radicals may also be stimulated by uncouplers of the cytochrome P-450 cycle (9). Finally, reactive oxygen species might also be released by activated macrophages and/or neutrophils after formation or deposition of immune complexes (10, 11).

*II.1.b. Reversible Molecular Interactions.* Some agents have high affinity to biomacromolecules and may as a result cause impairment of biochemical functions. Some nephrotoxic heavy metal ions, such as cadmium, mercury, lead, and silver, have a high affinity for sulfhydryl groups (12, 13). More or less reversible complexation of these metal ions with essential protein thiols can inactivate these proteins and is proposed as the mechanism ultimately leading to nephrotoxicity. The reactivity of metal ions with sulfhydryl groups may depend on pH: at pH 7.4, the nephrotoxic uranyl cation has high affinity for carboxyl, amino, and phosphate groups and only low reactivity toward sulfhydryl groups; at pH 6, however, it rapidly alkylates sulfhydryl groups (14). Reversible charge–charge interactions may also result in toxic responses. The glomerular structure is normally maintained by anion–anion repulsions. Interference with these charge–charge interactions due to the presence of cationic agents may

result in a collapse of the glomerular structure.  Charge–charge interactions between cationic agents and anionic phospholipids, which are essential constituents of membranes (15), or high-affinity binding interactions between several compounds and proteins [e.g., light hydrocarbons to $\alpha_{2\mu}$-globulin (16)] may result in interferences with the turnover of these macromolecules by formation of indigestible complexes.

A hypoxic cellular state can ultimately lead to various types of damage in highly oxygen-dependent tissues. The nephrotoxicants mercury chloride and cyclosporin A have been shown to decrease the supply of oxygen to the kidney by inducing ischemia due to vasoconstriction (17, 18).

Specific examples for the different types of initiation mechanisms, leading to selective toxicity to different segments of the renal tubule, will be discussed in detail in section IV.

*II.2. Biochemical Mechanisms Leading to Cell Death.* As mentioned in the previous section, chemically induced toxicity is generally a result of reversible or irreversible interactions between a toxicant and critical cellular macromolecules (proteins, DNA, RNA) and/or membranes. After a primary insult caused by a toxin, a sequence of secondary biochemical processes ultimately leads to cell necrosis. Some of the initially affected processes may be reversible, and recovery can occur if the injurious stimulus is removed. However, at a given point processes become irreversible ("point of no return"), and the cell is said to have entered the phase of necrosis. Seven stages characterized by ultrastructural changes have been described in the process of ischemia-induced nephrotoxicity; four of them are reversible, the others are irreversible (19).

Several biochemical mechanisms have been proposed to play a role in the pathogenesis of cytoxicity:

(a) *A sustained elevation of cytosolic free $Ca^{2+}$ concentration*, with subsequent activation of $Ca^{2+}$-dependent autolytic enzymes, alterations in the cytoskeleton, and surface blebbing (20). Activation of nonlysosomal proteases can cause alterations in the cytoskeleton and in integral membranal proteins (21). Also, the association of the cytoskeleton with the plasma membrane might be disturbed by dissociation of $\alpha$-actinin from actin filaments. Activation of a specific subset of phospholipases, collectively designated as phospholipase $A_2$, can also mediate membrane breakdown directly or indirectly by generating cytotoxic breakdown products. $Ca^{2+}$-mediated activation of endonucleases in the cell nucleus may lead to DNA fragmentation and appears to be responsible for hepatocyte killing by several chemical toxins (22). Recently, elevation of glycogen phosphorylase activity in the liver of rats treated with $CCl_4$, diquat, and paracetamol was found indicative for elevation of intracellular free $Ca^{2+}$ in vivo (23).

(b) *Depletion of Cellular ATP or Decreased ATP/ADP Ratio.* A depletion of cellular ATP or decreased ATP/ADP ratio has been shown to result in an inability of the cell to perform important energy-dependent activities. The decrease in $Na^+/K^+$- and $Ca^{2+}$-ATPase activity might result in elevation of cytosolic $Ca^{2+}$ and subsequent activation of autolytic enzymes, as described above. Since ATP is required for polymerization of microfilaments necessary for the maintenance of the cytoskeleton, ATP depletion can lead to cellular blebbing, a common event in the progress of toxic injury (24).

(c) *Modification of Protein Thiols and Disulfides.* Protein thiols and disulfides have multiple roles in protein structure and function, e.g., by stabilizing conformations essential for substrate binding and translocation or by

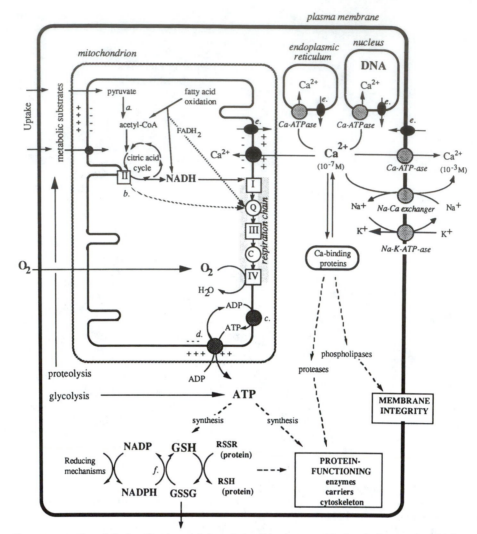

**Figure 2**. Schematic representation of the localization and the relationships between biochemical processes which at present are known to be affected by (nephro)toxicants. (a) Pyruvate dehydrogenase complex; (b) succinate dehydrogenase; (c)$F_1$-ATPase; (d) ADP/ATP antiporter; (e) calcium leak channels; (f) glutathione reductase. I, NADH dehydrogenase; Q, coenzyme Q; III, cytochrome $c$ reductase; C, cytochrome $c$; IV, cytochrome $c$ oxidase.

playing a dynamic role in catalysis. $Ca^{2+}$-ATPases, present in the plasma membrane and endoplasmic reticulum, are sulfhydryl-dependent proteins. Oxidation and/or alkylation of the sulfhydryl groups may lead to elevated cytoplasmic $Ca^{2+}$ levels (25, 26). Recently, oxidation of thiol groups in the cytoskeleton-related protein actin has also been associated with bleb formation in hepatocytes exposed to menadione (27).

Because of the multisite regulation of $Ca^{2+}$ homeostasis (see section II.3) disturbance of one of the processes involved often can be compensated by a relatively high capacity of other redistribution mechanisms. Therefore, most of the disturbances of $Ca^{2+}$ homeostasis per se may not necessarily be responsible for cytotoxicity, but they can nevertheless have a potentiating effect. A sustained elevated cytosolic $Ca^{2+}$ will be achieved when the $Ca^{2+}$-regulation processes are disturbed simultaneously at different levels. Recently, it was shown that a sustained high free $Ca^{2+}$ concentration induced by extracellular ATP did not immediately lead to cytotoxicity in hepatocytes (28). Both intracellular $Ca^{2+}$ and ATP were increased, which might argue against the concept of $Ca^{2+}$-induced activation of autolytic enzymes. Instead, it might point to a collapse of mitochondrial functioning and to subsequent ATP depletion as the final trigger to cell death. $Ca^{2+}$-induced

uncoupling of oxidative phosphorylation as a result of elevated cytosolic $Ca^{2+}$, however, may also contribute to the process of cytotoxicity.

Recently, it was shown that the synthesis of ATP by glycolysis in cytosol temporarily protects hepatocytes against menadione-induced mitochondrial dysfunctioning (29). Depletion of glycolytic substrates, such as glucose 6-phosphate, by fasting shortened the time course of both ATP depletion and cell death. Changes in redox status (NADPH/NADP$^+$) and thiol status (GSH/GSSG) did not correlate with the time course of cell death. Therefore, these two factors apparently are less essential in maintaining cellular integrity (29).

The inhibition of hepatotoxicity caused by inhibitors of phospholipase $A_2$ and proteases (30, 31) nevertheless points to the involvement of increased autolysis in some situations. Therefore, the relative importance of ATP depletion and autolytic processes probably depends on the type of toxicant and possibly the tissue involved.

**II.3. Localization of Early Pathogenic Biochemical Events.** Intracellular $Ca^{2+}$, ATP, and thiol homeostases are related and extraordinarily complex processes. A simplified schematic representation of the processes involved in these homeostases in mammalian cells is presented in Figure 2. Interference of a toxicant with any of the individual components of the scheme might contribute to some degree to the process of cell dysfunctioning or cell necrosis.

Events at the plasma membrane, inner mitochondrial membrane, endoplasmic reticulum, and nuclear membrane are all involved in regulating the intracellular $Ca^{2+}$ concentration. In addition to that, $Ca^{2+}$ is bound to cytosolic proteins (calmodulin). Toxicants have been shown to interact with these different sites of $Ca^{2+}$ homeostasis, resulting in an influx and/or redistribution of $Ca^{2+}$ in the cell.

*II.3.a. Effects of Toxicants on the Plasma Membrane.* The plasma membrane is an import barrier at which strong gradients are maintained: the intracellular $Ca^{2+}$ concentration is approximately 10 000 times lower than the extracellular concentration: $10^{-7}$ M vs $10^{-3}$ M. Despite a passive leak of calcium, via leak-channel proteins, into the cell, the calcium gradient is maintained by an ATP-dependent carrier, $Ca^{2+}$-ATPase, and a $Na^+/Ca^{2+}$ exchanger. The latter is driven by the sodium gradient at the plasma membrane (outside: 440 mM; inside: 50 mM), which in turn is maintained by the ATP-dependent $Na^+/K^+$-ATPase.

Toxicants can affect cellular $Ca^{2+}$ homeostasis by affecting plasma membrane bound transporters. Direct interactions with $Ca^{2+}$ transport are most commonly observed with divalent metals such as lead, ruthenium, cadmium, and others (32). Plasma membrane $Ca^{2+}$-ATPase, in the kidney located at the brush border membrane (33), is a sulfhydryl-dependent enzyme which can be inactivated by thiol oxidation [e.g., by menadione (34)], mixed disulfide formation [e.g., by cystamine (35)], or covalent binding [e.g., by paracetamol (36)]. $Na^+/K^+$-ATPase also has been shown to be inhibited by several nephrotoxicants: e.g., by vanadium (37), mercury chloride (38), and gentamicin (39). Increased net sodium influx resulting from inhibition of this enzyme secondarily results in elevation of $Ca^{2+}$ influx due to $Na^+/Ca^{2+}$ exchange. Both $Ca^{2+}$-ATPase and $Na^+/K^+$-ATPase depend on ATP for activity. ATP is produced in cytosol by glycolysis and in mitochondria by oxidative phosphorylation. Inhibition of these processes, therefore, will secondarily also result in disturbance of ATP-dependent electrolyte homeostases.

Toxicants can also act by increasing the permeability of the plasma membrane for ions, resulting in leakage of $Ca^{2+}$ and/or $Na^+$ along their respective gradients. The hepatotoxin $CCl_4$ has been shown to increase plasma membrane permeability to $Ca^{2+}$ (40). An increased influx of $Na^+$, if exceeding the capacity of $Na^+/K^+$-ATPase to antagonize it, will result in an increase of intracellular sodium and secondarily, by the action of the $Na^+/Ca^{2+}$ exchanger, intracellular $Ca^{2+}$ (41). An increased permeability toward these ions may result from lipid peroxidation, membrane protein thiol modification, or ionophores (e.g., nystatin) (19) and/or from interferences with membrane phospholipid turnover. The nephrotoxic silver ion ($Ag^+$) has been shown to increase the cell membrane $K^+$ and $Na^+$ permeability of the renal proximal tubule, while inhibiting the $Na^+/K^+$-ATPase (13). The reversal of these effects by thiol reagents points to the involvement of reversal of thiol modification (by complexation).

*II.3.b. Effects of Toxicants on the Mitochondrion.* The mitochondrion plays a very important role in maintaining $Ca^{2+}$ and ATP homeostasis, (a) because of its high capacity to sequester $Ca^{2+}$ and (b) because of its role in the synthesis of ATP, which is the metabolic fuel for the microsomal and plasmalemmal $Ca^{2+}$-ATPases and several important cellular processes (42). Because the proximal tu-

bular cell only has a very low glycolytic ATP-production activity (43), ATP production is almost exclusively mediated by mitochondrial oxidative phosphorylation in these cells. A prominent and early effect of many nephrotoxins and pathological conditions affecting the kidney consists of a dramatic decrease in ATP concentration (44). Recently, the ATP depletion in different parts of the proximal tubule was shown to correlate with the site of renal toxicity induced by mercuric chloride and ochratoxin A (45, 46).

$Ca^{2+}$ is actively carried into the mitochondria by a transport process that is driven by the membrane potential/proton gradient at the inner mitochondrial membrane. This gradient is generated by the process of mitochondrial respiration. The energy released during electron transport from oxidizable substrates [NADH, $FADH_2$ (free or enzyme bound) (e.g., succinate dehydrogenase)] to molecular oxygen is used to move protons ($H^+$) to the outside of the mitochondrion, thus creating a pH gradient as well as a voltage gradient (membrane potential). The formation of ATP by the $F_1$-ATPase is also driven by the membrane potential/proton gradient, but it is lower in hierarchy than $Ca^{2+}$ sequestration. When there is an increased need to accumulate $Ca^{2+}$ (due to elevation of cytosolic free $Ca^{2+}$), ATP synthesis will be inhibited. Several other processes also depend on the membrane potential [ADP/ATP exchange (47), aspartate/glutamate exchange (48), coenzyme A transport (49)] or proton gradient [phosphate transport (50), pyruvate and glutamate transport (51, 52), di- and tricarboxylate accumulation (53)].

(Nephro)toxicants can influence mitochondrial respiration and/or membrane potential at numerous levels. In the following, an overview is given of processes that have been identified or proposed as being important in this regard.

*Deficiency of Molecular Oxygen (Hypoxia).* Hypoxia or ischemia due to vasoconstriction has been proposed as a mechanism by which several toxicants induce nephrotoxicity. Vasoconstriction can be mediated by the renin–angiotensin system as a result of decreased tubular reabsorption of sodium [e.g., by mercury chloride (17)] or by increased $Ca^{2+}$ influx in renal smooth muscle cells [e.g., cyclosporin A (18)]. Hypoxia results in increased $K^+$ influx due to inhibition of $Na^+/K^+$-ATPase resulting from ATP depletion (54).

*Deficiency of Reductants (NADH, FADH₂).* NADH is able to release electrons into the respiratory chain via complex I (NADH dehydrogenase). NADH is produced (by reduction of $NAD^+$) by the citric acid cycle as well as by fatty acid β-oxidation (Figure 2). Inhibition of the citric acid cycle can be achieved by inhibition of the individual enzymes involved or by inhibition of uptake of necessary metabolic substrates. Fluoroacetate, for example, is a very potent rodent poison because it is metabolized to fluorocitrate which blocks the citric acid cycle at the level of the enzyme acetonitase (55). The nephrotoxin S-(1,2-dichlorovinyl)-L-cysteine (DCVC) has been shown to inhibit succinate:cyt *c* oxidoreductase and isocitrate dehydrogenase (56). Nephrotoxic cephalosporin antibiotics (cephaloridine, cephaloglycine) have been shown to inhibit by acylation the mitochondrial transporters for anionic metabolic substrates (e.g., succinate), which are necessary to maintain the citric acid cycle (57, 58). Cephaloridine also inhibits cellular uptake of metabolic substrates by competitive inhibition of the anion transport system (59, 60) and/or by causing lipid peroxidation (61). Recently, it has been shown that inhibition of glycolysis and proteolysis by menadione may lead to a deficiency of citric acid cycle intermediates, which secondarily results in impaired mitochondrial ATP production (62).

The citric acid cycle enzyme succinate dehydrogenase contains a FAD group which after reduction is regenerated by releasing its electrons via coenzyme Q to the respiratory chain. Inhibition of this enzyme, which has been observed with several nephrotoxins, such as DCVC, S-(1,2-dichlorovinyl)-L-homocysteine (DCVHC) (56), 2-bromohydroquinone (BHQ) (63), and S-(1,2,3,4,4-pentachlorobutadienyl)-L-cysteine (PCBC) (64), affects electron donation to the respiratory chain directly (via coenzyme Q) as well as indirectly (via inhibited NADH formation).

$FADH_2$, which is formed during fatty acid $\beta$-oxidation, also introduces electrons into the respiratory chain via coenzyme Q to complex III. However, inhibition of fatty acid oxidation as a cause of toxicity has not yet been demonstrated.

The pyruvate dehydrogenase complex converts pyruvate to acetyl coenzyme A (acetyl-CoA), which subsequently is converted to NADH by the citric acid cycle. Inhibition of this sulfhydryl-dependent enzyme has been observed with the nephrotoxicants DCVC (65) and 3-chloropyruvate formed from 3-chlorolactate (66). Synthesis of acetyl-CoA by pyruvate dehydrogenase and fatty acid $\beta$-oxidation requires mitochondrial CoA. Depletion of this thiol compound may contribute to mitochondrial toxicity. The nephrotoxicant 4-(dimethylamino)phenol has recently been shown to deplete mitochondrial CoA by covalent binding to the reactive intermediate $N,N'$-dimethylquinone imine (67).

*Inhibition of Electron Flow ($NADH \rightarrow I \rightarrow Q \rightarrow III \rightarrow Cyt\ c \rightarrow IV$).* NADH dehydrogenase (complex I) has been shown to be inhibited by the nephrotoxins S-(2-chloro-1,1,2-trifluoroethyl)-L-cysteine (CTFE-Cys) (68) and 1,2-dibromo-3-chloropropane (69). Accumulation of $Ca^{2+}$ and production of oxygen free radicals also inhibit mitochondrial respiration at this site (70). Cytochrome c oxidase (complex IV) is inhibited by the nephrotoxicants mercury chloride (38), BHQ (71), and PCBC (72), although for the latter two compounds it is a relatively late effect.

The electron flow via membrane-bound components can be inhibited by partitioning of electrons to foreign substrates. Menadione has thus been shown to compete with coenzyme Q for the respiratory electrons delivered by NADH by NADH:ubiquinone oxidoreductase and succinate–ubiquinone oxidoreductase (complex II) (73, 74).

Reduction of molecular oxygen by complex IV can be effectively inhibited by cyanide or azide, which compete for the oxygen-binding heme. Although cyanide toxicity generally is not directed to the kidney, it has been suggested to be involved in the mechanism of nephrotoxicity of several organonitriles. The nephrotoxicity of (2S)-1-cyano-2-hydroxy-3,4-epithiobutane, for instance, could be due to released cyanide (75). However, oxidation products of the aliphatic moiety, or the parent compound itself, may also play a role (75). Because cyanide release has not been quantified yet, its relative contribution to the nephrotoxicity of (2S)-1-cyano-2-hydroxy-3,4-epithiobutane remains to be established.

The nephrotoxic immunosuppressant cyclosporin A has a complicated effect on renal mitochondrial functioning. Cyclosporin A inhibits complex II of the electron transport chain (76) but stimulates NADH–complex Q reductase (complex I), resulting in a net stimulation of mitochondrial function in vivo (77). Stimulation of complex I was suggested to be a compensatory response to the ischemia resulting from the cyclosporin-induced decrease in renal blood flow. Inhibition of the respiratory chain was recently

suggested to be caused by inhibition of the synthesis of subunits of cytochrome oxidase and $F_1$-ATPase activity (78). The mechanism of inhibition of protein synthesis may involve binding to cyclophilin, a cytoplasmic protein active in folding proteins during ribosomal synthesis (78).

*Dissipation of Proton Gradient.* Most known uncouplers of oxidative phosphorylation act through dissipation of the proton gradient across the mitochondrial inner membrane (79). Several lipophilic weak acids (e.g., 3,5-dinitrophenol) are protonated at the more acidic outer side of the inner mitochondrial membrane and because of their lipophilicity diffuse across the inner membrane into the mitochondrial matrix. In the matrix they are deprotonated again due to the higher pH. The proton permeability of the inner membrane is also affected by interactions with membrane components regulating permeability, e.g., by alkylating thiol groups (80, 7). Recently, the nephrotoxicant PCBC has been shown to dissipate the proton gradient after metabolic activation to an alkylating reactive intermediate (81).

*Inhibition of ADP/ATP Antiporter.* The ADP/ATP antiporter provides the mitochondrion with ADP necessary to carry out oxidative phosphorylation, and it transports ATP to the outside of the mitochondrion where it is necessary for many vital cellular functions. Inhibition of this sulfhydryl-dependent carrier will increase the ATP/ADP ratio inside the mitochondrion, which leads to inhibition of respiration in coupled mitochondria, and to a decreased ATP/ADP ratio in cytosol. Inhibition of this carrier can be achieved by sulfhydryl-modifying agents (e.g., N-ethylmaleimide) (82). Inhibition of respiration of renal tubules by BHQ was proposed to result from inhibition of this transporter and/or inhibition of the $F_1$-ATPase (71).

*II.3.c. Effects of Toxicants on the Endoplasmic Reticulum.* Active (ATP-dependent) sequestration of $Ca^{2+}$ by the endoplasmic reticulum (ER) also participates in the regulation of the cytosolic free $Ca^{2+}$ concentration. The capacity of the ER to accumulate $Ca^{2+}$ is relatively low when compared to that of the mitochondria. However, an impairment of the microsomal $Ca^{2+}$ sequestration may also contribute to an imbalance of $Ca^{2+}$ homeostasis and as a consequence to cell death. Inactivation of the sulfhydryl-dependent $Ca^{2+}$-ATPase can be caused by alkylation [e.g., p-(chloromercuri)benzoate], oxidation of protein sulfhydryl groups (e.g., diamide), or formation of mixed disulfides (e.g., cystamine) (26). S-(2-Chloroethyl)-L-cysteine, chloroform, and 1,1-dichloroethylene (which are known nephrotoxicants) have been shown to inactivate the $Ca^{2+}$ transporter and $Ca^{2+}$-ATPase activity in hepatic smooth ER by covalent binding (83–85). The nephrotoxicant S-(1,2-dichlorovinyl)-L-homocysteine (DCVHC) inhibits both renal microsomal $Ca^{2+}$-ATPase and mitochondrial $Ca^{2+}$ sequestration; because of this it is a much more potent toxin than DCVC which only inhibits mitochondrial sequestration (56). The nephrotoxin cisplatin, however, causes an increase of the ER $Ca^{2+}$ pump (86). Whether this is a response to an increased cytosolic $Ca^{2+}$ concentration and/or whether this increased activity is responsible for disruption of the $Ca^{2+}$ homeostasis and the renal toxicity is not known.

Passive leakage of $Ca^{2+}$ from the ER to the cytosol also can be increased by toxicants. $CCl_4$ and diquat increase microsomal permeability by covalent binding and membrane lipid peroxidation, respectively (23). Both hepatotoxicants also destroy microsomal $Ca^{2+}$-ATPase.

*II.3.d. Effects of Toxicants on the Nuclear Membrane.* Recent studies have demonstrated that liver nuclei also possess an active $Ca^{2+}$-dependent uptake system, which appears to be calmodulin dependent and operates at physiological cytosolic $Ca^{2+}$ levels (87). Because the nucleus contains an endonuclease, which fragments DNA and which is activated by $Ca^{2+}$ at micromolar levels (88), this process cannot be regarded as a protection mechanism against elevated cytosolic free $Ca^{2+}$ concentrations. This mechanism might be involved in hepatocyte killing by chemical toxins (22). Involvement of this mechanism in nephrotoxicity, however, is not known yet.

*II.3.e. Effects of Toxicants on the Lysosome.* Biochemical mechanisms leading to cellular nephrotoxicity also may involve disturbances in lysosomal functioning. Cationic amphiphilic drugs (CAD's; molecules consisting of a hydrophobic portion and a hydrophilic portion containing a primary or substituted nitrogen group which is charged at physiological pH) induce the formation of concentric membranous structures, the so-called lysosomal lamellar bodies, which have been shown to be secondary lysosomes. These structures are formed by storage of polar lipids, primarily phospholipids, as a result of impairment in degradation of these substrates (reviewed in ref 89). The mechanism of impaired degradation may be formation of indigestible drug–(phospho)lipid complexes or direct action on lysosomal phospholipases. A labilization of the secondary lysosomes may result in the release of lysosomal hydrolytic enzymes into the cytosol. Such a process could lead to the attack on intracellular organelles, subsequent injury, and necrosis. Nephrotoxic aminoglycosides, such as gentamicin, and the anorectic agent chlorphentermine (90) are polycationic molecules capable of binding to phospholipids of the renal brush border and particularly the basolateral membranes, thus inhibiting the activity of lysosomal phospholipases. Binding of aminoglycosides to the brush border membrane also results in decreased membrane fluidity, and as a consequence in decreased activities of various transport processes (91). However, because gentamicin also interferes with $Ca^{2+}$ homeostasis, the relative contribution of both processes to gentamicin-induced nephrotoxicity is not clear.

A variety of (mixtures of) chemicals including unleaded gasoline, decalin, isophorone, 1,4-dichlorobenzene, tetrachloroethylene, and *d*-limonene have been shown to produce accumulation of $\alpha_{2\mu}$-globulin in lysosomes, subsequent degeneration of the proximal tubular cells, and nephrotoxicity in male rats (reviewed in ref 16). $\alpha_{2\mu}$-Globulin is a male rat specific low molecular weight protein synthesized in the liver. By reversible binding to $\alpha_{2\mu}$-globulin, the chemicals mentioned above cause indigestable complexes. Endocytosis of these complexes may lead to accumulation of $\alpha_{2\mu}$-globulin, to lysosomal overload, and to cytotoxicity. Subsequent cellular regeneration of proximal tubular cells might act as a tumor promotion mechanism by clonally expanding spontaneous initiated cells in the kidney.

## III. Factors Determining Site Selectivity of Nephrotoxicants

The kidney is a complex organ with a very heterogeneous nature. Anatomically, the kidney can be divided in two distinct areas, the cortex and the medulla (Figure 3). The medulla is subdivided into an inner and an outer medulla. The latter again is divided into an inner and an outer stripe. Within the cortex there are also medullary interdigitations, the medullary rays.

The functional subunit of the kidney is the nephron. The extent of differentiation between functionally distinct segments of the nephron has increased recently. Different nomenclatures to describe the heterogeneity of the kidney

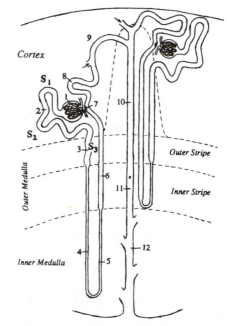

**Figure 3.** This scheme depicts the heterogeneity of the nephron. A short-looped and a long-looped nephron together with the collecting system are shown. A medullary ray is delineated by a dashed line. (1) Renal corpuscle including Bowman's capsule and the glomerulus; (2) proximal convoluted tubule; (3) proximal straight tubule; (4) descending thin limb; (5) ascending thin limb; (6) distal straight tubule (thick ascending limb); (7) macula densa; (8) distal convoluted tubule; (9) connecting tubule; (10) cortical collecting duct; (11) outer medullary collecting duct; (12) inner medullary duct. (Adapted from ref 92 with minor modifications.)

have been used throughout the last decades. This might lead to confusion when comparing different studies. Only very recently, the nomenclature to describe the various parts of the kidney and nephron was standardized (92). At present, a single nephron is divided into at least 10–12 different parts, which are structurally and functionally distinct. The major division of the nephron is: (a) glomerulus, (b) proximal tubule, (c) intermediate tubule, (d) distal tubule, and (e) collecting duct. Nephrotoxicity may occur at all different sites along the nephron. At present, many nephrotoxicants have been shown to be site-specific: the large majority appear to be proximal tubular toxins, while others are glomerular toxins or papillotoxins. Within the proximal tubular toxins, a distinction between $S_1$-, $S_2$-, and/or $S_3$-directing toxins can be made.

The site selectivity of nephrotoxicity will be determined by several factors, such as accumulation mechanisms and distribution of activating and deactivating enzymes. The studies describing renal distribution of toxicologically important transport processes and drug metabolizing enzymes often use different nomenclatures, however. The previously used name "outer medulla", for example, at present is subdivided into an outer and inner stripe, respectively containing the $S_3$ segment (proximal tubule) and the thick ascending limb of Henle (distal tubule). The frequently used name "pars recta" (or straight part) of the proximal tubule is not equivalent to the $S_3$ segment, but to the $S_3$ segment and part of the $S_2$ segment. Therefore, it is sometimes difficult to explain site selectivity of renal toxicity using literature data that uses different nomenclatures. In the following section, an attempt is made to describe and classify a number of factors, which determine site selectivity of nephrotoxicity.

*III.1. Localization of Transport and Accumulation Mechanisms.* In general, the glomerulus, acting as a filtration barrier, is susceptible to injury by deposition of high molecular weight circulating immune complexes. Subsequent activation of macrophages and/or neutrophiles, digesting these complexes, results in release of activated oxygen species in the glomerulus (10). Substances secreted from the neutrophils, which accumulated in the glomerulus, may also act on proximal tubular segments (93).

Many nephrotoxicants have their primary site of action on the proximal tubule, probably because most of the blood flow to the kidney is delivered to the cortex, which is predominantly proximal tubule, and because active secretion of potentially toxic organic anions and cations can lead there to several hundredfold higher intracellular concentrations than the corresponding plasma concentration (315). The activity of organic anion carriers with p-aminohippuric acid (PAH) as a substrate along the nephron was $S_2 > S_1 = S_3$ (94). In the rabbit, the $S_2$ region was 3–5 times as active in secretion of PAH as the $S_1$ and $S_3$ regions. The axial transition from $S_2$ to $S_3$ is probably gradual, rather than abrupt. The activity of the organic cation carriers along the nephron, however, was $S_1 > S_2 > S_3$ for superficial nephrons, and $S_1 = S_2 > S_3$ for juxtamedullary nephrons (94).

Cephaloridine is transported into proximal tubular cells by organic anion transport systems. However, release of cephaloridine from the tubular cells to the tubular lumen, carried out by organic cation transporters, is greatly restricted. This results in high intracellular cephaloridine concentrations in the proximal tubular cells (95). The relationship between transport and nephrotoxicity has been demonstrated by the reduction of toxicity by inhibitors of transport systems. The nephrotoxic effect of cephaloridine, citrinin, and several cysteine conjugates in rats and isolated renal cells can be blocked effectively by a probenecid, a competitive inhibitor of the organic anion transporter (95).

The high endocytosis/pinocytosis activity in the brush border membranes of the $S_1$ and $S_2$ segments of the proximal tubule makes these parts of the nephron vulnerable to nephrotoxic low molecular weight proteins [such as cadmium–metallothionein (96) and $\alpha_{2\mu}$-globulin complexes (16)] and to gentamicin-induced phospholipidosis (89). Glomerular filtration is required for delivering these toxicants to the tubular brush border membranes. Thus it was shown that, at a decreased glomerular filtration rate, the tubular uptake of gentamicin was increased because of a longer transition time (97).

Other compounds are concentrated passively along the nephron due to the continuing concentration of the urine. As a result, these compounds may reach toxic concentrations at distal parts (pappillae) of the nephron [e.g., 2-bromoethanamine (BEA)], or they may even precipitate [e.g., folic acid (98)]. The papillotoxicity of BEA does not appear in antidiuretic hormone deficient rats because of their inability to concentrate urine in the distal part of the tubule (99).

### III.2. Localization of Activating and Deactivating Enzymes.
Numerous nephrotoxicants require metabolic activation in the kidney before they can produce toxicity. Therefore, the presence of activating enzymes will be an important factor, especially in the case of short-lived reactive intermediates. The exact localization along the nephron of the numerous drug metabolizing enzymes, capable of toxifying and detoxifying xenobiotics, at present has only been partly elucidated. The most important activating and deactivating enzymes are discussed below briefly.

*Cytochrome P-450.* Although spectrophotometric (100) and immunohistochemical (101) determination revealed that cytochrome P-450 is only detectable in the cortex and outer stripe of outer medulla (OSOM), especially in $S_2$ and $S_3$ segments (101), activities of this enzyme system have also been demonstrated in the inner stripe of the outer medulla (ISOM) and the inner medulla. The distribution of the activities of cytochrome P-450 was strongly substrate-dependent. Aminopyrine N-demethylation, ethoxycoumarin O-deethylation, and benzo[a]pyrene 3-hydroxylation activities were highest in the cortex (102, 103). When referred to content of cytochrome P-450, the inner medulla showed the highest lauric acid hydroxylase activity (100), while the thick ascending limb of Henle (ISOM) had the highest cytochrome P-450 dependent arachidonic acid metabolism activity (102).

*NADPH–cytochrome P-450 reductase* only was detectable immunohistochemically in microsomes of the $S_2$ and $S_3$ segment (104); however, its activity decreased only gradually from the cortex to the inner medulla (100).

*Prostaglandin H Synthase.* The activity of prostaglandin H synthase (PHS), which is capable of activating xenobiotics by cooxidation by the prostaglandin hydroperoxidase component (105), decreased as follows: inner medulla ≫ outer medulla > cortex (104).

*Flavin-Containing Monooxygenase.* The FAD monooxygenase, which is active in the oxygenation of heteroatom (nitrogen, sulfur) containing compounds, had comparable activity in cortex and medulla of the pig kidney (106, 107).

*L- and D-Amino Acid Oxidase (L-α-Hydroxyacid Oxidase).* L-Amino acid oxidase, which has recently been shown to activate the nephrotoxin 3-chlorolactate (66), had its highest activity in the renal cortex (108). D-Amino acid oxidase is present in the kidney at higher levels than L-amino acid oxidase (109). This enzyme, which recently is suggested to be involved in the bioactivation of S-(1,2-dichlorovinyl)-D-cysteine (110), has its highest activity in the $S_3$ segment of the proximal tubule (111).

*NAD(P)H Quinone Reductase.* The activity of renal NAD(P)H quinone reductase, which is important in detoxification (by two-electron reduction) of reactive quinones, was found to be highest in the papillae and lowest in the cortex, with intermediary activity in the medulla (112).

*UDP-glucuronyl- and Sulfotransferases.* Activities of these enzymes were approximately 2-fold higher in the $S_2$ segment of the proximal tubule than in the cortex and outer stripe of the medulla (113).

*Glutathione S-Transferase(s).* The glutathione S-transferases are dimeric enzymes catalyzing the conjugation of GSH with a wide variety of potentially toxic electrophiles. Very recently, the distribution of glutathione S-transferase isozymes along the human nephron was studied (114). After 35 weeks of gestation, the so-called α isoenzymes (α, β, γ, δ, and ε) appear to be restricted to the proximal tubule. The π set, however, only appears in the distal and collecting tubules and in the loop of Henle. The physiological significance of this differential distribution is not clear yet. The cells of the proximal tubule contain numerous mitochondria, and this site of the nephron is the site of considerable metabolic activity and oxygen consumption. α but not π forms of GSH transferases are able to catalyze the detoxication of organic peroxides, and therefore, α isoenzymes may protect against the production of oxygen-derived free radicals and organic peroxides at this site of the nephron. In the rabbit kidney, the distribution of this so called non-selenium-dependent glutathione peroxidase activity was similar (115).

The activity of the *selenium-dependent glutathione peroxidase*, which is able to detoxify organic peroxides as well as hydrogen peroxide, is highest in the rabbit kidney cortex but also has relatively high activity in the inner and outer medulla (*115*).

The activity of *glutathione reductase*, which is active in the NADPH-dependent reduction of glutathione disulfide to GSH, has highest activity in the rabbit kidney cortex and decreases to half the activity in the inner medulla (*115*).

The concentration of *glutathione* (*GSH*), the cofactor of the glutathione transferases, glutathione peroxidases, and glutathione reductases, is highest in the rabbit kidney cortex (9.2 $\mu$mol/g of tissue wet weight), intermediary in the outer medulla (5.8 $\mu$mol/g), and lowest in the inner medulla (4.4 $\mu$mol/g) (*116*).

*Cytosolic Cysteine Conjugate β-Lyase*. This enzyme, which is responsible for the ultimate bioactivation of several nephrotoxic halogenated hydrocarbons, is present in rat kidney cytosol (*117*) and in mitochondria (*118, 119*). The cytosolic enzyme was detected by immunohistochemical detection using antibodies to a highly purified enzyme. The cytosolic enzyme was localized in the $S_3$ segment of the proximal tubule of the rat kidney (*120*). However, another study also reported the presence of this enzyme in both the $S_1$ and $S_2$ segments of the proximal tubule (*121*).

*γ-Glutamyltranspeptidase, Aminopeptidases, and Cysteine Conjugate N-Acetyltransferase*. Activities of these enzymes which participate in the conversion of GSH conjugates to mercapturic acids were highest in the outer stripe of the medulla, which corresponds to the proximal straight tubule (*122, 115*).

*Deacetylation* of paracetamol, which probably is important for its nephrotoxicity (*40*), was more active in proximal than in distal tubules (*103*).

### III.3. Stability of Reactive Intermediates.

For direct-acting nephrotoxicants or labile toxicants, which easily degrade chemically to reactive intermediates, the site selectivity is mainly determined by the availability of the toxic agent. Very short-lived reactive intermediates will only act within the cells of formation. Their toxicity will strongly depend on the distribution of the (proximate) precursor compound and on the balance of activating and deactivating systems (*123*). Reactive intermediates with intermediate lifetime may be able diffuse to neighboring cells, and their site selectivity will be determined by the vulnerability (due to insufficient protection mechanisms) of these cells. Long-lived reactive intermediates even might be formed in extrarenal tissues and be transported to the kidney where they exert their toxicity because of the concentrating ability or the vulnerability of the kidney. It has been proposed that the nephrotoxicity of 1,1-dichloroethylene may even be due to translocation of reactive intermediates formed by hepatic cytochrome P-450, from the liver to the kidney (*124*).

### III.4. Vulnerable Intracellular Targets.

In the proximal tubule of the kidney, the glycolytic enzymes only have a low activity (*43*). Therefore, in this part of the tubule, cells are highly dependent on mitochondria to perform the various energy-dependent activities necessary for normal cellular functioning. Dysfunctioning of mitochondria in this region, therefore, might have more serious consequences than in other parts of the nephron. Because the number of mitochondria is lowest in the $S_3$ segment, this part of the nephron will generally be most susceptible to mitochondrial toxicity. The convoluted part is more susceptible to hypoxia due to its higher oxygen demand

(*54*). However, results with the clamped ischemia model suggest that the convoluted ($S_1$ and $S_2$) part is only sublethally injured. In contrast, the straight ($S_3$) part sustains lethal injury such that upon reoxygenation the convoluted part quickly recovers, but the straight part becomes necrotic (*125*).

The $S_1$ and $S_2$ segments, which contain a very active endocytosis/lysosomal apparatus, are susceptible to nephrotoxicants which act by lysosomal dysfunctioning (*16, 89, 126*). The distal tubule, where the electrolyte homeostasis is finely regulated, is sensitive to high fluoride anion concentrations, which are responsible for the nephrotoxicity of the inhalation anesthetic methoxyflurane (*127*).

## IV.  Specific Examples of Nephrotoxicants

### IV.1. Glomerular Toxicants

*IV.1.a. Immune Toxicity.* The majority (70%) of cases of glomerulopathy are caused by circulating immune complexes, of which the antigens often are not of glomerular origin (*128*). Antigens which have been implicated as such include viruses and tumor antigens. After trapping in the glomeruli and binding of complement, attraction of neutrophils and phagocytosis of the complexes take place. With phagocytosis, some glomerular cells die and lysosomal enzymes are released, thus damaging the endothelial cells and the glomerular basement membrane (GBM). A second immunopathogenetic mechanism, which represents 5% of the cases, involves the formation of antibodies against the GBM. Once the anti-GBM antibodies are deposited, they damage the glomerulas by complement- and neutrophil-mediated processes. The mechanism responsible for this autoimmune disease is not known. Administration of mercury chloride ($HgCl_2$) to rabbits or certain strains of rats causes an autoantibody which immunolocalizes along the GBM (*129, 130*). Although the precise molecular mechanism of action of $HgCl_2$ is unknown, a direct effect on circulating lymphocytes may be involved (*131*).

Another mechanism of neutrophil-mediated toxicity may involve the release of reactive oxygen species, such as $H_2O_2$, superoxide anion radical, and $OCl^-$. Catalase prevented the glomerular damage induced by intrarenal injection of phorbol myristate acetate, a potent activator of neutrophils, whereas superoxide dismutase had no protective effect (*132*). Therefore, $H_2O_2$ from activated neutrophils, rather than the superoxide anion radical, was responsible for this type of glomerular damage. Neutrophils and macrophages are commonly seen within glomeruli in many types of acute, mostly immune complex mediated glomerulonephritis, and the generation of toxic oxygen species by these cell systems could well be an important mechanism leading to glomerular injury (*10, 11, 133, 134*).

*IV.1.b. Xanthine Oxidase Mediated Toxicity.* The aminonucleoside of puromycin (PA) produces glomerular damage resulting in proteinuria (*135*) which exhibits a biphasic pattern (*136*). The acute toxicity of PA appears to be directed toward the epithelial membrane on the outside of the GBM (*137*). Recently, the involvement of oxygen free radicals in PA-induced glomerular damage has been proposed (*138*). The observed protective effect of superoxide dismutase points to the involvement of the superoxide anion radical. PA has been shown to be metabolized in the liver via adenine and/or adenosine to hypoxanthine (Figure 4) (*139*). Hypoxanthine can serve as substrate for superoxide anion radical production via the xanthine oxidase system. Allopurinol, an inhibitor of xanthine oxidase, protects against PA-indiced glomerular damage. PA may be involved in the conversion of type

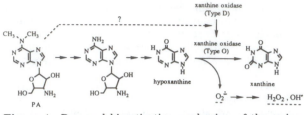

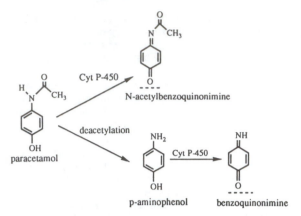

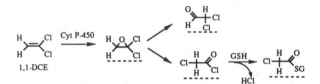

**Figure 4.** Proposed bioactivation mechanism of the aminonucleoside of puromycin (PA). PA is metabolized via adenosine to hypoxanthine in the liver. PA may be involved in conversion of type D to type O xanthine oxidase in the glomerulus, which may subsequently generate superoxide anion radicals. The superoxide anion radicals can produce hydrogen peroxide and hydroxyl radicals by the processes shown in Figure 1.

D xanthine oxidase (xanthine dehydrogenase) to the type O xanthine oxidase, which is the oxygen radical producing form (140). PA may also increase uptake of adenosine by increasing the membrane permeability of the glomerular endothelial cells, and thereby PA may provide the intracellular compartments with ample precursor for hypoxanthine formation and subsequent oxygen free radical generation via xanthine oxidase. Catalase and dimethyl sulfoxide did not protect against PA toxicity, which suggests that $H_2O_2$ and hydroxyl radicals, respectively, are less important in PA toxicity (138).

*IV.1.c. Direct-Acting Toxicants.* Recent evidence suggests that several nephrotoxicants can reduce the number of fixed anionic charges on glomerular elements by charge–charge interaction. The polyanionic aminoglycosides, such as gentamicin, can reduce the charge-selective filtration properties of the glomerulas as well as the maintenance of the glomerular ultrastructure, which appears to be dependent on charge–charge repulsion (141, 142).

### IV.2. Proximal Tubular Toxicants

#### IV.2.a. Oxidative Activation

IV.2.a.i. Activation by Renal Cytochrome P-450. Chloroform is both hepatotoxic and nephrotoxic (143). The mechanisms of both of these toxicities at present are well-defined and appear to be the result of cytochrome P-450 mediated bioactivation (1). Hydroxylation of chloroform by cytochrome P-450 is rapidly followed by loss of HCl to the highly reactive phosgene ($COCl_2$) (144, 145). Although the activity of cytochrome P-450 is higher in the liver, the nephrotoxicity of chloroform is nevertheless the result of bioactivation by cytochrome P-450 in situ in the kidney, because the half-life of phosgene is only very short (134). Renal fractions of rabbits have been shown to be able to bioactivate chloroform to phosgene, which is trapped as 2-oxothiazolidine-4-carboxylic acid in the presence of cysteine (146). Also, the different susceptibility of male and female rats correlated well with differences in the renal but not the hepatic contents of cytochrome P-450. High doses of paracetamol (acetaminophen) may result in hepatic centrilobular and also in renal cortical necrosis (147). At least two mechanisms of metabolic activation of paracetamol within renal cortical tissue have been demonstrated (Figure 5). One mechanism appears to be similar to the mechanism responsible for paracetamol-induced hepatotoxicity and involves activation by cytochrome P-450 to a reactive *N*-acetylquinone imine. The second mechanism involves initial deacetylation to *p*-aminophenol (148, 149). This compound, which has been identified as a urinary metabolite of paracetamol, produces a similar type of proximal tubular toxicity, due to cytochrome P-450 mediated activation to an arylating intermediate, presumably *p*-quinone imine (150, 151). The

**Figure 5.** Proposed bioactivation pathways, responsible for paracetamol-induced toxicity to the proximal renal tubule. Reactive intermediates capable of covalent binding to macromolecules are underlined by the dashed lines.

**Figure 6.** Proposed bioactivation mechanism responsible for 1,1-dichloroethylene-induced toxicity to the proximal renal tubule. Reactive intermediates capable of covalent binding to biomacromolecules are underlined by the dashed lines. The reactive GSH conjugate may be formed in the liver and subsequently be transported to the kidney.

activity for the deacetylation of paracetamol is highest in proximal tubules (103). Evidence for the contribution of the deacetylase pathway among others includes the following: (1) covalent binding of paracetamol was maximal when cytosol (deacetylase fraction) and microsomes (oxidation fraction) were combined, and (2) the deacetylase inhibitor *bis*(*p*-nitrophenyl) phosphate reduced covalent binding of paracetamol (152). The deacetylase-dependent metabolic activation of paracetamol appears to have substantial toxicological relevance as it was correlated with proximal tubular nephrotoxicity whereas cytochrome P-450 dependent arylation was not (153).

When compared to *p*-aminophenol, its N-methylated isomers 4-(methylamino)phenol and particularly 4-(dimethylamino)phenol are more potent proximal tubular nephrotoxins (154). It was postulated that the cytotoxicity of *p*-aminophenols might be directly correlated with steady-state levels of the *p*-quinone imines (150, 151).

IV.2.a.ii. Initial Activation by Hepatic Cytochrome P-450. 1,1-Dichlorethylene (DCE), a widely used monomeric intermediate in the production of plastics, may cause damage to liver, lung, and kidney (155–158). Administration of DCE to mice, however, indicated that the highest covalent binding occurred in the kidney (159). The metabolism of DCE in the liver has been shown to be NADPH-dependent and to involve the hepatic microsomal cytochrome P-450 system. Hepatic metabolism of DCE results in DCE oxide, 2-chloroacetyl chloride, its hydrolysis product 2-chloroacetic acid, and 2,2-dichloroacetaldehyde (160, 161). Renal microsomes, however, were incapable of bioactivating DCE to covalent binding species (162). Therefore, it was proposed that the nephrotoxicity of DCE may be due to translocation of reactive intermediates from the liver to the kidney. Acyl halides are extremely reactive intermediates which hydrolyze with rate constants exceeding 100 s$^{-1}$ at 22 °C (163). The DCE metabolite 2-

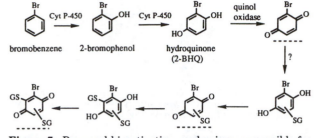

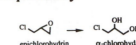

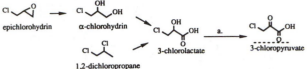

**Figure 7.** Proposed bioactivation mechanisms responsible for bromobenzene- and 2-bromophenol-induced nephrotoxicity in the rat. Reactive intermediates capable of covalent binding to biomacromolecules are underlined by the dashed lines. Contribution of the nephrotoxic mono- and bis-GSH-conjugated quinones to bromobenzene toxicity at present is uncertain as yet.

**Figure 8.** A proposed bioactivation pathway of 3-chloropropane compounds involving formation of the nephrotoxic metabolite 3-chloropyruvate. (a) Catalyzed by renal α-hydroxy acid oxidase. 3-Chloropyruvate may produce toxicity by inhibiting the mitochondrial citric acid cycle at the level of pyruvate dehydrogenase.

chloroacetyl chloride has been shown to alkylate GSH, producing S-(2-chloroacetyl)glutathione (Figure 6) (*164*). The 2-chloroacetyl moiety in this GSH conjugate, however, is still able to react covalently to thiol compounds. Therefore, S-(2-chloroacetyl)glutathione may interact with targets in distant locations inside or outside the hepatocyte. Proteins containing specific GSH-binding sites may interact specifically with this type of reactive GSH conjugate: hepatic glutathione S-transferases and hepatic canilicular membranes, which contain transport proteins extruding GSSG and GSH conjugates, have thus been shown to be damaged upon DCE intoxication (*165, 166*). The renal proximal tubules, which contain high γ-glutamyl transpeptidase (γ-GT) activity, therefore also might be target for this metabolite because of the GSH-binding site of γ-GT. However, evidence to support this hypothesis at present is not available.

Bromobenzene is also both hepatotoxic and nephrotoxic. The hepatotoxicity is believed to result from cytochrome P-450 mediated formation of reactive arene oxides in the mammalian liver. Because of the poor activity of renal cytochrome P-450 in bromobenzene metabolism, a different bioactivation mechanism is thought to be responsible for the nephrotoxicity. Mice treated with bromobenzene excreted *o*-, *m*-, and *p*-bromophenol as well as 4-bromocatechol in urine. All of these metabolites were more potent nephrotoxicants than bromobenzene itself both in vivo and in vitro (*167*). In rats, however, only 2-bromophenol was nephrotoxic (*168*). The current hypothesis is that 2-bromophenol is further bioactivated to 2-bromohydroquinone (BHQ) by hepatic cytochrome P-450 (*169*). BHQ is the major metabolite of both bromobenzene and 2-bromophenol in the rat. BHQ appears to be a potent toxin to proximal tubular cells and renal mitochondria (*170, 71*). Oxidation to 2-bromosemiquinone and/or 2-bromoquinone has been proposed as the bioactivation mechanism for BHQ. This oxidation step is catalyzed by a quinol oxidase associated with membranal and microsomal fractions from rat kidneys (*171*); cytochrome P-450 and prostaglandin H synthase apparently are not involved (*170*). As mentioned in section II.1, several semiquinones are capable of reducing oxygen to form superoxide anion radicals, which results in (re)generation of the quinone. This redox cycling results in oxidative stress and reactive oxygen species. Although BHQ is capable of redox cycling, nevertheless its toxicity to the proximal tubule appears not to be mediated by redox cycling and resulting oxidative stress (*172*). Covalent binding of the reactive (semi)quinone to proteins, instead, has been proposed to be responsible for mitochondrial dysfunctioning and tubular cell death (*173*). Microsomal incubation of BHQ in the presence of GSH results in the

formation of several isomeric monosubstituted and disubstituted GSH conjugates (Figure 7) (*174*). These GSH conjugates also are potent nephrotoxins. 2-Bromo-3,5- or 2-bromo-3,6-(diglutathionyl)hydroquinone [2-Br-(diG-Syl)HQ] even appears to be the most potent nephrotoxic metabolite of bromobenzene, causing toxicity at 0.3% of the dose of bromobenzene itself and at 4% of the dose of BHQ (*174*). Accumulation of 2-Br-(diGSyl)HQ or its cysteine derivative(s) in the proximal tubules, followed by oxidation of the hydroquinone moiety to a reactive quinone, may also result in selective nephrotoxicity of this compound. The observed inhibition of both renal accumulation and toxicity of 2-Br-(diGSyl)HQ by AT-125 suggests an imperative role of γ-GT, which was high activity in the proximal tubule, in directing toxicity to the kidney (*175, 176*). The inability of AT-125 to protect proximal tubules against BHQ-induced toxicity, however, questions again the sole participation of these conjugates in BHQ-induced toxicity (*173*).

IV.2.a.iii. Miscellaneous Oxidative Bioactivation Mechanisms. The selectivity of nephrotoxicity of hydroquinone–glutathione conjugates derived from 1,4-benzoquinone also may be explained by γ-GT-mediated accumulation in proximal tubules and subsequent oxidative activation to the corresponding quinone (*177*). Covalent binding of the quinone moiety is believed to be responsible for toxicity; the fully substituted tetraglutathion-S-yl conjugate is not toxic because of its inability to alkylate tissue compounds, although redox cycling still would be possible. In contrast, however, the nephrotoxicity of the GSH conjugate of menadione (thiodione) (*62*) has been attributed to γ-GT-mediated accumulation of this GSH conjugate in proximal tubules and to its ability to undergo redox cycling (*178*).

A number of chlorinated derivatives of propane have also been shown to produce nephrotoxicity (*179*). These compounds include the industrial solvent 3-chloropropane 1,2-oxide (ECH, epichlorohydrin), the rodenticide 3-chloro-1,2-propanediol (ACH, α-chlorohydrin), and 1,2-dichloropropane. The nephrotoxicity of these compounds appears to be primarily directed to the proximal tubule. A common metabolite (Figure 8), which may be derived from the *R* isomer of 3-chlorolactate, a potent nephrotoxin in the rat (*180*), recently has been proposed as the metabolite responsible for the chloropropane-induced proximal tubular toxicity (*66*). The oxidation of 3-chlorolactate to 3-chloropyruvate has been suggested to be catalyzed by rat kidney L-α-hydroxy aacid oxidase (L-amino acid oxidase). Another common oxidative metabolite of chloropropanes is oxalic acid. Precipitation of oxalic acid at the corticomedullary junction previously was believed to be responsible for the proximal tubular damage caused by ethylene glycol, glycine, and glycolic acid (*181*). Recently, however, the role of oxalic acid in the tubular damage caused by the chloropropane compounds (*182*) and ethylene glycol (*183*) has been questioned.

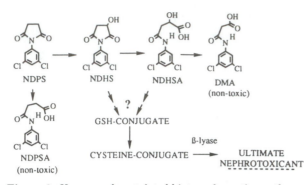

**Figure 9**. Known and postulated biotransformation pathways for N-(3,5-dichlorophenyl)succinimide (NDPS) in rats (169). NDHS and NDHSA produce nephrotoxicity in rats; inhibition of β-lyase blocks NDPS-induced nephrotoxicity, suggesting involvement of GSH conjugation. NDHS, N-(3,5-dichlorophenyl)-2-hydroxysuccinimide; NDHSA, N-(3,5-dichlorophenyl)-2-hydroxysuccinamic acid; DMA, N-(3,5-dichlorophenyl)malonamic acid; NDPSA, N-(3,5-dichlorophenyl)succinamic acid.

N-(3,5-Dichlorophenyl)succinimide (NDPS), an agricultural fungicide, produces proximal tubular necrosis (184). Hydroxylation of the succinimide ring by cytochrome P-450 appears to be an important step in the generation of nephrotoxic species (Figure 9) (185, 186). The 2-hydroxysuccinimide (NDHS) and 2-hydroxysuccinamic acid (NDPSA) are more potent nephrotoxins than NDPS and produce proximal tubular damage indistinguishable from that of NDPS (187). Because NDHSA is not nephrotoxic in vitro to cortical slices, it was suggested that it must undergo additional biotransformation and, furthermore, that the kidney does not convert NDHSA to its ultimate nephrotoxic species. The ability of (aminooxy)acetic acid, an inhibitor of pyridoxal phosphate dependent enzymes, to inhibit the nephrotoxicity of NDPS was suggested to be indicative for a role of a cysteine conjugate of NDPS or NDPS metabolite as a penultimate toxicant (188). However, the isolation and identification of a cysteine conjugate as toxic precursor still remains to be accomplished.

As discussed in section II.3e, a variety of compounds have been shown to cause proximal tubular toxicity in rats by reversible binding to the male-specific renal protein $\alpha_{2\mu}$-globulin. For some of these compounds, 2,4,4-trimethylpentane and d-limolene, the major binding species appears to be an oxidative metabolite, 2,4,4-trimethyl-2-pentanol (189) and d-limonene 1,2-oxide (190), respectively. In these cases, therefore, hepatic oxidative metabolism, also, might be regarded as a bioactivation mechanism for renal toxicity.

*IV.2.b. Activation by Glutathione Conjugation*

IV.2.b.i. Direct-Acting Glutathione-Derived Conjugates. 1,2-Dibromoethane (EDB), which is used as a pesticide and lead scavenger, produces both hepato- and nephrotoxicity and liver, lung, stomach, mammary, adrenal, skin, and kidney tumors in several species (191). Although the major portion of EDB metabolism is oxidative, as has been shown with deuterium labeling (192), EDB was also shown to form a sulfur half-mustard with glutathione (193–195). It is now clear that GSH conjugation with EDB occurs in both liver and kidney and that the GSH conjugate is reactive in situ. Due to its lability/reactivity, this compound has not as yet been isolated. The identification of S-[2-(N⁷-guanyl)ethyl]glutathione as a DNA adduct both in vivo and in vitro (196) is indicative for a cyclic episulfonium ion formed by internal displacement of the second halogen atom by the sulfur atom of the glutathione conjugate

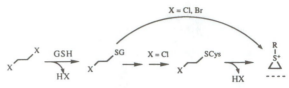

**Figure 10**. Formation of reactive episulfonium ions from GSH and cysteine S-conjugates, presumably responsible for the nephrotoxicity caused by 1,2-dihaloethanes.

(Figure 10). Because of the high reactivity of the glutathione conjugate of EDB, this compound probably plays an important role in the nephrotoxicity of EDB.

1,2-Dichloroethane (EDC) is used as a lead scavenger in gasoline, as an industrial solvent, and as a grain fumigant. EDC is both hepatotoxic and nephrotoxic and also is a suspected animal carcinogen. Although basically acting through the same mechanism of episulfonium ion formation, there appear to be subtle differences in the mechanism of renal toxicity of EDC and EDB. Conjugation of EDC to GSH occurs only in the liver. The GSH conjugate formed is relatively stable and is transported via the bile and gut to the kidney. In the bile, small intestine, and kidney, the GSH conjugate is hydrolyzed to the corresponding cysteine, conjugate. Both S-(2-chloroethyl)glutathione (CEG) and its breakdown product S-(2-chloroethyl)-L-cysteine (CEC) are direct-acting alkylating and therefore potentially nephrotoxic agents (Figure 10) (197). CEC has more alkylating power toward the N⁷-position of deoxyguanosine than the corresponding CEG conjugate. CEC but not CEG induces DNA strand breaks in plasmid pBR322 DNA (198). CEC has also been shown to deplete the intracellular GSH concentration in hepatocytes by covalent binding, forming S-(2-DL-cysteinylethyl)glutathione. CEC induced lipid peroxidation (LPO), as a consequence of GSH depletion, and inhibited microsomal Ca²⁺ sequestration in hepatocytes. Inhibition of LPO, however, did not protect against the cytotoxicity, and therefore LPO apparently is not a causal event in CEC-induced cytotoxicity. CEC, formed by further metabolism of CEG, therefore, has been implicated as the reactive intermediate involved in EDC-induced nephrotoxicity (199). The analogues S-ethyl-DL-cysteine, S-(2-hydroxyethyl)-DL-cysteine, and S-(3-chloropropyl)-DL-cysteine, which cannot form episulfonium ions, are not toxic in vivo or in vitro (83, 199).

Other nephrotoxic compounds with vicinal halogens are 1,2-dibromo-3-chloropropane (DBCP) and tris(2,3-dibromopropyl) phosphate (Tris-BP), both containing vicinal bromine atoms. The soil fumigant/nematocide DBCP has been shown to be an acute testicular toxicant and an acute nephrotoxicant (200, 201). The renal damage is characterized by necrosis of the proximal tubule in the outer medulla of the kidney (182). DBCP was found to concentrate in the rat kidney approximately 25 times relative to plasma 1 h after administration (202). Metabolism by cytochrome P-450 in the liver appears to protect against the nephrotoxicity of DBCP (203). The absence of any significant deuterium effect with the perdeuterio analogue indicates that breaking of a carbon–hydrogen bond, by oxygenation, is not the rate-limiting step in DBCP-induced nephrotoxicity (202). The mutagenicity of DBCP, however, was suggested to involve oxidative metabolism by cytochrome P-450 to 2-bromoacrolein (204). Alternatively, analogous to the bioactivation of 1,2-dibromoethane (EDB), conjugation of glutathione to this vicinally dibrominated compound has been suggested to play a role in DBCP mutagenicity and testicular toxicity (205). Whether this route contributes to the nephrotoxicity is not known, however. Probenecid, (aminooxy)acetic acid, and

AT-125, compounds known to protect against nephrotoxic GSH conjugates, did not affect the nephrotoxicity of DBCP (*202*). However, these compounds do not influence GSH conjugation in the kidney, which might be more important for bioactivation than GSH conjugation in the liver due to the high reactivity of the GSH conjugate. Recently, as an alternative route of bioactivation of DBCP, metabolism to the nephrotoxicant 2-chlorolactate has been proposed (*66*). This compound indeed has been shown to be an oxidative metabolite of DBCP (*206*). However, a casual relationship between this metabolite and DBCP-induced nephrotoxicity as yet remains to be proven.

The flame retardant Tris-BP has also been shown to be mutagenic and carcinogenic and to cause renal tubular necrosis (*207*). The major metabolite of Tris-BP, bis-(2,3-dibromopropyl) phosphate (Bis-BP) (*208*), is less mutagenic but more nephrotoxic. The mechanism responsible for mutagenicity, oxidative formation of the mutagen 2-bromoacrolein (*209*), therefore, appears to differ from the mechanism leading to nephrotoxicity. The mechanism responsible for nephrotoxicity at present remains to be established.

IV.2.b.ii. *β-Lyase-Mediated Bioactivation.* The first nephrotoxic cysteine conjugate identified was *S*-(1,2-dichlorovinyl)-L-cysteine (DCVC), which was formed nonenzymatically in trichloroethylene-extracted soybeans (*210*). When administered to calves, this compound produces anemia; however, to rats it was highly nephrotoxic (*211*). DCVC was shown to be bioactivated enzymatically by a β-lyase to reactive sulfur containing intermediates (*212*). The localization of DCVC-induced toxicity, the $S_3$ segment of the proximal tubule, correlates well with the localization of covalent binding of the reactive intermediates (*213*). The amount of DCVC formed in rats treated with trichloroethylene, however, appears to be extremely low, as indicated by extremely low urinary excretion of the corresponding mercapturic acid, *N*-acetyl-*S*-(1,2-dichlorovinyl)-L-cysteine (DCV-NAC) (*214*). A regioisomer of DCV-NAC, namely, *N*-acetyl(2,2-dichlorovinyl)-L-cysteine, also has only recently been detected in urine of trichloroethylene-treated rats, however, at even lower amounts (*215*). DCVC probably plays a more important role in the nephrotoxicity of dichloroacetylene, since exposure of rats to this compound results in much higher excretion of the corresponding mercapturic acid (*216*). Following the studies with DCVC a number of other haloalkenes and their cysteine conjugates have also been shown to damage the pars recta of the proximal tubule of the kidney. These include hexachlorobutadiene (HCBD) (*217*), tetrafluoroethylene (TFE) (*218*), hexafluoropropene (*219*), chlorotrifluoroethylene (CTFE) (*219*), 1,1-dichloro-2,2-difluoroethylene (DCDFE) (*220*), and 3,3,3-trifluoro-1,1,2-trichloroprop-1-ene (*221*).

The molecular bioactivation mechanism of these nephrotoxic chlorinated and fluorinated olfins is a complex route of metabolism involving different tissues (Figure 11). Data from different studies are indicative for the following processes. The initial GSH conjugates are formed predominantly in the liver and excreted in the blood and/or bile. The protection of bile canulation against nephrotoxicity induced by hexachlorobutadiene (HCBD) points to the importance of the biliary route (*222*). Peptidases present in bile and small intestine are active in degrading of the GSH conjugates to the corresponding cysteine conjugates, which are subsequently reabsorbed into the blood (*223*). Via the portal vein, the cysteine conjugate first passes the liver, where N-acetylation takes place to form the corresponding mercapturic acid. Via the blood,

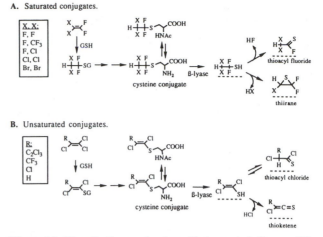

**Figure 11.** Bioactivation mechanism of halogenated alkenes. (A) β-Lyase-catalyzed bioactivation of saturated cysteine conjugates; (B) β-Lyase-catalyzed bioactivation of unsaturated cysteine conjugates. Proposed reactive intermediates are underlined.

the cysteine conjugates and mercapturic acids are transported to the kidney where they are secreted by different active transporters into the proximal tubular cells (*224*). In these cells, the cysteine conjugates can be N-acetylated or be bioactivated by β-lyase. The mercapturic acid can be excreted to urine; however, it also can be deacetylated again by renal acylases to the corresponding cysteine conjugates, which subsequently can be bioactivated by β-lyase. Apart from β-lyase, the activity of renal acylases is a crucial factor for the observed nephrotoxicity of a number of mercapturic acids both in vivo (*225*) and in vitro (*226*). Activities of *N*-acetyltransferase, acylase, and β-lyase toward nephrotoxic conjugates have been observed in both renal and hepatic fractions; however, activities were higher in the kidney (*227*).

In the kidney, the β-lyase-dependent toxicity of cysteine conjugates is directed to the $S_3$ segment of the proximal tubule. The specificity of toxicity to the kidney is probably the result of the following:

(*a*) *Accumulation of conjugates* by active transporters. Probenecid inhibits the nephrotoxicity of HCBD (*217*) and DCVC (*228*). Because probenecid does not inhibit the uptake of cysteine conjugates in rat kidney proximal tubules (*229*), it was concluded that uptake of mercapturates by the organic anion transporter and subsequent deacetylation are more important for providing toxic cysteine conjugates to the kidney than direct delivery of cysteine conjugates from the blood. However, in contrast to the study with proximal tubules, uptake of DCVC in isolated proximal tubular cells was inhibited by probenecid (*230*). The reason for this discrepancy might be than with isolated tubules only the basolateral membrane of the kidney cells are exposed to the cysteine conjugate, whereas in isolated kidney cells both basolateral and brush border membranes are exposed. Because transport of DCVC in brush border vesicles appears to be 6 times higher than in basolateral membrane vesicles (*231*), in isolated kidney cells transport might be carried out mainly by brush border membrane transporter(s). In proximal tubules transport might be carried out mainly by basolateral transporter(s). Therefore, probenecid may have different effects in these two models for kidney function. Nephrotoxic cysteine conjugates are transported in the kidney by multiple transporters; little is known about the nature, number, and tubular distribution of these transporters (*229*, *224*). Recently, it was shown that several nephrotoxic cysteine conjugates and GSH conjugates inhibited basolateral transport of *p*-aminohippuric acid in isolated rat proximal tubules (*232*). It is not known, however, whether these results reflect transport of the compounds by the organic anion transporter or binding of the lipophilic S-conjugates to the hydrophobic core of this transporter. As mentioned in section III.1, the proximal tubular $S_3$ segment has the lowest activity of organic anion transport. It is not known whether the microperfusion system used to localize the transport system is representative for the in vivo situation, because the organic anion transport appears to be controlled by exchange with dicarboxylates (*233*).

GSH conjugates are also nephrotoxic both in vivo to the rat (*228*) and in vitro to isolated rat renal cells (*234, 235*). Inhibition of toxicity by AT-125 points to involvement of $\gamma$-GT in the bioactivation mechanism. Almost complete (>97%) inhibition of renal $\gamma$-GT by AT-125, however, potentiated rather than diminished the nephrotoxicity of HCBD (*236*). This might indicate that for HCBD renal $\gamma$-GT does not contribute significantly to degradation of the GSH conjugate formed in the liver and excreted into the bile. The activity of $\gamma$-GT, dipeptidases, and *N*-acetyltransferases in the bile and small intestine may result in reabsorption of the lipophilic cysteinylglycine, cysteine, and *N*-acetylcysteine conjugates rather than reabsorption of the more hydrophilic, intact GSH conjugate into the blood. The cysteine conjugates and/or *N*-acetylcysteine conjugates, which may also be formed when cysteine conjugates pass the liver, most likely are the predominant compounds seen by the kidney, explaining the lack of inhibition of toxicity by AT-125.

(*b*) *The balance of deactivation/activation*, which in the kidney may be directed toward activation pathways by high deacetylase and $\beta$-lyase activities. The selective nephrotoxicity of a series of related haloethyl mercapturic acids to renal cells appeared to be strongly dependent on the high activity of $\beta$-lyase and acylase in the kidney, when compared to that in the liver (*226, 227*). A limited availability of acetyl coenzyme A, the cofactor for the deactivating N-acetylation step, also might direct biotransformation of this type of nephrotoxicants toward the bioactivation routes (*227*).

The reactive intermediates responsible for $\beta$-lyase-mediated nephrotoxicity have been partly elucidated. Fluorinated ethylenes, like tetrafluoroethylene (TFE), chlorotrifluoroethylene (CTFE), 1,1-dichloro-2,2-difluoroethylene (DCDFE), and 1,1-dibromo-2,2-difluoroethylene (DBDFE) are conjugated to GSH via an addition mechanism to saturated conjugates (*220, 237, 238*). The ethanethiol compounds formed from the respective cysteine conjugates may rearrange to form reactive thionoacyl fluorides or thiiranes. The cysteine conjugate of TFE is predominantly bioactivated to difluorothionoacyl fluoride, since covalent adducts derived from this reactive intermediate have been identified both in vivo and in vitro (*227*). The cysteine conjugate of CTFE has been demonstrated to be bioactivated to chlorofluorothionoacyl fluoride (*239*). The involvement of thiiranes in the nephrotoxicity has also been proposed for DCDFE and DBDFE (*225*). GSH conjugation of chlorinated olefins, such as trichloroethylene, tetrachloroethylene, and hexachlorobutadiene (HCBD), proceeds via addition/elimination, leading to the formation of unsaturated cysteine conjugates (*240, 241*). The corresponding thiols, formed by action of $\beta$-lyase, can probably tautomerize to reactive thionoacyl chlorides. Alternative reactive intermediates which have been proposed are thioketenes formed by spontaneous elimination of hydrogen chloride from the chlorinated enethiols. Both types of reactive intermediates are potent thioacylating

agents. Adducts which may be derived from both types of reactive intermediates are demonstrated after $\beta$-lyase-dependent degradation of chlorinated cysteine conjugates in the presence of selected nucleophiles (e.g., diethylamine) (*242, 243*).

Recently, S-(1,2-dichlorovinyl)-L-cysteine (DCVC) (*244*) and trichloroethylene also have been shown to produce lipid peroxidation in vivo by an as yet unknown mechanism. Hypoxia resulted in increased peroxidative membrane damage. Inhibitors of lipid peroxidation were reported to inhibit nephrotoxicity (*245*).

Cysteine conjugates of chlorinated ethylenes also have been shown to be mutagenic, while the conjugates of fluorinated ethylenes are not (*246*). Inhibition of mutagenicity by (aminooxy)acetic acid (AOA) points also to a $\beta$-lyase-dependent mechanism (*247*). Whether the difference between chlorinated and fluorinated conjugates might be attributed to the fact that different types of reactive intermediates are formed by $\beta$-lyase is not yet known. Recently, biotransformation of toxic cysteine conjugates by L-amino acid oxidase, thus producing hydrogen peroxide, has been shown to contribute significantly to the renal metabolism of these compounds (*248*). Both mutagenicity and cytotoxicity of hydrogen peroxide have been shown to be potentiated strongly by hydrogen sulfide (*249*). Hydrogen sulfide also has been identified as a product of $\beta$-lyase-catalyzed degradation of toxic cysteine conjugates (*68*). A possible contribution of hydrogen peroxide produced via this pathway to toxicity and/or mutagenicity, however, remains to be established.

*IV.2.c. Miscellaneous.* A number of the cephalosporin antibiotics produce acute proximal tubular necrosis and acute renal failure when given in large single doses (*250*). Several mechanisms underlying cephalosporin nephrotoxicity have been proposed, but only three of these have been supported by experimental evidence:

(*1*) *Concentrative uptake* into the tubular cell by the organic anion secretory carrier (*251*). Inhibitors of organic anion transport reduce both renal cortical accumulation and nephrotoxicity of cephaloridine. The $S_2$ segment of the proximal tubule, which represents the site of greatest activity of organic anion transport, is the primary target of cephaloridine-induced injury (*252*). Because the transport of cephaloridine from cell to lumen appears to be limited and dependent upon an organic cation carrier, inhibition of this transport by mepiphenidol increased intracellular concentrations of cephaloridine and nephrotoxicity (*253*).

(*2*) *Production of respiratory toxicity* through acylation and inactivation of mitochondrial transporters for anionic substrate uptake (*57, 254*). The irreversible inhibition of mitochondrial proteins in vivo was explained by the acylating properties of the $\beta$-lactam antibiotics. Cephaloridine, cephaloglycin, and the thienamycin imipenem are the most nephrotoxic $\beta$-lactams and the most reactive protein acylators, while cephalexin and penicillins, which have little or no nephrotoxic potential, are among the least reactive (*58*).

(*3*) *Production of Lipid Peroxidative Injury*. Cephaloridine has been shown to produce lipid peroxidation and oxidative stress in vivo (*255, 256*). It was speculated that one-electron reduction of the pyridinium ring of cephaloridine, catalyzed by renal NADPH–cytochrome P-450 reductase in mitochondria and endoplasmic reticulum, would lead to redox cycling, resulting in the production of superoxide anion radicals, which in turn would initiate lipid peroxidation (Figure 12). Diets deficient in selenium or vitamin E, which augment oxidative injury, indeed po-

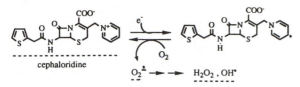

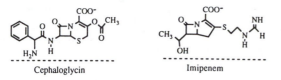

**Figure 12.** Proposed mechanism of cephaloridine nephrotoxicity (*30*). The reactive β-lactam ring of cephaloridine and reactive oxygen species derived from superoxide anion radical may be involved in initiating nephrotoxicity.

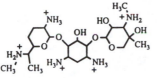

**Figure 13.** Structures of the nephrotoxic β-lactam antibiotics cephaloglycin and imipenem.

**Figure 14.** Polycationic character of gentamicin.

tentiated cephaloridine nephrotoxicity in rats. Cephaloglycin, however, which lacks a pyridinium ring (Figure 13), produced only a small amount of oxidative stress (measured as GSH oxidation) and lipid peroxidation, but was more nephrotoxic then cephaloridine in vivo (*58*). Imipenem (Figure 13) also exhibited peroxidative effects, albeit to a lesser extend than cephaloridine (*63*).

At least 10% of all cases of acute renal failure have been attributed to use of aminoglycoside antibiotics. The nephrotoxic potential of an aminoglycoside is related to the number of ionizable amino groups it contains: neomycin (six amino groups), gentamicin (five), tobramycin (five), netilmycin (five), kanamycin (four), amikacin (four), and streptomycin (three). Although aminoglycosides have glomerular effects, their primary target is the proximal tubular cell, in particularly the $S_1$ and $S_2$ segments. After charge–charge interaction with brush border membranes, they are taken up by endocytosis and stored in secondary lysosomes. By this mechanism, gentamicin (Figure 14) accumulates to levels 50–100 times the corresponding plasma concentrations. A good correlation was found between cortical accumulation and the nephrotoxic effect for aminoglycosides (*257*). As discussed previously, binding of aminoglycosides to acidic phospholipids results in decreased catabolism by phospholipases, and as a result in phospholipidosis. Binding of gentamicin to acidic phospholipids of the plasma membrane by charge–charge interaction results in changes in membrane fluidity and permeability; several membrane-bound processes are disturbed: $Na^+/K^+$-ATPase activity, adenylate cyclase activity, cation ($K^+$, $Ca^{2+}$, $Mg^{2+}$) transport, and D-glucose transport (*91, 258, 259*).

Next to its effect on phospholipid metabolism, gentamicin also appears to enhance the formation of reactive oxygen species. Gentamicin inhibits the mitochondrial state 3 respiration of isolated mitochondria, while it stimulated state 4 respiration. Because this effect was inhibitable with catalase, production of hydrogen peroxide might be involved in gentamicin nephrotoxicity (*260*). Scavengers of hydroxyl radicals (dimethyl sulfoxide, sodium benzoate, dimethylthiourea) or iron chelator (deferoxamine) inhibit Fenton reaction protected rats against

acute renal failure of gentamicin in vivo, which suggests a pivotal role for hydroxyl radicals (*261*).

Increase of malondialdehyde levels by gentamicin are indicative for lipid peroxidation. However, nhibition of the lipid peroxidation by vitamin E did not protect against gentamicin nephrotoxicity in the rat (*262*). The mechanism by which gentamicin increased formation of reactive oxygen species at present is unknown. Gentamicin is almost entirely excreted unchanged in the urine. A possible interference with calcium-dependent events in gentamicin-induced nephrotoxicity was suggested to explain the potentiating effect of the calcium antagonist diltiazem (*263*) and the protective effect of dietary calcium (*264*). The mechanism by which calcium interferes with gentamicin nephrotoxicity, however, at present remains unknown. Gentamicin has been shown to form a complex with pyridoxal 5′-phosphate (PLP), the active form of vitamin B6, which is exreted by renal tubular transport (*265*). Evidence suggests that depletion of PLP in the kidney may be a factor in the development of nephrotoxicity. Administration of PLP to rats protect against gentamicin nephrotoxicity (*266*). Gentamicin alone causes toxicity in the $S_1$ and $S_2$ segments (*44*). However, when a nontoxic dose of gentamicin was given during a period of renal hypoperfusion, severe toxicity was observed only at the $S_3$ segment (*97, 267*). Apparently, gentamicin strongly exacerbates renal ischemic injury.

Most heavy metals are potent nephrotoxins. The proximal tubule appears to be particularly sensitive to these compounds. The underlying mechanisms leading to the selective renal toxicity are poorly understood but appear to vary along the different heavy metals. Mercury may be introduced into the body as elemental mercury, as inorganic mercury, and as organic mercury (several diuretics, methylmercury chloride). Mercuric(II) chloride causes severe necrosis in the pars recta ($S_3$) of the proximal tubule, with extensive necrosis of the convoluted part at a later time point (*17, 268*). Proposed mechanisms involved in mercury-induced tubular toxicity are mitochondrial dysfunctioning due to angiotensin–renin-mediated ischemia (*269*) or inhibition of enzymes by direct binding to sulfhydryl groups. Methylmercury was reported to produce methyl radicals, due to homolytic scission, and to produce lipid peroxidation (*270*).

Acute exposure to cadmium produces hepatotoxicity but does not produce nephrotoxicity (*271*). With chronic administration, tolerance to hepatotoxicity is observed as a result of induction of metallothionein (MT) (*272*). MT binds cadmium and prevents reacting to critical macromolecules. With chronic administration of cadmium, the target organ of toxicity changes from liver to kidney (*273*). It has been suggested that the cadmium–MT complex might be the cause of renal injury. Due to minor hepatic injury, cadmium–MT is released from the liver into the bloodstream (*274*). Following glomerular filtration, this low molecular weight complex (MW 7000) is reabsorbed by endocytosis in the proximal tubule cells. The toxicity of cadmium, therefore, is primarily confined to $S_1$ and $S_2$. Lysosomal degradation of the complex yields free cadmium ions, which at high concentrations cause failure of secondary lysosome fusion and putative damage to systems involved in macromolecular synthesis (*126*). Although the kidney also is able to synthesize MT, the lower rate when compared to the liver may contribute to making the kidney the target organ of toxicity during chronic exposure (*275*).

Renal cortical toxicity is one of the important manifestations of plumbotoxicity and results from the accumulation of lead ($Pb^{2+}$) in the renal cortex. Lead interferes with the $Ca^{2+}$ homeostasis by competing with $Ca^{2+}$ for the mitochondrial $Ca^{2+}$ transporter (276) as well as for calmodulin (277). Accumulation of lead in mitchondria by the $Ca^{2+}$ transporter results in reduced mitochondrial respiration (276). Lead also may modify the plasma membrane to permit $Ca^{2+}$ to move more readily down its electrochemical gradient. Because lead has been shown to be able to substitute for $Ca^{2+}$ in the activation of calmodulin-dependent activities, lead may place the regulation of diverse cellular processes beyond the normal physiological controls of the $Ca^{2+}$ messenger system (278).

Potassium dichromate administered to rats affects the convoluted proximal tubule either by direct corrosive effect upon contact with this cell following filtration and peritubular blood flow delivery or by some specific (as yet unknown) functional activity located in this region that allows selective uptake (279).

Cisplatinum [cis-diamminedichloroplatinum(II)] is a cancer chemotherapeutic agent with a wide spectrum of antitumor activity. Renal toxicity is a major dose-limiting factor in its clinical use. The $S_3$ segment of the proximal tubule appears to be the primary target in cisplatin-induced renal failure. The renal uptake of cisplatinum at the peritubular side of the cell is oxygen-dependent and possible via cotransport and/or binding to components which belong to the organic base transport system of the tubulus (280). It is proposed that after entry into the cell, due to the low intracellular concentration of chloride, cisplatinum is hydrolyzed to reactive positively charged monoaquo- and diaquo species (281). The antitumor activity of cisplatinum has been attributed to formation of bifunctional adducts with DNA (282). Binding to critical protein sulfhydryl groups has been proposed as an explanation of cisplatinum nephrotoxicity (283). Diethyldithiocarbamate protects against nephrotoxicity but not the antitumor effect, because it selectively reverses platinum–thiol complexes without reversal of platinum–DNA cross-links (284). Recently, sodium selenite was also shown to provide effective protection against cisplatinum nephrotoxicity, without affecting its antineoplastic activity against several tumors (285). Renal brush border enzymes have been excluded as primary sites of toxicity in vivo (286), and changes in mitochondrial respiration and $Ca^{2+}$ sequestration do not occur until 3–5 days after cisplatinum (287). However, alteration of renal intracellular glutathione levels and depression of renal macromolecule synthesis (protein, DNA, RNA) were observed within hours after exposure (288).

### IV.3. Medullary/Papillary Toxicants

*IV.3.a. Bioactivation by Renal Prostaglandin H Synthase.* Chronic use of combination analgesics containing phenacetin, paracetamol, or aspirin is believed to be responsible for papillary necrosis observed in analgesic nephropathy (289). There still is controversy concerning the agent(s) which is (are) responsible for the nephrotoxicity and the mechanism(s) by which it (they) act. Use of a single analgesic substance seems to be less harmful then a combination of analgesics (290).

A high dose of aspirin results in papillary necrosis (291). It has been speculated that the ability of aspirin to inhibit renal medullary prostaglandin synthesis might remove an endogenous vasodilator prostaglandin, leading to localized vasoconstriction and finally to an ischemic injury to the papillae. However, experimental evidence to support this hypothesis at present is lacking.

Phenacetin is metabolized in the liver to two major metabolites, p-phenetidine and paracetamol (Figure 15). p-Phenetidine and paracetamol were shown to be con-

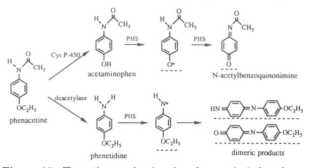
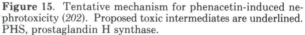
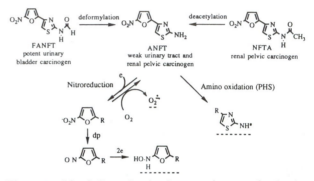

**Figure 15.** Tentative mechanism for phenacetin-induced nephrotoxicity (202). Proposed toxic intermediates are underlined. PHS, prostaglandin H synthase.

**Figure 16.** Metabolism of 5-nitrofurans and proposed reductive and oxidative routes of bioactivation to toxic and carcinogenic metabolites. dp, disproportionation; PHS, prostaglandin H synthase.

centrated from cortex to papilla after administration of phenacetin to dogs (292, 293). Prostaglandin H synthase (PHS), which has high activity in the medulla, has a very high affinity for both metabolites, and activation by this enzyme, therefore, may occur at therapeutic doses (294). Phenacetin itself is not concentrated in the kidney and is not a substrate for PHS. PHS activates both paracetamol and p-phenetidine to alkylating reactive intermediates. Paracetamol is believed to be oxidized by PHS to N-acetylbenzoquinone imine via a phenoxy radical intermediate (295). Phenetidine is believed to be oxidized to reactive nitrogen radical, which subsequently might produce reactive dimeric products (296). The fact that phenacetin is considerably more nephrotoxic than paracetamol, its major metabolite, may suggest that PHS-dependent oxidation of p-phenetidine is of more relevance for the toxicity of phenacetin (294).

Substituted 5-nitrofurans are a diverse group of compounds used as food additives, human medicines, and veterinary drugs (297). Several 5-nitrofurans both are nephrotoxic and acuase renal and urniary tract cancer. Bioactivation mechanisms that are thought to be essential for the biological activity are reduction of the 5-nitro moiety (298, 299) and/or cooxidation of the 2-aminothiazole moiety by PHS (300) (Figure 16). Nitroreduction has been demonstrated with a variety of enzymes including NADPH–cytochrome c reductase, xanthine oxidase, and aldehyde oxidase (301, 302). The nitro anion radical formed after one-electron reduction of nitrofurans can lead to redox cycling (303) or can be further reduced under hypoxic conditions to nitroso, hydronitroxide, and amine compounds (304).

5-Nitrofuran compounds differ strongly in inter- and intraorgan selectivity of toxicity and carcinogenicity. Next to metabolism, transport mechanisms appear to play an

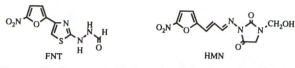

**Figure 17.** Structures of 1-formyl-2-[4-(5-nitro-2-furyl)-2-thia-zolyl]hydrazine (FNT) and 1-(hydroxymethyl)-3-{[3-(5-nitro-2-furyl)allylidene]amino}hydantoin (HMN).

important role. 1-Formyl-2-[4-(5-nitro-2-thiazo-lyl]hydrazine (FNT) and 1-(hydroxymethyl)-3-{[3-(5-nitro-2-furyl)allylidene]amino}hydantoin (HMN) (Figure 17) are thought to be activated by nitroreduction; however, they differ in their renal toxicity and carcinogenity (305, 306). HMN-fed rats showed renal papillary necrosis while FNT-fed rats did not. Both compounds caused interstitial nephritis. FNT but not HMN caused renal pelvic, mammary, and intestinal tumors. The different patterns of toxicity and carcinogenity have been attributed partly to different renal handling of these compounds. HMN appears to be excreted much more rapidly by the organic acid transport system than FNT, resulting in high intramedullary concentrations, causing the interstitial nephritis and papillary necrosis seen in toxicity studies. FNT appears to be more rapidly metabolized and/or taken up by the kidney and extrarenal tissues, resulting in the diverse and potent nonrenal toxicities of FNT (307).

Different renal handling also appears to be responsible for the different patterns of carcinogenicity of the related compounds *N*-[4-(5-nitro-2-furyl)-2-thiazolyl]formamide (FANFT), *N*-[4-(5-nitro-2-furyl)-2-thiazolyl]acetamide (NFTA), and 2-amino-4-(5-nitro-2-furyl)thiazole (ANFT). FANFT is a potent carcinogen causing transitional cell carcinoma of the urinary bladder (308). ANFT is of intermediate potency, causing bladder or renal pelvic malignancy, but it is a potent inducer of forestomach tumors. NFTA is a weak renal pelvis carcinogen but also is a mammary gland, salivary gland, and lung carcinogen in the rat. The biological activity of both FANFT and NFTA have been proposed to result from deformylation and deacetylation, respectively (Figure 16), to the putative proximate carcinogen ANFT (300, 301). FANFT is more rapidly taken up by the kidney than is ANFT (309). FANFT is then deformylated and rapidly excreted as ANFT. Higher urinary concentrations of ANFT are achieved with administration of FANFT than with ANFT, explaining its higher carcinogenicity in the bladder (310). Inhibition of bladder carcinogenesis by aspirin (311) can be explained by inhibition of PHS, since aspirin did not alter urinary excretion or half-life of FANFT or ANFT (308). NFTA has been shown to be deacetylated by kidney homogenates and slices (312, 300); however, in the isolated kidney no deacetylation to ANFT could be measured (310). Therefore, transport of NFTA might be located at a region of the nephron deficient of deacetylase.

*IV.3.b. Direct-Acting Toxicants.* 2-Bromoethylamine (BEA) is a model compound used in laboratory animals to simulate renal papillary necrosis (313). The toxicity caused by BEA is the result of medullary concentration of this direct-acting chemical. Studies of BEA's metabolism are hindered by its high chemical reactivity. The reactive intermediate responsible for toxicity is believed to be the ethylene imine, resulting from intramolecular cyclization (Figure 18). The concentration ability of the collecting ducts is an important determinant of toxicity of BEA. The high urine volume in rats deficient in antidiuretic hormone (ADH) production protected them against BEA-induced toxicity (99). Administration of ADH to these rats results in papillary nephrotoxicity.

**Figure 18.** Proposed activation of 2-haloethylamines by intramolecular rearrangement to reactive ethylene imines.

Substituting bromine with other halogens alters both reactivity and toxicity of 2-haloethylamines. The chlorinated analogue 2-chloroethylamine is a less potent papillotoxin due to the decreased likelihood of chloride ion dissociation (314). 2-Fluoroethylamine is acutely lethal at the doses at which BEA produced papillotoxicity; however, at a low dose mineralization of the $S_3$ segment of the proximal tubule was observed.

## V.  Conclusions

This review has dealt with the molecular and biochemical mechanisms responsible for the nephrotoxicity of a wide variety of xenobiotics. Although the process leading to nephrotoxicity has previously been subdivided into at least seven stages on the basis of ultrastructural changes (19), we have divided this process into three stages on the basis of the molecular and biochemical events. (1) The first ("initiation") phase involves the primary interaction of the toxic species with critical cellular constituents, such as proteins, DNA, and RNA. Nephrotoxicity may be initiated by xenobiotics by reversible interactions, such as complexation or charge–charge interaction, or by irreversible interactions, such as covalent binding by electrophiles or hydrogen abstraction by radical species. (2) In the second ("propagation") phase, cellular constituents are progressively affected by the initial events, resulting in protein and nonprotein thiol depletion, a shift of cellular redox couples to the oxidized state (oxidative stress), or lipid peroxidation. As a result of these interactions, essential biochemical processes, such as calcium homeostasis or oxidative phosphorylation, may be affected directly or secondarily. (3) When restorative mechanisms become insufficient or are impaired directly by the toxicant, cells may enter the third ("termination") phase, in which the disturbance of the biochemical processes has become irreversible and processes are leading to cell death. At present, it is not clear which disturbances of biochemical processes are responsible for cell death. A sustained elevation of intracellular free calcium, followed by activation of autolytic enzymes, and deprivation of cellular energy, followed by a collapse of energy-requiring processes have been proposed as critical toxicological events. In this review it is shown that known nephrotoxicants can interfere with processes localized on the plasma membrane, the endoplasmic reticulum, and the mitochondrion; depending on the nature of the nephrotoxicant, different types of interferences with these organelles have been identified. Also, interferences of nephrotoxicants with lysosomal functioning can result in release of lysosomal enzymes, and subsequently in the autolysis of the kidney cell.

The kidney is anatomically a highly complex organ, and the functional unit of the kidney, the nephron, can be subdivided into functionally different segments. Nephrotoxicants often appear to direct their toxicity to distinct parts of the nephron. This selective localization of nephrotoxicity is determined by a number of factors:

(a) *Concentrative Mechanisms.* Because of the high blood flow and the presence of concentrative mechanisms, such as active transport of organic anions or cations or efficient endocytosis, several xenobiotics may reach toxic concentrations in (cells of) the proximal tubular part of the nephron. Passive concentration of a xenobiotic, as a result of concentration of the primary urine, can lead to toxic concentrations at distal parts of the nephron.

(b) *Presence of Bioactivating Enzymes and Insufficient Protection Mechanisms.* Several nephrotoxicants are not toxic themselves but require bioactivation to reactive intermediates before causing toxicity. Therefore, the intrarenal distribution of bioactivating enzymes as well as intracellular protection systems also may determine the site of nephrotoxicity caused by toxicants. The relatively high concentration of activating enzymes, such as cytochrome P-450 and cysteine conjugate $\beta$-lyase, predispose the $S_3$ segment of the proximal tubule to several nephrotoxicants. The presence of prostaglandin H synthase in combination with a low GSH content makes the medullary part of the kidney vulnerable to certain toxicants.

(c) *Presence of Vulnerable Intracellular Targets.* Because the proximal tubule is completely dependent on mitochondrial respiration for energy, mitochondrial damage has important consequences, especially in this part of the nephron. Also, the presence of a high concentration of lysosomes predisposes the $S_1$ and $S_2$ segments of the proximal tubule to nephrotoxicants which act by interference with lysosomal functioning.

The elucidation of the molecular mechanisms and the factors contributing to the onset of nephrotoxicity may be of great importance in strategies aiming at the prediction of and the prevention against chemically induced nephrotoxicity. Structure–activity relationships may help to predict the toxicity of structurally related compounds. Considering the mechanism of nephrotoxicity, prevention against nephrotoxicity may involve the following: (a) interference with transport and concentration mechanisms, or (b) interference at the level of bioactivation, for example, by selective inhibition of activating enzymes or (c) by mechanism-based introduction of structural modifications preventing bioactivation. At present, the large majority of toxicological data are obtained from studies with animals exposed to a single agent. Only very limited data are available from exposure to combinations of drugs. Knowledge of common mechanisms (transport, bioactivation, or intracellular interactions) may help to explain the antagonism, potentiation, or synergism of nephrotoxicity which may occur after simultaneous exposure to several xenobiotics.

In this review, the molecular and biochemical mechanisms of several nephrotoxicants have been reviewed. For some of the examples given, the localization of toxicity could well be explained by the intrarenal distribution of concentration and/or bioactivating mechanisms. Because of the complex architecture of the kidney, however, the exact localization and activity of the different transport mechanisms and bioactivating and bioinactivating enzymes need to be further characterized.

## References

[1] Hook, J. B., and Hewitt, W. R. (1986) Toxic responses of the kidney. In *Casarett and Doull's Toxicology* (Klassen, C. D., Doull, J., and Amdur, M. O. Eds.) pp 310–329, MacMillan, New York.

[2] Vermeulen, N. P. E., Van der Straat, R., te Koppele, J. M., Baldew, G., Commandeur, J. N. M., Haenen, G. R. M. M., Koymans, L., and Van Welie, R. T. H. (1990) Molecular mechanisms in toxicology and drug design. In *Trends in Drug Research, Vol. 9, Proceedings of the Seventh Noordwijkerhout Symposium on Medicinal Chemistry* (Klaassen, V., Ed.) Elsevier, Amsterdam (in press).

[3] Coles, B. (1985) Effects of modifying structure on electrophilic reactions with biological nucleophiles. *Drug Metab. Rev.* 15, 1307–1334.

[4] Girotti, A. W. (1985) Mechanisms of lipid peroxidation. *J. Free Radicals Biol. Med.* 1, 87–95.

[5] Thornton, J. M. (1981) Disulphide bridges in globular proteins. *J. Mol. Biol.* 151, 261–287.

[6] Sokol, P. P., Holohan, P. D., and Ross, C. R. (1986) Essential disulfide and sulfhydryl groups for organic cation transport in renal brush-border membranes. *J. Biol. Chem.* 261, 3282–3287.

[7] Lê-Quôc, D., and Lê-Quôc, K. (1985) Crucial role of sulfhydryl groups in the mitochondrial inner membrane structure. *J. Biol. Chem.* 260, 7422–7428.

[8] Kappus, H. (1986) Overview of enzyme systems involved in bioreduction of drugs and in redox-cycling. *Biochem. Pharmacol.* 35, 1–6.

[9] Heinemeyer, G., Nigam, S., and Hildebrandt, A. G. (1980) Hexobarbital-binding, hydroxylation and hexobarbital-dependent hydrogen peroxide production in hepatic microsomes of guinea pig, rat and rabbit. *Naunyn-Schmiederberg's Arch. Pharmacol.* 314, 201–210.

[10] Shah, S. V. (1989) Role of reactive oxygen metabolites in experimental glomerular disease. *Kidney Int.* 35, 1093–1106.

[11] Boyce, N. W., Tipping, T. G., and Holdsworth, S. R. (1989) Glomerular macrophages produce reactive oxygen species in experimental glomerulonephritis. *Kidney Int.* 35, 778–782.

[12] Passow, H., Rothstein, A., and Clarkson, T. W. (1961) The general pharmacology of the heavy metals *Pharmacol. Rev.* 13, 185–224.

[13] Kone, B. C., Kaleta, M., and Gullans, S. R. (1988) Silver ion ($Ag^+$)-induced increases in cel membrane $K^+$ and $Na^+$-permeability in the renal proximal tubule: reversal by thiol reagents. *J. Membr. Biol.* 102, 11–19.

[14] Brady, H. R., Kone, B. C., Brenmer, R. M., and Gullans, S. R. (1989) Early effects on uranyl nitrate on respiration and $K^+$-transport in rabbit proximal tubule. *Kidney Int.* 36, 27–34.

[15] Brenner, B. N., Ichikawa, I., and Deen, W. M. (1981) Glomerular filtration. In *The kidney* (Brenner, B. M., and Rector, F. C., Jr., Eds.) Vol. 1, pp 289–327, Saunders Co., Philadelphia.

[16] Swenberg, J. A., Short, B., Borghoff, S., Strasser, J., and Charbonneau, M. (1989) The comparative pathobiology of $\alpha_{2\mu}$-globulin nephropathy. *Toxicol. Appl. Pharmacol.* 97, 35–46.

[17] Haagsma, B. H., and Pound, A. W. (1979) Mercuric chloride-induced renal tubular necrosis in the rat. *Br. J. Exp. Pathol.* 60, 341–352.

[18] Rossi, N. F., Churchill, P. C., McDonald, F. D., and Ellis, V. R. (1989) Mechanism of cyclosporin A-induced renal vasoconstriction in the rat. *J. Pharmac. Exp. Ther.* 250, 896–901.

[19] Trump, B. F., and Berezesky, I. K. (1984) The role of sodium and calcium regulation in toxic cell injury. In *Drug Metabolism and Drug Toxicity* (Mitchell, J. R., and Horning, M. G. Eds.) pp 261–300, Raven Press, New York.

[20] Orrenius, S., McConkey, D. J., Bellomo, G., and Nicotera, P. (1989) Role of $Ca^{2+}$ in toxic cell killing. *Trends Pharmacol. Sci.* 10, 281–285.

[21] Mirabelli, F. (1988) Cytoskeletal alterations in human platelets exposed to oxidative stress are mediated by oxidative and $Ca^{2+}$-dependent mechanisms. *Arch. Biochem. Biophys.* 270, 487–488.

[22] McConkey, D. J., Hartzell, P., Nicotera, P., Wyllie, A. H., and Orrenius, S. (1988) Stimulation of endogenous endonuclease activity in hepatocytes exposed to oxidative stress. *Toxicol. Lett.* 42, 123–130.

[23] Tsokos-Kuhn, J. O. (1989) Evidence in vivo for elevation of intracellular free $Ca^{2+}$ in the liver after diquat, acetaminophen, and $CCl_4$. *Biochem. Pharmacol.* 38, 3061–3065.

[24] Korn, E. D. (1982) Actin polymerization and its regulation by proteins from non muscle cells. *Physiol. Rev.* 62, 672–737.

[25] Bellomo, G., Jewell, S. A., Thor, H., and Orrenius, S. (1982) Regulation of intracellular calcium compartimentation: studies with isolated hepatocytes and t-butyl hydroperoxide. *Proc. Natl. Acad. Sci. U.S.A.* 79, 6842–6846.

[26] Thor, H., Hartzel, P., Svensson, S.-A., Orrenius, S., Mirabelli, F., Marinoni, V., and Bellomo, G. (1985) On the role of thiol groups in the inhibition of liver microsomal $Ca2+$-sequestration by toxic agents. *Biochem. Pharmacol.* 34, 3717–3723.

[27] Mirabelli, F., Salis, A., Marnoni, V., Finardi, G., Bellomo, G., Thor, H., and Orrenius, S. (1988) Menadione-induced bleb formation in hepatocytes is associated with oxidation of thiol groups in actin. *Arch. Biochem. Biophys.* 264, 261–269.

[28] Nagelkerke, J. F., Dogterom, P., de Bont, H. J. G. M., and Mulder, G. J. (1989) Prolonged high intracellular free calcium concentrations by ATP are not immediately cytotoxic in isolated rathepatocytes. *Biochem. J.* 263, 347–353.

[29] Redegeld, F. A. M., Moison, R. M. W., Koster, A. S., and Noordhoek, J. (1989) Alterations in energy status by menadione metabolism in hepatocytes isolated from fasted and fed rats. *Arch. Biochem. Biophys.* 273, 215–222.

[30] Chien, K. R., Pfau, R. G., and Farber, J. L. (1979) Ischemic myocardial cell injury. Prevention by chlorpromazine of an accelerated phospholipid degradation and associated membrane dysfunction. *Am. J. Pathol.* 97, 505–530.

[31] Nicotera, P., Hartzell, P., Baldi, C., Svensson, S.-A., Bellomo, G., and Orrenius, S. (1986) Cystamine induces toxicity in hepatocytes through the elevation of cytosolic $Ca^{2+}$ and the stimulation of a non-lysosomal proteolytic system. *J. Biol. Chem.* 261, 14628–14635.

[32] Pounds, J. G., and Rosen, J. F. (1988) Cellular $Ca^{2+}$ homeostasis and $Ca^{2+}$-mediated cell processes as critical targets for toxicant action: conceptual and methodological pitfalls. *Toxicol. Appl. Pharmacol.* 94, 331–341.

[33] Van Erum, M., Martens, L., Vanduffel, L., and Teuchy, H. (1988) The localization of ($Ca^{2+}$ or $Mg^{2+}$)-ATPase in plasma membranes of renal proximal tubular cells. *Biochim. Biophys. Acta* 937, 145–152.

[34] Nicotera, P., Moore, M., Mirabelli, F., Bellomo, G., and Orrenius, S. (1985) Inhibition of hepatocyte plasma membrane $Ca^{2+}$-ATPase activity by menadione and its restoration by thiols. *FEBS Lett.* 181, 149–153.

[35] Nicotera, P., Hartzell, P., Baldi, C., Svenson, S.-Å., Bellomo, G., and Orrenius, S. (1986) Cystamine induces toxicity in hepatocytes through the elevation of cytosolic $Ca^{2+}$ and the stimulation of a nonlysosomal proteolytic system. *J. Biol. Chem.* **261**, 14628–14635.

[36] Tsokos-Kuhn, J. O., Hughes, H., Smith, C. V., and Mitchell, J. R. (1988) Alkylation of the liver plasma membrane and inhibition of the $(Ca^{2+}, Mg^{2+})$-ATPase by acetaminophen. *Biochem. Pharmacol.* **37**, 2125–2131.

[37] Phillips, T. D., Nechay, B. R., and Heidelbaugh, N. D. (1983) Vanadium: chemistry and the kidney. *Fed. Proc. Fed. Am. Soc. Exp. Biol.* **42**, 2969–2973.

[38] Pfaller, W., and Rittinger, M. (1980) Quantitative morphology of the rat kidney. *Int. J. Biochem.* **12**, 17–22.

[39] Kacew, S. (1987) Cationic amphiphilic drug-induced renal cortical lysosomal phospholipidosis: an in vivo comparative study with gentamicin and chlorphentermine. *Toxicol. Appl. Pharmacol.* **91**, 469–476.

[40] Tsokos-Kuhn, J. O., Smith, C. V., Mitchell, J. R., Tate, C. A., and Entman, M. L. (1986) Evidence for increased membrane permeability of plasmalemmal vesicles from livers of $CCl_4$-intoxicated rats. *Mol. Pharmacol.* **30**, 444–451.

[41] Snowdowne, K. W., and Borle, A. B. (1985) Effects of low extracellular sodium on cytosolic ionized calcium. $Na^+$-$Ca^{2+}$-exchange as a major calcium influx pathway in kidney cells. *J. Biol. Chem.* **260**, 14998–15007.

[42] Alberts, B., Bray, D., Lewis, J., Raff, M., Roberts, K., and Watson, J. D., Eds. (1983) *Molecular Biology of the Cell*, Garland Publishing, New York and London.

[43] Guder, W. G., and Wirthensohn, G. (1985) Enzyme distribution and unique biochemical pathways in specific cells along the nephron. In *Renal heterogeneity and target cell toxicity* (Bach, P. H., and Lock, E. A., Eds.) pp 195–198, Wiley and Sons, Chinchester.

[44] Humes, H. D., and Weinberg, J. M. (1983) Cellular energetics in acute renal failure. In *Acute renal failure* (Brenner, B. M., and Lazarus, J. M., Eds.) pp 47–98, Saunders, Philadelphia, PA.

[45] Jung, K. Y., Uchida, S., and Endou, H. (1989) Nephrotoxicity assessment by measuring cellular ATP content. I. Substrate specificities in the maintainance of ATP content in isolated rat nephron segments. *Toxicol. Appl. Pharmacol.* **100**, 369–382.

[46] Jung, K. Y., and Endou, H. (1989) Nephrotoxicity assessment by measuring cellular ATP content. II. Intranephron site of ochratoxin A nephrotoxicity. *Toxicol. Appl. Pharmacol.* **100**, 383–390.

[47] Klingenberg, M. (1980) The ADP, ATP translocation in mitochondria: a membrane potential controlled transport. *J. Membr. Biol.* **56**, 97–105.

[48] LaNoue, K. F., Meijer, A. J., and Brouwer, A. (1974) Evidence for electrogenic aspartate transport in rat liver mitochondria. *Arch. Biochem. Biophys.* **161**, 544–550.

[49] Tahiliani, A. G. (1989) Dependence of mitochondrial coenzyme A uptake on the membrane electrical gradient. *J. Biol. Chem.* **264**, 18426–18432.

[50] McGivan, J. D., and Klingenberg, M. (1971) Correlation between $H^+$ and anion movement in mitochondria and the key role of the phosphate carrier. *Eur. J. Biochem.* **20**, 392–399.

[51] Hoek, J. B., and Njagu, R. M. (1976) Glutamate transport and the trans-membrane pH gradient in isolated rat-liver mitochondria. *FEBS Lett.* **71**, 341–346.

[52] Papa, S., Francavilla, A., Paradies, G., and Meduri, B. (1971) The transport of pyruvate in rat liver mitochondria. *FEBS Lett.* **12**, 285–288.

[53] Palmieri, L., Ronca, G., Cioni, L., and Puccini, R. (1984) Enzymuria as a marker of renal injury and disease. Studies of N-acetylglucosaminidase, alanine aminopeptidase and lysozyme in patients with renal disease. *Contrib. Nephrol.* **42**, 123–129.

[54] Ruegg, C. E., Gandolfi, A. J., Nagle, R. B., and Brendel, K. (1987) Differential patterns of injury to the proximal tubule of renal cortical slices following in vitro exposure to mercuric chloride, potassium dichromate or hypoxic condition. *Toxicol. Appl. Pharmacol.* **90**, 261–273.

[55] Metzler, D. E. (1977) Poisons: fluoroacetate and "lethal" synthesis. In *Biochemistry: the chemical reactions of living cells*, p 528, Academic Press, New York.

[56] Lash, L. H., and Anders, M. W. (1987) Mechanism of S-(1,2-dichlorovinyl)-l-cysteine- and S-(1,2-dichlorovinyl)-l-homocysteine-induced renal mitochondrial toxicity. *Mol. Pharmacol.* **32**, 549–556.

[57] Tune, B. M., Sibley, R. K., and Hsu, C.-Y. (1987) The mitochondrial respiratory toxicity of cephalosporin antibiotics. An inhibitory effect on substrate uptake. *J. Pharmacol. Exp. Ther.* **245**, 1054–1059.

[58] Tune, B. M., Fravert, D., and Hsu, C.-Y. (1989) The oxidative and mitochondrial toxic effects of cephalosporin antibiotics in the kidney. A comparative study of cephaloridine and cephaloglycin. *Biochem. Pharmacol.* **38**, 795–802.

[59] Tune, B. M. (1975) Relationship between the transport and toxicity of cephalosporins in the kidney. *J. Infect. Dis.* **132**, 189–194.

[60] Goldstein, R. S., Contardi, L. R., Pasino, D. A., and Hook, J. B. (1987) Mechanisms mediating cephaloridine inhibition of gluconeogenesis. *Toxicol. Appl. Pharmacol.* **87**, 297–305.

[61] Goldstein, R. S., Pasino, D. A., Hewitt, W. R., and Hook, J. R. (1986) Biochemical mechanisms of cephaloridine nephrotoxicity: Time and concentration-dependence of peroxidative injury. *Toxicol. Appl. Pharmacol.* **83**, 261–270.

[62] Redegeld, F. A. M. (1989) *Hepatic and renal toxicity of xenobiotics: the role of metabolism and transport*, Chapter 11, Academisch Proefschrift, Rijksuniversiteit Utrecht.

[63] Tune, B. M., Fravert, D., and Hsu, C.-Y. (1989) Thienamycin nephrotoxicity; mitochondrial injury and oxidative effects of imipenem in the rabbit kidney. *Biochem. Pharmacol.* **38**, 3779–3783.

[64] Wallin, A., Jones, T. W., Vercesi, A. E., Cotgreave, I., Ormstad, K., and Orrenius, S. (1987) Toxicity of S-pentachlorobutadienyl-l-cysteine studied with isolated rat renal cortical mitochondria. *Arch. Biochem. Biophys.* **258**, 365–372.

[65] Stonard, M. D. (1973) Further studies on the site and mechanism of action of S-(1,2-dichlorovinyl)-L-cysteine and S-(1,2-dichlorovinyl)-3-mercaptopropionic acid in rat liver. *Biochem. Pharmacol.* **22**, 1329–1335.

[66] Dobbie, M. S., Porter, K. E., and Jones, A. R. (1988) Is the nephrotoxicity of (R)-3-chlorolactate in the rat caused by 3-chloropyruvate? *Xenobiotica* **18**, 1389–1399.

[67] Eckert, K.-G., Elbers, F. R., and Eyer, P. (1989) Depletion of mitochondrial coenzyme A and glutathione by 4-dimethylaminophenol and formation of mixed thioethers. *Biochem. Pharmacol.* **38**, 3253–3259.

[68] Banki, K., Elfarra, A. A., Lash, L. H., and Anders, M. W. (1986) Metabolism of S-(2-chloro-1,1,2-trifluoroethyl)-L-cysteine to hydrogen sulfide and the role of hydrogen sulfide in S-(2-chloro-1,1,2-trifluoroethyl)-L-cysteine-induced mitochondrial toxicity. *Biochem. Biophys. Res. Commun.* **138**, 707–713.

[69] Greenwell, A., Tomaszewski, K. E., and Melnick, R. L. (1987) A biochemical basis for 1,2-dibromo-3-chloropropane-induced male infertility: inhibition of sperm mitochondrial electron transport activity. *Toxicol. Appl. Pharmacol.* **91**, 274–280.

[70] Malis, C. D., and Bonventre, J. V. (1986) Mechanism of calcium potentiation of oxygen free radical injury to renal mitochondria: a model for post-ischemic and toxic mitochondrial damage. *J. Biol. Chem.* **261**, 14201–14208.

[71] Schnellman, R. G., Ewell, F. P. Q., Sgambati, M., and Mandel, L. J. (1987) Mitochondrial toxicity of 2-bromohydroquinone in rabbit renal proximal tubules. *Toxicol. Appl. Pharmacol.* **90**, 420–426.

[72] Schnellman, R. G., Lock, E. A., and Mandel, L. J. (1987) A mechanism of S-(1,2,3,4,4-pentachlorobuta-1,3-dienyl)-l-cysteine to rabbit renal proximal tubules. *Toxicol. Appl. Pharmacol.* **90**, 513–521.

[73] De Haan, E. J., and Charles, R. (1969) The mechanism of uncoupling of oxidative phosphorylation by 2-methyl-1,4-naphtoquinone. *Biochim. Biophys. Acta* **180**, 417–419.

[74] Moore, G. A., O'Brien, P. J., and Orrenius, S. (1986) Menadione (2-methyl-1,4-naphtoquinone)-induced $Ca^{2+}$-release from rat-liver mitochondria is caused by NAD(P)H oxidation. *Xenobiotica* **16**, 873–882.

[75] Gould, D. H., Fettman, M. J., Daxenbichler, M. E., and Bartuska, B. M. (1985) Functional and structural alterations of the rat kidney induced by the naturally occuring organonitrile 2S-1-cyano-2-hydroxy-3,4-epithiobutane. *Toxicol. Appl. Pharmacol.* **78**, 190–201.

[76] Jung, K., Reinholdt, C., and Scholz, D. (1987) Inhibited efficiency of kidney mitochondria isolated from rats treated with cyclosporin A. *Nephron* **45**, 43–45.

[77] Lemmi, C. A. E., Pelikan, P. C. D., Sikka, S. C., Hirschberg, R., Geesaman, B., Miller, R. L., Park, K. S., Liu, S.-C., Koyle, M., and Rajfer, J. (1989) Cyclosporine augments renal mitochondrial function in vivo and reduces renal blood flow. *Am. J. Physiol.* **257**, F837–841.

[78] Buss, W. C., Stepanek, J., and Bennett, W. M. (1989) A new proposal for the mechanism of cyclosporine A nephrotoxicity. *Inhibition of renal microsomal protein chain elongation following in vivo cyclosporine A. Biochem. Pharmacol.* **38**, 4085–4093.

[79] Tzagoloff, A. (1982) Oxidative phosphorylation. In *Mitochondria*, p 132, Plenum Press, New York.

[80] Lê-Quôc, K., and Lê-Quôc, D. (1982) Control of the mitochondrial inner membrane permeability by sulfhydryl groups. *Arch. Biochem. Biophys.* **216**, 639–651.

[81] Schnellmann, R. G., Cross, T. J., and Lock, E. A. (1989) Pentachlorobutadienyl-l-cysteine uncouples oxidative phosphorylation by dissipating the proton gradient. *Toxicol. Appl. Pharmacol.* **100**, 498–505.

[82] Lê-Quôc, D., and Lê-Quôc, K. (1989) Relationships between the NAD(P) redox state, fatty acid oxidation, and inner membrane permeability in rat liver mitochondria. *Arch. Biochem. Biophys.* **273**, 466–478.

[83] Webb, W. W., Elfarra, A. A., Webster, K. D., Thom, R. E., and Anders, M. W. (1987) Role for an episulfonium ion in S-(2-chloroethyl)cysteine-induced cytotoxicity and its reaction with glutathione. *Biochemistry* **24**, 3017–3023.

[84] Moore, L. (1980) Inhibition of liver-microsome calcium pump by in vivo administration of $CCl_4$, $CHCl_3$ and 1,1-dichloroethylene (vinylidene chloride). *Biochem. Pharmacol.* **29**, 2505–2511.

[85] Moore, L. (1982) 1,1-Dichloroethylene inhibition of liver endoplasmic reticulum calcium pump function. *Biochem. Pharmacol.* **31**, 1463–1465.

[86] De Witt, L. M., Jones, T. W., and Moore, L. (1988) Stimulation of the renal endoplasmic reticulum calcium pump: a possible biomarker for platinate toxicity. *Toxicol. Appl. Pharmacol.* **92**, 157–169.

[87] Nicotera, P., McConkey, D. J., Jones, D. P., and Orrenius, S. (1989) ATP stimulates $Ca^{2+}$ uptake and increases the free $Ca^{2+}$ concentration in isolated rat liver nuclei. *Proc. Natl. Acad. Sci. U.S.A.* **86**, 453–457.

[88] Jones, D. P., McConkey, D. J., Nicotera, P., and Orrenius, S. (1989) Calcium-activated DNA fragmentation in rat liver nuclei. *J. Biol. Chem.* **264**, 6398–6403.

[89] Reasor, M. J. (1989) A review of the biology and toxicologic implications of the induction of lysosomal lamellar bodies by drugs. *Toxicol. Appl. Pharmacol.* **97**, 47–56.

[90] Lullmann, H., Lullmann-Rauch, R., and Mossinger, E. U. (1981) Impairment of renal function in rats with generalized lipidosis as induced by chlorphentermine. *Arzneim.-Forsch.* **31**, 795–799.

[91] Moriyama, T., Nakahama, H., Fukuhara, Y., Horio, M., Yanase, M., Orita, Y., Kamada, T., Kanashiro, M., and Miyake, Y. (1989) Decrease in the fluidity of brush-border membrane vesicles induced by gentamicin. *Biochem. Pharmacol.* **38**, 1169–1174.

[92] The Renal Commision of the International Union of Physiological Sciences (IUPS) (1988) A standard nomenclature for structures of the kidney. *Am. J. Physiol.* **254**, F1–F8.

[93] Hellberg, P. O. A. and Kållskog, T. Ø. K. (1989) Neutrophil-mediated post-ischemic tubular leakage in the rat kidney. *Kidney Int.* **36**, 555–561.

[94] McKinney, T. D. (1982) Heterogeneity of organic base secretion by proximal tubules. *Am. J. Physiol.* **243**, F404–F407.

[95] Berndt, W. O. (1989) Potential involvement of renal transport mechanisms in nephrotoxicity. *Toxicol. Lett.* **46**, 77–82.

[96] Squibb, K. S., Pritchard, J. B., and Fowler, B. A. (1984) Cadmium-metallothionein nephropathy: relationships between ultrastructural/biochemical alterations and intracellular cadmium binding. *J. Pharmacol. Exp. Ther.* **229**, 311–321.

[97] Zager, R. A. (1989) Gentamicin nephrotoxicity in the setting of acute renal hypoperfusion. *Am. J. Physiol.* **254**, F574–581.

[98] Taylor, D. M., Threlfall, G., and Buck, A. T. (1968) Chemically-induced renal hypertrophy in the rat. *Biochem. Pharmacol.* **17**, 1567–1574.

[99] Sabatini, S., Koppera, S., Manaligod, J., Arruda, J. A. L., and Kurtzman, N. A. (1983) Role of urinary concentrating ability in the generation of toxic papillary necrosis. *Kidney Int.* **23**, 705–710.

[100] Zenser, T. V., Mattammal, M. B., and Davis, B. B. (1978) Differential distribution of the mixed-function oxidase activities in rabbit kidney. *J. Pharmacol. Exp. Ther.* **207**, 719–725.

[101] Dees, J. H., Koseki, C., Hasumura, S., Kakuno, K., Hojo, K., and Johnson, E. F. (1982) Effect of 2,3,7,8-tetrachlorodibenz-p-dioxin and phenobarbital on the occurrence and distribution of four cytochrome P-450 isozymes in rabbit kidney, lung and liver. *Cancer Res.* **42**, 1423–1432.

[102] Schwartzman, M. L., Abraham, N. G., Carroll, M. A., Levere, R. D., and McGiff, J. C. (1986) Regulation of arachidonic acid metabolism by cytochrome P-450 in rabbit kidney. *Biochem. J.* **238**, 283–290.

[103] Cojocel, C., Maita, K., Pasino, D. A., Kuo, C. H., and Hook, J. B. (1983) Metabolic heterogeneity of the proximal and distal kidney tubules. *Life Sci.* **33**, 855–861.

[104] Zenser, T. Y., and Davis, B. B. (1984) Enzyme systems involved in the formation of reactive intermediates in the renal medulla: cooxidation via prostaglandin H synthetase. *Fundam. Appl. Toxicol.* **4**, 922–929.

[105] Krauss, R. S., and Eling, T. E. (1984) Arachidonic acid-dependent cooxidation. A potential pathway for the activation of chemical carcinogens in vivo. *Biochem. Pharmacol.* **33**, 3319–3324.

[106] Ziegler, D. M. (1980) In *Enzymic Basis of Detoxification* (Jacoby, W. B., Ed.) pp 201–227, Vol. I, Academic Press, New York.

[107] Dannan, G. A., and Guengerich, F. P. (1982) Immunochemical comparison and quantitation of microsomal flavin-containing monooxygenases in various hog, mouse, rat, rabbit, dog, and human tissues. *Mol. Pharmacol.* **22**, 787–794.

[108] Cromarty, T. H., and Walsh, C. T. (1975) Rat kidney l-α-hydroxyacid oxidase: isolation of enzyme with one flavin coenzyme per two subunits. *Biochemistry* **14**, 2588–2596.

[109] Hamilton, G. (1985) Peroxisomal oxidases. *Adv. Enzymol.* **57**, 86.

[110] Wolfgang, G. H. I., Gandolfi, A. J., Stevens, J. L., and Brendel, K. (1989) In vitro and in vivo nephrotoxicity of the L and D isomers of S-(1,2-dichlorovinyl)-cysteine. *Toxicology* **58**, 33–42.

[111] Waldman, R. H., and Burch, H. B. (1963) Rapid method for study of enzyme distribution in rat kidney. *Am. J. Physiol.* **204**, 749.

[112] Monks, T. J., Highet, R. J., Chu, P. S., and Lau, S. S. (1988) Synthesis and nephrotoxicity of 6-bromo-2,5-dihydroxythiophenol. *Mol. Pharmacol.* **34**, 15–22.

[113] Hjelle, J. T., Hazelton, G. A., Klaassen, C. D., and Hjelle, J. J. (1986) Glucuronidation and sulfation in rabbit kidney. *J. Pharmacol. Exp. Ther.* **236**, 150–156.

[114] Hiley, C., Bell, J., Hume, R., and Strange, R. (1989) Differential expression of alpha and pi isoenzymes of glutathione S-transferase in the developing human kidney. *Biochim. Biophys. Acta* **990**, 321–324.

[115] Mohandas, J., Marshall, J. J., Duggin, G. G., Horvath, J. S., and Tiller, D. J. (1984) Differential distribution of glutathione and glutathione-related enzymes in rabbit kidney. *Biochem. Pharmacol.* **33**, 1801–1807.

[116] Mohandas, J., Duggin, G. G., Horvath, J. S., and Tiller, D. J. (1981) Regional differences in peroxidatic activation of paracetamol (acetaminophen) mediated by cytochrome P-450 mixed function oxidase and prostaglandin endoperoxide synthetase in rabbit kidney. *Res. Commun. Chem. Pathol. Pharmacol.* **34**, 69–80.

[117] Stevens, J. L., Robbins, J. D., and Byrd, R. A. (1986) A purified cysteine conjugate β-lyase from rat kidney cytosol: requirement for an α-keto acid or an amino acid oxidase for activity, and identity with soluble glutamine transaminase K. *J. Biol. Chem.* **261**, 15529–15537.

[118] Lash, L. H., Elfarra, A. A., and Anders, M. W. (1986) Renal cysteine conjugate β-lyase: Bioactivation of nephrotoxic cysteine S-conjugates in mitochondrial outer membrane. *J. Biol. Chem.* **261**, 5930–5935.

[119] Stevens, J. L., Ayoubi, N., and Robbins, J. D. (1988) The role of mitochondrial matrix enzymes in the metabolism and toxicity of cysteine

conjugates. *J. Biol. Chem.* **263**, 3395–3401.

[120] MacFarlane, M., Foster, J. R., Gibson, G. G., King, L. J., and Lock, E. A. (1989) Cysteine conjugate β-lyase of rat kidney: characterization, immunocytochemical localization, and correlation with hexachlorobutadiene nephrotoxicity. *Toxicol. Appl. Pharmacol.* **98**, 185–197.

[121] Jones, T. W., Qin, C., Schaeffer, V. H., and Stevens, J. L. (1988) Immunohistochemical localization of glutamine transaminase K, a rat kidney cysteine conjugate β-lyase, and the relationship to the segment specificity of cysteine conjugate nephrotoxicity. *Mol. Pharmacol.* **34**, 621–627.

[122] Hughey, R. P., Rankin, B. B., Elce, J. S., and Curthoys, N. P. (1978) Specificity of a particulate rat renal peptidase and its localization with other enzymes of mercapturic acid synthesis. *Arch. Biochem. Biophys.* **1986**, 211–217.

[123] Jefcoate, C. R. (1983) In *The biological basis for detoxification* (Caldwell, J., and Jacoby, W. B., Eds.) Academic Press, New York.

[124] Okine, L. K., and Gram, T. E. (1986) In vitro studies on the metabolism and covalent binding of [$^{14}$C]-1,1-dichloroethylene by mouse liver, kidney and lung. *Biochem. Pharmacol.* **35**, 2789–2795.

[125] Venkatachalam, M. A., Bernard, D. B., Donohoe, J. F., and Levinsky, N. G. (1978) Ischemic damage and repair in the rat proximal tubule: Differences among the S1, S2, and S3 segments. *Kidney Int.* **14**, 31–49.

[126] Squibb, K. S., Pritchard, J. B., and Fowler, B. A. (1984) Cadmium-metallothionein nephropathy: relationships between ultrastructural/biochemical alterations and intracellular cadmium binding. *J. Pharmacol. Exp. Ther.* **229**, 311–321.

[127] Mazze, R. I. (1981) in *Toxicology of the Kidney* (Hook, J. B., Ed.) pp 135–149, Raven Press, New York.

[128] Robbins, S. L., Angell, M., and Kumor, M. D., Eds. (1981) The kidney and its collecting system. In *Basic Pathology*, pp 421–456, W. B. Saunders Co., Philadelphia.

[129] Sapin, C., Druet, E., and Druet, P. (1977) Induction of anti-glomerular basement membrane antibodies in the Brown-Norway rat by mercuric chloride. *Clin. Exp. Immunol.* **28**, 173–179.

[130] Fukatsu, A., Brentjens, J. R., Killen, P. D., Kleinman, H. K., Martin, G. R., and Andres, G. A. (1987) Studies on the formation of glomerular immune deposits in Brown Norway rats injected with mercuric chloride. *Clin. Immunol. Immunopathol.* **45**, 35–47.

[131] Hirsch, F., Couderc, J., Sapin, C., Fournie, G., and Druet, P. (1982) Polyclonal effect of mercury(II) chloride in the rat. Its possible role in an experimental autoimmune disease. *Eur. J. Immunol.* **12**, 620–625.

[132] Rehan, A., Johnson, K. J., Kunkel, R. G., and Wiggins, R. C. (1985) Role of oxygen radicals in phorbol myristate-induced glomerular injury. *Kidney Int.* **27**, 503–511.

[133] Weiss, S. J., and Ward, P. A. (1982) Immune complex induced generation of oxyfen metabolites by human neutrophiles. *J. Immunol.* **129**, 309–319.

[134] Fantone, J. C., and Ward, P. A. (1982) Role of oxygen derived free radicals and metabolites in leucocyte dependent inflammatory reactions. *Am. J. Pathol.* **107**, 395–418.

[135] Lannigan, R., Kark, R., and Pollak, V. E. (1962) The effect of a single intravenous injection of aminonucleoside of puromycin on the rat kidney. *J. Pathol. Bacteriol.* **83**, 357–362.

[136] Diamond, J. R., Pesek, I., Ruggieri, S., and Karnovski, M. J. (1989) Essential fatty acid deficiency during acute puromycin nephrosis ameliorates late renal injury. *Am. J. Physiol.* **257**, F798–807.

[137] Ryan, G. B., and Karnovsky, M. J. (1975) An ultrastructural study of the mechanisms of proteinuria in aminonucleoside nephrosis. *Kidney Int.* **8**, 219–232.

[138] Diamond, J. R., Bonventre, J. V., and Karnovsky, M. J. (1986) A role for oxygen free radicals in aminonucleoside nephrosis. *Kidney Int.* **29**, 478–483.

[139] Nagasawa, H. T., Swingle, K. F., and Alexander, C. S. (1967) Metabolism of aminonucleoside-8-14C in the rat and guinea pig. *Biochem. Pharmacol.* **16**, 2211–2219.

[140] Roy, R. S., and McCord, J. M. (1983) Superoxide and ischemia: conversion of xanthine dehydrogenase to xanthine oxidase. In *Oxy radicals and their scavenger systems. Cellular and medical aspects* (Greenwald, R. A., and Cohen, G., Eds.) Vol. II, p 145, Elsevier Science Publishers, New York.

[141] Bayliss, C., Rennke, H. R., and Brenner, B. M. (1977) Mechanism of the defect in glomerular ultrafiltration associated with gentamicin administration. *Kidney Int.* **12**, 344–353.

[142] Cojocel, C., Dociu, N., Maita, K., Sleight, S. D., and Hook, J. B. (1983) Effects of aminoglycosides on glomerular permeability, tubular reabsorption, and intracellular catabolism of the cationic low-molecular weight protein lysozyme. *Toxicol. Appl. Pharmacol.* **68**, 96–109.

[143] Ilett, K. F., Reid, W. D., Sipes, I. G., and Krishna, G. (1973) Chloroform toxicity in mice: Correlation of renal and hepatic necrosis with covalent binding of metabolites with tissue macromolecules. *Exp. Mol. Pathol.* **19**, 215–229.

[144] Pohl, L. R., Bhooshan, B., Whittaker, N. F., and Krishna, G. (1977) Phosgene: a metabolite of chloroform. *Biochem. Biophys. Res. Commun.* **79**, 684–691.

[145] Nash, T., and Pattle, R. E. (1971) The absorption of phosgene by aqueous solutions and its relation to toxicity. *Ann. Occup. Hyg.* **14**, 227–233.

[146] Bailie, M. B., Smith, J. H., Newton, J. F., and Hook, J. B. (1984) Mechanism of chloroform nephrotoxicity. IV. Phenobarbital potentiation of in vitro chloroform metabolism and toxicity in rabbit kidneys.

*Toxicol. Appl. Pharmacol.* **74**, 285–292.

[147] McMurtry, R. J., Snodgrass, W. R., and Mitchell, J. R. (1978) Renal necrosis, glutathione depletion, and covalent binding after acetaminophen. *Toxicol. Appl. Pharmacol.* **46**, 87–100.

[148] Newton, J. F., Kuo, C. H., Gemborys, M. W., Mudge, G. H., and Hook, J. B. (1982) Nephrotoxicity of p-aminophenol, a metabolite of acetaminophen, in the Fischer 344 rat. *Toxicol. Appl. Pharmacol.* **65**, 336–344.

[149] Newton, J. F., Yoshimoto, J., Bernstein, J., Rush, G. F., and Hook, J. B. (1983) Acetaminophen nephrotoxicity in the rat. I. Strain differences in nephrotoxicity and metabolism. *Toxicol. Appl. Pharmacol.* **69**, 291–306.

[150] Crowe, C. A., Young, A. C., Calder, I. C., Ham, K. N., and Tange, J. D. (1979) The nephrotoxicity of p-aminophenol. I. The effect on microsomal cytochromes, glutathione, and covalent binding in kidney and liver. *Chem.-Biol. Interact.* **27**, 235–243.

[151] Elbers, F. R., Eyer, P., Kampffmeyer, H., and Soboll, S. (1982) Organ toxicity and metabolic pathway of 4-dimethylaminophenol. In *Biological reactive intermediates II* (Snyder, R., Parke, D. V., Kocsis, J. J., Jollow, D. J., Gibson, G. G., and Witmer, C. M., eds.) pp 1173–1181, Plenum Press, New York.

[152] Newton, J. F., Bailie, M. B., and Hook, J. B. (1983) Acetaminophen nephrotoxicity in the rat. Renal metabolic activation in vitro. *Toxicol. Appl. Pharmacol.* **70**, 433–444.

[153] Newton, J. F., Pasino, D. A., and Hook, J. B. (1985) Acetaminophen nephrotoxicity in the rat: quantitation of renal metabolic activation in vivo. *Toxicol. Appl. Pharmacol.* **78**, 39–46.

[154] Elbers, F. R. (1982) Comparison of 4-dimethylaminophenol, 4-methylaminophenol, 4-aminophenol and acetaminophenol toxicity in the perfused rat kidney. *Arch. Pharmacol.* **319**, Suppl. R15.

[155] Reynolds, E. S., Moslen, M. T., Szabo, S., Jaeger, R. J., and Murphy, S. D. (1975) 1,1-Dichloroethylene hepatotoxicity. Time course of GSH changes and biochemical aberrations. *Am. J. Physiol.* **81**, 219–236.

[156] Krijgsheld, K. R., and Gram, T. E. (1984) Selective induction of renal microsomal cytochrome P-450-linked monooxygenases by 1,1-dichloroethylene in mice. *Biochem. Pharmacol.* **33**, 1951–1956.

[157] Jenkins, L. J., and Andersen, M. E. (1978) 1,1-Dichloroethylene nephrotoxicity in the rat. *Toxicol. Appl. Pharmacol.* **46**, 131.

[158] Jackson, N. M., and Conolly, R. B. (1985) Acute nephrotoxicity of 1,1-dichloroethylene in the rat after inhalation exposure. *Toxicol. Lett.* **29**, 191–199.

[159] Okine, L. K., Goochee, J. M., and Gram, T. E. (1985) Studies on the distribution and covalent binding of 1,1-dichloroethylene in the mouse. *Biochem. Pharmacol.* **34**, 4051–4057.

[160] Costa, A. K., and Ivanetich, K. M. (1982) Vinylidene chloride: its metabolism by hepatic microsomal cytochrome P-450 in vitro. *Biochem. Pharmacol.* **31**, 2083.

[161] Liebler, D. C., and Guengerich, F. P. (1983) Olefin oxidation by cytochrome P-450: evidence for group migration in catalytic intermediates formed with vinylidene chloride and trans-1-phenyl-1-butene. *Biochemistry* **22**, 5482–5489.

[162] Okine, L. K., and Gram, T. E. (1986) In vitro studies on the metabolism and covalent binding of [14C]-1,1-dichloroethylene by mouse liver, kidney and lung. *Biochem. Pharmacol.* **35**, 2789–2795.

[163] Palling, D. J., and Jencks, W. P. (1984) Nucleophilic reactivity toward acetyl chloride in water. *J. Am. Chem. Soc.* **106**, 4869–4876.

[164] Liebler, D. C., Latwesen, D. G., and Reeder, T. C. (1988) S-(2-chloroacetyl)glutathione, a reactive glutathione thiol ester and a putative metabolite of 1,1-dichloroethylene. *Biochemistry* **27**, 3652–3657.

[165] Moslen, M. T., and Reynolds, E. S. (1985) Rapid substrate-specific and dose-dependent deactivation of liver cytosolic glutathione-S-transferases in vivo by 1,1-dichloroethylene. *Res. Commun. Chem. Pathol. Pharmacol.* **47**, 59–72.

[166] Kanz, M. F., and Reynolds, E. S. (1986) Early effects of 1,1-dichloroethylene on canalicular and plasma membranes: ultrastructure and stereology. *Exp. Mol. Pathol.* **44**, 93–110.

[167] Rush, G. F., Newton, J. F., Maita, K., Kuo, C.-H., and Hook, J. B. (1984) Nephrotoxicity of phenolic bromobenzene metabolites in the mouse. *Toxicology* **30**, 259–272.

[168] Lau, S. S., Monks, T. J., Greene, K. E., and Gillette, J. R. (1984) The role of orthobromophenol in the nephrotoxicity of bromobenzene in rats. *Toxicol. Appl. Pharmacol.* **72**, 539–549.

[169] Lau, S. S., Monks, T. J., and Gillette, J. R. (1984) Identification of 2-bromohydroquinone as a metabolite of bromobenzene and 2-bromophenol: implications for bromobenzene-induced nephrotoxicity. *J. Pharmacol. Exp. Ther.* **230**, 360–366.

[170] Schnellmann, R. G., and Mandel, L. J. (1986) Cellular toxicity of S-(1,2,3,4,4-pentachloro-1,3-butadienyl)-l-cysteine toxicity to rabbit renal proximal tubules. *J. Pharmacol. Exp. Ther.* **237**, 456–461.

[171] Monks, T. J., and Lau, S. S. (1987) Characterization of a renal quinol oxidase. *Pharmacologist* **29**, 186.

[172] Schnellmann, R. G. (1989) 2-Bromohydroquinone-induced toxicity to rabbit renal proximal tubules: evidence against oxidative stress. *Toxicol. Appl. Pharmacol.* **99**, 11–18.

[173] Schnellmann, R. G., Monks, T. J., Mandel, L. J., and Lau, S. S. (1989) 2-Bromohydroquinone-induced toxicity to rabbit renal proximal tubules: the role of biotransformation, glutathione, and covalent binding. *Toxicol. Appl. Pharmacol.* **99**, 19–27.

[174] Monks, T. S., Lau, S. S., Highet, R. J., Gillette, J. R. (1985) Glutathione conjugates of 2-bromohydroquinone are nephrotoxic. *Drug.*

*Metab. Dispos.* **13**, 553–559.

[175] Lau, S. S., McMenamin, M. G., and Monks, T. J. (1988) Differential uptake of isomeric 2-bromohydroquinone glutathione conjugates into kidney slices. *Biochem. Biophys. Res. Commun.* **152**, 223–230.

[176] Monks, T. S., Highet, R. J., and Lau, S. S. (1988) 2-Bromo-(diglutathion-S-yl)hydroquinone nephrotoxicity: Physiological, biochemical, and electrochemical determinants. *Mol. Pharmacol.* **34**, 492–500.

[177] Lau, S. S., Hill, B. A., Highet, R. J., and Monks, T. J. (1988) Sequential oxidation and glutathione addition to 1,4-benzoquinone: correlation of toxicity with increased glutathione substitution. *Mol. Pharmacol.* **34**, 829–836.

[178] Wefers, H., and Sies, H. (1983) hepatic low-level chemiluminescence during redox cycling of menadione and the menadione–glutathione conjugate: relation to glutathione and NAD(P)H:quinone reductase (DT diaphorase) activity. *Arch. Biochem. Biophys.* **224**, 568–578.

[179] Jones, A. R., Porter, K., and Stevenson, D. (1981) The renal toxicity of some halogenated derivatives of propane in the rat. *Naturwissenschaften* **68**, 98.

[180] Porter, K. E., and Jones, A. R. (1987) The renal toxicity of (R)-3-chlorolactate in the rat. *Chem.-Biol. Interact.* **62**, 157–166.

[181] Khan, S. R., Finlayson, B., and Hackett, R. L. (1979) Histologic study of the early events in oxalate induced intranephronic calculosis. *Invest. Urol.* **17**, 199–202.

[182] Kluwe, W. M., Gupta, B. N., and Lamb, J. C., Iv. (1983) The comparative effects of 1,2-dibromo-3-chloropropane (DBCP) and its metabolites, 3-chloro-1,2-propaneoxide (epichlorohydrin), 3-chloro-1,2-propanediol (alphachlorohydrin), and oxalic acid, on urogenital system of male rats. *Toxicol. Appl. Pharmacol.* **70**, 67–86.

[183] Garella, S. (1989) Extracorporal techniques in the treatment of exogenous intoxications. *Kidney Int.* **33**, 735–754.

[184] Rankin, G. O. (1982) Nephrotoxicity of N-(3,5-dichlorophenyl)-succinimide in Sprague-Dawley rats. *Toxicology* **23**, 21–31.

[185] Rankin, G. O., Yang, D. J., Teets, V. J., and Brown, P. I. (1986) Deuterium isotope effect in acute N-(3,5-dichlorophenyl)succinimide-induced nephrotoxicity. *Life Sci.* **39**, 1291–1299.

[186] Rankin, G. O., Yang, D. J., Richmond, C. D., Teets, V. J., Wang, R. T., and Brown, P. I. (1987) Effect of microsomal enzyme activity modulation on N-(3,5-dichlorophenyl)succinimide-induced nephrotoxicity. *Toxicology* **45**, 269–289.

[187] Rankin, G. O., Shih, H. C., Yang, D. J., Richmond, C. D., Teets, V. J., and Brown, P. I. (1988) Nephrotoxicity of N-(3,5-dichlorophenyl)-succinimide metabolites in vivo and in vitro. *Toxicol. Appl. Pharmacol.* **96**, 405–416.

[188] Rankin, G. O., Teets, V., and Yang, D. J. (1986) Effect of cysteine conjugate β-lyase inhibition on N-(3,5-dichlorophenyl)succinimide-induced nephrotoxicity (abstract). *Fed. Proc., Fed. Am. Soc. Exp. Biol.* **45**, 571.

[189] Lock, E. A., Charbonneau, M., Strasser, J., Swenberg, J. A., and Bus, J. S. (1987) 2,2,4-Trimethylpentane-induced nephrotoxicity. II. The reversible binding of a TMP metabolite to a renal protein fraction containing $\alpha_{2\mu}$-globulin. *Toxicol. Appl. Pharmacol.* **91**, 182–192.

[190] Lehman-McKeeman, L. D., Rodriguez, P. A., Takigiku, R., Candill, D., and Fey, M. L. (1989) d-Limonene-induced male rat-specific nephrotoxicity: evaluation of the association between d-limonene and $\alpha_{2\mu}$-globulin. *Toxicol. Appl. Pharmacol.* **99**, 250–259.

[191] Sun, M. (1984) EDB Contamination kindles federal action. *Science (Washington, D.C.)* **223**, 464–466.

[192] Van Bladeren, B. J., Breimer, D. D., and Van Huijgevoort, J. A. T. C. M., Vermeulen, N. P. E., and Van der Gen, A. (1981) The metabolic formation of the N-acetyl-S-2-hydroxyethyl-l-cysteine from tetradeutero-1,2-dibromoethane. Relative importance of oxidation and glutathione conjugation in vivo. *Biochem. Pharmacol.* **30**, 2499–2502.

[193] Rannug, U., Sundvall, A., and Ramel, C. (1978) The mutagenic effect of 1,2-dichloroethane on *Salmonella typhimurium*. I. Activation through conjugation with glutathione in vitro. *Chem.-Biol. Interact.* **20**, 1–16.

[194] Van Bladeren, P. J., Breimer, D. D., Rotteveel-Smijs, G. M. T., De Jong, R. A. W., Buijs, W., Van der Gen, A., and Mohn, G. R. (1980) The role of glutathione conjugation in the mutagenicity of 1,2-dibromoethane. *Biochem. Pharmacol.* **29**, 2975–2982.

[195] Guengerich, F. P., Crawford, W. M., Domoradzki, J. Y., Macdonald, T. L., and Watanabe, P. G. (1980) In vitro activation of 1,2-dichloroethane by microsomal and cytosolic enzymes. *Toxicol. Appl. Pharmacol.* **55**, 303–317.

[196] Koga, N., Inskeep, P. B., Harris, T. M., and Guengerich, F. P. (1986) S-[2-(N7-guanyl)ethyl]glutathione, the major DNA adduct formed from 1,2-dibromoethane. *Biochemistry* **25**, 2192–2198.

[197] Foureman, G. L., and Reed, D. J. (1987) Formation of S-[2-(N7-guanyl)ethyl] adducts by the postulated S-(2-chloroethyl)cysteinyl and S-(2-chloroethyl)glutathionyl conjugates of 1,2-dichloroethane. *Biochemistry* **26**, 2028–2033.

[198] Vadi, H. V., Schasteen, C. S., and Reed, D. J. (1985) Interactions of S-(2-haloethyl)-mercapturic acid analogs with plasmid DNA. *Toxicol. Appl. Pharmacol.* **80**, 386–396.

[199] Elfarra, A. A., Baggs, R. B., and Anders, M. W. (1985) Structure-nephrotoxicity relationships of S-(2-chloroethyl)-DL-cysteine and analogs: Role for an episulfonium ion. *J. Pharmacol. Exp. Ther.* **233**, 512–516.

[200] Kluwe, W. M. (1981) Acute toxicity of 1,2-dibromo-3-chloropropane in the F344 male rat. I. Dose–response relationships and differences in routes of exposure. *Toxicol. Appl. Pharmacol.* **59**, 71–83.

[201] Kluwe, W. M. (1981) Acute toxicity of 1,2-dibromo-3-chloropropane in the F344 male rat. II. Development and repair of the renal, epididymal, testicular, and hepatic lesions. *Toxicol. Appl. Pharmacol.* **59**, 84–95.

[202] Omichinsky, J. G., Brunborg, G., Soderlund, E. J., Dahl, J. E., Bausano, J. A., Holme, J. A., Nelson, S. D., and Dybing, E. (1987) Renal necrosis and DNA damage caused by selectively deuterated and methylated analogs of 1,2-dibromo-3-chloropropane in the rat. *Toxicol. Appl. Pharmacol.* **91**, 358–370.

[203] Kluwe, W. M. (1983) Chemical modulation of 1,2-dibromo-3-chloropropane nephrotoxicity. *Toxicology* **27**, 287–299.

[204] Omichinski, J. G., Soderlund, E. J., Dybing, E., Pearson, P. G., and Nelson, S. D. (1988) Detection and mechanism of formation of the potent direct acting mutagen 2-bromoacrolein from 1,2-dibromo-3-chloropropane. *Toxicol. Appl. Pharmacol.* **92**, 286–194.

[205] Omichinski, J. G., Brunborg, G., Holme, Soderlund, E. J., Nelson, S. D., and Dybing, E. (1988) The role of oxidative and conjugative pathways in the activation of 1,2-dibromo-3-chloropropane to DNA-damaging products in rat testicular cells. *Mol. Pharmacol.* **34**, 74–79.

[206] Jones, A. R., Fakhouri, G., and Gadiel, P. (1979) The metabolism of the soil fumigant 1,2-dibromo-3-chloropropane in the rat. *Experienta* **35**, 1432–1434.

[207] Soderlund, E. J., Dybing, E., and Nelson, S. D. (1980) Nephrotoxicity and hepatotoxicity of tris(2,3-dibromopropyl)phosphate in the rat. *Toxicol. Appl. Pharmacol.* **56**, 171–186.

[208] Lynn, R. K., Garvie-Gould, C., Wong, K., and Kennish, J. M. (1982) Metabolism, distribution and excretion of the flame retardant tris(2,3-dibromopropyl)phosphate (Tris-BP) in the rat: identification of mutagenicand nephrotoxic metabolites. *Toxicol. Appl. Pharmacol.* **63**, 105–119.

[209] Soderlund, E. J., Gordon, W. P., Nelson, S. D., Omichinski, J. G., and Dybing, E. (1984) Metabolism in vitro of tris(2,3-dibromopropyl)phosphate: oxidative debromination and bis(2,3-dibromopropyl)phosphate formation as correlates of mutagenicity and covalent protein binding. *Biochem. Pharmacol.* **33**, 4017–4023.

[210] McKinney, L. L., Picken, J. C., Jr., Weakley, F. B., Eldridge, A. C., Campbell, R. E., Cowan, J. C., and Biester, H. E. (1959) Possible toxic factor of trichloroethylene-extracted soybean oil meal. *J. Am. Chem. Soc.* **81**, 909–915.

[211] Derr, R. F., and Schultze, M. O. (1963) The metabolism of S-(1,2-dichlorovinyl)-l-cysteine in the rat. *Biochem. Pharmacol.* **12**, 465–474.

[212] Anderson, P. M., and Schultze, M. O. (1965) Cleavage of S-(1,2-dichlorovinyl)-l-cysteine by an enzyme of bovine origin *Arch. Biochem. Biophys.* **111**, 593–602.

[213] Darnedud, P. O., Brandt, I., Feil, V. J., and Bakke, J. E. (1988) S-(1,2-dichloro-[14C]vinyl)-l-cysteine (DCVC) in the mouse kidney: correlation between tissue-binding and toxicity. *Toxicol. Appl. Pharmacol.* **95**, 423–434.

[214] Dekant, W., Metzler, M., and Henschler, D. (1986) Identification of S-1,2-dichlorovinyl-N-acetylcysteine as a urinary metabolite of trichloroethylene: a possible explanation of its nephrocarcinogenicity in male rats. *Biochem. Pharmacol.* **35**, 2455–2458.

[215] Commandeur, J. N. M., and Vermeulen, N. P. E. (1990) Identification of N-acetyl(2,2-dichlorovinyl)- and N-acetyl(1,2-dichlorovinyl)-L-cysteine as two regioisomeric mercapturic acids of trichloroethylene in the rat. *Chem. Res. Toxicol.* (companion paper in this issue).

[216] Kanhai, W., Dekant, W., and Henschler, D. (1989) Metabolism of the nephrotoxin dichloroacetylene by glutathione conjugation. *Chem. Res. Toxicol.* **2**, 51–56.

[217] Lock, E. A., and Ishmael, J. (1985) Effect of the organic acid transport inhibitor probenecid on renal cortical uptake and proximal tubular toxicity of hexachlorobutadiene and its conjugates. *Toxicol. Appl. Pharmacol.* **81**, 32–42.

[218] Dilley, J. V., Carter, V. L., and Harris, E. S. (1974) Fluoride excretion by male rats after inhalation of one of several fluoroethylenes or hexafluoropropane. *Toxicol. Appl. Pharmacol.* **27**, 582–590.

[219] Potter, C. L., Gandolfi, A. J., Nagle, R., and Clayton, J. W. (1981) Effects of inhaled chlorotrifluoroethylene and hexafluoropropene on the rat kidney. *Toxicol. Appl. Pharmacol.* **59**, 431.

[220] Commandeur, J. N. M., Oostendorp, R. A. J., Schoofs, P. R., Xu, B., and Vermeulen, N. P. E. (1987) Nephrotoxicity and hepatotoxicity of 1,1-dichloro-2,2-difluoroethylene in the rat. Indications for differential mechanisms of bioactivation. *Biochem. Pharmacol.* **36**, 4229–4237.

[221] Vamvakas, S., Kremling, E., and Dekant, W. (1989) Metabolic activation of the nephrotoxic haloalkene 1,1,2-trichloro-3,3,3-trifluoro-1-propene by glutathione conjugation. *Biochem. Pharmacol.* **38**, 2297–2304.

[222] Nash, J. A., King, L. J., Lock, E. A., and Green, T. (1984) The metabolism and disposition of hexachloro-1:3-butadiene in the rat and its relevance to nephrotoxicity. *Toxicol. Appl. Pharmacol.* **73**, 124–137.

[223] Tateishi, M. (1983) Methylthiolated metabolites. *Drug Metab. Rev.* **14**, 1207–1234.

[224] Monks, T. J., and Lau, S. L. (1987) Commentary: Renal transport processes and glutathione conjugate-mediated nephrotoxicity. *Drug Metab. Dispos.* **15**, 437–441.

[225] Commandeur, J. N. M., Brakenhof, J., de Kanter, F. J. J., and Vermeulen, N. P. E. (1988). Nephrotoxicity of mercapturic acids of three structurally related 2,2-difluoroethylenes in the rat. Indications for different bioactivation mechanisms. *Biochem. Pharmacol.* **37**, 4495–4505.

[226] Boogaard, P. J., Commandeur, J. N. M., Mulder, G. J., Vermeulen, N. P. E., and Nagelkerke, J. F. (1989) Toxicity of the cysteine-S-

conjugates and mercapturic acids of four structurally related difluoroethylenes in isolated proximal tubular cells from rat kidney. Uptake of the conjugates and activation to toxic metabolites. *Biochem. Pharmacol.* **38**, 3731–3741.

[227] Commandeur, J. N. M., de Kanter, F. J. J., and Vermeulen, N. P. E. (1989) Bioactivation of the cysteine-S-conjugate and mercapturic acid of tetrafluoroethylene to acylating reactive intermediates in the rat. Dependence of activation and deactivation activities on acetyl coenzyme A availability. *Mol. Pharmacol.* **36**, 654–663.

[228] Elfarra, A. A., Jacobson, I., and Anders, M. W. (1986) Mechanism of S-(1,2-dichlorovinyl)glutathione-induced nephrotoxicity. *Biochem. Pharmacol.* **35**, 283–288.

[229] Zhang, G., and Stevens, J. L. (1989) Transport and activation of S-(1,2-dichlorovinyl)-l-cysteine and N-acetyl-S-(1,2-dichlorovinyl)-l-cysteine in rat kidney proximal tubules. *Toxicol. Appl. Pharmacol.* **100**, 51–61.

[230] Lash, L. H., and Anders, M. W. (1989) Uptake of nephrotoxic S-conjugates by isolated rat renal proximal tubular cells. *J. Pharmacol. Exp. Ther.* **248**, 531–537.

[231] Schaeffer, V. H., and Stevens, J. L. (1987) Mechanism of transport for toxic cysteine conjugates in rat kidney cortex membrane vesicles. *Molec. Pharmacol.* **32**, 293–298.

[232] Ullrich, K. J., Rumrich, G., Wieland, Th., and Dekant, W. (1989) Contraluminal para-aminohippurate (PAH) transport in the proximal tubule of the rat kidney. VI. Specificity: amino acids, their N-methyl-, N-acetyl- and N-benzoylderivates; glutathione- and cysteine conjugates, di- and oligopeptides. *Pflüglers Arch.* **415**, 342–350.

[233] Ullrich, K. J., and Rumbich, G. (1988) Contraluminal transport systems in the proximal renal tubule involved in secretion of organic anions. *Am. J. Physiol.* **254**, F453–462.

[234] Dohn, D. R., Leininger, J. R., Lash, L. H., Quebbemann, A. J., and Anders M. W. (1985) Nephrotoxicity of S-(2-chloro-1,1,2-trifluoroethyl)glutathione and S-(2-chloro-1,1,2-trifluoroethyl)-L-cysteine, the glutathione and cysteine conjugates of chlorotrifluoroethylene. *J. Pharmacol. Exp. Ther.* **235**, 851–857.

[235] Lash, L. H., and Anders, M. W. (1986) Cytotoxicity of S-(1,2-dichlorovinyl)glutathione and S-(1,2-dichlorovinyl)-L-cysteine in isolated rat kidney cells. *J. Biol. Chem.* **261**, 13076–13081.

[236] Davis, M. E. (1988) Effect of AT-125 on the nephrotoxicity of hexachloro-1,3-butadiene in rats. *Toxicol. Appl. Pharmacol.* **95**, 44–52.

[237] Odum, J., and Green, T. (1988) The metabolism and nephrotoxicity of tetrafluoroethylene in the rat. *Toxicol. Appl. Pharmacol.* **76**, 306–318.

[238] Dohn, D. R., Quebbemann, A. J., Borch, R. F., and Anders, M. W. (1985) Enzymic reaction of chlorotrifluoroethylene with glutathione: 19F-NMR evidence for stereochemical control of the reaction. *Biochemistry* **24**, 5137–5143.

[239] Dekant, W., Lash, L. H., and Anders, M. W. (1987) Bioactivation Mechanisms of the cytotoxic and nephrotoxic S-conjugate S-(2-chloro-1,1,2-trifluoroethyl)-L-cysteine. *Proc. Natl. Acad. Sci. U.S.A.* **84**, 7443–7447.

[240] Dekant, W., Metzler, M., and Henschler, D. (1986) Identification of S-1,2-dichlorovinyl-N-acetylcysteine as a urinary metabolite of trichloroethylene: a possible explanation of its nephrocarcinogenicity in male rats. *Biochem. Pharmacol.* **35**, 2455–2458.

[241] Dekant, W., Metzler, M., and Henschler, D. (1986) Identification of S-1,1,2-trichlorovinyl-N-acetylcysteine as a urinary metabolite of trichloroethylene—Bioactivation through glutathione conjugation may explain its nephrocarcinogenity. *J. Biochem. Toxicol.* **1**, 57–71.

[242] Dekant, W., Berthold, K., Vamvakas, S., Henschler, D., and Anders, M. W. (1988) Thioacylating intermediates as metabolites of S-(1,2-dichlorovinyl)-l-cysteine and S-(1,2,2-trichlorovinyl)-l-cysteine formed by cysteine conjugate β-lyase. *Chem. Res. Toxicol.* **1**, 175–178.

[243] Dekant, W., Berthold, K., Vamvakas, S., and Henschler, D. (1988) Thioacylating agents as ultimate intermediates in the β-lyase catalysed metabolism of S-(pentachlorobutadienyl)-l-cysteine. *Chem.-Biol. Interact.* **67**, 139–148.

[244] Beuter, W., Cojocel, C., Mueller, W., Donaubauer, H. H., and Mayer, D. (1989) Peroxidative damage and nephrotoxicity of dichlorovinyl-cysteine in mice. *J. Appl. Toxicol.* **9**, 181–186.

[245] Cojocel, C., Beuter, W., Mueller, W., and Mayer, D. (1989) Lipid peroxidation: a possible mechanism of trichloroethylene induced nephrotoxicity. *Toxicology* **55**, 131–141.

[246] Green, T., and Odum, J. (1985) Structure activity studies of the nephrotoxic and mutagenic action of chloro- and fluoroalkenes. *Chem.-Biol. Interact.* **54**, 15–31.

[247] Vamvakas, S., Köchling, A., Berthold, K., and Dekant, W. (1989) Cytotoxicity of cysteine-S-conjugates: structure activity relationships. *Chem.-Biol. Interact.* **71**, 79–90.

[248] Stevens, J. L., Hatzinger, P. B., and Hayden, P. J. (1989) Quantitation of multiple pathways for the metabolism of nephrotoxic cysteine conjugates using selective inhibitors of l-α-hydroxy acid oxidase (l-amino acid oxidase) and cysteine conjugate β-lyase. *Drug Metab. Dispos.* **17**, 297–303.

[249] Berglin, E. H., and Carlsson, J. (1986) Effect of hydrogen sulphide on the mutagenicity of hydrogen peroxide in Salmonella Typhimurium strain TA102. *Mutat. Res.* **175**, 5–9.

[250] Tune, B. M. (1986) The nephrotoxicity of cephalosporin antibiotics—structure activity relationships. *Comments Toxicol.* **1**, 145–170.

[251] Tune, B. M., and Fernholt, M. (1973) Relationship between ce-

phaloridine and p-aminohippuric acid transport in the kidney. *Am. J. Physiol.* **225**, 1114–1117.

252 Silverblatt, F., Turck, M., and Bulger, R. (1970) Nephrotoxicity due to cephaloridine: light and electron-microscopic study in rabbits. *J. Infect. Dis.* **122**, 33.

253 Wold, J. S., and Turnipseed, S. A. (1980) The effect of renal cation transport inhibitors on the in vivo and in vitro accumulation and efflux of cephaloridine. *Life Sci.* **27**, 2559.

254 Tune, B. M., Wu, K. Y., Fravert, D., and Holtzman, D. (1979) Effect of cephaloridine on respiration by renal cortical mitochondria. *J. Pharmacol. Exp. Ther.* **210**, 98–100.

255 Kuo, C. H., Maita, K., Sleight, S. D., and Hook, J. B. (1983) Lipid peroxidation: a possible mechanism of cephaloridine-induced nephrotoxicity. *Toxicol. Appl. Pharmacol.* **67**, 78–88.

256 Cojocel, C., Hannemann, J., and Baumann, K. (1985) Cephaloridine-induced lipid peroxidation initiated by reactive oxygen species as a possible mechanism of cephaloridine nephrotoxicity. *Biochim. Biophys. Acta* **834**, 402–410.

257 Brier, R. A., Brier, M. E., and Mayer, P. R. (1985) Relationship between rat renal accumulation of gentamycin, tobramycin, and netilmicin and their nephrotoxicities. *Antimicrob. Agents Chemother.* **27**, 812–816.

258 Kaloyanides, G. J. (1984) Aminoglycoside-induced functional and biochemical defects in the renal cortex. *Fundam. Appl. Toxicol.* **4**, 930–943.

259 Humes, H. D., Weinberg, J. M., and Knauss, T. C. (1982) Clinical and pathophysiologic aspects of aminoglycoside nephrotoxicity. *Am. J. Kidney Dis.* **11**, 5–29.

260 Walker, P. D., and Shah, S. V. (1987) Gentamicin enhanced production of hydrogen peroxide by renal cortical mitochondria. *Am. J. Physiol.* **253**, C495–C499.

261 Walker, P. D., and Shah, S. V. (1988) Evidence suggesting a role for hydroxyl radical in gentamicin-induced acute renal failure in rats. *J. Clin. Invest.* **81**, 334–341.

262 Ramsammy, L. S., Josepovitz, C., King, K. Y., Lane, B. P., and Kaloyanides, G. J. (1987) Failure of inhibition of lipid peroxidation by vitamine E to protect against gentamicin nephrotoxicity in the rat. *Biochem. Pharmacol.* **36**, 2125–2132.

263 Gomez, A., Martos, F., Garcia, R., Perez, B., and Sanchez de la Cuesta, F. (1989) Diltiazem enhances gentamycin nephrotoxicity in rats. *Pharmacol. Toxicol. (Copenhagen)* **64**, 190–192.

264 Quarum, M. L., Houghton, D. C., Gilbert, D. N., McCarron, D. A., and Bennett, W. M. (1984) Increased dietary calcium moderates experimental gentamycin nephrotoxicity. *J. Lab. Clin. Med.* **103**, 104–114.

265 Keniston, R. C. (1979) Polyamine-pyridoxal-5'-phosphate interaction: effects of pH and phosphate concentration on Schiff's base formation. *Physiol. Chem. Phys.* **11**, 465–470.

266 Kacew, S. (1989) Inhibition of gentamicin-induced nephrotoxicity by pyridoxal-5'-phosphate in the rat. *J. Pharmacol. Exp. Ther.* **248**, 360–366.

267 Zager, R. A., and Sharma, H. M. (1983) Gentamicin increases renal susceptibility to an ischemic insult. *J. Lab. Clin. Med.* **101**, 670–678.

268 Gartland, K. P. R., Bonner, F. W., and Nicholson, J. K. (1989) Investigations into the biochemical effects of region-specific nephrotoxins. *Mol. Pharmacol.* **35**, 242–250.

269 Zalme, R. C., McDowell, E. M., Nagle, R. B., McNeil, J. S., Flamenbaum, W., and Trump, B. F. (1976) Studies on the pathophysiology of acute renal failure. II. A histochemical study of the proximal tubule following administration of mercuric chloride. *Virchows Arch. B Cell. Pathol.* **22**, 197–216.

270 Ganther, H. E. (1978) Methyl mercury-induced lipid peroxidation is thought to involve free radicals formed by homolytic fission of the carbon–metal bond. *Environ. Health Perspect.* **25**, 71.

271 Dudley, R. E., Svoboda, D. J., and Klaassen, C. D. (1982) Acute exposure to cadmium causes severe liver injury in rats. *Toxicol. Appl. Pharmacol.* **65**, 302–313.

272 Goering, P. K., and Klaassen, C. D. (1984) Zinc–induced tolerance to cadmium hepatoxicity. *Toxicol. Appl. Pharmacol.* **74**, 308–313.

273 Friberg, L., Piscator, M., Nordberg, G. F., and Kjellstrom, T. (1974) *Cadmium in the Environment*, 2nd ed., CRC Press Inc., Cleveland, OH.

274 Dudley, R. E., Gammal, L. M., and Klaassen, C. D. (1985) Cadmium-induced hepatic and renal injury in chronically exposed rats: likely role of hepatic cadmium–metallothionein in nephrotoxicity. *Toxicol. Appl. Pharmacol.* **77**, 414–426.

275 Sendelbach, L. E., and Klassen, C. D. (1988) Kidney synthesizes less metallothionein than liver in response to cadmium chloride and cadmium–metallothionein. *Toxicol. Appl. Pharmacol.* **92**, 95–102.

276 Kapoor, S. C., Van Rossum, G. D. V., O'Neill, K. J., and Mercolrella, I. (1985) Uptake of inorganic lead in vitro by isolated mitochondria and tissue slices of rat renal cortex. *Biochem. Pharmacol.* **34**, 1439–1448.

277 Fullmer, C. S., Edelstein, S., and Wasserman, R. H. (1985) Lead-binding properties of intestinal calcium-binding protein. *J. Biol. Chem.* **260**, 6816–6819.

278 Rosen, J. F., and Pounds, J. G. (1989) Quantitative interactions between $Pt^{2+}$ and $Ca^{2+}$ homeostasis in cultured osteoclastic bone cells. *Toxicol. Appl. Pharmacol.* **98**, 530–543.

279 Berndt, W. O. (1976) Renal chromium accumulation and its relationship to chromium-induced nephtotoxicity. *J. Toxicol. Environ. Health* **1**, 449–459.

280 Safirstein, R., Miller, P., and Guttenplan, J. B. (1984) Uptake and metabolism of cisplatin by rat kidney. *Kidney Int.* **25**, 753–758.

281 Goldstein, R. S., and Mayor, G. H. (1983) The nephrotoxicity of cisplatin. *Life Sci.* **32**, 685–690.

282 Roberts, J. J., and Thomson, A. J. (1979) The mechanism of action of antitumor platinum compounds. *Prog. Nucleic Acid Res. Mol. Biol.* **22**, 71–133.

283 Levi, J., Jacobs, C., Kalman, S. M., McTigue, M., and Weiner, M. W. (1980) Mechanism of *cis*-platinum nephrotoxicity: I. Effects of sulfhydryl groups in rat kidneys. *J. Pharmacol. Exp. Ther.* **213**, 545–550.

284 Bodenner, D. L., Dedon, P. C., Keng, P. C., and Borch, R. F. (1986) Effect of diethyldithiocarbamate on cis-diamminedichloroplatinum-(II)-induced cytotoxicity, DNA cross-linking and γ-glutamyl transpeptidase inhibition. *Cancer Res.* **46**, 2745–2750.

285 Baldew, G. S., Van den Hamer, C. J. A., Los, G., Vermeulen, N. P. E., De Goeij, J. J. M., and McVie, J. G. (1989) Selenium-induced protection against cis-diamminedichloroplatinum(II) nephrotoxicity in mice and rats. *Cancer Res.* **49**, 3020–3023.

286 Dedon, P. C., and Borch, R. F. (1987) Characterization of the reactions of platinum antitumor agents with biologic and nonbiologic sulfur-containing nucleophiles. *Biochem. Pharmacol.* **36**, 1955–1964.

287 Gordon, J. A., and Gattone, V. H. (1986) Mitochondrial alterations in cisplatin induced acute renal failure. *Am. J. Physiol.* **250**, F991–998.

288 Tay, L. K., Bregman, C. L., Masters, B. A., and Williams, P. D. (1988) Effects of cis-diamminedichloroplatinum(II) on rabbit kidney in vivo and on rabbit renal proximal tubule cells in culture. *Cancer Res.* **48**, 2538–2543.

289 Bach, P. H., and Bridges, J. W. (1985) Chemically induced renal papillary necrosis and upper urothelial carcinoma: parts 1 and 2. *CRC Crit. Rev. Toxicol.* **15**, 217–439.

290 Molland, E. A. (1976) Experimental renal papillary necrosis. *Kidney Int.* **13**, 5–14.

291 Nanra, R. S., and Kincaid-Smith, P. (1970) Papillary necrosis in rats caused by aspirin and aspirin containing mixtures. *Br. Med. J.* **3**, 559–561.

292 Barraclough, M. A., and Nilam, F. (1972) Effect of vasopressin on the renal tubular reabsorption and cortico-papillary concentration gradient of phenacetin and its metabolites. *Experimentia* **28**, 1065–1066.

293 Bluemle, L. W., Jr., and Goldberg, M. (1968) Renal accumulation of salicylate and phenacetin: possible mechanisms in the nephropathy of analgesic abuse. *J. Clin. Invest.* **47**, 2507–2514.

294 Larsson, R., Ross, D., Berlin, T., Olsson, L. I., and Moldeus, P. (1985) Prostaglandin synthetase catalyzed metabolic activation of p-phenetidine and acetaminophen by microsomes isolated from rabbit and human kidney. *J. Pharmacol. Exp. Ther.* **235**, 475–480.

295 West, P. R., Harman, L. S., Josephy, P. D., and Mason, R. P. (1984) Acetaminophen: enzymic formation of a transient phenoxyl free radical. *Biochem. Pharmacol.* **33**, 2933–2936.

296 Ross, D., Larsson, R., Norbeck, K., Ryhage, R., and Moldeus, P. (1985) Characterization and mechanism of formation of reactive products formed during peroxidase-catalyzed oxidation of p-phenetidine: trapping of reactive species by reduced glutathione and butylated hydroxyanisole. *Mol. Pharmacol.* **27**, 277–286.

297 Bryan, G. J. (1978) Occurrence, production, and uses of nitrofurans. In *Carcinogenesis: A Comprehensive Survey* (Bryan, G. T., Ed.) pp 1–11, Raven Press, New York.

298 Erturk, E., Morris, J. S., Cohen, S. M., Von Esch, A. M., Crovetti, A. J., Price, J. M., and Bryan, G. T. (1971) Comparative carcinogenicity of formic acid 2-[(4-(5-nitro-2-furyl)-2-thiazolyl]hydrazide and related chemicals in the rat. *J. Natl. Cancer Inst.* **47**, 437–445.

299 Zenser, T. V., Mattammal, M. B., Palmier, M. O., and Davies, B. N. (1981) Microsomal nitroreductase activity in rabbit kidney and bladder: implications in 5-nitrofuran-induced toxicity. *J. Pharmacol. Exp. Ther.* **219**, 735–740.

300 Zenser, T. V., Palmier, M. O., Mattammal, M. B., and Davies, B. B. (1984) Metabolic activation of the carcinogen N-[4-(5-nitro-2-furyl)-2-thiazolyl]acetamide by prostaglandin H synthetase. *Carcinogenesis* **5**, 1225–1230.

301 Swaminathan, S., and Lower, G. M., Jr. (1978) Biotransformations and excretion of nitrofurans. In *Carcinogenesis: A comprehensive Survey* (Bryan, G. T., Ed.) pp 59–97, Raven Press, New York.

302 Wang, C. Y., Behrens, B. C., Ichikawa, M., and Bryan, G. T. (1974) Nitroreduction of 5-nitrofuran derivatives by rat liver xanthine oxidase and reduced nicotinamide adenine dinucleotide phosphate–cytochrome c reductase. *Biochem. Pharmacol.* **23**, 3395–3404.

303 Docampo, R., Moreno, S. N. J., and Stoppani, A. O. M. (1981) Nitrofuran enhancement of microsomal electron transport, superoxide anion production and lipid peroxidation. *Arch. Biochem. Biophys.* **207**, 316.

304 Moreno, S. N. J., and Docampo, R. (1985) Mechanism of toxicity of nitro compounds used in the chemotherapy of trichomonias. *Environ. Health Perspect.* **64**, 199–208.

305 Erturk, E., Cohen, S. M., and Bryan, G. T. (1970) Induction, histogenesis and isotransplantability of renal tumors induced by 2-[4-(5-nitro-2-furyl)-2-thiazolyl]hydrazide in rats. *Cancer Res.* **30**, 2098–2106.

306 Tekele, S., Biava, C. G., and Price, J. M. (1973) The carcinogenic activity of 3-hydroxymethyl-1-(5-nitro-2-furyl)allydilene]aminohydantoin in rats. *Cancer Res.* **33**, 2894–2897.

307 Ballal, S., Spry, L. A., Zenser, T. V., and Davies, B. B. (1988) Renal handling of 5-nitrofuran nephrotoxins in the rat. *Drug Metab. Dispos.* **16**, 829–833.

[308] Cohen, S. M. (1978) Toxicity and carcinogenicity of nitrofurans. In *Carcinogenesis: A Comprehensive Survey* (Bryan, G. T., Ed.) pp 171–231, Raven Press, New York.

[309] Spry, L. A., Lakshmi, V. M., Zenser, T., and Davies, B. (1986) Metabolism and excretion of nitrofurothiazole bladder carcinogens. *J. Pharmacol. Exp. Ther.* **238**, 457–461.

[310] Spry, L. A., Zenser, T. V., Cohen, S. M., and Davies, B. B. (1985) Role of renal metabolism and excretion in 5-nitrofuran-induced uroepithelial cancer in the rat. *J. Clin. Invest.* **76**, 1025–1031.

[311] Murasaki, G., Zenser, T. V., Davies, B. B., and Cohen, S. M. (1984) Inhibition by aspirin of N-[4-(5-nitro-2-furyl)-2-thiazolyl]formamide-induced bladder carcinogenesis and enhancement of forestomach carcinogenesis. *Carcinogenesis* **5**, 53–55.

[312] Wang, C. Y., and Bryan, G. T. (1974) Deacylayion of carcinogenic 5-nitrofuran derivatives by mammalian tissues. *Chem.-Biol. Interact.* **9**, 423–428.

[313] Davies, D. J. (1969) The structural changes in the kidney and urinary tract caused by ethyleneimine. *J. Pathol.* **97**, 695–703.

[314] Powell, C. J., Tyobeka, T., Bach, P. H., and Bridges, J. W. (1985) The nephrotoxicity of structural analogues of 2-bromoethanamine. In *Renal heterogeneity and target cell toxicity* (Bach, P. H., and Lock, E. A., Eds.) pp 195–198, Wiley and Sons, Chichester.

[315] Grantham, J. J. (1982) Studies of organic anion and cation transport in isolated segments of the proximal tubule. *Kidney Int.* **22**, 519–525.

**Enzymes of Activation,
Inactivation, and Repair**

# Chapter 10

# Oxidation of Toxic and Carcinogenic Chemicals by Human Cytochrome P-450 Enzymes

F. Peter Guengerich*

*Department of Biochemistry and Center in Molecular Toxicology, Vanderbilt University School of Medicine, Nashville, Tennessee 37232-0146*

Tsutomu Shimada

*Osaka Prefectural Institute of Public Health, Nakamichi, Higashinari-ku, Osaka 537, Japan*

Reprinted from *Chemical Research in Toxicology*, Vol. 4, No. 4, July/August, 1991

The majority of toxic and carcinogenic chemicals do not produce their detrimental biological effects by themselves, except perhaps in acute doses. In most cases activation to electrophilic forms is necessary to produce molecules capable of reacting irreversibly with tissue nucleophiles. The pioneering work of the Millers demonstrated the need for such bioactivation in the formation of adducts involving proteins and nucleic acids (1–4). Such activation of chemicals is now known to be catalyzed by almost all of the enzymes involved in the biotransformation of xenobiotic chemicals (those not normally found in the body). Under the appropriate conditions, the so-called "phase II" enzymes epoxide hydrolase (5, 6), glutathione S-transferase (7, 8), UDP-glucuronosyltransferase (9, 10), cysteine conjugate β-lyase (8, 11, 12), γ-glutamyl transpeptidase (8), methylases (13), and others (14) may be involved in bioactivation as well as detoxication reactions. The majority of bioactivation reactions probably involve oxidation, and some of these detrimental oxidations are catalyzed by alcohol dehydrogenases (15), monoamine oxidase (16), microsomal flavin-containing monooxygenase (17), and peroxidases such as prostaglandin synthase (18, 19) and myeloperoxidase (20). However, the majority of oxidative bioactivation reactions can probably be attributed to P-450.[1]

There are several known models of activation of chemicals to electrophiles catalyzed by P-450s. Seven of these are shown in Scheme I and account for the bulk of possibilities, and these have been reviewed in more detail

**Scheme I. Some Modes of Activation of Procarcinogens and -toxicants by P-450 Enzymes**

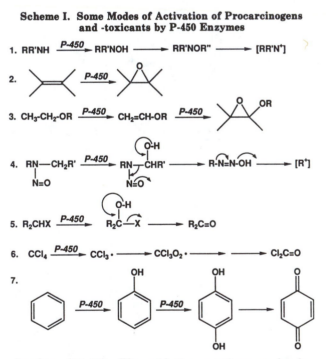

elsewhere (21–23). The oxidation reactions can be described in terms of a rather common chemistry which involves the abstraction of either a hydrogen atom or a nonbonded (or $\pi$) electron by a formal $[FeO]^{3+}$ iron–oxo porphyrin complex, followed by oxygen rebound (radical recombination) (24). Reaction 1 depicts the activation of

---

[1] Abbreviations: P-450, cytochrome P-450; GSH, glutathione; Trp P-1, 3-amino-1,4-dimethyl-5H-pyrido[4,3-b]indole.

2428–5/92/0098$06.00/0

arylamines and acetylamines; the resulting primary or secondary hydroxylamine is electrophilic in itself but esterification (by R″) often facilitates the formation of a (formal) nitrenium ion. Reaction 2, epoxidation, is common in bioactivation of olefins and arenes. In at least one case, that of ethyl carbamate, a P-450 catalyzes a desaturation of an alkane and then an epoxidation (25). The hydroxylation of N-nitrosodialkylamines leads to formation of alkanediazohydroxides, which can be thought of mechanistically as generating *formal* carbonium ions (reaction 4). Hydroxylation of some halogenated hydrocarbons produces *gem*-halohydrins that spontaneously dehydrohalogenate to reactive carbonyl compounds (reaction 5); e.g., ethylene dibromide is oxidized to 2-bromoacetaldehyde, and $CHCl_3$ is oxidized to phosgene. In an alternative mode of P-450 catalysis, P-450 (2E1) reduces $CCl_4$ to the trichloromethyl radical which can react with oxygen and ultimately generate phosgene and some electrophilic chlorine species (26) (reaction 6). In the final reaction (7) shown in Scheme I, hydroxylation of an aromatic compound can occur with or without the intermediacy of an epoxide (24). In the example of benzene shown, a second P-450 hydroxylation yields hydroquinone which can be oxidized by any of several mechanisms to benzoquinone (other modes of benzene activation have also been proposed) (27, 28). In other situations aliphatic (and particularly benzylic) hydroxyls formed by P-450 can be esterified to induce charge separation and thus generate electrophiles. There are additional modes of activation by P-450s, such as the formal dehydrogenation of pyrrolizidine alkaloids and acetaminophen (21).

P-450 is not a single enzyme but a generic term applied to a group of hemoproteins defined by unique spectral properties (near-UV $\lambda_{max}$ at 450 nm for the $Fe^{II}$–CO complex) imparted by the presence of an axial cysteinyl thiolate ligand to the heme iron (29). There are probably more than 30 different P-450 (Cyp) genes present in each animal species, and apparently a large fraction of these produce protein products at the same time (30, 31). P-450s are found principally in the liver but also at lower levels in most other tissues except erythrocytes and striated muscle—in some cases these same hepatic P-450s are expressed in other tissues (32, 33) while in other cases considerable tissue-selective expression occurs (i.e., a certain P-450 may be expressed in some extrahepatic tissues but not in liver) (34, 35). Ultimately, the inducibility and suppression of individual P-450s as well as the constitutive tissue specific expression may be understood in terms of the interaction of trans-acting factors with regulatory elements in the genes, but only in the case of P-450 1A1 have such genomic elements been extensively characterized (36–38). Most of the attention in this review will be given to the P-450s in the endoplasmic reticulum, since the microsomal fraction generally shows the highest level of oxidation of xenobiotics in vitro. The mitochondrial 11A and 11B P-450s function in anabolic reactions in steroidogenic tissues (39). Hepatic mitochondrial P-450s are known and have been shown to be capable of catalyzing the oxidation of carcinogens such as aflatoxin $B_1$ and benzo[a]pyrene (40–42); however, the overall significance of these P-450s in xenobiotic metabolism and their exact relationships to the microsomal enzymes remain to be further defined. Recently, other reviews on the topic of activation of chemical carcinogens have been published (43, 44).

What is the evidence that alteration of P-450 catalytic activities can be important in modulating the toxicity and carcinogenicity of chemicals? The subject has been reviewed at length elsewhere (43, 45, 46), but some of the pertinent information is presented here briefly. (i) In animal models where a particular P-450 is missing, the sensitivity to the toxic and tumorigenic effects of a particular chemical may be dramatically altered, either enhanced or attenuated (47, 48). Further, alterations in hepatic metabolism may shift sites of toxicity and tumor formation to or away from extrahepatic targets (45). (ii) The administration of P-450 inducers and inhibitors to experimental animals can also enhance and attenuate toxicity and tumorigenicity at certain sites (46). (iii) In vitro experiments can be done in which genotoxic events thought to be relevant to toxicity and carcinogenicity can be produced with P-450 enzymes, both in cell-free and in whole-cell systems (5, 49). In cellular environments the specificity of individual P-450s can be demonstrated (5, 50, 51). (iv) Humans are known to vary widely in terms of levels of individual P-450s (52) and in rates of metabolism of drugs (53–55). The variation can be up to 3 orders of magnitude, depending upon the parameter chosen for consideration (56, 57), and there are indications that rates of oxidation of drugs and nontherapeutic chemicals show considerable interindividual variation even when administered at very low doses (56). Thus, even though the molar dose of a particular chemical may be on the order of the amount of the P-450 that oxidizes it, considerable interindividual variation in metabolism may be seen (58, 59). (v) In the case of drugs, differences in levels of particular P-450s in humans can clearly be related to differences in pharmacological response and toxicity (60). In the case of tumor formation, intriguing epidemiological relationships have been reported (61–63), although mechanisms remain undefined.

### Background on Human P-450s

Although the existence of P-450 enzymes in human tissues has been known for many years (52), efforts to purify and characterize the proteins have been more recent. Kaschnitz and Coon (64) were able to partially separate P-450, NADPH-P-450 reductase, and phospholipid from human liver microsomes and demonstrated the need for all three fractions in the reconstitution of catalytic activity. In the late 1970s efforts in several laboratories led to the purification of human liver P-450s to a high degree of purity (65–67). In retrospect these proteins were probably P-450 3A7 (30, 66) and P-450 2C proteins (30, 67–70). Further efforts in this laboratory led to the separation of several proteins (69), which were also probably in the P-450 2C and 3A families. Since that time efforts have been directed primarily toward the purification of human P-450s on the basis of (a) immunochemical reactivity with animal P-450s (71–73) or (b) catalytic activity toward particular substrates (70, 74–79). The latter approach is technically difficult because of the need to do chromatography in the presence of detergents, remove the detergent from individual fractions, and estimate specific catalytic activities in the presence of NADPH–P-450 reductase and phospholipid. Nevertheless, such an approach has been used to purify human P-450 1A2 (74, 80), P-450 2C10 (76, 77), P-450$_{MP}$ (a P-450 2C enzyme) (70, 77), P-450 2D6 (74), P-450 3A4 (75), and certain other P-450s. The approach of utilizing immunochemical cross-reactivity with known rat P-450s has been used in the purification of human P-450 3A proteins (71, 81) and P-450 2E1 (73).

Two major approaches have been used to derive cDNAs coding for human P-450s. One has involved screening expression libraries with antibodies raised against purified human P-450s or their animal orthologues (82, 83). The

**Table I. Protoxicants and -carcinogens as Principal Substrates for Bioactivation by Individual Human P-450 Enzymes**

|  | refs |
|---|---|
| **P-450 1A1** | |
| benzo[a]pyrene and other polycyclic hydrocarbons | 92, 93 |
| **P-450 1A2** | |
| 2-(acetylamino)fluorene | 50, 93, 94 |
| 2-aminofluorene | 50, 94 |
| 2-aminoanthracene | 50, 94 |
| 2-amino-3-methylimidazo[4,5-f]quinoline (IQ) | 50, 51, 80, 93, 95, 96 |
| 2-amino-3,8-dimethylimidazo[4,5-f]quinoline (MeIQ) | 50, 51, 93 |
| 2-amino-3,8-dimethylimidazo[4,5-f]quinoxaline (MeIQx) | 50, 51, 93 |
| 2-amino-3,4,8-trimethylimidazo[4,5-f]quinoxaline (DiMeIQx) | 51, 93 |
| 2-amino-6-methyldipyrido[1,2-a:3',2'-d]imidazole (Glu P-1) | 50, 80, 95, 97 |
| 3-amino-1-methyl-5H-pyrido[4,3-b]indole (Trp P-2) | 50, 95 |
| 2-aminodipyrido[1,2-a:3',2'-d]imidazole (Glu P-2) | 50 |
| 2-naphthylamine | 80 |
| 4-aminobiphenyl | 50, 80, 94 |
| 2-amino-1-methyl-6-phenylimidazo[4,5-b]pyridine (PhIP) | 93 |
| (acetaminophen) | 98 |
| **P-450 2A6** | |
| N-nitrosodiethylamine | 99 |
| **P-450 2E1** | |
| N-nitrosodimethylamine | 73, 99–102 |
| benzene | 100 |
| carbon tetrachloride | 100, 103 |
| chloroform | 100 |
| methylene chloride | 100, 104 |
| trichloroethylene | 100 |
| ethylene dichloride | 100 |
| ethylene dibromide | 100 |
| 1,2-dichloropropane | 100 |
| styrene | 100 |
| vinyl chloride | 100 |
| vinyl bromide | 100 |
| acrylonitrile | 100 |
| ethyl carbamate | 25, 100 |
| vinyl carbamate | 25, 100 |
| (acetaminophen) | 98 |
| **P-450 3A4** | |
| aflatoxin $B_1$ | 96, 105–109 |
| aflatoxin $G_1$ | 105–107 |
| sterigmatocystin | 105–107 |
| senecionine | 110 |
| 7,8-dihydroxy-7,9-dihydrobenzo[a]pyrene | 50, 92, 107 |
| 9,10-dihydroxy-9,10-dihydrobenzo[b]fluoranthene | 02 |
| 3,4-dihydroxy-3,4-dihydro-7,12-dimethylbenz[a]anthracene | 92 |
| 6-aminochrysene | 50, 107 |
| tris(2,3-dibromopropyl) phosphate | 50, 107 |
| 1-nitropyrene | 111 |

**Table II. Procarcinogens Inactivated by Individual Human P-450 Enzymes**

|  | refs |
|---|---|
| **P-450 1A2** | |
| aflatoxin $B_1$ | 112, 113 |
| 2-furylfuramide | 114 |
| 1,3-dinitropyrene | 115 |
| **P-450 2E1** | |
| trichloroethylene | 100 |
| methylene chloride | 100, 104 |
| methyl chloride | 100 |
| ethylene dichloride | 100 |
| ethylene dibromide | 100 |
| 1,2-dichloropropane | 100 |
| 1,1,1-trichloroethane | 100 |
| **P-450 3A4** | |
| aflatoxin $B_1$ | 112 |
| senecionine | 110 |
| 1,6-dinitropyrene | 115 |

450s have been prepared in such a way (*31*). As pointed out below, this approach is particularly useful in evaluating which of several closely related sequences present in a system is involved in a particular reaction (*90*).

Features of human liver P-450s have been reviewed recently (*91*), and a current listing of protoxicants and -carcinogens activated and detoxicated predominantly by individual P-450 enzymes is presented in Tables I and II.

## Approaches to Elucidation of Catalytic Activities of P-450s

A number of in vitro approaches have been used in different laboratories, but most can be included in the general strategies we have used in our own work. It is best to first examine a number of different human liver samples (usually 5–15) for the activity under interest, in order to determine which can be practically used in further studies and to assess the extent of variation. With carcinogens, it is appropriate to utilize a *Salmonella typhimurium* tester strain to assay the production of genotoxic products. Often human tissue samples are not sterile—one approach we have used to remove contaminating bacteria is to expose frozen microsomes to γ-radiation. This procedure has been demonstrated not to affect any of several known human liver P-450s. Another approach we have utilized extensively involves a system in which the plasmid pSK1002 containing the *umuC''lacZ* chimera is harbored in *S. typhimurium* TA1535. Alkylation of bacterial DNA invokes the SOS response by activation of the recA protein, activating many genes including *umuC*. In this case activation of the chimera leads to β-galactosidase production, which can be measured colorimetrically after lysing the cells. The *S. typhimurium* host has the advantages of permeability to activated carcinogens and lack of DNA repair. The entire procedure takes only a few hours for one person to do ~100 assays, is insensitive to bacterial contamination, and is highly sensitive for many carcinogens (*50*). However, these bacterial systems give weak responses with some carcinogens (*50*), and appropriate radiometric, chromatographic, and other assays must be used (*100*), depending upon prior knowledge regarding relevant metabolic transformations. Even if the bacterial genotoxicity assays are used, it is desirable to define the chemistry of the reaction under consideration (*80, 106*). In certain cases mammalian cell toxicity and transformation can be used as end points (*49*), particularly when cytochrome P-450 cDNAs have been inserted (*116*).

With an assay in hand, some combination of the following approaches can be used.

other general approach has involved screening libraries with cDNA probes related to orthologous sequences in experimental animals (*84–86*). (Apparently the approach of using oligonucleotide probes derived from partial amino acid sequences for screening has not been used to isolate human P-450 cDNA clones.) It should be emphasized that in several cases very closely related genes exist (*30*) and the deduction of a sequence from partial-length, overlapping clones can generate chimeric artifacts. In principle, it should be possible to derive cDNAs from information regarding a partial N-terminal amino acid sequence and the poly(A)+ region using polymerase chain reaction technology. It should be pointed out that in many cases sequences may be highly similar to each other or to animal proteins. However, caution is advised in suggesting that certain sequences represent orthologues or that the proteins have similar catalytic activities just because of primary sequence similarity [incidentally, the appropriate term in making such comprisons is "similarity" or "identity" but *not* "homology" (*87–89*)].

It is now possible to express individual P-450 cDNAs in yeast and mammalian cells, and a number of human P-

**(i) Correlation.** The activity under consideration is measured in a series of different liver (microsomal) preparations and compared to (a) known marker activities for individual P-450s or (b) immunochemically determined levels of individual P-450s in the same samples. If the same enzyme is carrying out both of the reactions under consideration, then a high degree of correlation should be seen (*117*). The simplest analysis involves linear regression, and one way of considering the correlation coefficient $r^2$ is as the fraction of the variance that is accounted for by the relationship. More complex methods can also be applied, such as principal component analysis (*118, 119*).

**(ii) Selective Inhibition and Stimulation.** Certain chemicals are known to selectively inhibit and in at least one case (P-450 3A4) stimulate microsomal catalytic activity. Certain of these effects will be specifically discussed later.

**(iii) Immunoinhibition.** Specific antibodies can be used to inhibit P-450 reactions in microsomes. In principle, the degree of immunoinhibition indicates the fraction of a reaction attributable to a certain enzyme (as in the case of chemical inhibitors) (*120*). Monoclonal antibodies may offer some advantages in terms of increased specificity toward the individual enzymes (*121, 122*). However, it should be emphasized that the majority of monoclonal antibodies raised to P-450 enzymes are not inhibitory and only some are suitable for such purposes (*123*). Obviously, the interpretation of immunoinhibition experiments depends upon the specificity of the antibodies, and establishing specificty by various techniques is essential. In some cases P-450 proteins within a family have considerable similarity and distinguishing individual forms may not be trivial. Monoclonal antibodies may be useful (vide supra), or antibodies may be rendered more specific by adsorption against known cross-reactive proteins. It is also possible to raise antibodies against peptides, and antibodies can be directed against the distinct peptides in some highly related P-450 enzymes. These approaches require knowledge about how many proteins exist within the family and the availability of the related proteins, since cross-adsorption or screening will only eliminate certain proteins and not address other possibilities. In some cases the complete account of multiplicity within a gene family is not available—in such cases caveats may need to be drawn regarding the conclusions and other approaches may be needed to further refine specificity.

**(iv) Enzyme Purification.** If an assay to monitor the transformation of a substrate or an end point such as genotoxicity is available, it is possible to purify individual P-450 enzymes involved in particular reactions. This approach has been applied to the enzymes involved in the oxidation of drugs (*74*) and steroids (*124*) but can also be applied to procarcinogens and -toxicants. In cases where known proteins are not responsible for a reaction, this approach is one of the few available for identification of the involved enzyme(s). The procedure is not trivial because P-450 enzymes are intrinsic membrane proteins—chromatography procedures must be done in the presence of detergents, but detergents inhibit the coupling of the flavoprotein NADPH–P-450 reductase with P-450s and need to be removed from fractions prior to analysis of catalytic activity. This can make isolation of individual enzymes a painstaking procedure, but purifications can be accomplished (*70, 74, 75*). An alternative approach involves the use of a chemical surrogate oxygen donor instead of the reductase, NADPH, and molecular oxygen. Such systems often destroy the heme rapidly and are not particularly sensitive, but in at least two cases product for-

mation is linear (with respect to time) for at least 10 min (*107, 125, 126*). With such a system it is not necessary to eliminate the detergent before enzyme assay.

When an enzyme is purified and demonstrated to be homogeneous by a number of criteria, its catalytic activity is estimated in a system composed of NADPH–P-450 reductase and phospholipid, with the catalytic activity being expressed as a turnover number [nmol of product formed/(min·nmol of P-450), or min$^{-1}$]. Comparison of this rate is often made with rates measured in microsomes, on the basis of the total pool of P-450. It is somewhat difficult to compare such rates, although as a general rule the rate measured in the purified and reconstituted system should equal or exceed that measured in the microsomes. However, in some cases the reconstitution conditions are difficult to optimize and the rates observed in the reconstituted system are less than expected. The P-450 family 3A proteins are notorious in this regard (*75, 127, 128*), possibly due to deficiencies in rates of electron transfer from the reductase to P-450 (*107*). In some cases the presence of cytochrome $b_5$ or particular phospholipids may be beneficial (*129*). However, in the last analysis, interpretation of rates measured with the purified enzymes is a qualitative matter. Those P-450s showing the higher rates should be regarded as most likely to play the major roles in the reactions, with consideration given to how much of each enzyme is present. In one case involving rat liver P-450s it has been possible to qualitatively reconstruct a pattern of stereo- and regioselective warfarin oxidation from rates of individual product formation measured with purified enzymes, although there are some exceptions due to P-450 3A-linked reactions (*130*).

**(v) cDNA-Based Expression.** This approach has been made possible with the availability of particular human P-450 cDNAs and appropriate vectors for expression. It is now possible to express P-450s in bacteria (*131*), but NADPH–P-450 reductase is not present and direct activity measurements may be precluded unless there is an alternative source of electron equivalents. Several eukaryotic systems involving yeast (*132*) and mammalian cells have been used to express human P-450 enzymes. The mammalian expression systems include both transient and stable systems. The transient systems used most are COS-1 (monkey kidney) (*133*) and vaccinia systems (*134*)—the latter yield much higher levels of expression, and spectral measurements of P-450 can be performed, although there are a number of technical difficulties including the need for vaccination. Stable cell lines are also available—although these have very low levels of expression, the cells can be utilized in the measurement of mutations and cytotoxicity within the cell containing the P-450 enzyme (*135, 136*).

Such cDNA-based expression systems have an obvious advantage in being able to ascertain that a certain P-450 is capable of oxidizing a particular substrate. Thus they can help answer questions that arise about the catalytic capabilities of members of complex gene families containing closely related genes. Since the P-450 is inserted into the endoplasmic reticulum containing NADPH–P-450 reductase, cytochrome $b_5$, and the normal membranes, the rates of catalytic activity should tend to reflect those expected in liver or other tissues.

There are, however, some caveats that must be considered in the use of such systems. First, there have been several instances in which the activities of such cDNA-expressed P-450s have been unexplainedly low, in yeast microsomes and in mammalian microsomes and cells [see Brian et al. (*107*) and references therein]. Second, comparisons have sometimes been made without regard to how much P-450 is actually expressed—further, in the case of low-level expression systems quantitation cannot be done with regard to the amount of holoenzyme present. Finally, the same considerations regarding efforts to quantify respective roles of individual enzymes apply as in the case of the systems reconstituted with purified enzymes. It is difficult to evaluate a rate established in a cDNA-based expression system—the most rational guideline would be that the turnover number equal or exceed that measured in the tissue system which the vector system is intended to model. Such considerations have often been ignored in the literature. Of course, only those cDNAs that are available can be considered (as in the case of purified enzymes), and other enzymes will be overlooked.

Some dangers inherent in interpretation are well exemplified in the recent work of Lai and Chiang (137). Two major 3-methylcholanthrene-inducible hamster P-450s, P-450 1A2 and P-450 2A, were purified and shown to activate alfatoxin $B_1$, utilizing enzyme reconstitution systems containing the umu reporter gene. As expected, antibodies raised against these enzymes almost completely inhibited the respective reactions. However, when liver microsomes prepared from 3-methylcholanthrene-treated hamsters were used to activate aflatoxin $B_1$, anti-P-450 2A inhibited the reaction ~20% and anti-P-450 1A2 stimulated the activity by 175%! The results were appropriately interpreted to mean that P-450 1A2 converts aflatoxin $B_1$ to less toxic products in liver microsomes. This is probably the situation with P-450 1A2 enzymes as well—mouse P-450 1A2 is known to detoxicate aflatoxin $B_1$ by 9a-hydroxylation (to aflatoxin $M_1$) (138), and the observation that a cell line expressing only human P-450 1A2 can activate aflatoxin $B_1$ (116) must be considered in the context of the work of Lai and Chiang (137), since rates of aflatoxin $B_1$ 9a-hydroxylation in human liver microsomal preparations are well correlated with levels of P-450 1A2 (112, 113) and the P-450 1A2 inhibitor 7,8-benzoflavone (93) stimulates activation of aflatoxin $B_1$ (105, 139) [and also the formation of aflatoxin $B_1$ 8,9-oxide as judged by trapping of the glutathione conjugate (140)] in human liver microsomes. Similarly, the observation that cells infected with a vaccinia vector containing human P-450 1A2 are able to convert Trp P-1 to a mutagen (94) must be considered in light of the lack of ability of anti-P-450 1A2 or 7,8-benzoflavone to inhibit activation of the compound in human liver microsomes (50).

As we see, then, there are a number of means of obtaining information regarding the roles of individual P-450 enzymes in the oxidation of chemicals, including protoxicants and -carcinogens. All of these are useful, but caution is advised against reaching conclusions solely on the basis of one line of investigation. Our experience has been that the most useful studies have been based upon data from several approaches—when the results support each other, then a good case for the role of a particular enzyme can be made. Another point must be considered when a particular role is assigned to an enzyme—humans vary dramatically in the amounts of each enzyme they have, by at least 2 orders of magnitude in many cases (57). When a certain enzyme is assigned a major role in a particular reaction, it must be considered that some individuals will have very little of that enzyme and therefore the relative contribution of other enzymes may become greater. However, the point should be remembered that the rate is much lower in these individuals and that they will be less prone to the effects of the reaction, good or bad, than the individuals with higher levels. A case in point is that

of the oxidation of many arylamines—work from several laboratories has assigned a major role to P-450 1A2 in human liver (50, 51, 80, 93, 96). However, P-450 1A2 is expressed only at very low levels in fetal liver, and Kamataki and his associates have shown that the activation of arylamines can probably be attributed to P-450 3A7 in that case (96).

Another point to consider is that in vitro findings must be put into an in vivo context. Approaches to in vivo assignments of roles of enzymes parallel those used in vitro and include cosegregation of parameters (similar to the correlation analysis discussed above), use of diagnostic inhibitors, and induction studies (55, 141, 142). In many cases simple drugs can be administered and urinary metabolites can be measured to provide a guide to the phenotypic behavior of an individual. Since tissue samples are not required, such approaches are termed "noninvasive" and may offer considerable potential in studies related to what is termed molecular epidemiology. In considering the appropriateness and reliability of such noninvasive assays, which will be presented later, one must consider the tissue distribution of the enzyme under consideration and the contribution of extrahepatic metabolism, if hepatic metabolism is the object of consideration. In some cases, there is concern that intestinal enzymes may be of significance in the oxidation of certain drugs that are administered orally (143, 144).

These considerations are important in that interpretation of the available literature requires an understanding of the usefulness of each of the approaches used and its reliability. Further, work done largely in the past decade has set the stage for the widespread use of these approaches, and in the future the limit to understanding will not be the capability to obtain reagents but the ability to design and interpret experiments. With this background we will consider the available information regarding several of the known human P-450s and their substrates, inhibitors, and inducers.

## P-450 1A2

Human P-450 1A2 has a primary sequence 80% identical with that of P-450 1A1, the other member of the P-450 1A family (145) (it is possible that some species may contain more than these two P-450 1A genes). It was identified as an orthologue of the rat, mouse, and rabbit P-450 1A2 proteins (145–147) and isolated from liver on the basis of its phenacetin O-deethylation activity (74).

Correlation, immunoinhibition, and cDNA-directed expression studies all provide evidence that this enzyme is the major one involved in the oxidation of a variety of arylamines and arylacetamides to their respective hydroxylamines and hydroxamic acids (50, 51, 80, 93–96). Further, 7,8-benzoflavone ($\alpha$-naphthoflavone) is an excellent competitive inhibitor of these reactions, with a $K_i$ of <2 $\mu$M (50, 80) (Table III). 7,8-Benzoflavone also inhibits P-450 1A1 [to a lesser degree (93)], although several lines of investigation indicate that there are only low levels of this enzyme in human liver, even in smokers (74, 80, 156)—furafylline is reported to be a more specific inhibitor of P-450 1A2, although it is not readily available (148).

Not all arylamines are substrates for the enzyme. Evidence has been presented that 6-aminochrysene (50) and dapsone (157) are preferentially oxidized by P-450 3A4. Further, the food pyrolysate Trp P-1 appears to be oxidized by rat P-450 1A2 (158) but not by human P-450 1A2 (50). The enzyme appears to play a role in the detoxication of the nitro compounds 2-furylfuramide (114) and

**Table III. Diagnostic Chemical Inhibitors of Human P-450 Enzymes**

| | refs |
|---|---|
| P-450 1A1 | |
| 7,8-benzoflavone | 93 |
| P-450 1A2 | |
| 7,8-benzoflavone | 80, 93 |
| furafylline | 148 |
| P-450 2C10 | |
| sulfaphenazole[a] | 149 |
| P-450 2D6 | |
| quinidine[a] | 150 |
| ajmalicine | 150 |
| yohimbine | 150 |
| P-450 2E1 | |
| diethyldithiocarbamate (disulfiram)[a] | 100 |
| P-450 3A4 | |
| troleandomycin[a] | 151 |
| gestodene[a] | 152–154 |
| (naringenin) | 113 |
| (erythromycin)[a] | 155 |

[a] Shown to produce inhibition in vivo.

**Table IV. Potential Noninvasive Assays for Individual Human P-450 Enzymes**

| | refs |
|---|---|
| P-450 1A2 | |
| phenacetin O-deethylation[a] | 74, 80 |
| caffeine 3-demethylation | 80, 161 |
| P-450 2A6 | |
| coumarin 7-hydroxylation | 162 |
| P-450 2C10 (also 2C8, 2C9) | |
| tolbutamide hydroxylation | 149, 163 |
| P-450 2D6 | |
| debrisoquine 4-hydroxylation | 61 |
| sparteine $\Delta^2$- and $\Delta^6$-oxidation | 164 |
| dextromethorphan N-demethylation | 165 |
| P-450 2E1 | |
| chlorzoxazone 6-hydroxylation | 166 |
| P-450 3A4 | |
| nifedipine oxidation | 75, 167 |
| erythromycin N-demethylation | 107 |
| dapsone N-hydroxylation | 157 |
| lidocaine N-deethylation | 168, 169 |
| cortisol 6β-hydroxylation | 107, 170 |

[a] No longer possible to be used in the U.S. because of concern as a cancer suspect.

1,3-dinitropyrene (*115*) (Table II). The chemistry involved in these inactivations has not been definitively established but probably involves reduction of a nitro group to an amine (*159*). In addition, the 9a-hydroxylation of aflatoxin $B_1$ (to form aflatoxin $M_1$) is known to be catalyzed by the murine orthologue, and correlation studies support the view that human P-450 1A2 also catalyzes this detoxication reaction (*112, 113*).

Other substrates for human P-450 1A2 include phenacetin (O-deethylation) (*74*), 7-ethoxyresorufin (O-deethylation) (*160*), and caffeine (3-demethylation) (*80*). The enzyme also appears to be involved in the oxidation of acetaminophen to an iminoquinone, but P-450 2E1 also contributes extensively (*98*).

The role of P-450 1A2 in caffeine 3-demethylation, the major reaction in the metabolism of caffeine, offers the capability to estimate relative levels of the enzyme in different individuals (*80, 161*) (Table IV). The extents of variation measured in vitro (*80*) and in vivo (*161*) by the ratio of urinary 1,7-dimethylxanthine/caffeine are both ~50-fold. The enzyme appears to be largely restricted to the liver (*148, 156*). Previous work on phenacetin metabolism by Conney and his associates may be interpreted in terms of this enzyme (*171, 172*), but further studies using phenacetin are not possible in the United States because the drug has been implicated in cancers of the renal pelvis. The enzyme is inducible by cigarette smoking and by consumption of charbroiled food and certain cruciferous vegetables. However, the enzyme does not appear to be induced in all smokers, regardless of the level of smoking, and the basis of this observation is unclear (*161*). The drug omeprazole has been reported to induce P-450 1A2 in human liver, as judged by in vitro measurements with liver samples and cultured cells (*173*), but in vivo caffeine measurements argue against induction.[2] Among individuals using phenacetin (and the drug is still used in some countries), those with low rates of O-deethylation are at increased risk of methemoglobinemia (*174*). The hypothesis can be considered that those individuals with higher levels of P-450 1A2 are at increased risk for several cancers because of the increased capacity for bioactivation (*161*). Smoking and ingestion of charbroiled meat expose an individual to arylamines that are activated by P-450 1A2, in addition to inducing the enzyme (*171*). However, to date no firm epidemiological evidence is available to

support this conclusion. Recently, we have obtained evidence that P-450 1A2 appears to be the major enzyme in human and animal liver involved in the bioactivation of genotoxic components of cigarette smoke condensate, as judged by the *umu* assay.[3]

## P-450 2E1

The rat orthologue of human P-450 2E1, previously termed P-450j, was first isolated and characterized on the basis of its ability to oxidize ethanol to acetaldehyde (*175*). A number of studies on the nature of N-nitrosodimethylamine N-demethylation had shown that the enzyme responsible was a P-450 but that the "high-affinity" or "low-$K_m$" form of the enzyme was not induced by treatment of rats with the common inducers phenobarbital, 3-methylcholanthrene, and pregnenolone 16α-carbonitrile (*176–178*). Eventually it was understood that this rat P-450 (2E1) is suppressed by these treatments and that the enzyme is the principal catalyst of N-nitrosodimethylamine N-demethylation (a process that also generates a reactive methanediazohydroxide) at low substrate concentration (*179–181*). These findings were extended to human P-450 2E1 (*73, 101, 102*). Other studies with rat and rabbit P-450 2E1 have implicated it as a major catalyst involved in the oxidation of many primary alcohols to aldehydes, the N-oxygenation of pyridine, the hydroxylation of benzene (and subsequent hydroxylation of phenol to hydroquinone), and the one-electron reduction of $CCl_4$ (*180*). The enzyme is induced by these same substrates and also by isoniazid, acetone, and trichloroethylene (*182*).

In the past no drug substrates were available to serve as selective indicators of P-450 2E1 [the enzyme is able to oxidize acetaminophen to the iminoquinone but this reaction is also carried out by P-450 1A2 (*98*), and the (rat) P-450 2E1 substrate enflurane is complicated by its anesthetic properties and gaseous nature (*183*)]. Recently, the 6-hydroxylation of the muscle relaxant chlorzoxazone has been shown to be catalyzed almost exclusively by this enzyme in humans (*166*) and experimental animals.[4] Inhibitors of P-450 2E1 are also known. Indeed, essentially all organic solvents are competitive inhibitors and care

---

[2] T. Andersson, unpublished results.

[3] T. Shimada, and F. P. Guengerich, submitted for publication.
[4] R. Peter, R. B. Böcker, P. H. Beaune, T. D. Porter, and F. P. Guengerich, in preparation.

must be taken to avoid their presence in incubation systems (184). 3-Aminobenzotriazole and aminoacetonitrile are inhibitors, although they have rather low affinities (100, 185). Pyrazole derivatives have also been shown to be effective as inhibitors in vivo (186, 187). Evidence has existed for some time that both reduced and oxidized dialkyldithiocarbamates are mechanism-based inhibitors of P-450s, as first provided by Hunter and Neal in the case of benzphetamine N-demethylation (188). We have found that diethyldithiocarbamate and its oxidized form, disulfiram, selectively inhibit P-450 2E1 in human microsomes and that this inhibition is mechanism-based (100). Thus many in vitro and in vivo experiments can be interpreted in light of this information—for instance, disulfiram is a potent cocarcinogen in the case of ethylene dibromide (189), and the oxidation of ethyl carbamate to yield nucleic acid adducts is blocked (190). These two reactions involving experimental animals can now be understood in terms of P-450 2E1 action (vide infra).

Several studies have shown the role of rabbit and rat P-450 2E1 enzymes in the oxidation of potential carcinogens (180), but few of these had been extended to humans. Evidence has been provided through reconstitution, immunoinhibition, and correlation studies that human P-450 2E1 is the major catalyst of N-nitrosodimethylamine N-demethylation (73, 101, 102). Recently, we have provided evidence that human P-450 2E1 is a major catalyst in the oxidation of many low molecular weight cancer suspects, including ethyl carbamate, vinyl carbamate, acrylonitrile, vinyl chloride, vinyl bromide, benzene, styrene, carbon tetrachloride, chloroform, ethylene dibromide, trichloroethylene, ethylene dichloride, methyl chloride, methylene dichloride, 1,1,1-trichloropropane, and 1,2-dichloropropane (100). Most of these chemicals elicit only weak responses in the common assays of genotoxicity, and therefore systems involving the measurement of special oxidation products have been employed in many cases. Evidence has linked most of these reactions with genotoxicity—in other instances these oxidations represent detoxications (and other enzymes such as glutathione S-transferase are involved in the production of genotoxic species) (7, 191).

Major species differences in the catalytic specificity of P-450 2E1 enzymes have not been observed, and it may be expected that substrates found for the animal enzymes will also be substrates for the human enzyme. Thus, compounds such as azoxymethane, simple pyridine derivatives, and thiobenzamide may be substrates for human P-450 2E1 (192). Clearly, these compounds have little in the way of structural similarity (Table I). Some are quite hydrophilic (e.g., N-nitrosodimethylamine, ethyl carbamate), and most have little capability for forming hydrogen bonds. The one structural feature these do have in common is their small size. Thus, we can consider the active site of P-450 2E1 as a molecular sieve of sorts. The physical basis of the driving force underlying the selectivity of P-450 2E1 enzymes for these small substrates remains to be elucidated. This apparent selectivity of P-450 2E1 would argue that other chemicals such as vinylidene chloride, tetrachloroethylene, 1,2-dibromo-3-chloropropane, and naphthalene may be expected to be substrates. It should be pointed out that many of these compounds can be substrates for other P-450s at higher substrate concentrations, and what often distinguishes P-450 2E1 is not its $V_{max}$ but its lower $K_m$ [e.g., N-nitrosodimethylamine (178), trichloroethylene (193)].

It is interesting to note that compounds such as dialkyldithiocarbamates and pyrazoles inhibit such diverse proteins that oxidize alcohols, e.g., alcohol dehydrogenases

and P-450s. Although some physical resemblance of the active sites may exist, the mechanisms of the enzymes are quite distinct (194).

Recently, the role of human P-450 2E1 in the activation of ethyl carbamate (urethan) has been further clarified. The pathway of ethyl carbamate → vinyl carbamate → vinyl carbamate epoxide → (various) nucleic acid adducts proposed by the Millers (195) has been confirmed (Scheme I). P-450 2E1 catalyzes both the formal dehydrogenation of ethyl carbamate to vinyl carbamate (reaction 1) and the subsequent epoxidation (reaction 2) (25, 100). Reaction 1 has a high $K_m$ (>2 mM) and a low $V_{max}$ while reaction 2 has a relatively low $K_m$ (50 $\mu$M) and a high $V_{max}$ (25). Therefore, only low amounts of the intermediate vinyl carbamate accumulate in microsomal incubations. P-450 2E1 also apears to oxidize ethyl carbamate to 2-hydroxyethyl carbamate, but the 2-hydroxy compound is not an intermediate in the formation of vinyl carbamate. Instead, hydrogen atom abstraction is postulated to yield a transient methylene radical intermediate which partitions between oxygen rebound (to give 2-hydroxyethyl carbamate) and the abstraction of a proton/electron pair (to form the olefin).

Efforts to utilize chlorzoxazone 6-hydroxylation as a noninvasive in vivo assay are in progress. It is not clear yet whether or not the in vivo reaction can be ascribed to hepatic oxidation or not. Nevertheless, even if extrahepatic metabolism is a contributing factor, the noninvasive assay is useful if the regulation in extrahepatic tissues is similar to that in the liver. Recently, Song et al. (196) have been able to quantify P-450 2E1 in peripheral blood lymphocytes of some individuals and have shown the level to be considerably enhanced in diabetic patients who do not respond to insulin. The level of hepatic P-450 2E1 has been found to be elevated in alcoholics (197). Thus animal models may provide useful information regarding the regulation of this enzyme in humans (198, 199).

Whether or not alterations in the level of P-450 2E1 influence the health effects of the substrates indicated here is not yet known. However, maximum tolerated daily value limits exist for several of the compounds under consideration, and with the use of sensitive assays of reaction products it should be possible to modulate P-450 2E1 and carry out experiments directed toward such a goal.

## P-450 3A4

In this laboratory a P-450 protein was isolated from human liver microsomes on the basis of its ability to catalyze the oxidation of the dihydropyridine calcium channel blocker nifedipine to its pyridine derivative and termed "P-450$_{NF}$" (75). The isolated enzyme was found to be immunochemically related to some other human liver P-450 preparations previously isolated in this laboratory (69) and to rat liver P-450 3A proteins (75). The enzyme P-450$_{NF}$ is now defined as P-450 3A4 (30, 200, 201). The P-450 3A gene families are complex, and human P-450 3A4 is related to at least three or four rat P-450 3A proteins[5] (201–203) and one or more rabbit liver P-450 3A proteins (201). Three other P-450 3A proteins have been reported in human liver samples. P-450 3A7, one of the first P-450s purified (66), appears to be a major fetal P-450 (204), but selective hybridization studies with oligonucleotide probes indicate that P-450 3A7 mRNA is not often found in adult human liver (and that P-450 3A4 mRNA is not found in human fetal liver) (205). Immunochemical studies indicate that ~30% of adults express P-450 3A5 and that, when

---

[5] C. Kasper, personal communication.

expressed, levels of the protein are $\leq^1/_4$ those of other P-450 3A enzymes (*81, 206, 207*). Purified hepatic P-450 3A5 and cDNA-expressed P-450 3A5 oxidize some drugs and steroids but to date have been found to oxidize any carcinogens tested only at low rates (*90, 108, 206*). P-450 3A3, the last known member of the human P-450 3A family, represents a sequence derived from overlapping cDNA clones isolated from the same cDNA library as was the original P-450 3A4 clone (*82*), which was prepared from a single individual liver, thus arguing against the existence of allelic variants as a basis for the two cDNAs which differ only in 14 residues. Such a full-length clone has not yet been reported, and studies with two different oligo-nucleotide probes generated to the maximum differences between the reported P-450 3A3 and P-450 3A4 sequences indicated that the level of P-450 3A3 mRNA present in any of 12 different human liver samples was <8% of that related to P-450 3A4 or a related sequence (*208*); immuno-chemical analysis indicated that the results could not be attributed to the presence of P-450 3A5 in these samples (*206*). Further, similar studies with (i) polymerase chain reaction using selected oligonucleotides and human liver samples (*143*) and (ii) oligonucleotide probes and human hepatocyte cultures treated with rifampicin to induce P-450 3A enzymes (*209*) also provide evidence that significant amounts of P-450 3A3 mRNA are not found in human liver. Although the possibilities can be considered that protein levels may not reflect steady-state mRNA levels and that other related cDNAs may exist, the burden of proof argues against a major role for P-450 3A3. Another cDNA clone, NF-10, was isolated from the same cDNA library as P-450 3A4 (NF-25) (*208*). It differs in the addition of a 3-base insert, but the corresponding mRNA was not found at significant levels in human liver samples. However, the additional 3′ region found in this particular cDNA clone led to identification of the alternate poly-adenylation signals—the one yielding the longer mRNA is used ~10% of the time in most individuals (*208*).

There appear to be many substrates for the human P-450 3A enzymes, as shown by studies involving chemical inhibitors, immunoinhibition, and correlation (*50*). One of the features of P-450 3A4 is that many catalytic activities are stimulated (in vitro) by modifiers such as 7,8-benzoflavone. However, there are other P-450 enzymes in humans and experimental animals that can also be stimulated by flavones (*50, 92*) and not all of the catalytic activities of P-450 3A4 appear to be equally responsive to this modifier (*112*); it is not yet clear how much various catalytic activities differ in their response.[6] Nevertheless, stimulation of an activity in microsomes is suggestive of a role of P-450 3A enzymes, and the response to 7,8-benzoflavone can quickly provide a clear distinction between P-450 3A and P-450 1A enzymes (vide supra). Another point that should be made is that purified P-450 3A4 can be difficult to demonstrate activities with—this property is characteristic of some rat and rabbit liver P-450 3A enzymes as well (*75, 127, 128*) and can cause problems in interpretation of data regarding catalytic specificity. Even in some cDNA-directed expression systems, the enzyme has shown only low activity due to low rates of P-450 reduction (*107*). Catalytic activity may be improved by using particular mixtures of phospholipids, detergents, and

cytochrome $b_5$ (*129, 211*). In addition, this enzyme is relatively stable to cumene hydroperoxide (*107*), and this active oxygen surrogate can be used in studies regarding catalytic specificity (vide supra), although assays involving measurement of biological end points of genotoxicity are not practical.

P-450 3A4 appears to play a major role in the oxidations of many drugs, including the reactions nifedipine oxidation (*75*), (*R*)-warfarin 9,10-dehydrogenation, (*S*)-warfarin 10-hydroxylation (*107*), quinidine N-oxygenation and 3-hydroxylation (*107, 212*), erythromycin N-demethylation (*71, 107*), the conversion of troleandomycin to an inhibitory product (*213*), lidocaine N-deethylation (*168, 169*), dapsone N-oxygenation (*157*), midazolam 1′- and 4-hydroxylation (*214*), and the oxidation of the hypocholesterolemic agent lovastatin at three different positions (*215*). In addition, at least 19 other dihydropyridines structurally related to nifedipine have been shown to be primarily substrates for P-450 3A4 (*118*). Steroids hydroxylated by the enzyme (positions indicated in bold type) include testosterone (**6β**) (*75, 90, 107*), androstenedione (**6β**) (*90, 216*), progesterone (**6β**) (*90, 216*), cortisol (**6β**) (*107, 170*), dehydroepi-androstene 3-sulfate (**16α**) (*107*), estradiol (**2**) (*75, 107*), 17α-ethynylestradiol (**2**) (*107, 217*), and gestodene (mech-anism-based inactivation) (*152*). The ability of the enzyme to oxidize large substrates is dramatized by the case of the immunosuppressant cyclosporin A, $M_r$ 1201, which is hy-droxylated by P-450 3A4 at three different positions (*90, 218, 219*). The ability to hydroxylate so many large sub-strates might argue for a lack of specificity, but the striking regioselectivity should be emphasized—apparently these are critical features of the protein that influence the se-lectivity in a rather dramatic way, and P-450 3A5, with a primary sequence 88% identical, lacks some of the cata-lytic activities (of P-450 3A4) (*90, 206*). These reactions provide several substrates for noninvasive assays for P-450 3A4, including the oxidation of P-450 or other dihydro-pyridines (*220*), erythromycin N-demethylation (measured by isotopic $CO_2$ formation) (*167*), lidocaine N-deethylation (*168, 169, 221*), dapsone N-oxygenation (*157*), and the 6β-hydroxylation of the endogenous steroid cortisol (*170, 222–224*). These assays all have their advantages and disadvantages. P-450 3A4 is expressed in the intestine (*143*) and possibly other tissues; a potential concern is that extrahepatic P-450 3A4 contribution is reflected in non-invasive measurements made with drugs when they are administered orally. The reliability of the 6β-hydrocortisol measurements can be improved by using chromatography to separate 6α-hydroxycortisol and normalize for cortisol excretion.[7]

In vitro studies suggest that P-450 3A4 plays a major role in the bioactivation of aflatoxins $B_1$ and $G_1$, sterig-matocystin, 6-aminochrysene, tris(2,3-dibromopropyl) phosphate, and a number of polycyclic hydrocarbon di-hydrodiols that can be converted to a "bay-region" diol epoxides (Table I) (*50, 92, 105*). The pyrrolizidine alkaloid senecionine is oxidized to its pyrrolic derivative by this enzyme (*110*). The same enzyme has been implicated in the microsomal detoxication of 1,6-dinitropyrene (*115*). The enzyme has been implicated in two other detoxication reactions, the formation of the *N*-oxide of senecionine (*110*) and aflatoxin 3-hydroxylation (to form aflatoxin $Q_1$) (*112*). These latter two cases present dilemmas because one en-zyme appears to have the major roles in both bioactivation *and* detoxication pathways, and its is not directly clear as to whether a higher level of the enzyme would place a person at greater or less risk.

---

[6] The mechanistic basis of stimulation of certain P-450 activities by 7,8-benzoflavone and other compounds is not totally defined. Evidence has been presented that the affinity of the P-450 for both NADPH–P-450 reductase (*210*) and the substrate itself (*128*) is increased. Other effects may also be operative, but it appears clear that some type of allosteric phenomenon involving a distinct binding site on the P-450 must be invoked.

[7] J. D. Groopman, personal communication.

P-450 3A4 can be induced in cell culture by rifampicin (209, 225). The available literature on the increases in in vitro activities can be interpreted to indicate that P-450 3A4 is induced by rifampicin, dexamethasone, and macrolide antibiotics such as troleandomycin (71, 91, 167). The enzyme is also known to be inhibited by (nitroso) products generated by the oxidation of macrolide antibiotics containing amines, such as troleandomycin (213, 226). The acetylenic progestogen gestodene is a mechanism-based inactivator (152), and both troleandomycin and gestodene exert their effects in vivo (151, 153–155). Undoubtedly, much more remains to be learned about different modulators of P-450 3A4 enzyme levels and activities that are found in the diet. Of interest is the recent finding that in vivo oxidation of the dihydropyridines nifedipine and felodipine can be dramatically inhibited by the consumption of a single serving of grapefruit juice (224, 227). The flavone naringin may be responsible for this effect (113). Its aglycon product naringenin, known to be formed readily (228), inhibits both in vitro nifedipine oxidation and the bioactivation of the P-450 3A4 substrate aflatoxin $B_1$ to a genotoxic product (113).

Some preliminary data are available regarding epidemiology related to P-450 3A4. As mentioned above, the bioactivation of aflatoxin $B_1$ appears to be catalyzed primarily by P-450 3A4 in adult liver (105, 108, 109) and P-450 3A7 in fetal liver (229). Suggestions involving generally prominent roles of P-450 1A2, 2A3, 2B7, and 3A3 (108) have been dismissed (113), and indeed, P-450 1A2 should probably be considered a detoxicating enzyme in this instance (113, 137, 138). There does appear to be recent evidence, however, that P-450 3A4 is also involved in detoxication through formation of aflatoxin $Q_1$, which is not so readily converted to the epoxide (112). It is possible to quantify $N^7$-guanyl derived from aflatoxin–DNA adducts in the urines of people exposed to aflatoxin (230). The ratio of urinary 6β-hydroxycortisol:cortisol was used as a noninvasive index of P-450 3A4 and was found to be correlated to the level of the $N^7$-guanyl aflatoxins excreted,[7] and the results cannot be attributed to aflatoxin intake. There are obviously other contributing factors such a glutathione S-transferase inactivation of aflatoxin 8,9-oxide (140), but these preliminary findings suggest a potentially important relationship.

## P-450 1A1

Historically there has been considerable interest in this enzyme and its relationship to human cancer. Many studies have been done in experimental animals, and the work has been reviewed recently (46). Attempts have been made to link the inducibility of peripheral blood cell P-450 1A1, measured by benzo[a]pyrene (3-) hydroxylation activity, with the incidence of lung cancer in smokers (231, 232). No clear relationship has been discernible (233)—a correlation does appear to exist between the basal level of the enzyme activity (in lymphocytes) and lung cancer risk among smokers (234).

P-450 1A1 is known to be strongly induced by cigarette smoking in extrahepatic tissues such as placenta, lung, and peripheral blood cells and in cultured cells (122, 235). However, the level in liver appears to be extremely low and is not induced in cigarette smokers (122). Although the enzyme has never been purified from human tissue, the cDNA sequence is known and has been expressed in mammalian cells (93, 236). The enzyme catalyzes several oxidations of polycyclic hydrocarbons, forming both epoxides and phenols, and the N-oxygenation of some ary-

lamines (93). The latter reactions thus appear to be catalyzed primarily by P-450 1A2 in human liver and by P-450 1A1 in extrahepatic tissues. A low level of P-450 1A1 in human liver may be enough to account for the formation of benzo[a]pyrene 7,8-oxide there. P-450 3A4 appears to play a more dominant role in the conversion of 7,8-dihydroxy-7,8-dihydrobenzo[a]pyrene to the 9,10-epoxide, however, in human liver and possibly in lung (92). It should be emphasized that in human liver benzo[a]pyrene 3-hydroxylation is catalyzed primarily by other enzymes (92).

A restrictive fragment length polymorphism involving the P-450 1A1 gene has been associated with risk of lung cancer in one study (237) but not another (238). Clearly, the enzyme can activate some potential carcinogens and may be of significance in extrahepatic carcinogenesis. However, further studies will be required. There are no drugs known to be specific markers of this activity in humans.

## Other P-450 Enzymes

The list of human P-450s is extensive, with a total of more than 20 having been identified and characterized to a considerable degree. The question arises as to whether any of these will prove to also have significant roles in the biotransformation of chemical carcinogens. Those P-450s devoted primarily to steroid metabolism probably will not (i.e., P-450 7, P-450 11, P-450 17, P-450 19, P-450 21, P-450 26). However, P-450 19 has a major role in the formation of estrogens, involved in hormonal carcinogenesis, and is a primary target for therapeutic treatment of the breast and certain other tissues (239).

P-450 2A6 has been shown to have some inherent ability to activate aflatoxin $B_1$ (108) although there is no evidence that it plays a major role (113). This enzyme has been suggested to be more important than P-450 2E1 in the activation of N-nitrosodiethylamine (99). Recently P-450 2A6 has been purified from human liver in this laboratory.[8] The purified enzyme has coumarin 7-hydroxylation, 7-ethoxycoumarin O-deethylation, and 4,4′-methylenebis-(2-chloroaniline) (MOCA) N-hydroxylation activities and can activate aflatoxin $B_1$ and 6-aminochrysene at appreciable rates. However, the liver with the highest level of this enzyme was found to contain only ~10 pmol of P-450 2A6/mg of microsomal protein (1% of total P-450), and anti-P-450 2A6 only inhibited (>80%) coumarin 7-hydroxylation and 7-ethoxycoumarin O-deethylation activities of the list presented above; 4,4-methylenebis(2-chloroaniline) N-hydroxylation was only inhibited ~20%, and the activation of neither aflatoxin $B_1$ nor 6-amino-chrysene was blocked to a measurable degree.

In studies in this laboratory inhibitors selective for P-450 2C enzymes (sulfaphenazole and anti-P-450$_{MP}$) and P-450 2D6 (quinidine) were found not to block the activation of any of the carcinogens examined in human liver microsomal systems (50). In addition, other evidence against the activation of any carcinogens by P-450 2D6 has been presented (51, 108, 240, 241). These results are interesting in light of several reports of a tendency for more rapid debrisoquine 4-hydroxylation to be linked to a greater incidence of cancer in the lung (61, 63) and other tissues (62, 242), particularly in cigarette smokers (61). There are, however, some reports of failure to observe such a relationship (243, 244). However, in light of the lack of evidence for the role of this enzyme in carcinogen activation

---

[8] C.-H. Yun, T. Shimada, and F. P. Guengerich, submitted for publication.

(vide supra) and its expression in extrahepatic tissues, it should be remembered that early studies showed a general but not perfect cosegregation of expression of P-450 2D6 activity with that of P-450 1A2 (phenacetin O-deethylation) (*245*). The basis of this relationship is not yet clear, since the location of the P-450 2D6 gene (human chromosome 22) (*201, 246*) differs from that of the P-450 1A2 gene (chromosome 15). In mice the P-450 2D orthologues are located on chromosome 15 (*247*) and the *Ah* receptor has been assigned to chromosome 12 (*248, 249*). However, the possibility should be considered that the relationship seen between cancer and P-450 2D6 activity is actually one with P-450 1A2, an enzyme known to activate arylamines found in tobacco smoke (*50*).[3]

## Some Other Carcinogens for Which a Dominant Role for a Known P-450 Has Not Been Found

It should not be expected that a single P-450 will predominate in the oxidation of every chemical. For instance, 3-methoxy-4-aminoazobenzene appears to be activated to a genotoxic derivative by a number of different P-450s, none of which is dominant (*250*). In our own studies we found that the activation of 6-nitrochrysene and Trp P-1 could not be ascribed to a known human P-450 (*50*). The difficulty may be the result of a series of steps in bioactivation catalyzed primarily by different P-450s. For instance, it would be difficult to ascertain the nature of activation of polycyclic aromatic hydrocarbons such as benzo[*a*]pyrene by focusing on experiments designed to implicate a single enzyme (*92*), and in such cases individual steps must be analyzed. Alternatively, P-450s other than those currently characterized may be involved, and in such cases purification studies are required. To date, efforts to identify the human P-450(s) involved in fluoranthene or 4,4'-methylenebis(2-chloroaniline) oxidation have not been successful but are continuing in this laboratory. Although benzo[*a*]pyrene 3-hydroxylation is the major reaction measured by the so-called "aryl hydrocarbon hydroxylase" fluorescence assay, there is substantial evidence which argues that this reaction is not catalyzed by P-450 1A1 in human liver. Although suggestions have been made that P-450 2C (*117, 251*) and P-450 3A (*93*) enzymes are involved in human liver, several lines of investigation in our own laboratory have not supported these conclusions (*70, 149*). Although P-450 2E1 appears to play a dominant role in the oxidation of *N*-nitrosodimethylamine (*73, 101, 102*) and can contribute to the oxidation of other methyl-substituted nitrosamines (*102*), it does not appear to contribute substantially to the oxidation of larger tobacco-specific and other nitrosamines (*252, 253*). There are, of course, additional classes of carcinogens that have not been studied with regard to which human P-450s are involved in bioactivation and detoxication (*254*).

## Future Prospects

The view that variations in levels of individual human P-450s play a significant role in cancer risk differences is a hypothesis, and considerable speculation can be made regarding the future of research in this area. We will highlight only a few of what we feel are major needs and areas in which future success will be possible.

Our ability to understand the catalysis of oxidation reactions by P-450s is limited by our lack of knowledge of the three-dimensional structures of the proteins. Ultimately, the ability of individual P-450s to catalyze new reactions may be predicted if structural models become available. The structures can be determined by X-ray diffraction methods, although currently, difficulties with intrinsic membrane proteins have prevented the crystallization of any eukaryotic P-450s. Knowledge about the three-dimensional structure of a bacterial P-450 (*255*) and the similarity of regions of primary sequences has been used to suggest structures, but it is not clear that this approach will ever provide the detail necessary for meaningful insight into protein–substrate docking. Affinity labeling of active site residues and site-directed mutagenesis will also not, in themselves, lead to complete pictures of substrate binding sites. One approach that may be of use in predicting the nature of reactions of P-450s with new substrates is that of developing active site diagrams on the basis of computer modeling using the structures of known substrates and inhibitors (*256*). Such an approach is in progress with P-450 2D6, where most of the protein ligands have similar features such as a basic nitrogen (*240*), and has also been used with rat P-450 1A1 and polycyclic aromatic hydrocarbons (*22, 257*). However, such an approach will probably not be useful with P-450s whose substrates have few general chemical features (e.g., P-450 2E1) or have potentially large active sites (e.g., P-450 3A4).

More information needs to be obtained regarding which DNA adducts are most relevant to cancer. Defining details of biotransformation is limited in its usefulness if no information regarding their biological relevance is available. Studies on mechanisms of mutagenesis are required, as well as animal studies and molecular epidemiology in humans. Approaches such as site-specific mutagenesis and physical studies on oligonucleotide interaction offer great potential toward understanding the contributions of individual adducts when viewed in the context of cancer biology.

As pointed out earlier, there are more carcinogens that need to be studied with regard to which of the P-450s have the most prominent roles in bioactivation and detoxication. Several chemicals for consideration have already been mentioned above. In addition, there are a number in which studies with experimental animal models are available and inferences regarding the catalytic specificity of the human P-450s can be suggested. Finally, a need exists to extend work to the conjugation of oxidation products by the so-called "phase II" enzymes. In some cases specific enzymes may have particular roles in detoxication (and bioactivation) which will ultimately influence our understanding of in vivo situations.

Epidemiology needs to be developed to utilize basic information gathered about roles of individual enzymes in order to determine the significance of such studies. The most relevant assays for each P-450 enzyme need to be used, and there is also a great need to develop appropriate noninvasive methods for the estimation of the conjugating enzymes in humans. Adducts that relate to the cancer risk need to be established, along with methods for their measurement. Better ways of utilizing cancer as an end point would be helpful. The matter of tissue specificity of bioactivation, adduct formation, and cancer development needs to be carefully considered.

Finally, as information is gathered regarding the biochemical basis of carcinogen biotransformation and its relevance to cancer risk through studies in epidemiology, rational schemes for cancer prevention and intervention need to be formulated. Basic information regarding the enzymes involved in bioactivation and detoxication should be important in this regard, for the use of appropriate inhibitors and inducers may be possible. It would be of great interest to identify such compounds in foods, for these could be used most safely and would offer few obstacles in their use. Indeed, much is already known regarding the effects of certain vegetables and other foods on cancer, and it may be possible to understand some of these effects in terms of specific alterations of enzymes that generate and detoxicate tumor initiators.

**Acknowledgment.** This research was supported in part by U.S. Public Health Service Grants CA 44353 and ES 00267 and by a Grant-in-Aid for Cancer Research from the Ministry of Education, Science, and Culture of Japan.

**Registry No.** P-450, 9035-51-2.

## References

(1) Miller, J. A. (1970) Carcinogenesis by chemicals: an overview—G. H. A. Clowes memorial lecture. *Cancer Res.* **30**, 559–576.

(2) Mueller, G. C., and Miller, J. A. (1948) The metabolism of 4-dimethylaminoazobenzene by rat liver homogenates. *J. Biol. Chem.* **176**, 535–544.

(3) Miller, E. C. (1951) Studies on the formation of protein-bound derivatives of 3,4-benzpyrene in the epidermal fraction of mouse skin. *Cancer Res.* **11**, 100–108.

(4) Miller, E. C., and Miller, J. A. (1981) Searches for ultimate chemical carcinogens and their reactions with cellular macromolecules. *Cancer* **47**, 2327–2345.

(5) Wood, A. W., Levin, W., Lu, A. Y. H., Yagi, H., Hernandez, O., Jerina, D. M., and Conney, A. H. (1976) Metabolism of benzo[a]pyrene and benzo[a]pyrene derivatives to mutagenic products by highly purified hepatic microsomal enzymes. *J. Biol. Chem.* **251**, 4882–4890.

(6) Oesch, F. (1979) Epoxide hydratase. In *Progress in Drug Metabolism* (Bridges, J. W., and Chasseaud, L. F., Eds.) pp 253–301, John Wiley, Chichester, England.

(7) Peterson, L. A., and Guengerich, F. P. (1988) Comparison of and relationships between glutathione S-transferase and cytochrome P-450 systems. In *Glutathione Conjugation: Its Mechanisms and Biological Significance* (Sies, H., and Ketterer, B., Eds.) pp 193–233, Academic Press, London.

(8) Anders, M. W., Lash, L., Dekant, W., Elfarra, A. A., and Dohn, D. R. (1988) Biosynthesis and biotransformation of glutathione S-conjugates to toxic metabolites. *CRC Crit. Rev. Toxicol.* **18**, 311–341.

(9) Green, M. D., and Tephly, T. R. (1987) N-Glucuronidation of carcinogenic aromatic amines catalyzed by rat hepatic microsomal preparations and purified rat liver uridine 5′-diphosphate-glucuronosyltransferases. *Cancer Res.* **47**, 2028–2031.

(10) Tephley, T. R. (1990) Isolation and purification of UDP-glucuronosyltransferases. *Chem. Res. Toxicol.* **3**, 509–516.

(11) Tateishi, M., Suzuki, S., and Shimizu, H. (1978) Cysteine conjugate β-lyase in rat liver: a novel enzyme catalyzing formation of thiol-containing metabolites of drugs. *J. Biol. Chem.* **253**, 8854–8859.

(12) Boogaard, P. J., Commandeur, J. N. M., Mulder, G. J., Vermeulen, N. P. E., and Nagelkerke, J. F. (1989) Toxicity of the cysteine-S-conjugates and mercapturic acids of four structurally related difluoroethylenes in isolated proximal tubular cells from rat kidney: uptake of the conjugates and activation to toxic metabolites. *Biochem. Pharmacol.* **38**, 3731–3741.

(13) Ziegler, D. M., Ansher, S. S., Nagata, T., Kadlubar, F. F., and Jakoby, W. B. (1988) N-Methylation: potential mechanism for metabolic activation of carcinogenic primary arylamines. *Proc. Natl. Acad. Sci. U.S.A.* **85**, 2514–2517.

(14) Jakoby, W. B., Ed. (1980) *Enzymatic Basis of Detoxication*, Vols. 1 and 2, Academic Press, New York.

(15) Ohno, Y., Ormstad, K., Ross, D., and Orrenius, S. (1985) Mechanism of allyl alcohol toxicity and protective effects of low-molecular-weight thiols studied with isolated rat hepatocytes. *Toxicol. Appl. Pharmacol.* **78**, 169–179.

(16) Chiba, K., Peterson, L. A., Castagnoli, K. P., Trevor, A. J., and Castagnoli, N., Jr. (1985) Studies on the molecular mechanism of bioactivation of the selective nitrostriatal toxin 1-methyl-4-phenyl-1,2,3,6-tetrahydropyridine. *Drug Metab. Dispos.* **13**, 342–347.

(17) Ziegler, D. M. (1988) Flavin-containing monooxygenases: catalytic mechanism and substrate specificities. *Drug Metab. Rev.* **19**, 1–32.

(18) Reed, G. A., and Marnett, L. J. (1982) Metabolism and activation of 7,8-dihydrobenzo[a]pyrene during prostaglandin bio-synthesis: intermediacy of a bay-region epoxide. *J. Biol. Chem.* **257**, 11368–11376.

(19) Kadlubar, F. F., Miller, J. A., and Miller, E. C. (1976) Microsomal N-oxidation of the hepatocarcinogen N-methyl-4-aminoazobenzene and the reactivity of N-hydroxy-N-methyl-4-aminoazobenzene. *Cancer Res.* **36**, 1196–1206.

(20) Shen, J. H., Wegenke, M., and Wolff, T. (1990) Capability of human blood cells to form the DNA adduct, C8-($N^2$-aminofluorenyl)-deoxyguanosine-3′,5′-diphosphate from 2-aminofluorene. *Carcinogenesis* **11**, 1441–1444.

(21) Guengerich, F. P., and Liebler, D. C. (1985) Enzymatic activation of chemicals to toxic metabolites. *CRC Crit. Rev. Toxicol.* **14**, 259–307.

(22) Kadlubar, F. F., and Hammons, G. J. (1987) The role of cytochrome P-450 in the metabolism of chemical carcinogens. In *Mammalian Cytochromes P-450* (Guengerich, F. P., Ed.) Vol. 2, pp 81–130, CRC Press, Boca Raton, FL.

(23) Guengerich, F. P. (1990) Enzymatic oxidation of xenobiotic chemicals. *Crit. Rev. Biochem. Mol. Biol.* **25**, 97–153.

(24) Guengerich, F. P., and Macdonald, T. L. (1990) Mechanisms of cytochrome P-450 catalysis. *FASEB J.* **4**, 2453–2459.

(25) Guengerich, F. P., and Kim, D.-H. (1991) Enzymatic oxidation of ethyl carbamate to vinyl carbamate and its role as an intermediate in the formation of 1,$N^6$-ethenoadenosine. *Chem. Res. Toxicol.* (in press).

(26) Mico, B. A., and Pohl, L. R. (1983) Reductive oxygenation of carbon tetrachloride: trichloromethylperoxyl radical as a possible intermediate in the conversion of carbon tetrachloride to electrophilic chlorine. *Arch. Biochem. Biophys.* **225**, 596–609.

(27) Busby, W. F., Jr., Wang, J. S., Stevens, E. K., Padykula, R. E., Aleksejczyk, R. A., and Berchtold, G. A. (1990) Lung tumorigenicity of benzene oxide, benzene dihydrodiols and benzene diolepoxides in the BLU:Ha newborn mouse assay. *Carcinogenesis* **11**, 1473–1478.

(28) Witz, G., Latriano, L., and Goldstein, B. D. (1989) Metabolism and toxicity of trans,trans-muconaldehyde, an open-ring microsomal metabolite of benzene. *Environ. Health Perspect.* **82**, 19–22.

(29) Ortiz de Montellano, P. R., Ed. (1986) *Cytochrome P-450*, Plenum Press, New York.

(30) Nebert, D. W., Nelson, D. R., Adesnik, M., Coon, M. J., Estabrook, R. W., Gonzalez, F. J., Guengerich, F. P., Gunsalus, I. C., Johnson, E. F., Kemper, B., Levin, W., Phillips, I., Sato, R., and Waterman, M. (1989) The P450 superfamily: updated listing of all genes and recommended nomenclature for the chromosomal loci. *DNA* **8**, 1–13.

(31) Gonzalez, F. J. (1990) Molecular genetics of the P-450 superfamily. *Pharmacol. Ther.* **45**, 1–38.

(32) Parandoosh, Z., Fujita, V. S., Coon, M. J., and Philpot, R. M. (1987) Cytochrome P-450 isozymes 2 and 5 in rabbit lung and liver: comparisons of structure and inducibility. *Drug Metab. Dispos.* **15**, 59–67.

(33) Vanderslice, R. R., Domin, B. A., Carver, G. T., and Philpot, R. M. (1987) Species-dependent expression and induction of homologues of rabbit cytochrome P-450 isozyme 5 in liver and lung. *Mol. Pharmacol.* **31**, 320–325.

(34) Masters, B. S. S., Muerhoff, A. S., and Okita, R. T. (1987) Enzymology of extrahepatic cytochromes P-450. In *Mammalian Cytochromes P-450* (Guengerich, F. P., Ed.) pp 107–131, CRC Press, Boca Raton, FL.

(35) Johnson, E. F., Walker, D. L., Griffin, K. J., Clark, J. E., Okita, R. T., Muerhoff, A. S., and Masters, B. S. (1990) Cloning and expression of three rabbit kidney cDNAs encoding lauric acid ω-hydroxylases. *Biochemistry* **29**, 873–879.

(36) Fujisawa-Sehara, A., Yamane, M., and Fujii-Kuriyama, Y. (1988) A DNA-binding factor specific for xenobiotic responsive elements of P-450c gene exists as a cryptic form in cytoplasm: its possible translocation to nucleus. *Proc. Natl. Acad. Sci. U.S.A.* **85**, 5859–5863.

(37) Whitlock, J. P., Jr. (1990) Genetic and molecular aspects of 2,3,7,8-tetrachlorodibenzo-p-dioxin action. *Annu. Rev. Pharmacol. Toxicol.* **30**, 251–277.

(38) Fisher, J. M., Wu, L., Denison, M. S., and Whitlock, J. P., Jr. (1990) Organization and function of a dioxin-responsive enhancer. *J. Biol. Chem.* **265**, 9676–9681.

(39) Waterman, M. R., Mason, J. I., Zuber, M. X., John, M. E., Rodgers, R. J., and Simpson, E. R. (1986) Control of gene expression of adrenal steroid hydroxylases and related enzymes. *Endocr. Res.* **12**, 393–408.

(40) Niranjan, B. G., Avadhani, N. G., and DiGiovanni, J. (1985) Formation of benzo(*a*)pyrene metabolites and DNA adducts catalyzed by a rat liver mitochondrial monooxygenase system. *Biochem. Biophys. Res. Commun.* **131**, 935–942.

(41) Niranjan, B. G., Wilson, N. M., Jefcoate, C. R., and Avadhani, N. G. (1984) Hepatic mitochondrial cytochrome P-450 system: distinctive features of cytochrome P-450 involved in the activation of aflatoxin $B_1$ and benzo(a)pyrene. *J. Biol. Chem.* **259**, 12495–12501.

(42) Shayiq, R. M., and Avadhani, N. G. (1989) Purification and characterization of a hepatic mitochondrial cytochrome P-450 active in aflatoxin $B_1$ metabolism. *Biochemistry* **28**, 7546–7554.

(43) Guengerich, F. P. (1988) Roles of cytochrome P-450 enzymes in chemical carcinogenesis and cancer chemotherapy. *Cancer Res.* **48**, 2946–2954.

(44) Guengerich, F. P. (1990) Characterization of roles of human cytochrome P-450 enzymes in carcinogen metabolism. *Asia Pacific J. Pharmacol.* **5**, 327–345.

(45) Nebert, D. W. (1978) Genetic control of carcinogen metabolism leading to individual differences in cancer risk. *Biochimie* **60**, 1019–1029.

(46) Nebert, D. W. (1989) The *Ah* locus: genetic differences in toxicity, cancer, mutation, and birth defects. *Chem. Res. Toxicol.* **2**, 153–174.

(47) Nebert, D. W. (1981) Genetic differences in susceptibility to chemically induced myelotoxicity and leukemia. *Environ. Health Perspect.* **39**, 11–22.

(48) Levitt, R. C., Fysh, J. M., Jensen, N. M., and Nebert, D. W. (1979) The Ah locus: biochemical basis for genetic differences in brain tumor formation in mice. *Genetics* **92**, 1205–1210.

(49) Bradley, M. O., Bhuyan, B., Francis, M. C., Langenbach, R., Peterson, A., and Huberman, E. (1981) Mutagenesis by chemical agents in V79 Chinese hamster cells: a review and analysis of the literature. *Mutat. Res.* **87**, 81–142.

(50) Shimada, T., Iwasaki, M., Martin, M. V., and Guengerich, F. P. (1989) Human liver microsomal cytochrome P-450 enzymes involved in the bioactivation of procarcinogens detected by *umu* gene response in *Salmonella typhimurium* TA1535/pSK1002. *Cancer Res.* **49**, 3218–3228.

(51) Aoyama, T., Gelboin, H. V., and Gonzalez, F. J. (1990) Mutagenic activation of 2-amino-3-methylimidazo[4,5-*f*]quinoline by complementary DNA-expressed human liver P-450. *Cancer Res.* **50**, 2060–2063.

(52) Distlerath, L. M., and Guengerich, F. P. (1987) Enzymology of human liver cytochromes P-450. In *Mammalian Cytochromes P-450* (Guengerich, F. P., Ed.) Vol. 1, pp 133–198, CRC Press, Boca Raton, FL.

(53) Küpfer, A., and Preisig, R. (1983) Inherited defects of hepatic drug metabolism. *Semin. Liver Dis.* **3**, 341–354.

(54) Eichelbaum, M. (1984) Polymorphic drug oxidation in humans. *Fed. Proc.* **43**, 2298–2302.

(55) Smith, R. L. (1988) The role of metabolism and disposition studies in the safety assessment of pharmaceuticals. *Xenobiotica* **18**, 89–96.

(56) Idle, J. R., and Smith, R. L. (1979) Polymorphisms of oxidation at carbon centers of drugs and their clinical significance. *Drug Metab. Rev.* **9**, 301–317.

(57) Guengerich, F. P., and Turvy, C. G. (1991) Comparison of levels of several human microsomal cytochrome P-450 enzymes and epoxide hydrolase in normal and disease states using immunochemical analysis of surgical liver samples. *J. Pharmacol. Exp. Ther.* **256**, 1189–1194.

(58) Guengerich, F. P. (1990) Metabolism of 17α-ethynylestradiol in humans. *Life Sci.* **47**, 1981–1988.

(59) Peterson, L. A., Carmella, S. G., and Hecht, S. S. (1990) Investigations of metabolic precursors to hemoglobin and DNA adducts of 4-(methylnitrosamino)-1-(3-pyridyl)-1-butanone. *Carcinogenesis* **11**, 1329–1333.

(60) Lennard, M. S., Ramsay, L. E., Silas, J. H., Tucker, G. T., and Woods, H. F. (1983) Protecting the poor metabolizer: clinical consequences of genetic polymorphism of drug oxidation. *Pharm. Int.* **4**, 61–65.

(61) Ayesh, R., Idle, J. R., Ritchie, J. C., Crothers, M. J., and Hetzel, M. R. (1984) Metabolic oxidation phenotypes as markers for susceptibility to lung cancer. *Nature* **312**, 169–170.

(62) Kaisary, A., Smith, P., Jaczq, E., McAllister, C. B., Wilkinson, G. R., Ray, W. A., and Branch, R. A. (1987) Genetic predisposition to bladder cancer: ability to hydroxylate debrisoquine and mephenytoin as risk factors. *Cancer Res.* **47**, 5488–5493.

(63) Caporaso, N., Hayes, R. B., Dosemeci, M., Hoover, R., Ayesh, R., Hetzel, M., and Idle, J. (1989) Lung cancer risk, occupational exposure, and the debrisoquine metabolic phenotype. *Cancer Res.* **49**, 3675–3679.

(64) Kaschnitz, R. M., and Coon, M. J. (1975) Drug and fatty acid hydroxylation by solubilized human liver microsomal cytochrome P-450 phospholipid requirement. *Biochem. Pharmacol.* **24**, 295–297.

(65) Beaune, P., Dansette, P., Flinois, J. P., Columelli, S., Mansuy, D., and Leroux, J. P. (1979) Partial purification of human liver cytochrome P-450. *Biochem. Biophys. Res. Commun.* **88**, 826–832.

(66) Kitada, M., and Kamataki, T. (1979) Partial purification and properties of cytochrome P450 from homogenates of human fetal livers. *Biochem. Pharmacol.* **28**, 793–797.

(67) Wang, P., Mason, P. S., and Guengerich, F. P. (1980) Purification of human liver cytochrome P-450 and comparison to the enzyme isolated from rat liver. *Arch. Biochem. Biophys.* **199**, 206–219.

(68) Guengerich, F. P., Wang, P., Mason, P. S., and Mitchell, M. B. (1981) Immunological comparison of rat, rabbit, and human microsomal cytochromes P-450. *Biochemistry* **20**, 2370–2378.

(69) Wang, P. P., Beaune, P., Kaminsky, L. S., Dannan, G. A., Kadlubar, F. F., Larrey, D., and Guengerich, F. P. (1983) Purification and characterization of six cytochrome P-450 isozymes from human liver microsomes. *Biochemistry* **22**, 5375–5383.

(70) Shimada, T., Misono, K. S., and Guengerich, F. P. (1986) Human liver microsomal cytochrome P-450 mephenytoin 4-hydroxylase, a prototype of genetic polymorphism in oxidative drug metabolism. Purification and characterization of two similar forms involved in the reaction. *J. Biol. Chem.* **261**, 909–921.

(71) Watkins, P. B., Wrighton, S. A., Maurel, P., Schuetz, E. G., Mendez-Picon, G., Parker, G. A., and Guzelian, P. S. (1985) Identification of an inducible form of cytochrome P-450 in human liver. *Proc. Natl. Acad. Sci. U.S.A.* **82**, 6310–6314.

(72) Wrighton, S. A., Thomas, P. E., Willis, P., Maines, S. L., Watkins, P. B., Levin, W., and Guzelian, P. S. (1987) Purification of a human liver cytochrome P-450 immunochemically related to several cytochromes P-450 purified from untreated rats. *J. Clin. Invest.* **80**, 1017–1022.

(73) Wrighton, S. A., Thomas, P. E., Ryan, D. E., and Levin, W. (1987) Purification and characterization of ethanol-inducible human hepatic cytochrome P-450HLj. *Arch. Biochem. Biophys.* **258**, 292–297.

(74) Distlerath, L. M., Reilly, P. E. B., Martin, M. V., Davis, G. G., Wilkinson, G. R., and Guengerich, F. P. (1985) Purification and characterization of the human liver cytochromes P-450 involved in debrisoquine 4-hydroxylation and phenacetin O-deethylation, two prototypes for genetic polymorphism in oxidative drug metabolism. *J. Biol. Chem.* **260**, 9057–9067.

(75) Guengerich, F. P., Martin, M. V., Beaune, P. H., Kremers, P., Wolff, T., and Waxman, D. J. (1986) Characterization of rat and human liver microsomal cytochrome P-450 forms involved in nifedipine oxidation, a prototype for genetic polymorphism in oxidative drug metabolism. *J. Biol. Chem.* **261**, 5051–5060.

(76) Ged, C., Umbenhauer, D. R., Bellew, T. M., Bork, R. W., Srivastava, P. K., Shinriki, N., Lloyd, R. S., and Guengerich, F. P. (1988) Characterization of cDNAs, mRNAs, and proteins related to human liver microsomal cytochrome P-450(*S*)-mephenytoin 4'-hydroxylase. *Biochemistry* **27**, 6929–6940.

(77) Srivastava, P. K., Yun, C.-H., Beaune, P. H., Ged, C., and Guengerich, F. P. (1991) Separation of human liver tolbutamine hydroxylase and (S)-mephenytoin 4'-hydroxylase cytochrome P-450 enzymes. *Mol. Pharmacol.* (in press).

(78) Gut, J., Meier, U. T., Catin, T., and Meyer, U. A. (1986) Mephenytoin-type polymorphism drug oxidation: purification and characterization of a human liver cytochrome P-450 isozyme catalyzing microsomal mephenytoin hydroxylation. *Biochim. Biophys. Acta* **884**, 435–447.

(79) Gut, J., Catin, T., Dayer, P., Kronbach, T., Zanger, U., and Meyer, U. A. (1986) Debrisoquine/sparteine-type polymorphism of drug oxidation: purification and characterization of two functionally different human liver cytochrome P-450 isozymes involved in impaired hydroxylation of the prototype substrate bufuralol. *J. Biol. Chem.* **261**, 11734–11743.

(80) Butler, M. A., Iwasaki, M., Guengerich, F. P., and Kadlubar, F. F. (1989) Human cytochrome P-450PA (P-450IA2), the phenacetin O-deethylase, is primarily responsible for the hepatic 3-demethylation of caffeine and N-oxidation of carcinogenic arylamines. *Proc. Natl. Acad. Sci. U.S.A.* **86**, 7696–7700.

(81) Wrighton, S. A., Ring, B. J., Watkins, P. B., and Vandenbranden, M. (1989) Identification of a polymorphically expressed member of the human cytochrome P-450III family. *Mol. Pharmacol.* **86**, 97–105.

(82) Beaune, P. H., Umbenhauer, D. R., Bork, R. W., Lloyd, R. S., and Guengerich, F. P. (1986) Isolation and sequence determination of a cDNA clone related to human cytochrome P-450 nifedipine oxidase. *Proc. Natl. Acad. Sci. U.S.A.* **83**, 8064–8068.

(83) Umbenhauer, D. R., Martin, M. V., Lloyd, R. S., and Guengerich, F. P. (1987) Cloning and sequence determination of a complementary DNA related to human liver microsomal cytochrome P-450 S-mephenytoin 4-hydroxylase. *Biochemistry* **26**, 1094–1099.

(84) Kimura, S., Pastewka, J., Gelboin, H. V., and Gonzalez, F. J. (1987) cDNA and amino acid sequences of two members of the human P450IIC gene subfamily. *Nucleic Acids Res.* **15**, 10053.

(85) Yamano, S., Nagata, K., Yamazoe, Y., Kato, R., Gelboin, H. V., and Gonzalez, F. J. (1989) cDNA and deduced amino acid sequences of human P450 IIA3 (CYP2A3). *Nucleic Acids Res.* **17**, 4888.

(86) Nhamburo, P. T., Gonzalez, F. J., McBride, O. W., Gelboin, H. V., and Kimura, S. (1989) Identification of a new P450 expressed in human lung: complete cDNA sequence, cDNA-directed expression, and chromosome mapping. *Biochemistry* **28**, 8060–8066.

(87) Reeck, G. R., de Haën, C., Teller, D. C., Doolittle, R. F., Fitch, W. M., Dickerson, R. E., Chambon, P., McLachlan, A. D., Margoliash, E., Jukes, T. H., and Zuckerkandl, E. (1987) "Homology" in proteins and nucleic acids: a terminology muddle and a way out of it. *Cell* **50**, 667.

(88) Lewin, R. (1987) When does homologoy means something else? *Science* **237**, 1570.

(89) Petrella, R. J., and Yokoyama, M. M. (1990) Ubihomologous homology usage. *Nature* **343**, 518.

(90) Aoyama, T., Yamano, S., Waxman, D. J., Lapenson, D. P., Meyer, U. A., Fischer, V., Tyndale, R., Inaba, T., Kalow, W., Gelboin, H. V., and Gonzalez, F. J. (1989) Cytochrome P-450 hPCN3, a novel cytochrome P-450 IIIA gene product that is differentially expressed in adult human liver. *J. Biol. Chem.* **264**, 10388–10395.

(91) Guengerich, F. P. (1989) Characterization of human microsomal cytochrome P-450 enzymes. *Annu. Rev. Pharmacol. Toxicol.* **29**, 241–264.

(92) Shimada, T., Martin, M. V., Pruess-Schwartz, D., Marnett, L. J., and Guengerich, F. P. (1989) Roles of individual human cytochrome P-450 enzymes in the bioactivation of benzo(a)pyrene, 7,8-dihydroxy-7,8-dihydrobenzo(a)pyrene, and other dihydrodiol derivatives of polycyclic aromatic hydrocarbons. *Cancer Res.* **49**, 6304–6312.

(93) McManus, M. E., Burgess, W. M., Veronese, M. E., Huggett, A., Quattrochi, L. C., and Tukey, R. H. (1990) Metabolism of 2-acetylaminofluorene and benzo(a)pyrene and activation of food-derived heterocyclic amine mutagens by human cytochromes P-450. *Cancer Res.* **50**, 3367–3376.

(94) Gonzalez, F. J., Aoyama, T., and Gelboin, H. V. (1990) Activation of promutagens by human cDNA-expressed cytochrome P450s. In *Mutation and the Environment, Part B: Metabolism, Testing Methods, and Chromosomes* (Mendelshon, M. L., and Albertini, R. J., Eds.) pp 77–86, Wiley-Liss, New York.

(95) Ohta, K., Kitada, M., Ohi, H., Komori, M., Nagashima, K., Sato, N., Muroya, K., Kodama, T., Nagao, M., and Kamataki, T. (1989) Interspecies homology of cytochrome P-450: toxicological significance of cytochrome P-450 cross-reactive with anti-rat P-448-H antibodies in liver microsomes from dogs, monkeys and humans. *Mutat. Res.* **226**, 163–167.

(96) Kitada, M., Taneda, M., Ohta, K., Nagashima, K., Itahashi, K., and Kamataki, T. (1990) Metabolic activation of aflatoxin B₁ and 2-amino-3-methylimidazo[4,5-*f*]-quinoline by human adult and fetal livers. *Cancer Res.* **50**, 2641–2645.

(97) Yamazoe, Y., Abu-Zeid, M., Yamauchi, K., and Kato, R. (1988) Metabolic activation of pyrolysate arylamines by human liver microsomes: possible involvement of P-448-H type cytochrome P-450. *Jpn. J. Cancer Res.* **79**, 1159–1167.

(98) Raucy, J. L., Lasker, J. M., Lieber, C. S., and Black, M. (1989) Acetaminophen activation by human liver cytochrome P450IIE1 and P450IA2. *Arch. Biochem. Biophys.* **271**, 270–283.

(99) Crespi, C. L., Penman, B. W., Leakey, J. A., Arlotto, M. P., Stark, A., Parkinson, A., Turner, T., Steimel, D. T., Rudo, K., Davies, R. L., and Langenbach, R. (1990) Human cytochrome P450IIA3:cDNA sequence, role of the enzyme in the metabolic activation of promutagens, comparison to nitrosamine activation by human cytochrome P450IIE1. *Carcinogenesis* **11**, 1293–1300.

(100) Guengerich, F. P., Kim, D.-H., and Iwasaki, M. (1991) Role of human cytochrome P-450 IIE1 in the oxidation of several low molecular weight cancer suspects. *Chem. Res. Toxicol.* **4**, 168–179.

(101) Wrighton, S. A., Thomas, P. E., Molowa, D. T., Haniu, M., Shively, J. E., Maines, S. L., Watkins, P. B., Parker, G., Mendez-Picon, G., Levin, W., and Guzelian, P. S. (1986) Characterization of ethanol-inducible human liver N-nitrosodimethylamine demethylase. *Biochemistry* **25**, 6731–6735.

(102) Yoo, J. S. H., Guengerich, F. P., and Yang, C. S. (1988) Metabolism of N-nitrosodialkylamines by human liver microsomes. *Cancer Res.* **48**, 1499–1504.

(103) Ekström, G., von Bahr, C., and Ingelman-Sundberg, M. (1989) Human liver microsomal cytochrome P-450IIIE1. Immunological evaluation of its contribution to microsomal ethanol oxidation, carbon tetrachloride reduction and NADPH oxidase activity. *Biochem. Pharmacol.* **38**, 689–693.

(104) Reitz, R. H., Mendrala, A., and Guengerich, F. P. (1989) In vitro metabolism of methylene chloride in human and animal tissues: use in physiologically-based pharmacokinetic models. *Toxicol. Appl. Pharmacol.* **97**, 230–246.

(105) Shimada, T., and Guengerich, F. P. (1989) Evidence for cytochrome P-450$_{NF}$, the nifedipine oxidase, being the principal enzyme involved in the bioactivation of aflatoxins in human liver. *Proc. Natl. Acad. Sci. U.S.A.* **86**, 462–465.

(106) Baertschi, S. W., Raney, K. D., Shimada, T., Harris, T. M., and Guengerich, F. P. (1989) Comparison of rates of enzymatic oxidation of aflatoxin B₁, aflatoxin G₁, and sterigmatocystin and activities of the epoxides in forming guanyl-$N^7$ adducts and inducing different genetic responses. *Chem. Res. Toxicol.* **2**, 114–122.

(107) Brian, W. R., Sari, M.-A., Iwasaki, M., Shimada, T., Kaminsky, L. S., and Guengerich, F. P. (1990) Catalytic activities of human liver cytochrome P-450 IIIA4 expressed in *Saccharomyces cerevisiae*. *Biochemistry* **29**, 11280–11292.

(108) Aoyama, T., Yamano, S., Guzelian, P. S., Gelboin, H. V., and Gonzalez, F. J. (1990) 5 of 12 forms of vaccinia virus-expressed human hepatic cytochrome-P450 metabolically activate aflatoxin B₁. *Proc. Natl. Acad. Sci. U.S.A.* **87**, 4790–4793.

(109) Forrester, L. M., Neal, G. E., Judah, D. J., Glancey, M. J., and Wolf, C. R. (1990) Evidence for involvement of multiple forms of cytochrome P-450 in aflatoxin B₁ metabolism in human liver. *Proc. Natl. Acad. Sci. U.S.A.* **87**, 8306–8310.

(110) Miranda, C. L., Reed, R. L., Guengerich, F. P., and Buhler, D. R. (1991) Role of cytochrome P450IIIA4 in the metabolism of the pyrrolizidine alkaloid senecionine in human liver. *Carcinogenesis* **12**, 515–519.

(111) Howard, P. C., Aoyama, T., Bauer, S. L., Gelboin, H. V., and Gonzalez, F. J. (1990) The metabolism of 1-nitropyrene by human cytochromes P450. *Carcinogenesis* **11**, 1539–1542.

(112) Raney, K. D., Shimada, T., Kim, D.-H., Groopman, J. D., Harris, T. M., and Guengerich, F. P. (1991) Oxidation of aflatoxin B₁ and related dihydrofurans by human liver microsomes: role of aflatoxin Q₁ as a detoxication product. *Chem. Res. Toxicol.* (in press).

(113) Guengerich, F. P., and Kim, D.-H. (1990) *In vitro* inhibition of dihydropyridine oxidation and aflatoxin B₁ activation in human liver microsomes by naringenin and other flavonoids. *Carcinogenesis* **11**, 2275–2279.

(114) Shimada, T., Yamazaki, H., Shimura, H., Tanaka, R., and Guengerich, F. P. (1990) Metabolic deactivation of furylfuramide by cytochrome P450 in human and rat liver microsomes. *Carcinogenesis* **11**, 103–110.

(115) Shimada, T., and Guengerich, F. P. (1990) Inactivation of 1,3-, 1,6-, and 1,8-dinitropyrene by human and rat microsomes. *Cancer Res.* **50**, 2036–2043.

(116) Crespi, C. L., Steimel, D. T., Aoyama, T., Gelboin, H. V., and Gonzalez, F. J. (1990) Stable expression of human cytochrome P450IA2 cDNA in a human lymphoblastoid cell line: role of the enzyme in the metabolic activation of aflatoxin B₁. *Mol. Carcinog.* **3**, 5–8.

(117) Beaune, P., Kremers, P. G., Kaminsky, L. S., de Graeve, J., and Guengerich, F. P. (1986) Comparison of monooxygenase activities and cytochrome P-450 isozyme concentrations in human liver microsomes. *Drug Metab. Dispos.* **14**, 437–442.

(118) Guengerich, F. P., Brian, W. R., Iwasaki, M., Sari, M.-A., Bäärnhielm, C., and Berntsson, P. (1991) Oxidation of dihydropyridine calcium channel blockers and analogs by human liver cytochrome P-450 IIA4. *J. Med. Chem.* (in press).

(119) Wold, S., Albano, C., Dunn, W. J., III, Edlund, U., Esbensen, K., Geladi, P., Hellberg, S., Johansson, E., Lindberg, W., and

Sjöström, M. (1984) Multivariate data analysis in chemistry. In *Chemometrics, Mathematics and Statistics in Chemistry* (Kowalski, B. R., Ed.) pp 17–95, D. Reidel Publishing Co., Dordrecht, Holland.

(120) Thomas, P. E., Lu, A. Y. H., West, S. B., Ryan, D., Miwa, G. T., and Levin, W. (1977) Accessibility of cytochrome P450 in microsomal membranes: inhibition of metabolism by antibodies to cytochrome P450. *Mol. Pharmacol.* **13**, 819–831.

(121) Park, S. S., Fujino, T., West, D., Guengerich, F. P., and Gelboin, H. V. (1982) Monoclonal antibodies that inhibit enzyme activity of 3-methylcholanthrene-induced cytochrome P-450. *Cancer Res.* **42**, 1798–1808.

(122) Fujino, T., Park, S. S., West, D., and Gelboin, H. V. (1982) Phenotyping of cytochromes P-450 in human tissues with monoclonal antibodies. *Proc. Natl. Acad. Sci. U.S.A.* **79**, 3682–3686.

(123) Park, S. S., Waxman, D. J., Miller, H., Robinson, R., Attisano, C., Guengerich, F. P., and Gelboin, H. V. (1986) Preparation and characterization of monoclonal antibodies to pregnenolone-16α-carbonitrile inducible rat liver cytochrome P-450. *Biochem. Pharmacol.* **35**, 2859–2867.

(124) Ogishima, T., Deguchi, S., and Okuda, K. (1987) Purification and characterization of cholesterol 7α-hydroxylase from rat liver microsomes. *J. Biol. Chem.* **262**, 7646–7650.

(125) Zanger, U. M., Vibois, F., Hardwick, J. P., and Meyer, U. A. (1988) Absence of hepatic cytochrome P450bufI causes genetically deficient debrisoquine oxidation in man. *Biochemistry* **27**, 5447–5454.

(126) Pompon, D., and Nicolas, A. (1989) Protein engineering by cDNA recombination in yeasts: shuffling of mammalian cytochrome P-450 functions. *Gene* **83**, 15–24.

(127) Elshourbagy, N. A., and Guzelian, P. S. (1980) Separation, purification, and characterization of a novel form of hepatic cytochrome P-450 from rats treated with pregnenolone-16α-carbonitrile. *J. Biol. Chem.* **255**, 1279–1285.

(128) Schwab, G. E., Raucy, J. L., and Johnson, E. F. (1988) Modulation of rabbit and human hepatic cytochrome P-450-catalyzed steroid hydroxylations by α-naphthoflavone. *Mol. Pharmacol.* **33**, 493–499.

(129) Halvorson, M., Greenway, D., Eberhart, D., Fitzgerald, K., and Parkinson, A. (1990) Reconstitution of testosterone oxidation by purified rat cytochrome P450$_p$ (IIIA1). *Arch. Biochem. Biophys.* **277**, 166–180.

(130) Kaminsky, L. S., Guengerich, F. P., Dannan, G. A., and Aust, S. D. (1983) Comparisons of warfarin metabolism by liver microsomes of rats treated with a series of polybrominated biphenyl congeners and by the component-purified cytochrome P-450 isozymes. *Arch. Biochem. Biophys.* **225**, 398–404.

(131) Porter, T. D., Pernecky, S. J., Larson, J. R., Fujita, V. S., and Coon, M. J. (1990) Expression of cytochrome P-450 in yeast and *Escherichia coli*. In *Drug Metabolizing Enzymes: Genetics, Regulation and Toxicology, Proceedings of the Eighth International Symposium on Microsomes and Drug Oxidations* (Ingelman-Sundberg, M., Gustafsson, J.-Å., and Orrenius, S., Eds.) p 20, Karolinska Institute, Stockholm.

(132) Guengerich, F. P., Brian, W. R., Sari, M.-A., and Ross, J. T. (1991) Expression of mammalian cytochrome P-450 enzymes using yeast-based vectors. *Methods Enzymol.* (in press).

(133) Zuber, M. X., Simpson, E. R., and Waterman, M. R. (1986) Expression of bovine 17α-hydroxylase cytochrome P-450 cDNA in nonsteroidogenic (COS 1) cells. *Science* **234**, 1258–1261.

(134) Battula, N., Sagara, J., and Gelboin, H. V. (1987) Expression of P$_1$-450 and P$_3$-450 DNA coding sequences as enzymatically active cytochromes P-450 in mammalian cells. *Proc. Natl. Acad. Sci. U.S.A.* **84**, 4073–4077.

(135) Crespi, C. L., Langenbach, R., Rudo, K., Chen, Y. T., and Davies, R. L. (1989) Transfection of a human cytochrome P-450 gene into the human lymphoblastoid cell line, AHH-1, and use of the recombinant cell line in gene mutation assays. *Carcinogenesis* **10**, 295–301.

(136) Crespi, C. L., Langenbach, R., and Penman, B. W. (1990) The development of a panel of human cell lines expressing specific human cytochrome P450 cDNAs. In *Mutation and the Environment, Part B: Metabolism, Testing Methods, and Chromosomes* (Mendelsohn, M. L., and Albertini, R. J., Eds.) pp 97–106, Wiley-Liss, New York.

(137) Lai, T. S., and Chiang, J. Y. L. (1990) Aflatoxin B$_1$ metabolism by 3-methylcholanthrene-induced hamster hepatic cytochrome P-450s. *J. Biochem. Toxicol.* **5**, 147–153.

(138) Faletto, M. B., Koser, P. L., Battula, N., Townsend, G. K., Maccubbin, A. E., Gelboin, H. V., and Gurtoo, H. L. (1988) Cytochrome P$_3$-450 cDNA encodes aflatoxin B$_1$-4-hydroxylase. *J. Biol. Chem.* **263**, 12187–12189.

(139) Buening, M. K., Chang, R. L., Huang, M. R., Fortner, J. G., Wood, A. W., and Conney, A. H. (1981) Activation and inhibition of benzo(*a*)pyrene and aflatoxin B$_1$ metabolism in human liver microsomes by naturally occurring flavonoids. *Cancer Res.* **41**, 67–72.

(140) Raney, K. D., Meyer, D. J., Ketterer, B., Harris, T. M., and Guengerich, F. P. (1991) Conjugation of aflatoxin B$_1$ 8,9-oxide with glutathione by rat and human glutathione S-transferases. *Chem. Res. Toxicol.* (in press).

(141) Beaune, P. H., and Guengerich, F. P. (1988) Human drug metabolism *in vitro*. *Pharmacol. Ther.* **37**, 193–211.

(142) Wilkinson, G. R., Guengerich, F. P., and Branch, R. A. (1989) Genetic polymorphism of S-mephenytoin hydroxylation. *Pharmacol. Ther.* **43**, 53–76.

(143) Kolars, J., Schmiedelin-Ren, P., Dobbins, W., Merion, R., Wrighton, S., and Watkins, P. (1990) Heterogeneity of P-450 IIIA expression in human gut epithelia. *FASEB J.* **4**, A2242.

(144) Watkins, P. B., Hamilton, T. A., Annesley, T. M., Ellis, C. N., Kolars, J. C., and Voorhees, J. J. (1990) The erythromycin breath test as a predictor of cyclosporine blood levels. *Clin. Pharmacol. Ther.* **48**, 120–129.

(145) Quattrochi, L. C., Okino, S. T., Pendurthi, U. R., and Tukey, R. H. (1985) Cloning and isolation of human cytochrome P-450 cDNAs homologous to dioxin-inducible rabbit mRNAs encoding P-450 4 and P-450 6. *DNA* **4**, 395–400.

(146) Wrighton, S. A., Campanile, C., Thomas, P. E., Maines, S. L., Watkins, P. B., Parker, G., Mendez-Picon, G., Haniu, M., Shively, J. E., Levin, W., and Guzelian, P. S. (1986) Identification of a human liver cytochrome P-450 homologous to the major isosafrole-inducible cytochrome P-450 in the rat. *Mol. Pharmacol.* **29**, 405–410.

(147) Jaiswal, A. K., Nebert, D. W., McBride, O. W., and Gonzalez, F. J. (1987) Human P$_3$-450: cDNA and complete protein sequence, repetitive *Alu* sequences in the 3′ nontranslated region, and localization of gene to chromosome 15. *J. Exp. Pathol.* **3**, 1–17.

(148) Sesardic, D., Boobis, A. R., and Davies, D. S. (1989) Tissue specific inducibility of members of the cytochrome P450IA gene sub-family in rat and man. In *Xenobiotic Metabolism and Disposition, Second International Meeting, International Society for the Study of Xenobiotics* (Cayen, R., Estabrook, R. W., and Kato, R., Eds.) pp 45–52, Taylor and Francis, London.

(149) Brian, W. R., Srivastava, P. K., Umbenhauer, D. R., Lloyd, R. S., and Guengerich, F. P. (1989) Expression of a human liver cytochrome P-450 protein with tolbutamide hydroxylase activity in *Saccharomyces cerevisiae*. *Biochemistry* **28**, 4993–4999.

(150) Inaba, T., Jurima, M., Mahon, W. A., and Kalow, W. (1985) In vitro inhibition studies of two isozymes of human liver cytochrome P-450: mephenytoin p-hydroxylase and sparteine monooxygenase. *Drug Metab. Dispos.* **13**, 443–448.

(151) Pessayre, D., Tinel, M., Larrey, D., Cobert, B., Funck-Brentano, C., and Babany, G. (1983) Inactivation of cytochrome P-450 by a troleandomycin metabolite. Protective role of glutathione. *J. Pharmacol. Exp. Ther.* **224**, 685–691.

(152) Guengerich, F. P. (1990) Mechanism-based inactivation of human liver cytochrome P-450 IIIA4 by gestodene. *Chem. Res. Toxicol.* **3**, 363–371.

(153) Kuhl, H., Jung-Hoffmann, C., and Heidt, F. (1988) Alterations in the serum levels of gestodene and SHBG during 12 cycles of treatment with 30 μg ethinylestradiol and 75 μg gestodene. *Contraception* **38**, 477–486.

(154) Jung-Hoffmann, C., and Kuhl, H. (1990) Interaction with the pharmacokinetics of ethinylestradiol of progestogens contained in oral contraceptives. *Contraception* **40**, 299–312.

(155) Tinel, M., Descatoire, V., Larrey, D., Loeper, J., Labbe, G., Letteron, P., and Pessayre, D. (1989) Effects of clarithromycin on cytochrome P-450. Comparison with other macrolides. *J. Pharmacol. Exp. Ther.* **250**, 746–751.

(156) Sesardic, D., Boobis, A. R., Edwards, R. J., and Davies, D. S. (1988) A form of cytochrome P450 in man, orthologous to form d in the rat, catalyses the O-deethylation of phenacetin and is inducible by cigarette smoking. *Br. J. Clin. Pharmacol.* **26**, 363–372.

(157) Fleming, C. M., Branch, R. A., Wilkinson, G. R., and Guengerich, F. P. (1990) Human liver microsomal N-hydroxylation of dapsone by cytochrome P-450 IIIA4. *Pharmacologist* **32**, 140.

(158) Yamazoe, Y., Shimada, M., Maeda, K., Kamataki, T., and Kato, R. (1984) Specificity of four forms of cytochrome P-450 in the metabolic activation of several aromatic amines and benzo-[a]pyrene. *Xenobiotica* 14, 549–552.

(159) Djuric, Z., Potter, D. W., Heflich, R. H., and Beland, F. A. (1986) Aerobic and anaerobic reduction of nitrated pyrenes in vitro. *Chem.-Biol. Interact.* 59, 309–324.

(160) Bourdi, M., Larrey, D., Nataf, J., Berunau, J., Pessayre, D., Iwasaki, M., Guengerich, F. P., and Beaune, P. H. (1990) A new anti-liver endoplasmic reticulum antibody directed against human cytochrome P-450 IA2: a specific marker of dihydralazine-induced hepatitis. *J. Clin. Invest.* 85, 1967–1973.

(161) Kadlubar, F. F., Talaska, G., Butler, M. A., Teitel, C. H., Massengill, J. P., and Lang, N. P. (1990) Determination of carcinogenic arylamine N-oxidation phenotype in humans by analysis of caffeine urinary metabolites. In *Mutation and the Environment, Part B: Metabolism, Testing Methods, and Chromosomes* (Mendelsohn, M. L., and Albertini, R. J., Eds.) pp 107–114, Wiley-Liss, New York.

(162) Yamano, S., Tatsuno, J., and Gonzalez, F. J. (1990) The CYP2A3 gene product catalyzes coumarin 7-hydroxylation in human liver microsomes. *Biochemistry* 29, 1322–1329.

(163) Relling, M. V., Aoyama, T., Gonzalez, F. J., and Meyer, U. A. (1990) Tolbutamide and mephenytoin hydroxylation by human cytochrome P450s in the CYP2C subfamily. *J. Pharmacol. Exp. Ther.* 252, 442–447.

(164) Eichelbaum, M., Spannbrucker, N., Steincke, B., and Dengler, H. J. (1979) Defective N-oxidation of sparteine in man: a new pharmacogenetic defect. *Eur. J. Clin. Pharmacol.* 16, 183–187.

(165) Küpfer, A., Schmid, B., Preisig, R., and Pfaff, G. (1984) Dextromethorphan as a safe probe for debrisoquine hydroxylation polymorphism. *Lancet*, 517–518.

(166) Peter, R., Böcker, R. G., Beaune, P. H., Iwasaki, M., Guengerich, F. P., and Yang, C.-S. (1990) Hydroxylation of chlorzoxazone as a specific probe for human liver cytochrome P-450 IIE1. *Chem. Res. Toxicol.* 3, 566–573.

(167) Watkins, P. B., Murray, S. A., Winkelman, L. G., Heuman, D. M., Wrighton, S. A., and Guzelian, P. S. (1989) Erythromycin breath test as an assay of glucocorticoid-inducible liver cytochrome P-450: studies in rats and patients. *J. Clin. Invest.* 83, 688–697.

(168) Bargetzi, M. J., Aoyama, T., Gonzalez, F. J., and Meyer, U. A. (1989) Lidocaine metabolism in human liver microsomes by cytochrome P450IIIA4. *Clin. Pharmacol. Ther.* 46, 521–527.

(169) Imaoka, S., Enomoto, K., Oda, Y., Asada, A., Fujimori, M., Shimada, T., Fujita, S., Guengerich, F. P., and Funae, Y. (1990) Lidocaine metabolism by human cytochrome P-450s purified from hepatic microsomes: comparison of those with rat hepatic cytochrome P-450s. *J. Pharmacol. Exp. Ther.* 255, 1385–1391.

(170) Ged, C., Rouillon, J. M., Pichard, L., Combalbert, J., Bressot, N., Bories, P., Michel, H., Beaune, P., and Maurel, P. (1989) The increase in urinary excretion of 6β-hydroxycortisol as a marker of human hepatic cytochrome P450IIIA induction. *Br. J. Clin. Pharmacol.* 28, 373–387.

(171) Conney, A. H. (1982) Induction of microsomal enzymes by foreign chemicals and carcinogenesis by polycyclic aromatic hydrocarbons: G. H. A. Clowes memorial lecture. *Cancer Res.* 42, 4875–4917.

(172) Pantuck, E. J., Hsiao, K.-C., Maggio, A., Nakamura, K., Kuntzman, R., and Conney, A. H. (1974) Effect of cigarette smoking on phenacetin metabolism. *Clin. Pharmacol. Ther.* 15, 9–17.

(173) Diaz, D., Fabre, I., Daujat, M., Saintaubert, B., Bories, P., Michel, H., and Maurel, P. (1990) Omeprazole is an aryl hydrocarbon-like inducer of human hepatic cytochrome-P450. *Gastroenterology* 99, 737–747.

(174) Veronese, M. E., McLean, S., D'Souza, C. A., and Davies, N. W. (1985) Formation of reactive metabolites of phenacetin in humans and rats. *Xenobiotica* 15, 929–940.

(175) Ryan, D. E., Ramanathan, L., Iida, S., Thomas, P. E., Haniu, M., Shively, J. E., Lieber, C. S., and Levin, W. (1985) Characterization of a major form of rat hepatic microsomal cytochrome P-450 induced by isoniazid. *J. Biol. Chem.* 260, 6385–6393.

(176) Yang, C. S., Koop, D. R., Wang, T., and Coon, M. J. (1985) Immunochemical studies on the metabolism of nitrosamines by ethanol-inducible cytochrome P-450. *Biochem. Biophys. Res. Commun.* 128, 1007–1013.

(177) Lai, D. Y., Myers, S. C., Woo, Y. T., Greene, E. J., Friedman, M. A., Argus, M. F., and Arcos, J. C. (1979) Role of dimethylnitrosamine-demethylase in the metabolic activation of di-

methylnitrosamine. *Chem.-Biol. Interact.* 28, 107–126.

(178) Levin, W., Thomas, P. E., Oldfield, N., and Ryan, D. E. (1986) N-Demethylation of N-nitrosodimethylamine catalyzed by purified rat hepatic microsomal cytochrome P-450: isozyme specificity and role of cytochrome $b_5$. *Arch. Biochem. Biophys.* 248, 158–165.

(179) Thomas, P. E., Bandiera, S., Maines, S. L., Ryan, D. E., and Levin, W. (1987) Regulation of cytochrome P-450j, a high-affinity N-nitrosodimethylamine demethylase, in rat hepatic microsomes. *Biochemistry* 26, 2280–2289.

(180) Yang, C. S., Yoo, J. S. H., Ishizaki, H., and Hong, J. (1990) Cytochrome P450IIE1: roles in nitrosamine metabolism and mechanisms of regulation. *Drug Metab. Rev.* 22, 147–159.

(181) Yoo, J. S. H., Ishizaki, H., and Yang, C. S. (1990) Roles of cytochrome P450IIE1 in the dealkylation and denitrosation of N-nitrosodimethylamine and N-nitrosodiethylamine in rat liver microsomes. *Carcinogenesis* 11, 2239–2243.

(182) Ryan, D. E., Koop, D. R., Thomas, P. E., Coon, M. J., and Levin, W. (1986) Evidence that isoniazid and ethanol induce the same microsomal cytochrome P-450 in rat liver, an isozyme homologous to rabbit liver cytochrome P-450 isozyme 3a. *Arch. Biochem. Biophys.* 246, 633–644.

(183) Pantuck, E. J., Pantuck, C. B., and Conney, A. H. (1987) Effect of streptozotocin-induced diabetes in the rat on the metabolism of fluorinated volatile anesthetics. *Anesthesiology* 66, 24–28.

(184) Yoo, J. S. H., Cheung, R. J., Patten, C. J., Wade, D., and Yang, C. S. (1987) Nature of N-nitrosodimethylamine demethylase and its inhibitors. *Cancer Res.* 47, 3378–3383.

(185) Koop, D. R. (1990) Inhibition of ethanol-inducible cytochrome P450IIE1 by 3-amino-1,2,4-triazole. *Chem. Res. Toxicol.* 3, 377–383.

(186) Phillips, J. C., Lake, B. G., Gangolli, S. D., Grasso, P., and Lloyd, A. G. (1977) Effects of pyrazole and 3-amino-1,2,4-triazole on the metabolism and toxicity of dimethylnitrosamine in the rat. *J. Natl. Cancer Inst.* 58, 629–633.

(187) Fiala, E. S., Kulakis, C., Christiansen, G., and Weisburger, J. H. (1978) Inhibition of the metabolism of the colon carcinogen, azoxymethane, by pyrazole. *Cancer Res.* 38, 4515–4521.

(188) Hunter, A. L., and Neal, R. A. (1975) Inhibition of hepatic mixed-function oxidase activity in vitro and in vivo by various thiono-sulfur-containing compounds. *Biochem. Pharmacol.* 24, 2199–2205.

(189) Wong, L. C. K., Winston, J. M., Hong, C. B., and Plotnick, H. (1982) Carcinogenicity and toxicity of 1,2-dibromoethane in the rat. *Toxicol. Appl. Pharmacol.* 63, 155–165.

(190) Leithauser, M. T., Liem, A., Stewart, B. C., Miller, E. C., and Miller, J. A. (1990) 1,$N^6$-Ethenoadenosine formation, mutagenicity and murine tumor induction as indicators of the generation of an electrophilic epoxide metabolite of the closely related carcinogens ethyl carbamate (urethane) and vinyl carbamate. *Carcinogenesis* 11, 463–473.

(191) Kim, D. H., and Guengerich, F. P. (1990) Formation of the DNA adduct S-[2-($N^7$-guanyl)ethyl]glutathione from ethylene dibromide: effects of modulation of glutathione and glutathione S-transferase levels and lack of a role for sulfation. *Carcinogenesis* 11, 419–424.

(192) Chieli, E., Saviozzi, M., Puccini, P., Longo, V., and Gervasi, P. G. (1990) Possible role of the acetone-inducible cytochrome P-450IIE1 in the metabolism and hepatotoxicity of thiobenzamide. *Arch. Toxicol.* 64, 122–127.

(193) Nakajima, T., Wang, R. S., Murayama, N., and Sato, A. (1990) Three forms of trichloroethylene-metabolizing enzymes in rat liver induced by ethanol, phenobarbital, and 3-methylcholanthrene. *Toxicol. Appl. Pharmacol.* 102, 546–552.

(194) Walsh, C. (1979) *Enzymatic Reaction Mechanisms*, W. H. Freeman, San Francisco.

(195) Dahl, G. A., Miller, J. A., and Miller, E. C. (1978) Vinyl carbamate as a promutagen and a more carcinogenic analog of ethyl carbamate. *Cancer Res.* 38, 3793–3804.

(196) Song, B. J., Veech, R. L., and Saenger, P. (1990) Cytochrome P450IIE1 is elevated in lymphocytes from poorly controlled insulin-dependent diabetics. *J. Clin. Endocrinol. Metab.* 71, 1036–1040.

(197) Perrot, N., Nalpas, B., Yang, C. S., and Beaune, P. H. (1989) Modulation of cytochrome P450 isozymes in human liver, by ethanol and drug intake. *Eur. J. Clin. Invest.* 19, 549–555.

(198) Song, B. J., Veech, R. L., Park, S. S., Gelboin, H. V., and Gonzalez, F. J. (1989) Induction of rat hepatic N-nitrosodimethylamine demethylase by acetone is due to protein stabilization. *J. Biol. Chem.* 264, 3568–3572.

(199) Kim, S. G., and Novak, R. F. (1990) Induction of rat hepatic P450IIIE1 (CYP 2E1) by pyridine: evidence for a role of protein synthesis in the absence of transcriptional activation. *Biochem. Biophys. Res. Commun.* **166**, 1072–1079.

(200) Nebert, D. W., Adesnik, M., Coon, M. J., Estabrook, R. W., Gonzalez, F. J., Guengerich, F. P., Gunsalus, I. C., Johnson, E. F., Kemper, B., Levin, W., Phillips, I. R., Sato, R., and Waterman, M. R. (1987) The P450 gene superfamily: recommended nomenclature. *DNA* **6**, 1–11.

(201) Nebert, D. W., Nelson, D. R., Coon, M. J., Estabrook, R. W., Feyereisen, R., Fujii-Kuriyama, Y., Gonzalez, F. J., Guengerich, F. P., Gunsalus, I. C., Johnson, E. F., Loper, J. C., Sato, R., Waterman, M. R., and Waxman, D. J. (1991) The P450 superfamily: update on new sequences, gene mapping, and recommended nomenclature. *DNA Cell Biol.* **10**, 1–14.

(202) Graves, P. E., Kaminsky, L. S., and Halpert, J. (1987) Evidence for functional and structural multiplicity of pregnenolone-16α-carbonitrile-inducible cytochrome P-450 isozymes in rat liver microsomes. *Biochemistry* **26**, 3887–3894.

(203) Shimada, M., Nagata, K., Murayama, N., Yamazoe, Y., and Kato, R. (1989) Role of growth hormone in modulating the constitutive and phenobarbital-induced levels of two P-450$_{6\beta}$ (testosterone 6β-hydroxylase) mRNAs in rat livers. *J. Biochem.* **106**, 1030–1034.

(204) Kitada, M., Kamataki, T., Itahashi, K., Rikihisa, T., Kato, R., and Kanakubo, Y. (1985) Purification and properties of cytochrome P-450 from homogenates of human fetal livers. *Arch. Biochem. Biophys.* **241**, 275–280.

(205) Komori, M., Nishio, K., Kitada, M., Shiramatsu, K., Muroya, K., Soma, M., Nagashima, K., and Kamataki, T. (1990) Fetus-specific expression of a form of cytochrome P-450 in human livers. *Biochemistry* **29**, 4430–4433.

(206) Wrighton, S. A., Brian, W. R., Sari, M. A., Iwasaki, M., Guengerich, F. P., Raucy, J. L., Molowa, D. T., and Vandenbranden, M. (1990) Studies on the expression and metabolic capabilities of human liver cytochrome P450IIIA5 (HLp3). *Mol. Pharmacol.* **38**, 207–213.

(207) Wrighton, S. A., and Vandenbranden, M. (1989) Isolation and characterization of human fetal liver cytochrome P450HLp2: a third member of the P450III gene family. *Arch. Biochem. Biophys.* **268**, 144–151.

(208) Bork, R. W., Muto, T., Beaune, P. H., Srivastava, P. K., Lloyd, R. S., and Guengerich, F. P. (1989) Characterization of mRNA species related to human liver cytochrome P-450 nifedipine oxidase and the regulation of catalytic activity. *J. Biol. Chem.* **264**, 910–919.

(209) Daujat, M., Pichard, L., Fabre, I., Diaz, D., Maurice, M., Pineau, T., Blanc, P., Fabre, G., Fabre, J. M., Saint Aubert, B., and Maurel, P. (1990) Human P450IA and IIIA subfamilies: regulation of expression and inducibility in primary cultures of human hepatocytes. In *Drug Metabolizing Enzymes: Genetics, Regulation and Toxicology, Proceedings of the Eighth International Symposium on Microsomes and Drug Oxidations* (Ingelman-Sundberg, M., Gustafsson, J.-Å., and Orrenius, S., Eds.) p 16, Karolinska Institute, Stockholm.

(210) Huang, M. T., Johnson, E. F., Muller-Eberhard, U., Koop, D. R., Coon, M. J., and Conney, A. H. (1981) Specificity in the activation and inhibition by flavonoids of benzo[a]pyrene hydroxylation by cytochrome P-450 isozymes from rabbit liver microsomes. *J. Biol. Chem.* **256**, 10897–10901.

(211) Yamazoe, Y., Murayama, N., Shimada, M., Yamauchi, K., Nagata, K., Imaoka, S., Funae, Y., and Kato, R. (1988) A sex-specific form of cytochrome P-450 catalyzing propoxycoumarin O-depropylation and its identity with testosterone 6β-hydroxylase in untreated rat livers: reconstitution of the activity with microsomal lipids. *J. Biochem.* **104**, 785–790.

(212) Guengerich, F. P., Müller-Enoch, D., and Blair, I. A. (1986) Oxidation of quinidine by human liver cytochrome P-450. *Mol. Pharmacol.* **30**, 287–295.

(213) Renaud, J.-P., Cullin, C., Pompon, D., Beaune, P., and Mansuy, D. (1990) Expression of human liver cytochrome P450 IIIA4 in yeast: a functional model for the hepatic enzyme. *Eur. J. Biochem.* **194**, 889–896.

(214) Kronbach, T., Mathys, D., Umeno, M., Gonzalez, F. J., and Meyer, U. A. (1989) Oxidation of midazolam and triazolam by human liver cytochrome P450IIIA4. *Mol. Pharmacol.* **36**, 89–96.

(215) Wang, R. W., Kari, P. H., Lu, A. Y. H., Thomas, P. E., Guengerich, F. P., and Vyas, K. P. (1991) Identification of cytochrome P-450 3A proteins as the major enzymes responsible for the oxidative metabolism of lovastatin in rat and human liver microsomes. *Arch. Biochem. Biophys.* (in press).

(216) Waxman, D. J., Attisano, C., Guengerich, F. P., and Lapenson, D. P. (1988) Cytochrome P-450 steroid hormone metabolism catalyzed by human liver microsomes. *Arch. Biochem. Biophys.* **263**, 424–436.

(217) Guengerich, F. P. (1988) Oxidation of 17α-ethynylestradiol by human liver cytochrome P-450. *Mol. Pharmacol.* **33**, 500–508.

(218) Combalbert, J., Fabre, I., Fabre, G., Dalet, I., Derancourt, J., Cano, J. P., and Maurel, P. (1989) Metabolism of cyclosporin A. IV. Purification and identification of the rifampicin-inducible human liver cytochrome P-450 (cyclosporin A oxidase) as a product of P450IIIA gene subfamily. *Drug. Metab. Dispos.* **17**, 197–207.

(219) Kronbach, T., Fischer, V., and Meyer, U. A. (1988) Cyclosporine metabolism in human liver: identification of a cytochrome P-450III gene family as the major cyclosporine-metabolizing enzyme explains interactions of cyclosporine with other drugs. *Clin. Pharmacol. Ther.* **43**, 630–635.

(220) Schellens, J. H. M., Soons, P. A., and Breimer, D. D. (1988) Lack of bimodality in nifedipine plasma kinetics in a large population of healthy subjects. *Biochem. Pharmacol.* **37**, 2507–2510.

(221) Thomson, P. D., Melmon, K. L., Richardson, J. A., Cohn, K., Steinbrunn, W., Cudihee, R., and Rowland, M. (1973) Lidocaine pharmacokinetics in advanced heart failure, liver disease, and renal failure in humans. *Ann. Intern. Med.* **78**, 499–508.

(222) Park, B. K. (1981) Assessment of urinary 6β-hydroxycortisol as an *in vivo* index of mixed function oxidase activity. *Br. J. Clin. Pharmacol.* **12**, 97–102.

(223) Saenger, P., Forster, E., and Kream, J. (1981) 6β-Hydroxycortisol: a noninvasive indicator of enzyme induction. *J. Clin. Endocrinol. Metab.* **52**, 381–384.

(224) Edgar, B., Bailey, D. G., Bergstrand, R., Johnsson, G., and Lurje, L. (1990) Formulation dependent interaction between felodipine and grapefruit juice. *Clin. Pharmacol. Ther.* **47**, 181.

(225) Morel, F., Beaune, P. H., Ratanasavanh, D., Flinois, J.-P., Yang, C.-S., Guengerich, F. P., and Guillouzo, A. (1990) Expression of cytochrome P-450 enzymes in cultured human hepatocytes. *Eur. J. Biochem.* **191**, 437–444.

(226) Delaforge, M., Sartori, E., and Mansuy, D. (1988) In vivo and in vitro effects of a new macrolide antibiotic roxithromycin on rat liver cytochrome P-450: comparison with troleandomycin and erythromycin. *Chem.-Biol. Interact.* **68**, 179–188.

(227) Bailey, D. G., Edgar, B., Spence, J. D., Munzo, C., and Arnold, J. M. O. (1990) Felodipine and nifedipine interactions with grapefruit juice. *Clin. Pharmacol. Ther.* **47**, 180.

(228) Booth, A. N., Jones, F. T., and De Eds, F. (1958) Metabolic and glucosuria studies on naringin and phloridzin. *J. Biol. Chem.* **233**, 280–282.

(229) Kitada, M., Taneda, M., Ohi, H., Komori, M., Itahashi, K., Nagao, M., and Kamataki, T. (1989) Mutagenic activation of aflatoxin B$_1$ by P-450 HFLa in human fetal livers. *Mutat. Res.* **227**, 53–58.

(230) Groopman, J. D., Donahue, P. R., Zhu, J., Chen, J., and Wogan, G. N. (1985) Aflatoxin metabolism in humans: detection of metabolites and nucleic acid adducts in urine by affinity chromatography. *Proc. Natl. Acad. Sci. U.S.A.* **82**, 6492–6496.

(231) Kellerman, G., Luyten-Kellerman, M., and Shaw, C. R. (1973) Genetic variation of aryl hydrocarbon hydroxylase in human lymphocytes. *Am. J. Hum. Genet.* **25**, 327–331.

(232) Kellerman, G., Shaw, C. R., and Luyten-Kellerman, M. (1973) Aryl hydrocarbon hydroxylase inducibility and bronchogenic carcinoma. *N. Engl. J. Med.* **298**, 934–937.

(233) Paigen, B., Ward, E., Reilly, A., Houten, L., Gurtoo, H. L., Minowada, J., Steenland, K., Havens, M. B., and Sartori, P. (1981) Seasonal variation of aryl hydrocarbon hydroxylase activity in human lymphocytes. *Cancer Res.* **41**, 2757–2761.

(234) Kouri, R. E., McKinney, C. E., Slomiany, D. J., Snodgrass, D. R., Wray, N. P., and McLemore, T. L. (1982) Positive correlation between high aryl hydrocarbon hydroxylase activity and primary lung cancer as analyzed in cryopreserved lymphocytes. *Cancer Res.* **42**, 5030–5037.

(235) McLemore, T. L., Adelberg, S., Czerwinski, M., Hubbard, W. C., Yu, S. J., Storeng, R., Wood, T. G., Hines, R. N., and Boyd, M. R. (1989) Altered regulation of the cytochrome P4501A1 gene: novel inducer-independent gene expression in pulmonary carcinoma cell lines. *J. Natl. Cancer Inst.* **81**, 1787–1794.

(236) Jaiswal, A. K., Gonzalez, F. J., and Nebert, D. W. (1985) Human dioxin-inducible cytochrome $P_1$-450: complementary DNA and amino acid sequence. *Science* **228**, 80–83.

(237) Kawajiri, K., Nakachi, K., Imai, K., Yoshii, A., Shinoda, N., and Watanabe, J. (1990) Identification of genetically high risk individuals to lung cancer by DNA polymorphisms of the cytochrome P450IA1 gene. *FEBS Lett.* **263**, 131–133.

(238) Tefre, T., Børresen, A.-L., Ryberg, D., Haugen, A., and Brøgger, A. (1990) Allele association studies of the P450IA1 and P450IID1 DNA-polymorphisms in lung cancer patients and controls. In *Drug Metabolizing Enzymes: Genetics, Regulation and Toxicology, Proceedings of the Eighth International Symposium on Microsomes and Drug Oxidations* (Ingelman-Sundberg, M., Gustaffson, J.-Å., and Orrenius, S., Eds.) p 51, Karolinska Institute, Stockholm.

(239) Brodie, A. M. H. (1985) Aromatase inhibition and its pharmacologic implications. *Biochem. Pharmacol.* **34**, 3213–3219.

(240) Wolff, T., Distlerath, L. M., Worthington, M. T., Groopman, J. D., Hammons, G. J., Kadlubar, F. F., Prough, R. A., Martin, M. V., and Guengerich, F. P. (1985) Substrate specificity of human liver cytochrome P-450 debrisoquine 4-hydroxylase probed using immunochemical inhibition and chemical modeling. *Cancer Res.* **45**, 2116–2122.

(241) Plummer, S., Boobis, A. R., and Davies, D. S. (1986) Is the activation of aflatoxin $B_1$ catalysed by the same form of cytochrome P-450 as that 4-hydroxylating debrisoquine in rat and/or man? *Arch. Toxicol.* **58**, 165–170.

(242) Ritter, J., Somasundaram, R., Heinemeyer, G., and Roots, I. (1986) The debrisoquine hydroxylation phenotype and the acetylator phenotype as genetic risk factors for the occurrence of larynx and pharynx carcinoma. *Acta Pharmacol. Toxicol.* **59** (Suppl. 5), 221.

(243) Drakoulis, N., Minks, T., Ploch, M., Otte, F., Heinemeyer, G., Kampf, D., Loddenkemper, R., and Roots, I. (1986) Questionable association of debrisoquine hydroxilator phenotype and risk for bronchial carcinoma. *Acta Pharmacol. Toxicol.* **59** (Suppl. 5), 220.

(244) Speirs, C. J., Murray, S., Davies, D. S., Mabadeje, A. F. B., and Boobis, A. R. (1990) Debrisoquine oxidation phenotype and susceptibility to lung cancer. *Br. J. Clin. Pharmacol.* **29**, 101–109.

(245) Kahn, G. C., Boobis, A. R., Brodie, M. J., Toverud, E. L., Murray, S., and Davies, D. S. (1985) Phenacetin O-deethylase: an activity of a cytochrome P-450 showing genetic linkage with that catalysing the 4-hydroxylation of debrisoquine? *Br. J. Clin. Pharmacol.* **20**, 67–76.

(246) Eichelbaum, M., Baur, M. P., Dengler, H. J., Osikowska-Evers, B., Tieves, G., Zekorn, C., and Rittner, C. (1987) Chromosomal assignment of human cytochrome P-450 (debrisoquine/sparteine type) to chromosome 22. *Br. J. Clin. Pharmacol.* **23**, 455–458.

(247) Wong, G., Itakura, T., Kawajiri, K., Skow, L., and Negishi, M. (1989) Gene family of male-specific testosterone 16α-hydroxylase (C-P-450$_{16\alpha}$) in mice: organization, differential regulation, and chromosome localization. *J. Biol. Chem.* **264**, 2920–2927.

(248) Poland, A., Glover, E., and Taylor, B. A. (1987) The murine Ah locus: a new allele and mapping to chromosome 12. *Mol. Pharmacol.* **32**, 471–478.

(249) Cobb, R. R., Stomming, T. A., and Whitney, J. B., III (1987) The aryl hydrocarbon hydroxylase Ah locus and a novel restriction-fragment length polymorphism (RFLP) are located on mouse chromosome 12. *Biochem. Genet.* **25**, 401–413.

(250) Yamazaki, H., Degawa, M., Funae, Y., Imaoka, S., Inui, Y., Guengerich, F. P., and Shimada, T. (1990) Roles of different cytochrome P450 enzymes in bioactivation of the potent hepatocarcinogen 3-methoxy-4-aminoazobenzene by rat and human liver microsomes. *Carcinogenesis* **12**, 133–139.

(251) Yasumori, T., Murayama, N., Yamazoe, Y., Nogi, Y., Fukasawa, T., and Kato, R. (1989) Expression of a human P-450IIC gene in yeast cells using galactose-inducible expression system. *Mol. Pharmacol.* **35**, 443–449.

(252) Lin, D. X., Malaveille, C., Park, S. S., Gelboin, H. V., and Bartsch, H. (1990) Contribution of DNA methylation and benzylation to N-nitroso-N-methylamine-induced mutagenesis in bacteria: effects of rat liver cytochrome P450 isozymes and glutathione transferases. *Carcinogenesis* **11**, 1653–1658.

(253) Lee, M., Ishizaki, H., Brady, J. F., and Yang, C. S. (1989) Substrate specificity and alkyl group selectivity in the metabolism of N-nitrosodialkylamines. *Cancer Res.* **49**, 1470–1474.

(254) Searle, C. E., Ed. (1984) *Chemical Carcinogens*, Vols. 1 and 2, American Chemical Society, Washington, DC.

(255) Poulos, T. L., Finzel, B. C., Gunsalus, I. C., Wagner, G. C., and Kraut, J. (1985) The 2.6-Å crystal structure of *Pseudomonas putida* cytochrome P-450. *J. Biol. Chem.* **260**, 16122–16130.

(256) Lewis, D. F. V., Ioannides, C., and Parke, D. V. (1987) Structural requirements for substrates of cytochromes P-450 and P-448. *Chem.-Biol. Interact.* **64**, 39–60.

(257) van Bladeren, P. J., Armstrong, R. N., Cobb, D., Thakker, D. R., Ryan, D. E., Thomas, P. E., Sharma, N. D., Boyd, D. R., Levin, W., and Jerina, D. M. (1982) Stereoselective formation of benz-[a]anthracene (+)-(5S,6R)-oxide and (+)-(8R,9S)-oxide by a highly purified and reconstituted system containing cytochrome P-450c. *Biochem. Biophys. Res. Commun.* **106**, 602–609.

## Chapter 11

# Glutathione *S*-Transferases:  Reaction Mechanism, Structure, and Function

Richard N. Armstrong

*Department of Chemistry and Biochemistry, University of Maryland,*
*College Park, Maryland 20742*

Reprinted from *Chemical Research in Toxicology,* Vol. 4, No. 2, March/April, 1991

## Introduction

The glutathione *S*-transferases (EC 2.5.1.18) catalyze the nucleophilic addition of the tripeptide glutathione (GSH; see Chart I) to substrates that have electrophilic functional groups.  The enzyme is found in most aerobic microorganisms, plants, and animals.  The primary function of the enzyme, particularly in higher organisms, is generally considered to be the detoxication of both endogenous and xenobiotic alkylating agents such as epoxides, $\alpha,\beta$-unsaturated aldehydes and ketones, alkyl and aryl halides, and others.  It is probably fair to say that this family of proteins is the single most important group of enzymes involved in the metabolism of electrophilic compounds.  Certainly one of the most fascinating aspects of the biochemistry of the GSH transferases is their ability to catalyze reactions toward such a large number of structurally diverse substrates, a characteristic shared with most other detoxication enzymes.  This ability is a consequence of the existence of several isoenzymes each of which has a unique substrate selectivity and the remarkable tolerance of each isoenzyme for both the type of electrophilic functional group and the structure of the molecule to which it is appended.  It is clear that a structural and mechanistic understanding of the substrate preferences of the GSH transferases is essential for the elucidation of their influence on the toxicology of alkylating agents.

Although the GSH transferases occur in a large number of organisms, most of the mechanistic and structural work has been carried out on enzymes from higher organisms such as rats, mice, and humans.  The bulk of the GSH transferase activity in these species is due to a collection of cytosolic isoenzymes all of which are dimeric proteins

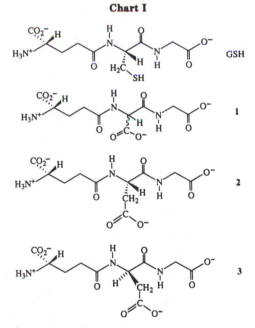

Chart I

with subunit molecular weights of about 25K.  In the last 15 years considerable attention has been given to the classification of the large number of isoenzymes from various species (*1–3*).  The early classifications, based primarily on substrate specificity and subunit molecular weight, have given way more recently to the development of a genetic catalogue of the proteins.  The advances being made in the genetic classification of the isoenzymes and the regulation of gene expression are of tremendous importance in unraveling the role of the GSH transferases in toxicology. However, they are not the subject of this review. These particular topics have been the subject of recent reviews (4, 5).

A very brief discussion of the classification of GSH transferases is, however, warranted by way of introduction. In higher organisms there is very clear evidence for at least three gene classes of the cytosolic enzymes. They have been designated $\alpha$, $\mu$, and $\pi$. Each gene class consists of two or more genes that encode different subunit types. For example, the polypeptides encoded by $\alpha$ class genes in rat, designated subunits 1, 2, 8, and 10, combine to form active homodimeric or heterodimeric holoenzymes 1-1, 2-2, 1-2, etc. $\mu$ class genes encode subunit types 3, 4, 6, 9, and 11 which similarly combine into homodimeric and heterodimeric holoenzymes. Inter-gene-class heterodimers are not known, a fact that suggests the interfacial regions for subunit interactions are unique, in this respect, for each gene class. Primary sequence homologies for subunits within a gene class are quite high, typically $\geq 70\%$ sequence identity. Not surprisingly, homologies between subunits of different gene classes are significant but considerably less, usually about 40%. In addition to the cytosolic isoenzymes, there is a microsomal enzyme that bears little resemblance to any of the former except that it catalyzes the same reaction (6, 7). The coverage of this review is limited to the cytosolic enzymes.

Although the actual reaction catalyzed by the GSH transferases is quite straightforward, relatively little is known concerning the details of how the enzyme actually accelerates the addition of GSH to the electrophilic substrate. The general reaction that is catalyzed is illustrated in eq 1. The central features of the chemical and kinetic

$$E + GSH \rightleftharpoons E{\cdot}GS^- + H^+ \underset{\longleftarrow}{\overset{R-X}{\rightleftharpoons}} E{\cdot}GS^-{\cdot}R{-}X \rightleftharpoons$$
$$E{\cdot}GSR{\cdot}X^- \rightleftharpoons E + GSR + X^- \quad (1)$$

mechanisms are that the nucleophilic species in the active site of the enzyme is the thiolate anion of GSH and that the reaction is sequential; that is, the addition reaction takes place in the ternary complex of enzyme, GSH, and electrophile. Over the last decade a considerable amount of information has been gathered concerning the kinetic mechanisms and substrate specificities of a number of different isoenzymes. Most of the detailed kinetic investigations of the enzyme-catalyzed reactions suggest that the addition of substrates is random. Equation 1 is misleading in this regard. However, given that the typical dissociation constant for GSH and the enzyme (10–30 $\mu$M) is 2–3 orders of magnitude lower than the normal concentration of GSH (1–5 mM) in healthy cells, the kinetic mechanism of the enzyme under physiological conditions is, for all practical purposes, ordered with GSH adding first. Much of the work on the kinetic mechanism and the more descriptive investigations of substrate specificity have been summarized in recent reviews (4, 5, 8) and will not be reiterated here in any detail.

To fully appreciate the participation of the GSH transferases in the metabolism of xenobiotics, it is necessary to understand the chemical mechanism of catalysis as well as the influence of protein structure on substrate selectivity. It is the progress that has been made, particularly during the last three or four years, toward an understanding of these general issues that is the topic of this review. It is convenient, if not entirely logical, to segregate the discussion of the chemical mechanism of catalysis into the issues that are somewhat more specific. First to be considered is the interaction of the enzyme with

$$\overset{\delta-}{GS}{-}\overset{\delta+}{H}{\cdots}B \qquad \qquad \overset{-}{GS} {\cdot} \overset{+}{HB}$$

**Figure 1.** Two possible scenarios for the removal of the proton from the thiol of enzyme-bound GSH in catalysis. Polarization of the S–H bond by a general base (B) in the active site is illustrated on the left. Alternatively, the bound peptide may exist as the thiolate anion at neutral pH in the binary complex if the thiol is positioned in a positively charged electrostatic field represented on the right as BH+.

the physiological substrate GSH and the influence of the protein on the chemical properties of the tripeptide. Second, the experimental evidence for the putative involvement of particular amino acid residues in binding and catalysis will be evaluated. The third topic to be discussed is the nature of transition states and intermediates on the reaction coordinates of specific substrates. This latter point is intimately related to differences in the substrate selectivity observed with different isoenzymes. Finally, the current approaches to the elucidation of structure–function relationships of the various isoenzymes and the design of catalysts with altered catalytic properties will be discussed.

## Thiolate Ion and Catalysis

It is generally accepted that the addition of thiols to electrophiles occurs by way of the thiolate anion. In fact, a recent estimate of the relative reactivities of simple thiols and thiolates in aqueous solution suggests that the thiolate anion is up to $10^9$ times more reactive than its conjugate acid (9). It therefore stands to reason that one role of the enzyme is to remove the proton from the thiol of GSH and generate the more reactive thiolate, GS$^-$. Although there is no doubt that the proton must be removed at some point on the reaction coordinate, there are essentially two views as to how this might occur (10, 11). The two views differ with respect to the nature of the nucleophilic species in the active site. The species are essentially distinguished by the position of the proton in the transition state for the reaction as illustrated in Figure 1. The nucleophile in the general-base-catalyzed reaction shown on the left is a thiol with the S–H bond polarized by hydrogen bonding to a base in the active site. The alternative is preionization of the thiol to the thiolate assisted by a positively charged electrostatic field in the active site as shown on the right in Figure 1. The properties of the ground-state binary complex of the enzyme and GSH can help to differentiate these possibilities. That is to say, evidence for the existence of a significant fraction of the enzyme-bound peptide in the thiolate ionization state would be a strong indication for the preionization scenario.

Several lines of evidence now suggest that, in fact, the predominate ionization state of the thiol of GSH in the binary complex is the thiolate (i.e., E·GS$^-$). Thiolate anions have a relatively strong absorption band in the ultraviolet at approximately 235 nm ($\epsilon$ = 4000–6000 M$^{-1}$ cm$^{-1}$) which is attributable to an n $\rightarrow$ $\sigma^*$ electronic transition of a lone pair on the sulfur. A similar transition for the protonated thiol occurs at higher energy (195 nm). In spite of the considerable backbone absorption of the protein in the 235–240-nm region, it is possible to observe the thiolate anion of bound GSH by UV difference spectroscopy. Difference spectroscopy of the binary complex of isoenzyme 4-4 of rat liver GSH transferase with GSH and the enzyme alone or as the binary complex with the oxygen analogue ($\gamma$-L-glutamyl-L-serylglycine) of GSH at neutral pH has revealed an absorption band at 239 nm ($\epsilon$ = 5200 M$^{-1}$ cm$^{-1}$) that is most reasonably assigned to E·GS$^-$, the thiolate anion of the bound tripeptide (11). Furthermore,

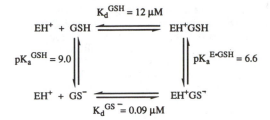

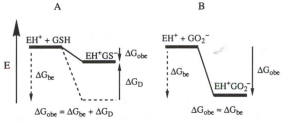

**Figure 2.** Proposed binding and acid–base equilibria for GSH and isoenzyme 4-4. The dissociation constant for the thiolate anion from the enzyme is calculated as described in the text from the known $pK_a$s of the enzyme-bound and free GSH and the observed dissociation constant of GSH measured at low pH.

**Figure 3.** (A) Utilization of binding energy $\Delta G_{be}$ for the deprotonation of GSH in the active site of the enzyme. The actual observed binding energy for the protonated peptide $\Delta G_{obe}$ is smaller than $\Delta G_{be}$ by the energy required to destabilize the thiol, $\Delta G_D$. (B) Realization of $\Delta G_D$ as observed binding energy for the resonance-stabilized carboxylate analogues of GSH. The amount of additional binding energy $\delta\Delta G_{obe}$ actually realized with the carboxylate analogues can be calculated from $\delta\Delta G_{obe} = -RT \ln (K_d^{GSH}/K_d^{analogue}) = -2.4$ kcal/mol for **1**.

titration of the difference absorption bands over the pH range of 5–8 indicates that the $pK_a$ of the thiol of bound GSH is in the range 6.4–6.7. This is about 2.5 $pK$ units lower than the acid dissociation constant of GSH in aqueous solution ($pK_a^{GSH} = 9.0$) and is quite close to the $pK_a$ of 6.6 estimated for E·GSH from the pH dependence of $k_c/K_m^{CDNB}$ in the enzyme-catalyzed addition of GSH to 1-chloro-2,4-dinitrobenzene (CDNB) under conditions of saturating GSH (*12*).

The evidence cited above suggests that one role of the enzyme is to lower the $pK_a$ of the bound thiol so that a significant fraction is ionized at physiological pH. Given the $pK_a$ of both the enzyme-bound and free thiol and the dissociation constant ($K_d^{GSH}$) of the enzyme and GSH at low pH, it is possible to complete the thermodynamic box shown in Figure 2 by calculation of the dissociation constant for the thiolate; $K_d^{GS^-} = K_d^{GSH}(K_a^{GSH}/K_a^{E·GSH}) = 9 \times 10^{-8}$ M. The analysis reveals that the thiolate is bound much more tightly than is the conjugate acid. The enzyme can be viewed as utilizing some of the binding energy of the peptide to destabilize the thiol as is illustrated in Figure 3A. The fraction of the binding energy of GSH utilized to destabilize the bound thiol is given by $\Delta G_D = RT \ln (K_a^{E·GSH}/K_a^{GSH}) = 3.3$ kcal/mol. This analysis predicts that a good structural mimic of GS⁻, that is, a peptide that has a stabilized anionic group replacing the side chain of cysteine, should bind to the active site with much higher affinity than does GSH since some of the binding energy normally used to destabilize the thiol should be translated into observed binding energy. This particular situation is illustrated in Figure 3B.

Two such molecules have been prepared, and their properties illustrate this point quite convincingly (*11*). Replacement of the cysteine side chain with a carboxylate group as in γ-L-glutamyl-(D,L-2-aminomalonyl)glycine (**1**) results in a mixture of diastereomeric peptides at least one

of which binds to the enzyme much more tightly ($K_d = 7 \times 10^{-7}$ M at pH 6.5) than does GSH ($K_d = 2.2 \times 10^{-5}$ M). The facile epimerization about the α-carbon of the aminomalonyl residue precludes a direct determination of the dissociation constant for the correctly configured isomer. However, if it is assumed that only the L-configured epimer is responsible for the inhibition, then the carboxylate analogue binds about 60-fold more tightly than does GSH. This corresponds to a realization of an additional −2.4 kcal/mol of binding energy ($\delta\Delta G_{obe}$) as observed binding energy in the carboxylate-containing peptide. That $\delta\Delta G_{obe} \neq \Delta G_D$ is simply a manifestation of the fact that the carboxylate analogue is not a perfect mimic of the thiolate. Just about the same result is obtained with the configurationally stable carboxylate analogue γ-L-Glu-L-AspGly (**2**), which binds with a $K_d$ of $9 \times 10^{-7}$ M. The incorrectly configured analogue γ-L-Glu-D-AspGly (**3**) binds somewhat less tightly ($K_d = 4.7 \times 10^{-5}$ M) than does GSH.

The rather pronounced affinity of the enzyme for the side-chain anion is an obvious indication of a positively charged electrostatic field in the active site of the enzyme, the function of which is to destabilize or lower the $pK_a$ of the thiol. The rest of the complementary surfaces of the enzyme and peptide presumably function to position the thiol in that electrostatic field. In the absence of a crystal structure of the enzyme several laboratories have resorted to what might be called "site-directed substrate mutagenesis", the alteration of the peptide substrate structure to probe enzyme–substrate interactions. A fairly large number of substrate analogues of GSH have been synthesized for this purpose (*12–17*), and a good deal of valuable information has been catalogued concerning the likes and dislikes of various isoenzymes with respect to specific modifications of the peptide structure (*14–17*). In some instances it is obvious that alteration of the peptide structure changes the relative orientation of the peptide and electrophilic substrates in the active site. For example, the stereoselectivity of the addition of the retro–inverso isomer of GSH (*13*) to phenanthrene 9,10-oxide catalyzed by isoenzyme 4-4 is quite distinct from that shown with the natural substrate (*12*). Although conjecture about the specific orientation of peptide and protein functional groups from these types of studies is entertaining, it is also potentially misleading without some notion of the mechanistic consequences or a knowledge of the protein structure. Unfortunately, only a couple of the alternative substrates have been pursued in any mechanistic detail (*12*), and virtually nothing is known about the protein structure at this time.

Two alternative substrates, $N$-acetylGSH and γ-L-Glu-L-Cys, which are reasonably good substrates for isoenzyme 4-4, have been shown, in one very important respect, to behave similarly to GSH in catalysis (*12*). The enzyme is capable of significantly lowering the apparent $pK_a$ of the thiol group of both of these alternative substrates when bound in the active site. Kinetically determined $pK_a$ values, which obviously should be viewed with considerable caution, suggest that the enzyme reduces the $pK_a$ of the thiol by as much as 2 $pK$ units in the ternary complex of enzyme, analogue, and CDNB. But there is a difference; it is that the limiting values of $k_c$ for the two analogues at high pH are smaller by an order of magnitude than that for GSH. The apparent $pK_a$ of enzyme-bound $N$-acetylGSH and γ-L-Glu-L-Cys (7.5 and 7.7, respectively) are somewhat higher than that for E·GSH. What is unusual is that the nucleophiles (E·analogues) with the higher $pK_a$ are less reactive at high pH, where they are completely deprotonated, than is the less basic nucleophile, E·GS⁻.

This is contrary to the expected Brønsted behavior where the nucleophiles of higher $pK_a$ should be at least as reactive as GSH (Brønsted $\beta_{nuc} = 0$) or more reactive ($\beta_{nuc} > 0$). It is certainly possible, even likely, that the kinetically determined $pK_a$s are not precise. However, another interesting possibility is that the analogues are not optimally oriented in the active site such that the thiolate is coordinated to another functional group on the protein or is more heavily solvated than it is in E·GS⁻. Both scenarios would have the effect of reducing the nucleophilicity of the thiolate analogues, though to what extent is difficult to predict. Thus another role of the active site of the enzyme may be to shield the thiolate anion from solvent and enhance its reactivity.

Other considerations suggest that desolvation of the nucleophile may be a key feature in catalysis. As suggested above, lowering the $pK_a$ of the thiol in the active site will, depending on the sensitivity of the transition state, actually decrease to a greater or lesser extent the effective nucleophilicity of the anion. The fact that transition states for addition of thiolate anions to electrophiles are relatively insensitive to the basicity of the nucleophile ($\beta_{nuc}$ typically $< 0.5$) suggests that a decrease in the $pK_a$ will only marginally decrease the reactivity of the anion and will be more than offset by the increased fraction of the anion (11, 12, 18). Another way in which the enzyme might enhance the reactivity of the thiolate is to shield it from solvent. A recent investigation by Huskey et al. (19) has directly addressed this problem through the use of kinetic solvent isotope effects on both the specific-base- (spontaneous) and enzyme-catalyzed additions of GSH to CDNB. The inverse kinetic solvent isotope effect of 0.84 on the spontaneous reaction was used to estimate that the extent of desolvation in the transition state is a modest 34%, an observation that is consistent with the low sensitivity ($\beta_{nuc} = 0.16$) of the reaction to the basicity of the nucleophile (12). Fractionation factors estimated for E·GS⁻, and the ternary complex E·GS⁻·CDNB from the kinetic solvent isotope effects of near unity for $k_c$ and $k_c/K_m^{GSH}$ and of 0.79 for $k_c/K_m^{CDNB}$ in the reactions catalyzed by isoenzyme 1-1 suggest that a change in solvation of the thiolate ion occurs in proceeding from E·GS⁻ to the transition state, a change similar in nature to that observed in the spontaneous reaction (19). More importantly, it appears that, in the enzyme-catalyzed reaction, the desolvation in the transition state is almost complete as compared to the partial desolvation apparent in the specific-base-catalyzed reaction.

### Identification of Essential Catalytic Residues

One of the first questions posed by mechanistic enzymologists in attempting to describe the catalytic mechanism of an enzyme is the following: What are the identities of the amino acid side chains that are essential for catalysis? For instance, apropos to the discussion above is the identity of the contributors to the electrostatic field that lowers the $pK_a$ of enzyme-bound GSH. There has been speculation that an imidazolium side chain of a histidine residue in the active site might be partially responsible for the electrostatic field that lowers the $pK_a$ of GSH (4, 12). This would be an intermolecular example of the type of thiol activation seen in the active site of papain, for example. There are a few reports that chemical modification of specific residues which might be involved in the deprotonation of GSH, namely, histidine (20) and arginine (21), leads to substantial inactivation of particular GSH transferases. Some investigators have interpreted this to mean that these residues are "essential" for catalysis. If

the history of chemical modification of proteins and the more recent lessons of site-specific mutagenesis have taught us anything, it is that such conclusions should not be so quickly drawn. Van Bladeren and co-workers have presented evidence that modification of a cysteine residue at or near the active site by tetrachloro-1,4-benzoquinone and related compounds leads to irreversible inactivation of the $\mu$ class isoenzymes (22, 23). These workers have suggested that even though cysteine is probably not an essential catalytic residue, there is at least one side chain near enough to the active site to be a target for active site directed reagents.

The availability of vectors for the heterologous expression of the mammalian enzymes in *Escherichia coli* (24–29) has encouraged the more the recent application of site-specific mutagenesis to this problem. Inasmuch as most of this work has yet to appear in the primary literature, only a few highlights will be summarized here. Board, Mannervik, and co-workers (30) have prepared and characterized to some degree over a dozen site-specific mutants of the human isoenzyme $\epsilon$. The residues that were chosen as targets were among those found to be conserved among the various isoenzymes and for which some catalytic role might be imagined. For example, two highly conserved arginine residues appear to contribute to the binding of GSH since both the R13A and R20A mutants show diminished affinity for GSH and S-hexylGSH. It is possible that arginine side chains help to position GSH in the active site or constitute part of the electrostatic field that destabilizes the thiol of bound GSH. Further work on the mechanistic consequences of these and other mutations should be enlightening.

The case for involvement of a histidine residue as a general feature in catalysis by the GSH transferases is not good. In the first place there is no highly conserved histidine in the sequences of the mammalian enzymes when all the gene classes are compared. There are, however, conserved histidines within a gene class. For example histidine residues 14 and 83 are conserved in the $\mu$ class isoenzymes, so it is conceivable that one or both might be crucial to the catalytic mechanism. Zhang et al.[1] have mutated each of the four histidine residues to asparagine in isoenzyme 3-3 from rat. None of the four mutants, H14N, H83N, H84N, and H167N, show any particular gross deficiency in catalyzing the addition of GSH to CDNB, phenanthrene 9,10-oxide, or 4-phenyl-3-buten-2-one. In addition, $^1H$ and $^{13}C$ NMR studies[1] of the imidazole side chains of isoenzyme 3-3 show absolutely no perturbation in their $pK_a$s upon binding of GSH, an observation that makes it highly unlikely that histidine side chains are involved in catalysis (31). Histidine is not an essential residues for isoenzyme 3-3, and it can be safely said that it does not play an immutable role in GSH transferase catalysis in general though it may participate in some ancillary fashion in other isoenzymes. To further emphasize the point, Lu and co-workers[2] have also found that the H8V, H143V, and H159V mutants of the $\alpha$ class isoenzyme 1-1 from rat are quite active in the CDNB assay.

Unpublished evidence[3] also suggests that the cysteine residues in the $\mu$ class isoenzymes, which are probably the targets for the chemical modification by tetrachloro-1,4-benzoquinones, are not essential for catalytic activity. The single cysteine mutants C86S, C114S, and C173S and the double mutant N198K/C199S of isoenzyme 3-3, all of

[1] P. Zhang, G. F. Graminski, and R. N. Armstrong, unpublished results.
[2] A. Y. H. Lu, private communication.
[3] P. Zhang, and R. N. Armstrong, unpublished results.

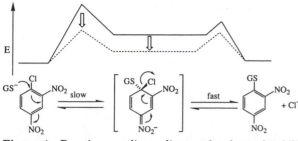

**Figure 4.** Reaction coordinate diagram for the nucleophilic addition of GS⁻ to 1-chloro-2,4-dinitrobenzene. Catalysis by the enzyme beyond simple deprotonation the thiol must involve stabilization of the transition state for formation of the σ-complex, the slow step. Such stabilization of the transition state is likely also to be apparent in stabilization of the intermediate itself, particularly if it resembles the transition state.

which were made for crystallographic purposes, are good catalysts of the CDNB reaction. The inhibition observed after chemical modification is probably due to the introduction of a bulky group that either fills or perturbs the structure of the active site.

## Chemical Mechanism and Substrate Specificity

**Nucleophilic Aromatic Substitution.** The substrate specificity of any enzyme is a function of how effectively the enzyme can lower the activation barrier(s) for a particular chemical process. To truly understand the diversity in the catalytic specificity of the GSH transferase, it is essential to come to grips with the nature of the intermediates and transition states on the reaction coordinate. These can vary widely with the type of electrophilic functional group as well as the general topology of the substrate.

Some progress has been made in sorting out the process of nucleophilic aromatic substitution reactions catalyzed by the enzyme. This reaction is illustrated in Figure 4 for the most common spectrophotometric substrate for the GSH transferases, 1-chloro-2,4-dinitrobenzene. In solution, this general class of reaction is known to follow an addition–elimination sequence with the formation of σ-complex or Meisenheimer complex intermediates. There is strong evidence with many nucleophiles and leaving groups that the rate-limiting step in the spontaneous reaction is the formation of the intermediate. If the rate acceleration provided by the enzyme involves something other than ionization and desolvation of the nucleophile, then the best place to look is at the transition state for σ-complex formation. Does the enzyme, as illustrated in Figure 4, stabilize the transition state for formation of the σ-complex? Several lines of evidence suggest that isoenzymes 3-3 and 4-4 can do just that.

The most commonly cited evidence for rate-limiting formation of the σ-complex in nucleophilic aromatic substitution reactions is the increased reactivity of the substrate upon substitution of F⁻ for Cl⁻. The more electronegative leaving group is anticipated to facilitate formation of the intermediate but to have just the opposite effect on its decomposition. The second-order rate constant for the spontaneous reaction of GSH with 1-chloro-2-nitro-4-(trifluoromethyl)benzene is about 50-fold smaller than that for the 1-fluoro compound. The kinetic constants $k_c$ and $k_c/K_m$ for these two substrates with isoenzyme 4-4 differ by factors of 40 and 10, respectively, suggesting that the enzyme-catalyzed reaction has the same rate-limiting step as the base-catalyzed reaction (*12*). A similar observation has been made by Blanchard and co-workers with isoenzyme 6-6.[4] The differences between the kinetic con-

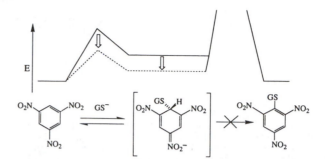

**Figure 5.** Reaction coordinate diagram for the equilibrium formation of the dead-end σ-complex between GS⁻ and 1,3,5-trinitrobenzene. The enzyme stabilizes the σ-complex by a factor of about 10³ and may also lower the barrier to its formation as indicated by the dashed line.

stants for the same two substrates in reactions catalyzed by isoenzyme 3-3 are more modest, with the fluoro substrate being better by factor of only 4 (*32*). The lower sensitivity of isoenzyme 3-3 to the identity of the leaving group is perhaps significant since this isoenzyme is better than isoenzyme 4-4 at catalyzing nucleophilic aromatic substitutions.

The sensitivity of the transition state in enzyme-catalyzed aromatic substitution reactions to electronic effects of the para substituent is instructive particularly when compared to the specific-base-catalyzed reaction (*12*). A Hammett plot of log $k_c$ vs σ⁻ for a series of 4-substituted 1-chloro-2-nitrobenzene substrates and isoenzyme 4- *gives* a ρ value of 1.2, which is substantially smaller than the value of 3.4 found in the specific-base-catalyzed reaction. This finding is consistent with rate-limiting formation of the σ-complex and perhaps an earlier or much later transition state for the enzyme-catalyzed reaction (*12*). A smaller substituent effect in the enzyme-catalyzed reaction is certainly compatible with the view that the thiolate in the active site is more reactive, thus requiring less electronic stabilization of the transition state. It is interesting that the substituent effect in the enzyme-catalyzed reactions with the two peptide analogues *N*-acetylGSH (ρ = 1.9) and γ-L-Glu-L-Cys (ρ = 2.4) is greater than that for GSH. This observation is perhaps a reflection of a greater degree of solvation in the transition states for the two less reactive analogues.

Direct evidence for the enzyme-catalyzed formation of the σ-complex has been obtained from an investigation of the interaction of 1,3,5-trinitrobenzene with binary complexes of GSH with isoenzymes 3-3 and 4-4 of the GSH transferases from rat liver (*31, 32*). This electron-deficient arene, which does not posses a good leaving group, favors reversible accumulation of the σ-complex in the active site of the enzyme at relatively low substrate concentration as illustrated in Figure 5. Isoenzyme 3-3 appears to have a particular capacity to stabilize the σ-complex between GSH and trinitrobenzene (TNB). A comparison of the equilibrium constant, $K_f$, for formation [GS⁻$_{(aq)}$ + TNB ⇌ M] of the Meisenheimer complex, M, to that of the enzyme-bound dead-end intermediate (E·GS⁻ + TNB ⇌ E·M) can be made from the pH independent formation constants. The two formation constants of 28 M⁻¹ for M and 9.6 × 10⁴ M⁻¹ for E·M suggest that the enzyme stabilizes the intermediate by as much as 4.8 kcal/mol. It is not known whether the much more favorable equilibrium in the active site is due to a larger rate constant for the forward reaction or a smaller rate constant for decomposition of E·M.

---

[4] K. K. Wong, G. Banks, and J. S. Blanchard, private communication.

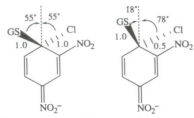

**Figure 6.** Two possible transition states for the enzyme-catalyzed addition of GSH to CDNB derived from BEBOVIB calculations and the observed inverse $^{35}$S and $^{14}$C primary kinetic isotope effects on the reaction. A late transition state for $\sigma$-complex formation with unit C–S and C–Cl bond orders is shown on the left. Angles of the incoming nucleophile and the leaving group with respect to the plane of the cyclohexadienate ring are indicated. The alternative of a late (product-like) transition state for decomposition of the intermediate with complete C–S bond formation and substantial C–Cl bond cleavage is shown on the right. The figure is courtesy of K. K. Wong, G. Banks, and J. S. Blanchard.[4]

Preliminary studies (*33*) of the preequilibrium kinetics of formation suggest that the enzyme does accelerate the forward reaction with a $k_1$ which is much larger than $k_1$ = 2.9 × 10$^3$ M$^{-1}$ s$^{-1}$ (*34*) observed in aqueous solution. It therefore seems that the enzyme can also stabilize the transition state for formation of a $\sigma$-complex and that the intermediate must resemble, to a first approximation, the structure of the transition state.

The ability of a particular isoenzyme to stabilize M should be related to its catalytic efficiency toward nucleophilic aromatic substitution reactions if the analysis above is correct. Comparison of $K_{f(app)}$ at pH 7.5 reveals that isoenzyme 3-3 stabilizes the 1-(*S*-glutathionyl)-2,4,6-trinitrocyclohexadienate anion about 70-fold more effectively than does isoenzyme 4-4. In most instances isoenzyme 3-3 is roughly an order of magnitude more efficient (judged by $k_c/K_m$) at aromatic substitutions than isoenzyme 4-4 so that the relationship appears to hold. Trinitrobenzene is, as expected, a very potent reversible inhibitor of the enzyme since the dead-end intermediate has a quite high affinity for the active site (*32, 35*).

Evidence that the transition state for the isoenzyme 6-6 catalyzed addition of GSH to CDNB resembles the structure of the $\sigma$-complex intermediate has been obtained through other means by Blanchard and co-workers.[4] Both $^{35}$S ($^{35}$S-GSH) and $^{14}$C ($^{14}$C-CDNB) primary kinetic isotope effects on $k_c/K_m$ for this reaction are inverse ($^{35}$S, $k_c/K_m$ = 0.971 ± 0.005; $^{14}$C, $k_c/K_m$ = 0.964 ± 0.010). These values, when matched to the isotope effects predicted for various transition-state structures modeled by BEBOVIB calculations, suggest two possible alternatives for the transition-state structure as shown in Figure 6. The isotope effects are compatible with a transition-state structure that resembles either the structure of the intermediate (a late transition state for $\sigma$-complex formation) or the structure of the product (a late transition state for decomposition of the $\sigma$-complex). Wong, Banks, and Blanchard[4] have concluded that the first possibility is more likely given the preponderance of evidence that formation of the intermediate is rate-limiting.

The dead-end intermediate complex formed with trinitrobenzene is also a useful tool with which to study the ionization behavior of GSH and perhaps other functional groups in the active site of the enzyme. The pH dependence of complex formation on the enzyme is not simple. The dependence of log $K_f$ on pH clearly suggests that the equilibrium with isoenzyme 3-3 is affected by multiple ionizations (*32*). Our original analysis of the pH dependence of $K_f$ concluded that $\sigma$-complex formation occurred in a simple collision complex between TNB and E·GS$^-$, was

**Scheme I**

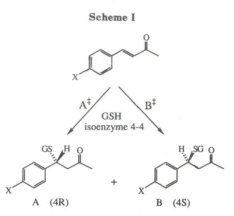

dependent on two ionizations in the active site, and ignored the rather curious sigmoidal dependence of the log $\epsilon$ on pH. In this analysis the p$K_a$ of enzyme-bound GSH was estimated to be 5.7 with an additional ionization of another group with p$K_a$ of 7.6. Cleland subsequently derived a model that incorporates the dependence of $\epsilon$ (maximum color formation) on pH assuming the formation of a colorless collision complex followed by a unimolecular step for formation of the $\sigma$-complex and a pH dependence in the internal equilibrium between the colorless complex and the $\sigma$-complex.[5] At high pH the internal equilibrium constant is very large, but at low pH it is assumed to be 0.61, the factor by which $\epsilon$ changes going from high to low pH. Cleland's model predicts a p$K_a^{E-GSH}$ = 6.7, which is in good agreement with the p$K_a$ measured by other techniques with isoenzyme 4-4 (*9, 10*). Preequilibrium kinetic studies which should help to sort out this intriguing system are in progress.

**Michael Additions and Epoxide Ring Openings.** The evolutionary development of catalytic specificity in the GSH transferases is an interesting matter on which to speculate. Different electrophilic functional groups do have different requirements with respect to catalysis. However, the reactions involving the addition of GSH to $\alpha,\beta$-unsaturated aldehydes and ketones and epoxides appear wholly unrelated to nucleophilic aromatic substitution, yet all share a common need for stabilization of anionic intermediates or transition states. The evolutionary pressure for the detoxication of Michael acceptors and epoxides can be readily imagined. So, have specific isoenzymes evolved to accommodate particular functional groups and transition states? Or does each isoenzyme just possess some general but slightly modified ability to stabilize anionic transition states? Mechanistic investigations can perhaps help to define such questions if not answer them directly.

The Michael addition of GSH to 4-phenyl-3-buten-2-one (Scheme I, X = H) catalyzed by isoenzyme 4-4 from rat is stereoselective such that a 9:1 ratio of the two possible diastereomeric products is observed (*36*). Clearly, the transition state for addition of GSH to one prochiral face of the enone is preferred. The question arises as to whether this stereochemical preference is mechanism based (i.e., stereoelectronic in nature) or is simply a consequence of some steric interference with one possible mode of binding. Were the latter the case, one might anticipate essentially the same stereoselectivity of the enzyme toward a series of sterically similar substrates regardless of their reactivity. Just the opposite has been found for a series

---

[5] W. W. Cleland, private communication. The alternative model derived by Cleland gives a considerably better statistical analysis than does our original calculations.

**Table I. X-ray Diffraction Quality Single Crystals of GSH Transferase**

| species isoenzyme, gene family | space group | unit cell dimensions, Å | unit cell volume, nm³ | $V_m$, Å³/dalton | asymmetric unit | resolution, Å | ref |
|---|---|---|---|---|---|---|---|
| human, GST2, $\alpha$ | monoclinic C2 | $a = 100.8$, $b = 95.4$, $c = 105.2$, $\beta = 92.4°$ | 1012 | 2.49 | two dimers | 2.7 | 43 |
| rat, 3-3, $\mu$ | monoclinic C2 | $a = 88.4$, $b = 69.4$, $c = 81.3$, $\beta = 106.0°$ | 478.5 | 2.3 | one dimer | <2.0 | 41 |
| bovine, neutral, $\pi$ | tetragonal, $P4_12_12$ or $P4_32_12$ | $a = b = 61$, $c = 237$ | 882 | 2.4 | one dimer | 2.6 | 42 |
| human, acidic, $\pi$ | tetragonal, $P4_12_12$ or $P4_32_12$ | $a = b = 60.1$, $c = 244$ | 881 | 2.45 | one dimer | <2.7 | 44 |

of para-substituted 4-phenyl-3-buten-2-ones in which the reactivity of the enone was varied by changing the electron-withdrawing ability of the para substituent. The degree of stereoselectivity actually decreases with increasing reactivity of the substrate in a manner that is consistent with differential electronic stabilization of the two possible transition states. The enzyme exhibits little stereoselectivity toward more reactive substrates (X = NO₂, 66% product A) and a much higher stereoselectivity toward the least reactive substrate (X = OCH₃, 95% product A).

The log of $(k_c/K_m)_{obs}$ for the addition of GSH to a series of para-substituted 4-phenyl-3-buten-2-ones follows the linear free energy relationship given in eq 2 (*36*), where

$$\log (k_c/K_m)_{obs} - c_{app}\pi = \rho_{app}\sigma \qquad (2)$$

$\sigma$ is the electronic substituent constant, $\pi$ is the Hansch hydrophobic substituent constant (*37*), and $\rho_{app}$ and $c_{app}$ are measures of the sensitivity of the rate constant to the electronic and hydrophobic nature of the substituent. Of course, as illustrated in Scheme I, $(k_c/K_m)_{obs}$ is a composite rate constant that represents a weighted average for the reaction proceeding through the two possible parallel transition states and is dominated by the transition state of lowest activation energy. The apparent substituent effects $\rho_{app}$ and $c_{app}$ are therefore a linear combination of the substituent effects for each transition state A⁺ and B⁺. Measurement of the stereoselectivity (product ratio) allows a measurement of the individual substituent constants such that $\rho_A = 0.78$, $c_A = 0.44$, $\rho_B = 1.74$, and $c_B = 0.51$. It can be readily seen that the sensitivity of the two transition states to the hydrophobicity of the substituent is roughly the same, $c_A \approx c_B$, but that there is a substantial difference in the electronic sensitivity, $\rho_A < \rho_B$. Thus the transition state for the reaction which is *least* effectively catalyzed by the enzyme is considerably *more* sensitive to the presence of an electron-withdrawing group. Furthermore, since the log of the ratio of the two products reflects the difference in free energies of activation for the two diastereomeric transition states, it can be readily shown that a linear free energy relationship exists between this quantity and $\sigma$ as expressed in

$$\log ([A]/[B]) - (c_A - c_B)\pi = (\rho_A - \rho_B)\sigma \qquad (3)$$

where $\rho_{obs} = \rho_A - \rho_B = -0.94$ and $c_{obs} = c_A - c_B = -0.07$.

One possible interpretation of these results shown in Figure 7 suggests that one transition state, A⁺, is much less sensitive to the nature of the substituent because the enzyme provides additional electrostatic stabilization of the enolate, perhaps through a specific functional group (BH⁺) in the active site. This stabilization is not available in transition state B⁺ so that formation of product B is much more sensitive to the reactivity of the substrate itself. This type of mechanistic observation tends to support the notion that specific isoenzymes may have evolved to deal specifically and stereospecifically with particular functional groups.

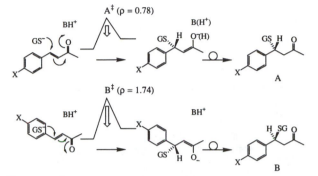

**Figure 7.** Two possible transition states for the isoenzyme 4-4 catalyzed addition to substituted phenylbutenones. Transition state A⁺, which has the lower activation barrier, due to enzymic assistance in stabilization of the enolate, is less sensitive to the electronic character of X as indicated by the lower $\rho$ value. The transition state for attack of GSH on the opposite face of the molecule does not derive assistance from BH⁺; hence B⁺ has a higher activation energy and is much more sensitive to the nature of X. Note that the absolute configuration of the products is not known.

It is interesting to note that isoenzyme 4-4 which is a very good catalyst for Michael additions is also quite good in catalyzing the addition of GSH to epoxides. Both reactions benefit from stabilization of oxyanions on the reaction coordinate. Some time ago we found that this isoenzyme was stereospecific in the addition of GSH to many K-region arene oxide substrates and that the stereoselective preference was lost with azaarene oxide substrates (*38, 39*). This behavior was interpreted to suggest that the hydrophobic character of the substrate was a major determinant in the stereoselectivity. The extent to which stereoselective protonation of the oxyanion may contribute to the observed stereochemistry in this reaction is not known. It is now known that the enzyme loses its stereoselectivity toward certain very reactive substrates, for example, the enantiomers of 3,4,5,6-tetramethylphenanthrene 9,10-oxide (*40*). It should be possible to determine the electronic contribution to the stereoselectivity toward epoxides with the appropriate substrates.

## Protein Structure and Catalytic Specificity

**Crystal Structure of the Enzyme.** Crystals, which are suitable for X-ray diffraction studies, have been described in the literature for four different isoenzymes representative of the three major gene classes of the GSH transferases (*41–44*). The characteristics of each are summarized in Table I. Given the level of interest in the three-dimensional structure of this class of proteins and the number of laboratories pursuing its solution, it is quite likely that a refined, high-resolution structure of at least one isoenzyme will be available in the near future. Indeed, it is probable that a structure of an example from each gene class will be forthcoming in a short period of time.

In one instance the protein crystals have been shown to be catalytically competent. Crystals of isoenzyme 3-3 are

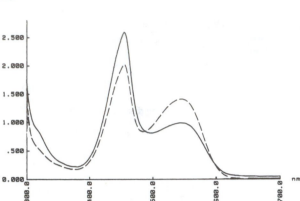

**Figure 8.** Polarized adsorption spectra of the 1-(S-glutathionyl)-2,4,6-trinitrocyclohexadiene anion bound at the active site of crystalline isoenzyme 3-3. The two spectra were recorded with the plane of polarization parallel to the $b$ (—) and $c$ (---) axes of the unit cell after soaking the crystals in the presence of 1 mM trinitrobenzene and 1 mM GSH for 1.5 h.[6]

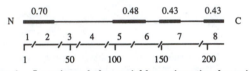

**Figure 9.** Location of the variable regions in the primary structure of the class $\mu$ glutathione transferases. The four variable regions are indicated by the bold line. The fraction of identical positions in each region for subunits 3, 4, and 6 is indicated at the top. The conserved regions indicated by the narrow lines exhibit ≥90% sequence identity. The bottom line shows the relationship of the exon/intron interfaces to the residue number of the protein. Exons are numbered 1–8, with the bottom scale indicating the amino acid residue number.

capable of forming the Meisenheimer complex between GSH and 1,3,5-trinitrobenzene (*32*). After incubation for a short period of time in the presence of moderate concentrations of GSH and TNB the crystals turn a deep red-orange color indicative of complex formation. In addition, the chromophore in the crystal exhibits a distinct polarized adsorption spectra when the plane of the polarized light is aligned along two of the principal axes as shown in Figure 8. The chromophore therefore has a specific orientation with respect to the crystallographic axes.[6] Single-crystal spectroscopy should be valuable in characterizing the properties and orientation of the complex in the crystal. Ultimately, of course, it should be possible to obtain a high-resolution picture of exactly how the enzyme stabilizes the dead-end intermediate through X-ray crystallography.

**Chimeric Glutathione Transferases.** Inspection of the primary sequence information available from cDNA clones (*25*, *45–48*) of the class $\mu$ isoenzymes from rat indicates that there are several variable regions (<70% amino acid identity) in the primary structure connected by highly conserved regions with >90% identity as shown in the schematic of Figure 9. Interesting questions arise as to the importance of these variable regions to the different catalytic properties of individual isoenzymes and to the evolution of catalytic function. Is one region more important than another? Can the substrate specificity of one isoenzyme be conferred upon another related isoenzyme by transplantation of one or more of the variable regions? Can new catalysts be designed by focusing on the redesign of sequence within variable regions? These types of

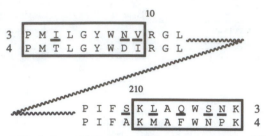

**Figure 10.** Amino acid changes made in the construction of three chimeric GSH transferases listed in Table II. The regions swapped from subunit 4 into subunit 3 are indicated in the boxes. The nonconserved residues are underlined.

**Table II. Catalytic Properties of Native and Chimeric Class $\mu$ GSH Transferases toward 4-Phenyl-3-buten-2-one at pH 6.6 (*28*)**

| iso-enzyme or chimera | $K_d^{GSH}$, $\mu M^a$ | $k_c$, $s^{-1}$ | $k_c/K_m^s$, $M^{-1}\,s^{-1}$ | % isomer A | $\delta\Delta G^{\ddagger}$, kcal/mol$^b$ |
|---|---|---|---|---|---|
| 3-3 | 21 | 0.70 | 3900 | 58 | 0.2 |
| 4-4 | 24 | 8.5 | 47000 | 90 | 1.3 |
| $(4^9 3^{208})_2$ | 42 | 1.4 | 6100 | 93 | 1.5 |
| $(3^{209} 4^8)_2$ | 180 | 0.91 | 3500 | 67 | 0.5 |
| $(4^9 3^{200} 4^8)_2$ | 20 | 0.95 | 3700 | 75 | 0.6 |

$^a$ Determined by fluorescence titration. $^b$ Calculated from $RT \ln$ ([A]/[B]).

questions are just beginning attract the attention of biochemists interested in the enzymes of detoxication (*49*). It is likely that some of these questions can be answered by construction and characterization of chimeric enzymes in which the variable region of one isoenzyme is used to replace the analogous region of another.

A set of first-generation chimeric GSH transferases have been constructed in order to test the feasibility of this approach for sorting structure–function relationships in the class $\mu$ isoenzymes (*28*, *31*). The class $\mu$ subunits 3, 4, and 6 share about 77% sequence identity, with most of the sequence differences clustered in four regions (Figure 9). The variable regions roughly encompass residues 1–34, 98–135, 150–172, and 198–217. Our initial focus was on the variable regions at the ends of the polypeptide chains. The expression vector for subunit 3 (*28*) was constructed and manipulated to give three chimeric genes that encode hybrid polypeptides that contain short sequences derived from the gene encoding the amino- and carboxyl-terminal domains of subunit 4. The first three hybrid polypeptides successfully expressed were designated $4^9 3^{208}$, $3^{209} 4^8$, and $4^9 3^{200} 4^8$ to indicate, for example, in the first instance that the sequence of first nine amino acids is derived from subunit 4 and that the remaining 208 are from subunit 3. The actual mutations are illustrated in Figure 10.

The three active hybrid dimers are quite different in catalytic character with respect both to each other and to the parent proteins from which they were derived. Their catalytic properties toward one substrate shown in Table II serve to illustrate the point. It is obvious from the data in Table II that alterations in both the amino- and carboxyl-terminal regions of the polypeptide influence to some degree the structure of the active site. Although the changes in the kinetic constants are generally modest, the relative $k_c/K_m$ for the two diastereomeric transition states changes by a factor of almost 10 with $(4^9 3^{208})_2$ as reflected in the substantial increase in the stereoselectively compared to the native isoenzyme 3-3. That the C-terminal domain of the protein may be part of the active site has independent support from the work of Boyer and co-

---

[6] A. Mozzarelli and R. N. Armstrong, unpublished results. The polarized absorption spectra were obtained and kindly provided by Prof. A. Mozzarelli at the University of Parma.

workers (*50*), who have reported the photoaffinity labeling of the C-terminal region of isoenzymes 1-1 and 2-2 with an active site directed photoprobe. The region labeled corresponds to the region altered in the $\mu$ class chimeras.

Even more interesting is the apparent complementation of changes introduced in the C-terminal region by comodification of the N-terminus. The quadruple mutation in the C-terminus of $(3^{209}4^8)_2$ introduces a minor defect in the binding of GSH as evidenced by the 9-fold increase in $K_d^{GSH}$. This defect appears to be corrected in the tripartite chimera $(4^93^{200}4^8)_2$ which has a normal $K_d^{GSH}$. This is the first indication that the N- and C-termini of the protein, which are as distant in sequence space as they can be, may actually be close in three dimensions. Whether this is an inter- or intrasubunit interaction in the active dimeric protein is not known. Even though such results are quite intriguing, it is clear that considerable caution should be exercised when interpreting the changes in catalytic character of hybrid enzymes.

## Conclusions

We are rapidly approaching the point at which we can say that we have a basic understanding of the mechanism of the GSH transferases. Although it will require a crystal structure or six to put many of the pieces together with regard to structure–function relationships and substrate specificity, all is not lost in the meantime. Mechanistic investigations are of the utmost importance if we are going to understand what the structures can tell us. In addition, the applications of molecular biology through random mutagenesis and the construction of libraries of chimeric proteins should allow us to generate and screen new GSH transferases with altered, interesting, and perhaps even useful catalytic properties.

**Acknowledgment.** I am particularly indebted to the students and postdoctoral associates and colleagues in my laboratory including John Darnow, Helen Fung (Chen), Gerard Graminski, Pinghui Zhang, Muctarr Sesay, Prof. Yasuo Kubo, and Prof. Herman Ammon, whose hard work and good humor are responsible for the results cited from the University of Maryland. Work in my laboratory was supported by grants from the NIH (GM 30910) and the American Cancer Society (BC-632). I also express my appreciation to Prof. John Blanchard, Dr. Kenny Wong, and Dr. Anthony Lu for generously providing preprints and unpublished data from their most recent work, to Prof. W. W. Cleland for his interest and insight, and to Andrea Mozzarelli for the polarized absorption spectra. I also thank the University of Maryland Center for Advanced Research in Biotechnology for their very generous accommodations during a sabbatical leave in which this review was written.

**Registry No.**  GSH transferase, 50812-37-8.

## References

(1) Jakoby, W. B. (1978) The glutathion S-transferases: a group of multifunctional detoxification proteins. *Adv. Enzymol. Relat. Areas Mol. Biol. 46*, 383–414.

(2) Mannervik, B. (1985) The isoenzymes of glutathione transferase. *Adv. Enzymol. Relat. Areas Mol. Biol. 57*, 357–417.

(3) Jakoby, W. B., Ketterer, B., and Mannervik, B. (1984) Glutathione transferases: nomenclature. *Biochem. Pharmacol. 33*, 2539–2540.

(4) Mannervik, B., and Danielson, U. H. (1988) Glutathione transferases—Structure and catalytic activity. *CRC Crit. Rev. Biochem. 23*, 283–337.

(5) Pickett, C. B., and Lu, A. Y. H. (1989) Glutathione S-transferases: gene structure, regulation and biological function. *Annu. Rev. Biochem. 58*, 743–764.

(6) Morgenstern, R., and DePierre, J. W. (1983) Microsomal glutathione transferase. Purification in unactivated form and further characterization of the activation process, substrate specificity and amino acid composition. *Eur. J. Biochem. 134*, 591–597.

(7) Morgenstern, R., DePierre, J. W., and Jornvall, H. (1985) Microsomal glutathione transferase. Primary structure. *J. Biol. Chem. 260*, 13976–13983.

(8) Armstrong, R. N. (1987) Enzyme-catalyzed detoxication reactions: Mechanisms and stereochemistry *CRC Crit. Rev. Biochem. 22*, 39–88.

(9) Roberts, D. D., Lewis, S. D., Ballou, D. P., Olson, S. T., and Shafer, J. A. (1986) Reactivity of small thionate anions and cysteine-25 in papain toward methylmethanethiosulfonate. *Biochemistry 25*, 5595–5601.

(10) Douglas, K. T. (1987) Mechanism of action of glutathione-dependent enzymes. *Adv. Enzymol. Relat. Areas Mol. Biol. 59*, 103–167.

(11) Graminski, G. F., Kubo, Y., and Armstrong, R. N. (1989) Spectroscopic and kinetic evidence for the thiolate anion of glutathione at the active site of glutathione S-transferase. *Biochemistry 28*, 3562–3568.

(12) Chen, W.-J., Graminski, G. F., and Armstrong, R. N. (1988) Dissection of the catalytic mechanism of isozyme 4-4 of glutathione S-transferase with alternative substrates. *Biochemistry 27*, 647–654.

(13) Chen, W.-J., Lee, D. Y., and Armstrong, R. N. (1986) 4N-(Malonyl-D-cysteinyl)-L-2,4-diaminobutyrate: The end-group modified retro-inverso isomer of glutathione. *J. Org. Chem. 51*, 2848–2850.

(14) Adang, A. E. P., Duindam, A. J. G., Brussee, J., Mulder, G. J., and Van der Gen, A. (1988) Synthesis and nucleophilic reactivity of a series of glutathione analogues modified at the $\gamma$-glutamyl moiety. *Biochem. J. 255*, 715–720.

(15) Adang, A. E. P., Brussee, J., Meyer, D. J., Coles, B., Ketterer, B., Van der Gen, A., and Mulder, G. J. (1988) Substrate specificity of rat liver glutathione S-transferase for a series of glutathione analogues modified at the $\gamma$-glutamyl moiety. *Biochem. J. 255*, 721–724.

(16) Adang, A. E. P., Meyer, D. J., Brussee, J., Van der Gen, A., Ketterer, B., and Mulder, G. J. (1989) Interaction of rat glutathione S-transferases 7-7 and 8-8 with $\gamma$-glutamyl or glycyl modified glutathione analogues. *Biochem. J. 264*, 759–764.

(17) Adang, A. E. P., Brussee, J., van der Gen, A., and Mulder, G. J. (1990) The glutathione binding site in glutathione S-transferases: Investigation of the cysteinyl, glycyl and $\gamma$-glutamyl domains. *Biochem. J. 269*, 47–54.

(18) Bruice, P. Y., Bruice, T. C., Yagi, H., and Jerina, D. M. (1976) Nucleophilic displacement on the arene oxides of phenanthrene. *J. Am. Chem. Soc. 98*, 2973–2981.

(19) Huskey, S.-E., Huskey, W. P., and Lu, A. Y. H. (1990) Contributions of thiolate "desolvation" to catalysis by glutathione-S-transferase isozymes 1-1 and 2-2: Evidence from kinetic solvent isotope effects. *J. Am. Chem. Soc.* (in press).

(20) Awasthi, Y. C., Bhatnager, A., and Singh, S. V. (1987) Evidence for the involvement of histidine at the active site of glutathione S-transferase $\psi$ from human liver. *Biochem. Biophys. Res. Commun. 143*, 965–970.

(21) Asaoka, K., and Takahashi, K. (1989) Inactivation of bovine liver glutathione S-transferase by specific modification of arginine residues with phenylglyoxal. *J. Enzyme Inhibition 3*, 77–80.

(22) Van Ommen, B., den Besten, C., Rutten, A. L. M., Ploemen, J. H. T. M., Vos, R. M. E., Muller, F., and Van Bladeren, P. J. (1988) Active site directed irreversible inhibition of glutathione S-transferases by the glutathione conjugate of tetrachloro-1,4-benzoquinone. *J. Biol. Chem. 263*, 12939–12942.

(23) Van Ommen, B., Ploemen, J. H. T. M., Ruven, H. J., Vos, R. M. E., Bogaards, J. J. P., and Van Bladeren, P. J. (1989) Studies on the active site of rat glutathione S-transferase 4-4-chemical modification by tetrachloro-1,4-benzoquinone and its glutathione conjugate. *Eur. J. Biochem. 181*, 423–429.

(24) Board, P. G., and Pierce, K. (1987) Expression of human glutathione S-transferase 2 in *Escherichia coli*. Immunological comparison with the basic glutathione S-transferases from human liver. *Biochem. J. 248*, 937–941.

(25) Lai, H.-C. J., Qian, B., Grove, G., and Tu, C.-P. D. (1988) Gene expression of rat glutathione S-transferases. Evidence for gene conversion in the evolution of the Y$_b$ multigene family. *J. Biol. Chem. 263*, 11389–11395.

(26) Abramovitz, M., Ishigaki, S., Felix, A. M., and Listowsky, I. (1988) Expression of an enzymatically active Y$_{b3}$ glutathione S-

transferase in *Escherichia coli* and identification of its natural form in rat brain. *J. Biol. Chem.* **263**, 17627–17631.

(27) Wang, R. W., Pickett, C. B., and Lu, A. Y. H. (1989) Expression of a cDNA encoding a rat liver glutathione S-transferase $Y_a$ subunit in *Escherichia coli*. *Arch. Biochem. Biophys.* **269**, 536–543.

(28) Zhang, P., and Armstrong, R. N. (1990) Construction, expression, and preliminary characterization of chimeric class $\mu$ glutathione S-transferases with altered catalytic properties. *Biopolymers* **29**, 159–169.

(29) Husky, S. E. W., Wang, R. W., Linemeyer, D. L., Pickett, C. B., and Lu, A. Y. H. (1990) Expression in *Escherichia coli* of rat liver cytosolic glutathione S-transferase Yc cDNA. *Arch. Biochem. Biophys.* **279**, 116–121.

(30) Mannervik, B., Board, P. G., Berhane, K., Bjornestedt, R., Castro, V. M., Danielson, U. H., Hao, X.-Y., Kolm, R., Olin, B., Principato, G. B., Ridderstrom, M., Stenberg, G., and Widersten, M. (1990) Classes of glutathione transferases: Structural and catalytic properties of the enzymes. In *Glutathione S-transferases and drug resistance* (Hayes, J. D., Pickett, C. B., and Mantle, T. J., Eds.) pp 35–46, Taylor & Francis, London.

(31) Armstrong, R. N., Ammon, H. L., Graminski, G. F., Zhang, P., Sesay, M. A., Kubo, Y., and Dickert, L. A. (1990) Mechanistic and structural investigations of glutathione S-transferase. In *Glutathione S-transferases and drug resistance* (Hayes, J. D., Pickett, C. B., and Mantle, T. J., Eds.) pp 65–74, Taylor & Francis, London.

(32) Graminski, G. F., Zhang, P., Sesay, M. A., Ammon, H. L., and Armstrong, R. N. (1989) Formation of the 1-(S-glutathionyl)-2,4,6-trinitrocyclohexadienate anion at the active site of glutathione S-transferase: Evidence for enzymic stabilization of $\sigma$-complex intermediates in nucleophilic aromatic substitution reactions. *Biochemistry* **28**, 6252–6258.

(33) Graminski, G. F. (1990) *Mechanistic studies of glutathione S-transferase*. Ph.D. Dissertation, University of Maryland, College Park, MD.

(34) Gan, L.-H. (1977) Kinetic studies of the reacton of glutathione with aromatic compounds. *Aust. J. Chem.* **30**, 1475–1479.

(35) Clark, A. G., and Sinclair, M. (1988) The Meisenheimer complex of glutathione and trinitrobenzene. A potent inhibitor of the glutathione S-transferase from *Galleria mellonella*. *Biochem. Pharmacol.* **37**, 259–263.

(36) Kubo, Y., and Armstrong, R. N. (1989) Observation of a substituent effect on the stereoselectivity of glutathione *S*-transferase toward para-substituted 4-phenyl-3-buten-2-ones. *Chem. Res. Toxicol.* **2**, 144–145.

(37) Hansch, C., and Leo, A. (1979) *Substituent Constants for Correlation Analysis in Chemistry and Biology*, Wiley, New York.

(38) Cobb, D., Boehlert, C., Lewis, D., and Armstrong, R. N. (1983) Stereoselectivity of isozyme C of glutathione S-transferase toward arene and azaarene oxides. *Biochemistry* **22**, 805–812.

(39) Boehlert, C. C., and Armstrong, R. N. (1984) Investigation of the kinetic and recognition of arene and azaarene oxides by isozymes A2 and C2 of glutathione S-transferase. *Biochem. Biophys. Res. Commun.* **121**, 980–986.

(40) Darnow, J. N., and Armstrong, R. N. (1990) Conformational enantiomers of 3,4,5,6-tetramethylphenanthrene 9,10-oxide. A novel axially chiral arene oxide with unusual conformational stability and reactivity. *J. Am. Chem. Soc.* **112**, 6725–6726.

(41) Sesay, M. A., Ammon, H. L., and Armstrong, R. N. (1987) Crystallization and a preliminary X-ray diffraction study of isozyme 3-3 of glutathione S-transferase from rat liver. *J. Mol. Biol.* **197**, 377–378.

(42) Schaffer, J., Galley, O., and Ladenstein, R. (1988) Glutathione transferase from bovine placenta. Preparation, biochemical characterization, crystallization, and preliminary crystallography analysis of a neutral class pi enzyme. *J. Biol. Chem.* **263**, 17405–17411.

(43) Cowan, S. W., Bergfors, T., Jones, T. A., Tibbelin, G., Olin, B., Board, P. G., and Mannervik, B. (1989) Crystallization of GST2, a human class alpha glutathione transferase. *J. Mol. Biol.* **208**, 369–370.

(44) Parker, M. W., Lo Bello, M., and Federici, G. (1990) Crystallization of glutathione S-transferase from human placenta. *J. Mol. Biol.* **213**, 221–222.

(45) Ding, G. J.-F., Ding, V. D.-H., Rodkey, J. A., Bennett, C. D., Lu, A. Y. H., and Pickett, C. B. (1986) Rat liver glutathione S-transferases. DNA sequence analysis of a $Y_{b2}$ cDNA clone and regulation of the $Y_{b1}$ and $Y_{b2}$ RNAs by phenobarbital. *J. Biol. Chem.* **261**, 7952–7957.

(46) Lai, H.-C., Grove, G., and Tu, C.-P. D. (1986) Cloning and sequence analysis of a cDNA for a rat liver glutathione S-transferase $Y_b$ subunit. *Nucleic Acids Res.* **14**, 6101–6114.

(47) Lai, H.-C. J., and Tu, C.-P. D. (1986) Rat glutathione S-transferase supergene family. Characterization of an anionic Yb subunit cDNA clone. *J. Biol. Chem.* **261**, 13793–13799.

(48) Abramovitz, M., and Listowsky, I. (1987) Selective expression of a unique glutathione S-transferase $Y_{b3}$ gene in rat brain. *J. Biol. Chem.* **262**, 7770–7773.

(49) Armstrong, R. N. (1990) Structure–function relationships in enzymic catalysis. Can chimeric enzymes contribute? *Chem. Rev.* **90**, 1309–1325.

(50) Hoesch, R. M., and Boyer, T. D. (1989) Localization of a portion of the active site of two rat liver glutathione S-transferases using a photoaffinity label. *J. Biol. Chem.* **263**, 17712–17717.

# Chapter 12

# Isolation and Purification of UDP-Glucuronosyltransferases

Thomas R. Tephly

*Department of Pharmacology, University of Iowa, Iowa City, Iowa 52242*

Reprinted from *Chemical Research in Toxicology,* Vol. 3, No. 6, November/December, 1990

## Introduction

The study of the metabolic fate of xenobiotics in mammals has been of interest for about 200 years, ever since benzoic acid conjugation to glycine was discovered (*1*). Glucuronide and sulfate conjugation was revealed in work dating back to the 1870s as was the oxidative transformation of benzene to phenol (*1*). Over the last 40 years, the biological role of the hepatic endoplasmic reticulum in the metabolism of xenobiotics by oxidation, reduction, and conjugation has been studied extensively.

Over the last 15 years, we have begun to appreciate the wide diversity of enzymes involved in glucuronide synthesis. This review will indicate progress made in the isolation and purification of UDP-glucuronosyltransferases (UDPGTs) since 1980; research prior to 1981 has been reviewed by Dutton (*2*), Kasper and Henton (*3*), and Burchell (*4*). Recent reviews (*5, 6*) have also appeared which deal with numerous aspects of UDPGTs, including purification. Although this review will emphasize our knowledge concerning the functional and physical properties of highly purified UDPGTs, some molecular biological research will also be referred to. Isolation and expression of cDNAs for UDPGTs have been carried out in parallel with studies on enzyme purification, and each approach has enhanced our understanding of the nature and variety of UDPGTs which mediate the glucuronidation of xenobiotics and endobiotics.

In this review the terminology used to denote the various UDPGTs employs the guidelines of Bock et al. (*7*), where a UDPGT is named after an endogenous substrate (bilirubin UDPGT) or a class of endogenous substrates (17$\beta$- or 3$\alpha$-hydroxysteroid UDPGTs) if a specific site on the molecule is linked to glucuronic acid. Where no endogenous substrate has been identified, the UDPGT is named for a highly reactive xenobiotic substrate (*p*-nitrophenol UDPGT, digitoxigenin monodigitoxoside UDPGT).

Glucuronides are generally more water soluble than the parent aglycon and are readily excreted into the urine or bile. It should be noted, however, that in certain cases glucuronide metabolites may be more active pharmacologically than the parent drug or may possess toxic potential. Morphine 6-*O*-glucuronide has been shown to be a more potent analgesic than morphine (*8*). Whereas this metabolite is not formed to any great extent in rats, it is found in humans (*9*) after morphine administration and can be generated, in vitro, in human liver microsomes (*10*). Adverse reactions produced by glucuronides have also been reported. Glucuronidation of the *N*-hydroxy group of *N*-hydroxyphenacetin resulted in a glucuronide conjugate that was covalently bound to protein (*11*). Estrogen D-ring glucuronides have been shown to produce an inhibition of canalicular bile flow and bile acid secretory rate in rats and to decrease hepatic excretory function in monkeys (*12, 13*). Acyl glucuronides formed from a number of nonsteroidal antiinflammatory drugs have been shown to bind covalently to proteins in vivo, and it has been suggested that the adduct acts to produce an immunological response which may result in anaphylactic reactions (*14, 15*). Thus, although glucuronidation is a major route of drug elimination in animals, potential toxicities may result from elevated levels of certain glucuronides given the right xenobiotic or even with certain endobiotic substances.

Assessment of rates of glucuronidation in tissue preparations or in vivo depends largely, but not exclusively, upon the type and amount of UDP-glucuronosyltransferase (UDPGT) present. Understanding the properties of UDPGTs, such as their substrate specificity and kinetics of reaction, has therefore been an important area of research. In addition, recent work has also revealed important information on their abundance in untreated and induced animals, qualitative and quantitative differences in lower animals and humans, and genetic regulation in a number of animal species (5, 6).

UDPGTs are membrane-bound enzymes of the endoplasmic reticulum and are found primarily in liver, a major source of tissue for work on purification. They catalyze the linkage of glucuronic acid from UDP-glucuronic acid to substances possessing OH, COOH, $NH_2$, SH, and C moieties (2, 3). These proteins are extremely labile, especially when purification procedures are employed to resolve them from other microsomal proteins and from each other. Overcoming the problems of lability in order to obtain preparations of UDPGTs retaining catalytic function has led many investigators to molecular biological work where expression of cDNAs has become a popular approach. On the other hand, our laboratory has been fortunate to ultimately purify some UDPGTs to homogeneity from rabbit, rat, and human liver microsomes. Each UDPGT, in general, has required a separate series of experimental procedures, and for each, basic questions have directed the work. Questions addressed have included the following: which detergent should be used, what concentration of detergent should be used for solubilization, what substances should be used for protection of activity, and what is the rate of decay in enzymic activity throughout a given procedure?

In the early 1970s, Dr. Emilio Sanchez and Dr. Eugenia del Villar began work on morphine glucuronidation in this laboratory. Studies were performed to determine whether p-nitrophenol or p-nitrophenol glucuronide might competitively inhibit morphine glucuronidation in microsomal preparations. The reason for doing this experiment came from the work of Zakim et al. (16), who had shown that p-nitrophenol glucuronide inhibited competitively the glucuronidation of p-nitrophenol and that p-aminophenol glucuronide did not inhibit p-nitrophenol glucuronidation. Morphine and morphine 3-glucuronide were tested as inhibitors of p-nitrophenol glucuronidation, and no inhibition of morphine glucuronidation by p-nitrophenol or its glucuronide was observed. p-Nitrophenol glucuronide competitively inhibited p-nitrophenol glucoronidation (17) as had previously been reported by Zakim et al. (16). Morphine 3-glucuronide inhibited morphine glucuronidation, but neither morphine nor its glucuronide inhibited p-nitrophenol glucuronidation. Interestingly, bilirubin stimulated (18) the rate of glucuronidation of both p-nitrophenol and morphine and produced no inhibition at any concentration. On the basis of this work one might suppose that at least three UDPGTs existed in rat liver microsomes.

Next, experiments were performed to solubilize and separate p-nitrophenol and morphine UDPGTs from both rat and rabbit liver microsomes by using Emulgen 911 as detergent (19, 20). This represented the first separation of UDPGTs from hepatic microsomes. We noted that the morphine UDPGT was extremely labile and allowed for only limited purification because the enzyme activity decreased rapidly once resolution was achieved. This problem retarded progress on study with morphine UDPGT, and it took a decade to finally find a partial solution to the problem, which resulted in the ultimate purification of the enzyme to homogeneity.

By the early 1980s it was clear that a number of hepatic microsomal UDPGTs existed in rabbits and rats. Reports from our laboratory, Bock and his colleagues (21–23), and Burchell's laboratory (24) led to this inescapable conclu-

sion. However, evidence that rat hepatic microsomal UDPGTs could be isolated to homogeneity was still of major concern. From our point of view, three major breakthroughs occurred which allowed for improvements in UDPGT purification. Gorski and Kasper (25) used a UDP-hexanolamine Sepharose 4B affinity procedure which facilitated the purification of a rat liver microsomal UDPGT reacting with p-nitrophenol. This procedure was useful but did not allow for separation of UDPGTs if several UDPGTs were present when applied to the resin. The procedure depended largely on binding of the UDPGT through the UDPGA binding site, and since this site was common to all UDPGTs, elution of the proteins by high concentrations of UDPGA generally yielded a mixture of UDPGTs. A second advance was the finding that UDPGTs could be separated on the basis of different isoelectric points (26). Lastly, we found that certain UDPGTs could be separated on the UDP-hexanolamine Sepharose 4B affinity columns by adjustment of salt and UDPGA concentrations (27).

## Studies on Rabbit Liver UDP-Glucuronosyltransferases

The first highly purified preparations of rabbit liver microsomal UDPGTs were described by Billings et al. (26) and Tukey et al. (28). In this work, highly purified p-nitrophenol and estrone UDPGTs were isolated by using anion-exchange, isoelectric focusing, and affinity chromatographic procedures. This report was also one of the first to show the dependence of activity and reconstitution of several purified UDPGTs on phospholipids (29).

Shortly after, Tukey and Tephly (30) reported on the purification to homogeneity of estrone and p-nitrophenol UDPGTs from rabbit liver microsomes. This work clearly established that the two proteins possessed different functional properties. Peptide mapping experiments clearly indicated different peptide patterns, supporting the notion that these UDPGTs were products of different genes. Isoelectric points and amino acid composition were different, but surprisingly, these proteins had identical subunit molecular weights, 57 000. This observation was, in fact, an ominous sign and would in the future lead to more than the usual difficulty in proving homogeneity. Other instances of UDPGTs with identical monomeric molecular weights will be noted later. A procedure for the rapid separation and purification of estrone and p-nitrophenol UDPGTs from rabbit liver microsomes has also been published (31).

Substrate specificities of these proteins have been studied. Falany et al. (32) found that rabbit liver estrone UDPGT catalyzed the glucuronidation of estrogens almost exclusively at the 3-position of the steroid A ring, and more recently (33), we have shown the rank order of reactivity for estrogens to be estrone > 17β-estradiol > estriol (approximately 11:5:1, respectively). When the A ring of the steroid is not aromatic, as with androsterone, no activity is observed, and testosterone with a 17β-hydroxyl group also does not serve as a substrate. However, rabbit liver estrone UDPGT catalyzes the N-glucuronidation of α-naphthylamine, β-napthylamine, and 4-aminobiphenyl at respectable rates (33) but does not react with morphine or 4-hydroxybiphenyl.

Further studies with highly purified rabbit liver microsomal p-nitrophenol UDPGT have also been performed (33). p-Nitrophenol and 1-naphthol are excellent substrates as is 2-aminophenol. This is interesting since rat liver p-nitrophenol UDPGT does not react with 2-

aminophenol (unpublished results). Although slight activity was observed with estrone and β-estradiol, no activity with estriol was observed. Furthermore, it does not react with androgens, naphthylamines, morphine, or 4-hydroxybiphenyl (33).

An antibody raised in sheep against rabbit liver *p*-nitrophenol UDPGT has been very useful. This antibody preparation inhibits and immunoprecipitates *p*-nitrophenol UDPGT activity in solubilized rabbit liver microsomes. It also immunoprecipitates certain human liver UDPGT activities (34).

Both estrone and *p*-nitrophenol UDPGTs have been shown to be glycosylated (35). Each reacts with endoglycosidases to yield peptides on SDS–PAGE with lower monomeric molecular weights than the untreated proteins. However, deglycosylation of the UDPGTs had no effect on the catalytic activity of these proteins. Thus, glycosylation appears not to play a role in the catalytic function of these enzymes.

## Purification of Rat Liver UDP-Glucuronosyltransferases

A practical extension of the principle of enzyme separation based on isoelectric focusing was the use of chromatofocusing chromatography, whereby UDPGTs could be separated on the basis of their pI values. Subsequent application and resolution of certain UDPGTs with low and high concentrations of UDPGA on UDP-hexanolamine Sepharose 4B affinity resin allowed for three rat hepatic microsomal UDPGTs to be purified to homogeneity (27). These UDPGTs were identified by different substrate specificities, different subunit molecular weights on SDS–PAGE, and unambiguous NH$_2$-terminal amino acid sequence. A *p*-nitrophenol UDPGT with a monomeric molecular weight of 56 000 and pI value of about 9.0 was shown to employ *p*-nitrophenol, 4-methylumbelliferone, and 1-naphthol as substrates but showed no reactivity with steroids or morphine (27). It was induced by pretreating rats with 3-methylcholanthrene (3-MC). Two other UDPGTs, not induced by 3-MC treatment, were shown to react with steroid substrates and named for their specific reactivity with either 17β- or 3α-hydroxyl substituents of testosterone and androsterone, respectively (27). The 17β-hydroxysteroid UDPGT reacted with testosterone and the 17β-hydroxyl position of estradiol. Interestingly, this UDPGT, $M_r$ 50 000, reacted well with *p*-nitrophenol as well as with 17β-hydroxylated steroids. In livers from untreated rats, this enzyme is probably the major form reacting with *p*-nitrophenol since the *p*-nitrophenol UDPGT in untreated rats is in very low abundance. These studies alerted us to the unsuitability of comparing reaction rates of *p*-nitrophenol in hepatic microsomal preparations from untreated and 3-MC-pretreated rats. The 17β-hydroxysteroid UDPGT does not react with 4-methylumbelliferone, morphine, or 3α-hydroxylated steroids.

A 3α-hydroxysteroid UDPGT, $M_r$ 52 000, was shown to react with steroids possessing a 3α-hydroxyl moiety (27). This protein reacts well with androsterone and etiocholanolone but does not react with *p*-nitrophenol, 4-methylumbelliferone, 1-naphthol, morphine, or β-estradiol. Later, it was shown that this protein was capable of reacting with bile acids if they had a 3α-hydroxyl substituent. Kirkpatrick et al. (36) showed that lithocholic acid and the 24-methyl ester of lithocholate were excellent substrates but that chenodeoxycholate was a marginal substrate. More recently, studies of the reactivity of this protein with bile acids of different chain lengths have been published

(37, 38). No glucuronidation of the terminal carboxyl moiety occurs in the reaction of bile acids with this protein (38).

Another interesting and important observation relevant to 3α-hydroxysteroid UDPGT was made by Matsui and Hakozaki (39), who found that 50% of Wistar rats possess a deficiency of this enzyme in liver microsomes. Wistar rat liver microsomal preparations (LA) devoid of 3α-hydroxysteroid UDPGT have been extremely useful. Hepatic microsomes prepared from these LA Wistar rats have been shown to possess low glucuronidation activity toward substrates reacting with this enzyme (40), i.e., androsterone and bile acids. We have shown that 4-aminobiphenyl glucuronidation is mediated by 3α-hydroxysteroid UDPGT by using hepatic microsomes from LA Wistar rat livers and with the purified enzyme. Furthermore, purification of other UDPGTs from LA Wistar rat liver microsomes has had the advantage of lacking a major interfering UDPGT. The use of these livers was a major factor in allowing us to purify morphine and 4-hydroxybiphenyl UDPGTs.

In 1979, Bock et al. (23) reported on the separation and purification of two UDPGTs from rat liver microsomes, a GT-1 which catalyzed *p*-nitrophenol glucuronidation and a GT-2 which catalyzed morphine, 4-hydroxybiphenyl, and chloramphenicol glucuronidation. The GT-1 was induced by 3-MC and the GT-2 was induced by phenobarbital. The GT-1, *p*-nitrophenol UDPGT, has been discussed above (*p*-nitrophenol UDPGT). The GT-2 has turned out to be at least two and, possibly, four other UDPGTs.

Ten years after the separation of morphine UDPGT, Puig and Tephly were able to purify this enzyme to appear homogeneity from rat liver (41). We took advantage of three strategies. Livers from LA Wistar rats were employed. Phenobarbital was used to pretreat these rats providing a marked induction of morphine UDPGT activity. Lastly, a relatively high concentration of exogenous phosphatidylcholine was employed throughout most of the purification procedures, a feature that provided for a substantial degree of stabilization of the morphine UDPGT. Although many detergents were tried over the years, Emulgen 911 still provided the best means of solubilization. The morphine UDPGT had a monomeric molecular weight of 56 000, a value identical with that of the *p*-nitrophenol UDPGT, but by use of successive chromatofocusing procedures, a separation of morphine UDPGT from *p*-nitrophenol UDPGT and 17β-hydroxysteroid UDPGT was possible. The enzyme catalyzed the glucuronidation of morphine and was competitively inhibited by codeine, but no reactivity toward steroids, bilirubin, *p*-nitrophenol, 4-aminobiphenyl, or α-naphthylamine was observed. Of major importance was the observation that this enzyme did not react with 4-hydroxybiphenyl. Another UDPGT (41) was resolved by chromatofocusing which mediated 4-hydroxybiphenyl but not morphine glucuronidation.

More recently, we have been able to photoaffinity label the morphine UDPGT of rat liver microsomes. It had been suspected in studies performed in the 1970s that *N*-alkyl moieties on certain morphine agonists and antagonists determined their ability to react with morphine UDPGT (42). del Villar et al. (43) showed that cyproheptadine, a tertiary amine that is not glucuronidated in rat hepatic microsomes, was a potent and competitive inhibitor of morphine UDPGT activity ($K_i = 80$ μM). When desmethylcycloheptadine, the secondary amine, was studied, inhibitory potency was greatly decreased ($K_i = 400$ μM).

del Villar and Sanchez and their colleagues later showed that benzodiazepines which are not glucuronidated but which have tertiary amine structure were potent competitive inhibitors of morphine glucuronidation (44–46). Furthermore, they demonstrated that these benzodiazepines did not inhibit p-nitrophenol, testosterone, or estrone glucuronidation. More recently, competitive inhibition by benzodiazepines of human liver morphine glucuronidation has been demonstrated by Rane et al. (47). Of the benzodiazepines, flunitrazepam (FNZ) was shown to be one of the most potent (45) in inhibiting morphine UDPGT.

We have confirmed the findings of Vega et al. (45) that FNZ produces a competitive inhibition of morphine glucuronidation ($K_i$ = 130 $\mu$M) in solubilized rat hepatic microsomes (48). As expected, FNZ had no effect on 4-hydroxybiphenyl glucuronidation. Since flunitrazepam (FNZ) has been used as a photoaffinity probe for peripheral benzodiazepine receptors (49) and since it competitively inhibits morphine glucuronidation, we have studied its potential in photolabeling morphine binding sites on morphine UDPGT. Under UV irradiation, the concentration of FNZ that produced 50% inhibition of morphine glucuronidation (using 10 mM morphine) was reduced from 2.5 mM to 100 $\mu$M. A time-dependent increase in [3H]FNZ binding to microsomal protein and a coincident decrease in morphine UDPGT activity was observed. Morphine antagonized both binding of [3H]FNZ and the inhibition of morphine UDPGT activity (48). No inhibition of testosterone, androsterone, or 4-hydroxybiphenyl glucuronidation occurred when FNZ was used in either ambient or UV light. Most importantly, UDPGA did not protect against light-enhanced FNZ inhibition of morphine glucuronidation or [3H]FNZ binding to solubilized microsomal protein. p-Nitrophenol glucuronidation is also unaffected by FNZ after light irradiation.

When purification procedures were carried out as described by Puig and Tephly (41), only fractions of protein containing morphine UDPGT were labeled by [3H]FNZ. Protein fractions with other UDPGTs present did not bind [3H]FNZ. A comparison of morphine UDPGT specific activity and [3H]FNZ binding throughout a series of purification steps gave an excellent correlation. Fluorography of an SDS–PAGE gel of solubilized rat liver microsomes which had been photolabeled with [3H]FNZ showed on band appearing between 54 and 58 kDa which corresponded to the reported subunit molecular weight of homogeneous morphine UDPGT, 56 000.

It appears that the photoaffinity labeling of morphine UDPGT may be relatively specific and may lead to many new avenues of research with this enzyme. An important observation is that FNZ only binds to active morphine UDPGT and, as activity is lost, FNZ binding is diminished. Thus, photolabeling has recently been carried out early in the purification process, and radioactivity is followed through the procedure rather than the more labor-intensive enzymic activity measurement. Also, we have performed preliminary studies with human liver microsomal preparation where morphine UDPGT activity is relatively low (50). So far, results have been similar to those of rat liver microsomal preparations.

Recently, Coughtrie et al. (10) have reported on the use of (+)-morphine and (–)-morphine with rat liver microsomal preparations. Although natural (–)-morphine formed only the 3-O-glucuronide, (+)-morphine formed glucuronides at both the 3-OH and 6-OH positions. Using a series of induction, ontogenic, selective inhibition (1-

naphthylacetic acid), and genetic deficiency parameters, they determined that there may be two isoenzymes responsible for morphine glucuronidation in rat liver. Indeed, they suggest that bilirubin UDPGT may be responsible for (+)-morphine glucuronidation. Further studies will be needed to resolve this issue.

Digitoxin (DT-3) is a cardiac glycoside containing three digitoxose sugar residues and is known to be metabolized in animals, including humans, through a series of stepwise cleavages to the bis- (DT-2) and monodigitoxoside (DT-1). The first two steps appear to be mediated by hepatic microsomal P-450 species (51). Previous studies by Schmoldt and colleagues (52, 53) and Castle (54) demonstrated that, once DT-1 is formed, a UDPGT-dependent reaction leading to DT-1 glucuronide functions for ultimate elimination of this substance. Schmoldt and Promies (52) proposed that a specific UDPGT might be responsible for DT-1 glucuronidation. Watkins and Klaassen (55) suggested that there may be a separation UDPGT which catalyzed DT-1 glucuronidation on the basis of development of DT-1 glucuronidation in rats. Watkins et al. (56) had previously suggested that DT-1 and bilirubin glucuronidation might be mediated by the same UDPGT.

The separation and purification of a DT-1 UDPGT from rat liver microsomes was accomplished by von Meyerink et al. (57). Spironolactone and pregnenolone-16$\alpha$-carbonitrile have been shown to induce DT-1 glucuronidation in rats (51–53). Rats pretreated with spironolactone were employed since Schmoldt and Promies (52) had previously shown that steroid, bile acid, bilirubin, p-nitrophenol, and 4-hydroxybiphenyl glucuronidation was not increased by spironolactone treatment. After many preliminary studies exploring various detergents and conditions for stabilization it was found that Emulgen 911 was the detergent of choice for solubilization and that high thiol concentrations were required for stabilization of DT-1 UDPGT throughout purification. DT-1 UDPGT eluted at an extremely high pH on chromatofocusing columns (p$I$ approximately 10) which allowed for a separation from p-nitrophenol UDPGT which eluted at about pH 9.0. After affinity column procedures were employed, a protein displaying DT-1 glucuronidation was obtained and studied for its substrate specificity. DT-1 UDPGT displayed a very narrow substrate specificity; it reacted only with DT-1 and DT-2. No reactivity with DT-3 was observed. Also, no reaction with bilirubin, morphine, p-nitrophenol, 4-methylumbelliferone, 4-hydroxybiphenyl, carcinogenic amines, steroids, or bile acids was observed. Furthermore, none of these xenobiotics or endobiotics inhibited DT-1 UDPGT catalyzed glucuronidation.

Recently, we have purified a 4-hydroxybiphenyl UDPGT from rat liver microsomes from animals treated with phenobarbital (58, 59). Procedures used were similar to those described by Puig and Tephy (41) except that an affinity column was used for purification of this enzyme to apparent homogeneity. This protein has a monomeric molecular weight of about 52 500. It does not react with morphine or chloramphenicol (preliminary data). It does not react with steroids but possesses a high reactivity with 4-methylumbelliferone and p-nitrophenol.

Mackenzie (60) has reported on the expression of a cDNA (UDPGT$_r$-2) which demonstrated activity toward testosterone, chloramphenicol, 4-hydroxybiphenyl, and 4-methylumbelliferone. Since the substrate specificity of the protein expressed in his work is different from that of the protein we isolated, one can suggest that there are two UDPGTs capable of catalyzing 17$\beta$-hydroxysteroids, one

**Table I. Rat Hepatic UDP-Glucuronosyltransferases**[a]

| UDPGT | $M_r$ | substrate specificity endogenous | substrate specificity exogenous |
|---|---|---|---|
| 17$\beta$-OH steroid | 50 000 | testosterone, $\beta$-estradiol | p-nitrophenol, 1-naphthol, $\alpha$- and $\beta$-naphthylamine |
| 3$\alpha$-OH steroid | 52 000 | androsterone, 3-OH bile acids | 4-aminobiphenyl, $\alpha$- and $\beta$-naphthylamine |
| p-nitrophenol | 56 000 | ? | p-nitrophenol, 1-naphthol, 4-methylumbelliferone |
| bilirubin | 54 000 | bilirubin, bilirubin monoglucuronide | 4-hydroxydimethylaminoazobenzene |
| morphine | 56 000 | morphine | naloxone |
| DT-1 | ? | ? | digitoxigenin monodigitoxoside, digitoxigenin bisdigitoxoside |
| 4-OH-biphenyl-1 | 52 500 | ? | 4-OH-biphenyl, 4-methylumbelliferone, p-nitrophenol |
| 4-OH-biphenyl-2 | 52 000 | testosterone, $\beta$-estradiol | 4-OH-biphenyl, chloramphenicol |
| serotonin | 54 000 | serotonin | p-nitrophenol, 1-naphthol, eugenol |

[a] For details see refs 27, 60–62, 41, 57–59, and 64.

of which would have broader substrate specificity to include chloramphenicol and 4-hydroxybiphenyl. Recent work in our laboratory with chloramphenicol glucuronidation suggests that a protein corresponding to Mackenzie's cDNA may be resolvable from rat liver microsomes obtained from phenobarbital-treated animals. Thus, there appears to be a separate UDPGT that reacts with testosterone, 4-hydroxybiphenyl, and chloramphenicol.

The subject of bilirubin glucuronidation has been of major importance due to the possible toxicity of bilirubin in states where relative or absolute deficiency of bilirubin UDPGT occur. The isolation of apparently homogeneous bilirubin UDPGT from rat liver microsomes of phenobarbital-treated rats was reported by Burchell and Blackaert in 1984 (61). They found a Coomassie Blue staining band on SDS–PAGE with a subunit molecular weight of 53 000 although a recent review indicates the $M_r$ to be 54 000 (5). Roy Chowdhury et al. (62) have reported purification of a rat hepatic bilirubin UDPGT and a Coomassie Blue stained protein band with a subunit molecular weight of 53 000. Although both groups found UDP-glucose and UDP-xylose to serve as substrates, no extensive aglycon substrate specificity has yet been reported. Roy Chowdhury et al. (62) have shown reactivity of bilirubin UDPGT with 4-hydroxydimethylaminoazobenzene. This protein must receive more study considering its importance.

Recently, Sato et al. (63) have reported on the isolation of a cDNA for rat liver bilirubin UDPGT. The cDNA was transfected into COS7 cells, where glucuronidation of bilirubin was observed in cell homogenates. Interestingly, this cDNA shares an identical 913-bp sequence with 3-MC-inducible p-nitrophenol UDPGT corresponding to 247 amino acid residues from the carboxy-terminal portion of the protein.

A UDPGT has recently been purified from hepatic microsomes of 3-methylcholanthrenez-treated rats which reacts with serotonin, p-nitrophenol, 1-naphthol, and eugenol (64). It has a subunit molecular weight of 54 000 on SDS–PAGE gels and differs from other UDPGTs in that it has a threonine residue at the NH$_2$ terminus. All other UDPGTs analyzed to date have had either a glycine or aspartate residue at the NH$_2$ terminus.

Table I summarizes information concerning substrate specificities for purified rat hepatic UDPGTs or for cDNAs which have been isolated and expressed. It can be seen that certain substrates such as p-nitrophenol or 1-naphthol are capable of reacting with more than one isoenzyme in rat hepatic microsomes. On the other hand, several UDPGTs have relatively narrow substrate specificities (morphine and DT-1 UDPGTs).

## Rat Kidney UDPGTs

Coughtrie et al. (65) have recently purified two transferases from rat kidney, a phenol and a bilirubin UDPGT. The subunit molecular weights on SDS–PAGE were 54 000 (phenol) and 55 000 (bilirubin). However, when Western blot analysis was performed using specific antibodies, subunit molecular weights of 53 000 (phenol) and 54 000 (bilirubin) were observed.

Yokota et al. (66) have isolated a phenol-reactive UDPGT from kidney microsomes of rats treated with $\beta$-naphthoflavone. This protein has a subunit molecular weight of 54 000 on SDS–PAGE gels and reacts with p-nitrophenol, 1-naphthol, 4-methylumbelliferone, and serotonin but not with 4-hydroxybiphenyl or chloramphenicol. Activity toward bilirubin was minimal. The NH$_2$-terminal amino acid sequence was similar to that of rat hepatic microsomal p-nitrophenol UDPGT (67) but different from that reported by Yokota et al. (64) for the serotonin UDPGT from rat liver microsomes.

## Purification of Human Liver UDPGTs

Two human liver microsomal UDPGTs have been purified to apparent homogeneity (68). One protein present in relatively high abundance was eluted from chromatofocusing chromatographic columns at pH 7.4 and one with lesser abundance that eluted at pH 6.2. Both were purified on UDP-hexanolamine Sepharose 4B affinity columns. The p$I$ 7.4 UDPGT was found to react with 4-methyumbelliferone, p-nitrophenol, $\alpha$-naphthylamine, and estriol. No activity with 4-aminobiphenyl, estrone, testosterone, androsterone, 17$\beta$-estradiol, or 5$\alpha$-androstane-3$\alpha$,17$\beta$-diol was observed. The p$I$ 6.2 UDPGT possessed the same substrate specificity as the p$I$ 7.4 UDPGT except that it dit not react with estriol. It was distinguished functionally from the p$I$ 7.4 UDPGT on the basis of its reactivity with 4-aminobiphenyl, which was not a substrate for the p$I$ 7.4 UDPGT. We have called the p$I$ 7.4 UDPGT by the name estriol UDPGT. The p$I$ 6.2 UPDGT could be termed a phenol UDPGT.

The estriol UDPGT has a monomeric molecular weight on SDS–PAGE of 53 000 and required a relatively high concentration of phosphatidylcholine for reconstitution of maximal activity. The product of estriol glucuronidation was estriol 16$\alpha$-glucuronide and not the 3- or 17$\beta$-glucuronide. Thus, this protein is characteristic of the steroid type of UDPGT but possesses a distinctive reaction with the 16$\alpha$-hydroxyl position of the steroid structure. This type of reactivity has not been noted for UDPGTs from hepatic microsomes of other species.

Recent studies with purified human liver estriol UDPGT have shown that it does not react with morphine, 4-hydroxybiphenyl, or tertiary amines such as tripelennamine (69). It demonstrates immunoreactivity with antibodies raised against rat hepatic microsomal 3α- and 17β-hydroxysteroid UDPGTs but not with antibodies raised against rat hepatic microsomal p-nitrophenol UDPGT. NH$_2$-terminal analysis yielded an amino acid sequence which aligned to the deduced amino acid sequence of a cDNA cloned from a human liver library in λgt 11(HLUG4). Sequence analysis showed that HLUG4 is 2094 base pairs in length, encoding a protein of 523 amino acids which includes a 16 amino acid leader sequence.

Recently, Ritter et al. (70) isolated a human liver cDNA clone encoding a UDPGT which expressed glucuronidation in COS cells toward estriol as well as toward 3,4-catechol estrogens such as 4-hydroxyestrone, 2-hydroxyestradiol, and 4-hydroxyestradiol. The deduced amino acid sequence shows that it is different from the estriol UDPGT isolated by Irshaid and Tephly (68) and the HLUG4 which corresponds to the estriol UDPGT. Thus, there are two estriol UDPGTs in human liver.

The pI 6.2 UDPGT has a subunit molecular weight of 54 000 and reacts with 4-aminobiphenyl. This UDPGT has been shown to immunoprecipitate with sheep anti-rabbit p-nitrophenol UDPGT (34) and to be immunoreactive on Western blots with anti-rabbit and anti-rat p-nitrophenol UDPGTs. It does not demonstrate immunoreactivity with anti-rat 3α- or 17β-hydroxysteroid UDPGTs.

One might be tempted to say that the pI 6.2 UDPGT protein corresponds to a human hepatic UDPGT cDNA (termed HLUGP1) which has been expressed by Harding et al. (71) in COS cells and which expressed activity toward small phenolic compounds. However, NH$_2$-terminal amino acid sequence analysis of this protein shows that it does not possess the same sequence as the deduced amino acid sequence of HLUGP1 (preliminary results). Thus, it appears that there are a number of UDPGTs in human liver capable of reacting with small phenolic compounds.

Another human liver UDPGT cDNA (HLUG25) has been expressed in COS cells (72). This cDNA expressed activity toward hyodeoxycholic acid, producing a 6-O-β-glucuronide, a reaction previously shown by Radominska-Pyrek et al. (73) to occur in human liver microsomal fractions. It is interesting to note that the estriol UDPGT (HLUG4) has about 82% identity with HLUG25, suggesting that these proteins are members of the same gene subfamily. The protein corresponding to HLUG25 has not yet been purified from human liver microsomes.

There are at least three other UDPGTs that are known to be present in human liver microsomes which catalyze the glucuronidation of bilirubin, morphine, and tertiary amines other than morphine (74). These proteins have not yet been purified from human liver, but their presence is indicated by a number of studies. They are very labile proteins, and their abundance appears to be low.

Table II summarizes information on UDPGTs of human liver which have either been purified to homogeneity or whose cDNAs have been expressed.

## Summary

UDPGTs are members of a class of enzymes located in the endoplasmic reticulum and are encoded by a multigene family. These proteins are responsible for the glucuronidation of hundreds of xenobiotics of many chemical classes and many endogenous substances such as steroid hormones, bile acids, and bilirubin. There are a number of UDPGTs which have been identified by purification and characterization studies and a significant number which have been characterized by expression of cDNAs. On the basis of the primary structures elucidated they appear to have marked similarities (5) and are highly conserved. However, key differences in their functional properties appear to depend primarily on differences in amino acid sequences at or about the NH$_2$-terminal area of the protein (5).

Many of the UDPGTs have an extraordinarily broad substrate specificity; a few, however, are relatively specific for a given class of substrate (morphine, DT-1 UDPGTs). This places a burden on investigators to clearly identify which substrate and how many UDPGTs will be involved in any analysis of rates of glucuronidation in microsomal preparations. Caution should also be advised for extrapolation of data from hepatic microsomes of experimental animals to human hepatic microsomal preparations because human liver microsomes possess UDPGTs which are qualitatively different and, in certain cases, UDPGTs are present in human liver which are not present in lower animals.

**Table II. Human Liver UDP-Glucuronosyltransferases[a]**

|  | $M_r$ | endogenous | exogenous |
|---|---|---|---|
| estriol (pI 7.4) | 53 000 | estriol (16α-OH) | p-nitrophenol, 4-methylumbelliferone, α-naphthylamine |
| estriol (UDPGT$_h$-2) | 52 000 | estriol, catechol estrogens | ? |
| phenol (4-aminobiphenyl) | 54 000 | ? | phenols, 4-aminobiphenyl |
| HLUGP1 | 53 000 | ? | phenols |
| HLUG25 | 52 000 | 6-OH bile acids | ? |

[a] For details see refs 68–72.

**Acknowledgment.** I gratefully acknowledge the participation of many colleagues who have made much of the work reported here possible. These include Drs. Sanchez, del Villar, Billings, Tukey, Falany, Puig, Ishaid, and Thomassin. The research contributions of Mitchell Green and Birgit Coffman have been invaluable. The research was supported by NIH Grant GM 26221.

**Registry No.** UDPGT, 9030-08-4.

## References

(1) Williams, R. T. (1959) Introductory and Historical. In *Detoxication Mechanisms*, pp 13–22, Chapman & Hall, London.

(2) Dutton, G. J. (1980) *Glucuronidation of Drugs and Other Compounds*, CRC Press, Boca Raton, FL.

(3) Kasper, C. B., and Henton, D. (1980) Glucuronidation. In *Enzymatic Basis of Detoxication* W. B., (Jacoby, Ed.) Vol. 2, pp 3–26, Academic Press, New York.

(4) Burchell, B. (1981) Identification and purification of multiple forms of UDP-glucuronosyltransferase. *Rev. Biochem. Toxicol.* **3**, 1–39.

(5) Burchell, B., and Coughtrie, M. W. H. (1989) UDP-glucuronosyltransferases. *Pharmacol. Ther.* **43**, 261–289.

(6) Mackenzie, P. I., Roy Chowdhury, N., and Roy Chowdhury, J. (1989) Characterization and regulation of rat liver UDP-glucuronosyltransferases. *Clin. Exp. Pharmacol.* **16**, 501–504.

(7) Bock, K. W., Burchell, B., Dutton, G. J., Hanninen, O., Mulder, G. J., Owens, I. S., Siest, G., and Tephly, T. R. (1983) UDP-Glucuronosyltransferase activities. Guidelines for consistent interim terminology and assay conditions. *Biochem. Pharmacol.* **32**, 953–955.

(8) Abbott, F. V., and Palmour, R. M. (1988) Morphine-6-glucuronide: Analgesic effects and receptor binding profile in rats. *Life Sci.* **43**, 1685–1695.

(9) Yeh, S. V. H. (1975) Urinary excretion of morphine and its metabolites in morphine-dependent subjects. *J. Pharmacol. Exp. Ther.* **192**, 201–210.

(10) Coughtrie, M. W. H., Ask, B., Rane, A., Burchell, B., and Hume, R. (1989) The enantioselective glucuronidation of morphine in rats and humans: Evidence for the involvement of more than one UDP-glucuronosyltransferase isoenzyme. *Biochem. Pharmacol.* **38**, 3272–3280.

(11) Mulder, G. J., Hinson, J. A., and Gillette, J. R. (1977) Generation of reactive metabolites of N-hydroxy-phenacetin by glucuronidation and sulfation. *Biochem. Pharmacol.* 26, 189–196.

(12) Meyers, M., Slikker, W., and Vore, M. (1981) Steroid D-ring glucuronides: Characterization of a new class of cholestatic agents in the rat. *J. Pharmacol. Exp. Ther.* 218, 63–73.

(13) Slikker, W., Vore, M., Bailey, J. R., Meyers, M., and Montgomery, C. (1983) Hepatotoxic effects of estradiol-17β-D-glucuronide in the rat and monkey. *J. Pharmacol. Exp. Ther.* 225, 138–143.

(14) Smith, P. C., McDonagh, A. F., and Benet, L. Z. (1986) Irreversible binding of zomepirac to plasma protein *in vitro* and *in vivo*. *J. Clin. Invest.* 77, 934–939.

(15) Hyneck, M. L., Smith, P. C., Munafo, A., McDonagh, A. F., and Benet, L. Z. (1988) Disposition and irreversible plasma protein binding of tolmetin in humans. *Clin. Pharmacol. Ther.* 44, 107–114.

(16) Zakim, D., Goldenberg, J., and Vessey, D. A. (1973) Differentation of homologous forms of UDP-glucuronyltransferase. I. Evidence for the glucoronidation of o-aminophenol and p-nitrophenol by separate enzymes. *Biochem. Pharmacol.* 309, 67–74.

(17) Sanchez, E., and Tephly, T. R. (1974) Morphine metabolism. I. Evidence for separate enzymes in the glucuronidation of morphine and p-nitrophenol by rat hepatic microsomes. *Drug Metab. Dispos.* 2, 247–253.

(18) Sanchez, E., and Tephly, T. R. (1973) Activation of hepatic microsomal glucuronyltransferase by bilirubin. *Life. Sci.* 13, 1488–1490.

(19) del Villar, E., Sanchez, E., Autor, A. P., and Tephly, T. R. (1975) Morphine metabolism. III. Solubilization and separation of morphine and p-nitrophenol uridine diphosphoglucuronyltransferases. *Mol. Pharmacol.* 11, 236–240.

(20) del Villar, E., Sanchez, E., and Tephly, T. R. (1977) Morphine metabolism. V. Isolation of separate glucuronyltransferase activities for morphine and p-nitrophenol from rabbit liver microsomes. *Drug Metab. Dispos.* 5, 273–278.

(21) Bock, K. W., Clausbruch, U. C., Kaufmann, R., Lilienblum, W., Oesch, F., Pfeil, H., and Platt, K. L. (1980) Functional heterogeneity of UDP-glucuronyltransferase in rat tissues. *Biochem. Pharmacol.* 29, 495–500.

(22) Lilienblum, W., Walli, A. K., and Bock, K. W. (1982) Differential induction of rat liver microsomal UDP-glucuronosyltransferase activities by various inducing agents. *Biochem. Pharmacol.* 31, 907–913.

(23) Bock, K. W., Josting, D., Lilienblum, W., and Pfeil, H. (1979) Purification of rat-liver microsomal UDP-glucuronyltransferase. Separation of two enzyme forms inducible by 3-methylcholanthrene or phenobarbital. *Eur. J. Biochem.* 98, 19–26.

(24) Weatherill, P. J., and Burchell, B. (1980) The separation and purification of rat liver UDP-glucuronyltransferase activities towards testosterone and oestrone. *Biochem. J.* 189, 377–390.

(25) Gorski, J. P., and Kasper, C. B. (1977) Purification and properties of microsomal UDP-glucuronyltransferase from rat liver. *J. Biol. Chem.* 252, 1336–1343.

(26) Billings, R. E., Tephly, T. R., and Tukey, R. H. (1978) The separation and purification of estrone and p-nitrophenol UDP-glucuronyltransferase activities. In *Conjugation Reactions in Drug Biotransformation* (Aitio, A., Ed.) pp 365–376, Elsevier/North-Holland Press, Amsterdam.

(27) Falany, C. N., and Tephly, T. R. (1983) Separation, purification and characterization of three isoenzymes of UDP-glucuronyltransferase from rat liver microsomes. *Arch. Biochem. Biophys.* 227, 248–258.

(28) Tukey, R. H., Billings, R. E., and Tephly, T. R. (1978) Separation of oestrone UDP-glucuronyltransferase and p-nitrophenol UDP-glucuronyltransferase activities. *Biochem. J.* 171, 659–663.

(29) Tukey, R. H., Billings, R. E., Autor, A. P., and Tephly, T. R. (1979) Phospholipid-dependence of oestrone UDP-glucuronyltransferase and p-nitrophenol UDP-glucuronyltransferase. *Biochem. J.* 179, 59–65.

(30) Tukey, R. H., and Tephly, T. R. (1981) Purification and properties of rabbit liver estrone and p-nitrophenol UDP-glucuronyltransferases. *Arch. Biochem. Biophys.* 209, 565–578.

(31) Tukey, R. H., Robinson, R., Holm, B., Falany, C. N., and Tephly, T. R. (1982) A procedure for the rapid separation and purification of UDP-glucuronyltransferases from rabbit liver microsomes. *Drug Metab. Dispos.* 10, 97–101.

(32) Falany, C. N., Chowdhury, J. R., Chowdhury, N. R., and Tephly, T. R. (1983) Steroid 3- and 17-OH UDP-glucuronyltransferase activities in rat and rabbit liver microsomes. *Drug Metab. Dispos.* 11, 426–432.

(33) Tephly, T. R., Green, M., Puig, J., and Irshaid, Y. (1988) Endogenous substrates for UDP-glucuronosyltransferases. *Xenobiotica* 18, 1201–1210.

(34) Green, M. D., Coffman, B. L., Irshaid, Y. M., and Tephly, T. R. (1988) Characterization of antibodies to a rabbit hepatic UDP-glucuronosyltransferase and the identification of an immunologically similar enzyme in human liver. *Arch. Biochem. Biophys.* 262, 367–374.

(35) Green, M. D., and Tephly, T. R. (1989) N-glycosylation of purified rat and rabbit hepatic UDP-glucuronosyltransferases. *Arch. Biochem. Biophys.* 273, 72–78.

(36) Kirkpatrick, R. B., Falany, C. N., and Tephly, T. R. (1984) Glucuronidation of bile acids by rat liver 3-OH androgen UDP-glucuronyltransferase. *J. Biol. Chem.* 259, 6176–6180.

(37) Kirkpatrick, R. B., Green, M. D., Hagey, L. R., Hofmann, A. F., and Tephly, T. R. (1988) Effect of side chain length on bile acid conjugation: Glucuronidation, sulfation and coenzyme A formation of nor-bile acids and their natural $C_{24}$ homologs by human and rat liver fractions. *Hepatology* 8, 353–357.

(38) Radominska, A., Green, M. D., Zimniak, P., Lester, R., and Tephly, T. R. (1988) Biosynthesis of hydroxy-linked glucuronides of short-chain bile acids by rat liver 3-hydroxysteroid UDP-glucuronosyltransferase. *J. Lipid Res.* 29, 501–508.

(39) Matsui, M., and Hakozaki, M. (1979) Discontinuous variation in hepatic uridine diphosphate glucuronyltransferase toward androsterone in Wistar rats. A regulatory factor for *in vivo* metabolism of androsterone. *Biochem. Pharmacol.* 28, 411–415.

(40) Green, M. D., Falany, C. N., Kirkpatrick, R. B., and Tephly, T. R. (1985) Strain differences in purified rat hepatic 3α-hydroxysteroid UDP-glucuronosyltransferase. *Biochem. J.* 230, 403–409.

(41) Puig, J. F., and Tephly, T. R. (1986) Isolation and purification of rat liver morphine UDP-glucuronosyltransferase. *Mol. Pharmacol.* 30, 558–565.

(42) Sanchez, E., del Villar, E., & Tephly, T. R. (1978) Structural requirements in the reaction of morphine uridine diphosphate glucuronyltransferase with opioid substances. *Biochem. J.* 169, 173–177.

(43) del Villar, E., Sanchez, E., and Tephly, T. R. (1977) The inhibition of morphine: UDP-glucuronyltransferase in rabbit liver microsomes by cyproheptadine. *Life Sci.* 21, 1801–1806.

(44) del Villar, E., Sanchez, E., Letelier, M. E., and Vega, P. (1984) Differential inhibition by diazepam and nitrazepam of UDP-glucuronosyltransferase activity in rats. *Res. Commun. Chem. Pathol. Pharmacol.* 33, 433–447.

(45) Vega, P., Carrasco, M., Sanchez, E., and del Villar, E. (1984) Structure activity relationship in the effect of 1,4-benzodiazepines on morphine, aminopyrine and oestrone metabolism. *Res. Commun. Chem. Pathol. Pharmacol.* 44, 179–198.

(46) Vega, P., Gaule, G., Sanchez, E., and del Villar, E. (1986) Inhibition and activation of UDP-glucuronosyltransferase in alloxan diabetic rats. *Gen. Pharmacol.* 17, 641–645.

(47) Rane, A., Sawe, J., Pacifici, G. M., Svenson, J. O., and Kager, L. (1986) Regioselective glucuronidation of morphine and interaction with benzodiazepines in human liver. *Adv. Pain Res. Ther.* 8, 57–64.

(48) Thomassin, J., and Tephly, T. R. (1990) Photoaffinity labeling of rat liver microsomal morpine UDP-glucuronosyltransferase by [³H]flunitrazepam. *Mol. Pharmacol.* (in press).

(49) Snyder, S. H., Verma, A., and Trifiletti, R. R. (1987) The peripheral-type benzodiazepine receptor: A protein of mitochondrial outer membranes utilizing porphyrins as endogenous ligands. *FASEB J.* 1, 282–288.

(50) Thomassin, J., Styczyski, P., and Tephly, T. R. (1989) UV-light-activated inhibition of human liver morphine UDP-glucuronosyltransferase (UDPGT) activity by flunitrazepam (FNZ). *Pharmacologist* 31, 174.

(51) Schmoldt, A., and Roholoff, C. (1978) Dehydro-digitoxosides of digitoxigenin: Formation and importance for the digitoxin metabolism in the rat. *Naunyn Schmiedeberg's Arch. Pharmacol.* 305, 167–172.

(52) Schmoldt, A., and Promies, J. (1982) On the substrate specificity of the digitoxigenin monodigitoxoside conjugating UDP-glucuronosyltransferase in rat liver. *Biochem. Pharmacol.* 31, 2285–2289.

(53) Schmoldt, A. (1978) Increased digitoxin cleavage by liver microsomes of spironolactone-pretreated rats. *Naunyn-Schmiedeberg's Arch. Pharmacol.* **305**, 261–263.

(54) Castle, M. C. (1980) Glucuronidation of digitalis glycosides by rat liver microsomes: Stimulation by spironolactone and pregnenolone-16α-carbonitrile. *Biochem. Pharmacol.* **29**, 1497–1502.

(55) Watkins, J. B., and Klaassen, C. D. (1985) Development of UDP-glucuronosyltransferase activity toward digitoxigenin-monodigitoxoside in neonatal rats. *Drug. Metab. Dispos.* **13**, 186–191.

(56) Watkins, J. B., Gregus, Z., Thompson, T. N., and Klaassen, C. D. (1982) Induction studies on the functional heterogeneity of rat liver UDP-glucuronosyltransferases. *Toxicol. Appl. Pharmacol.* **64**, 439–446.

(57) von Meyerinck, L., Coffman, B. L., Green, M. D., Kirkpatrick, R. B., Schmoldt, A., and Tephly, T. R. (1985) Separation, purification and characterization of digitoxigenin-monodigitoxoside UDP-glucuronosyltransferase activity. *Drug Metab. Dispos.* **13**, 700–704.

(58) Tephly, T. R., Townsend, M., Coffman, B., Puig, J., and Green, M. (1988) Characterization of UDP-glucuronosyltransferases from animal and human liver. In *Cellular and Molecular Aspects of Glucuronidation* (Siest, G., Magdalou, J., and Burchell, B., Eds.) Vol. 173, pp 37–42, Colloques INSERM/John Libbey Eurotext, London.

(59) Tephly, T. R., Townsend, M., and Green, M. D. (1989) UDP-glucuronosyltransferases in the metabolic disposition of xenobiotics. In *Drug Metabolism Reviews* (Di Carlo, F. J., Ed.) Vol. 20, pp 689–695, Marcel Dekker, New York.

(60) Mackenzie, P. I. (1984) Rat liver UDP-glucuronosyltransferase: identification of cDNAs encoding two enzymes which glucuronidate testosterone, dihydrotestosterone and β-estradiol. *J. Biol. Chem.* **262**, 9744–9749.

(61) Burchell, B., and Blanckaert, N. (1984) Bilirubin mono- and diglucuronide formation by purified rat liver microsomal bilirubin UDP-glucuronotransferase. *Biochem. J.* **223**, 461–465.

(62) Chowdhury, N. R., Arias, I. M., Lederstein, M., and Chowdhury, J. R. (1986) Substrates and products of purified rat liver bilirubin UDP-glucuronosyltransferase. *Hepatology* **6**, 123–128.

(63) Sato, H., Koiwai, O., Tanabe, K., and Kashiwamata, S. (1990) Isolation and sequencing of rat liver bilirubin UDP-glucuronosyltransferase cDNA: Possible alternate splicing of a common primary transcript. *Biochem. Biophys. Res. Common.* **169**, 260–264.

(64) Yokota, H., Yuasa, A., and Sato, R. (1988) Purification and properties of a form of UDP-glucuronosyltransferase from liver microsomes of 3-methylcholanthrene-treated rats. *J. Biochem.* **104**, 531–536.

(65) Coughtrie, M. W. H., Burchell, B., and Bend, J. R. (1987) Purification and properties of rat kidney UDP-glucuronosyltransferase. *Biochem. Pharmacol.* **36**, 245–251.

(66) Yokota, H., Ohgiua, N., Ishihara, G., Ohta, K., and Yuasa, A. (1989) Purification and properties of UDP-glucuronosyltransferse from kidney microsomes of β-naphtho-flavone-treated rat. *J. Biochem.* **106**, 248–252.

(67) Iyanagi, T., Haniv, M., Sagawa, K., Fujii-Kuriyama, Y., Watanabe, S., Shively, J. E., and Anan, K. F. (1987) Cloning and characterization of cDNA encoding of 3-methylcholanthrene-inducible rat mRNA for UDP-glucuronosyltransferase. *J. Biol. Chem.* **261**, 15607–15614.

(68) Irshaid, Y. M., and Tephly, T. R. (1987) Isolation and purification of two human liver UDP-glucuronosyltransferases. *Mol. Pharmacol.* **31**, 27–34.

(69) Coffman, B. L., Tephly, T. R., Irshaid, Y. M., Green, M. D., Smith, C., Jackson, M. R., Wooster, R., and Burchell, B. (1990) Characterization and primary sequence of human hepatic microsomal estriol UDP-glucuronosyltransferase. *Arch. Biochem. Biophys.* **281**, 170–175.

(70) Ritter, J. K., Sheen, Y. Y., and Owens, I. S. (1990) Cloning and expression of human liver UDP-glucuronosyltransferase in COS-1 cells: 3,4-catechol estrogens and estriol as primary substrates. *J. Biol. Chem.* **265**, 7900–7906.

(71) Harding, D., Fournel-Gigleux, S., Jackson, M. R., and Burchell, B. (1988) Cloning and substrate specificity of a human phenol UDP-glucuronosyltransferase expressed in COS-7 cells. *Proc. Natl. Acad. Sci. U.S.A.* **85**, 8381–8385.

(72) Fournel-Gigleux, S., Jackson, M. R., Wooster, R., and Burchell, B. (1989) Expression of a human liver cDNA encoding UDP-glucuronosyltransferase catalyzing the glucuronidation of hyodeoxycholic acid in cell culture. *FEBS Lett.* **243**, 119–112.

(73) Radominska-Pyrek, A., Zimniak, P., Irshaid, Y. M., Lester, R., Tephly, T. R., and Pyrek, J., St. (1987) Glucuronidation of 6α-hydroxy bile acids by human liver microsomes. *J. Clin. Invest.* **80**, 234–241.

(74) Stycynski, P. B., Coffman, B. L., Green, M. D., and Tephly, T. R. (1989) Studies on quaternary ammonium-linked glucuronidation in human liver microsomes. *Pharmacologist* **31**, 131.

**Chapter 13**

# DNA Repair

Gary M. Myles and Aziz Sancar*

*Department of Biochemistry, School of Medicine, University of North Carolina, Chapel Hill, North Carolina 27599*

Reprinted from *Chemical Research in Toxicology*, Vol. 2, No. 4, July/August, 1989

## Contents

* Corresponding author.

## Introduction

DNA is a dynamic molecule; it undergoes structural changes in the process of carrying out its physiological functions: replication, transcription, recombination and transposition. While the mutations resulting from these structural changes may be helpful on an evolutionary time scale (by creating diversity within the species), on an individual level they are often harmful. Some structural changes, often called DNA damage, can block replication and/or transcription, or they may cause mutations which, in turn, may result in an unfavorable phenotype, cell death, or cancer. It is of the utmost in importance, therefore, that the level of genetic changes be kept in check by correcting the structural changes caused by intrinsic and extrinsic agents. This is accomplished by molecular mechanisms of varying complexities referred to as DNA repair pathways.

In recent years, important progress has been made in many areas of the DNA repair field. Although most of these developments have been reviewed recently (1–5), these reviews are, by design or necessity, limited in scope. In the present review, we will survey all of the major repair pathways (i.e., direct repair, base excision, nucleotide excision, recombination, and cross-link repair) as well as mismatch correction which is mechanistically and teleologically related to DNA repair. In addition, we will review the biochemical evidence for preferential DNA repair in the transcriptionally active genes of mammalian cells. Finally, we will review the molecular mechanisms involved in regulating the expression of DNA repair genes as well as mechanisms of damage avoidance and survival without repair.

## I.  Direct Reversal of DNA Damage

*A. DNA Photolyases.* Photoreactivation is the reversal of the effects of far-UV (200–300 nm) by exposure of an organism to near-UV (300–500 nm). Although there are several molecular mechanisms for this biological phenomenon (6), it is caused principally by the repair of far-UV-induced pyrimidine dimers by the photoreactivating enzyme DNA photolyase. Photolyases bind to pyrimidine dimers in a light-independent manner, absorb a photon, reverse the dimer, and dissociate from the DNA (7). Photolyases from *Escherichia coli* (8), *Saccharomyces cerevisiae* (9), *Anacystis nidulans* (Kiener and Walsh, personal communication), *Methanobacterium thermoautotrophicum* (10), and *Scenedesmus acutus* (11) have been purified to near-homogeneity, and all have molecular weights from 50 000 to 60 000. The genes for the *E. coli* (12), yeast (13), and *A. nidulans* (14) enzymes have been cloned and sequenced.

Photolyases bind to pyrimidine dimers in DNA with $K_D = 10^{-8}$–$10^{-9}$ M (15, 16, 10) and to nondamaged DNA with very low affinity, i.e., $K_{NS} = 10^{-3}$–$10^{-4}$ M (16). The contact sites of *E. coli* (17), *M. thermoautotrophicum* (10), and *S. cerevisiae* (18) photolyases have been determined. All three enzymes interact with DNA in essentially the same manner; the phosphodiester bond 5′ and the three or four phosphodiester bonds 3′ to the dimer are contacted. There is also a weak contact with the phosphodiester bond across

the minor groove from the intradimer phosphodiester bond. The thymine dimer bends DNA by 29° and unwinds the helix by 20° (19). Molecular modeling and molecular dynamics studies indicate that the damaged strand has essentially the same conformation in both single- and double-stranded DNA (20, 21). Not surprisingly, all the significant contacts by the three photolyases studied are on the damaged strand, and all three photolyases repair single- and double-stranded DNAs with equal efficiency. Methylation protection and interference experiments indicate that in dsDNA these photolyases straddle the major groove containing the cyclobutane ring (17, 10). Photolyases are thought to achieve the high specificity typical of DNA binding proteins by making hydrogen bonds with phosphates in the unique configuration forced upon the DNA backbone by the pyrimidine dimer. However, the pyrimidine dimer itself must play a significant role in damage recognition because the *E. coli* photolyase breaks the cyclobutane ring of an isolated thymine dimer, albeit at a much slower rate compared to a dimer incorporated in DNA. Binding of photolyase to pyrimidine dimers in DNA makes these lesions more accessible to *E. coli* ABC excinuclease (22) and increases the efficiency of excision repair in *E. coli* (23). Similarly, yeast photolyase appears to stimulate yeast nucleotide excision repair in vivo (24).

All photolyases studied to date contain two chromophores (7), and each has been categorized into one of two classes according to its chromophore composition. The folate class enzymes, including those from *E. coli* and *S. cerevisiae* (25), contain $FADH_2$ and folate, whereas the deazaflavin enzymes, of which *A. nidulans*, *Streptomyces griseus* (26), *S. acutus* (11), and *M. thermoautotrophicum* (10) are members, contain $FADH_2$ and deazaflavin. The latter is usually in the form of F420 which like folic acid has a polyglutamate tail of varying length.

Photolyases have remote similarities to photosynthetic systems. It appears that one of the chromophores, folate or deazaflavin [which are also referred to as "the second chromophore" (27, 11)], functions as a light harvester while the $FADH_2$ provides the catalytic center similar to the reaction centers (RC) of the photosynthetic systems (7, 28).

The flavin cofactor of the purified *E. coli* enzyme is in the form of an $FADH^0$ neutral blue radical (27). This appears to be a purification artifact, however, as the $FADH^0$ form of the enzyme is incapable of repairing pyrimidine dimers (28, 29). EPR studies on cells overproducing the enzyme show that in vivo the cofactor is fully reduced (i.e., in the $E$–$FADH_2$ form) (28). The neutral radical form $E$–$FADH^0$ is active in vitro because illumination of the enzyme in buffers that contain electron donors such as dithiothreitol or EDTA photoreduces $FADH^0$ to the catalytically active $FADH_2$ form (30, 31). The near-UV absorption maximum of the *E. coli* enzyme is at 384 nm, and in enzyme purified from an overproducing strain (which contains only 30% of its folate complement) two-thirds of this absorption is contributed by the folate (25). The enzyme can be saturated with folate by supplementing with additional chromophore, methenyltetrahydrofolate, or its open-ringed precursor, 10-formyltetrahydrofolate, which is in equilibrium with the enzyme's cofactor, methenyltetrahydrofolate (32). The native *E. coli* photolyase has $E_{385} = 35\,000$ M$^{-1}$ cm$^{-1}$, of which 30 000 is contributed by the folate and 5000 by the flavin. Thus, since folate absorbs most of the light, it is said to be the "antenna" of photolyase.

A likely scenario for the photochemical half of the photolyase reaction is as follows (see Figure 1): a photon is absorbed by the folate which transfers energy (not necessarily by dipole–dipole interaction) to $FADH_2$ which

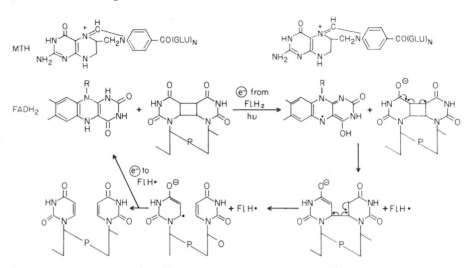

**Figure 1.** Reaction mechanism for the folate class photolyases. A photon is absorbed by methenyltetrahydrofolate (MTH), which transfers energy to the $FADH_2$ cofactor with relatively high efficiency by an, as yet, unknown mechanism. The $FADH_2$ that is excited by this energy transfer, or by direct absorption of a photon, donates an electron to the pyrimidine dimer. The resulting dimer anion is unstable; the cyclobutane ring is cleaved by a nonsynchronous mechanism and the electron is taken up by the flavin radical. The enzyme can now enter new rounds of repair. The overall quantum yield of repair is about 0.5 (*28*). Reproduced, with permission, from ref 1. Copyright 1988 Annual Reviews, Inc.

in turn transfers an electron to the dimer (presumably C4 of one of the pyrimidines). The dimer undergoes an electronic rearrangement resulting in the breakage of both bonds connecting the two pyrimidines, thus generating a pyrimidine and a pyrimidine anion radical the latter of which donates an electron to the $FADH^0$ to regenerate $FADH_2$.

The quantum yields for photolyases are quite high but not 1.0 as previously suggested (*29, 33*). The *E. coli* enzyme has a quantum yield of 0.5 (*28*) while *M. thermoautotrophicum* works with a quantum efficiency of 0.2 (*10*).

Interestingly, under saturating light, the *E. coli* enzyme containing only $FADH_2$ is as active as the enzyme containing both chromophores (i.e., E–$FADH_2$–MTHF) (*34, 38*), the major distinction being the lesser probability of the E–$FADH_2$ form absorbing an incident photon because of its lower extinction coefficient (*28, 32*). That E–$FADH_2$ is catalytically equivalent to the holoenzyme is indicative of the central role of $FADH_2$ in catalysis. Consistent with this view is the observation that E–MTHF is catalytically inert (Hamm–Alvarez, unpublished). Hence, the folate most likely serves as an antenna molecule while the flavin is the "cofactor" essential for catalysis. It is likely that deazaflavin acts as the antenna in that class of enzymes while the $FADH_2$ is in the active center of both classes.

**B. DNA Alkyltransferases (*2, 35*).** DNA is a strong electrophile and, therefore, is subject to modification by many alkylating agents. Experimentally, nitrosoureas and nitrosoguanidine are used to alkylate DNA. Also, *S*-adenosylmethionine has been shown to alkylate DNA in vitro (*36*); however, whether it alkylates DNA at a rate that is physiologically relevant in vivo is not known (*37*).

Alkylation damage is repaired by direct removal of the alkyl group (dealkylation) as well as by the excision of the modified base or neighboring nucleotides (*38*). In *E. coli*, there are two enzymes that repair DNA by direct alkyl transfer, the Ada and Ogt proteins. For about a decade, Ada was the only known alkyltransferase in this organism. Interestingly, cells mutated in the *ada* gene do retain some residual methyltransferase activity. It has recently been discovered that the residual activity is due to a second alkyltransferase (*39–41*). The gene for this enzyme has been cloned and sequenced (*39*). The gene product is named Ogt for its $O^6$-methylguanine DNA methyl-

transferase activity. In contrast to Ada, which removes methyl groups from $O^4$-mThy in addition to $O^6$-mGua, Ogt is specific for $O^6$-mGua. It has been proposed that the Ogt protein be called alkyltransferase I (ATI) and the Ada protein alkyltransferase II (ATII) by analogy to 3-mAde DNA glycosylases I and II (*39*); however, Rebeck et al. (*40*) have suggested the opposite designations for the two alkyltransferases. Here, we will refer to these two proteins by their gene symbols.

**1. The Ada Protein (*2, 35*).** The *ada* gene has been cloned and sequenced and its product greatly overproduced (*42–44*); the protein is a monomer of $M_r = 38\,000$ which has no cofactor requirement, it is functional in the presence of EDTA, and it has a strong preference for dsDNA. The enzyme has two active sites: (1) cysteine 69 is the acceptor of methyl groups from the $S_p$ stereoisomer of methyl phosphotriesters (*45, 46*) and (2) cysteine 321 is the acceptor site for the methyl from $O^6$-methylguanine ($O^6$-mGua; *42*) and to a lesser extent from $O^4$-methylthymine ($O^4$-mThy). Changing cysteine 321 to any other residue by site-specific mutagenesis abolishes $O^6$-mGua methyltransferase activity (*47*). The methylations at either site are irreversible, and therefore, there is no catalytic turnover of the enzyme (suizyme; *35*). The active site cysteine in the $O^6$-mGua methyltransferase function is embedded in the Pro-Cys-His amino acid triplet (*42*).

Two models have been proposed for the reaction mechanism based on the active site sequences of Ada and Ogt (*35*) and on the crystal structure of $O^6$-mGua (*48*): the imino hydrogen of the active site histidine makes a hydrogen bond with the $N^7$ of $O^6$-mGua. The other nitrogen of the histidine forms a hydrogen bond with the sulfhydryl group of cysteine, thereby increasing the sulfur's $S^-$ character and enabling it to act as a strong methyl acceptor. Such an ion pair is known to be involved in proteolysis by the cysteine protease papain. Alternatively, the SH group of Cys in the active site may be activated into a nucleophile by a basic residue which is not adjacent to the Cys in the primary sequence. Such a mechanism appears to be operative in thymidylate synthase (see Figure 2).

Although the most efficient substrate for Ada is $O^6$-mGua, it has been demonstrated that $O^6$-ethyl-Gua is a moderate substrate and that alkyltransferases will inhibit interstrand cross-link formation by (chloroethyl)nitroso-

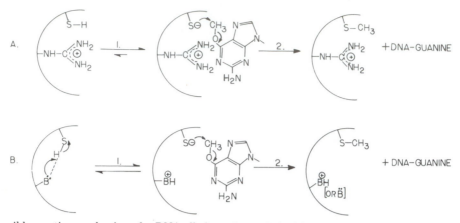

**Figure 2.** Two possible reaction mechanisms for DNA alkyltransferase (35). Mechanism A is based on the structure and reaction mechanism of thymidylate synthase. The active site cysteine thiol is converted into a strong nucleophile by interacting with an arginine residue which is not adjacent to this cysteine in the primary structure but is in its proximity in the active site pocket. Mechanism B is based on the known three-dimensional structure and mechanism of the cysteine protease papain. Here the thiolate–imidazolium ion pair in the active site is in equilibrium with the inactive nonionized form. The thioalte ion generated by either mechanism attacks the oxygen-linked alkyl group in DNA, resulting in the irreversible transfer of the alkyl group to cysteine. Reprinted with permission from ref 35. Copyright 1988 CRC Press, Inc.

ureas; however, the cross-link caused by these agents is between $N^1$ of guanine and $N^3$ of cytosine—there is no evidence that alkyltransferases repair $N^1$ of guanine. A mechanism has been proposed that explains how alkyltransferases may prevent interstrand cross-link formation as well as formation of DNA–alkyltransferase cross-links as side products of this reaction (49). The (chloroethyl)-nitrosoureas react with the $O^6$ of guanine to produce (chloroethyl)-$O^6$-Gua; the chloroethyl may be removed by the alkyltransferase by the usual alkyl-transfer reaction. More frequently, however, the adduct undergoes an intramolecular rearrangement to yield $O^6,N^1$-ethanoguanine. This intermediate may react with $N^3$ of cytidine in the complementary strand to yield $N^1$-Gua-ethano-$N^3$-Cyt. Alternatively, alkyltransferase may attack the $O^6$-ethanol bridge of the cyclic intermediate and link the ethano group to the cysteine in the active site in which case a DNA–protein cross-link is formed.

Ada is inactive on $O^6$-mGua in Z-DNA (50), and its activity on B-DNA seems to be strongly influenced by neighboring sequences (51).

The Ada protein has been of interest as much for its transcriptional regulatory activity as for its methyltransferase activity. The protein is the positive regulator of the genes involved in the adaptive response in *E. coli*; it binds to the Ada box upstream of the *ada*, the *alkA* (52, 53), and presumably the *alkB* and *aidB* genes (54, 55) and activates their transcription. The unmethylated protein binds weakly to the Ada box adjacent to the *ada* gene and binds with moderate affinity to the Ada box next to the *alkA* gene.

Methylation of Cys-69 by methyl transfer from a methyl phosphotriester results in a conformational change that converts the enzyme into a strong positive regulator (52). Recently, it was shown that the protein can also be activated by direct methylation with methyl methanesulfonate (56). The transcriptional activator function of Ada can also be modulated by mutations in the carboxyl domain of the protein. Interestingly, fusion proteins containing the N-terminus of Ada but C-termini from the cloning vectors act differently on the *ada* and *alkA* genes, suggesting that the enzyme may not be interacting with the upstream sequences of the two genes in the same fashion (57).

Another interesting aspect of structure–function of the Ada protein is its proteolytic cleavage. Ada is rapidly

cleaved in cell-free extracts. In fact, when it was first purified (by assaying for methyltransferase activity) the C-terminal domain of 19 kDa was obtained (58). For several years following the discovery of the methyltransferase activity, the Ada protein was presumed to be a monomer of 19 kDa. Cloning the gene helped identify the gene product as a 39-kDa protein. This protein is cleaved at Lys-129–Ala-130 and Lys-178–Glu-179, resulting in a 19-kDa C-terminal domain with the guanine methyltransferase activity and a 15-kDa N-terminal domain with the regulatory activity (59). Interestingly, the same protease cleaves the UvrB protein at a homologous region (60). Much speculation has been made regarding the regulatory significance of this cleavage in both the adaptive response (61) and the SOS response of ABC excinuclease (62) although there is no evidence that the cleavage occurs in vivo (63). Indeed, the protease responsible for the specific cleavage of Ada has been identified as the outer membrane protein OmpT (64) which also cleaves T7 RNA polymerase when it is overproduced in *E. coli*. Thus, cleavage of Ada and UvrB by this protease is an experimental artifact and, therefore, has no bearing on the regulation of the adaptive or SOS responses (64).

**2. The Ogt Protein.** This protein is a monomer of 171 amino acids with a molecular weight of 19 000. The *ogt* gene has been cloned and sequenced, but no mutants defective in this gene have been described (39). Its activities are similar to that of the 19-kDa C-terminal fragment of Ada. This protein shows about 40% identity over its C-terminal half with the corresponding region of the C-terminal fragment of the Ada protein. This homologous sequence includes a five amino acid stretch around the cysteine residue known to be methylated in Ada. Ogt has $O^6$-mGua methyltransferase activity and perhaps $O^4$-mThy methyltransferase activity but no methyl phosphotriester methyltransferase function. The *ogt* gene appears to be expressed constitutively, and the activity of the enzyme does not increase upon treatment with alkylating agents.

**3. DNA Alkyltransferases in Other Organisms.** The mechanism of alkyl repair in *Bacillus subtilis* is interesting; it appears that this bacterium encodes three proteins which have functions similar to the N- and C-termini of Ada and Ogt protein (65). The first two are inducible while the Ogt analogue is not. It, thus, appears that Ada may have evolved by fusion of a regulatory protein with an enzyme whose sole function is DNA repair.

The same may be true for *Micrococcus luteus*; this organism has three methyltransferase activities associated with 31-, 22-, and 13-kDa proteins (*66*).

$O^6$-Methylguanine DNA methyltransferases have been purified from various mammalian cells (*67–69*); these enzymes do not have the methyl phosphotriester methyltransferase activity. Some transformed cell lines are deficient in methyltransferase (MT) and are said to be Mer- or Mex- (*67*). The cause and consequence of this phenotypic change are not known.

## II. Base Excision Repair

In base excision repair, the damaged or unusual base is removed by a DNA glycosylase, and the resulting abasic sugar is excised by apurinic/apyrimidinic (AP) endonucleases. In some instances both glycosylase and AP endonuclease activities are on a single enzyme while in other cases the two activities are provided by independent proteins.

**A. DNA Glycosylases.** These enzymes cleave the base–sugar phosphodiester bond by hydrolysis and not by phosphorolysis. Glycosylases are unique among other repair enzymes in that they have very narrow substrate specificities.

**1. Uracil DNA *N*-Glycosylase.** Uracil is incorporated into DNA by DNA polymerases at approximately $10^{-3}$–$10^{-4}$ the frequency of thymine. Uracil is also generated by the deamination of cytosine which occurs either spontaneously or by treatment with sodium bisulfite or nitrous acid. Enzymes from *E. coli*, yeast, and human tissues (*70*) have been isolated that cleave the glycosylic bond linking uracil to DNA (*1*). The genes for uracil *N*-glycosylases (*ung*) from *E. coli* (*71, 72*) and yeast (*73*) have been cloned and sequenced; the gene products are about 30% identical. The NH$_2$-terminal amino acid sequence of the glycosylase from human placenta has been recently reported (*74*), which should facilitate cloning of the gene. Null mutants of *E. coli* (*75*) and yeast (*76*) uracil *N*-glycosylases are viable, indicating that the enzyme is not essential. Uracil *N*-glycosylases are monomeric proteins of 25–30 000 Da; they have no cofactors and no requirement for divalent ions. The enzymes prefer single-stranded DNA, have some sequence specificity (*77*), and are inhibited by high concentrations of uracil as well as by apurinic/apyrimidinic (AP) sites.

Eukaryotic cells apparently have two separate genes encoding nuclear (*74*) and mitochondrial uracil *N*-glycosylases (*78*). In addition, Herpes simplex virus type 2 (HSV-2) encodes its own uracil *N*-glycosylase (*79*). The *B. subtilis* phage PBS2, which contains uracil instead of thymine in its DNA, synthesizes an inhibitor of uracil *N*-glycosylase upon infection. Interestingly, this protein inhibits all uracil *N*-glycosylases tested. The inhibitor gene (*ugi*) has been cloned and sequenced (*80*). It encodes an acidic protein of calculated $M_r = 9477$ which behaves like a protein twice this size by gel filtration chromatography and migrates with an apparent molecular weight of 3500 on SDS–polyacrylamide gels. The inhibitor makes a complex of $M_r = 36 000$ with the *E. coli* enzyme and apparently inhibits all other uracil glycosylases by a similar mechanism, that is, by physical association with the enzymes (*81*). By expressing the cloned inhibitor in an *ung*$^+$ cell line, phenotypically Ung$^-$ cells can be obtained. This has enabled the investigation of the physiological consequences of the *ung* deficiency in cell lines in which it had previously not been possible to produce *ung*$^-$ mutants (*80*).

(Hydroxymethyl)uracil DNA glycosylase, which removes (hydroxymethyl)urea generated in DNA by oxidative damage of thymine, is a separate enzyme from uracil DNA glycosylase. It has been found in humans but has not been purified (*82*). The other spontaneously generated abnormal base is hypoxanthine; glycosylases removing this base have been identified but have not been characterized in any detail.

**2. 3-Methyladenine DNA Glycosylases.** 3-Methyladenine is one of the major products of DNA methylation. It is a lethal DNA adduct by virtue of its interference with DNA replication. Enzymes that remove this adduct have been identified in *E. coli* and *M. luteus* as well as in various eukaryotic cells.

*a. Tag I and Tag II of E. coli.* This bacterium has two genes which encode 3-mAde glycosylases, *tag* and *alkA*. The *tag* gene encodes a protein of $M_r = 21 000$ (Tag I) which is produced constitutively and is specific for 3-methyladenine (*83, 84*). AlkA encodes Tag II, a protein of $M_r = 31 400$ (*85, 86*). The Tag II glycosylase is induced during the adaptive response and removes, in addition to 3-mAde, 3-mGua, 7-mGua, $O^2$-mThy, and $O^2$-mCyt from DNA. Both genes have been cloned and sequenced (*83–85*); the Tag II protein has been crystallized (*87*). Both enzymes are specific for double-stranded DNA, but only Tag I is inhibited by low concentrations (1–2 mM) of 3-mAde. The enzymes from *M. luteus* are quite similar to those from *E. coli* (*88*) even though the relationship between the cloned *M. luteus* genes (*89*) and Tag I and Tag II type enzymes is not clear at present.

It was originally proposed (*90*) that the Ada protein recognizes and repairs methyl groups in the major groove ($O^4$-mThy and $O^6$-mGua) while Tag II patrols the minor groove for bases with methyl groups (i.e., 3-mGua, 3-mAda, and $O^2$-mThy). Although the discovery that Tag II repairs 7-mGua (*85*) is suggestive that Tag II can enter both major and minor grooves, the idea of an enzyme patrolling the minor groove is still attractive. Clearly, the change in the base's charge distribution upon methylation must also play an important role in its recognition and removal.

*b. 3-mAde DNA Glycosylases from Mammalian Cells.* Enzymes that release 3-mAde from DNA have been identified in mammalian tissues and cell lines. These enzymes generally have the biochemical properties of *E. coli* Tag II; that is, they remove 3-mAde, 3-mGua, $O^2$-mThy, and 7-mGua and are not inhibited by the product 3-mAde. The calf thymus 3-mAde DNA glycosylase is the best characterized eukaryotic enzyme; two species of $M_r = 42 000$ and 27 000 were identified after several chromatographic steps (*91*). Both species release 3-mAde, 3-mGua, and 7-mGua; are inactive on ssDNA; and are not product-inhibited. The low molecular weight species has been investigated in some detail (*92*); it releases methylated bases at relative rates of 3-mAde > 7-mGua > 3-mGua even though the $K_m$ values for 3-mAde and 7-mGua are similar.

It appears that the best substrates are those adopting an A-DNA conformation such as poly[d(A-T)] and poly-[d(G-C)]. The sequence context greatly influences the repair rate; in certain environments, some adducts are totally resistant to the enzyme. However, no general pattern has emerged for the sequence-dependent discriminating effect of the enzyme on various adducts (*92*).

3-mAde DNA glycosylases are specific for DNA; however, there is an interesting case of an RNA *N*-glycosylase. The Ricin A-chain (the cytotoxic protein isolated from castor beans) cleaves the glycosylic bond of A$^{4324}$ of 28S rRNA specifically and inactivates eukaryotic ribosomes (*93*).

### 3. Formamidopyrimidine–DNA Glycosylase.

7-mGua is the major adduct of methylating agents such as dimethyl sulfate, methyl methanesulfonate, and methylnitrosourea. This adduct is not lethal nor is it mutagenic; however, methylation does facilitate cleavage of the glycosylic bond, thus generating AP sites. Methylation also leads to cleavage of the imidazole ring. The product of cleavage of the imidazole ring is 2,6-diamino-4-oxy-5-(methylformamido)pyrimidine (FAPY). This adduct occurs in vivo (*94*); in vitro it inhibits DNA replication (*95*) and, therefore, is thought to be lethal. DNA glycosylases in both mammalian cells and *E. coli* specifically release FAPY from DNA (*96, 95*). The *E. coli* gene encoding the glycosylase (*fpg*) has been cloned and sequenced (*97*). The product of this gene is a monomer of 30 200 Da, and it has been purified to homogeneity from an overproducing strain. The enzyme appears to be specific for FAPY and does not release other alkylated bases but apparently also releases imidazole ring opened adenine—4,6-diamino-5-formamidopyrimidine—generated in DNA by γ-irradiation under hypoxic conditions (*96*). The enzyme is inactive against FAPY in Z-DNA (*94*). By use of the cloned gene, *E. coli* mutants with insertions in *fpg* have been isolated (*98*). These mutants have no detectable change in their phenotype; they have normal sensitivity to alkylating agents and to ionizing radiation.

### B. DNA Glycosylase–AP Endonucleases (*99*).

Considerable confusion exists over the nomenclature of these enzymes. The term UV endonuclease has been used to describe the enzymes from *M. luteus* and T4-infected *E. coli* cells that specifically incise UV-irradiated DNA at pyrimidine dimers. Subsequently, enzymes from various sources have been identified which also nick UV-irradiated DNA and have, therefore, also been termed UV endonucleases even though there is no evidence that these enzymes act on pyrimidine dimers—the major UV photoproduct. Nuclease activities specific for DNA irradiated with X-rays or γ-rays have also been identified and named X-ray and γ-endonuclease, respectively. Finally, there exists a class of enzymes that act on thymine glycols or urea residues (pyrimidine saturation and fragmentation products) referred to as thymine glycol endonucleases or urea endonucleases. Now, with a better understanding of the molecular mechanisms of these enzymes some order has been restored to the nomenclature. Perhaps the UV endonuclease term should be avoided as it has been used for enzymes acting both on pyrimidine dimers and on nondimer photoproducts. Some so-called UV endonucleases have turned out to be X-ray endonucleases (e.g., endonuclease III of *E. coli*).

### 1. Pyrimidine Dimer (PD) DNA Glycosylase–AP Endonucleases.

There are only two known examples of this class of enzymes. The T4 endonuclease V (T4 UV endonuclease) and the *M. luteus* "UV endonuclease". Both enzymes are small molecular weight monomeric proteins with no cofactor requirements, and both are active in the presence of EDTA. The enzymes work by the same mechanism (see Figure 3; *99a*): (1) cleavage of glycosylic bond of the 5′ pyrimidine of the dimer and (2) cleavage of the phosphodiester bond 3′ to the apyrimidinic sugar. A model has been proposed whereby the cleavage reactions occur by two sequential β-elimination reactions involving the same enzyme active site (*62*). However, the two cleavage reactions are frequently uncoupled (perhaps always in the case of T4 endonuclease V) (*100, 101*)—a result inconsistent with a mechanism involving sequential β-elimination reactions. In fact, on the basis of competition experiments it has been suggested that cleavage of the

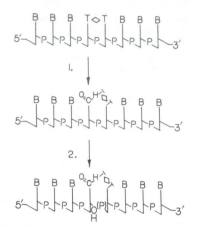

**Figure 3.** Reaction mechanism of pyrimidine dimer DNA glycosylase–AP endonuclease. The enzyme cleaves the glycosylic bond of the 5′ pyrimidine of the dimer (1) and then the intradimer phosphodiester bond (2) by a sequential mechanism. The cleavage of the phosphodiester bond is by a β-elimination reaction and not by hydrolysis. There is significant uncoupling of reactions 1 and 2 carried out by the enzyme from T4 phage both in vivo and in vitro but less so with the enzyme from *M. luteus*.

glycosylic and phosphodiester bonds requires two independent encounters of the enzyme and substrate (*101*).

T4 endonuclease V has been characterized in great detail. The gene has been cloned and sequenced, and overproducing strains have been constructed (*102–104*). The gene has also been synthesized and this gene used for protein overproduction (*105*). Finally, T4 endonuclease V has been crystallized, and some preliminary X-ray diffraction studies have been conducted.

T4 endonuclease V is approximately 100-fold more effective on dsDNA than on ssDNA (*105*). Even though its contact sites on DNA are not expected to extend beyond five nucleotides on either side of the dimer, the enzyme is not very effective on a dodecamer with a centrally located thymine dimer although it cleaves dimer containing 14- and 18-nucleotide duplexes quite efficiently with $K_m$ = $10^{-8}$ M and $k_{cat}$ = 7.7–1.2 min$^{-1}$ (*105*). Binding of the enzyme to these synthetic substrates in the absence of cleavage can be studied by conducting the binding reactions at 2 °C and filtering the reaction mixture in a high-salt buffer (15 × SSC) (*106*). By use of the filter binding assay, the dissociation constant $K_D$ = $10^{-9}$ M has been obtained for both the 14- and 18-mer. When the glycosylase and AP endonuclease activities are studied independently, the glycosylase function is essentially sequence-independent over a 10–18-base-pair range; however, a 10-mer duplex is a 5–10-fold poorer substrate as compared to the two other (14- and 18-mer) oligomer duplexes (*105*).

The two enzymatic activities of both T4 and *M. luteus* PD–DNA glycosylase–AP endonucleases can be easily uncoupled (*107*). When the enzyme reaction is carried out at high pH (8.0–8.5), substantial uncoupling occurs (*100*); that is, molecules accumulate with a cleaved glycosylic bond and intact AP site. The AP endonuclease activity can be completely inhibited by modification of the AP site with methoxyamine (*108*). The glycosylase and AP endonuclease activities can also be uncoupled by mutations; thus, the uvs13 mutation (trp128ser in a su⁺ background) has approximately wild-type glycosylase but no AP endonuclease activity. In contrast, the double mutant tyr129asn and tyr131asn has no glycosylase activity but nearly wild-type AP endonuclease activity (*109–111*). These results support the notion proposed by Valerie et al. (*102*) that the region trp128-tyr-lys-tyr-tyr132, which resembles the artificial AP endonuclease activity of lys-trp-lys (*112, 113*), is in the active site of T4 endonuclease V and is involved not only in AP endonuclease activity but in glycosylase function as well. Recently, Kim and Linn (*114*) have shown that the abasic sugar residue produced following cleavage of the 3' phosphodiester bond is not a normal deoxyribose but has the properties of 4,5 unsaturated deoxyribose which is expected to be generated by a β-elimination reaction. Since similar products have been identified with all other glycosylase–AP endonucleases tested, it is probable that cleavage of the phosphodiester backbone occurs through β-elimination (*115*). T4 endonuclease V acts processively in low salt in vitro (*116, 117*) and apparently also in vivo (*118, 119*). The role of the *M. luteus* pyrimidine dimer DNA glycosylase–AP endonuclease in cell survival is not known as this organism has a nucleotide excision repair system comparable to that of *E. coli* and mutations in the nucleotide excision repair genes render cells sensitive to UV (*120, 121*).

**2. Redoxyendonucleases (*122*).** Although this nomenclature has not been widely accepted and technically it may not be the most appropriate name, it is preferable to other terms such as UV endonuclease, γ-endonuclease, X-ray endonuclease, urea–DNA glycosylase, and thymine glycol–DNA glycosylase. All these enzymes release damaged bases (generally by oxidation–reduction reactions) by DNA–glycosylase action and then cleave the phosphodiester bond 3' to the abasic sugar. Enzymes from *E. coli*, yeast, mouse, and man have similar properties and similar catalytic mechanisms (*123–125*). Redoxyendonucleases are monomeric proteins of $M_r = 30\,000–40\,000$ and have no requirements for cofactors or divalent cations. Substrates for these enzymes are generated by UV, ionizing radiation, and chemical oxidants such as $H_2O_2$, potassium permanganate, and osmium tetroxide (*123*) and include thymine glycols, 5-hydroxy-5-methylhydantoin, tartronyl-urea methyl ester, urea, and a stable cytosine photohydrate of unknown structure (*126*), as well as possible guanine radiolysis products. Thymine glycols have been shown to block DNA replication and, thus, cause lethality (*127–129*). All redoxyendonucleases studied act on a "clean" AP site as well, incising 3' to the abasic sugar. The following are the best characterized redoxyendonucleases:

*a. E. coli Endonuclease III.* This enzyme is encoded by the *nth* gene, which has been cloned and sequenced and the protein overproduced (*130, 131, 131a*). An insertion mutation completely eliminates the thymine glycol glycosylase activity in this bacterium although the mutants are not more sensitive to $H_2O_2$ or ionizing radiation; however, the mutant has a high spontaneous mutation rate (*130*). The gene product has a $M_r = 23\,546$ and comprises approximately 10% of the total AP endonuclease activity in *E. coli* and 70–80% of the EDTA-resistant AP endonuclease in this bacterium; the remaining EDTA-resistant activity is due to endonuclease IV. The purified protein has two absorption peaks at 280 and 410 nm. The latter is due to the presence of [4Fe-4S]$^{2+}$ clusters in the enzyme (*131b*).

Although the enzyme is often called thymine glycol–DNA glycosylase or urea–DNA glycosylase, this term is inappropriate because when terminally labeled DNA (either irradiated with high UV dose or ionizing radiation, or treated with chemical oxidants) is used as substrate, the major incision sites are at cytosines—presumably due to the presence of as yet unidentified cytosine oxidation–reduction products (*124*). The enzyme does, however, quantitatively release thymine glycols, urea, and other thymine ring contraction and opening products from DNA

(*132*). It has long been suspected that endonuclease III, and indeed many other AP endonucleases, may not cleave the phosphodiester bond by hydrolysis but rather by β-elimination. Recently, unambiguous evidence has been obtained that cleavage does in fact occur by β-elimination: (1) the enzyme releases from [5'-$^{32}$P]pdT$_8$d(–)dTn a radioactive product which migrates as a doublet similar to the doublet generated by alkaline treatment (*133*); (2) after endonuclease III treatment followed by cleavage with a class II AP endonuclease, the released sugar comigrates with the β-elimination product (2,3-didehydro-2,3-dideoxyribose 5'-phosphate) and not 5-deoxyribose phosphate (*114*); (3) reduction of the C-1' aldehyde with cyanoborohydride or methoxyamine completely inhibits endonuclease III action (*134*); and (4) 2-deoxyribose is removed by the 3'–5' exonuclease activity of polymerase I whereas 2,3-didehydro-2,3-dideoxyribose is not and the product of endonuclease III is a poor substrate for polymerase I (*114*). It has, thus, been suggested that endonuclease III is a phosphoric monoester lyase belonging in EC4 and is not a phosphoric diester hydrolase which is in EC3 and, therefore, should be called an AP lyase rather than AP endonuclease (*133*). However, the name will likely be retained since nuclease does not necessarily mean cleavage of the phosphodiester bond by hydrolysis (*114*).

In contrast with pyrimidine dimer DNA glycosylase–AP endonucleases, the glycosylase and AP endonuclease functions of endonuclease III are tightly coupled; AP sites do not accumulate during the reaction of the enzyme with DNA containing thymine glycols or urea. The tight coupling as well as the inhibition of the enzyme by pyridoxal 5'-phosphate has led to the suggestion that an amino group on the enzyme may catalyze a series of β-elimination reactions which involve the formation of a Schiff base between the enzyme and C-1' of the deoxyribose (*134*). The γ-endonuclease of *M. luteus* is very similar in all aspects to endonuclease III of *E. coli* (*135*).

*b. Yeast Redoxyendonuclease (*125*).* An enzyme has been partially purified from *S. cerevisiae* that cleaves OsO$_4$-treated DNA at sites of thymine glycols and heavily irradiated DNA at sites of cytosine, thymine, and guanine photoproducts. The enzyme has an apparent $M_r = 40\,000$, is active in the presence of EDTA, and is very similar to *E. coli* endonuclease III. Yeast mutants known to be sensitive to X-ray (rad 6 and rad 52) have normal levels of this enzyme.

*c. Murine Redoxyendonucleases (UV Endonucleases I and II) (*136*).* Three endonucleases that nick UV-irradiated DNA have been identified in murine plasmacytoma cells, and two of these (I and II) have been purified to apparent homogeneity and characterized. Because of their functional similarities to other thymine glycol glycosylase/AP endonucleases, we will refer to these enzymes as murine redoxyendonucleases I and II. Redoxyendonuclease I has a $M_r = 47\,000$, and redoxyendonuclease II is 28\,000 Da. Redoxyendonuclease I is marginally more effective on relaxed as compared to supercoiled DNA while redoxyendonuclease II has a 6-fold higher affinity for supercoiled DNA. Both enzymes are active on simple AP sites in addition to their thymine DNA glycosylase activities. With OsO$_4$-damaged DNA as substrate, both enzymes release a thymine glycol as does *E. coli* endonuclease III, but redoxyendonuclease I also releases an unidentified base not seen with the *E. coli* enzyme, and endonuclease III releases a base as a major product that is not seen with the murine enzyme. These and a number of other observations indicate that redoxyendonucleases have overlapping but not identical substrate specificities. The redoxyendonuclease purified from calf thymus (123, 124) which has been reported to incise on both the 3′ and 5′ sides of the AP site appears to have been contaminated with a type II AP endonuclease. The mammalian endonucleases also appear to cleave the phosphodiester bond by β-elimination as the nick produced by these enzymes does not serve as a primer for nick translation by E. coli DNA polymerase I (114).

Another general characteristic of these enzymes is that the glycosylase and AP endonuclease functions are tightly coupled even though the enzymes are capable of carrying out the second half of the reaction (AP endonuclease) when presented with AP site containing DNA. Redoxyendonucleases incise DNA containing AP sites by a type I AP endonuclease mechanism. It will be interesting to determine whether any uncoupling occurs with thymine glycol substrates when the reaction is carried out in the presence of methoxyamine. Human redoxyendonuclease from HeLa cells releases thymine glycols from DNA (137) and has the general characteristics of other redoxyendonucleases.

### C. AP Endonucleases.

Apurinic/apyrimidinic sites in DNA are generated by several mechanisms. The glycosylic bond of purines is hydrolyzed spontaneously at physiologically relevant rates; these rates are accelerated by low pH and high temperature. AP sites are also generated by spontaneous cleavage of the glycosylic bond of alkylated bases (especially purines) whose glycosylic bond becomes unstable upon alkylation. Ionizing radiation and chemical oxidants, among other oxidation products, produce AP sites or, alternatively, deoxyribose linked to a remnant of the base (such as urea). AP sites are also the immediate product of DNA glycosylases. There are four possible incision sites for AP endonucleases, and the enzymes have been classified according to their sites of incision: class I and class II AP endonucleases incise 3′ and 5′, respectively, to the AP site, both leaving 3′-OH and 5′-P termini; class III and IV enzymes cleave 3′ and 5′, respectively, to the AP site, generating 3′-P and 5′-OH termini. All known class I enzymes are glycosylase–AP endonucleases, and these have been discussed above. Most AP endonucleases are of class II. There is only one known example of class III AP endonucleases, the Drosophila AP endonuclease, and no known examples of class IV enzymes. The following are some of the most extensively investigated AP endonucleases.

### 1. E. coli AP Endonucleases.

The two major class II AP endonucleases of E. coli are exonuclease III and endonuclease IV, although recently additional AP endonucleases have been reported.

a. Exonuclease III. This enzyme is encoded by the xth gene, which has been cloned and sequenced (138). Point, insertion, and deletion mutants of the gene have been isolated. Mutants are extremely sensitive to $H_2O_2$ but not to X-rays or alkylating agents (139, 140). Exonuclease III is the major E. coli AP endonuclease, accounting for 85–90% of total AP endonuclease activity in cell-free extracts. The enzyme is a monomeric protein of $M_r = 31\,000$ and requires $Mg^{2+}$. It has several enzymatic activities (141): (1) double-strand-specific 3′ → 5′ exonuclease; (2) 3′-phosphatase; (3) class II type AP endonuclease and the related function, 3′-phosphoglycolaldehyde diesterase; and (4) cleavage of the phosphodiester bond 5′ to a urea–deoxyribose. The latter function may be of special significance regarding repair of $H_2O_2$ damage of DNA. $H_2O_2$ generates breaks that have 3′-phosphoglycolaldehyde esters which cannot be removed by DNA polymerase I nor can they act as primers. Thus, exonuclease III converts a nonrepairable strand scission to a "clean" break which can be repaired (140). Similarly, the 2,3-didehydro-2,3-dideoxyribose 5′-phosphate product at the 3′ terminus of the class I AP endonuclease scission site is efficiently removed by exonuclease III (133). It has previously been suggested that exonuclease III needs "empty space" 3′ to the hydrolyzed phosphodiester bond; however, recent work by Kow has shown that the enzyme can efficiently hydrolyze the phosphodiester bond 3′ to all alkylhydroxylamine residues (i.e. O-Me-, O-Et-, O-alkyl-, and O-benzylhydroxylamine N-glycosides). In the presence of O-alkylhydroxylamines, the damaged base is believed to shift from the β-conformation to one that is coplanar with the deoxyribose and thus creates the necessary "empty space" in the inter-DNA strand area (141a).

Recently, the Streptococcus pneumonia exoA gene (which corresponds to exoIII of E. coli) was cloned and sequenced. The enzyme $M_r = 31\,263$ and is 26% identical with exonuclease III. Additionally, all of its enzymatic activities appear to be the same as for exonuclease III; however, in contrast to E. coli, null exoA mutants in S. pneumonia appear to be inviable (141b).

b. Endonuclease IV. The E. coli nfo gene that encodes this enzyme has been cloned and sequenced (142). Gene expression is induced by superoxide radicals (143). Insertion and deletion mutants of nfo have been isolated (144). The mutants are moderately sensitive to the alkylating agents N-methyl-N′-nitro-N-nitrosoguanidine and mitomycin C (145) and are markedly sensitive to the oxidants tert-butyl hydroperoxide and bleomycin. The nfo or xth single mutants are only marginally sensitive to UV and ionizing radiation; however, the combination of xth and nfo mutations makes cells extremely sensitive to these agents, indicating that the two enzymes constitute a backup system for dealing with damage caused by free radicals (144). The combination xth⁻, nfo⁻, and uvrA⁻ is lethal (145a).

Endonuclease IV has been purified to homogeneity and found to be a monomeric protein of 31 000 Da (146). It is functional in the presence of EDTA, but it apparently contains an intrinsic zinc and can be inactivated in the absence of substrate by incubating with EDTA or 1,10-phenanthroline (146). The enzyme is a simple class II AP endonuclease, but it also releases fragments of sugars remaining on the 3′ side of a nick generated by free radicals or class I AP endonucleases (140).

Recently, in addition to exonuclease III and endonuclease IV, two additional 3′-phosphoglycolaldehyde diesterases have been identified in E. coli (147). These enzymes also act as class II AP endonucleases.

### 2. Yeast AP Endonucleases.

Five AP endonucleases of class II (D1, D2, D3, D4, and E) have been partially purified from yeast (148, 149). These proteins, which differ in their molecular weights (10 000–37 000 Da), all require $Mg^{2+}$ and, at least the AP endonuclease E, incise 5′ to urea residues in DNA in a manner similar to that of exonuclease III of E. coli. A yeast enzyme has been purified to homogeneity by testing fractions for the 3′-phosphoglycolaldehyde diesterase activity (150). This enzyme of $M_r = 40\,500$ may be the yeast AP endonuclease E; it is very similar in all of its enzymatic properties of E. coli endonuclease IV. This protein is the major yeast enzyme which activates 3′-termini generated by bleomycin, γ-rays, and $H_2O_2$. It contains a tightly bound $Co^{2+}$ and can be inactivated by removal of the metal. Even though its biochemical properties suggest that it must be an important enzyme in yeast's defense against ionizing radiation, none of the radiation-sensitive mutants (rad 6 and rad 52 groups) are defective in this activity.

**3. *Drosophila* AP Endonucleases.** Two AP endonucleases have been identified in *Drosophila* embryos and have been partially purified (*151*). AP endonucleases I ($M_r$ = 66 000) appears to function as a class III AP endonuclease. Like AP endonuclease I, AP endonuclease II also cleaves 3' to the AP site; however, its cleavage is in the form of class I AP endonucleases, i.e., it produces 3'-deoxyribose and 5'-phosphomonoester termini. It is not known whether the enzyme has an associated DNA glycosylase activity as do all the other class I AP endonucleases.

**4. Human AP Endonucleases.** AP endonucleases from HeLa cells, fibroblasts, and placenta have been purified. The HeLa cell AP endonuclease (*152*) is a monomeric enzyme of $M_r$ = 41 000, it has an absolute requirement for $Mg^{2+}$ and acts as a class II AP endonuclease. Interestingly, the enzyme is inhibited by adenine, hypoxanthine, adenosine, and $NAD^+$ but not by caffeine. Antibodies to HeLa AP endonucleases react with AP endonucleases from various sources, and indeed, these antibodies were recently used as probes to clone the *Drosophila* AP endonuclease from a λ-gt11 library (*152a*). The placental AP endonuclease (*153*) has an apparent $M_r$ of 37 000 and requires $Mg^{2+}$ for activity. The enzyme cleaves at AP sites either 3' (class I) or 5' (class II) to the abasic sugar in a 40:60 ratio. Apparently, however, it does not carry out both reactions on the same substrate, and therefore, a deoxyribose 5-phosphate or its α,β-unsaturated derivative is not released. The latter apparently is released in small quantities by a nonenzymatic β-elimination reaction on 5'-incised molecules. Interestingly, the enzyme cleaves the phosphodiester bonds in the same proportion but at 20% efficiency when the AP sites are reduced with $NaBH_4$. It is unclear, therefore, whether the 3'-incision with the nonreduced substrate occurs by β-elimination. It would be expected that if 3'-cleavage were by β-elimination, reduction of the AP site would completely inhibit the incision reaction. Two AP endonucleases have been partially purified from human fibroblasts which have the characteristics of class I and class II AP endonucleases, respectively. Both the placental and fibroblast AP endonucleases cross-react with antibodies against HeLa AP endonucleases, and the relationship between the enzymes from these various human sources is a matter that needs further investigation.

The bovine AP endonuclease analogous to the HeLa cell enzyme has been purified to homogeneity from calf thymus and characterized in detail; its properties are similar to those of the HeLa enzyme. In a recent study, it has been shown that the bovine AP endonuclease cleaves 5' to abasic or anucleolytic sites containing either ethylene glycol, propanediol, or tetrahydrofuran interphosphate linkages. However, it is inhibited by reduction of C1' or the presence of urea of alkoxyamine at that position (*153a*).

**D. *3'-Phosphoglycolaldehyde Diesterase and DNA Deoxyribophosphodiesterase*.** AP endonucleases generate nicks in DNA which are flanked by a nucleotide on one side and a deoxyribose or a dideoxydihydrodeoxyribose on the other. Similar types of structures may be generated by direct action of ionizing radiation and free radicals generated by oxidizing agents. For repair of the nick the deoxyribose must be removed. It appears that there may exist specific enzymes to remove the deoxyribose or the deoxyribose fragments (3'-phosphoglycolaldehyde) from the termini in preparation for DNA polymerases. Two such activities have been described:

**1. 3'-Phosphoglycolaldehyde Diesterase (*147*).** In *E. coli*, the 2,3-dideoxyribose remaining at the 3'-terminus

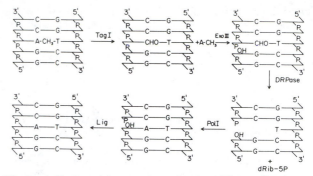

**Figure 4.** Base excision repair mechanism. The modified nucleotide is removed by a specific glycosylase—in the specific example here, 3-mAde by 3-mAde glycosylase I of *E. coli* (TagI). An AP endonuclease (ExoIII) incises 5' to the AP site; the abasic sugar is released by DNA deoxyribose phosphodiesterase (dRPase) to generate a one-nucleotide gap which is filled by DNA polymerase I (PolI) and sealed by DNA ligase (Lig). Reproduced, with permission, from ref 1. Copyright 1988 Annual Reviews, Inc.

after cleavage by class I AP endonucleases can be liberated by exonuclease III and endonuclease IV. These enzymes are equally effective in releasing the 3'-phosphoglycolaldehyde esters generated at nicks caused by reactive oxygen species ($O_2^{\bullet -}$ and $OH^{\bullet}$). In addition to these enzymes, there are two activities in *E. coli*, one of $M_r$ = 50 000 and one of $M_r$ = 25 000, both of which repair 3'-damaged termini. Whether these enzymes are specific for 3'-termini or are bona fide AP endonucleases that can also release phosphoglycolaldehyde remains to be determined. A yeast enzyme (*150*) purified to homogeneous 3'-phosphoglycolaldehyde diesterase activity is an AP endonuclease similar in many ways to *E. coli* endonuclease IV and was discussed in the previous section.

**2. DNA Deoxyribophosphodiesterase (dRPase) (*154*).** The 5'-deoxyribose termini generated by cleavage of class II AP endonucleases cannot be released in *E. coli* by the major class I AP endonuclease, endonuclease III, nor is it removed efficiently by the 5' → 3' exonuclease function of DNA polymerase I. An enzyme has been purified from *E. coli* that liberates the 5'-deoxyribose very efficiently. The partially purified enzyme has $M_r$ = 50 000, requires $Mg^{2+}$, and has no AP endonuclease or exonuclease activities. It does not nick depurinated DNA and does not release nucleotides from a nick. However, the partially purified enzyme liberates the 3'-dideoxydihydroribose residue remaining at the nick site of endonuclease III (3'-phosphoglycolaldehyde diesterase activity). It is unclear at present whether it is the same enzyme as the 50 000-Da protein originally identified by its 3'-phosphoglycoaldehyde diesterase function. A dRpase has been identified and partially purified from human cells. Figure 4 shows how base excision repair is accomplished by sequential actions of a glycosylase, AP endonuclease, dRPase, and finally DNA polymerase and ligase.

## III. Nucleotide Excision Repair

Adducts and DNA lesions that presumably cause major helical distortion and which are lethal and/or mutagenic are removed by an enzyme complex that incises DNA on both sides of the adduct. Although nucleotide excision repair is the primary mechanism for repair of such DNA adducts as pyrimidine dimers, 6–4 photoproducts, and psoralen–thymine adducts, recent research suggests that it may play a backup role in repairing DNA lesions which do not distort the helix extensively and which are ordinarily repaired by methyltransferases (*155*) and glycosylases (*156*). It does not, however, repair mismatches or extrahelical loops (157). This section describes recent advances in our understanding of nucleotide excision repair in E. coli, yeast, and man.

**A. Nucleotide Excision Repair in E. coli.** Biochemically, this is the most extensively characterized nucleotide excision repair system. The incision complex is encoded by three genes, uvrA, uvrB, and uvrC; mutations in any of the three genes abolishes excision repair in vivo (1). The genes have been cloned and sequenced, and the corresponding proteins have been purified (63).

The UvrA protein has been crystallized (63) although no structural data have been obtained. The UvrA protein is a dimer (monomer $M_r$ = 103874; 158), is an ATPase, has two "zinc finger" DNA binding motifs (159), as confirmed by EXAFS and site-specific mutagenesis (160), and binds to damaged DNA with $K_D$ = $10^{-8}$ M but has relatively high nonspecific binding affinity as well, $K_{NS}$ = $10^{-5}$ M (161, 162).

The UvrB protein ($M_r$ = 76118; 60) also possesses the consensus ATPase sequence but has no detectable ATPase activity and does not bind to DNA. However, UvrB makes a $(UvrA)_2(UvrB)_1$ complex with UvrA in an ATP-dependent reaction (163). UvrB also causes a 2–3-fold increase in ATPase activity when added to a UvrA–DNA mixture (63, 164). Whether this is caused by an allosteric effect on UvrA's ATPase activity or an unmasking of UvrB's cryptic ATPase is unknown. The OmpT protease that cleaves Ada (64) also removes 40 amino acids from the C-terminus of UvrB, generating UvrB* (60). The UvrB* protein is inactive in forming the nuclease complex but is a relatively potent ATPase (164, 165). Generation of UvrB* by proteolysis and the consequent unmasking of its ATPase activity have no biological significance as the cleavage does not occur in vivo (63).

UvrC ($M_r$ = 66038; 166) has no ATPase activity and binds to single- and double-stranded DNA (both damaged and undamaged) with low affinity ($K_D$ = $10^{-5}$ M). UvrC, in the absence of DNA, does not interact with UvrA, UvrB, or the $(UvrA)_2(UvrB)_1$ complex (163).

UvrA, UvrB, and UvrC together incise damaged DNA on both sides of the lesion in an ATP-dependent reaction, and together, these proteins are referred to as ABC excinuclease (167). The "Uvr" has been dropped from the name to emphasize that UV damage is not the only biological substrate for the enzyme in E. coli, and excinuclease (excision nuclease) has been used instead of endonuclease to emphasize its unique action mechanism, i.e., specific release of a single-stranded oligomer containing damaged nucleotide(s). The construction by Hearst and colleagues of substrates containing psoralen adducts at defined positions (167a) has made possible the study of the sequential assembly of the enzyme and the determination of the contact sites of the enzyme with the substrate. It has been shown that UvrA dimer contacts 33 bp on both strands surrounding the psoralen monoadduct. Addition of UvrB results in the "shrinking" of the footprint on the nonadducted strand to 19 bp and the appearance of a single nuclease-hypersensitive site 11 nucleotides 5′ to the psoralen-modified thymine on the damaged strand. In light of our current findings concerning the assembly of the enzyme, the smaller footprint appears to be that of UvrB alone. Addition of UvrC does not result in any observable change in the footprint on the nondamaged strand and the ABC excinuclease specific double incision on the modified strand (161). Addition of proteins known to be involved in catalytic turnover of ABC excinuclease (i.e., helicase II and DNA polymerase I) fails to reveal additional footprints at the incision sites, indicating that

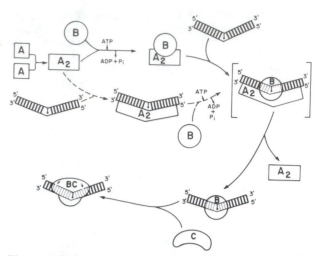

**Figure 5.** Reaction mechanism of E. coli ABC excinuclease. The damage recognition subunit, UvrA, forms a $(UvrA)_2(UvrB)_1$ complex in an ATP-driven reaction, delivers UvrB to the damage site, and dissociates from the DNA. The resulting UvrB–DNA complex is extremely stable and has high affinity for UvrC. Binding of UvrC to the UvrB–DNA complex results in immediate cleavage of the eighth phosphodiester bond 5′ and the fifth phosphodiester bond 3′ to the damaged nucleotide. ATP is probably required in other steps in addition to those indicated in the figure. Reprinted with permission from ref 163. Copyright 1989 National Academy Press.

the six proteins involved in nucleotide excision repair (i.e., UvrA, UvrB, UvrC, UvrD, DNA polymerase I, and DNA ligase) do not make a supramolecular complex ("repairosome") during the repair process. By use of the same system, it was also shown that DNA polymerase I fills in the excision gap in the presence or absence of helicase II and that 90% of the repair patches are 12 nucleotides long. This length is not affected by helicase II (162).

The current model for the incision reaction by ABC excinuclease is as follows (163; see Figure 5): UvrA associates with UvrB to form a $(UvrA)_2(UvrB)_1$ complex in a process requiring ATP hydrolysis. The complex, guided by UvrA's affinity for damaged DNA, binds to the damage site, a UvrB–DNA complex is formed, and $(UvrA)_2$ dissociates from the complex. The UvrB–DNA complex is extremely stable ($k_{off}$ = $1.1 \times 10^{-4}$ $s^{-1}$). UvrC, which has no affinity for UvrB and only marginal affinity for damaged DNA, interacts with the damaged DNA–UvrB complex and mediates the dual incisions. This three-step delivery mechanism is a unique macromolecular recognition process that ensures high specificity of damage recognition by employing three steps with limited specificities. The incision complex is composed of only the UvrB and UvrC subunits, and thus, UvrA functions as a damage recognition and UvrB delivery protein. Therefore, it has recently been proposed that the nuclease be renamed (A)BC excinuclease. The main incision pattern of the enzyme is the hydrolysis of the eighth phosphodiester bond 5′ and the fourth phosphodiester bond 3′ to the adducted nucleotide(s), which results in the liberation of a 12-mer containing the modified nucleotide (167). Minor modifications of this general pattern are observed depending upon whether the enzyme is excising a nucleotide mono- or diadduct. The cutting pattern is partially influenced by the type of adduct as well as its sequence context (168). Under suboptimal reaction conditions, and perhaps more frequently with certain DNA adducts, uncoupled incisions are observed (169, 170). It is not known if uncoupling has any physiological significance.

(A)BC excinuclease acts stoichiometrically under certain experimental conditions (*171, 172*). However, in light of our present understanding of the stepwise recognition process, it appears that only UvrB and UvrC must function in a stoichiometric fashion. The mixture of the three proteins can be made to act catalytically by including DNA polymerase I and helicase II to the reaction mixture. While the precise mechanism by which these enzymes promote the turnover of (A)BC excinuclease is not known, the following is a likely scenario in vivo. Uninduced *E. coli* has about 20 molecules of UvrA, 200 of UvrB, and 10–20 of UvrC (*1*). The 20 UvrA molecules make 10 (UvrA)$_2$ dimers which deliver the 200 UvrB molecules to the damage sites. UvrC binds to the stable UvrB–DNA complexes mediating the dual incision of the damaged DNA. The UvrB–UvrC–DNA complex is recognized by helicase II and polymerase I and is dissociated from the incision site along with the excised oligomer (*171, 172*). The released UvrC can now engage in new rounds of incisions.

**B. Nucleotide Excision Repair in Yeast.** DNA repair in the yeast *S. cerevisiae* requires the participation of at least 95 genetic loci (for reviews, see refs 173, 3, and 174). Mutations in DNA repair have been classified into epistasis groups; two mutants are defined as being members of the same epistasis group if a strain carrying a mutation in both alleles is no more sensitive to UV irradiation, etc., than the most sensitive of the single mutant strains. Consequently, yeast mutants sensitive to UV light have been classified into three epistasis groups: (1) RAD3 mutants are defective in excision repair, (2) RAD6 mutants are abnormal with regard to mutagenesis, and (3) RAD52 mutants are defective in recombination. Accordingly, it has been suggested that each epistasis group corresponds to a specific biochemical pathway. Here, we will consider only members of the RAD3 epistasis group. More specifically, we will describe current advances with regard to the RAD1, RAD2, RAD3, RAD4, and RAD10 loci as mutants in these are particularly sensitive to UV irradiation and do not carry out detectable DNA incision during postirradiation incubation. Mutants in the other loci, RAD7, RAD14, RAD16, RAD23, and MMS19, have significant incision activity and, hence, may only provide accessory functions in nucleotide excision repair. Each of the genes has been cloned—principally by restoring UV resistance to mutant yeast strains after transforming with DNA from a yeast genomic library.

The RAD1 gene has been sequenced (*175*); it encodes a protein of 1100 amino acids with a molecular weiight of 126 360. Comparison of the amino acid sequence with other DNA repair proteins or protein sequences in the National Biomedical Research Foundation Library data bank has not revealed any sequence homologies. RAD1–lacZ fusion experiments indicate that the gene is weakly expressed (*3*) and that its expression is not increased by exposure to DNA damage (*176, 177*). Characterization of this protein has been hampered by its limited solubility when overexpressed in *S. cerevisiae* (*3*). Recently, it has been reported that, in addition to its role in nucleotide excision repair, Rad1 plays a role in recombination after the formation of the recombinogenic substrate. This activity has not been found in Rad2, Rad3, and Rad4 and apparently participates in a pathway distinct from that reported for Rad52 (*178*).

The RAD2 gene encodes a protein of 117 700 Da. Studies with RAD2–lacZ fusions have demonstrated that RAD2 is a DIN (damage-inducible) gene. β-Galactosidase synthesis is stimulated by UV irradiation, 4NQO, mitomycin C, and a variety of other DNA-damaging agents

(*177, 179*). Unfortunately, like Rad1, Rad2 is mainly insoluble when overexpressed in yeast cells and has not been purified to date.

In addition to its role in nucleotide excision repair, the RAD3 gene is essential for cell viability (*180, 181*) although the nature of this essential function is unknown. The RAD3 gene has been sequenced (*182*); it encodes a protein of $M_r = 89\,700$. The amino terminus of Rad3 has a region of homology with other known ATPases (*183*) including *E. coli* UvrA, UvrB, UvrD, RecA, RecC, and RecD proteins. Rad3 also has, in its C-terminal half, a region of homology to the DNA binding helix–turn–helix motif. Naumovski and Friedberg (*184*) have constructed mutants in both of these regions and demonstrated that each is required for the excision repair function of this gene. The Rad3 protein has been purified and partially characterized; it is a single-strand DNA-dependent ATPase (*185*) and a 5′ to 3′ DNA helicase (*186*). These data are consistent with the prior reports that the rem1 mutation in the RAD3 locus increases the spontaneous mutation and mitotic recombination rates (*187, 188*), both of which are suggestive of a deficiency in DNA replication. Sung and co-workers suggest that this activity may be analogous to that described for the UvrD (helicase II) protein of *E. coli* (*171, 172*) by catalyzing the strand displacement repair synthesis at the site of incision in DNA. Site-directed mutagenesis of the conserved lysine residue in the Walker A "GKT" sequence abolishes both the ATPase and helicase activities of the protein although it still retains approximately 50% of its ATP binding activity (*189*). Interestingly, yeast mutants carrying this mutation (lys48arg) are still able to incise DNA but are apparently defective in a postincision step again reminiscent of *E. coli* UvrD mutants (*190*). RAD3 expression is not inducible by DNA-damaging agents (*191*).

The RAD4 gene has been cloned, and it has been shown that its expression is not induced by DNA-damaging agents (*192*). Also, its nucleotide sequence has been determined (*193, 194*); RAD4 encodes a protein of 87 173 Da and has a short stretch of 24 amino acids which is homologous with an amino-terminal portion of Rad10 (*3*). It is interesting that Rad4 is toxic to *E. coli*; transformation with the RAD4 gene results in mutational inactivation (*192*). However, a small number of *E. coli* transformants have been identified that propagate mutant RAD4 plasmids (*194*). Characterization of these plasmids has revealed that the spontaneous mutations map to a 372-bp region in the central one-third of the RAD4 gene. Coincidentally, yeast chromosomal RAD4 mutations also map to this region, and both the *E. coli* and yeast mutations encode truncated polypeptides. These observations suggest that the C-terminal one-third of the protein may be required for both the yeast nucleotide excision repair and *E. coli* lethality functions. The C-terminus of the protein is predominantly acidic whereas the N-terminal half of the protein possess numerous basic patches and includes a putative helix–turn–helix DNA binding motif (*193, 194*).

Rad10 ($M_r = 24\,200$) is highly homologous to the human excision repair protein ERCC-1 (*195, 196*), one particular domain of which possesses a putative DNA binding helix–turn–helix motif. This homology is particularly significant in light of the recent report describing the partial complementation of CHO complementation group 1 by the RAD10 (*197*). This observation represents direct experimental evidence for the close evolutionary relationship between such diverse eukaryotic excision repair pathways.

**C. Nucleotide Excision Repair in Humans.** Recent characterization of nucleotide excision repair in mammalian cells has progressed principally in five areas: (1) cloning of human genes complementing UV-sensitive rodent cell lines, (2) characterization of the phenomenon of preferential DNA repair in transcriptionally active gene segments, (3) reconstitution of nucleotide excision repair in cell-free extracts, (4) complementation of mutant cell lines with bacterial genes, and (5) characterization of human syndromes caused by defects in DNA repair. The following describes recent advances in each of these areas.

**1. Cloning of Excision Repair Genes.** Patients with the disease xeroderma pigmentosum have long been known to be deficient in their repair of UV-induced DNA damage (198). A minority of XP cases (variants) have been suggested to be deficient in "postreplication repair" (199). Somatic hybridizations of cells from patients with this disease have revealed the presence of at least nine complementation groups (200, 201), and it has been found that members of these complementation groups are unable to perform the initial nicking step that facilitates replacement of the damaged DNA segment (202).

A similar level of complexity has been demonstrated for nucleotide excision repair in rodent cells as eight complementation groups have been described to date (203–206). Groups 1–5 are highly UV sensitive and deficient in the incision step (207). Complementation group six is characterized by an apparently normal removal of 6–4 photoproducts but is defective in pyrimidine dimer removal. Complementation groups 7 and 8 have only recently been described (206); these authors tested complementation of CHO mutants with mouse mutants from complementation groups II and III and a UV-sensitive mutant of V79 cells (V-B11) (208). They found that V-B11 complemented each of the six CHO mutants whereas the mouse mutant in group III complemented the six CHO groups as well as the V79 mutant. It is striking that no overlap of complementation groups between CHO and XP have been identified in cell fusion experiments (although all combinations have not been reported; 209, 210).

Cell fusion techniques have been used to assign ERCC genes to their respective chromosomes. ERCC-1 has been mapped to the long arm of human chromosome 19 (211). ERCC-2, -3, -4, and -5 have been assigned to chromosomes 19, 2, 16, and 13, respectively (212, 213).

Isolation of human excision repair genes by complementation of XP cells has not been successful (reviewed by Lehmann; 214) presumably due to the limited amount of DNA that becomes integrated into the genome after genomic transfections. Using a cDNA library, however, Teitz et al. (215) have been able to complement XP-A cell lines although the clone responsible for this complementation has not been isolated. Between 20- and 100-fold more DNA is incorporated into CHO cells as compared to XP cells (216). For this reason, cloning efforts have been concentrated primarily on transfecting human DNA into CHO cells. Transformants corrected by human genes have been reported for several CHO complementation groups (217, 218, 210). Westerveld et al. (219) were the first to describe the cloning of a human nucleotide excision repair gene— ERCC-1. These authors transfected a CHO complementation group 1 mutant (43-3B; 220) with human DNA which had been digested with a restriction enzyme and ligated into a plasmid containing the dominant gene marker Ecogpt (which renders transformed cells resistant to mycophenolic acid). After selection of mycophenolic acid resistant clones, cells were tested for resistance to UV irradiation and mitomycin C. DNA from resistant colonies was retransfected into the 43-3B cells to eliminate cointegrated sequences, reisolated, digested with a re-

striction enzyme, ligated into a cosmid cloning vector, and transformed into *E. coli*. Clones were identified by hybridization with a gpt probe. These clones were shown to confer resistance to both UV and mitomycin C.

The cDNA corresponding to the ERCC-1 gene has been cloned by screening a cDNA library (196). Northern analysis of poly(A) RNA from HeLa cells has revealed two ERCC-1 transcripts, one of 1.0 kb and the other 1.1 kb. Only the cDNA from the larger transcript confers UV and MMC resistance to 43-3B cells. Zdzienicka et al. (208) have further shown correction of sensitivity to 4NQO, ENU, and N-Ac-AAF. Further, analysis of poly(A) RNA at various times after UV irradiation showed no evidence for UV inducibility of the ERCC-1 gene. ERCC-1 complementation is restricted to complementation group 1, suggesting that correction is specific.

DNA sequence analysis of the ERCC-1 cDNA has revealed that the protein is comprised of 297 amino acids with a molecular weight of 32 562. Numerous regions of sequence homology exist between ERCC-1 and a variety of other proteins. The entire RAD10 gene of yeast is homologous to the N-terminus of ERCC-1; ERCC-1 is longer by a tail of approximately 110 amino acids. This homology is particularly striking in light of the recent observation that RAD10 partially complements CHO cell lines from complementation group 1 (197). Also, a region of 42 amino acids is homologous to UvrA (31% identity). ERCC-1 possesses a nuclear localization signal although van Duin (196) have demonstrated that it is not required for CHO cell complementation. The stretch homologous with Rad10 is also homologous to the helix–turn–helix motif found in numerous DNA binding proteins. ERCC-1 also has a sequence homologous to the ADP ribosylation site of several G-proteins (see Hoeijmakers for a review; 216), and its C-terminus is homologous to *E. coli* UvrC (159, 221). Because of the strong homologies between repair proteins of such diverse species, it has been suggested that another strategy for isolating human genes may be to hybridize with DNA from evolutionary intermediates (222).

By using the same approach as did Westerveld et al. (219) for the cloning of ERCC-1, Weber et al. (223) have obtained a cosmid clone of ERCC-2. The gene has been sequenced; it encodes a protein of 760 amino acids which is 50% identical with the yeast RAD3 gene (223a). Complementation of CHO cells is specific for group 2 consistent with the report that CHO complementation groups 3, 4, and 5 are each corrected by a human chromosome other than 19. Although, chromosome 19 restores UV resistance to both CHO groups 1 and 2, the phenotypic differences between these groups (i.e., group 2 is hypersensitive to cross-linking agents whereas group one is not; 224) correlate with the specificity of complementation of CHO group 1 with ERCC-1 and group 2 with ERCC-2. Likewise, Hoeijmakers and his collegues have cloned and sequenced the ERCC-3 gene. It encodes a protein of 782 amino acids with a Walker nucleotide binding consensus sequence (223a).

To date, none of the proteins involved in the human nucleotide excision repair complex have been purified to homogeneity although a variety of methods have been proposed. One of the most promising, albeit tedious, methods utilizes the microinjection of cell extracts at various stages of fractionation and measures the transient correction of members of the XP complementation groups (reviewed by Hoeijmakers; 216). Initially developed for the human repair system by de Jonge and co-workers (225), correction of repair deficiencies in each of the nine XP complementation groups has been demonstrated by microinjection of extracts from wild-type cells (226). Using this procedure, Hoeijmakers has reported the partial purification of the XPA correcting factor from calf thymus. Its molecular weight is approximately 40000–45000 (223a). Studies with XP-A cells have revealed that the correcting factor binds with high affinity to both single-stranded and UV-irradiated double-stranded DNA, whereas this protein has reduced affinity for unirradiated double-stranded DNA (216). These findings suggest that this factor contributes to the binding of the complex to UV-induced DNA lesions.

Microinjection has also been used to demonstrate activities of purified repair proteins from a variety of species. Nonspecific correction of all XP complementation groups has been achieved by injecting purified *M. luteus* UV endonuclease (227). Similarly, Zwetsloot et al. (228) have shown reversal of pyrimidine dimers by yeast photoreactivating enzyme in competition with unscheduled DNA synthesis in normal fibroblasts. Further studies (229) demonstrated photoreactivation of pyrimidine dimers in a variety of XP cell lines. Interestingly, XP-C, -F, and -I show reduced levels of UDS as compared to noninjected XP cells, whereas XP-A, -D, -E, and -H cells show no reduction in UDS with DNA photolyase. These authors conclude that photoproducts in different XP cells are not equally accessible. They speculate that dimers in some cells may be protected from photoreactivating enzyme by defective repair enzymes. In a similar study with purified *E. coli* UvrA, -B, -C, and -D proteins, Zwetsloot et al. (230) were not able to show increased UDS in either XP-A, -C, or -H cells. They suggest that the chromatin structure may inhibit DNA repair.

**2. Gene- and Strand-Specific Repair.** There are approximately $10^5$ coding sequences in the DNA of mammals comprising only 1% of the entire genome. It is, therefore, not surprising that repair of DNA damage has been found to occur preferentially in these coding sequences. Evidence for such preferential repair has accumulated over the years. Many carcinogen-induced DNA adducts are removed with biphasic kinetics from mammalian cells (reviewed by Hanawalt et al.; 231); two different repair rates have also been observed for pyrimidine dimers in human fibroblasts (232). Mayne and Lehman (233) reported rapid recovery of RNA synthesis in UV-irradiated mammalian cells before any detectable genomic DNA repair, implying that DNA repair is somehow directed to transcriptionally active chromatin. Interestingly, these authors observed that cells from patients with Cockayne's syndrome (which are known to be hypersensitive to UV; 234) are deficient in their recovery of RNA synthesis following UV exposure although repair of their genomic DNA appears normal.

The chromatin structure of expressed genes causes these sequences to be highly sensitive to digestion by DNAse I while nonexpressed DNA sequences exist in tightly condensed chromatin structures. Accordingly, it has been suggested that the accessibility of certain genomic regions to repair enzymes may in part be responsible for their preferential repair. It has also been proposed that the repair complex proper may participate in the regulation of such open chromatin structure.

While UV light induced pyrimidine dimers are distributed evenly over the mammalian genome (235), this is not the case for many chemically induced DNA adducts. Lesions produced by aflatoxin B1, furocoumarins, and *N*-acetoxy-2-(acetylamino)fluorene reside principally in open chromatin "linker" regions coincident with their accessibility to repair enzymes. Unrepaired DNA damage

has been correlated with the activation of protooncogenes leading to carcinogenesis. Patients with xeroderma pigmentosum, ataxia telangiectasia, Fanconi's anemia, Bloom's syndrome, Cockayne's syndrome, and hereditary retinoblastoma are subject to cancer and are also deficient in DNA repair (198, 101, 236).

Recent advances in our understanding of preferential DNA repair in mammalian cells have been made possible by assaying repair in specific genomic segments whether coding on noncoding. Bohr et al. (237) have described a method for quantifying the repair of UV-induced pyrimidine dimers in a desired gene segment. Briefly, total genomic DNA is isolated from cells at various times after irradiation with UV light. The purified DNA is reacted with a pyrimidine dimer specific endonuclease (T4 endonuclease V), digested with a restriction endonuclease, electrophoresed on an alkaline agarose gel, and transferred to a nitrocellulose membrane. Radiolabeled probes consisting of desired chromosomal DNA segments are hybridized to the immobilized restriction fragments. DNA repair is measured as the increase in intensity of a full-length restriction fragment.

It has been a long-standing paradox that rodent cells in culture are deficient in removing pyrimidine dimers compared to human cells, yet they are equally resistant to UV radiation. Typically, only 10–20% of pyrimidine dimers are removed after 24 h, yet UV survival is comparable to human cells which remove 80% of dimers. Repair measurements have, however, been made over the entire genome. It has been speculated that some regions of the genome might be repaired preferentially to others.

In their initial study, Bohr et al. (237) irradiated Chinese hamster ovary cells with UV light and compared repair of pyrimidine dimers in the actively transcribed dihydrofolate reductase (DHFR) gene to that in a nontranscribed flanking sequence. They found that whereas 70% of pyrimidine dimers were removed from the DHFR gene, 10–15% were eliminated from the upstream sequence. They concluded that the resistance of rodent cells to UV irradiation is due to selective repair of transcribed sequences.

Analysis of DNA repair in the human DHFR gene (238) revealed that while after 2 h repair in the coding and noncoding sequences is the same, after 4 h, more than twice as many pyrimidine dimers are eliminated from the DHFR gene as compared to the flanking DNA. Hence, DNA repair cannot be accurately characterized by overall repair measurements, but the repair must be studied in specific gene segments. Further, Mellon et al. (238) have suggested that the major distinction between rodent and primate cells in repair of UV damage in transcriptionally active DNA appears to be one of restricted repair versus rate of repair, respectively.

It is interesting that Mansbridge and Hanawalt (239) showed that while XPC cells repair only 10–15% of pyrimidine dimers (240) as compared to normal human cells, the repair is limited to a portion of the genome. Mulleanders et al. (241) suggested that this repair is localized to the nuclear matrix. Whereas the extent of repair in XPC cells is equivalent to that in CHO cells, the resistance of XPC cells to UV light is only 1% of that for CHO cells (242). A comparison of cell survival, overall genome repair, and repair in the DHFR gene showed that repair in the DHFR gene correlates more closely to UV survival than does overall repair in the genome (243). These authors speculate that the defect in XPC "may reflect an inability of the cell to direct the repair mechanism to the vital regions of the genome".

A CHO cell line has been described that expresses the T4 *denV* gene (which encodes the pyrimidine dimer specific endonuclease V) (*244*). Bohr and Hanawalt (*245*) studied dimer repair in three CHO cell lines: (1) *denV*⁺ CHO; (2) wild-type CHO, and (3) a UV-sensitive cell line. Each cell line was irradiated with equal doses of UV light. In the wild-type cells, 49% of the pyrimidine dimers were repaired after 8 h, whereas none were repaired in the flanking regions. In the UV-sensitive cells, no repair was detected in either the gene or downstream segment. In the *denV*⁺ cells, 70% of the DHFR gene and 66% of the downstream segment were repaired, i.e., the same in both. Thus, repair by endonuclease V is not directed to transcriptionally active regions.

Is DNA repair selective for actively transcribed gene segments or merely for coding sequences? Madhani et al. (*246*) addressed this question by comparing the repair efficiencies in two protooncogenes in Swiss mouse 3T3 cells, the actively transcribed *c-abl* gene and the transcriptionally silent *c-mos* gene. While 85% of the pyrimidine dimers were removed in 24 h from the *c-abl* gene, only 10–20% were eliminated from the *c-mos* coding segment, suggesting that only transcriptionally active gene segments are repaired faster than nontranscribed genes.

One conclusion that cannot be drawn from the previous study is whether repair efficiency correlates with transcriptional activity or, alternatively, with the propensity toward active transcription. By comparing the repair of the metallothionein gene both under normal growth conditions and in the presence of the transcriptional inducer $ZnCl_2$, Okumoto and Bohr (*247*) found that, when not transcribed, repair of the metallothionein gene is indistinguishable from the overall genomic repair. In contrast, addition of $ZnCl_2$ increases repair activity by a factor of 2. These authors speculate that these results reflect a change in the chromatin structure associated with the increased transcriptional rate.

Mellon et al. (*248*) have reported a striking difference in the efficiency with which pyrimidine dimers are removed from the transcribed and nontranscribed strands of the DHFR gene in both CHO cells as well as human fibroblasts. In the hamster cells, 80% of the dimers are removed from the transcribed strand in 4 h, but little repair occurs in the nontranscribed strand—even after 24 h. Similar findings were reported in the human cells. These results challenge the simplicity of the model that the open chromatin conformation permits accessibility of repair enzymes. Strand selection must in some way be coupled to transcription; it has been suggested that the RNA polymerase complex may be involved in the repair process.

Thomas et al. (*249*) have described a modification of the assay described above which permits the measurement of the repair of a broad range of DNA adducts. These investigators replaced the pyrimidine dimer specific T4 endonuclease V with the ABC excinuclease from *E. coli*. As described previously, this enzyme binds to and excises a vast array of bulky DNA lesions. Thomas et al. (*250*) have studied the preferential repair of UV-induced 6–4 photoproducts by preincubating the cellular DNA with *E. coli* DNA photolyase to reverse all remaining pyrimidine dimers. They have found that, in 4 h, CHO cells repair 46% of 6–4 photoproducts in the DHFR gene as compared to only 21% in noncoding sequences. After 24 h, 66% of gene-specific lesions are repaired as compared to 40% in the noncoding region.

Regarding the mechanism of strand specificity, an observation of Vrieling et al. (*250a*) is perhaps significant. They found that, in CHO V79 cells, 11 of 17 mutational

base-pair changes were caused by photoproducts in the nontranscribed strand of the *hprt* gene whereas in the UV-sensitive derivative (V-H1), 10 out of 11 base-pair changes were caused by photoproducts in the transcribed strand. They suggest that strand specificity in V-H1 cells is due to differences in fidelity of DNA replication of leading and lagging strands. (They also propose that in the normal V79 cells, two processes, preferential repair of transcribed strands and higher fidelity of DNA replication in nontranscribed strands, were the sources of strand-specific UV-induced mutations.)

Leadon et al. (*251*) have approached many of these questions by an alternative approach. They have developed an immunological method to isolate DNA fragments containing bromouracil in repair patches by using monoclonal antibodies that recognize bromouracil. Excision repair of damage in DNA fragments containing the integrated and transcribed *E. coli* gpt gene was compared to that in the genome overall. A more rapid repair of both UV and AFB1 damage was observed in the DNA fragments containing the *E. coli* gpt genes. Leadon (*252*) studied the repair of UV and aflatoxin B1 induced damage in the human metallothionein gene family. DNA damage was initially repaired faster in transcribed genes than in the genome overall. After 6 h, there was twice as much repair in these genes as in the rest of the genome. A nontranscribed pseudogene was not repaired more efficiently than the genome. Induction of transcription of three of the expressed genes stimulated their repair 2-fold over the basal level. They concluded that efficiency of repair of damage in a DNA sequence is dependent on the level of transcriptional activity associated with that sequence.

Vos and Hanawalt (*252a*) have utilized the chemical properties of psoralens to develop another method for investigating gene-specific repair. Psoralen forms interstrand cross-links with DNA, which results in retarded electrophoretic migration following denaturation–renaturation. Using this method, these authors have found that, in the DHFR gene, the interstrand cross-links are repaired at a faster rate than the psoralen monoadducts.

Recent studies indicate, however, that preferential repair may not be the rule for all DNA adducts. Specifically, Scicchitano and Hanawalt have reported the absence of strand specificity in the repair of *N*-methylpurines in specific DNA sequences in the DHFR gene of CHO cells (*252b*).

**3. In Vitro Systems for Nucleotide Excision Repair.** Permeabilized cells and cell-free extracts have recently been used to study nucleotide excision in human cells. Cells permeabilized by osmotic shock have been used to demonstrate that UV-induced repair synthesis is ATP dependent, that ATP is required for the incision reaction, and that DNA polymerase δ is the enzyme responsible for repair synthesis (*253, 254*). The permeabilized cells are currently being used to isolate and purify the XP factors by using the proper permeabilized XP cells lines as "receptors" in the complementation assay.

Recently, cell-free extracts prepared by the method of Manley and originally developed for transcription by RNA polymerase II have been used to study nucleotide excision repair. Wood et al. (*255*) found that Manley extract is capable of carrying out repair synthesis on UV-irradiated DNA that presumably contains only pyrimidine dimers and 6–4 photoproducts. The repair synthesis activity requires ATP and is absent in all XP cell lines tested, i.e., XP-A, XP-C, XP-D, and XP-variant. Sibghat-Ullah et al. (*256*) confirmed these results and furthermore showed directly that about 50% of the repair synthesis was due to pyrimidine dimers; photoreactivation of the DNA with either *E. coli* or yeast DNA photolyase reduces the repair synthesis by half. Interestingly, when these enzymes are added in molar excess to irradiated DNA and incubated in the absence of photoreactivating light prior to repair synthesis, the same level of inhibition is observed. This further supports the conclusion that 50% of repair synthesis on UV-irradiated DNA is due to pyrimidine dimers. Photolyases bind to pyrimidine dimers and in the absence of photoreactivating light remain attached for approximately 1 min. Because *E. coli* photolyase and (A)BC excinuclease bind to opposite faces of DNA, they do not interfere with one another's binding, and in fact binding of *E. coli* photolyase stimulates incision by ABC excinuclease, facilitating either its binding or turnover (22). Similarly, yeast photolyase apparently stimulates yeast excision repair in vivo. However, yeast photolyase inhibits *E. coli* ABC excinuclease presumably because of the subtle differences in its binding mode compared to that of *E. coli* photolyase. Thus, it appears that photolyases stimulate excision repair in homologous systems and are inhibitory in heterologous systems (24)—consistent with the inhibition observed in human cells by the *E. coli* and yeast enzymes. The inhibitory effect of *E. coli* and yeast photolyases is specific for pyrimidine dimers, and the enzymes have no effect on repair synthesis elicited by other agents.

Interestingly, in vitro photoreactivation by photolyases of oligonucleotides containing the excised thymine dimers causes release of thymidine or thymidine monophosphate, indicating that human cells have a nuclease that incises the intradimer phosphodiester bond (257). DNA damaged by cisplatinum or psoralen is an equally good substrate for the repair synthesis. Extracts from XP cell lines are defective in carrying out repair synthesis on DNA damaged by any of these agents, and therefore, it is concluded that a human excision nuclease initiates the repair synthesis observed with UV, psoralen, and cisplatinum.

It is expected that the three repair synthesis assay systems, microinjection (226, 258, 230), permeabilized cells, and cell-free extracts, will eventually make it possible to purify the repair proteins absent from each XP complementation group. This will not only be the first step toward reconstituting human nucleotide excision in vitro but will also provide proteins for making antibodies and designing oligonucleotide probes to be used in cloning the repair genes.

In an exciting recent report, Chu and Chang (259) claim to have identified the damage recognition subunit by gel retardation. It appears that, of the xeroderma pigmentosum (XP) cell lines tested, only XP-E is deficient in this protein.

**4. Complementation of Human Repair Deficiencies with Bacterial Repair Genes.** As mentioned previously, T4 endonuclease V has been transfected into XP cell lines, repair-deficient CHO cell lines (244), rad mutants of *S. cerevisiae* (260, 104), and mei-9 and mus201 mutants of *Drosophila melanogaster* (260a). The transfected cells express the gene and gain partial resistance to UV. Similarly the *E. coli* ada gene has been transfected into various *Mer⁻* (*Mex⁻*) cell lines, and as a result, the transfected cells attain the Mer⁺ phenotype (261–264). The long-term biochemical and practical implications of these interesting experiments remain to be seen. Transfection of XP cells with the *uvrA* gene resulted in expression of the gene but no increase in the repair activity (265). Similarly, microinjection of UvrA, UvrB, UvrC, and UvrD, singularly or in combination, into XP cell lines does not restore repair

synthesis as measured by unscheduled DNA synthesis (230).

**5. Human Syndromes Caused by Defective Repair.** There are a number of human diseases associated with increased sensitivity to radiation. The biochemically best characterized of these is xeroderma pigmentosum. It is well established now that this syndrome is due to defective incision of UV-irradiated DNA and thus to a lack of incision (excision) nuclease. The molecular basis of another repair-related human syndrome has been recently elucidated. Bloom's syndrome, which manifests itself by telengiectic erythema and mental and developmental retardation, is known to be associated with high frequency of chromosome rearrangement, high frequency of sister chromatid exchange, and high incidence of cancers of all types. It has been recently demonstrated that Bloom's syndrome is associated with a missense mutation in DNA ligase I (266–268). The enzyme appears to be made in normal quantities, but its reactivity with ATP is decreased and as a result only low levels of the ligase–AMP intermediate is formed in in vitro assays (269). These observations are consistent with those of Runger and Kraemer (269a), who have reported a deficiency in in vivo DNA ligase activity in Bloom's syndrome cells in addition to error-prone DNA end joining and high levels of spontaneous mutation. However, other enzymatic defects have also been reported to be associated with Bloom's syndrome, notably abnormal hypoxanthine DNA glycosylase activity (270), and therefore, a cause and effect relationship between defective DNA ligase I and Bloom's syndrome remains to be established. In fact, Mezzina et al. have indicated that Bloom's syndrome cells have a level of DNA ligase activity equal to or higher than that in control cells (270a).

## IV. Recombination–Postreplication Repair

The molecular mechanisms of the recombination and postreplication repair pathways are relatively well-defined in *E. coli*. In other organisms molecular processes that ensure cell survival without removal of the DNA damage have been at times included in this category. In view of the lack of convincing models for the biochemical survival mechanisms in eukaryotes we will discuss the molecular mechanism of this pathway of repair/survival in *E. coli*. In this organism two separate repair events employ some of the same proteins and share many of the same intermediates: recombinational repair and cross-link repair. In both cases, genetic information in a homologous duplex is required to restore the integrity of a duplex that has a lesion involving both strands.

**A. Recombination Repair** (1; *See Figure 6*). When the *E. coli* replication fork reaches a DNA adduct that blocks replication, synthesis of DNA on the undamaged strand continues. However, synthesis on the damaged template strand resumes only after leaving a gap of several hundred nucleotides. Thus, a gap is created in the strand complementary to the damaged one. This gap is repaired by the strand-transfer activity of RecA protein in conjunction with other proteins involved in genetic recombination. The sister duplex is the source of the strand used to fill in the "postreplication gap." The gap thus generated in the sister duplex is restored by DNA polymerase I using the intact strand as a template. It is not known whether the nature of the gap formed and the mechanism of postreplication gap filling are the same when the lesion is in the leading or lagging strand template. Recombinational repair has not been reconstituted in vitro. For that to occur, *E. coli* DNA replication and recombination must

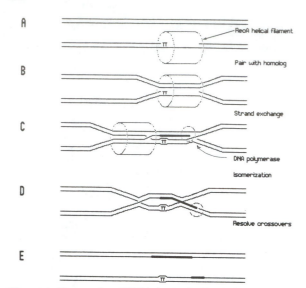

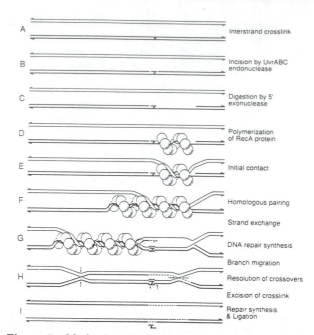

**Figure 6.** Restoration of DNA duplex by recombinational repair in *E. coli*. Bulky DNA adducts, in this case a pyrimidine dimer, constitute replication blocks and lead to the formation of gaps in replicating DNA. The RecA helical filaments form on these gaps and initiate a strand-transfer reaction with the sister duplex. The reciprocal strand exchange is presumably associated with partial DNA synthesis by a copy choice mechanism. The end product of the reaction is two duplexes connected to one another by two Holliday crossovers which are resolved by a resolvase producing two continuous DNA duplexes. Reproduced, with permission, from ref 1. Copyright 1988 Annual Reviews, Inc.

be taking place, neither of which has been achieved in its entirety in vitro. It should be noted that postreplication gap filling by recombination is not, strictly speaking, a repair process. Two intact duplexes are generated but the lesion still remains. However, in a duplex, the DNA adduct is once again substrate for ABC excinuclease or an appropriate glycosylase. Even in the absence of base or nucleotide excision repair, recombinational repair contributes to cell survival; it maintains the integrity of the duplex and if the lesion is not in an essential gene, it is eventually diluted out by successive rounds of replication and cell division.

**B. Repair of Cross-Links.** Psoralen, mitomycin C, cisplatinum, nitrous acid, bifunctional alkylating agents, and a number of other chemicals cross-link the two DNA strands. The repair of cross-links involves the same mechanical steps as recombinational repair, and it, too, is absolutely dependent on the presence of two homologous duplexes in the cell. A likely mechanisms is as follows (see Figure 7): ABC excinuclease incises a single strand on both sides of the damaged nucleotides; in the case of psoralen this is the furan-side adducted strand (271). The incised strand is invaded by the intact duplex in a reaction mediated by the strand-transfer activity of RecA protein. With the aid of other recombinational proteins, such as RecBCD nuclease (exonuclease V) and DNA polymerase I, single-stranded fragments of undetermined length are covalently linked into the "gap" and a three-strand intermediate is generated. This new structure is again recognized by ABC excinuclease, which excises the 11-mer crosslinked to the second strand as a monoadduct; ABC excinuclease incises on both sides of the cross-linked base, and the cross-link is released. The gaps in the sister duplex and in the second strand of the cross-linked duplex are filled in by repair synthesis (272). The first (the initial incision of cross-link) and the third (incision of triple-stranded intermediate to release the cross-link as an 11-mer-X-12-mer) steps have been accomplished in vitro

**Figure 7.** Mechanism of the repair of psoralen interstrand cross-links in *E. coli*. ABC excinuclease incises the ninth phosphodiester bond 5′ and the third phosphodiester bond 3′ to the thymine adducted to psoralen via the furan ring. Polymerase I binds to the nicks on the 3′ side and generates a gap by its 5′–3′ exonuclease action. The RecA protein binds to the single-strand region and polymerizes in a 5′–3′ direction and then interacts with the major groove of the homologous duplex and initiates the strand exchange reaction, which is accomplished by repair synthesis and branch migration. A double Holliday crossover is generated and resolved to produce an intact duplex and a triple-stranded intermediate. The short third strand in the triple-strand intermediate is recognized as a monoadduct by ABC excinuclease and removed as such. The resulting gap is filled by polymerase I and sealed by DNA ligase. Reprinted with permission from ref 272a. Copyright 1989 Journal of Biological Chemistry.

(271–272), and work is in progress to generate the triple-strand intermediate from an intact duplex and a homologous cross-linked duplex with the aid of RecA protein and DNA polymerase I. It has been suggested that the 5′–3′ exonuclease activity of DNA polymerase I creates a single-strand region following the initial biincision at the cross-link and that RecA initiates strand exchange from this gap (272a). Symmetrical cross-links are repaired similarly except that the strand that is incised initially is thought to be selected at random.

## V. Mismatch Repair

Mispairs in DNA arise as a result of at least three cellular events: (1) uncorrected errors in DNA synthesis, (2) recombination between two nonidentical alleles, and (3) spontaneous deamination of 5-methylcytosine in a guanine–Me-cytosine base pair. These premutagenic lesions are subject to correction by a variety of distinct molecular mechanisms. Errors in DNA synthesis are corrected in *E. coli* by the best characterized system, i.e., the methyl-directed mismatch repair pathway, whereas mismatches resulting from recombination between two nonidentical alleles may in part be reversed in a methyl-independent fashion. The unique problem of spontaneous deamination of 5-methylcytosine is addressed by the very short patch mismatch repair mechanism. Each of these pathways will be considered in turn.

**A. Methyl-Directed Mismatch Repair (See Figure 8).** Mismatches in newly synthesized DNA that escape

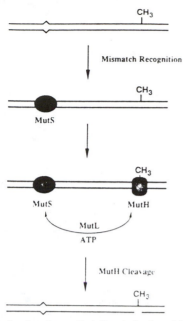

**Figure 8.** Mechanism for methyl-directed mismatch correction in *E. coli*. MutS protein binds to the region of mismatches, and MutH binds to hemimethylated d(GATC) sequences. MutL is involved in signal transduction between MutH and MutS. MutH nicks the DNA 5′ to the dG of the unmethylated d(GATC) sequence, and in the presence of ATP, SSB, helicase II, and polymerase III, the mispaired base in the unmethylated strand and the region between that base and the MutH generated nick are removed and replaced with the correct sequence. Reprinted with permission from ref 272b. Copyright 1989 Journal of Biological Chemistry.

the editing function of DNA polymerase are most frequently restored by the methyl-directed mismatch repair mechanism (reviewed by Modrich; 5, 272b); this mechanism is thought to increase replication fidelity 100–1000-fold (273). A newly synthesized daughter strand is identified by its transient undermethylation at the adenine residues of dam sites [d(GATC)] thus, $N^6$-adenine methylation in the parental strand serves as the basis for strand discrimination (274, 275) and ensures that the coding strand remains unaltered. Strains of *E. coli* deficient in dam methylase exhibit a mutator phenotype (276–278); likewise, those that overproduce the methylase are hypermutable presumably as a result of the inhibitory effect of an increased rate of substrate methylation coupled with the reduced efficiency of correction on symmetrically modified regions.

In all, the methyl-directed mismatch repair system requires the participation of at least seven proteins: MutH, MutL, MutS, MutU (UvrD or helicase II), SSB, DNA ligase, and the DNA polymerase holoenzyme. Also, the dam methylase is required as previously described. Each of these proteins has been isolated in pure form for in vitro studies.

MutH ($M_r$ = 15000; 279, 280) possesses a $Mg^{2+}$-dependent endonuclease activity which responds to the state of methylation of d(GATC) sequences (281). The introduction of single-strand breaks does not depend on the presence of a base-pair mismatch within the DNA substrate but responds differentially to the state of methylation of d(GATC) sequences present. The endonuclease cleaves 5′ to the dG of d(GATC) sequences to generate 5′-phosphoryl and 3′-hydroxyl termini. Symmetrically methylated d(GATC) sites are resistant to the endonuclease, hemimethylated sequences are cleaved on the unmethylated strand, and unmethylated d(GATC) sites

are usually subject to scission on only one DNA strand. Interestingly, the MutH-associated d(GATC) endonuclease is activated in the presence of MutL, MutS, and ATP in a manner that is dependent on the presence of a base-pair mismatch within the DNA substrate. Apparently, MutH acts at the strand discrimination stage of mismatch correction with incision at a d(GATC), providing a simple mechanism by which methylation may dictate strandedness of the excision event associated with mismatch repair.

MutS ($M_r$ = 97 000) binds specifically to DNA regions containing a single base-pair mismatch (282). DNase I footprinting studies revealed that the protein protects about 20 base pairs from hydrolysis. Also, the protein binds in an asymmetric manner relative to each mispair although the orientation of the asymmetry varies with each mispair.

An in vitro activity has only recently been described for the MutL gene product (283). DNase I footprinting studies have revealed that MutL ($M_r$ = 70 000, dimer = 140 000 Da) binds in an ATP-dependent manner to a MutS–mismatch complex. In contrast to the approximately 20 base pairs protected by MutS alone, the MutL–MutS–mismatch ternary complex covers virtually the entire 143-bp substrate used for the footprinting experiments. MutL does not interact specifically with the mispair (G-T) either in the presence or in the absence of ATP, and MutL does not affect the MutS–mismatch complex without ATP (or ATP-γ-S). Grilley et al. propose that MutL may stabilize the MutS–mismatch complex or mediate the interactions between the MutS–mismatch complex and other proteins or DNA sites [i.e., d(GATC)]. That is, MutL may have a central role in signal transfer.

Numerous aspects of the methyl-directed mismatch repair pathway have been elucidated by studying the fate of artificial substrates treated with *E. coli* cell-free extracts. Such substrates typically possess a defined number of d(GATC) sequences separated from a specified base–base mismatch. These closed circular heteroduplexes may be methylated on both strands, either one or the other strand, or neither strand. Hemimethylated heteroduplexes are repaired principally on the nonmethylated strand. Heteroduplexes containing one or more d(GATC) sites are efficiently corrected in a Mut-dependent manner whereas DNA devoid of such sequences is unrepaired (284, 285). Su et al. (286) have addressed the substrate specificity of methyl-directed mismatch repair in *E. coli* extracts. They have determined that transition mutations (i.e., G-T and A-T) are better substrates than transversion mutations (i.e., C-T, A-A, T-T, and G-G). Interestingly, C-C and A-G are repaired in a different manner; both C-C and A-G mismatches are poor substrates for mismatch correction, and each shows little bias for the unmethylated strand. A-G to A·T occurs by a MutHLS-dependent process whereas A-G to C·G occurs in a methylation- and MutHLS-independent manner (287). DNase I footprinting studies revealed that MutS binds to each of the eight mismatches with an affinity that roughly correlates with repair efficiency (286): highest affinity was for G-T and A-C transitions, while A-A, T-T, and G-G bound with intermediate affinity and A-G, C-T, and C-C with low affinity. Hence, the repair efficiency correlates approximately with the MutS substrate affinity. It has been suggested that, with the possible exception of T or C insertions, all mispairs can adopt an intrahelical form.

Several questions remain to be addressed: (1) What are the sites of incision and directionalities of excision and resynthesis? (2) What is the nature of signal transduction between a d(GATC) site and the mismatch; i.e., does communication arise from the binding of a protein (proteins) to a mismatch, followed by the transport of DNA past the protein complex in search of a d(GATC) site, or, alternatively, does mismatch recognition trigger the polymerization of some protein along the DNA helix until a d(GATC) site is encountered? (5).

Mismatch repair has been detected in several organisms; in fact, it was first described in *S. pneumonia*. The Hex system of *S. pneumonia* is functionally similar to the Mut system of *E. coli*. HexA is homologous to the MutS proteins of both *E. coli* and *Salmonella typhimurium* (288, 289). The Hex system is unique, however, in its method of strand targeting; i.e., *S. pneumonia* uses single-strand breaks as opposed to the methylation-directed nature of the *E. coli* Mut system (290–293). Lahue et al. (284) have observed, however, that closed circular heteroduplexes lacking a d(GATC) sequence are subject to strand-specific repair if they contain a strand-specific single-strand break. Interestingly, this correction reaction does not require MutH. Both Hex and Mut remove a large portion (i.e., >1 kb) of the targeted strand (294). Pang (295) has demonstrated that *mutS* from *S. typhimurium* fully complements *mutS* of *E. coli*; however, *hexA* does not complement *mutS* of *S. typhimurium* (289). *MutS* is required in two other less efficient mismatch repair processes. One of these acts on symmetrically methylated DNA and may serve to repair mismatches during recombination (296, 297). Another system corrects C to T transitions at the internal C of the Dcm methylase sequence d(CCA/TGG) and also requires the MutL protein and Dcm methylase (298).

**B. Methyl-Independent Mismatch Repair.** In contrast to the methyl-directed mismatch repair pathway, there exists in *E. coli* a mismatch repair pathway that (1) is specific for A-G mispairs, (2) converts A-G exclusively to C·G, (3) operates independently of d(GATC) methylation, and (4) is independent of MutH, -L, and -S activity (299, 286). Two complementary mutator loci in *E. coli* have been described, *mutT* and *mutY*, which result in an elevated frequency of T·A to G·C and G·C to T·A transversions, respectively (300, 301). It is believed that these loci may be involved in the methyl-independent repair of A-G mispairs.

Unlike *mutY*, the *mutT* phenotype is expressed only during DNA replication (302, 303) and presumably acts to exclude A-G mispairs where adenine resides in the template strand. The recent discovery of Bhatnager and Bessman (304) that the *mutT* gene product has a strong dGTPase activity suggests that this protein may operate by precluding the incorporation of guanine residues opposite the template adenine, presumably by hydrolyzing the syn form of dGTP (304a). Interestingly, MutT has no DNA binding or endonuclease activity (304).

Au et al. (305) have demonstrated that *mutY⁻ E. coli* are defective in methyl-independent repair of A-G mispairs to C·Gs but are otherwise fully active with respect to methyl-directed mismatch repair. In contrast to MutT, the MutY activity acts independently of DNA replication; it is thought that MutY operates on DNA after MutT and that it works exclusively on A-G mismatches where the guanine is in the template strand. Because MutT has eliminated A-G mispairs with adenine in the template strand, there is no requirement for strand discrimination in the case of MutY (305). Lu and Chang (306) have recently reported the identification of a novel excision repair protein specific for A-G mismatches. This endonuclease, termed AGP for A-G mismatch binding protein, binds preferentially to A-G mismatches and makes dual

**Figure 9.** Repair of an A-G mismatch in *E. coli*. An A-G-specific excision nuclease incises the second phosphodiester bond 5′ and the first phosphodiester bond 3′ to dA of the A-G mismatch. The resulting two-nucleotide gap is filled in with the correct bases. The incision mode shown in the figure was observed in crude extracts, and therefore, it is not clear at present whether the two incisions are made by a single protein, nor is it clear that phosphodiester bond incision is not preceded by the removal of the mismatched base by a glycosylase.

incisions in DNA one phosphodiester bond 3′ and the other two phosphodiester bonds 5′ to the dA of the mismatch, thereby leaving a two nucleotide patch (see Figure 9). Further, incisions occur exclusively on the A-containing strand regardless of the orientation of the mismatch. It will be interesting to learn if this activity exists in *mutY⁻ E. coli*.

**C. Very Short Patch Mismatch Repair.** Methylation of cytosine residues is the most common form of DNA modification (307). In prokaryotes, methylation at the second cytosine in the dcm sequence CC(A/T)GG (308) serves to protect DNA from degradation by restriction endonucleases, whereas in eukaryotes, cytosine methylation has been implicated in suppression of gene transcription (309); decreases in 5-methylcytosine levels are thought to be involved in tumor formation as a result of gene modulation (310). It has been suggested that loss of certain 5-methylcytosines may "affect cell differentiation and behavior" (311).

Deamination of 5-methylcytosine results in the formation of a T-G base pair. If left unrepaired, this mismatch will be mutagenic in 50% of the daughter cells. Likewise, random reversal will also lead to a 50% mutation frequency. This is different from deamination of cytosine, which forms a U-G mispair repairable by a specific uracil glycosylase (312) followed by excision of the resulting AP site (313).

The very short patch (VSP) mismatch repair pathway is specialized to process T-G mismatches in the context of the recognition sequence of the *E. coli* dcm-methyltransferase. The term was coined by Lieb because of its ability to cause excess recombination between very closely spaced markers (314, 315). It restores sites that have undergone spontaneous hydrolytic deamination at the inner (5-methylated) cytosine residue. This is the only known spontaneous intracellular reaction that converts a naturally occurring DNA component (m⁵C) into another (T) and consequently leads to a base–base mismatch in nonreplicating, nonrecombining DNA molecules. Because hydrolytic deamination of m⁵C occurs mainly in quiescent, hence fully dam-methylated, DNA, no methyl-directed mismatch is expected to occur. The only hemimethylation occurs at the location of the mismatch.

VSP repair has not been characterized at the molecular level; however, it is known that the process requires the *E. coli* dcm methylase and some components of the methyl-directed repair pathway. This pathway is specific for T-G mismatches and restores C methylation specifically. Jones et al. (316) have demonstrated that sequence-specific repair occurs only at CC(A/T)GG sequences, only at sites of G-m⁵C, and only at G-T mismatches, replaces only T, and requires MutL and -S of the methyl-directed mismatch repair pathway; MutH and MutU are not required. These points have been confirmed by Zell and Fritz (317) and Lieb (318). Recently, it has been shown that recognition and methylation of CC(A/T)GG sites are not sufficient for the participation of a protein in VSP repair, i.e., EcoRII methylase restores the cytosine methylase activity to *dcm⁻* cells but does not restore mismatch repair (*319*).

VSP repair has been demonstrated in mammalian cells. Hare and Taylor (*320*) have shown that T-G mismatches are restored to C·G pairs in SV40-transfected African queen monkey kidney cells. Brown and Jiricny (*311*) have extended these findings by showing that T-G mispairs are converted to C·Gs 90% of the time in African green monkey cells (CV1). Recently, Jiricny et al. (*321*) have identified a 200-kDa protein from HeLa cells that binds specifically to oligonucleotide heteroduplexes possessing G-T mismatches. Methylation of the adjacent guanine residue inhibits binding of this factor. They propose that this protein may carry out a role analogous to that of the *E. coli* MutS protein in the VSP repair pathway. Very interestingly, preliminary evidence of Wiebauer and Jiricny (*321a*) indicate that G-T to G·C mispair correction in human cells is initiated by a glycosylase that removes the mismatched thymine residue.

## VI.  Regulation of DNA Repair

Cellular mechanisms that protect DNA from damage and that repair the damage once it has occurred, like most other metabolic pathways, respond to external stimuli. Agents that damage DNA or are potentially mutagenic induce cellular responses. The molecular mechanisms of these cellular responses are best characterized in *E. coli*. However, evidence is accumulating that, in eukaryotes as well, DNA damage induces protective and/or life-saving mechanisms.

*A. Inducible Responses in E. coli*. Three damage-inducible responses have been characterized in this organism (*322*).

**1. The SOS Response.** DNA damage by UV and chemical agents that make bulky adducts such as mitomycin C, psoralen, and 4-nitroquinoline oxide, in addition to conditions such as thymine starvation that stop DNA replication, induces the SOS response. The response results in increased resistance to bulky adducts as well as to certain oxidative base damages such as thymine glycols (*322a*), increased mutagenesis, and induction of lysogenic phages such as λ and phage 80. The molecular mechanism is as follows: LexA protein is a repressor of about 20 genes (including *lexA*, *recA*, *uvrA*, *uvrB*, and *uvrD*) that are involved in cell division, recombination, DNA repair, and mutagenesis (*323*). Upon DNA damage, the RecA protein binds to the resulting single-stranded DNA produced as a direct result of the damage (or, alternatively, during processing the damage or as a result of a halt in replication) and becomes "activated". The activated RecA binds to the LexA repressor, promoting LexA's proteolysis. The LexA protein is cleaved autocatalytically (*324*) as follows: Lys-156 activates Ser-159 by a charge relay system, and the activated Ser-159 becomes a strong nucleophile, attacks the Ala-84–Gly-85 bond, and cleaves the protein in half (*325*). Even though the $NH_2$-terminal half of the cleaved protein has some affinity for the operators (CTG–$N_{10}$–CAG), this affinity is reduced by 2–3 orders of magnitude, thereby effectively inactivating LexA (*326*). Upon inactivation of LexA, the so-called SOS genes are expressed with the consequent increase in excision repair (*uvrA* and *uvrB* but not *uvrC* are inducible) and increased recombinational repair (*recA*, *recN*, *recQ*, *recD*) (see Figure 10; *326a*). A report claiming induction of the photolyase gene *phr* (*327*) by the SOS response has not been confirmed (*328*). Likewise, a claim of the regulation of the *ssb* gene by SOS has not been confirmed (*329*). Genes involved in

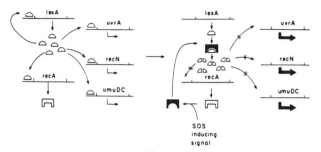

**Figure 10.** Mechanism of the SOS response. Lex A is the repressor of approximately 20 *E. coli* genes involved in DNA repair, replication, recombination, mutagenesis, and cell division. Upon DNA damage or stalling of the replication fork, an inducing signal (presumably single-stranded DNA) is produced, which activates RecA. Activated RecA binds to LexA, stimulates the rate of cleavage of LexA by autocatalysis, and derepresses the constituent genes of the regulon. In addition to LexA, RecA also stimulates the autocatalytic cleavage of UmuD, promoting mutagenesis. Repressors of the lambdoid phages are also cleaved autocatalytically, resulting in the initiation of lytic growth. Reproduced, with permission, from ref 326a. Copyright 1985 Annual Reviews, Inc.

cellular division are also turned on and cell division is inhibited, allowing the cell more time to recover before dividing. Upon recovery and disappearance of the inducing signal, LexA (which itself is induced as part of the response) accumulates and binds to the constituent operators, and the repressed state is reestablished.

The repressors of phages λ and 80 have limited sequence homology to LexA but do have significant homology around the residues that correspond to the "active site" of LexA protease, Lys-156 and Ser-159, and are cleaved at an Ala–Gly bond as is the case for LexA.

Discounting the phage repressors, until recently it was assumed that LexA protein was the only *E. coli* protein that undergoes RecA-promoted self-cleavage. However, it has now been shown that SOS induction results in the cleavage of UmuD protein as well. UmuD (15-kDa) and UmuC (48-kDa) proteins are encoded within a single operon and are absolutely required for UV mutagenesis. The UmuD protein is "cleaved" by RecA in vitro and in vivo, and this cleavage at the Cys-24–Gly-25 bond seems essential for its function in mutagenesis (*330*, *331*). Changing the glycine residue at the cleaved peptide bond or the Ser-60 and Lys-97 residues, which presumably constitute the serine protease active site, results in either nonmutability or great reduction in mutagenesis by UV light (*332*). It has been suggested that the cleaved UmuD may stimulate ssDNA-dependent ATPase activity of UmuC and enable it to serve as a polymerase (replicase) accessory factor, increasing the processivity of polymerase and consequently the frequency of translesion synthesis and, hence, mutation (*332*). A correlation of these results to a recent finding that *E. coli* DNA polymerase II (which has been an oddity until recently) is induced 7-fold during the SOS response and incorporates nucleotides efficiently across AP sites on the template (*333*) is unknown. An interesting new development in unraveling the intricate network of reactions called "the SOS response" is the discovery of the *psiB* (plasmid SOS inhibitor) gene. Conjugative plasmids R100 and R6-5 have a gene *psiB* near the transfer origin (*oriT*) which encodes a 12-kDa that inhibits the SOS response. A similar gene has been found on the F sex factor which inhibits SOS response only when amplified. The PsiB protein of R6-5 inhibits the SOS induction function of RecA but not its recombinogenic function (*334*). The exact action mechanism of PsiB protein is not known although it is likely to be by binding

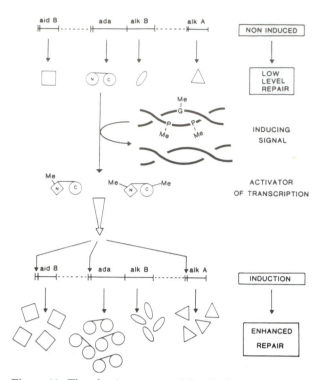

**Figure 11.** The adaptive response of *E. coli*. Cysteine 69 of Ada accepts the methyl (alkyl) group from the $S_p$ stereoisomer of methyl phosphotriester. Thus modified, Ada acts as a positive regulator, binding upstream to promoters of the constitutive genes of the regulon. Gene products of this regulon detoxify alkylating agents and repair the DNA damage caused by these agents. Reproduced, with permission, from ref 2. Copyright 1988 Annual Reviews, Inc.

to RecA and interfering with its "coprotease" function in cleavage of LexA repressor and UmuD protein.

**2. Adaptive Response to Alkylation Damage (2).** *E. coli* exposed to low levels of a methylating agent such as *N*-methyl-*N'*-nitro-*N*-nitrosoguanidine become more resistant to lethal and mutagenic effects of higher doses of the same or similar alkylating agents (335). The molecular mechanism of this response is well understood (see Figure 11). Upon alkylation of DNA the Ada protein transfers methyl groups from one of the two stereoisomers ($S_p$ form) of the methyl phosphotriester to the Cys-69 residue and becomes a positive regulator. It binds to the "Ada box" $A_3N_2A_3GCGCA$ upstream of promoters of *ada*, *alkA*, and *aidB* genes and induces their transcrpition. The increased levels of Ada repairs $O^6$-mGua and thus prevents mutagenesis while the increase in *alkA* gene product (3-mAde DNA glycosylase II) results in increased capacity to repair 3-mAde and 3-mGua which block DNA replication and are lethal. Recent studies have shown that a number of methylating agents can directly alkylate Cys-69 of Ada and activate it (56). Whether activation by this route or by methyl transfer from methyl phosphotriester is the physiologically important route is not known. Interestingly, the activated Ada protein does not interact with all Ada boxes in the same fashion, and it has been reported that even unmethylated Ada can activate the *alkA* promoter but not that of *ada* (53). The NH$_2$-terminal half of Ada seems to be sufficient for binding; however the carboxy-terminal half must play some role as the effect of fusion proteins which contain the N-terminus of Ada depends on the sequences in the carboxy-terminal half (57). There is some fine tuning of the Ada regulon. For example, the *aidB* gene is induced by anaerobiosis in addition to its induction by alkylation damage (55). The function of AidB

is not known; however, it is probably involved in inactivation of methylation agents rather than in repair.

**3. Adaptive Response to Oxidative Damage.** Oxidative metabolism and chemical oxidants such as paraquat and ionizing radiation generate activated oxygen species in the form of $H_2O_2$, $O_2^{\bullet-}$, and hydroxyl radicals (336). These species damage DNA extensively (337) but also induce a number of proteins which eliminate the active oxygen species and repair the damaged DNA. Thus, *E. coli*, exposed to low level of $H_2O_2$, becomes resistant to toxic levels of the oxidant (338, 339). The molecular mechanisms of the adaptive response to oxidative stress are not fully understood. Oxidative stress results in the synthesis of AppppA and related dinucleotides which have been called alarmones (340). How alarmones function to turn on the adaptive response in not known. In addition, there appears to be considerable overlap between the adaptive response to oxidative damage and the heat shock response (341, 342) as well as between the adaptive response to near-UV (343, 344). Genetic studies in *E. coli* indicate that there are three separate adaptive mechanisms to oxidative stress:

*a. The OxyR Regulon.* The *oxyR* gene was identified by a mutation that increases the cell's resistance to $H_2O_2$ (345). It appears that *oxyR* is a positive regulator of eight genes, four of which code for catalase (*catG*), manganese superoxide dismutase (Mn-SoD), glutathione reductase, and NADPH-dependent alkyl hydroperoxide reductase (346). The other proteins have not been identified as yet. The *oxyR* gene has been cloned and sequenced; it encodes a protein of 34 400 Da which shows homology to a number of positive regulatory proteins including LysR in *E. coli* and NadD in *Rhizobium*. It is a positive regulator of genes in the *oxyR* regulon but is negatively autoregulated. OxyR2 is a missence mutant (ala → val), and this mutation converts the protein to a permanently active state (346a). Deletion of the *oxyR* gene eliminates the induction of these eight genes by oxidative stress.

*b. The H$_2$O$_2$ Regulon.* Exposure of cells to $H_2O_2$ induces 21 proteins in addition to the 8 which are also induced by $H_2O_2$ but whose induction is dependent on the *oxyR* gene (345). The molecular mechanism of induction by $H_2O_2$ of the *oxyR*-independent genes is not known. However, overproduction of peroxide-scavenging enzymes suppresses the sensitivity of *oxyR*⁻ *E. coli* to redoxy-cycling agents (347). In addition, it has recently been reported that the *katF* gene product is a 44-kDa positive regulator which induces the synthesis of exonuclease III and hydroperoxidase II, whose products are involved in eliminating reactive oxygen species or repairing the damage caused by these species. Interestingly, mutations in *katF* result in nearly complete loss of exonuclease III activity while overproduction of *katF* leads to an approximately 3-fold increase of exonuclease III activity in the cell (347a).

*c. The SoX Regulon.* Superoxide radicals induce 26 proteins independent of those induced by the *oxyR* gene product (348, 349). These genes are induced by superoxide-generating conditions but not by $H_2O_2$. One of these proteins is endonuclease IV; the identity of the others is not known. Endonuclease IV is the only repair enzyme that is induced by this oxidative stress. Other enzymes known to play important roles in survival to oxidative damage, RecA, RecBCD (ExoV), and DNA polymerases I and III (339), are not significantly induced by oxidative damage. An interesting feature of the SoX response is the finding that insertional inactivation of three member genes investigated results in constitutive expression of some other members of the region (349).

**B. Inducible Responses in S. cerevisiae: The DIN and DDR Genes.** There is a wealth of information concerning damage-inducible responses in yeast (*350, 351*); however, the biochemical mechanisms underlying these stress response reactions is not known, nor is it understood whether these responses are beneficial or harmful to the cell.

Ruby and Szostak (*352*) fused random yeast genes to *E. coli lacZ* gene, inserted them back into yeast as plasmids or into the chromosomes, and searched for cells whose β-galactosidase activity increased upon DNA damage by various agents. Six genes that were inducible by DNA damage were identified. Five of these were induced by all damaging treatments tested (UV, 4NQO, MMS, MNNG), and DIN5 was induced by UV and thymine starvation but not by alkylating agents. The way these experiments were conducted did not disrupt the resident DIN genes, and therefore, it is not known at present whether the DIN genes are involved in DNA repair. However, it is known that DIN genes are separate from RAD genes.

McClanahan and McEntee (*353*) isolated DNA damage responsive (DDR) genes by using differential hybridization to cDNA prepared from transcripts of UV- or 4NQO-induced cells. Six DDR genes were identified that were inducible by MNNG and UV. These genes were shown to be different from the DIN genes. Interestingly, two of the DDR genes DDRA2 and DDR48 are also inducible by heat shock. RAD3, RAD6, and RAD52 genes appear to play indirect roles in DDR gene expression (*354*).

UV irradiation, 4NQO, and MMS also induce the gene encoding the small subunit of ribonucleotide reductase (*355, 356*). The CDC8 gene, which encodes thymidylate kinase (*355*), and the CDC9 gene, which encodes DNA ligase (*357*), are also induced by DNA damage. It has also been found that UV irradiation increases the transcription of the Ty element; whether this transcription is associated with increased transposition is not known (*358*).

Of the DNA repair genes, RAD2, which is involved in nucleotide excision (*179, 177*), and PHR1, which encodes DNA photolyase (J. Sebastian and G. Sancar), are induced by UV and other DNA-damaging agents. The molecular mechanisms of induction are not known.

An interesting case of an SOS-like response in plants is the transcriptional activation of pigment biosynthetic genes by UV light. This results in the increased production of pigments that absorb UV and presumably shield the DNA (*359*).

**C. Damage-Inducible Responses in Mammalian Cells.** The presence and physiological significance of damage-induced repair responses in mammalian cells have been the source of much speculation and controversy. Dasgupta and Summers (*360*) reported that reactivation of UV-irradiated HSV1 by human cells was inducible by pretreatment of cells with UV in a manner analogous to Weigle reactivation in *E. coli*. It was also shown that pretreatment of monkey COS cells with DNA-damaging agents caused an increase in mutation frequency of a UV-irradiated shuttle vector (*361*). Recent developments in damage-inducible responses in mammalian cells may be summarized as follows:

**1. UV-Inducible Proteins.** By comparison of [$^{35}$S]-methionine incorporation into nonirradiated and irradiated cells, a number of proteins have been found to be UV inducible (*362*). Furthermore, "conditioned medium" from UV-induced cells can elicit induction in nonirradiated cells, suggesting a signal transmission mechanism from irradiated cells to nondamaged cells. There have been no further reports on this intercellular signaling factor. However, some of the UV-inducible genes have been identified and cloned (*363*). In humans three of these are metallothionein II and related genes (*364*), the plasminogen activator gene, and the collagenase gene. In Chinese hamster ovary cells, metallothionein I and II genes are inducible (*365*). In humans, the UV-inducible metallothionein II and collagenase genes are also inducible with phorbol esters, and conversely, two genes that were known to be inducible by phorbol esters were found to be inducible by UV (*363*). The upstream sequence elements responsible for UV induction have been identified in collagenase, HIV, and c-fos. These 8–20 base pair long sequences have no apparent homology. Removal of these sequences abolishes the UV inducibility. This has led to the conclusion that induction is at the transcriptional level and that induction occurs by positive regulation as is most often the case in eukaryotes. The induction of c-fos, HIV-1 promoter, and collagen after UV irradiation is very rapid, and it has been suggested that it is mediated by a protein kinase. There is also some evidence that the amount of the trans-acting factors (positive regulators) that binds to these UV-responsive elements (URE) increases after UV irradiation of cells. Recently, Herrlich has suggested that *fos* plays a key role in the UV-induced response by cooperating with other transcription factors (*223a*). The induction, however, is not abolished by inhibitors of protein synthesis, suggesting that the increase in the amount of URE binding proteins is due to posttranslational modification.

Glazer et al. (*366*) used a different approach to find UV-induced proteins. Cells were irradiated with UV, and proteins from these cells were separated on SDS gels, transferred to nitrocellulose filters, and then probed for DNA binding by using radioactive DNA labels. They identified seven proteins whose DNA binding is induced by UV. Most of these proteins have higher affinity for dsDNA than ssDNA, but none has higher affinity for damaged DNA. Interestingly, synthesis of these proteins is not increased by UV irradiation as determined by postirradiation labeling with [$^{35}$S]methionine. The DNA binding proteins are also inducible by serum starvation; induction by either treatment does not require de novo protein synthesis. Thus, it appears that preexisting DNA binding proteins are activated for binding by posttranslational modification.

**2. DNA Damage Inducible Transcripts.** Using the low ratio–hybridization subtraction method, Fornace et al. (*367*) isolated the cDNA of UV-induced transcripts from CHO cells. Induction of these genes is inhibited by actinomycin D, indicating a requirement for transcription in the relative increase in the observed RNA levels. The induction of these transcripts is not mediated by a secreted protein factor (probably interleukin 2), nor is it elicited by phorbol esters. The UV-inducible transcripts fall into two groups: (1) class I (10 transcripts) are inducible by UV and AAAF but not by alkylating agents, and (2) class II (14 transcripts) are induced by both UV and MMS but not by X-rays. Only one of the class II transcripts is also induced by heat shock, while the others appear to be specific for UV damage.

**3. Adaptive Response in Mammalian Cells.** Increased $O^6$-mGua methyltransferase activity, after treatment of whole animals or rodent and human cell lines, has been reported, but these reports either have not been confirmed or have been ascribed to trivial causes (such as apparent increase due to liver cell proliferation). There is still much controversy; however, it is doubtful that mammalian cells have an adaptive response similar to that in *E. coli* (*2*).

In summary, there is no doubt that DNA-damaging agents result in increased expression of certain proteins in mammalian cells—as a result of either transcriptional activation or posttranslational modification. However, there is as yet no evidence that these inducible proteins participate in processes (repair, replication, cell division) that enhance cell survival.

**Acknowledgment.** We are grateful to Dr. Bruce Demple for making available preprints of recent work as well as for useful discussions. We also thank Dr. David Orren for his criticism of the manuscript. Financial support for research in our laboratory is provided by NIH Grants GM31082 and GM32833 and partly from The Council for Tobacco Research U.S.A., Inc., Grant CTR1872.

## References

(1) Sancar, A., and Sancar, G. B. (1988) DNA repair enzymes. *Annu. Rev. Biochem.* **57**, 29–67.

(2) Lindahl, T., Sedgwick, B., Sekiguchi, M., and Nakabeppu, Y. (1988) Regulation and expression of the adaptive response to alkylating agents. *Annu. Rev. Biochem.* **57**, 133–157.

(3) Friedberg, E. C. (1988) Deoxyribonucleic acid repair in the yeast *Saccharomyces cerevisiae*. *Microbiol. Rev.* **52**, 70–102.

(4) Bohr, V. A., and Hanawalt, P. C. (1988) DNA repair in genes. *Pharmacol. Ther.* **38**, 305–319.

(5) Modrich, P. (1987) DNA Mismatch Correction. *Annu. Rev. Biochem.* **56**, 435–466.

(6) Husain, I., and Sancar, A. (1987) Photoreactivation in *phr* mutants of *Escherichia coli* K-12. *J. Bacteriol.* **169**, 2367–2372.

(7) Sancar, G. B., and Sancar, A. (1987) Structure and function of DNA photolyases. *Trends Biochem. Sci.* **12**, 259–261.

(8) Sancar, A., and Sancar, G. B. (1984) *Escherichia coli* DNA photolyase is a flavoprotein. *J. Mol. Biol.* **172**, 223–227.

(9) Sancar, G. B., Smith, F. W., and Heelis, P. F. (1987) Purification of the yeast PHR1 photolyase from an *Escherichia coli* overproducing strain and characterization of the intrinsic chromophores of the enzyme. *J. Biol. Chem.* **262**, 15457–15465.

(10) Kiener, A., Husain, I., Sancar, A., and Walsh, C. (1989) Purification and properties of *Methanobacterium thermoautotrophicum* DNA photolyase. *J. Biol. Chem.* (in press).

(11) Eker, A. P. M., Hessels, J. K. C., and van de Velde, J. (1988) Photoreactivating enzyme from the green alga *Scenedesmus acutus*. Evidence for the presence of two different flavin chromophores. *Biochemistry* **27**, 1758–1765.

(12) Sancar, G. B., Smith, F. W., Lorence, M. C., Rupert, C. S., and Sancar, A. (1984) Sequences of the *Escherichia coli* photolyase gene and protein. *J. Biol. Chem.* **259**, 6033–6038.

(13) Sancar, G. B. (1985) Sequence of the *Saccharomyces cerevisiae phrA* gene and homology of the PHR1 photolyase to *E. coli* photolyase. *Nucleic Acids Res.* **13**, 8231–8246.

(14) Yasui, A., Oikawa, A., Kiener, A., Walsh, C., and Eker, A. P. M. (1988) Cloning and characterization of a photolyase gene from the cyanobacterium *Anacystis nidulans*. *Nucleic Acids Res.* **16**, 4447–4463.

(15) Sancar, G. B., Smith, F. W., Reid, R., Payne, G., Levy, M., and Sancar, A. (1987) Action mechanism of *Escherichia coli* DNA photolyase. *J. Biol. Chem.* **262**, 478–485.

(16) Husain, I., and Sancar, A. (1987) Binding of *E. coli* DNA photolyase to a defined substrate containing a single T<>T dimer. *Nucleic Acids Res.* **15**, 1109–1120.

(17) Husain, I., Sancar, G. B., Holbrook, S. R., and Sancar, A. (1987) Mechanism of damage recognition by *Escherichia coli* DNA photolyase. *J. Biol. Chem.* **262**, 13188–13197.

(18) Baer, M., and Sancar, G. B. (1989) Photolyases from *Saccharomyces cerevisiae* and *Escherichia coli* recognize common binding determinants in DNA containing pyrimidine dimers. *Mol. Cell. Biol.* (in press).

(19) Husain, I., Griffith, J., and Sancar, A. (1988) Thymine dimers bend DNA. *Proc. Natl. Acad. Sci. U.S.A.* **85**, 2558–2562.

(20) Broyde, S., Stellman, S., and Hingerty, B. (1980) DNA backbone conformations in cis-syn pyrimidine[]pyrimidine cyclobutane dimers. *Biopolymers* **19**, 1695–1701.

(21) Pearlman, D. A., Holbrook, S. R., Pirkle, D. H., and Kim, S.-H. (1985) Molecular models for DNA damaged by photoreaction. *Science* **227**, 1304–1308.

(22) Sancar, A., Franklin, K. A., and Sancar, G. B. (1984) *Escherichia coli* DNA photolyase stimulates uvrABC excision nuclease *in vitro*. *Proc. Natl. Acad. Sci. U.S.A.* **81**, 7397–7401.

(23) Yamamoto, K., Satake, M., Shinagawa, H., and Fujiwara, Y. (1983) Amelioration of the ultraviolet sensitivity of an *Escherichia coli recA* mutant in the dark by photoreactivating enzyme. *Mol. Gen. Genet.* **190**, 511–515.

(24) Sancar, G. B., and Smith, F. W. (1989) Interactions between yeast photolyase and nucleotide excision repair proteins in *Saccharomyces cerevisiae* and *E. coli*. *Mol. Cell. Biol.* (in press).

(25) Johnson, J. L., Hamm-Alvarez, S., Payne, G., Sancar, G. B., Rajagopalan, K. V., and Sancar, A. (1988) Identification of the second chromophore of *Escherichia coli* and yeast DNA photolyases as 5,10-methenyltetrahydrofolate. *Proc. Natl. Acad. Sci. U.S.A.* **85**, 2046–2050.

(26) Eker, A. P. M., Dekker, R. H., and Berends, W. (1981) Photoreactivating enzyme from *Streptomyces griseus*—IV. On the nature of the chromophoric cofactor in *Streptomyces griseus* photoreactivating enzyme. *Photochem. Photobiol.* **33**, 65–72.

(27) Jorns, M. S., Sancar, G. B., and Sancar, A. (1984) Identification of a neutral flavin radical and characterization of a second chromophore in *Escherichia coli* DNA photolyase. *Biochemistry* **23**, 2673–2679.

(28) Payne, G., Heelis, P. F., Rohrs, B. R., and Sancar, A. (1987) The active form of *Escherichia coli* DNA photolyase contains a fully reduced flavin and not a flavin radical, both *in vivo* and *in vitro*. *Biochemistry* **26**, 7121–7127.

(29) Sancar, G. B., Jorns, M. S., Payne, G., Fluke, D. J., Rupert, C. S., and Sancar, A. (1987b) Action mechanism of *Escherichia coli* DNA photolyase. *J. Biol. Chem.* **262**, 492–498.

(30) Heelis, P. F., and Sancar, A. (1986) Photochemical properties of *Escherichia coli* DNA photolyase: A flash photolysis study. *Biochemistry* **25**, 8163–8166.

(31) Heelis, P. F., Payne, G., and Sancar, A. (1987) Photochemical properties of *Escherichia coli* DNA photolyase: selective photodecomposition of the second chromophore. *Biochemistry* **26**, 4634–4640.

(32) Hamm-Alvarez, S., Sancar, A., and Rajagopalan, K. V. (1989) The role of the enzyme-bound 5,10-methenyltetrahydropteroylpolyglutamate in catalysis by *Escherichia coli* DNA photolyase. *J. Biol. Chem.* **264**, 9649–9656.

(33) Eker, A. P. M., Hessels, J. K. C., and Dekker, R. H. (1986) Photoreactivating enzyme from *Streptomyces griseus*—VI. Action spectrum and kinetics of photoreactivation. *Photochem. Photobiol.* **44**, 197–205.

(34) Jorns, M. S., Baldwin, E. T., Sancar, G. B., and Sancar, A. (1987) Action mechanism of *Escherichia coli* DNA photolyase. *J. Biol. Chem.* **262**, 486–491.

(35) Demple, B. (1988) Self-methylation by suicide DNA repair enzymes. In *Protein Methylation* (Paikard, W. K., and Kim, S., Eds.) CRC Press: Boca Raton, FL (in press).

(36) Barrows, L. R., and Magee, P. N. (1982) Nonenzymatic methylation of DNA by S-adenosylmethionine *in vitro*. *Carcinogenesis* **3**, 349–351.

(37) Rydberg, B., and Lindahl, T. (1982) Nonenzymatic methylation of DNA by the intracellular methyl group donor S-adenosyl-L-methionine is a potentially mutagenic reaction. *EMBO J.* **1**, 211–216.

(38) Samson, L., Thomale, J., and Rajewsky, M. F. (1988) Alternative pathways for the *in vivo* repair of $O^6$-alkylguanine and $O^4$-alkylthymine in *Escherichia coli*: the adaptive response and nucleotide excision repair. *EMBO J.* **7**, 2261–2267.

(39) Potter, P. M., Wilkinson, M. C., Fitton, J., Carr, F. J., Brennand, J., Cooper, D. P., and Margison, G. P. (1987) Characterization and nucleotide sequence of *ogt*, the $O^6$-alkylguanine-DNA-alkyltransferase gene of *E. coli*. *Nucleic Acids Res.* **15**, 9177–9193.

(40) Rebeck, G. W., Coons, S., Carroll, P., and Samson, L. (1988) A second DNA methyltransferase repair enzyme in *Escherichia coli*. *Proc. Natl. Acad. Sci. U.S.A.* **85**, 3039–3043.

(41) Shevell, D. E., Abou-Zamzam, A. M., Demple, B., and Walker, G. C. (1988) Construction of an *Escherichia coli* K-12 *ada* deletion by gene replacement in a *recD* strain reveals a second methyltransferase that repairs alkylated DNA. *J. Bacteriol.* **170**, 3294–3296.

(42) Demple, B., Sedgwick, B., Robins, P., Totty, N., Waterfield, M. D., and Lindahl, T. (1985) Active site and complete sequence of the suicidal methyltransferase that counters alkylation mutagenesis. *Proc. Natl. Acad. Sci. U.S.A.* **82**, 2688–2692.

(43) Nakabeppu, Y., Kondo, H., Kawabata, S.-I., Iwanaga, S., and Sekiguchi, M. (1985) Purification and structure of the intact Ada regulatory protein of *Escherichia coli* K12, $O^6$-Methylguanine-DNA methyltransferase. *J. Biol. Chem.* **260**, 7281–7288.

(44) Bhattacharyya, D., Tano, K., Bunick, G. J., Uberbacher, E. C., Behnke, W. D., and Mitra, S. (1988) Rapid, large-scale purification and characterization of "Ada protein" ($O^6$-methylguanine-DNA methyltransferase) of *E. coli*. *Nucleic Acids Res.* **16**, 6397–6410.

(45) Weinfeld, M., Drake, A. F., Saunders, J. K., and Paterson, M. C. (1985) Stereospecific removal of methyl phosphotriesters from DNA by an *Escherichia coli ada+* extract. *Nucleic Acids Res.* **13**, 7067–7077.

(46) McCarthy, T. V., and Lindahl, T. (1985) Methyl phosphotriesters in alkylated DNA are repaired by the Ada regulatory protyein of *E. coli*. *Nucleic Acids Res.* **13**, 2683–2698.

(47) Tano, K., Bhattacharyya, D., Foote, R. S., Mural, R. J., and Mitra, S. (1989) Site-directed mutation of the *Escherichia coli ada* gene: effects of substitution of methyl acceptor cysteine-321 by histidine in Ada protein. *J. Bacteriol.* **171**, 1535–1543.

(48) Yamagata, Y., Kohda, K., and Tomita, K.-I. (1988) Structural studies of $O^6$-methyldeoxyguanosine and related compounds: a premutagenic DNA lesion by methylating carcinogens. *Nucleic Acids Res.* **16**, 9307–9321.

(49) Brent, T. P., and Remack, J. S. (1988) Formation of covalent complexes between human $O^6$-alkylguanine-DNA alkyltransferases and BCNU-treated defined length synthetic oligodeoxynucleotides. *Nucleic Acids Res.* **16**, 6779–6788.

(50) Boiteux, S., de Oliveira, R. C., and Laval, J. (1985) The *Escherichia coli* $O^6$-methylguanine-DNA methyltransferase does not repair premutagenic $O^6$-methylguanine residues when present in Z-DNA. *J. Biol. Chem.* **260**, 8711–8715.

(51) Topal, M. D., Eadie, J. S., and Conrad, M. (1986) $O^6$-methylguanine mutation and repair is non-uniform. Selection for DNA most interactive with $O^6$-methyl-guanine. *J. Biol. Chem.* **261**, 9879–9887.

(52) Teo, I., Sedgwick, B., Kilpatrick, M. W., McCarthy, T. V., and Lindahl, T. (1986) The intracellular signal for induction of resistance to alkylating agents in *E. coli*. *Cell* **45**, 315–324.

(53) Nakabeppu, Y., and Sekiguchi, M. (1986) Regulatory mechanisms for induction of synthesis of repair enzymes in response to alkylating agents: Ada protein acts as a transcriptional regulator. *Proc. Natl. Acad. Sci. U.S.A.* **83**, 6297–6301.

(54) Kataoka, H., and Sekiguchi, M. (1985) Molecular cloning and characterization of the *alkB* gene of *Escherichia coli*. *Mol. Gen. Genet.* **198**, 263–269.

(55) Volkert, M. R., Hajec, L. I., and Nguyen, D. C. (1989) Induction of the alkylation-inducible *aidB* gene of *Escherichia coli* by anaerobiosis. *J. Bacteriol.* **171**, 1196–1198.

(56) Takahashi, K., Kawazoe, Y., Sakumi, K., Nakabeppu, Y., and Sekiguchi, M. (1988) Activation of Ada protein as a transcriptional regulator by direct alkylation with methylating agents. *J. Biol. Chem.* **263**, 13490–13492.

(57) Shevell, D. E., LeMotte, P. K., and Walker, G. C. (1988) Alteration of the carboxyl-terminal domain of Ada protein influences its inducibility, specificity, and strength as a transcriptional activator. *J. Bacteriol.* **170**, 5263–5271.

(58) Demple, B., Jacobsson, A., Olsson, M., Robins, P., and Lindahl, T. (1982) Repair of alkylated DNA in *E. coli*: physical properties of $O^6$-methyl guanine-DNA-methyltransferase. *J. Biol. Chem.* **252**, 13776–13782.

(59) Teo, I. A. (1987) Proteolytic processing of the Ada protein that repairs DNA $O^6$-methylguanine residues in *E. coli*. *Mutat. Res.* **183**, 123–127.

(60) Arikan, E., Kulkarni, M. S., Thomas, D. C., and Sancar, A. (1986) Sequences of the *E. coli uvrB* gene and protein. *Nucleic Acids Res.* **14**, 2637–2650.

(61) Yoshikai, T., Nakabeppu, Y., and Sekiguchi, M. (1988) Proteolytic cleavage of Ada protein that carries methyltransferase and transcriptional regulator activities. *J. Biol. Chem.* **263**, 19174–19180.

(62) Grossman, L., Caron, P., Mazur, S. J., and Oh, E. Y. (1988) Repair of DNA containing pyrimidine dimers. *FASEB J.* **2**, 2696–2701.

(63) Thomas, D. C., Levy, M., and Sancar, A. (1985) Amplification and purification of UvrA, UvrB and UvrC proteins of *Escherichia coli*. *J. Biol. Chem.* **260**, 9875–9883.

(64) Sedgwick, B. (1989) *In vitro* proteolytic cleavage of the *Escherichia coli* Ada protein by the *ompT* gene product. *J. Bacteriol.* **171**, 2249–2251.

(65) Morohoshi, F., and Munakata, N. (1987) Multiple species of *Bacillus subtilis* DNA alkyltransferases involved in the adaptive response to simple alkylating agents. *J. Bacteriol.* **169**, 587–592.

(66) Riazuddin, S., Athar, A., and Sohail, A. (1987) Methyl transferases induced during chemical adaptation of *M. luteus*. *Nucleic Acids Res.* **15**, 9471–9486.

(67) Yarosh, D. B. (1985) The role of $O^6$-methylguanine-DNA methyltransferase in cell survival, mutagenesis and carcinogenesis. *Mutat. Res.* **145**, 1–16.

(68) Brent, T. P., Dolan, M. E., Fraenkel-Conrat, H., Hall, J., Karran, P., Laval, F., Margison, G. P., Montesano, R., Pegg, A. E., Potter, P. M., Singer, B., Swenberg, J. A., and Yarosh, D. B. (1988) Repair of O-alkylpyrimidines in mammalian cells: a present consensus. *Proc. Natl. Acad. Sci. U.S.A.* **85**, 1759–1762.

(69) Satoh, M. S., Huh, H.-H., Rajewsky, M. F., and Kuroki, T. (1988) Enzymatic removal of $O^6$-ethylguanine from mitochondrial DNA in rat tissues exposed to *N*-ethyl-*N*-nitrosourea *in vivo*. *J. Biol. Chem.* **263**, 6854–6856.

(70) Seal, G., Arenaz, P., and Sirover, M. A. (1987) Purification and properties of the human placental uracil DNA glycosylase. *Biochim. Biophys. Acta* **925**, 226–233.

(71) Varshney, U., Hutcheon, T., and van de Sande, J. H. (1988) Sequence analysis, expression, and conservation of *Escherichia coli* uracil DNA glycosylase and its gene (*ung*). *J. Biol. Chem.* **263**, 7776–7784.

(72) Varshney, U., and van de Sande, J. H. (1989) Characterization of the *ung1* mutation of *Escherichia coli*. *Nucleic Acids Res.* **17**, 813.

(73) Percival, K. J., Klein, M. B., and Burgers, P. M. J. (1989) Molecular cloning and primary structure of the uracil-DNA-glycosylase gene from *Saccharomyces cerevisiae*. *J. Biol. Chem.* **264**, 2593–2598.

(74) Wittwer, C. U., Bauw, G., and Krokan, H. E. (1989) Purification and determination of the $NH_2$-terminal amino acid sequence of uracil-DNA glycosylase from human placenta. *Biochemistry* **28**, 780–784.

(75) Duncan, B. K. (1985) Isolation of insertion, deletion, and nonsense mutations of the uracil-DNA glycosylase (*ung*) gene of *Escherichia coli* K-12. *J. Bacteriol.* **164**, 689–695.

(76) Burgers, P. M. J., and Klein, M. B. (1986) Selection by genetic transformation of a *Saccharomyces cerevisiae* mutant defective for the nuclear uracil-DNA-glycosylase. *J. Bacteriol.* **166**, 905–913.

(77) Delort, A.-M., Duplaa, A.-M., Molko, D., and Teoule, R. (1985) Excision of uracil residues in DNA: mechanism of action of *Escherichia coli* and *Micrococcus luteus* uracil-DNA glycosylases. *Nucleic Acids Res.* **13**, 319–335.

(78) Domena, J. D., Timmer, R. T., Dicharry, S. A., and Mosbaugh, D. W. (1988) Purification and properties of mitochondrial uracil DNA glycosylase from rat liver. *Biochemistry* **27**, 6742–6751.

(79) Caradonna, S., Worrad, D., and Lirette, R. (1987) Isolation of a Herpes Simplex Virus cDNA encoding the DNA repair enzyme uracil-DNA glycosylase. *J. Virol.* **61**, 3040–3047.

(80) Wang, Z., and Mosbaugh, D. W. (1988) Uracil-DNA glycosylase inhibitor of bacteriophage PBS2: cloning and effects of expression of the inhibitor gene in *Escherichia coli*. *J. Bacteriol.* **170**, 1082–1091.

(81) Wang, Z., and Mosbaugh, D. W. (1989) Uracil-DNA glycosylase inhibitor gene of bacteriophage PBS2 encodes a binding protein specific for uracil-DNA glycosylase. *J. Biol. Chem.* **264**, 1163–1171.

(82) Hollstein, M. C., Brooks, P., Linn, S., and Ames, B. N. (1984) Hydroxymethyluracil DNA glycosylase in mammalian cells. *Proc. Natl. Acad. Sci. U.S.A.* **81**, 4003–4007.

(83) Steinum, A.-L., and Seeberg, E. (1986) Nucleotide sequence of the *tag* gene from *Escherichia coli*. *Nucleic Acids Res.* **14**, 3763–3772.

(84) Sakumi, K., Nakabeppu, Y., Yamamoto, Y., Kawabata, S., Iwanaga, S., and Sekiguchi, M. (1986) Purification and structure of 3-methyladenine-DNA glycosylase I of *Escherichia coli*. *J. Biol. Chem.* **261**, 15761–15766.

(85) Nakabeppu, Y., Kondo, H., and Sekiguchi, M. (1984) Cloning and characterization of the *alkA* gene of *Escherichia coli* that encodes 3-methyladenine DNA glycosylase II. *J. Biol. Chem.* **259**, 13723–13729.

(86) Kaasen, I., Evensen, G., and Seeberg, E. (1986) Amplified expression of the *tag+* and *alk+* genes in *Escherichia coli*: identification of gene products and effects on alkylation resistance. *J. Bacteriol.* **168**, 642–647.

(87) Yamagata, Y., Odawara, K., Tomita, K., Nakabeppu, Y., and Sekiguchi, M. (1988) Crystallization and preliminary X-ray diffraction studies of 3-methyladenine-DNA glycosylase II from *Escherichia coli*. *J. Mol. Biol.* **204**, 1055–1056.

(88) Riazuddin, S., Athar, A., Ahmed, Z., Lali, S. M., and Sohail, A. (1987) DNA glycosylase enzymes induced during chemical adaptation of *M. luteus*. *Nucleic Acids Res.* **15**, 6607–6624.

(89) Pierre, J., and Laval, J. (1986) Cloning of *Micrococcus luteus* 3-methyladenine-DNA glycosylase genes in *Escherichia coli*. *Gene* **43**, 139–146.

(90) McCarthy, T. V., Karran, P., and Lindahl, T. (1984) Inducible repair of O-alkylated DNA pyrimidines in *Escherichia coli*. *EMBO J.* **3**, 545–550.

(91) Male, R., Helland, D. E., and Kleppe, K. (1985) Purification and characterization of 3-methyladenine-DNA glycosylase from calf thymus. *J. Biol. Chem.* **260**, 1623–1629.

(92) Male, R., Haukanes, B. I., Helland, D. E., and Kleppe, K. (1987) Substrate specificity of 3-methyladenine-DNA glycosylase from calf thymus. *Eur. J. Biochem.* **165**, 13–19.

(93) Endo, Y., and Tsurugi, K. (1987) RNA *N*-Glycosidase activity of Ricin A-chain. *J. Biol. Chem.* **262**, 8128–8130.

(94) Lagravere, C., Malfoy, B., Leng, M., and Laval, J. (1984) Ring-opened alkylated guanine is not repaired in Z-DNA. *Nature (London)* **310**, 798–800.

(95) Boiteux, S., Belleney, J., Roques, B. P., and Laval, J. (1984) Two rotameric forms of open ring 7-methylguanine are present in alkylated polynucleotides. *Nucleic Acids Res.* **12**, 5429–5439.

(96) Breimer, L. H. (1984) Enzymatic excision from γ-irradiated polydeoxyribonucleotides of adenine residues whose imidazole rings have been ruptured. *Nucleic Acids Res.* **12**, 6359–6367.

(97) Boiteux, S., O'Connor, T. R., and Laval, J. (1987) Formamidopyrimidine-DNA glycosylase of *Escherichia coli*: cloning and sequencing of the *fpg* structural gene and overproduction of the protein. *EMBO J.* **6**, 3177–3183.

(98) Boiteux, S., and Huisman, O. (1989) Isolation of a formamidopyrimidine-DNA glycosylase (*fpg*) mutant of *Escherichia coli*. *Mol. Gen. Genet.* **215**, 300–305.

(99) Demple, B., and Linn, S. (1980) DNA *N*-glycosylases and UV repair. *Nature (London)* **287**, 203–208. (a) Haseltine, W. A., Gordon, L. K., Lindan, C. P., Grafstrom, R. H., Shaper, N. L., and Grossman, L. (1980) Cleavage of pyrimidine dimers in specific DNA sequences by a pyrimidine dimer DNA-glycosylase of *M. luteus*. *Nature (London)* **285**, 634–641.

(100) Nakabeppu, Y., and Sekiguchi, M. (1981) Physical association of pyrimidine dimer DNA glycosylase and apurinic/apyrimidinic DNA endonuclease essential for repair of ultraviolet-damaged DNA. *Proc. Natl. Acad. Sci. U.S.A.* **78**, 2742–2746.

(101) Friedberg, E. C. (1985) *DNA Repair*, W. H. Freeman and Company, New York.

(102) Valerie, K., Henderson, E. E., and de Riel, J. K. (1984) Identification, physical map location and sequence of the *denV* gene from bacteriophage T4. *Nucleic Acids Res.* **12**, 8085–8096.

(103) Recinos, A., III, Augustine, M. L., Kiggins, K. M., and Lloyd, R. S. (1986) Expression of the bacteriophage T4 *denV* structural gene in *Escherichia coli*. *J. Bacteriol.* **168**, 1014–1018.

(104) Chenevert, J. M., Naumovski, L., Schultz, R. A., and Friedberg, E. C. (1986) Partial complementation of the UV sensitivity of *E. coli* and yeast excision repair mutants by the cloned *denV* gene of bacteriophage T4. *Mol. Gen. Genet.* **203**, 163–171.

(105) Inaoka, T., Ishida, M., and Ohtsuka, E. (1989) Affinity of single- or double-stranded oligodeoxyribonucleotides containing a thymine photodimer for T4 endonuclease V. *J. Biol. Chem.* **264**, 2609–2614.

(106) Seawell, P. C., Simon, T. J., and Ganesan, A. K. (1980) Binding of T4 endonuclease V to deoxyribonucleic acid irradiated with ultraviolet light. *Biochemistry* **19**, 1685–1691.

(107) Liuzzi, M., Weinfeld, M., and Paterson, M. C. (1987) Selective inhibition by methoxyamine of the apurinic/apyrimidinic endonuclease activity associated with pyrimidine dimer-DNA glycosylases from *Micrococcus luteus* and Bacteriophage T4. *Biochemistry* **26**, 3315–3321.

(108) Liuzzi, M., and Talpaert-Borle, M. (1985) A new approach to the study of the base-excision repair pathway using methoxyamine. *J. Biol. Chem.* **260**, 5252–5258.

(109) Nakabeppu, Y., Yamashita, K., and Sekiguchi, M. (1982) Purification and characterization of normal and mutant forms of T4 endonuclease V. *J. Biol. Chem.* **257**, 2556–2562.

(110) Recinos, A., III, and Lloyd, R. S. (1988) Site-directed mutagenesis of the T4 endonuclease V gene: role of lysine-130. *Biochemistry* **27**, 1832–1838.

(111) Stump, D. G., and Lloyd, R. S. (1988) Site-directed mutagenesis of the T4 endonuclease V gene: role of tyrosine-129 and -131 in pyrimidine dimer-specific binding. *Biochemistry* **27**, 1839–1843.

(112) Pierre, J., and Laval, J. (1981) Specific nicking of DNA at apurinic sites by peptides containing aromatic residues. *J. Biol. Chem.* **256**, 10217–10220.

(113) Behmoaras, T., Toulme, J.-J., and Helene, C. (1981) A tryptophan-containing peptide recognizes and cleaves DNA at apurinic sites. *Nature (London)* **292**, 858–859.

(114) Kim, J., and Linn, S. (1988) The mechanism of action of E. coli endonuclease III and T4 UV endonuclease (endonuclease V). *Nucleic Acids Res.* **16**, 1135–1141.

(115) Manoharan, M., Mazumder, A., Ransom, S. C., Gerlt, J. A., and Bolton, P. H. (1988) Mechanism of UV endonuclease V cleavage of abasic sites in DNA determined by $^{13}$C labeling. *J. Am. Chem. Soc.* **110**, 2690–2691.

(116) Gruskin, E. A., and Lloyd, R. S. (1986) The DNA scanning mechanism of T4 endonuclease V. *J. Biol. Chem.* **261**, 9607–9613.

(117) Ganesan, A. K., Seawell, P. C., Lewis, R. J., and Hanawalt, P. C. (1986) Processivity of T4 endonuclease V is sensitive to NaCl concentration. *Biochemistry* **25**, 5751–5755.

(118) Gruskin, E. A., and Lloyd, R. S. (1988) Molecular analysis of plasmid DNA repair within ultraviolet-irradiated *Escherichia coli*. *J. Biol. Chem.* **263**, 12728–12737.

(119) Gruskin, E. A., and Lloyd, R. S. (1988) Molecular analysis of plasmid DNA repair within ultraviolet-irradiated *Escherichia coli*. *J. Biol. Chem.* **263**, 12738–12743.

(120) Tao, K., Noda, A., and Yonei, S. (1987) The roles of different excision-repair mechanisms in the resistance of *Micrococcus luteus* to UV and chemical mutagens. *Mutat. Res.* **183**, 231–239.

(121) Shiota, S., and Nakayama, H. (1988) Evidence for a *Micrococcus luteus* gene homologous to *uvrB* of *Escherichia coli*. *Mol. Gen. Genet.* **213**, 21–29.

(122) Doetsch, P. W., Helland, D. E., and Haseltine, W. A. (1986) Mechanism of action of a mammalian DNA repair endonuclease. *Biochemistry* **25**, 2212–2220.

(123) Helland, D. E., Doetsch, P. W., and Haseltine, W. A. (1986) Substrate specificity of a mammalian DNA repair endonuclease that recognizes oxidative base damage. *Mol. Cell. Biol.* **6**, 1983–1990.

(124) Doetsch, P. W., Henner, W. D., Cunningham, R. P., Toney, J. H., and Helland, D. E. (1987) A highly conserved endonuclease activity present in *Escherichia coli*, bovine, and human cells recognizes oxidative DNA damage at sites of pyrimidines. *Mol. Cell. Biol.* **7**, 26–32.

(125) Gossett, J., Lee, K., Cunningham, R. P., and Doetsch, P. W. (1988) Yeast Redoxyendonuclease, a DNA repair enzyme similar to *Escherichia coli* endonuclease III. *Biochemistry* **27**, 2629–2634.

(126) Weiss, R. B., Gallagher, P. E., Brent, T. P., and Duker, N. J. (1989) Cytosine photoproduct-DNA glycosylase in *Escherichia coli* and cultured human cells. *Biochemistry* **28**, 1488–1492.

(127) Ide, H., Kow, Y. W., and Wallace, S. S. (1985) Thymine glycols and urea residues in M13 DNA constitute replicative blocks *in vitro*. *Nucleic Acids Res.* **13**, 8035–8052.

(128) Clark, J. M., and Beardsley, G. P. (1986) Thymine glycol lesions terminate chain elongation by DNA polymerase I *in vitro*. *Nucleic Acids Res.* **14**, 737–749.

(129) Clark, J. M., and Beardsley, G. P. (1989) Template length, sequence context, and 3'-5' exonuclease activity modulate replicative bypass of thymine glycol lesions *in vitro*. *Biochemistry* **28**, 775–779.

(130) Cunningham, R. P., and Weiss, B. (1985) Endonuclease III (*nth*) mutants of *Escherichia coli*. *Proc. Natl. Acad. Sci. U.S.A.* **82**, 474–478.

(131) Weiss, B., and Cunningham, R. P. (1985) Genetic mapping of *nth*, a gene affecting endonuclease III (thymine glycol-DNA glycosylase) in *Escherichia coli* K12. *J. Bacteriol.* **162**, 607–610. (a) Asahara, H., Wistort, P. M., Bank, J. F., Bakerian, R. H., and Cunningham, R. P. (1989) Purification and characterization of *Escherichia coli* endonuclease III from the cloned *nth* gene. *Biochemistry* **28**, 4444–4449. (b) Cunningham, R. P., Asahara, H., Bank, J. F., Scholes, C. P., Salerno, J. C., Surerus, K., Munck, E., McCracken, J., Peisach, J., and Emptage, M. H. (1989) Endonuclease III is an iron–sulfur protein. *Biochemistry* **28**, 4450–4455.

(132) Breimer, L. H., and Lindahl, T. (1985) Thymine lesions produced by ionizing radiation in double-stranded DNA. *Biochemistry* **24**, 4018–4022.

(133) Bailly, V., and Verly, W. G. (1987) *Escherichia coli* endonuclease III is not an endonuclease but a β-elimination catalyst. *Biochem. J.* **242**, 565–572.

(134) Kow, Y. W., and Wallace, S. S. (1987) Mechanism of action of *Escherichia coli* endonuclease III. *Biochemistry* **26**, 8200–8206.

(135) Jorgensen, T. J., Kow, Y. W., Wallace, S. S., and Henner, W. D. (1987) Mechanism of action of *Micrococcus luteus* AP-endonuclease. *Biochemistry* **26**, 6436–6443.

(136) Kim, J., and Linn, S. (1989) Purification and characterization of UV endonuclease I and II from murine plasmacytoma cells. *J. Biol. Chem.* **264**, 2739–2745.

(137) Higgins, S. A., Frenkel, K., Cummings, A., and Teebor, G. W. (1987) Definitive characterization of human thymine glycol *N*-glycosylase activity. *Biochemistry* **26**, 1683–1688.

(138) Saporito, S. M., Smith-White, B. J., and Cunningham, R. P. (1988) Nucleotide sequence of the *xth* gene of *Escherichia coli* K-12. *J. Bacteriol.* **170**, 4542–4547.

(139) Kow, Y. W., and Wallace, S. S. (1985) Exonuclease III recognizes urea residues in oxidized DNA. *Proc. Natl. Acad. Sci. U.S.A.* **82**, 8354–8358.

(140) Demple, B., Johnson, A., and Fung, D. (1986) Exonuclease III and endonuclease IV remove 3′ blocks from DNA synthesis primers in $H_2O_2$-damaged *Escherichia coli*. *Proc. Natl. Acad. Sci. U.S.A.* **83**, 7731–7735.

(141) Weiss, B., and Grossman, L. (1987) Phosphodiesterases involved in DNA repair. *Adv. Enzymol. Relat. Areas Mol. Biol.* **60**, 1–34. (a) Kow, Y. W. (1989) Mechanism of action of *Escherichia coli* Exonuclease III. *Biochemistry* **28**, 3280–3287. (b) Puyet, A., Greenberg, B., and Lacks, S. A. (1989) The *exoA* gene of *Streptococcus pneumonia* and its product, a DNA exonuclease with apurinic endonuclease activity. *J. Bacteriol.* **171**, 2278–2286.

(142) Saporito, S. M., and Cunningham, R. P. (1988) Nucleotide sequence of the *nfo* gene of *Escherichia coli* K-12. *J. Bacteriol.* **170**, 5141–5145.

(143) Chan, E., and Weiss, B. (1987) Endonuclease IV of *Escherichia coli* is induced by paraquat. *Proc. Natl. Acad. Sci. U.S.A.* **84**, 3189–3193.

(144) Cunningham, R. P., Saporito, S. M., Spitzer, S. G., and Weiss, B. (1986) Endonuclease IV (*nfo*) mutant of *Escherichia coli*. *J. Bacteriol.* **168**, 1120–1127.

(145) Foster, P. L., and Davis, E. F. (1987) Loss of apurinic/apyrimidinic site endonuclease increase the mutagenicity of *N*-methyl-*N′*-nitro-*N*-nitrosoguanidine to *Escherichia coli*. *Proc. Natl. Acad. Sci. U.S.A.* **84**, 2891–2895. (a) Saporito, S. M., Gedenk, M., and Cunningham, R. P. (1989) Role of exonuclease III and endonuclease IV in repair of pyrimidine dimers initiated by bacteriophage T4 pyrimidine dimer-DNA glycosylase. *J. Bacteriol.* **171**, 2542–2546.

(146) Levin, J. D., Johnson, A. W., and Demple, B. (1988) Homogenous *Escherichia coli* endonuclease IV. *J. Biol. Chem.* **263**, 8066–8071.

(147) Bernelot-Moens, C., and Demple, B. (1989) Multiple activities for 3′-deoxyribose fragments in *Escherichia coli*. *Nucleic Acids Res.* **17**, 587–600.

(148) Armel, P. R., and Wallace, S. S. (1984) DNA repair in *Saccharomyces cerevisiae*: purification and characterization of apurinic endonucleases. *J. Bacteriol.* **160**, 895–902.

(149) Chang, C.-C., Kow, Y. W., and Wallace, S. S. (1987) Apurinic endonucleases from *Saccharomyces cerevisiae* also recognize urea residues in oxidized DNA. *J. Bacteriol.* **169**, 180–183.

(150) Johnson, A. W., and Demple, B. (1988) Yeast DNA diesterase for 3′-fragments of deoxyribose: purification and physical properties of a repair enzyme for oxidative DNA damage. *J. Biol. Chem.* **263**, 18009–18022.

(151) Spiering, A. L., and Deutsch, W. A. (1986) *Drosophila* apurinic/apyrimidinic DNA endonucleases. *J. Biol. Chem.* **261**, 3222–3228.

(152) Kane, C. M., and Linn, S. (1981) Purification and characterization of an apurinic/apyrimidinic endonuclease from HeLa cells. *J. Biol. Chem.* **256**, 3405–3414. (a) Kelley, M. R., Venugopal, S., Harless, J., and Deutsch, W. A. (1989) Antibody to a human DNA repair protein allows for cloning of a *Drosophila* cDNA that encodes an apurinic endonuclease. *Mol. Cell. Biol.* **9**, 965–973.

(153) Grafstrom, R. H., Shaper, N. L., and Grossman, L. (1982) Human placental apurinic/apyrimidinic endonuclease. *J. Biol. Chem.* **257**, 13459–13464. (a) Sanderson, B. J. S., Chang, C.-N., Grollman, A. P., and Henner, W. D. (1989) Mechanism of DNA cleavage and substrate recognition by a bovine apurinic endonuclease. *Biochemistry* **28**, 3894–3901.

(154) Franklin, W. A., and Lindahl, T. (1988) DNA deoxyribophosphodiesterase. *EMBO J.* **7**, 3617–3622.

(155) Voigt, J. M., Van Houten, B., Sancar, A., and Topal, M. D. (1989) Repair of $O^6$-methylguanine by ABC excinuclease of *Es-*

(156) Van Houten, B., and Sancar, A. (1987) Repair of *N*-methyl-*N*-nitro-nitrosoguanidine DNA damage by ABC excision nuclease. *J. Bacteriol.* **169**, 540–545.

(157) Thomas, D. C., Kunkel, T. A., Casna, N. J., Ford, J. P., and Sancar, A. (1986) Activities and incision patterns of ABC excinuclease on modified DNA containing single-base mismatches and extrahelical bases. *J. Biol. Chem.* **261**, 14496–14505.

(158) Husain, I., Van Houten, B., Thomas, D. C., and Sancar, A. (1986) Sequences of the *uvrA* gene and protein reveal two potential ATP binding sites. *J. Biol. Chem.* **261**, 4895–4901.

(159) Doolittle, R. F., Johnson, M. S., Husain, I., Van Houten, B., Thomas, D. C., Sancar, A. (1986) Domainal evolution of a prokaryotic DNA-repair protein: relationship to active transport proteins. *Nature (London)* **323**, 451–453.

(160) Navaratnam, S., Myles, G. M., Strange, R. W., and Sancar, A. (1989) Evidences from EXAFS and site-specific mutagenesis for zinc fingers in UvrA protein of *E. coli*. *J. Biol. Chem.* (in press).

(161) Van Houten, B., Gamper, H., Sancar, A., and Hearst, J. E. (1987) DNAse I footprint of ABC excinuclease. *J. Biol. Chem.* **262**, 13180–13187.

(162) Van Houten, B., Gamper, H., Hearst, J. E., and Sancar, A. (1988) Analysis of sequential steps of nucleotide excision repair in *Escherichia coli* using synthetic substrates containing single psoralen adducts. *J. Biol. Chem.* **263**, 16553–16560.

(163) Orren, D. K., and Sancar, A. (1989) The (A)BC excinuclease of *E. coli* has only the UvrB and UvrC subunits in the incision complex. *Proc. Natl. Acad. Sci. U.S.A.* **86**, 5237–5241.

(164) Caron, P. R., and Grossman, L. (1988) Involvement of a cryptic ATPase activity of UvrB and its proteolysis product, UvrB*, in DNA repair. *Nucleic Acids Res.* **16**, 9651–9662.

(165) Caron, P. R., and Grossman, L. (1988) Potential role of proteolysis in the control of UvrABC incision. *Nucleic Acids Res.* **16**, 9641–9650.

(166) Sancar, G. B., Sancar, A., and Rupp, W. D. (1984) Sequences of the *E. coli uvrC* gene and protein. *Nucleic Acids Res.* **11**, 4593–4608.

(167) Sancar, A., and Rupp, W. D. (1983) A novel repair enzyme: UvrABC excision nuclease of *Escherichia coli* cuts a DNA strand on both sides of the damaged region. *Cell* **33**, 249–260. (a) Van Houten, B., Gamper, H., Hearst, J. E., and Sancar, A. (1986) Construction of DNA substrates modified with psoralen at a unique site and study of the action mechanism of ABC excinuclease on these uniformly modified substrates. *J. Biol. Chem.* **261**, 14135–14141.

(168) Myles, G. M., Van Houten, B., and Sancar, A. (1987) Utilization of DNA photolyase, pyrimidine dimer endonucleases, and alkali hydrolysis in the analysis of aberrant ABC excinuclease incisions adjacent to UV-induced DNA photoproducts. *Nucleic Acids Res.* **15**, 1227–1243.

(169) Selby, C. S., and Sancar, A. (1988) ABC excinuclease incises both 5′ and 3′ to the cc-1065-DNA adduct and its incision activity is stimulated by DNA helicase II and DNA polymerase I. *Biochemistry* **27**, 7184–7188.

(170) Tang, M.-S., Lee, C.-S., Doisy, R., Ross, L., Needham–Van Devanter, D. R., and Hurley, L. H. (1988) Recognition and repair of the cc-1065-($N_3$-adenine)-DNA adduct by UVRABC nuclease. *Biochemistry* **27**, 893–901.

(171) Husain, I., Van Houten, B., Thomas, D. C., Abdel-Monem, M., and Sancar, A. (1985) The effect of DNA polymerase I and helicase II on the turnover rate of ABC excinuclease. *Proc. Natl. Acad. Sci. U.S.A.* **82**, 6774–6778.

(172) Caron, P. R., Kushner, S. R., and Grossman, L. (1985) Involvement of helicase II (*uvrD* gene product) and DNA polymerase I in excision mediated by the uvrABC protein complex. *Proc. Natl. Acad. Sci. U.S.A.* **82**, 4925–4929.

(173) Rubin, J. S. (1988) The molecular genetics of the incision step in the DNA excision repair process. *Int. J. Radiat. Biol.* **54**, 309–365.

(174) Haynes, R. H., and Kunz, B. A. (1981) DNA Repair and mutagenesis in yeast. In *The molecular biology of the yeast Saccharomyces* (Strathern, J., Jones, E. W., and Broach, J. R., Eds.) Cold Spring Harbor Laboratory, Cold Spring Harbor, N.Y.

(175) Reynolds, P. L., Prakash, D. L., and Prakash, S. (1987) Nucleotide sequence and functional analysis of the RAD1 gene of *Saccharomyces cerevisiae*. *Mol. Cell. Biol.* **7**, 1012–1020.

(176) Nagpal, M. L., Higgins, D. R., and Prakash, S. (1985) Expression on the RAD1 and RAD3 genes of *Saccharomyces cerevisiae* is not affected by DNA damage or during cell division. *Mol. Gen. Genet.* **199**, 59–63.

(177) Robinson, G. W., Nicolet, C. M., Kalainov, D., and Friedberg, E. C. (1986) A yeast excision-repair gene is inducible by DNA-damaging agents. *Proc. Natl. Acad. Sci. U.S.A.* **83**, 1842–1846.

(178) Schiestl, R. H., and Prakash, S. (1988) RAD1, an excision repair gene of *Saccharomyces cerevisiae*, is also involved in recombination *Mol. Cell. Biol.* **8**, 3619–3626.

(179) Madura, K., and Prakash, S. (1986) Nucleotide sequence, transcript mapping, and regulation of the RAD2 gene of *S. cerevisiae*. *J. Bacteriol.* **166**, 914–923.

(180) Naumovski, L., and Friedberg, E. C. (1983) A DNA repair gene required for the incision of damaged DNA is essential for viability in *Saccharomyces cerevisiae*. *Proc. Natl. Acad. Sci. U.S.A.* **80**, 4818–4821.

(181) Higgins, D. R., Prakash, L., Reynolds, P., Polakowska, R., Weber, S., and Prakash, L. (1983) Isolation and characterization of the RAD3 gene of *Saccharomyces cerevisiae* and inviability of *rad3* deletion mutants. *Proc. Natl. Acad. Sci. U.S.A.* **80**, 5680–5684.

(182) Reynolds, P., Higgins, D. R., Prakash, L., and Prakash, S. (1985) The nucleotide sequence of the RAD3 gene of *Saccharomyces cerevisiae*: a potential adenine nucleotide binding amino acid sequence and a nonessential carboxyl terminal domain. *Nucleic Acids Res.* **13**, 2357–2372.

(183) Walker, J. E., Saraste, M., Runswick, M. J., and Gay, N. J. (1982) Distantly related sequences in the α- and β-subunits of ATP synthase, myosin, kinases, and other ATP-requiring enzymes and a common nucleotide binding fold. *EMBO J.* **1**, 945–951.

(184) Naumovski, L., and Friedberg, E. C. (1986) Analysis of the essential and excision repair functions of the RAD3 gene of *Saccharomyces cerevisiae* by mutagenesis. *Mol. Cell. Biol.* **6**, 1218–1227.

(185) Sung, P. L., Prakash, S., Weber, S., and Prakash, S. (1987) The RAD3 gene of *Saccharomyces cerevisiae* encodes a DNA-dependent ATPase. *Proc. Natl. Acad. Sci. U.S.A.* **84**, 6045–6049.

(186) Sung, P., Prakash, L., Matson, S. W., and Prakash, S. (1987) RAD3 protein of *Saccharomyces cerevisiae* is a DNA helicase. *Proc. Natl. Acad. Sci. U.S.A.* **84**, 8951–8955.

(187) Golin, J. E., and Esposito, M. S. (1977) Evidence for joint genic control of spontaneous mutation and genetic recombination during mitosis in *Saccharomyces*. *Mol. Gen. Genet.* **150**, 127–135.

(188) Montelone, B. A., Hoekstra, M. F., and Malone, R. E. (1988) Spontaneous mitotic recombination in yeast: the hyperrecombinational rem1 mutations are alleles of the RAD3 gene. *Genetics* **119**, 289–301.

(189) Sung, P., Higgins, D., Prakash, L., and Prakash, S. (1988) Mutation of lysine-48 to arginine in the yeast RAD3 protein abolishes its ATPase and DNA helicase activities but not the ability to bind ATP. *EMBO J.* **7**, 3263–3269.

(190) Kuemmerle, N. B., and Masker, W. E. (1980) Effect of the *uvrD* mutations of excision repair. *J. Bacteriol.* **142**, 535–546.

(191) Naumovski, L., Chu, G., Berg, P., and Friedberg, E. C. (1985) RAD3 gene of *Saccharomyces cerevisiae*: nucleotide sequence of wild-type and mutant alleles, transcript mapping, and aspects of gene regulation. *Mol. Cell. Biol.* **5**, 17–26.

(192) Fleer, R., Nicolet, C. M., Pure, G. A., and Friedberg, E. C. (1987) RAD4 gene of *Saccharomyces cerevisiae*: molecular cloning and partial characterization of a gene that is inactivated in *Escherichia coli*. *Mol. Cell. Biol.* **7**, 1180–1192.

(193) Gietz, R. D., and Prakash, S. (1988) Cloning and nucleotide sequence analysis of the *Saccharomyces cerevisiae* RAD4 gene required for excision repair of UV-damaged DNA. *Gene* **74**, 535–541.

(194) Couto, L. B., and Friedberg, E. C. (1989) Nucleotide sequence of the wild-type RAD4 gene of *Saccharomyces cerevisiae* and characterization of mutant *rad4* alleles. *J. Bacteriol.* **171**, 1862–1869.

(195) Reynolds, P., Prakash, L., Dumais, D., Perozzi, G., and Prakash, S. (1985) Nucleotide sequence of the RAD10 gene of *Saccharomyces cerevisiae*. *EMBO J.* **4**, 3549–3552.

(196) van Duin, M., de Wit, J., Odijk, H., Westerveld, A., Yasui, A., Koken, M. H. M., Hoeijmakers, J. H. J., and Bootsma, D. (1986) Molecular characterization of the human excision repair gene ERCC-1: cDNA cloning and amino acid homology with the yeast DNA repair gene RAD10. *Cell* **44**, 913–923.

(197) Lambert, C., Couto, L. B., Weiss, W. A., Schultz, R. A., Thompson, L. H., and Friedberg, E. C. (1988) A yeast DNA repair gene partially complements defective excision repair in mammalian cells. *EMBO J.* **7**, 3245–3253.

(198) Cleaver, J. E. (1968) Defective repair replication of DNA in xeroderma pigmentosum. *Nature (London)* **218**, 652–656.

(199) Lehman, A., Kirk-Bell, S., Arlett, C., Paterson, M. C., Lohman, P. H. M., de Weerd-Kastelein, E. A., and Bootsma, D. (1975) Xeroderma pigmentosum cells with normal levels of excision repair have a defect in DNA synthesis after UV-irradiation. *Proc. Natl. Acad. Sci. U.S.A.* **72**, 219–223.

(200) de Weerd-Kastelein, E. A., Keijzer, W., and Bootsma, D. (1972) Genetic heterogeneity of xeroderma pigmentosum demonstrated by somatic cell hybridization. *Nature (London)* **238**, 80–83.

(201) Fisher, E., Keijzer, W., Thielmann, H. W., Popanda, O., Bohnert, E., Edler, L., Jung, E. G., and Bootsma, D. (1985) A ninth complementation group in xeroderma pigmentosum, XP-I. *Mutat. Res.* **145**, 217–225.

(202) Cleaver, J. E. (1983) *Xeroderma pigmentosum*. In *The Metabolic Basis of Inherited Disease*, 5th ed. (Stanbury, J. B., Wyngaarden, J. B., Fredrickson, D. S., Goldstein, J. L., and Brown, M. S., Eds.) McGraw-Hill, New York.

(203) Thompson, L. H., Busch, D. B., Brookman, K., Mooney, C. L., and Glaser, P. A. (1981) Genetic diversity of UV-sensitive DNA-repair mutants of Chinese hamster ovary cells. *Proc. Natl. Acad. Sci. U.S.A.* **78**, 3734.

(204) Thompson, L. H., Salazar, E. P., Brookman, K. W., Collins, C. C., Stewart, S. A., Busch, D. B., and Weber, C. A. (1987) Recent progress with the DNA repair mutants of Chinese hamster ovary cells. *J. Cell Sci.* **6** (Suppl.), 97–110.

(205) Thomas, L. H., and Carrano, A. V. (1983) Analysis of mammalian cell mutagenesis and DNA repair using *in vitro* selected CHO cell mutants. *UCLA Symp. Mol. Cell. Biol.* **11**, 125–143.

(206) Thompson, L. H., Shiomi, T., Salazar, E. P., and Stewart, S. A. (1988) An eighth complementation group of rodent cells hypersensitive to ultraviolet irradiation. *Somat. Cell Mol. Genet.* **14**, 605–612.

(207) Thompson, L. H., Brookman, K. W., Dillehay, L. E., Mooney, C. L., and Carrano, A. V. (1982) Hypersensitivity to mutation and sister-chromatid exchange induction in CHO cell mutants defective in incising DNA containing UV-lesions. *Somatic Cell Genet.* **8**, 759–773.

(208) Zdzienicka, M. Z., Roza, L., Westerveld, A., Bootsma, D., and Simons, J. W. I. M. (1987) Biological and biochemical consequences of the human ERCC-1 repair gene after transfection into a repair-deficient CHO cell line. *Mutat. Res.* **183**, 69–74.

(209) Stefanini, M., Keijzer, W., Westerveld, A., and Bootsma, D. (1985) Interspecies complementation analysis of xeroderma pigmentosum and UV-sensitive Chinese hamster cells. *Exp. Cell Res.* **161**, 373–380.

(210) Thompson, L. H., Mooney, C. L., and Brookman, K. W. (1985) Genetic complementation between UV-sensitive CHO mutants and xeroderma pigmentosum fibroblasts. *Mutat. Res.* **150**, 423–429.

(211) Rubin, J. S., Prideaux, V. R., Huntington, F. W., Dulhanty, A. M., Whitmore, G. F., and Bernstein, A. (1985) Molecular cloning and chromosomal localization of DNA sequences associated with a human DNA repair gene. *Mol. Cell. Biol.* **5**, 398–405.

(212) Thompson, L. H., Carrano, A. V., Sato, K., Salazar, E. P., White, B. F., Stewart, S. A., Minkler, J. L., and Siciliano, M. J. (1987) Identification of nucleotide-excision-repair genes on human chromosomes 2 and 13 by functional complementation in hamster-human hybrids. *Somatic Cell Mol. Genet.* **13**, 539–551.

(213) Siciliano, M. J., Carrano, A. V., and Thompson, L. H. (1986) Assignment of a human DNA-repair gene associated with sister-chromatid exchange to chromosome 19. *Mutat. Res.* **174**, 303–308.

(214) Lehman, A. R. (1985) Use of recombinant DNA techniques in cloning DNA repair genes and in the study of mutagenesis in human cells. *Mutat. Res.* **150**, 61.

(215) Teitz, T., Naiman, T., Avissar, S. S., Bar, S., Okayama, H., and Canaani, D. (1987) Complementation of the UV-sensitive phenotype of a xeroderma pigmentosum human cell line by transfection with a cDNA clone library. *Proc. Natl. Acad. Sci. U.S.A.* **84**, 8801–8804.

(216) Hoeijmakers, J. H. J. (1987) Characterization of genes and proteins involved in excision repair of human cells. *J. Cell Sci.* **6** (Suppl.), 111–125.

(217) Rubin, J. S., Joyner, A. L., Bernstein, A., and Whitmore, G. F. (1983) Molecular identification of a human DNA repair gene following DNA-mediated gene transfer. *Nature (London)* **306**, 206–208.

(218) MacInnes, M. A., Bingham, J. D., Thompson, L. H., and strniste, G. F. (1984) DNA-mediated cotransfer of excision repair capacity and drug resistance into Chinese hamster ovary cell line UV-135. *Mol. Cell. Biol.* **4**, 1152–1158.

(219) Westerveld, A., Hoeijmakers, J. H. J., van Duin, M., de Wit, J., Odijk, H., Pastink, A., Wood, R. D., and Bootsma, D. (1984) Molecular cloning of a human DNA repair gene. *Nature (London)* **310**, 425–429.

(220) Wood, R. D., and Burki, H. J. (1982) Repair capability and the cellular age response for killing and mutation induction after UV. *Mutat. Res.* **95**, 505–514.

(221) van Duin, M., van den Tol, J., Warmerdam, P., Odijk, H., Meijer, D., Westerveld, A., Bootsma, D., and Hoeijmakers, J. H. J. (1988) Evolution and mutagenesis of the mammalian excision repair gene ERCC-1. *Nucleic Acids Res.* **16**, 5305–5322.

(222) Freidberg, E. C. (1987) The molecular biology of nucleotide excision repair of DNA: recent progress. *J. Cell Sci.* **6** (Suppl.), 1–23.

(223) Weber, C. A., Salazar, E. P., Stewart, S. A., and Thompson, L. H. (1988) Molecular cloning and biological characterization of a human gene, ERCC2, that corrects the nucleotide excision repair defect in CHO UV5 cells. *Mol. Cell. Biol.* **8**, 1137–1146. (a) Hanawalt, P. C. (1989) DNA Repair Mechanisms and their biological implications in mammalian cells. *Mutat. Res.* **217**, 173–184.

(224) Hoy, C. A., Thompson, L. H., Mooney, C. L., and Salazar, E. P. (1985) Defective DNA cross-link removal in Chinese hamster cell mutants hypersensitive to bifunctional alkylating agents. *Cancer Res.* **45**, 1737–1743.

(225) de Jonge, A. J. R., Vermeulen, W., Klein, B., and Hoeijmakers, J. H. J. (1983) Microinjection of human cell extracts corrects xeroderma pigmentosum defect. *EMBO J.* **2**, 637–641.

(226) Vermeulen, W., Osseweijer, P., de Jonge, A. J. R., and Hoeijmakers, J. H. J. (1986) Transient correction of excision repair defects in fibroblasts of 9 xeroderma pigmentosum complementation groups by microinjection of crude human cell extracts. *Mutat. Res.* **165**, 199–206.

(227) de Jonge, A. J. R., Vermeulen, W., Keijzer, W., Hoeijmakers, J. H. J., and Bootsma, D. (1985) Microinjection of *Micrococcus luteus* UV-endonuclease restores UV-induced unscheduled DNA synthesis in cells of 9 xeroderma pigmentosum complementation groups. *Mutat. Res.* **150**, 99–105.

(228) Zwetsloot, J. C. M., Vermeulen, W., Hoeijmakers, J. H. J., Yasui, A., Eker, A. P. M., and Bootsma, D. (1985) Microinjected photoreactivating enzymes from *Anacystes* and *Saccharomyces* monomerize dimers in chromatin of human cells. *Mutat. Res.* **146**, 71–77.

(229) Zwetsloot, J. C. M., Hoeijmakers, J. H. J., Vermeulen, W., Eker, A. P. M., and Bootsma, D. (1986) Unscheduled DNA synthesis in xeroderma pigmentosum cells after microinjection of yeast photoreactivating enzyme. *Mutat. Res.* **165**, 109–115.

(230) Zwetsloot, J. C. M., Barbeiro, A. P., Vermeulen, W., Arthur, H. M., Hoeijmakers, J. H. J., and Backendorf, C. (1986) Microinjection of *Escherichia coli* UvrA, B, C and D proteins into fibroblasts of xeroderma pigmentosum complementation groups A and C does not result in restoration of UV-induced unscheduled DNA synthesis. *Mutat. Res.* **166**, 89–98.

(231) Hanawalt, P. C., Cooper, P. K., Ganesan, A. K., and Smith, C. A. (1979) DNA repair in bacterial and mammalian cells. *Annu. Rev. Biochem.* **48**, 783–836.

(232) Kantor, G. J., and Setlow, R. B. (1981) Rate and extent of DNA repair in non-dividing human diploid fibroblasts. *Cancer Res.* **41**, 819–825.

(233) Mayne, L. V., and Lehman, A. R. (1982) Failure of RNA synthesis to recover after UV-irradiation: an early defect in cells from individuals with Cockaynes syndrome and xeroderma pigmentosum. *Cancer Res.* **42**, 1473–1478.

(234) Schmickel, R. D., Chu, E. H., Trosko, J. E., and Chang, C. C. (1977) Cockayne syndrome: a cellular sensitivity to ultraviolet light. *Pediatrics* **60**, 135–139.

(235) Williams, J., and Friedberg, E. C. (1979) DNA excision repair in chromatin after UV irradiation of human fibroblasts in culture. *Biochemistry* **18**, 3965–3976.

(236) Hanawalt, P. C., and Sarasin, A. (1986) Cancer-prone hereditary diseases with DNA processing abnormalities. *Trends Genet.* **2**, 124–129.

(237) Bohr, V. A., Smith, C. A., Okumoto, D. S., and Hanawalt, P. C. (1985) DNA repair in an active gene: removal of pyrimidine dimers from the DHFR gene of CHO cells is much more efficient than in the genome overall. *Cell* **40**, 359–369.

(238) Mellon, I., Bohr, V. A., Smith, C. A., and Hanawalt, P. C. (1986) Preferential DNA repair of an active gene in human cells. *Proc. Natl. Acad. Sci. U.S.A.* **83**, 8878–8882.

(239) Mansbridge, J. N., and Hanawalt, P. C. (1983) Domain-limited repair of DNA in ultraviolet irradiated fibroblasts from xeroderma pigmentosum group C. *UCLA Symp. Mol. Cell. Biol.* **2**, 195–207.

(240) Kleijer, W. J., Weerd-Kastelein, D., Sluyter, M. L., Kleijzer, N., De Wit, J., and Bootsma, D. (1973) UV induced DNA repair synthesis in cells of patients with different forms of xeroderma pigmentosum and of heterozygotes. *Mutat. Res.* **20**, 417–428.

(241) Mulleanders, L. H. F., van Kesteren, A. C., Bussman, C. J. M., van Zeeland, A. A., and Natarajan, A. T. (1984) Preferential repair of nuclear matrix associated DNA in xeroderma pigmentosum complementation group C. *Mutat. Res.* **141**, 75–82.

(242) Andrews, A. D., Barett, S. F., and Robbins, J. H. (1978) Xeroderma pigmentosum neurological abnormalities correlate with colony forming ability after UV irradiation. *Proc. Natl. Acad. Sci. U.S.A.* **75**, 1984–1988.

(243) Bohr, V. A., Okumoto, D. S., and Hanawalt, P. C. (1986) Survival of UV-irradiated mammalian cells correlates with efficient DNA repair in an essential gene. *Proc. Natl. Acad. Sci. U.S.A.* **83**, 3830–3833.

(244) Valerie, K., de Riel, J. K., and Henderson, E. E. (1985) Genetic complementation of UV-induced DNA repair in Chinese hamster ovary cells by the *denV* gene of phage T4. *Proc. Natl. Acad. Sci. U.S.A.* **82**, 7656–7660.

(245) Bohr, V. A., and Hanawalt, P. C. (1987) Enhanced repair of pyrimidine dimers in coding and non-coding genomic sequences in CHO cells expressing a prokaryotic DNA repair gene. *Carcinogenesis* **8**, 1333–1336.

(246) Madhani, H. D., Bohr, V. A., and Hanawalt, P. C. (1986) Differential DNA repair in transcriptionally active and inactive proto-oncogenes: *c-abl* and *c-mos*. *Cell* **45**, 417–423.

(247) Okumoto, D. S., and Bohr, V. A. (1987) DNA repair in the metallothionein gene increases with transcriptional activation. *Nucleic Acids Res.* **15**, 10021–10030.

(248) Mellon, I., Spivak, G., and Hanawalt, P. C. (1987) Selective removal of transcription-blocking DNA damage from the transcribed strand of the mammalian DHFR gene. *Cell* **51**, 241–249.

(249) Thomas, D. C., Morton, A. G., Bohr, V. A., and Sancar, A. (1988) General method for quantifying base adducts in specific mammalian genes. *Proc. Natl. Acad. Sci. U.S.A.* **85**, 3723–3727.

(250) Thomas, D. C., Okumoto, D. S., Sancar, A., and Bohr, V. A. (1989) Preferential DNA repair of (6–4) photoproducts in the DHFR gene of CHO cells (submitted for publication). (a) Vrieling, H., Van Rooijen, M. L., Groen, N. A., Zdzienicka, M. Z., Simons, J. W. I. M., Lohman, P. H. M., and van Zeeland, A. A. (1989) DNA strand specificity for UV-induced mutations in mammalian cells. *Mol. Cell. Biol.* **9**, 1277–1283.

(251) Leadon, S. A. (1986) Differential repair of DNA damage in specific nucleotide sequences in monkey cells. *Nucleic Acids Res.* **14**, 8979–8995.

(252) Leadon, S. A., and Snowden, M. M. (1988) Differential repair of DNA damage in the human metallothionein gene family. *Mol. Cell. Biol.* **8**, 5331–5338. (a) Vos, J.-M. H., and Hanawalt, P. C. (1987) Processing of psoralen adducts in an active human gene: Repair and replication of DNA containing monoadducts and interstrand cross-links. *Cell* **50**, 789–799. (b) Scicchitano, D. A., and Hanawalt, P. C. (1989) Repair of N-methylpurines in specific DNA sequences in Chinese hamster ovary cells: Absence of strand specificity in the dihydrofolate reductase gene. *Proc. Natl. Acad. Sci. U.S.A.* **86**, 3050–3054.

(253) Nishida, C., Reinhard, P., and Linn, S. (1988) DNA repair synthesis in human fibroblasts requires DNA polymerase δ. *J. Biol. Chem.* **263**, 501–510.

(254) Dresler, S. L., Gowans, B. J., Robinson-Hill, R. M., and Hunting, D. J. (1988) Involvement of DNA polymerase δ in DNA repair synthesis in human fibroblasts at late times after ultraviolet irradiation. *Biochemistry* **27**, 6379–6383.

(255) Wood, R. D., Robins, P., and Lindahl, T. (1988) Complementation of the xeroderma pigmentosum DNA repair defect in cell-free extracts. *Cell* **53**, 97–106.

(256) Sibghat-Ullah, Husain, I., Carlton, W., and Sancar, A. (1989) Human nucleotide excision repair *in vitro*: repair of pyrimidine dimers, psoralen, and cisplatin adducts by HeLa cell-free extracts. *Nucleic Acids Res.* (in press).

(257) Weinfeld, M., Genter, N. E., Johnson, L. D., and Paterson, M. C. (1986) Photoreversal-dependent release of thymidine and thymidine monophosphate from pyrimidine dimer-containing DNA excision fragments isolated from ultraviolet-damaged human fibroblasts. *Biochemistry* **25**, 2656–2664.

(258) Yamaizumi, M., Sugano, T., Asahina, H., Okada, Y., and Uchida, T. (1986) Microinjection of partially purified protein

factor restores DNA damage specifically in group A of xeroderma pigmentosum cells. *Proc. Natl. Acad. Sci. U.S.A.* **83**, 1476–1479.

(259) Chu, G., and Chang, E. (1988) Xeroderma pigmentosum group E cells lack a nuclear factor that binds to damaged DNA. *Science* **242**, 564–567.

(260) Valerie, K., Fronko, G., Henderson, E. E., and de Riel, J. K. (1986) Expression of the *denV* gene of coliphage T4 in UV-sensitive *rad* mutants of *Saccharomyces cerevisiae. Mol. Cell. Biol.* **6**, 3559–3562. (a) Banga, S. S., Boyd, J. B., Valerie, K., Harris, P. V., Kurz, E. M., and de Riel, J. K. (1989) *denV* gene of bacteriophage T4 restores DNA excision repair to *mei-9* and *mus201* mutants of *Drosophila melanogaster. Proc. Natl. Acad. Sci. U.S.A.* **86**, 3227–3231.

(261) Ding, R., Ghosh, K., Eastman, A., and Bresnick, E. (1985) DNA-mediated transfer and expression of a human DNA repair gene that demethylates $O^6$-methylguanine. *Mol. Cell. Biol.* **5**, 3293–3296.

(262) Samson, L., Derfler, B., and Waldstein, E. A. (1986) Suppression of human DNA alkylation-repair defects by *Escherichia coli* DNA-repair genes. *Proc. Natl. Acad. Sci. U.S.A.* **83**, 5607–5610.

(263) Brennand, J., and Margison, G. P. (1986) Reduction of the toxicity and mutagenicity of alkylating agents in mammalian cells harboring the *Escherichia coli* alkyltransferase gene. *Proc. Natl. Acad. Sci. U.S.A.* **83**, 6292–6296.

(264) Kataoka, H., Hall, J., and Karran, P. (1986) Complementation of sensitivity to alkylating agents in *Escherichia coli* and Chinese hamster ovary cells by expression of a cloned bacterial DNA repair gene. *EMBO J.* **5**, 3195–3200.

(265) Dickstein, R., Huh, N. D., Sandlie, I., and Grossman, L. (1988) The expression of the *Escherichia coli urvA* gene in human cells. *Mutat. Res.* **193**, 75–86.

(266) Willis, A. E., and Lindahl, T. (1987) DNA ligase I deficiency in Bloom's syndrome. *Nature* (London) **325**, 355–357.

(267) Chan, J. Y. H., Becker, F. F., German, J., and Ray, J. H. (1987) Altered DNA ligase I activity in Bloom's syndrom cells. *Nature* (*London*) **325**, 357–359.

(268) Willis, A. E., Weksberg, R., Tomlinson, S., and Lindahl, T. (1987) Structural alterations of DNA ligase I in Bloom's syndrome. *Proc. Natl. Acad. Sci. U.S.A.* **84**, 8016–8020.

(269) Chan, J. Y.-H., and Becker, F. F. (1988) Defective DNA ligase I in Bloom's syndrome cells. *J. Biol. Chem.* **263**, 18231–18235. (a) Runger, T. M., and Kraemer, K. H. (1989) Joining of linear plasmid DNA is reduced and error-prone in Bloom's syndrome cells. *EMBO J.* **8**, 1419–1425.

(270) Dezaya, P., and Sirover, M. A. (1986) Regulation of hypoxanthine DNA glycosylase in normal human and Bloom's syndrome fibroblasts. *Cancer Res.* **46**, 3756–3761. (a) Mezzina, M., Nardelli, J., Nocentini, S., Renault, G., and Sarastin, A. (1989) DNA ligase activity in human cell lines from normal donors and Bloom's syndrome patients. *Nucleic Acids Res.* **17**, 3091–3106.

(271) Van Houten, B., Gamper, H., Hearst, J. E., and Sancar, A. (1986) Action mechanism of ABC excision nuclease on a DNA substrate containing a psoralen crosslink at a defined position. *Proc. Natl. Acad. Sci. U.S.A.* **83**, 8077–8083.

(272) Cheng, S., Van Houten, B., Gamper, H., Hearst, J. E., and Sancar, A. (1988) Use of psoralen-modified oligonucleotides to trap three-standard RecA-DNA complexes and repair of these crosslinked complexes by ABC excinuclease. *J. Biol. Chem.* **263**, 11451–11460. (a) Sladek, F. M., Munn, M. M., Rupp, W. D., and Howard-Flanders, P. (1989) *In vitro* repair of psoralen–DNA cross-links by RecA, UvrABC, and the 5'-exonuclease of DNA polymerase I. *J. Biol. Chem.* **264**, 6755–6765. (b) Modrich, P. (1989) Methyl-directed DNA mismatch correction. *J. Biol. Chem.* **264**, 6597–6600.

(273) Glickman, B. W., and Radman, M. (1980) *Escherichia coli* mutator mutants deficient in methylation-instructured DNA mismatch correction. *Proc. Natl. Acad. Sci. U.S.A.* **77**, 1063–1067.

(274) Pukkila, P. J., Peterson, J., Herman, G., Modrich, P., and Meselson, M. (1983) Effect of high levels of DNA adenine methylation on methyl-directed mismatch repair in *Escherichia coli. Genetics* **104**, 571–582.

(275) Lu, A.-L., Clark, S., and Modrich, P. (1983) Methyl-directed repair of DNA base pair mismatches *in vitro. Proc. Natl. Acad. Sci. U.S.A.* **80**, 4639–4643.

(276) Marinus, M. G., and Morris, N. R. (1974) Biological function for 6-methyladenine residues in the DNA of *Escherichia coli* K12. *J. Mol. Biol.* **85**, 309–322.

(277) Herman, G. E., and Modrich, P. (1981) *Escherichia coli* K12 clones that overproduce *dam* methylase are hypermutable. *J. Bacteriol.* **145**, 644–646.

(278) Marinus, M. G. (1984) Methylation of prokaryotic DNA. In *DNA Methylation* (Razin, A., Cedar, H., and Riggs, A. D., Eds.) Springer-Verlag, New York.

(279) Grafstrom, R. H., and Hoess, R. H. (1983) Cloning of *mutH* and identification of the gene product. *Gene* **22**, 245–253.

(280) Grafstrom, R. H., and Hoess, R. H. (1987) Nucleotide sequence of the *Escherichia coli mutH* gene. *Nucleic Acids Res.* **15**, 3073–3084.

(281) Welsh, K. M., Lu, A.-L., Clark, S., and Modrich, P. (1987) Isolation and characterization of the *Escherichia coli mutH* gene product. *J. Biol. Chem.* **262**, 15624–15629.

(282) Su, S., and Modrich, P. (1986) *Escherichia coli* mutS-encoded protein binds to mismatched DNA base pairs. *Proc. Natl. Acad. Sci. U.S.A.* **83**, 5057–5061.

(283) Grilley, M., Welsh, K. M., Su, S.-S., and Modrich, P. (1989) Isolation and characterization of the *Escherichia coli mutL* gene product. *J. Biol. Chem.* **264**, 1000–1004.

(284) Lahue, R. S., Su, S.-S., and Modrich, P. (1987) Requirement for d(GATC) sequences in *Escherichia coli mutHLS* mismatch correction. *Proc. Natl. Acad. Sci. U.S.A.* **84**, 1482–1486.

(285) Langle-Rouault, F., Maenhault-Michel, G., and Radman, M. (1986) GATC sequence and mismatch repair in *Escherichia coli. EMBO J.* **5**, 2009–2013.

(286) Su, S.-S., Lahue, R. S., Au, K. G., and Modrich, P. (1988) Mismatch specificity of methyl-directed DNA mismatch correction *in vitro. J. Biol. Chem.* **263**, 6829–6835.

(287) Lu, A.-L., and Chang, D.-Y. (1988) Repair of single base-pair transversion mismatches of *Escherichia coli in vitro*: correction of certain A/G mismatches is independent of *dam* methylation and host *mutHLS* gene functions. *Genetics* **118**, 593–600.

(288) Priebe, S. D., Hadi, S. M., Greenberg, B., and Lacks, S. A. (1988) Nucleotide sequence of the *hexA* gene for DNA mismatch repair in *Streptococcus pneumonia* and homology of *hexA* and *mutS* of *Escherichia coli* and *Salmonella typhimurium. J. Bacteriol.* **170**, 190–196.

(289) Haber, L. T., Pang, P. P., Sobell, D. I., Mankovich, J. A., and Walker, G. C. (1988) Nucleotide sequence of the *Salmonella typhimurium mutS* gene required for mismatch repair: homology of MutS and HexA of *Streptococcus pneumonia. J. Bacteriol.* **170**, 197–202.

(290) Claverys, J. P., Roger, M., and Sicard, A. M. (1980) Excision and repair of mismatched base pairs in transformation of *Streptococcus pneumoniae. Mol. Gen. Genet.* **178**, 191–201.

(291) Claverys, J.-P., and Lacks, S. A. (1986) Heteroduplex deoxyribonucleic acid base mismatch repair in bacteria. *Microbiol. Rev.* **50**, 133–165.

(292) Guild, W. R., and Shoemaker, N. B. (1976) Mismatch correction in pneumococcal transformation: donor length and *hex*-dependent marker efficiency. *J. Bacteriol.* **125**, 125–135.

(293) Lacks, S. A., Dunn, J. J., and Greenberg, B. (1982) Identification of base mismatches recognized by the heteroduplex-DNA-repair system of *Streptococcus pneumonia. Cell* **31**, 327–336.

(294) Wagner, R., Jr., and Meselson, M. (1976) Repair tracts in mismatched DNA heteroduplexes. *Proc. Natl. Acad. Sci. U.S.A.* **73**, 4135–4139.

(295) Pang, P. P., Tsen, S. D., Lundberg, A. S., and Walker, G. C. (1984) The *mutH, mutL, mutS*, and *uvrD* genes of *Salmonella typhimurium* LT2. *Cold Spring Harbor Symp. Quant. Biol.* **49**, 597–602.

(296) Fishel, R. A., and Kolodner, R. (1984) An *Escherichia coli* cell-free system that catalyses the repair of symmetrically methylated heteroduplex DNA. *Cold Spring Harbor Symp. Quant. Biol.* **49**, 603–609.

(297) Fishel, R. A., Siegel, E. C., and Kolodner, R. (1986) Gene conversion in *Escherichia coli*. Resolution of heteroallelic mismatched nucleotides by co-repair. *J. Mol. Biol.* **188(2)**, 147–158.

(298) Lieb, M., Allen, E., and Read, D. (1986) Very short patch mismatch repair in phage lambda: repair sites and length of repair tracts. *Genetics* **114**, 1041–1060.

(299) Lu, A.-L., and Chang, D.-Y. (1988) Repair of single base-pair transversion mismatches of *Escherichia coli in vitro*: correction of certain A/G mismatches is independent of *dam* methylation and host *mutHLS* gene functions. *Genetics* **118**, 593–600.

(300) Cox, E. C. (1976) Bacterial mutator genes and the control of spontaneous mutation. *Annu. Rev. Genet.* **10**, 135–156.

(301) Nghiem, Y., Cabrera, M., Cupples, C. G., and Miller, J. H. (1988) The *mutY* gene: mutator locus in *Escherichia coli* that generates G·C → T·A transversions. *Proc. Natl. Acad. Sci. U.S.A.* **85**, 2709–2713.

(302) Cox, E. C. (1973) Mutator gene studies in *Escherichia coli*: the *mutT* gene. *Genetics* **73** (Suppl.), 67–80.

(303) Schaaper, R. M., and Dunn, R. L. (1987) *Escherichia coli mutT* mutator effect during *in vitro* DNA synthesis. *J. Biol. Chem.* **262**, 16267–16270.

(304) Bhatnagar, S. K., and Bessman, M. J. (1988) Studies on the mutator gene, *mutT* of *Escherichia coli*. *J. Biol. Chem.* **263**, 8953–8957. (a) Akiyama, M., Maki, H., Sekiguchi, M., and Horiuchi, T. (1989) A specific role of MutT protein: to prevent dG·dA mispairing in DNA replication. *Proc. Natl. Acad. Sci. U.S.A.* **86**, 3949–3952.

(305) Au, K. G., Cabrera, M., Miller, J. H., and Modrich, P. (1988) *Escherichia coli mutY* gene product is required for specific A·G → C·G mismatch correction. *Proc. Natl. Acad. Sci. U.S.A.* **85**, 9163–9166.

(306) Lu, A.-L., and Chang, D.-Y. (1988) A novel nucleotide excision repair for the conversion of an A/G mismatch for C/G base pair in *E. coli*. *Cell* **54**, 805–812.

(307) Razin, A., and Riggs, A. (1980) DNA methylation and gene function. *Science* **210**, 604–610.

(308) May, M. S., and Hattman, S. (1975) Deoxyribonucleic acid-cytosine methylation by host- and plasmid-controlled enzymes. *J. Bacteriol.* **122**, 129–138.

(309) Razin, A., Cedar, H., and Riggs, A. D. (1984) *DNA Methylation: Biochemistry and Biological Significance*, Springer, New York.

(310) Barr, F. G., Rajagopalan, S., MacArthur, C. A., and Lieberman, M. W. (1986) Genomic hypomethylation and far 5′ sequence alterations are associated with carcinogen-induced activation of the hamster thymidine kinase gene. *Mol. Cell. Biol.* **6**, 3023–3033.

(311) Brown, T. C., and Jiricny, J. (1987) A specific mismatch repair event protects mammalian cells from loss of 5-methylcytosine. *Cell* **50**, 945–950.

(312) Lindahl, T. (1982) DNA repair enzymes. *Annu. Rev. Biochem.* **51**, 61–87.

(313) Loeb, L. A., and Preston, B. D. (1986) Mutagenesis by apurinic/apyrimidinic sites. *Annu. Rev. Genet.* **20**, 201–230.

(314) Lieb, M. (1983) Specific mismatch correction in bacteriophage λ crosses by very short patch repair. *Mol. Gen. Genet.* **191**, 118–125.

(315) Lieb, M. (1985) Recombination in the λ repressor gene: evidence that very short patch (VSP) mismatch correction restores a specific sequence. *Mol. Gen. Genet.* **199**, 465–470.

(316) Jones, M., Wagner, R., and Radman, M. (1987) Mismatch repair of deaminated 5-methyl-cytosine. *J. Mol. Biol.* **194**, 155–159.

(317) Zell, R., and Fritz, H.-J. (1987) DNA mismatch-repair in *Escherichia coli* counteracting the hydrolytic deamination of 5-methyl-cytosine residues. *EMBO J.* **6**, 1809–1815.

(318) Lieb, M. (1987) Bacterial genes *mutL*, *mutS*, and *dcm* participate in repair of mismatches at 5-methylcytosine sites. *J. Bacteriol.* **169**, 5241–5246.

(319) Lieb, M., and Bhagway, A. S. (1988) Very short patch mismatch repair activity associated with gene *dcm* is not conferred by a plasmid coding for EcoRII methylase. *J. Bacteriol.* **170**, 4967–4968.

(320) Hare, J. T., and Taylor, J. H. (1985) One role for DNA methylation in vertebrate cells in strand discrimination in mismatch repair. *Proc. Natl. Acad. Sci.* **82**, 7350–7354.

(321) Jiricny, J., Hughes, M., Corman, N., and Rudkin, B. B. (1988) A human 200-kDa protein binds selectively to DNA fragments containing G·T mismatches. *Proc. Natl. Acad. Sci. U.S.A.* **85**, 8860–8864. (a) Wiebauer, K., and Jiricny, J. (1989) *In vitro* correction of G-T mispairs to G·C pairs in nuclear extracts from human cells. *Nature* **339**, 234–236.

(322) Walker, G. C. (1984) Mutagenesis and inducible responses to deoxyribonucleic acid damage in *Escherichia coli*. *Microbiol. Rev.* **48**, 60–93. (a) Laspia, M. F., and Wallace, S. S. (1989) SOS processing of unique oxidative DNA damages in *Escherichia coli*. *J. Mol. Biol.* **207**, 53–60.

(323) Sedgwick, S. G. (1986) Inducible DNA repair in microbes. *Microbiol. Rev.* **3**, 76–83.

(324) Little, J. W. (1984) Autodigestion of *lexA* and phage λ repressors. *Proc. Natl. Acad. Sci. U.S.A.* **81**, 1375–1379.

(325) Slilaty, A. N., and Little, J. W. (1987) Lysine-156 and serine-119 are required for LexA repressor cleavage: a possible mechanism. *Proc. Natl. Acad. Sci. U.S.A.* **84**, 3987–3991.

(326) Granger-Schnarr, M., Schnarr, M., and van Sluis, C. A. (1986) *In vitro* study of the interaction of the LexA repressor and the UvrC protein with a *uvrC* regulatory region. *FEBS Lett.* **198**, 61–65. (a) Walker, G. C. (1985) Inducible DNA repair systems. *Annu. Rev. Biochem.* **54**, 425–457.

(327) Ihara, M., Yamamoto, K., and Ohnishi, T. (1987) Induction of *phr* gene expression by irradiation of ultraviolet light in *Escherichia coli*. *Mol. Gen. Genet.* **209**, 200–202.

(328) Payne, N., and Sancar, A. (1989) The LexA protein does not bind specifically to the two SOS box-like sequences immediately 5′ to the *phr* gene. *Mutat. Res.* (in press).

(329) Perrino, F. W., Rein, D. C., Bobst, A. M., and Meyer, R. R. (1987) The relative rate of synthesis and levels of single-stranded DNA binding protein during induction of SOS repair in *Escherichia coli*. *Mol. Gen. Genet.* **209**, 612–614.

(330) Shinagawa, H., Iwasaki, h., Kato, T., and Nakata, A. (1988) RecA protein-dependent cleavage of UmuD protein and SOS mutagenesis. *Proc. Natl. Acad. Sci. U.S.A.* **85**, 1806–1810.

(331) Burckhadt, S. E., Woodgate, R., Scheurmann, R. H., and Echols, H. (1988) The UmuD mutagenesis protein of *E. coli*: overproduction, purification and cleavage by RecA. *Proc. Natl. Acad. Sci. U.S.A.* **85**, 1811–1815.

(332) Nohmi, T., Battista, J. R., Dodson, L. A., and Walker, G. C. (1988) RecA-mediated cleavage activates UmuD for mutagenesis: mechanistic relationship between transcriptional derepression and posttranslational activation. *Proc. Natl. Acad. Sci. U.S.A.* **85**, 1816–1820.

(333) Bonner, C. A., Randall, S. K., Rayssiguier, C., Radman, M., Eritja, R., Kaplan, B. E., McEntee, K., and Goodman, M. F. (1988) Purification and characterization of an inducible *Escherichia coli* DNA polymerase capable of insertion and bypass at abasic lesions in DNA. *J. Biol. Chem.* **263**, 18946–18952.

(334) Golub, E., Bailone, A., and Devoret, R. (1988) A gene encoding an SOS inhibitor is present in different conjugative plasmids. *J. Bacteriol.* **170**, 4392–4394.

(335) Samson, L., and Cairns, J. (1977) A new pathway for DNA repair in *Escherichia coli*. *Nature* (*London*) **267**, 281–283.

(336) Hassan, H. M., and Fridovich, I. (1979) Intracellular production of superoxide radical and of hydrogen peroxide by redox active compounds. *Arch. Biochem. Biophys.* **196**, 385–395.

(337) Richter, C., Park, J.-W., and Ames, B. N. (1988) Normal oxidative damage to mitochondrial and nuclear DNA is extensive. *Proc. Natl. Acad. Sci. U.S.A.* **85**, 6465–6467.

(338) Demple, B., and Holbrook, J. (1983) Inducible repair of oxidative DNA damage in *E. coli*. *Nature* (*London*) **304**, 446–448.

(339) Imlay, J. A., and Linn, S. (1986) Bimodal pattern of killing of DNA repair-defective of anoxically grown *E. coli* by hydrogen peroxide. *J. Bacteriol.* **166**, 797–799.

(340) Bochner, B. R., Lee, P. C., Wilson, S. W., Cutler, C. W., and Ames, B. N. (1984) ApppA and related adenylated nucleotides are synthesized as a consequence of oxidation stress. *Cell* **37**, 225–232.

(341) Morgan, R. W., Christman, M. F., Jacobson, F. S., Storz, G., and Ames, B. N. (1986) Hydrogen peroxide-inducible proteins in *Salmonella typhimurium* overlap with heat shock and other stress proteins. *Proc. Natl. Acad. Sci. U.S.A.* **83**, 8059–8063.

(342) Privalle, C. T., and Fridovich, I. (1987) Induction of superoxide dismutase in *Escherichia coli* by heat shock. *Proc. Natl. Acad. Sci. U.S.A.* **84**, 2723–2726.

(343) Sammartano, L. J., Tuveson, R. W., Davenport, R. (1986) control of sensitvity to inactivation by $H_2O_2$ and broad-spectrum near-UV radiation by the *Escherichia coli katF* locus. *J. Bacteriol.* **168**, 13–21.

(344) Kramer, G. F., and Ames, B. N. (1987) Oxidative mechanisms of toxicity of low-intensity near-UV light in *Salmonella typhimurium*. *J. Bacteriol.* **169**, 2259–2266.

(345) Christman, M. F., Morgan, R. W., Jacobson, F. S., and Ames, B. N. (1985) Positive control of a regulon for defenses against oxidative stress and some heat-shock proteins in *Salmonella typhimurium*. *Cell* **41**, 753–762.

(346) Jacobson, F. S., Morgan, R. W., Christman, M. F., and Ames, B. N. (1989) An Alkyl hydroperoxide reductase from *Salmonella typhimurium* involved in the defense of DNA against oxidative damage. *J. Biol. Chem.* **264**, 1488–1496. (a) Christman, M. F., Storz, G., and Ames, B. N. (1989) OxyR, a positive regulator of hydrogen peroxide-inducible genes in *Escherichia coli* and *Salmonella typhimurium*, is homologous to a family of bacterial

regulatory proteins. *Proc. Natl. Acad. Sci. U.S.A.* **86**, 3484–3488.

(347) Greenberg, J. T., and Demple, B. (1988) Overproduction of peroxide-scavenging enzymes in *Escherichia coli* supresses spontaneous mutagenesis and sensitivity to redox-cycling agents in oxyR- mutants. *EMBO J.* **7**, 2611–2617. (a) Sak, B. D., Eisenstark, A., and Touati, D. (1989) Exonuclease III and the catalase hydroperoxidase II in *Escherichia coli* are both regulated by the *KatF* gene product. *Proc. Natl. Acad. Sci. U.S.A.* **86**, 3271–3275.

(348) Van Bogelen, R. A., Kelley, P. M., and Neidhardt, F. C. (1987) Differential induction of heat shock, SOS, and oxidation stress regulons and accumulation of nucleotides in *Escherichia coli*. *J. Bacteriol.* **169**, 26–32.

(349) Walkup, L. K. B., and Kogoma, T. (1989) *Escherichia coli* proteins inducible by oxidative stress mediated by the superoxide radical. *J. Bacteriol.* **171**, 1476–1484.

(350) Mitchel, R. E. J., and Morrison, D. P. (1987) Inducible DNA repair systems in yeast: competition for lesions. *Mutat. Res.* **183**, 149–159.

(351) Burtscher, H. J., Cooper, A. J., and Couto, L. (1988) Cellular responses to DNA damage in the yeast *Saccharomyces cerevisiae*. *Mutat. Res.* **194**, 1–8.

(352) Ruby, S. W., and Szostak, J. W. (1985) Specific *Saccharomyces* genes are expressed in response to DNA-damaging agents. *Mol. Cell. Biol.* **5**, 75–84.

(353) McClanahan, T., and McEntee, K. (1986) DNA damage and heat shock dually regulate genes in *Saccharomyces cerevisiae*. *Mol. Cell. Biol.* **6**, 90–96.

(354) Maga, J. A., McClanahan, T. A., and McEntee, K. (1986) Transcriptional regulation of DDR genes in different *rad* mutant strains of *S. cerevisiae*. *Mol. Gen. Genet.* **205**, 276–284.

(355) Elledge, S. J., and Davis, R. W. (1987) Identification and isolation of the gene encoding the small subunit of ribonucleotide reductase from *S. cerevisiae*: DNA damage-inducible gene required for mitotic viability. *Mol. Cell. Biol.* **7**, 2783–2793.

(356) Hurd, H. K., Roberts, C. W., and Roberts, J. W. (1987) Identification of the gene for the yeast ribonucleotide reductase small subunit and its inducibility by methyl methanesulfonate. *Mol. Cell. Biol.* **7**, 3673–3677.

(357) Johnson, A. L., Barker, D. G., and Johnson, L. H. (1986) Induction of yeast DNA ligase in exponential and stationary phase cultures in response to DNA damaging agents. *Curr. Genet.* **11**, 107–112.

(358) Rolfe, M., Spanos, A., and Banks, G. (1986) Induction of yeast Ty element transcription by ultraviolet light. *Nature (London)* **319**, 339–340.

(359) Schulze-Lefert, P., Dangl, J. L., Becker-Andre, M., Hahlbrock, K., and Schulz, W. (1989) Inducible *in vivo* DNA footprints define sequences necessary for UV light activation of the parsley chalcone synthase gene. *EMBO J.* **8**, 651–656.

(360) Dasgupta, U., and Summers, W. C. (1978) UV-reactivation of herpes simplex virus is mutagenic and inducible in mammalian cells. *Proc. Natl. Acad. Sci. U.S.A.* **75**, 2378–2381.

(361) Sarkar, S. N., Dasgupta, U. B., and Summers, W. C. (1984) Error-prone mutagenesis detected in mammalian cells by a shuttle vector containing the *supF* gene of *Escherichia coli*. *Mol. Cell. Biol.* **4**, 2227–2230.

(362) Schorpp. M., Mallick, U., Rahmsdorf, H. J., and Herrlich, P. (1984) UV induced extracellular factor from human fibroblasts communicates the UV response to non-irradiated cells. *Cell* **37**, 861–868.

(363) Stein, B., Rahmsdorf. H. J., Schunthal, A., Buscher, M., Ponta, H., and Herrlich, P. (1988) The UV induced signal transduction pathway to specific genes. *UCLA Symp. Mol. Cell. Biol.* **83**, 557–570.

(364) Kartasova, T., and van de Putte, P. (1988) Isolation, characterization, and UV-stimulated expression of two families of genes encoding polypeptides of related structure in human epidermal keratinocytes. *Mol. Cell. Biol.* **8**, 2195–2203.

(365) Fornace, A. J., Jr., Schalch, H., and Alamo, I., Jr. (1988) Coordinate induction of metallothioneins I and II in rodent cells by UV irradiation. *Mol. Cell. Biol.* **8**, 4716–4720.

(366) Glazer, P. M., Greggio, N. A., Metherall, J. E., and Summers, W. C. (1989) UV-induced DNA-binding proteins in human cells. *Proc. Natl. Acad. Sci. U.S.A.* **86**, 1163–1167.

(367) Fornace, A. J., Jr., Alamo, I., Jr., and Hollander, M. C. (1988) DNA damage-inducible transcripts in mammalian cells. *Proc. Natl. Acad. Sci. U.S.A.* **85**, 8800–8804.

# Physical Methods

# Chapter 14

# Applications of NMR Spectroscopy to Studies of Reactive Intermediates and Their Interactions with Nucleic Acids

Thomas M. Harris,* Michael P. Stone,* and Constance M. Harris

*Department of Chemistry and Center in Molecular Toxicology, Vanderbilt University, Nashville, Tennessee 37235*

Reprinted from *Chemical Research in Toxicology*, Vol. 1, No. 2, March/April, 1988

The origins of NMR[1] spectroscopy lie in the work of Bloch and Hahn in the late 1940s. The value of the technique quickly became evident and within a decade spectrometers were commercially available. The impact of NMR was initially limited to the chemical sciences; chemists turned the resonance phenomenon into a technique with enormous value for structure determination. The use of NMR in the biological sciences including toxicology was minimal for some time, primarily because major developments in theory and instrumentation were required before NMR could be employed to attack biological problems. The large size, scarcity and spectral complexity of many molecules of biological interest still presents constant problems in the application of NMR methods. Nevertheless, increasingly frequent mention of NMR in biochemical journals testifies to the growing importance of this technique in elucidation of complex macromolecular structures, conformations, molecular dynamics, and reaction mechanisms.

One of the key advances has been the development of pulsed spectrometers. Early spectrometers employed continuous wave (CW) techniques in which the spectrum was slowly scanned, i.e., ~5 min, with a low-power transmitter. Advances in computer technology and radio

frequency electronics have made possible pulsed spectrometers in which the spectrum is created by a high-power radio frequency pulse of a few $\mu$seconds duration. The effect is to excite simultaneously all the protons in the sample; the individual nuclei then relax to the ground state by reemitting radio frequency signals, each at its characteristic frequency. The resulting cacophony of RF signals, called a free induction decay (or fid), is converted to a frequency domain NMR spectrum by Fourier transformation. In this process the complex wave pattern of the fid is separated into its individual frequencies which become the signals in the resulting NMR spectrum; the intensities and shapes of the signals are determined by the intensities and rates of decay of the individual frequencies in the fid. An important advantage of the FT-NMR technique is that a spectrum can be acquired rapidly, typically in ~2–3 seconds for high-resolution $^1$H spectra. A further advantage is that the method is well-suited for signal averaging; multiple fids can be added together to improve signal-to-noise in samples of low concentration.

A second major development has been in magnet technology. The frequencies at which the signals are observed are directly proportional to the strength of the magnetic field. For protons and most other nuclei, two major benefits accrue from increasing the observation frequency: increased sensitivity and increased spectral dispersion. Iron core permanent magnets and electromagnets are limited in field strength to approximately 2.3 T. Thus, 100 MHz was the highest accessible frequency for proton spectroscopy until the development of superconducting magnets. With superconducting magnets, spectrometers with operating frequencies up to 600 MHz for proton spectroscopy are now commercially available and those in the 300–500-MHz range are commonplace in university and industrial laboratories. Sensitivity improvements have

---

[1] Abbreviations: NMR, nuclear magnetic resonance; FT, Fourier transformation; COSY, two-dimensional *J*-correlated spectroscopy; 1D, one-dimensional; 2D, two dimensional; NOESY, two-dimensional nuclear Overhauser enhancement spectroscopy; DNA, deoxyribonucleic acid; NOE, nuclear Overhauser enhancement; CD, circular dichroism; O6meG, *O*$^6$-methylguanine; O4meT, *O*$^4$-methylthymine; AAF, *N*-acetyl-2-amino-fluorene. The internucleotide linkage in oligodeoxynucleotides has been eliminated for brevity, except for the references to dinucleotides; there are no terminal phosphate groups except where indicated. Several carcinogen–DNA adducts are referred to in the literature with multiple nomenclatures; we have attempted to include structures where there is potential for confusion.

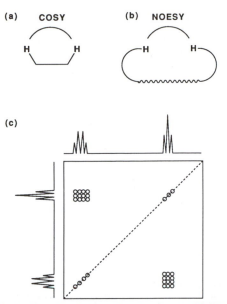

**Figure 1.** (a) Coupling interactions leading to COSY spectra; (b) cross relaxation leading to NOESY spectra; (c) a hypothetical COSY spectrum showing coupling between $CH_2$ and $CH_3$ in an ethyl group.

made it possible to acquire $^1H$ spectra on small compounds (e.g., carcinogen–nucleoside adducts) when only a few micrograms of material is available.

A third major advance of crucial importance to biochemical NMR has been the development of the theory and application of multiple pulse techniques which lead to two-dimensional spectra. These techniques are an outgrowth of the marriage of computers to the instruments; computers not only store data and carry out Fourier transformations but also control data acquisition at the microsecond time level. In the current generation of spectrometers the computer has full control over the transmitter, decoupler, and receiver and can create transmitter and decoupler pulses in which the frequency, duration, intensity, and phase are controlled with exactitude. Multiple pulses with different characteristics can be produced at precisely determined intervals. The existence of these sophisticated pulse programmers has made possible the development of two-dimensional spectroscopy.

In 2D spectroscopy, spectra are created in which the two axes represent spectral characteristics of the sample; most commonly each of them is chemical shift. The spectra are in actuality three dimensional with the third dimension being intensity. The most common method of displaying the data is as a contour plot in which the contours represent the intensity of the signals. The simplest of the 2D techniques is proton correlation spectroscopy (i.e., COSY spectra) which reveals scalar coupling between signals and hence through-bond connectivities (Figure 1a). In COSY spectra the chemical shifts are displayed along both axes and the normal 1D spectrum appears along the diagonal of the spectrum (where the chemical shifts are identical for the two axes) as shown in Figure 1c. Any protons which are mutually coupled will produce additional signals off the diagonal at the chemical shifts of their coupling partners. Many variants of this technique have been developed. A second very important class of 2D spectra are NOESY spectra, which are based on nuclear Overhauser effects. Their appearance is similar to COSY spectra but the cross-peaks in NOESY spectra are determined by dipolar coupling interactions between nuclei. Dipolar interactions will be seen between protons which lie within

5 Å of one another irrespective of the presence of chemical bonds (Figure 1b). This is a powerful technique for establishing conformational details of structure and for mapping substrate–receptor interactions. For further discussion of these valuable techniques the reader is referred to several excellent monographs which have appeared recently (1–3).

NMR techniques involving other nuclei have found fewer applications in the biological sciences because to varying degrees the other nuclei, with the exception of $^{19}F$, have significantly decreased sensitivity or occur in low natural abundance. However, $^{31}P$ and $^{19}F$ NMR and, with isotopically enriched samples, $^{13}C$, $^2H$, and $^{15}N$ NMR have all played important roles in some cases. The usefulness of low sensitivity nuclei such as $^{15}N$ is being enhanced by 2D methods in which the spectrum of the heteroatom is observed via the high sensitivity proton spectrum.

This survey of the use of NMR spectroscopy in the area of toxicology will concentrate on interactions of carcinogens with nucleic acids. Many of the most interesting carcinogens are intrinsically difficult to study because of high reactivity and low solubility. It should be noted that in many cases the methods employed for analysis of carcinogen/DNA interactions parallel those being used to study drug/DNA interactions.

Four different areas in which NMR has been employed for the study of carcinogen interactions with nucleic acids will be discussed. The first is the structural characterization of ultimate carcinogens. The second is characterization of carcinogen–nucleoside adducts. Usually these adducts have been formed by reactions on high molecular weight DNA and analyzed after excision from the phosphate backbone. These studies define the site(s) on the DNA at which reaction has occurred and in some cases reveal configurational details of the reaction. The third section involves studies of the conformation of adducts formed with mono- and dinucleotides. The final section involves similar investigations employing oligonucleotides; these studies, although more complex, offer far more insight into the actual structure of carcinogen adducts with polymeric DNA.

## I. Structures of Ultimate Carcinogens

Many compounds have now been identified as mutagens by the Ames assay; some of these have also been shown to be carcinogens, and the remainder are suspected of being carcinogens. The majority of these compounds act by electrophilic attack on DNA, causing structural alterations which fail to be removed by repair mechanisms and are subsequently replicated. Some of the compounds, such as dimethyl sulfate, act directly, but many of them are inert until activated by enzyme-mediated processes which exist in living organisms to detoxify xenobiotics. In many cases this activation involves oxidation by microsomal cytochrome P-450 or by flavin-containing monooxygenases. A third oxidative pathway involves prostaglandin synthetase-dependent peroxidative activation. Not all activation mechanisms are oxidative; condensation of 1,2-dihalides with glutathione gives sulfur mustards which readily alkylate DNA. NMR has played a very important role in the assignment of structures of many of the reactive species.

**Reactive Epoxides.** Epoxides formed by oxidation of olefins represent a particularly prominent class of ultimate carcinogens. In many cases the unactivated olefins have little or no toxicity. The epoxides of polycyclic aromatic hydrocarbons provide a prime example where activation involves epoxidation, hydration to the diol, and then

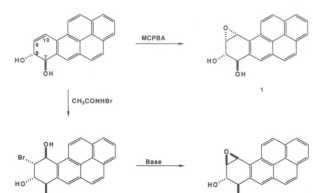

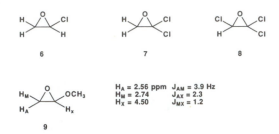

**Figure 2.** Synthesis of epimeric diol epoxides 1 and 2 of benzo[a]pyrene.

**Figure 3.** Valence bond tautomerism between benzene oxide (4) and oxepin (5).

$H_A = 2.56$ ppm    $J_{AM} = 3.9$ Hz
$H_M = 2.74$         $J_{AX} = 2.3$
$H_X = 4.50$         $J_{MX} = 1.2$

**Figure 4.** Epoxides of vinyl chloride, vinylidene chloride, trichloroethylene, and methyl vinyl ether.

further epoxidation. 7β,8α-Dihydroxy-9α,10α-epoxy-7,8,9,10-tetrahydrobenzo[a]pyrene (1) was prepared (4) by treatment of 7β,8α-dihydroxy-7,8-dihydrobenzo[a]pyrene with m-chloroperbenzoic acid; the diastereomeric 9β,10β-epoxide 2 was prepared by addition of HOBr to 7,8-dihydroxy-7,8-dihydrobenzo[a]pyrene to form bromo triol 3 followed by dehydrohalogenation (Figure 2). The C10 hydroxyl group in 3 assumes a pseudoaxial position to avoid severe steric interactions in the bay region. The relative configurations of the other substituents in 3 were established from the NMR spectrum; H7 and H8 are pseudoaxial as reflected by the strong coupling ($J$ = 10.0 Hz) between them. In diol epoxide 1, a larger coupling between H7 and H8 (9.0 vs 6.0 Hz) is an indication that the hydroxyl groups of 1 are pseudoequatorial whereas those of 2 are pseudoaxial. Trimethylsilylation of 2 causes substantial changes in conformation such that the hydroxyl groups shift to more pseudoequatorial positions with the vicinal coupling constant increasing to 9.0 Hz. Hydrogen bonding between the 7-hydroxyl group and the epoxide in 2 has been postulated to account for the difference in conformation of the two compounds. This conclusion is of relevance to the reactivity of the two compounds; 2 is attacked by nucleophiles ∼2 orders of magnitude faster than 1, presumably because of anchimeric assistance by the 7-hydroxyl group.

The toxicity of benzene, the simplest of the aromatic hydrocarbons, and the requirement that it must undergo metabolic activation are well recognized, but the nature of the ultimate carcinogen remains obscure; possibilities include the epoxide and benzoquinone. The chemistry of benzene oxide (4) and related arene oxides has been studied in great detail with contributions from many laboratories (For reviews, see ref 5 and 6). The first synthesis (7, 8) involved bromination of 1,3-cyclohexadiene followed by epoxidation with m-chloroperbenzoic acid and finally elimination of HBr with sodium methoxide. Later syntheses involved thermolysis of 3-oxaquadricyclane and thermal (and photochemical) isomerization of the monoepoxide of Dewar benzene. The pure product is a reasonably stable liquid which is rearranged to phenol by acid catalysts. Reactions of the product gave confusing results with some reactions supporting the benzene oxide structure, others an isomer, oxepin 5. NMR studies revealed that the compound was a rapidly equilibrating mixture of the two species (Figure 3). At room temperature the spectrum was a weighted average of the spectra of the individual structures. The chemical shifts of the three types of protons were sensitive to solvent and temperature reflecting changes in the mole ratio of the benzene oxide

and oxepin. A variable-temperature study carried out on a 60-MHz NMR spectrometer showed that equilibration slowed sufficiently at low temperature to allow resolution of the spectra of the individual species. The equilibrium constant for the benzene oxide–oxepin interconversion could be established below the coalescence temperature (−113 °C) by the ratio of the signal areas for the two species and above that temperature by the resonance position of the α-proton signal.

The isolation of benzene oxide came after many unsuccessful attempts because a synthetic strategy was required which would generate the epoxide under conditions where the compound is stable; the epoxide although presumably formed by cytochrome P-450 oxidation of benzene has never been detected in biological systems. There has been considerable interest in the synthesis and characterization of other reactive epoxides believed to be oxidative metabolites of olefins. The epoxides of chlorinated ethylenes are a case in point (Figure 4). The epoxide (6) of vinyl chloride has been prepared (9, 10). Its structure was indicated by the proton spectrum, which shows a doublet of doublets for the C1 proton at 5.00 ppm with coupling constants of 1.5 and 2.5 Hz and signals at 2.96 ppm for the two protons on the other carbon. The epoxide (7) of vinylidene chloride gives a singlet at 3.2 ppm in the $^1$H spectrum. A $^{13}$C spectrum of the epoxide has been obtained at −50 °C in $CD_2Cl_2$ at 75 MHz (11). A signal was observed at 58.32 ppm for the $CH_2$ group; the $CCl_2$ group was not observed due to the low concentration of the sample and inherently poor sensitivity of the unprotonated carbon caused by its slow relaxation and lack of nuclear Overhauser enhancement. The epoxide (8) of trichloroethylene gives a $^1$H signal at 5.31 ppm (11). All three of these epoxides are several orders of magnitude more labile than ethylene oxide (10).

The epoxides of enol ethers represent another very labile class of reactive epoxides. The epoxide (9) of methyl vinyl ether was first prepared by the reaction of diazomethane with methyl formate (13) and later by reaction of ground-state ($^3$P) oxygen atoms with the vinyl ether (14). The compound is sufficiently stable that NMR spectra can be run in $CDCl_3$ at ambient temperature but it decomposes during short-path distillation at atmospheric pressure. The NMR spectrum was assigned as shown in Figure 4 and is consistent with the proposed oxirane structure.

The pursuit of highly electrophilic carcinogens continues with increasing emphasis on labile species. The epoxide (11) of aflatoxin $B_1$ (10) is one of the putative carcinogens which has eluded detection. Attempts to prepare it by

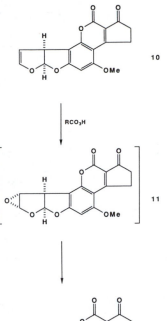

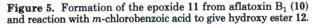

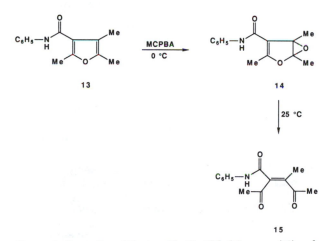

**Figure 5.** Formation of the epoxide **11** from aflatoxin B$_1$ (**10**) and reaction with *m*-chlorobenzoic acid to give hydroxy ester **12**.

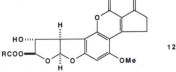

**Figure 6.** Formation of the epoxide **14** of Methfuroxam (**13**) and thermal rearrangement to diketone **15**.

oxidation with peroxy acids have failed due to the ease with which the epoxide is reopened by the carboxylic acid by-product to give the 8-acyloxy-9-hydroxy derivative **12** (Figure 5) (*15–17*). Possibly other reaction conditions would permit the isolation or detection of **11**; recently Ruzo et al. (*18*) prepared the 4,5-epoxide (**14**) of the fungicide Methfuroxam (**13**) by reaction with *m*-chloroperbenzoic acid at 0 °C (Figure 6). The epoxide was detected by $^1$H NMR. The epoxide could not be detected in reactions carried out at 25 °C due to rapid rearrangement to diketone **15**.

Havel and Chan (*14*) using ground-state oxygen atoms have successfully prepared the epoxide of 2,3-dihydropyran in sufficient quantity and purity to obtain a $^1$H NMR spectrum but the corresponding synthesis of the epoxide of 2,3-dihydrofuran failed, apparently due to facile fragmentation of the diradical precursor of the epoxide.

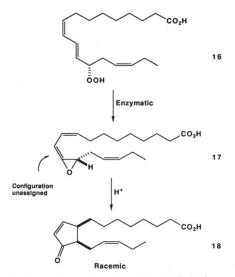

**Figure 7.** Enzymatic conversion of the 13(*S*)-hydroperoxide **16** of linolenic acid to allene oxide **17** and acid-catalyzed rearrangement of **17** to racemic *cis*-12-oxophytodienoic acid (**18**).

The prospect still exists for preparation of the epoxide of aflatoxin B$_1$ but a synthetic strategy is required which will yield the epoxide faster than it decomposes, an isolation scheme which segregates the epoxide from the other components of the reaction mixture, and finally storage conditions which permit the epoxide to be held long enough to obtain spectra and carry out chemical characterization. Recently, these problems were overcome for isolation of another very labile epoxide. The formation of *cis*-12-oxophytodienoic acid (**18**) from the 13(*S*)-hydroperoxide (**16**) of α-linolenic acid by plants had been postulated to occur via an allene oxide (**17**) (*19–21*). Isolation of enzymically prepared epoxide has been achieved by employing 5-s incubations with a very active enzyme preparation at 0 °C followed by rapid extraction into organic solvent, esterification with CH$_2$N$_2$, and HPLC on a straight-phase (silica gel) column below 0 °C (Figure 7) (*22*). Although the epoxide was completely destroyed on standing in the 0 °C incubation mixture for as little as 60 s, the ester was sufficiently stable in hexane-$d_{14}$ solution that $^1$H COSY spectra could be obtained at −40 °C; assignment of structure was made primarily on the basis of the $^1$H NMR spectra.

**Reactive Species Other than Epoxides.** Electrophiles other than epoxides can also alter DNA structure. Many alkyl halides probably alkylate directly; however, ethylene dibromide (EDB) acts by a circuitous mechanism of initially forming a β-bromoethyl adduct with glutathione by an S$_N$2 process catalyzed by glutathione *S*-transferase. The adduct then undergoes internal displacement to give an episulfonium ion which is the putative ultimate carcinogen. The analogous episulfonium ion derived from cysteine has recently been synthesized from *S*-(2-hydroxyethyl)cysteine by treatment with "super acid" (trifluoromethanesulfonic acid diluted with SO$_2$) (Figure 8) and is sufficiently stable in that medium for $^1$H and $^{13}$C spectra to be obtained (*23*). In the $^{13}$C spectrum the two methylene carbon signals are nonequivalent due to the chirality of the amino acid *and* the nonplanarity of the trisubstituted sulfur atom. The episulfonium ion prepared from *S*-(2-hydroxyethyl)cysteine-2,2-$d_2$ gave spectra showing that scrambling of the CD$_2$ group had occurred between the two carbon atoms of the thiiranium ion. The $^1$H spectrum showed splitting of the signal for the thiiranium protons with an overall reduction in area of 50%. It should be noted that the four protons

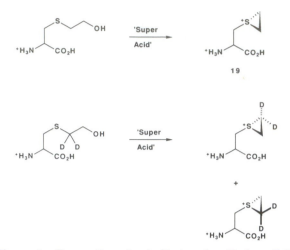

**Figure 8.** Preparation of episulfonium ion **19** from S-(2-hydroxyethyl)cysteine.

in the thiiranium ion ring are nonequivalent due to the asymmetric environment. The situation may be more complex than it at first appears; McManus and co-workers (24) have found evidence of facile ion-pair return during solvolysis of specifically deuterated 2-(methylthio)ethyl dinitrophenylate in 50% water/acetone and 50% water/acetonitrile. Dohn and Casida (23) question whether return could occur from episulfonium ions, which should react preferentially with water due to their high electrophilicity; they suggest that a trigonal bipyramidal sulfurane may be formed initially (in a reversible process) followed by isomerization to the episulfonium ion.

## II. Carcinogen–Nucleoside Adducts

Evidence for the sites of linkage between carcinogens and polymeric DNA has been derived from studies in which polymeric DNA was reacted with specific carcinogen molecules in vitro or in vivo, followed by controlled enzymatic and/or chemical degradation to release adducts with mononucleosides or with the heterocyclic bases. Structural assignments have then been made by NMR and mass spectrometry. The parameters of interest are the identity of the nucleotide base, the site of reaction on the nucleotide, the site of reaction in the electrophile, and the stereochemistry of the linkage in the resulting adduct. The risk in this process is that some adducts may be destroyed or undergo structural rearrangements during the degradation procedure. For convenience in discussing spectra, structures and numbering of the common DNA and RNA bases and sugars are shown in Figure 9.

**Polycyclic Aromatic Hydrocarbon Adducts.** The nucleic acid adducts of epoxides derived from polycyclic aromatic hydrocarbons have been studied intensively. Initially attention was focused on K-region epoxides because of the belief that the K region was the primary locus of the mutagenic and carcinogenic activity. The reaction of racemic 7,12-dimethylbenz[a]anthracene 5,6-oxide (**20**) with polyguanylic acid followed by hydrolysis to the nucleoside–hydrocarbon stage gave four adducts (**21–24**) as shown in Figure 10 (25, 26). Deglycosylation of the four compounds gave two pairs of enantiomers. The site of attachment to the guanine nucleus was deduced to be the 2-amino group from the mass spectra. Analysis of chemical shifts for H5 and H6 in the hydrocarbon portion of the adducts showed one pair to be linked via H5 and the other via H6. Furthermore the spectra provided stereochemical details; both pairs of adducts had been formed by $S_N2$ opening of the epoxide to give trans adducts. It is note-

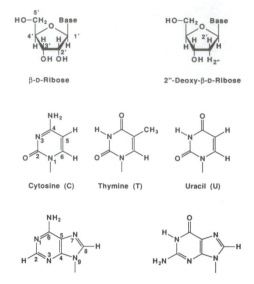

**Figure 9.** Structures and numbering of the sugars and bases of DNA and RNA.

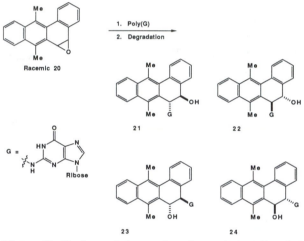

**Figure 10.** Products of the reaction of racemic 7,12-dimethylbenz[a]anthracene 5,6-epoxide (**20**) with polyguanylic acid.

worthy that none of these products corresponded to the in vivo adducts.

Subsequently, guanosine was treated with epoxide **20**; six new products were isolated. Four were diastereomeric pairs of cis and trans adducts resulting from attack of guanosine C8 on the 5-position of the two enantiomers of the racemic epoxide (27). The remaining two were unusual adducts resulting from attack by the 2′-ribosyl hydroxyl group at C5 and C6 of the arene oxide with trans opening of the epoxide in each case (28). It is noteworthy that only the 5R,6S enantiomer of the racemic epoxide participated in formation of these ribosyl adducts. Three of the six products (**25–27**), all arising from the 5R,6S epoxide, corresponded to in vivo adducts (Figure 11).

Other investigations pointed to the diol epoxides as being more important physiological electrophiles (29, 30). A major in vivo adduct of [³H]benzo[a]pyrene with RNA was found to be identical with one of the products of the reaction of racemic diol epoxide 1 with polyguanylic acid after both had been degraded to the nucleoside stage (Figure 12) (31). The structure of adduct **28** was determined from the ¹H NMR spectrum. Analysis of the NMR spectrum was complicated by overlap of signals for several

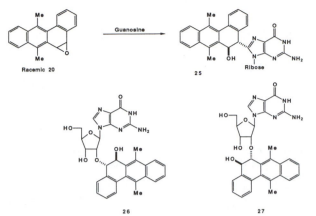

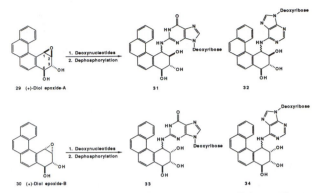

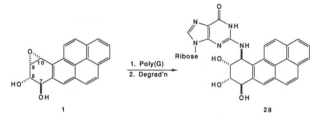

Figure 11. Products of the reaction of racemic 7,12-dimethyl-benz[a]anthracene 5,6-epoxide (20) with guanosine which have also been found in vivo (27, 28).

Figure 13. Adducts formed by the reaction of diol epoxides 29 and 30 of benzo[c]phenanthrene with deoxyadenylic and deoxyguanylic acids.

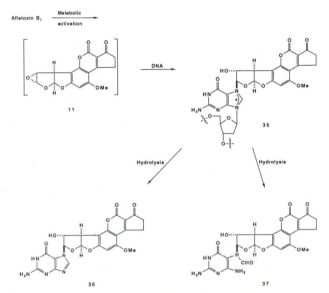

Figure 12. Formation of adduct 28 from the reaction of one of the enantiomers of diol epoxide 1 with polyguanylic acid.

Figure 14. DNA adducts 35–37 formed from aflatoxin epoxide (11) after metabolic activation of aflatoxin B₁.

of the key protons with signals for residual protons in the D₂O and CD₃OD solvent mixture. The spectrum was improved by suppression of the solvent signals by a preliminary 180° pulse, taking advantage of the fact that the solvent protons relax more slowly than those of the adduct. However, further manipulation of the chemical shifts of the signals of interest versus the conflicting solvent signals was required and was achieved by addition of benzene-$d_6$ and by elevation of the temperature of the sample. Spectral analysis indicated that trans opening of the epoxide had occurred by attack at C10 of the epoxide. The site of reaction of guanine was assigned to the amino group at C2 from the NMR spectrum. The spectrum when measured in DMSO-$d_6$ showed an exchangeable doublet ($J = 8$ Hz) for the remaining NH proton which collapsed on irradiation of H10. The absolute configuration was assigned from the circular dichroism spectrum of the dibenzoate derivative (32).

The racemic diol epoxide of benzo[a]pyrene having the epimeric orientation of the epoxide 2 was also found to react with the 2-amino group of polyguanylic acid (33). Degradation of the poly(G) including cleavage of the ribose left two arene–guanine adducts, both linked from the 2-amino group to C10. The major isomer was 9,10-trans and the minor one cis.

Characterization has been reported for 16 adducts formed by the reactions of the four configurational isomers of 3,4-trans-diol 1,2-epoxides of benzo[c]phenanthrene with deoxyadenylic and deoxyguanylic acids (34). Figure 13 shows the cis and trans forms of dephosphorylated adducts 31–34 formed from the (+) enantiomers 29 and 30; the (−) enantiomers reacted analogously. The exocyclic amino groups of both adenine and guanine reacted with the epoxides at the C1 position with the attack occurring by both back-side and front-side mechanisms to give trans and cis adducts. The stereochemistry of epoxide opening was deduced in each case by careful analysis of the ¹H spectrum. The linkage in the deoxyguanosine adducts was

established by decoupling experiments on pentaacetate esters. The NMR signals for the exocyclic NH groups in the adenine adducts could not be detected in the NMR spectrum presumably because of exchange broadening and chemical methods had to be employed to assign the linkage.

**Aflatoxin.** Although the epoxide (11) of aflatoxin B₁ has not been isolated, adduct 35 of the epoxide with DNA is readily prepared by in situ oxidation of aflatoxin B₁ by $m$-chloroperbenzoic acid or by microsomal oxidation (Figure 14). The adduct bears a positive charge on the imidazole ring and consequently is vulnerable to hydrolysis. Three reaction courses are observed: (1) solvolytic loss of the aflatoxin residue to give back unaltered DNA; (2) cleavage of the glycosyl linkage to give adduct 36 and leave an apurinic site in the chain; and (3) hydrolytic attack at C8 of the guanine residue leading to ring opening to give formamidopyrimidine derivative 37 (17, 35, 36). NMR spectra of adduct 36 showed that (1) chemical and enzymatic epoxidation of aflatoxin had occurred stereospecifically to give the exo epoxide and (2) opening of the epoxide ring had given exclusively the trans adduct. The related mycotoxin sterigmatocystin also contains the dihydrofuran ring system and requires oxidative activation prior to formation of covalent adducts with DNA; sterigmatocystin yields a similar adduct with guanine at N7 (37). These adducts have also been prepared by reaction of in

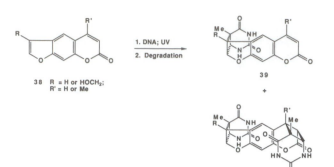

**Figure 15.** Mono and bisthymine photoadducts **39** and **40** formed from psoralens **38**.

situ generated epoxides with 3',5'-di-O-butyryldeoxy-guanosine (*17*).

**Psoralens.** Psoralens (**38**) are furocoumarin natural products which undergo covalent binding to DNA in the presence of light (Figure 15). The process involves (1) noncovalent, intercalative association with the DNA, (2) formation of a photoadduct upon irradiation at 365 nm, and (3) subsequent absorption of a second quantum to yield a bis adduct which results in interstrand cross-linking (*38, 39*). Thymines are the preferred site for monoadduct formation (*40*); cross-link formation occurs at 5'-TpA-3' sites in DNA (*41–43*). The structures of the photoadducts have been established by excision from the DNA and characterization by ¹H NMR and mass spectroscopy (*43, 44*). The adducts are mono- and biscyclobutanes arising from 2 + 2 cycloadditions first to the 3,4-double bond (in the pyrone ring) and then the 4',5'-double bond (in the furan ring). The stereochemistry of addition in the mono and bis adducts with thymidine was established as cis–syn by homonuclear spin decoupling and by nuclear Overhauser effects. The adducts were isolated as mixtures of two nucleoside diastereomers, having opposite configurations at each of the sites in the cyclobutane moieties but identical configurations in the chiral centers of the deoxyribose. Hydrolytic cleavage of the deoxyribose converted the diastereomers to enantiomers (Figure 15); **39** is one of the enantiomers of the mono adduct, and **40** is one of the bis adduct enantiomers. The structures of the adducts are consistent with intercalation into both dA-dT and dT-dA sequences.

**Aromatic Amines.** Aromatic amines undergo activation by oxidation and acetylation to give N-arylhydroxamic acids. O-Acylation of the hydroxamates or N→O rearrangement of the acyl group gives reactive N-(acyloxy)-arylamines which dissociate to give nitrenium ions which react with nucleic acids and proteins. A well-studied member of this class is N-acetyl-2-aminofluorene (AAF, **41**), which is a potent chemical mutagen and carcinogen. It is activated in vivo to the N-hydroxy species **42**, which forms nitrenium ions (either with or without N→O transfer of the acetyl group) which react with DNA primrily at C8 of guanine. The in vivo products are 1-(deoxyguanosin-8-yl)-N-acetyl-2-aminofluorene **44** and 1-(deoxyguanosin-8-yl)-2-aminofluorene **45**. These guanine adducts may be obtained in vitro by treating guanine nucleotides or guanine-containing oligodeoxynucleotides with N-acetoxy-N-acetyl-2-aminofluorene (AAAF, **43**). The AF adduct (**45**) can be prepared from the AAF adduct (**44**) by treatment with ammonia at elevated temperature (*45*). One of the early examples of the use of NMR for solving structural problems in chemical toxicology was a study of adduct **44** (Figure 16) (*46*). Whereas guanosine showed a signal for

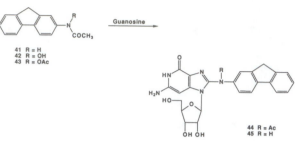

**Figure 16.** C8 adducts **44** and **45** formed by reaction of guanosine with N-acetoxy-N-acetyl-2-aminofluorene.

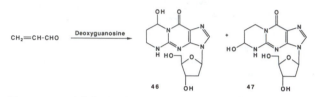

**Figure 17.** 1,N²-Cycloadducts **46** and **47** formed from reaction of acrolein with deoxyguanosine.

the C8 proton at 8.38 ppm, the adduct showed no absorption in the region between 8 and 10 ppm suggesting that substitution had occurred at C8.

**Aldehydes.** Aldehydes can react with DNA without activation. Many of them arise from lipids by enzymatic or non-enzymatic routes. Acrolein, which is ubiquitous in the environment, is a direct mutagen; it reacts with DNA to form adducts with deoxyguanosine which have been shown to be 1,N²-cycloadducts **46** and **47** involving both the possible orientations; i.e., Michael addition to either the exocyclic amino group or to N¹ followed by cyclization (Figure 17) (*47–49*). The adducts were characterized by a combination of chemical and spectroscopic techniques. With less reactive α,β-unsaturated compounds, such as crotonaldehyde, the adducts that have been isolated arise only by the first process. Other examples of reactive aldehydes include trans-4-hydroxy-2-alkenals arising through lipid peroxidation (*50*) and by microsomal metabolism of the pyrrolizidine alkaloid senecionine (*51*); these also react by conjugate attack by N² followed by carbinolamine formation with N¹ (*52*).

Malondialdehyde represents a special case. It can arise by enzymic and nonenzymatic degradation of prostaglandin endoperoxides (*53–55*). It, like acrolein, is a potent mutagen; however, acrolein and malondialdehyde act by different mechanisms. Acrolein induces base-pair substitutions in *Salmonella* whereas malondialdehyde not only induces base-pair substitutions but also causes frame shifts (additions and deletions) (*56, 57*). Marnett and co-workers have studied the condensations of malondialdehyde with guanosine. A 1,N²-cycloadduct (**48**) has been isolated; additionally a complex adduct (**49**) was found arising from condensation of 2 equiv of malondialdehyde with the nucleoside (Figure 18) (*58, 59*). The structure of the 2:1 adduct was established by X-ray crystallography after NMR failed to yield unambiguously a single structure. The mechanism by which the 2:1 adduct arises has not been investigated in detail. A reasonable scenario involves the carbinolamine precursor of **48** reverting to aldehyde and undergoing aldol condensation with the second equivalent of malondialdehyde. This 2:1 adduct then undergoes two ring closures to give **49**. Nair and co-workers have isolated additional adducts of adenosine and cytidine (*60*). These include uncyclized compounds **50** and **51** arising from addition at the exocyclic amino groups plus two 3:1 adducts, which have been assigned extraordinary

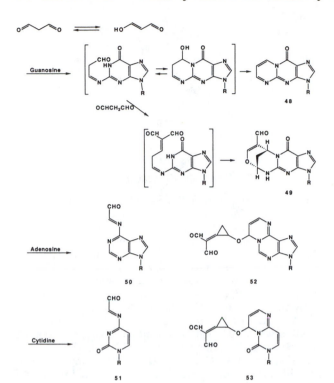

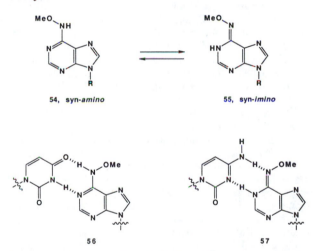

**Figure 18.** Nucleoside adducts 48–53 formed with malondialdehyde.

**Figure 19.** Tautomers 54 and 55 of the $N^6$-methoxy derivative of adenosine and their base pairing with uracil (56) and cytidine (57), respectively.

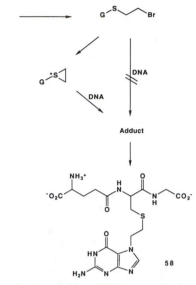

**Figure 20.** Formation of ternary adduct 58 between ethylene dibromide, glutathione, and guanine.

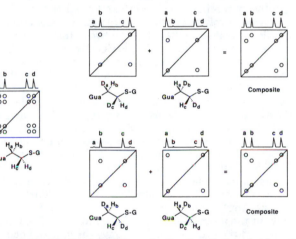

**Figure 21.** Idealized COSY spectra for the ethylene bridge of ternary ethylene dibromide adduct 58 and for the two pairs of 1,2-dideuterio adducts which would result from use of threo and erythro dideuteriated ethylene dibromides, if the reaction follows a single stereospecific pathway. The relative chemical shifts are arbitrarily assigned.

methylenecyclopropane structures **52** and **53** on the basis of NMR studies. On mechanistic and spectroscopic grounds these structures should be regarded as provisional until confirmed by X-ray crystallography.

**Hydroxylamines.** Hydroxylamine and methoxyamine are mutagens which react with adenosine to give $N^6$-hydroxyl and $N^6$-methoxyl derivatives (*61–63*) and with cytidine to give the corresponding $N^4$ derivatives (*64, 65*). $^1$H NMR spectroscopy indicates that tautomerism occurs between amino (**54**) and imino (**55**) forms of $N^6$-methoxyadenosine (Figure 19). Whereas the amino tautomer forms a base pair (**56**) with uridine (*66*), the imino tautomer forms a base pair (**57**) with cytidine (*67*).

**Ethylene Dibromide.** Studies performed by Guengerich and co-workers on the carcinogen ethylene dibromide have employed NMR for both determination of structure and study of mechanism. The ternary adduct of ethylene

dibromide with glutathione and DNA (formed both in vivo and in vitro) was degraded to the modified purine (**58**) to demonstrate that reaction had occurred at N7 of guanine (Figure 20). The $^1$H NMR spectrum of the adduct was assigned by means of a COSY spectrum in comparison with the spectra of glutathione and its dimer (*68*). On account of the asymmetry of the amino acids of glutathione, both sets of geminal protons in the ethylene bridge of the adduct showed nonequivalence in the $^1$H NMR spectrum. The existence of this nonequivalence was subsequently exploited to show that the initially formed S-(2-bromoethyl)glutathione forms the DNA adduct via an episulfonium ion rather than by direct displacement (*69*). The experiment involved utilization of stereospecifically 1,2-dideuteriated ethylene dibromide for adduct formation to ascertain whether the adduct is formed by an odd, even, or nonintegral number of inversion processes on the methylene carbons. Whereas the undeuteriated ternary adduct shows six cross-peaks for the ethylene bridge, reflecting two geminal and four vicinal couplings, each dideuterio species will show only one, a vicinal homonuclear

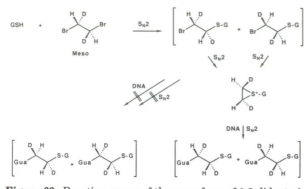

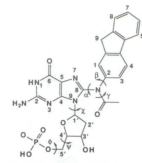

**Figure 22.** Reaction course of the meso form of 1,2-dideuterio ethylene dibromide with guanyl residues in DNA showing stereochemical consequences of the $S_N2$ displacements.

**Figure 23.** The 8-AAF–dGMP adduct.

coupling as shown in the drawing of a COSY spectrum (Figure 21). The experiment was complicated by the fact that the threo and erythro [1,2-$^2$H$_2$]EDBs which were used were racemic and meso forms, respectively. As a consequence, adduct formation with the glutathione which is a single enantiomer would lead in both cases to two diastereomers at each stage in the reaction sequence, irrespective of mechanism if an integral number of displacements was involved; i.e., the ternary product would contain two diastereomers and would yield two of the four possible vicinal off-diagonal signals in the COSY spectrum (Figure 21). The meso (i.e., erythro) dideuteriated EDB yielded one pair of signals, and the racemic (threo) yielded the other. Therefore, an integral number of displacements had, in fact, occurred in the reaction sequence reflecting the existence of a single mechanism which did not involve carbocations or other intermediates possessing a plane of symmetry (Figure 22) (23). Analysis of coupling and NOE data failed to yield unambiguous assignments of the individual diastereomers and the final conclusion that formation of the ternary complex involves a total of three $S_N2$ displacements rested on synthesis of the adduct by an independent route involving only two displacements without the intermediacy of a thiiranium ion. The COSY spectra of the synthetic adducts gave the opposite pairing of cross-peaks from that observed with the biosynthetic adducts.

Whereas the preceding sections have dealt with the use of NMR for the determination of primary structure and stereochemistry of carcinogen–nucleic acid adducts, the remaining sections discuss applications of NMR in evaluation of solution conformations of carcinogen–nucleotide adducts.

## III. Characterization of Adducts Formed with Mono- and Dinucleotides by NMR Spectroscopy. Model Studies

Commercially available mono- and dinucleotides have been frequently utilized as binding substrates for NMR investigations of various carcinogen–DNA adducts. Although mono- and dinucleotides are not entirely adequate models for polymeric DNA, a substantial amount of information may be obtained from these studies. The following discussion will focus on the nucleotide adducts of the arylamine N-acetoxy-N-acetyl-2-aminofluorene (AAAF, **43**) which was mentioned in the previous section.

$^1$H NMR spectra of AAAF adducts with 5'-GMP, ApG, and GpA were reported by Nelson et al. (70). The information obtainable from these low-field studies was limited due to poor resolution of the spectrum. However, substantial upfield chemical shifts of the adenine H8 and H2

protons in the modified dinucleotides as compared to the unmodified dinucleotides were used to support the proposal that in the adduct, the planar fluorenyl ring of the carcinogen was stacked above the adenine ring; the guanine base was displaced. This model came to be known as the "base-displacement" model (70, 71); a similar model known as the "insertion–denaturation" model was independently proposed (72–76). Support for these related models comes also from the observation that DNA, when modified with AAAF, exhibits greater sensitivity to $S_1$ nuclease, suggestion regions of single-stranded or denatured DNA are present at the locations of bound AAF moieties (77). $^1$H NMR spectra of UpG, GpU, and the corresponding dimers in which the guanosine residue was modified with AAF failed to indicate a stacking interaction between the fluorene moiety and the adjacent residue (78), in contrast to the studies where the fluorene moiety was adjacent to adenine residues. These results provided an early indication of the sequence specificity of carcinogen–DNA interactions. $^1$H NMR studies of the deoxydinucleotide d(ApG) were subsequently reported (45). The observations for the deoxydinucleotide–AAF adduct were similar to and supported the results previously obtained (70) for the corresponding ribodinucleotide. The deacetylated AF adduct analogous to **45** was also examined as it had been reported to account for 80% of the C8 adduct in vivo (79–81). The results suggested that in the AF adduct the fluorene ring is less stacked on the adjacent adenine than in the AAF adduct. It was proposed that this was due to steric factors arising from the presence of the relatively large acetyl group (45). Thus it appears that the functionality of the central nitrogen atom in the adduct also determines conformation.

Substantial contributions to the understanding of the conformational and dynamic properties of the dGMP–AAF adduct have been made by Evans and co-workers utilizing $^1$H, $^{13}$C, $^{15}$N, and $^{31}$P NMR (82–84, 88). Rotational barriers exist about the guanyl nitrogen ($\alpha$) and amide ($\gamma$) bonds of the dGMP–AAF adduct (see Figure 23). The magnitude of the barrier is reported to be 12–15 kcal/mol for both bonds. Four separate subspectra of the dGMP–AAF adduct are observable both in methanolic and aqueous solution at low temperature (83). This is illustrated in Figure 24, where signals arising from individual subspectra are labeled I–IV. Analysis of chemical shift data and saturation transfer experiments led to the conclusion that the four torisonal diastereomers are interrelated via a cyclic interconversion pathway, as illustrated in Figure 25. Analysis of $^{13}$C–$^1$H vicinal coupling constants indicated that the glycosyl torsion angle of the dGMP was in the syn conformation. In the aqueous solution experiments (84), the additional effect of self-association of the dGMP–AAF adduct was observed. Subsequent investigations of the dinucleotides ApG, d(ApG), GpA, and d(GpA) using $^{13}$C and $^{31}$P NMR supported observations obtained from $^1$H

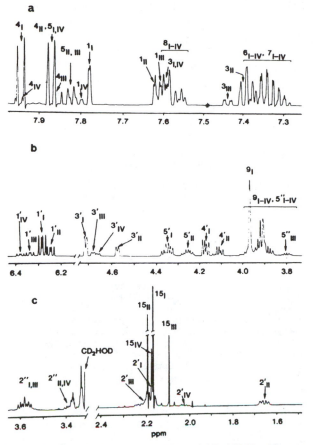

**Figure 24.** $^1$H NMR spectrum of the 8-AAF–dGMP adduct at −50 °C: observation of four distinct conformational isomers. Subspectra I–IV are indicated by subscripts in the figure: (A) the aromatic region of the spectrum; (B) the aliphatic region of the spectrum between 3.8 and 6.4 ppm; (C) the aliphatic region of the spectrum between 1.5 and 3.6 ppm. Reprinted with permission from ref 83. Copyright 1984 American Chemical Society.

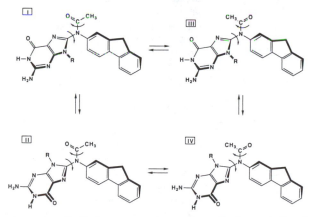

**Figure 25.** Cyclic interconversion pathway for four rotational isomers of 8-AAF–dGMP as determined from analysis of NMR data. The rotational isomers result from hindered rotation about the α and γ bonds, as shown in Figure 23.

NMR (*85–87*). Evans and Levine (*88*) utilized an analysis of one-bond $^{15}$N–$^{13}$C coupling constants to examine the configuration of the central nitrogen atom of the 5′-dGMP–AF adduct. By comparison to various model compounds and theoretically computed coupling constants as a function of torisonal angle, it was concluded that the amine nitrogen in the AF adduct is nearly planar. Modeling studies indicated that in this conformation the

fluorene ring could be accommodated within the major groove of the DNA.

## IV. Characterization of Specific Sequence Oligodeoxynucleotide–Carcinogen Interactions by NMR Spectroscopy

Utilization of mono- and dinucleotides to study carcinogen–DNA interactions is ultimately limited by the inadequacy of these systems as models for polymeric DNA. With the possible exception of the simple alkylating agents, the binding of carcinogenic agents to DNA is dictated both by sequence and by conformation at potential binding sites. Native DNA polymers are heterogeneous in sequence; commercial synthetic DNA polymers are available in a limited number of repeating sequences. Unfortunately, neither synthetic nor native polymeric DNA molecules are readily amenable to detailed high-field NMR studies. Their size results in slow reorientation in solution and resulting line broadening of NMR spectra. The information obtainable from NMR spectra of polymeric DNA is somewhat limited. In contrast, oligodeoxynucleotides of length 4–15 base pairs tumble rapidly in solution and yield high-resolution NMR spectra. Furthermore, they can be constructed to have a specific sequence. Oligodeoxynucleotides of this length appear to be reasonable models for polymeric DNA; although when working with shorter oligodeoxynucleotides especially, end effects must be considered. The development of automated oligodeoxynucleotide synthesis chemistry is one major development that has promoted the utilization of NMR methods beyond the mono- and dinucleotide level. It is now routinely possible to obtain NMR quantities of oligodeoxynucleotides having defined sequence and conformation (*89*).

When utilizing oligodeoxynucleotides, it is important to distinguish between self-complementary and non-self-complementary sequences. Oligodeoxynucleotides which are self-complementary will self-associate into Watson–Crick duplexes. Self-complementary sequences have been utilized extensively since only a single strand of DNA must be chemically synthesized. Furthermore, antiparallel self-complementary sequences contain a pseudodyad axis of symmetry, which simplifies spectral interpretation. In the analysis of carcinogen–DNA interactions, the chemistry of the specific system being examined is often the determining factor in deciding whether a self-complementary or a non-self-complementary oligodeoxynucleotide is best suited to the problem at hand.

A second major development which has contributed to the utilization of NMR methods with longer oligodeoxynucleotides is the ability to assign the complex spectra of these macromolecules. It is now possible to make partial, and in many cases complete, assignments of the $^1$H signals arising from oligodeoxynucleotides up to 15 base pairs in length. This is due principally to the development of two-dimensional NMR analysis techniques, especially the homonuclear autocorrelated experiment (COSY) and the 2D-NOE experiment (NOESY), as discussed in a previous section of this review.

A major difficulty in the utilization of oligodeoxynucleotides to examine carcinogen–DNA adducts is that more than one binding site can be present in an oligodeoxynucleotide; in principle, upon reacting the carcinogen with the oligodeoxynucleotide, a number of discrete adducts corresponding to various binding sites may be obtained. This presents formidable purification problems in obtaining sufficient amounts of chemically pure species for detailed spectroscopic characterization of carcinogen–

**Figure 26.** Sequence design of an oligodeoxynucleotide to limit reaction of the carcinogen AAAF to a single site. The two strands were synthesized separately; the sequence d(CCACGCACC) has a single reactive guanosine residue. After reaction of d-(CCACGCACC) with AAAF, it was combined with the complementary strand d(GGTGCGTGG) for NMR analysis.

oligodeoxynucleotide adducts. Major approaches which have been utilized to solve this problem include the following: (a) design of oligodeoxynucleotides to limit reaction to a single site and (b) construction of oligodeoxynucleotides using carcinogen-modified mononucleotide units. Both of these strategies have limitations which will be discussed in the following sections.

**Design of Oligodeoxynucleotide Sequences to Limit Binding to a Specific Site.** An example of this method is illustrated by the study of the interactions of *N*-acetoxy-*N*-acetyl-2-aminofluorene (AAAF) (**43**) with the non-self-complementary oligodeoxynucleotide sequence d-(CCACGCACC):d(GGTGCGTGG) by Krugh and co-workers (*90*). The utilization of a non-self-complementary oligodeoxynucleotide sequence required separate synthesis of two different strands, in contrast to a self-complementary sequence which is capable of self-association into a Watson–Crick duplex. This strategy avoided the potential problem of disproportionation which could occur if a self-complementary sequence were used, should the carcinogen–oligodeoxynucleotide adduct form a less stable duplex structure than the unreacted self-complementary duplex. This particular sequence was designed such that one chain contained only a single guanine nucleotide and no adenine nucleotides. Thus, when the d(CTCCGTTCC) strand was reacted with AAAF, the site of reactivity was limited to the single guanine, making purification of the bound adduct a much simpler problem. After reaction with AAAF, the adducted chain d(CTCCG*TTCC) was combined stoichiometrically with the complement strand for ¹H NMR spectroscopy. This scheme is shown in Figure 26.

The base-displacement model (*70, 71*) and related insertion–denaturation model (*72–76*) for AAF–DNA adducts both propose that the guanine base at the site of covalent attachment is displaced from the helix. The base-displacement model should be accompanied by substantial disruption of the DNA helix. To address this question, a study of the hydrogen bonded imino protons (thymine N3 and guanine N1) of the AAF-modified oligodeoxynucleotide duplex was carried out (*90*). The imino protons of oligodeoxynucleotides are normally only observable under conditions such as base pairing where the exchange rate with solvent is slow. Inspection of the modified oligodeoxynucleotide showed that for this sequence, all nine imino protons were visible; two imino protons were shifted significantly upfield from their location in the unmodified duplex. This is shown in Figure 27. A series of NOE experiments were used to assign the various imino resonances. These results argue against the base displacement model for this particular sequence, which would predict a loss of one or more imino resonances at the modification site, due to rapid exchange with solvent.

The polymer poly(dGdC)·poly(dGdC) is able to assume the left-handed (*Z*) conformation (*91, 92*). In the left-handed conformation, the guanine nucleotides are rotated

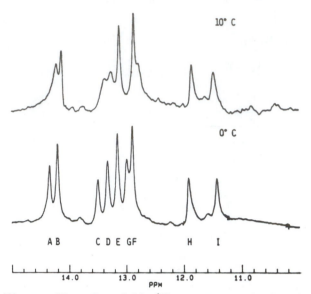

**Figure 27.** Low-field ¹H spectrum of the d-(CCACG$^{AAF}$CACC)·d(GGTGCTGG) complex showing signals arising from hydrogen bonding interactions of the guanosine N1 and thymine N3 protons in the modified duplex. Two signals, designated as H and I, are shifted upfield in the modified duplex. These signals reflect the distortion of the helix upon modification with the carcinogen. The presence of all nine signals in the spectrum argues against disruption of base pairing in the modified duplex. Reprinted with permission from ref 90. Copyright 1987 Adenine Press.

about the glycosyl bond from the anti to the syn conformation. When modified with AAF, this polymer was not sensitive to digestion by nuclease S₁ (*77*). Furthermore, the CD spectrum of poly(dGdC)·poly(dGdC) when highly modified with AAAF was similar to the CD spectrum of the left handed form of this polymer. The oligodeoxynucleotide sequence studied by Krugh and co-workers (*90*) consists of an alternating purine–pyrimidine sequence at the binding site; CD studies of the AAF adduct in this oligodeoxynucleotide (*93*) indicate that the modified duplex has at least a portion of the helix in an altered conformation, presumably left-handed. This study is illustrative of the potential utilization of high-field NMR methods to study carcinogen-DNA adducts. It appears that the binding of AAAF to DNA will prove to depend on a variety of factors, which include both sequence and conformation.

**Synthesis of Oligodeoxynucleotides Using Modified Mononucleotide Units.** The obvious limitation of specific design of the oligodeoxynucleotide sequence is that many sequences potentially of interest cannot easily be examined. An alternative approach is the preparation of modified nucleotide units which are then incorporated into an oligodeoxynucleotide as it is synthesized. This has been successfully used to examine several methylated oligodeoxynucleotides. The primary limitation of this approach is that the modified nucleotide units must be stable under the conditions utilized in automated DNA synthesis (i.e., phosphoramidite or phosphotriester chemistry) and in subsequent deblocking of the oligodeoxynucleotides. Thus, studies of methylated oligomers are ideally suited to this approach; $O^6$-methylguanine is now commercially available as the phosphoramidite reagent.

Alkylation of guanine at the O6 position, and of thymine at the O4 position is believed to be important in the action of *N*-nitroso compounds. Both $O^6$-methylguanine (*94–96*) and $O^4$-methylthymine (*97*) have been successfully incorporated into oligodeoxynucleotides. Substantial infor-

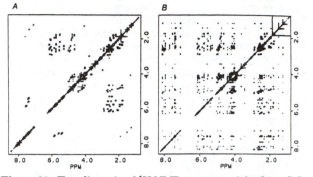

**Figure 28.** Two-dimensional $^1H$ NMR spectrum of the O6meG·C dodecadeoxynucleotide investigated by Patel and co-workers (*98*). A comparison of the $^1H$ COSY and NOESY spectrum is shown in the contour plot representation. In the COSY spectrum (A), only cross-peaks representative of scalar (through-bond) coupling interactions are observed. The corresponding NOESY experiment (B) shows cross-peaks arising from dipolar (through-space) coupling interactions. The pattern of dipolar couplings which are observed is used to assign the sequence of the oligodeoxynucleotide and to answer questions about its conformation. Reprinted with permission from ref 98. Copyright 1986 American Chemical Society.

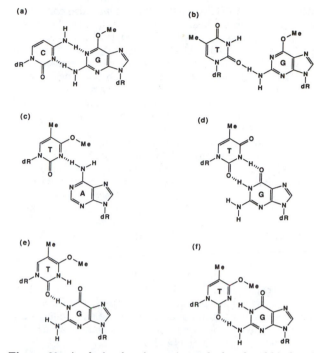

**Figure 29.** Analysis of various mismatched and wobble base pairing schemes involving C,T, G, A, O6meG, and O4meT. Base pairing involving C·O6meG (b), T·O6meG (b), A·O4meT (c), G·T (d), G·O4meT (e, f). From various $^1H$ NMR studies of Patel and co-workers (*102–104, 106, 107, 109*). The C·O6meG (a) and G·T (d) pairs are stabilized by two hydrogen bonds and are accommodated within the helix. Analysis of the low-field $^1H$ spectrum and NOE experiments led to the proposal that the G·O4meT pair had the configuration as shown in f rather than e, which would have been analogous to the G·T mismatch (d). These mismatch and wobble pairs apparently are accommodated within the DNA helix, although localized distortion of the helix occurs.

mation on the base pairing interactions of these alkylated nucleotides has been derived from NMR studies performed by Patel. In a set of two papers, Patel and co-workers (*98, 99*) compared $O^6$-methylguanine when it was paired with cytosine and when paired with thymine; in previous studies (*100–102*) the nonmethylated oligodeoxynucleotides of the same sequence had been examined. The sequences used were (a) d(CGCGAATTCG\*CG) and (b) d-(CGTGAATTCG\*CG); the second of these contains a G·T mismatch and the asterisk represents the site where $O^6$-methylguanine was incorporated. A series of 1D NOE and 2D COSY and NOESY experiments were utilized to study the conformations of the O6meG·C and O6meG·T mismatch-containing oligodeoxynucleotides (Figure 28). The results indicated that in both instances, the methylated guanine base and its hydrogen bonded partner (either C or T) were incorporated into the helix with minor disruption as compared to the corresponding nonmethylated helix. In the case of the O6meG·C-containing oligomer, $^{31}P$ NMR spectroscopy suggested the existence of an altered phosphodiester backbone at the modification site. Since analysis of the NMR data led to the conclusion that the modified G·C base pair was accommodated within the helix, the authors proposed a model for O6meG·C base pairing involving two hydrogen bonds: between the guanine exocyclic amino group and $N^3$ of cytosine and between $N^1$ of guanine and the exocyclic amino group of cytosine (Figure 29a). Likewise, on the basis that helical perturbations for the O6meG·T mismatch pair were minimal, it was proposed that the mismatch pair contains a single hydrogen bond between the exocyclic amino group of guanine and the C2 carbonyl moiety of thymine (Figure 29b).

The incorporation of $O^4$-methylthymine into oligodeoxynucleotides via chemical synthesis is somewhat more difficult than $O^6$-methylguanine, due to the reactivity of the methylated thymine. $O^4$-Methylthymine is sensitive to both nucleophilic attack and to dealkylation by protic acids. Thus, modification of the deblocking procedures to remove the protecting groups in the phosphotriester chemistry for oligodeoxynucleotide synthesis was necessary (*97*). In a second set of papers, Patel and co-workers compared A·O4meT and G·O4meT base-pairing interactions with the corresponding nonmethylated A·C and G·T

mismatch pairs (*103, 104*). The A·C mismatch forms a wobble base pair, as was determined from X-ray (*105*) and NMR studies (*106, 107*); wobble pairing is also predicted by theoretical studies (*108*). The comparative NMR study on the A·C mismatch and A·O4meT mismatch [utilizing the self-complementary oligodeoxynucleotide sequences d(CGCAAGCTC\*GCG) and d(CGCAAGCTT\*GCG), where the asterisk represents the site of incorporation of O4meT or the A·C mismatch] indicates similar wobble pairing for the A·O4meT lesion (*103*). From an analysis of NOE intensities, these workers determined that the $OCH_3$ group of O4meT is in the syn orientation with respect to the thymine N3 position. It should be noted that the $^1H$ NMR spectra of both of these sequences were found to be sensitive to pH; thus the data were reported at pH 5.5 where higher quality spectra were obtained. It was proposed on the basis of the NMR studies that the A·O4meT pair forms a single hydrogen bond between the adenine exocyclic amino group and the N3 position of thymine (Figure 29c). The possibility of an additional hydrogen bond between adenine N1 and the C2 carbonyl group of thymine was postulated, although such a resonance was not detected in the $^1H$ NMR spectrum (the low pH of these studies leads to the possibility of protonation of adenine $N^1$).

Wobble pairing for the G·T mismatch interaction in the self-complementary oligodeoxynucleotide d-(CGTGAATTCG\*CG) containing one mismatch (\*) was reported to involve formation of two imino proton–carbonyl hydrogen bonds between guanine and thymine (*109*), as shown in Figure 29d. Examination of d-

(CGCG*AGCTTGCG) G·T mismatch was also consistent with formation of these two hydrogen bonds. However, when T was replaced by O4meT (*104*), the guanosine imino proton was observed to be shifted upfield to 8.67 ppm from its normal position (12–13 ppm). This large upfield shift was interpreted to mean that hydrogen bonding between the guanine imino proton and O4meT as shown in Figure 29e (analogous to the G·T wobble pair) is weak or absent. Analysis of NOESY data showed that both the guanine and O4meT at the modification site have anti glycosyl torsion angles and are stacked into the duplex. In addition, observation of a strong NOE between the OCH$_3$ group of O4meT and the imino proton of G led to the conclusion that the OCH$_3$ group was in the syn orientation with respect to the N$^3$ position of O4meT. Because of the large change in chemical shift for the guanine imino proton in the G·O4meT mismatch pair as compared to the G·T mismatch pair and the observation of a strong NOE between the guanine imino proton and the O4 methoxy group of O4meT, it was concluded that wobble pairing (Figure 29e) analogous to that observed for G·T (Figure 29d) (*109*) is not formed for G·O4meT. Instead it was proposed that an alternative base-pairing mode existed, stabilized by a single short hydrogen bond between the exocyclic amino group at the C2 position of guanine and the C2 carbonyl of O4meT (Figure 29f). In these studies of methylated guanine and thymine nucleotides incorporated into the DNA double helix, the modified bases are apparently incorporated (stacked) into the helix. These alkylated species do not appear to change the gross conformational properties of the helix, although significant perturbations of the helical geometry are observed at the site of modification.

A related area of investigation is a study of oligodeoxynucleotides containing apurinic sites. Abasic sites in DNA are inherently unstable; NMR studies of abasic sites in oligodeoxynucleotides require utilization of stable analogs of deoxyribose. In one approach, the O1′ of the deoxyribose was methylated (*110*). Two oligodeoxynucleotides containing methyl 2′-deoxy-α-D-ribofuranoside were synthesized using phosphotriester chemistry: d-(CSCG) and d(CSCGCG), where the symbol S refers to the methyl 2′-deoxy-α-D-ribofuranoside unit. Whereas the shorter sequence did not form a duplex under the conditions examined, it was concluded from 1D and 2D $^1$H NMR spectroscopy that the hexamer sequence containing the apurinic site occurs at low temperature as a right-handed "staggered" DNA duplex where the apurinic site is located in a dangling end sequence. In a second approach, 3-hydroxy-2-(hydroxymethyl)tetrahydrofuran was utilized as a deoxyribose analogue (*111, 112*). Patel and co-workers (*113*) synthesized the complementary nonanucleotide duplex d(CATGAGTAC)·d(GTACFCATG), where F represents the site of incorporation of the 3-hydroxy-2-(hydroxymethyl)tetrahydrofuran analogue into the duplex. NOESY experiments at low temperature indicated that the adenine base opposite the apurinic site is inserted into the helix and stacked between the neighboring Watson–Crick base pairs. Two of the 16 phosphodiester groups in the altered nonadeoxynucleotide have altered conformation, as is evidenced by inspection of $^{31}$P NMR spectra.

The incorporation of the aminofluorene adduct into an oligodeoxynucleotide starting from a protected guanosine–AAF nucleotide via phosphotriester chemistry has also been reported (*114*). The AAF adduct is converted to the AF adduct during deblocking under basic conditions. In addition, the deblocking reaction was performed in the presence of a trace of mercaptoethanol to prevent oxidative loss of the fluorene moiety from the product.

**NMR-Derived Solution Structures of Carcinogen–Oligodeoxynucleotide Adducts.** Rapid advances in the development of NMR spectrometers and pulse programs are augmented by concurrent advances in data-processing capabilities. It is increasingly clear that NMR methodology will become a major method for elucidation of solution conformations of large biomolecules, including nucleic acids and proteins. The intensity of the NOE signal observed between two discrete nuclei in the NOE experiment is directly related to the internuclear distance between them. The NOE effect is operative over short range only; typically two nuclei having an observable NOE must be within 5 Å of each other. The 2D NOESY experiment yields a complete set of NOE interactions between all pairs of nuclei in the molecule. In principle, it should be possible by analysis of these data, in combination with a known set of distance and conformational restraints for particular atoms in the structure, to calculate the solution conformation of the molecule. Algorithms for conformational analysis of NOE data have been independently developed in several laboratories [a detailed review of this field and specific programs in use, such as DISMAN, CONFOR, and DISGEO, can be found in the volume by Wüthrich (*2*)]. These methods require not only the hardware to obtain NMR data at the highest possible field strength, but also substantial data processing capability.[2]

One illustration of the utilization of distance geometry methods in combination with NOE data is the NMR-derived structure of a psoralen–oligodeoxynucleotide adduct. As was previously discussed, psoralens are capable of crosslinking duplex DNA in a multiple-step process. Kim, Wemmer, and Tomic (*117*) designed the self-complementary oligodeoxynucleotide sequence d(GGGTACCC)$_2$ to examine the $^1$H NMR spectrum of a photo-cross-linked psoralen–DNA adduct. This sequence contains a single psoralen photocrosslinking site. Upon modification with psoralen the symmetry of the self-complementary sequence is lost, and in the resulting NMR spectrum of the psoralen adduct there are six distinct G·C and two distinct A·T base pairs.

The conformation of the psoralen adduct with the oligodeoxynucleotide d(GGGTACCC)$_2$ has been calculated by using a distance geometry algorithm (*117*). 2D NOESY experiments were used to assign the spectrum and estimate distances between 171 proton pairs in the cross-linked oligodeoxynucleotide. These estimated distances were then evaluated in combination with a set of known distance and geometry constraints to arrive at a NMR-derived solution structure for the psoralen adduct. The derived model showed a 53° bend into the major groove at the site of modification, and a 56° unwinding, spanning the 8-mer duplex. Figure 30 shows a stereoview of the solution structure of the oligodeoxynucleotide–psoralen adduct.

**Further Developments in Spectral Assignment.** Development of spectral assignment strategies for oligodeoxynucleotides plays a major role in the utilization of these macromolecules in the study of carcinogen–DNA interactions. To date, signal assignments in oligodeoxynucleotides have relied primarily on 2D NOESY methods. These methods depend on the oligodeoxynucleotide

---

[2] Another technique which shows great promise is maximum entropy Fourier spectral deconvolution (MEFSD). As recently developed (*115*), the MEFSD technique can be exploited to simultaneously obtain resolution enhancement and noise suppression. Thus it is ideally suited to problems encountered in biological NMR. Recently, this technique has been extended to evaluation of NOESY NMR spectra, such as a 25 nucleotide RNA molecule containing a hairpin loop (*116*).

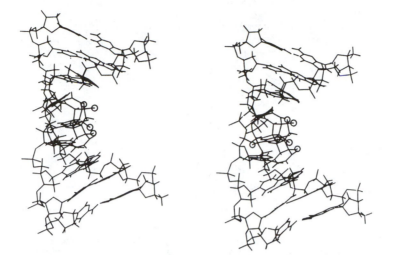

**Figure 30.** The psoralen–octadeoxynucleotide complex as determined from distance–geometry computations. Stereo picture of the NMR-derived model. The methyl groups and the amino group of AMT are indicated by circles. Reprinted with permission from ref 117. Copyright 1987 American Association for the Advancement of Science.

maintaining a sufficiently ordered conformation that internucleotide NOEs can be observed. Furthermore, the NOESY sequential assignment methods are based upon assumption of known conformation for the oligodeoxynucleotide in question (generally the right-handed B-DNA conformation). The NOESY methods may fail for oligodeoxynucleotides in the single-stranded conformation or having single-stranded regions (e.g., the insertion–denaturation model for AAF adducts). Binding of a carcinogen to DNA may result in substantial alteration of the DNA conformation at the site of modification in which case prior knowledge of the DNA conformation cannot be assumed. Thus, alternative methods of signal assignment will undoubtedly be required in many instances. We discuss two methods of approaching this problem: (a) selective isotopic substitution methods and (b) NMR assignment techniques which rely on scalar rather than dipolar interactions between nuclei.

**Specific Deuteriation as an Aid to Spectral Assignment.** Although the use of deuteriation in the spectral assignment of oligodeoxynucleotides is not a new idea and has been demonstrated to be a powerful method for assignment of complex spectra, it has not been extensively utilized due to the difficulty and expense usually associated with the preparation of specifically labeled molecules. Selective deuteriation through incorporation of specifically deuteriated mononucleotide units into an oligodeoxynucleotide provides an avenue of approach toward spectral simplification and also a means to strategically locate deuterium atoms in specific target locations in the macromolecule. Potentially, the ability to construct deuteriated oligodeoxynucleotides will allow the examination of substantially longer sequences than is now feasible even with the highest available magnetic field strengths. Dramatic spectral simplification of two-dimensional NOESY spectra of non-self-complementary oligodeoxynucleotides was demonstrated (*118*) using relatively simple deuteriation procedures. In these experiments the NOESY spectrum of the dodecadeoxynucleotide d(CGTTA-TAATGCG)·d(CGCATTATAACG) was simplified by selectively deuteriating the purine H8 and pyrimidine H5 protons of one strand at a time and combining with the opposite strand. The resulting NOESY spectra are shown in Figures 31 and 32. This facile deuteriation was completed on the intact oligodeoxynucleotides through heating in $D_2O$ in the presence of $(ND_4)_2SO_3$ at pD 7.8.

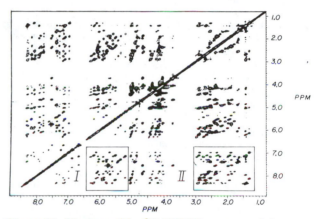

**Figure 31.** Phase-sensitive $^1H$ NOESY spectrum of the non-self-complementary dodecadeoxynucleotide d(CGTTA-TAATGCG)·d(CGCATTATAACG). A contour plot representation of the data; cross-peaks arising from dipolar interactions between the purine H8 and pyrimidine H6 protons and the deoxyribose anomeric protons are located in box I, as are the scalar coupling interactions between cytosine H5 and H6 protons (intense cross-relaxation between the cytosine H5 and H6 protons in the NOESY experiments contributes to the intensity of the latter cross-peaks). The sequence of the dodecanucleotide can be determined from analysis of the pattern of cross-peaks in this region of the NOESY spectrum; characteristic cross-peaks are observed between each nucleotide unit in the sequence and its 5'-neighbor nucleotide unit when the DNA is in the B conformation. Cross-peaks arising from dipolar interactions between the purine H8 and pyrimidine H6 protons and the deoxyribose H2' and H2'' protons are located in box II. Reprinted with permission from ref 118. Copyright 1988 American Chemical Society.

Sequence assignments for oligodeoxynucleotides which do not form ordered B-DNA structures in solution could be determined in a single experiment if an oligomer were synthesized with different defined isotopic ratios at a single site at each individual position in the sequence. Such an approach has been previously utilized for assignment of $^{31}P$ spectra of oligonucleotides through synthetic $^{17}O/^{18}O$ labeling procedures (*119–122*). The tetramer sequence TTTT forms a disordered random coil configuration in solution. Relatively simple deuteriation procedures allow the four thymidines of this sequence to be assigned with confidence (*123*). Examination of sequences longer than 15 base pairs may require the use of selective deuteriation. A possible strategy would be construct specific protonated

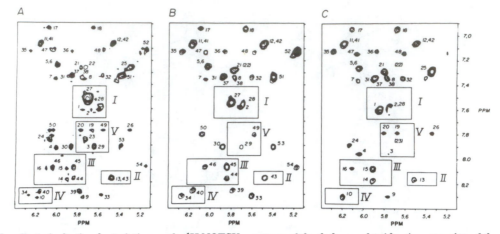

**Figure 32.** The effect of selective deuteriation on the $^1$H NOESY spectrum of the dodecanucleotide. An expansion of the region enclosed by box I in the contour plot of the NOESY spectrum shown in Figure 31. NOE cross-peaks between the purine H8 and deoxyribose anomeric protons in the fully protonated duplex (A), the duplex where strand A has been deuteriated (B), and the duplex where strand B has been deuteriated (C). Numbers next to the cross-peaks identify specific NOE interactions in the duplex. Boxes I–V contain specific sets of NOE cross-peaks which are affected by deuteriation. Reprinted with permission from ref 118. Copyright 1988 American Chemical Society.

binding regions (relatively simple NMR interpretation using existing methods) flanked by deuteriated sequences which would provide stability for the binding interaction but be invisible in the NMR spectrum.

**NMR Spectral Assignments Derived from Scalar Coupling Interactions.** Utilization of NOESY sequential assignment techniques requires that the DNA structure be sufficiently ordered to observe specific internucleotide dipolar coupling interactions. An alternative to the NOESY methods is to use scalar coupling interactions to arrive at internucleotide connectivities. The advantage of utilizing scalar rather than dipolar couplings to obtain sequence assignments in oligodeoxynucleotides is that the scalar methods do not require prior assumption of any particular geometry; thus they are ideally suited to problems in analyzing nonstandard conformations, such as may be observed for a variety of carcinogen–DNA interactions.

The relay method, extensively utilized for analysis of protein spectra (2), allows one to follow magnetization transfer between two nuclei mutually coupled to a third nucleus. The method which appears to be most applicable to oligodeoxynucleotide analysis is $^{31}$P-relayed $^1$H–$^1$H correlation spectroscopy, where magnetization is followed from the H3′ of one nucleotide unit to the H4′ of the following unit, relayed via the mutual coupling of the two protons to the $^{31}$P nucleus in the phosphodiester linkage (124–126). Potential limitations of this method appear to be resolution of the H4′ spectral region, the need to repeat the relay experiment for a range of coupling constants, and loss of signal intensity due to $T_2$ relaxation (125). Nevertheless, this method has been demonstrated for a number of short oligomers and has the advantage of not requiring resolution of the $^{31}$P spectrum, often a difficult or impossible task.

## Conclusions

High-field NMR spectroscopy will play an increasingly important role in the field of chemical toxicology, particularly in the investigation of the interactions between a variety of toxic agents (mutagens, carcinogens) with DNA. This is a result of several factors, including the development of commercial spectrometers with superconducting high-field magnets, development of two-dimensional NMR spectroscopy (particularly COSY and NOESY spectroscopy), development of strategies for the assignment of

complex NMR spectra, and development of chemical techniques, such as oligodeoxynucleotide synthesis, making it possible to obtain sufficient quantities of these compounds for NMR analysis.

In the future, it is likely that NMR data will be able to define the solution conformation between a number of carcinogens and oligodeoxynucleotides. Whereas 500–600 MHz will likely be the limit on magnetic field strength in NMR spectrometers for the immediate future, the next 5–10 years should witness substantial progress in the area of spectral data deconvolution, data processing, and structure determination.

**Acknowledgment.** Financial support for research in this laboratory on carcinogen interactions with DNA is derived from NIH Project Grant ES03755 and Center Grant ES00267.

## References

(1) Sanders, J. K. M., and Hunter, B. K. (1987) *Modern NMR Spectroscopy*, Oxford University Press, Oxford, UK.
(2) Wüthrich, K. (1986) *NMR of Proteins and Nucleic Acids*, Wiley, New York.
(3) Croasmun, W. R., and Carlson, R. M. K., Eds. (1987) *Two-Dimensional NMR Spectroscopy Applications for Chemists and Biochemists*, VCH, New York.
(4) Yagi, H.; Hernandez, O., and Jerina, D. (1975) "Synthesis of (±)-7β,8α-dihydroxy-9β,10α-epoxy-7,8,9,10-tetrahydrobenzo[a]-pyrene, a potential metabolite of the carcinogen benzo[a]pyrene with stereochemistry related to the antileukemic triptolides". *J. Am. Chem. Soc.* **97**, 6881–6883.
(5) Vogel, E., and Günther, H. (1967) "Benzene oxide–oxepin valence tautomerism". *Angew. Chem., Int. Ed. Engl.* **6**, 385–401.
(6) Boyd, D. R., and Jerina, D. M. (1985) "Arene oxides–oxepines". In *Small Ring Heterocycles, Part 3. Oxiranes, Arene Oxides, Oxaziridines, Thietanes, Thietes and Thiazetes* (Hassner, A., Ed.) pp 197–281, Wiley, New York.
(7) Vogel, E., Schubart, R., and Böll, W. A. (1964) "Synthesis of an oxepin derivative". *Angew. Chem., Int. Ed. Engl.* **3**, 510.
(8) Vogel, E., Böll, W. A., and Günther, H. (1965) "Oxepin–benzoloxyd-valenztautomerie". *Tetrahedron Lett.* 609–615.
(9) Walling, C., and Fredricks, P. S. (1962) "Positive halogen compounds. IV. Radical reactions of chlorine and t-butyl hypochlorite with some small ring compounds". *J. Am. Chem. Soc.* **84**, 3326–3332.
(10) Kline, S. A., Solomon, J. J., and Van Duuren, B. L. (1978) "Synthesis and reactions of chloroalkene epoxides". *J. Org. Chem.* **43**, 3596–3600.
(11) Liebler, D. C., and Guengerich, F. P. (1983) "Olefin oxidation by cytochrome P-450: evidence for group migration in catalytic

intermediates formed with vinylidene chloride and *trans*-1-phenyl-1-butene". *Biochemistry* **22**, 5482–5489.

(12) Kline, S. A., and Van Duuren, B. L. (1977) "Reactions of epoxy-1,1,2-trichloroethane with nucleophiles". *J. Heterocycl. Chem.* **14**, 455–458.

(13) Meerwein, H., Disselnkötter, H., Rappen, F., v. Rintelen, H., and van de Vloed, H. (1957) "Über die Einwirkung von Diazomethan auf organische Verbindung im Licht II". *Justus Liebigs Ann. Chem.* **604**, 151–167.

(14) Havel, J. J., and Chan, K. H. (1976) "Atomic oxygen. V. Reactions of enol ethers with oxygen ($^3$P) atoms". *J. Org. Chem.* **41**, 513–516.

(15) Coles, B. F., Lindsay Smith, J. R., and Garner, R. C. (1979) "The halogenation and attempted epoxidation of 3a,8a-dihydrofuro[2,3-*b*]benzofuran and aflatoxin B$_1$". *J. Chem. Soc., Perkin Trans. 1* 2664–2671.

(16) Gorst-Allman, C. P., Steyn, P. S., and Wessels, P. L. (1977) "Oxidation of the bisdihydrofuran moieties of aflatoxin B$_1$ and sterigmatocystin; conformation of tetrahydrofurobenzofurans". *J. Chem. Soc., Perkin Trans. 1* 1360–1364.

(17) Büchi, G., Fowler, K. W., and Nadzan, A. M. (1982) "Photochemical epoxidation of alfatoxin B$_1$ and sterigmatocystin: synthesis of guanine-containing adducts". *J. Am. Chem. Soc.* **104**, 544–547.

(18) Ruzo, L. O., Casida, J. E., and Holden, I. (1985) "Direct N.M.R. detection of an epoxyfuran intermediate in peracid oxidation of the fungicide methfuroxam". *J. Chem. Soc., Chem. Commun.* 1642–1643.

(19) Corey, E. J., d'Alarcao, M., Matsuda, S. P., and Lansbury, P. T., Jr. (1987) "Intermediacy of 8(*R*)-HPETE in the conversion of arachidonic acid to pre-clavulone A by *Clavularis viridis*. Implications for the biosynthesis of marine prostanoids". *J. Am. Chem. Soc.* **109**, 289–290.

(20) Hamberg, M. (1987) "Mechanism of corn hydroperoxide isomerase: detection of 12,13(S)-oxido-9(Z),11-octadecadienoic acid". *Biochim. Biophys. Acta* **920**, 76–84.

(21) Baertschi, S. W., Ingram, C. D., Harris, T. M., and Brash, A. R. (1988) "Absolute configuration of *cis*-12-oxophytodienoic acid of flaxseed: implications for the mechanism of biosynthesis from the 13(S)-hydroperoxide of linolenic acid". *Biochemistry* **27**, 18–24.

(22) Brash, A. R., Baertschi, S. W., Ingram, C. D., and Harris, T. M. "Isolation and characterization of natural allene oxides". *Proc. Natl. Sci. U.S.A.* (in press).

(23) Dohn, D. R., and Casida, J. E. (1987) "Thiiranium ion intermediates in the formation and reactions of S-(2-Haloethyl)-L-cysteines". *Bioorg. Chem.* **15**, 115–124.

(24) McManus, S. P., Maemati-Mazraeh, N., Hovanes, B. A., Paley, M. S., and Harris, J. M. (1985) "Hydrolysis of mustard derivatives. Failure of the Raber–Harris probe in predicting nucleophilic assistance". *J. Am. Chem. Soc.* **107**, 3393–3395.

(25) Blobstein, S. H., Weinstein, I. B., Grunberger, D., Weisgras, J., and Harvey, R. G. (1975) "Products obtained after in vitro reaction of 7,12-dimethylbenz[*a*]anthracene 5,6-oxide with nucleic acids". *Biochemistry* **14**, 3451–3458.

(26) Jeffrey, A. M., Blobstein, S. H., Weinstein, I. B., Beland, F. A., Harvey, R. G., Kasai, H., and Nakanishi, K. (1976) "Structure of 7,12-dimethylbenz[*a*]anthracene–guanosine adducts". *Biochemistry* **73**, 2311–2315.

(27) Nakanishi, K., Komura, H., Miura, I., Kasai, H., Frenkel, K., and Grunberger, D. (1980) "Structure of a 7,12-dimethylbenz[*a*]anthracene 5,6-oxide derivative bound to C-8 of guanosine". *J. Chem. Soc., Chem. Commun.* 82–83.

(28) Kasai, H., Nakanishi, K., Frenkel, K., and Grunberger, D. (1977) "Structures of 7,12-dimethylbenz[*a*]anthracene 5,6-oxide derivatives linked to the ribose moiety of guanosine". *J. Am. Chem. Soc.* **99**, 8500–8502.

(29) Borgen, A., Darvey, H., Castagnoli, N., Crocker, T. T., Rasmussen, R. E., and Wang, I. Y. (1973) "Metabolic conversion of benzo[*a*]pyrene by syrian hamster liver microsomes and binding of metabolites to deoxyribonucleic acid". *J. Med. Chem.* **16**, 502–506.

(30) Swaisland, A. J., Grover, P. L., and Sims, P. (1974) "Reactions of polycyclic hydrocarbon epoxides with RNA and polyribonucleotides". *Chem.-Biol. Interact.* **9**, 317–326.

(31) Jeffrey, A. M., Jennette, K. W., Blobstein, S. H., Weinstein, I. B., Beland, F. A., Harvey, R. G., Kasai, H., Miura, I., and Nakanishi, K. (1976) "Benzo[*a*]pyrene-nucleic acid derivative found in vivo: structure of a benzo[*a*]pyrenetetrahydrodiol epoxide–guanosine adduct". *J. Am. Chem. Soc.* **98**, 5714–5715.

(32) Nakanishi, K., Kasai, H., Cho, H., Harvey, R. G., Jeffrey, A. M., Jennette, K. W., and Weinstein, I. B. (1977) "Absolute configuration of a ribonucleic acid adduct formed in vivo by metabolism of benzo[*a*]pyrene". *J. Am. Chem. Soc.* **99**, 258–260.

(33) Koreeda, M., Moore, P. D., Yagi, H., Yeh, H. Y. C., and Jerina, D. M. (1976) "Alkylation of polyguanylic acid at the 2-amino group and phosphate by the potent mutagen (±)-7β,8α-dihydroxy-9β,10β-epoxy-7,8,9,10-tetrahydrobenzo[*a*]pyrene". *J. Am. Chem. Soc.* **98**, 6770–6722.

(34) Agarwal, S. K., Sayer, J. M., Yeh, H. J. C., Pannell, L. K., Hilton, B. D., Pigott, M. A., Dipple, A., Yagi, H., and Jerina, D. M. (1987) "Chemical characterization of DNA adducts derived from the configurationally isomeric benzo[*c*]phenanthrene-3,4-diol 1,2-epoxides". *J. Am. Chem. Soc.* **109**, 2497–2504.

(35) Essigmann, J. M., Croy, R. G., Nadzan, A. M., Busby, W. F., Jr., Reinhold, V. N., Büchi, G., and Wogan, G. N. (1977) "Structural identification of the major DNA adduct formed by aflatoxin B$_1$ *in vitro*". *Proc. Natl. Acad. Sci. U.S.A.* **74**, 1870–1874.

(36) Essigmann, J. M., Green, C. L., Croy, R. G., Fowler, K. W., Büchi, G. H., and Wogan, G. N. (1983) "Interactions of aflatoxin B$_1$ and alkylating agents with DNA: structural and functional studies". *Cold Spring Harbor Symp. Quant. Biol.* **47**, 327–337.

(37) Essigmann, J. M., Barker, L. J., Fowler, K. W., Francisco, M. A., Reinhold, V. N., and Wogan, G. N. (1979) "Sterigmatocystin–DNA interactions: identification of a major adduct formed after metabolic activation in vitro". *Proc. Natl. Acad. Sci. U.S.A.* **76**, 179–183.

(38) Song, P.-S., and Tapley, K. J. (1979) "Photochemistry and photobiology of psoralens". *Photochem. Photobiol.* **29**, 1177–1197.

(39) Scott, D. R., Pathak, M. A., and Mohn, G. R. (1976) "Molecular and genetic basis of furocoumarin reactions". *Muta. Res.* **39**, 29–74.

(40) Kanne, D., Straub, K., Rapoport, H., and Hearst, J. E. (1982) "Psoralen–deoxyribonucleic acid photoreaction. Characterization of the monoaddition products from 8-methoxypsoralen and 4,5′,8-trimethylpsoralen". *Biochemistry* **21**, 861–871.

(41) Gamper, H., Piette, J., and Hearst, J. E. (1984) "Efficient formation of a crosslinkable HMT monoadduct at the Kpn I recognition site". *Photochem. Photobiol.* **40**, 29–30.

(42) Zhen, W., Buchardt, O., and Nielsen, P. E. (1986) "Site-specificity of psoralen–DNA interstrand cross-linking determined by nuclease BaI31 digestion". *Biochemistry* **25**, 6598–6603.

(43) Straub, K., Kanne, D., Hearst, J. E., and Rapoport, H. (1981) "Isolation and characterization of pyrimidine–psoralen photoadducts from DNA". *J. Am. Chem. Soc.* **103**, 2347–2355.

(44) Kanne, D., Straub, K., Hearst, J. E., and Rapoport, H. (1982) "Isolation and characterization of pyrimidine–psoralen–pyrimidine photoadducts from DNA". *J. Am. Chem. Soc.* **104**, 6754–6764.

(45) Santella, R. M.; Kriek, E.; Grunberger, D. (1980) "Circular dichroism and proton magnetic resonance studies of dApdG modified with 2-aminofluorene and 2-acetylaminofluorene". *Carcinogenesis* **1**, 897–902.

(46) Kriek, E., Miller, J. A., Juhl, U., and Miller, E. C. (1967) "8-(N-2-Fluorenylacetamide)guanosine, an arylamidation reaction product of guanosine and the carcinogen N-acetoxy-N-2-fluorenylacetamide in neutral solution". *Biochemistry* **6**, 177–182.

(47) Galliani, G., and Pantarotto, C. (1983) "The reaction of guanosine and 2″-deoxyguanosine with acrolein". *Tetrahedron Lett.* **24**, 4491–4492.

(48) Chung, F. L., and Hecht, S. S. (1983) "Formation of cyclic 1,N$^2$-propanodeoxyguanosine adducts in DNA upon reaction with acrolein or crotonaldehyde". *Cancer Res.* **43**, 1230–1235.

(49) Chung, F. L., Young, R., and Hecht, S. S. (1984) "Formation of cyclic 1,N$^2$-propanodeoxyguanosine adducts in DNA upon reaction with acrolein or crotonaldehyde". *Cancer Res.* **44**, 990–995.

(50) Benedetti, A., Comporti, M., and Esterbauer, H. (1980) "Identification of 4-hydroxynonenal as a cytotoxic product originating from the peroxidation of liver microsomal lipids". *Biochim. Biophys. Acta* **620**, 281–296.

(51) Segall, H. J., Wilson, D. W., Dallas, J. L., and Haddon, W. F. (1985) "Trans-4-hydroxy-2-hexanal: a reactive metabolite from the macrocyclic pyrrolizidine alkaloid senecionine". *Science (Washington, D.C.)* **229**, 472–475.

(52) Winter, C. K., Segall, H. J., and Haddon, W. F. (1986) "Formation of cyclic adducts of deoxyguanosine with the aldehydes trans-4-hydroxy-2-hexenal and trans-4-hydroxy-2-nonenal in vitro". *Cancer Res.* **46**, 5682–5686.

(53) Bernheim, F., Bernheim, M. L. C., and Wilbur, K. M. (1948) "The reaction between thiobarbituric acid and the oxidation

products of certain lipids". *J. Biol. Chem.* **174**, 257–264.

(54) Hamberg, M., and Samuelsson, B. (1967) "Oxygenation of unsaturated fatty acids by the vesicular gland of sheep". *J. Biol. Chem.* **242**, 5344–5354.

(55) Diczfalusy, U., Falardeau, P., and Hammarström, S. (1977) "Conversion of prostaglandin endoperoxides to $C_{17}$-hydroxy acids by human platelet thromboxane synthase". *FEBS Lett.* **84**, 271–274.

(56) Makai, F. H., and Goldstein, B. D. (1976) "Mutagenicity of malondialdehyde, a decomposition product of peroxidized polyunsaturated fatty acid". *Science (Washington, D.C.)* **191**, 868–869.

(57) Marnett, L. J., Hurd, H., Hollstein, M., Levin, D. E., Esterbauer, H., and Ames, B. N. (1984) "Naturally occurring carbonyl compounds are mutagens in Salmonella tester strain TA104". *Mutat. Res.* **148**, 25–34.

(58) Marnett, L. J., Basu, A. K. O'Hara, S. M., Weller, P. E., Maqsudur Rahman, A. F. M., and Oliver, J. P. (1986) "Reaction of malondialdehyde with guanine nucleosides: formation of adducts containing oxadiazabicyclononene residues in the basepairing region". *J. Am. Chem. Soc.* **108**, 1348–1350.

(59) Basu, A. K., O'Hara, S. M., Valladier, P., Stone, K., Mols, O., and Marnett, L. J. (1988) "Identification of adducts formed by reaction of guanine nucleosides with malondialdehyde and structurally related aldehydes". *Chem. Res. Toxicol.* **1**, 53–59.

(60) Nair, V., Turner, G. A., and Offerman, R. J. (1984) "Novel adducts from the modification of nucleic acid bases by malondialdehyde". *J. Am. Chem. Soc.* **106**, 3370–3371.

(61) Budowsky, E. I. (1976) "The mechanism of the mutagenic action of hydroxylamines". *Prog. Nucleic Acid Res. Mol. Biol.* **16**, 125–188.

(62) Marfey, F., and Robinson, E. (1981) "The genetic toxicology of hydroxylamines". *Mutat. Res.* **86**, 155–191.

(63) Shugar, D., and Kierdaszuk, B. (1985) "New light on tautomerism of purines and pyrimidines and its biological and genetic implications". *Proc. Int. Symp. Biomol. Struct. Interact. Suppl. J. Biosci.* **8**, 657–668.

(64) Budowsky, E. I., Sverdlov, E. D., and Spasokukotskaya, T. N. (1971) "Tautomerism of nucleic bases and the template synthesis of polynucleotides". *FEBS Lett.* **17**, 336–338.

(65) Flavell, R. A., Sabo, D. L., Bandle, E. F., and Weissmann, C. (1974) "Site-directed mutagenesis: generation of an extracistronic mutation in bacteriophage Q$\beta$RNA". *J. Mol. Biol.* **89**, 255–272.

(66) Stolarski, R., Kierdaszuk, B., Hagberg, C.-E., and Shugar, D. (1984) "Hydroxylamine and methoxyamine mutagenesis: displacement of the tautomeric equilibrium of the promutagen $N^6$-methoxyadenosine by complementary base-pairing". *Biochemistry* **23**, 2906–2913.

(67) Stolarski, R., Kierdaszuk, R., Hagberg, C.-E., and Shugar, D. (1987) "Mechanism of hydroxylamine mutagenesis: tautomeric shifts and proton exchange between the promutagen $N^6$-methoxyadenosine and cytidine". *Biochemistry* **26**, 4332–4337.

(68) Koga, N., Inskeep, P. B., Harris, T. M., and Guengerich, F. P. (1986) "$S$-[2-($N^7$-Guanyl)ethyl]glutathione, the major DNA adduct formed from 1,2-dibromoethane". *Biochemistry* **25**, 2192–2198.

(69) Peterson, L. A., Harris, T. M., and Guengerich, F. P. "Evidence for an episulfonium ion intermediate in the formation of $S$-[2-($N^7$-guanyl)ethyl]glutathione in DNA". *J. Am. Chem. Soc.* (in press).

(70) Nelson, J. H., Grunberger, D., Cantor, C. R., and Weinstein, I. B. (1971) "Modification of ribonucleic acid by chemical carcinogens. IV. Circular dichroism and proton magnetic resonance studies of oligonucleotides modified with N-2-acetylaminofluorene". *J. Mol. Biol.* **62**, 331–346.

(71) Grunberger, D., Nelson, J. H., Cantor, C. R., and Weinstein, I. B. (1970) "Coding and conformational properties of oligonucleotides modified with the carcinogen N-2-acetylaminofluorene". *Proc. Natl. Acad. Sci. U.S.A.* **66**, 488–494.

(72) Fuchs, R., and Daune, M. (1971) "Changes of stability and conformation of DNA following the covalent binding of a carcinogen". *FEBS Lett.* **14**, 206–208.

(73) Fuchs, R., and Duane, M. (1972) "Physical studies on deoxyribonucleic acid after covalent binding of a carcinogen". *Biochemistry* **11**, 2659–2666.

(74) Fuchs, R. P. P., and Duane, M. P. (1974) "Dynamic structure of DNA modified with the carcinogen N-acetoxy-N-2-acetylaminofluorene". *Biochemistry* **13**, 4435–4440.

(75) Fuchs, R. P. P. (1975) "*In vitro* recognition of carcinogen-induced local denaturation sites in native DNA by S$_1$ endonuclease from *Aspergillus oryzae*". *Nature (London)* **257**, 151–152.

(76) Fuchs, R. P. P., Lefebre, J.-F., Pouyet, J., and Daune, M. P. (1976) "Comparative orientation of the fluorene residue in native DNA modified by N-acetoxy-N-2-acetylaminofluorene and two 7-halogeno derivatives". *Biochemistry* **15**, 3347–3351.

(77) Santella, R. M., Grunberger, D., Weinstein, I. B., and Rich, A. (1981) "Induction of the Z conformation in poly(dGdC)·poly(dGdC) by binding of N-2-acetylaminofluorene to guanine residues". *Proc. Natl. Acad. Sci. U.S.A.* **106**, 1451–1455.

(78) Grunberger, D., Blobstein, S. H., and Weinstein, I. B. (1974) "Modification of ribonucleic acid by chemical carcinogens. VI. Effect of N-2-acetylaminofluorene modification of guanosine on the codon function of adjacent nucleosides in oligonucleotides". *J. Mol. Biol.* **82**, 459–468.

(79) Kriek, E. (1972) "Persistent binding of a new reaction product of the carcinogen N-hydroxy-N-2-acetylaminofluorene with guanine in rat liver DNA in vivo". *Cancer Res.* **32**, 2042–2048.

(80) Irving, C. C. (1966) "Enzymatic deacetylation of N-hydroxy-2-acetylaminofluorene by liver microsomes". *Cancer Res.* **26**, 1390–1396.

(81) King, C. M., and Phillips, B. (1969) "N-Hydroxy-2-fluorenylacetamide". *J. Biol. Chem.* **244**, 6209–6216.

(82) Evans, F. E., Miller, D. W., and Beland, F. A. (1980) "Sensitivity of the conformation of deoxyguanosine to binding at the C-8 position by N-acetylated and unacetylated 2-aminofluorene". *Carcinogenesis* **1**, 955–959.

(83) Evans, F. E., Miller, D. W., and Levine, R. A. (1984) "Conformation and dynamics of the 8-substituted deoxyguanosine 5'-monophosphate adduct of the carcinogen 2-(acetylamino)fluorene". *J. Am. Chem. Soc.* **106**, 396–401.

(84) Evans, F. E., Miller, D. W., and Levine, R. A. (1986) "$^1$H NMR study of self-association and restricted internal rotation of the C8-substituted deoxyguanosine 5'-monophosphate adduct of the carcinogen 2-(acetylamino)fluorene". *J. Biomol. Struct. Dyn.* **3**, 935–948.

(85) Alderfer, J. L., Lilga, K. T., French, J. B., and Box, H. G. (1984) "$^{13}$C NMR studies of the effects of the carcinogen acetylaminofluorene on the conformation of dinucleoside monophosphate". *Chem.-Biol. Interact.* **48**, 69–80.

(86) Box, H. C., Lilga, K. T., French, J., and Alderfer, J. L. (1984) "$^{13}$C- and $^{31}$P-NMR studies of the conformation of carcinogen-modified nucleic acid dimers". *Chem.-Biol. Interact.* **52**, 93–102.

(87) Box, H. C., Sharma, M., and Alderfer, J. L. (1985) "NMR studies of carcinogen-modified DNA model compounds". In *Molecular Basis of Cancer, Part A: Macromolecular Structure, Carcinogens, and Oncogenes* (Rein, R., Ed.) pp 199–206, Alan R. Liss, Inc., New York.

(88) Evans, F. E., and Levine, R. A. (1986) "Conformation and configuration at the central amine nitrogen of a nucleotide adduct of the carcinogen 2-(acetylamino)fluorene as studied by $^{13}$C and $^{15}$N NMR spectroscopy". *J. Biomol. Struct. Dyn.* **3**, 923–935.

(89) This subject is extensively reviewed in the volume: Gait, M. J., Ed. (1984) *Oligonucleotide Synthesis: a Practical Approach*, IRL, Oxford.

(90) Krugh, T. R., Sanford, D. G., Walker, G. T., and Huang, G. (1987) "Drug and carcinogen complexes with left-handed and right-handed DNAs". In *Molecular Mechanisms of Carcinogenic and Antitumor Activity, Pontificiae Academiae Scientiarum Scripta Varia* (Pullman, B., and Chagas, C., Eds.) Vol. 70, pp 147–167, Adenine Press, Schenectady, NY.

(91) Pohl, F. M., and Jovin, T. M. (1972) "Salt-induced co-operative conformation change of a synthetic DNA: equilibrium and kinetic studies with Poly(dG-dC)". *J. Mol. Biol.* **67**, 375–396.

(92) Wang, A. H.-J., Quigley, G. J., Kolpak, F. J., Crawford, J. L., van Boom, J. H., van der Marel, G. A., and Rich, A. (1979) "Molecular structure of a left-handed double helical DNA fragment at atomic resolution". *Nature (London)* **282**, 680–686.

(93) Sanford, D. G., and Krugh, T. R. (1985) "N-Acetoxy-2-acetylaminofluorene modification of a deoxyoligonucleotide duplex". *Nucleic Acids Res.* **13**, 5907–5917.

(94) Fowler, K. W., Büchi, G., and Essigmann, J. M. (1982) "Synthesis and characterization of an oligonucleotide containing a carcinogen-modified base: $O^6$-methylguanine". *J. Am. Chem. Soc.* **104**, 1050–1054.

(95) Kuzmich, S., Marky, L. A., and Jones, R. A. (1983) "Specifically alkylated DNA fragments. Synthesis and physical characterization of d[CGC($O^6$Me)GCG] and d[CGT($O^6$Me)GCG]". *Nucleic Acids Res.* **11**, 3393–3404.

(96) Gaffney, B. L., Marky, L. A., and Jones, R. A. (1984) "Synthesis and characterization of a set of four dodecadeoxyribonucleoside undecaphosphates containing $O^6$-methylguanine opposite adenine,

cytosine, guanine, and thymine". *Biochemistry* **23**, 5686–5691.

(97) Li, B. F., Reese, C. B., and Swann, P. F. (1987) "Synthesis and characterization of oligodeoxynucleotides containing 4-*O*-methylthymine". *Biochemistry* **26**, 1086–1093.

(98) Patel, D. J., Shapiro, L., Kozlowski, S. A., Gaffney, B. L., and Jones, R. A. (1986) "Structural studies of the O⁶meG·C interaction in the d(C-G-C-G-A-A-T-T-C-O⁶meG-C-G) duplex". *Biochemistry* **25**, 1027–1036.

(99) Patel, D. J., Shapiro, L., Kozlowski, S. A., Gaffney, B. L., and Jones, R. A. (1986) "Structural studies of the O⁶meG·T interaction in the d(C-G-T-G-A-A-T-T-C-O⁶meG-C-G) duplex". *Biochemistry* **25**, 1036–1042.

(100) Patel, D. J., Kozlowski, S. A., Marky, L. A., Broka, C., Rice, J. A., Itakura, K., and Breslauer, K. J. (1982) "Premelting and melting transitions in the d(CGCGAATTCGCG) duplex". *Biochemistry* **21**, 428–436.

(101) Hare, D. R., Wemmer, D. E., Chou, S. H., Drobny, G., and Reid, B. R. (1983) "Assignment of the non-exchangeable proton resonances of d(C-G-C-G-A-A-T-T-C-G-C-G) using two-dimensional nuclear magnetic resonance methods". *J. Mol. Biol.* **171**, 319–336.

(102) Patel, D. J., Kozlowski, S. A., Marky, L. A., Rice, J. A., Broka, C., Dallas, J., Itakura, K., and Breslauer, K. J. (1982) "Structure, dynamics, and energetics of deoxyguanosine–thymidine wobble base pair formation in the self-complementary d-(CGTGAATTCGCG) duplex in solution". *Biochemistry* **21**, 437–444.

(103) Kalnick, M. W., Kouchakdjian, M., Li, B. F., Swann, P., and Patel, D. J. (1988) "Base pair mismatches and carcinogen-modified bases in DNA: an NMR study of A·C and A·O⁴meT pairing in dodecanucleotide duplexes". *Biochemistry* **27**, 100–108.

(104) Kalnick, M. W., Kouchakdjian, M., Li, B. F., Swann, P., and Patel, D. J. (1988) "Base pair mismatches and carcinogen-modified bases in DNA: an NMR study of G·T and G·O⁴meT pairing in dodecanucleotide duplexes". *Biochemistry* **27**, 108–115.

(105) Hunter, W. N., Brown, T., Anand, N. N., and Kennard, O. (1986) "Structure of an adenine–cytosine base pair in DNA and its implications for mismatch repair". *Nature (London)* **320**, 552–555.

(106) Patel, D. J., Kozlowski, S. A., Ikuta, S., and Itakura, K. (1984) "Deoxyadenosine–deoxycytidine pairing in the d(C-G-C-G-A-A-T-T-C-A-C-G) duplex: conformation and dynamics at and adjacent to the dA·dC mismatch site". *Biochemistry* **23**, 3218–3226.

(107) Patel, D. J., Kozlowski, S. A., Ikuta, S., and Itakura, K. (1984) "Dynamics of DNA duplexes containing internal G:T, G:A, A:C, and T:C pairs: hydrogen exchange at and adjacent to mismatch sites". *Fed. Proc. Fed. Am. Soc. Exp. Biol.* **43**, 2663–2670.

(108) Keepers, J. W., Schmidt, P., James, T. L., and Kollman, P. A. (1984) "Molecular-mechanical studies of the mismatched base analogs of d(CGCGAATTCGCG)₂: d(CGTGAATTCGCG)₂, d-(CGAGAATTCGCG)₂, d(CGCGAATTCACG)₂, d-(CGCGAATTCTCG)₂, and d(CGCAGAATTCGCG)₂: d-(CGCGAATTCGCG)₂". *Biopolymers* **23**, 2901–2929.

(109) Hare, D. R., Shapiro, L., and Patel, D. J. (1986) "Wobble dG·dT pairing in right-handed DNA: solution conformation of the d(C-G-T-G-A-A-T-T-C-G-C-G) duplex deduced from distance–geometry analysis of nuclear Overhauser effect spectra". *Biochemistry* **25**, 7445–7456.

(110) Raap, J., Dreef, C. E., van der Marel, G. A., van Boom, J. H., and Hilbers, C. W. (1987) "Synthesis and proton-NMR studies of oligonucleotides containing an apurinic (AP) site". *J. Biomol. Struct. Dyn.* **5**, 219–247.

(111) Millican, T. A., Mock, G. A., Chauncey, M. A., Patel, T. P., Eaton, M. A. W., Gunning, J., Cutbush, S. D., Neidle, S., and Mann, J. (1984) "Synthesis and biophysical studies of short oligodeoxynucleotides with novel modifications: a possible approach to the problem of mixed base oligodeoxynucleotide synthesis". *Nucleic Acids Res.* **12**, 7435–7453.

(112) Takeshita, M., Chang, C.-N., Johnson, F., Will, S., and Grollman, A. P. (1987) "Oligodeoxynucleotides containing synthetic abasic sites model substrates for DNA polymerases and apurinic/apyrimidinic endonucleases". *J. Biol. Chem.* **262**, 10171–10179.

(113) Kalnick, M. W., Chang, C.-.N., Grossman, A. P., and Patel, D. J. (1988) "NMR studies of abasic sites in DNA duplexes: deoxyadenosine stacks into the helix opposite the cyclic analogue of 2-deoxyribose". *Biochemistry* **27**, 924–931.

(114) Stöhrer, G., Osband, J. A., and Alvarado-Urbina, G. (1983) "Site-specific modification of the lactose operator with acetyl-aminofluorene". *Nucleic Acids Res.* **11**, 5093–5102.

(115) Ni, F., Levy, G. C., and Scheraga, H. A. (1986) "Simultaneous resolution enhancement and noise suppression in NMR signal processing by combined use of maximum entropy and fourier self-deconvolution methods". *J. Magn. Reson.* **66**, 385–390.

(116) Levy, G. C., Delaglio, F., Macur, A., and Begemann, J. (1986) "NMR2: a powerful software system for processing multi-dimensional NMR data". *Comput. Enhanced Spectrosc.* **3**, 1–12.

(117) Tomic, M. T., Wemmer, D. E., and Kim, S.-H. (1987) "Structure of a psoralen cross-linked DNA in solution by nuclear magnetic resonance". *Science (Washington, D.C.)* **238**, 1722–1725.

(118) Brush, C. K., Stone, M. P., and Harris, T. M. (1988) "Selective reversible deuteration of oligodeoxynucleotides: simplification of two-dimensional nuclear Overhauser effect NMR spectral assignment of a non-self-complementary dodecamer duplex". *Biochemistry* **27**, 115–122.

(119) Connolly, B. A., and Eckstein, F. (1984) "Assignment of resonances in the ³¹P NMR spectra of d(GGAATTCC) by regiospecific labeling with oxygen-17". *Biochemistry* **23**, 5523–5527.

(120) Gorenstein, D. G., Lai, K., and Shah, D. O. (1984) "³¹P and two-dimensional ³¹P/¹H correlated NMR spectra of duplex d-(Ap[¹⁷O]Gp[¹⁸O]Cp[¹⁶O]T) and assignment of ³¹P signals in d-(ApGpCpT)₂–actinomycin D complex". *Biochemistry* **23**, 6717–6723.

(121) Shah, D. O., Lai, K., and Gorenstein, D. G. (1984) "Facile synthesis and ³¹P NMR spectra of a double-labeled oligonucleotide d(Ap[¹⁷O]Gp[¹⁸O]Cp[¹⁶O]T)". *J. Am. Chem. Soc.* **106**, 4302–4303.

(122) Petersheim, M., Mehdi, S., and Gerlt, J. A. (1984) "A general procedure for assigning the ³¹P spectra of nucleic acids". *J. Am. Chem. Soc.* **106**, 439–440.

(123) Brush, C. K., Stone, M. P., and Harris, T. M. "Selective deuteriation as an aid in the assignment of ¹H NMR spectra of single-stranded oligodeoxynucleotides". *J. Am. Chem. Soc.* (in press).

(124) Byrd, R. A., Summers, M. F., Zon, G. A., Spellmeyer Fouts, C., and Marzilli, L. G. (1986) "A new approach for assigning ³¹P NMR signals and correlating adjacent nucleotide deoxyribose moieties via ¹H-detected multiple quantum NMR. Application to the adduct of d(TGGT) with the anticancer agent (ethylenediamine)dichloroplatinum". *J. Am. Chem. Soc.* **108**, 504–505.

(125) Frey, M. H., Leupin, W., Sorensen, O. W., Denny, W. A., Ernst, R. R., and Wüthrich, K. (1985) "Sequence-specific assignment of the backbone ¹H and ³¹P-NMR lines in a short DNA duplex with homo- and heteronuclear correlated spectroscopy". *Biopolymers* **24**, 2371–2380.

(126) Pardi, A., Walker, R., Rapoport, H., Wider, G., and Wüthrich, K. (1983) "Sequential assignments for the ¹H and ³¹P atoms in the backbone of oligonucleotides by two-dimensional nuclear magnetic resonance". *J. Am. Chem. Soc.* **105**, 1652–1653.

# Chapter 15

# Fluorescence Line Narrowing: A High-Resolution Window on DNA and Protein Damage from Chemical Carcinogens

Ryszard Jankowiak and Gerald J. Small*

*Department of Chemistry and Ames Laboratory—USDOE, Iowa State University, Ames, Iowa 50011*

Reprinted from *Chemical Research in Toxicology*, Vol. 4, No. 3, May/June, 1991

## I. Introduction

Covalent binding of chemical carcinogens to DNA is believed to be the first critical step in the initiation of the tumor formation process (1–3). The polycyclic aromatic hydrocarbons (PAH) constitute a class of carcinogens for which metabolic activation to electrophilic intermediates is a necessary condition for DNA and other macromolecular damage. By virtue of the rich heterogeneous distribution of adducts they produce, the PAH provide a full spectrum of the complexity associated with understanding the initial phase of chemical carcinogenesis. There are several types of heterogeneity that one needs to consider. They stem from the existence of different metabolic pathways [for PAH the monooxygenation or diol epoxide (4, 5) and one-electron oxidation or radical cation (6, 7) pathways have been quite extensively studied], reaction

of metabolites with different bases, reaction of a metabolite with different nucleophilic centers of a given base, different DNA sites for a *given* chemical adduct, stereospecific reactions, and base sequence specificity. In addition to the problem of heterogeneity, one needs to be concerned with the persistence of macromolecular adducts, which depends on factors such as enzyme repair efficiency, chemical stability, and DNA replication rate.

During the past several years we have been attempting to develop bioanalytical spectroscopic techniques that possess the selectivity and sensitivity required to address the problem of heterogeneity. Of particular interest are techniques that can be used for the characterization and determination of macromolecular adducts. The primary purpose of this paper is to review the progress that has been made with fluorescence line narrowing spectroscopy (FLNS).[1]

Fluorescence line narrowing is a low-temperature solid-state technique that can eliminate or greatly reduce the contribution of site-inhomogeneous line broadening ($\Gamma_{inh}$) to vibronic fluorescence bandwidths. For a molecule imbedded in an amorphous host, such as a glass or polymer or protein, this broadening is very large ($\sim$100–300 cm$^{-1}$) and reflects the intrinsic structural disorder of the host. In FLNS, narrow-line laser excitation into the inhomogeneously broadened vibronic absorption bands of the $S_1 \leftarrow S_0$ absorption system leads to *site excitation energy selectivity*. Fluorescence vibronic bandwidths of 1–5 cm$^{-1}$ are readily achievable (low temperatures near 4 K are required to eliminate the thermal broadening contribution to the bandwidth). This represents an improvement in spectral resolution of about 2 orders of magnitude when measured relative to $\Gamma_{inh}$ of the vibronic absorption widths in amorphous hosts. Site-inhomogeneous line broadening

---

[1] Abbreviations: ACR, acrylamide; ALA-BP, 7,8,9-trihydroxy-*r*-7,*t*-8,*t*-9,*c*-10-tetrahydrobenzo[*a*]pyren-10-yl *N*-*t*-BOC-L-alaninate; BP, benzo[*a*]pyrene; BP-C8-dG, one-electron oxidation adduct formed by covalent bonding between C-6 of BP and the N-7 position of deoxyguanosine; BP-N7-Gua, one-electron oxidation adduct formed by covalent bonding between C-6 of BP and the N-7 position of guanine; BPDE, benzo[*a*]pyrene diol epoxide(s); BPDE–DNA, covalent adduct of BPDE and DNA; BPDE-N2-dG, monooxygenation adduct formed by covalent bonding between C-10 of the BP-diol epoxide and the 2-amino group of dG; BPT, benzo[*a*]pyrene tetraol; (dG-dC)$_2$, poly(dG-dC)·poly(dG-dC); dG, deoxyguanosine; ELISA, enzyme-linked immunosorbent assay; FLNS, fluorescence line narrowing spectroscopy; FWHM, full width at half-maximum; GMP, guanine monophosphate; HB, hole burning; HPLC, high-pressure liquid chromatography; NPHB, nonphotochemical hole burning; PAH, polycyclic aromatic hydrocarbon(s); PHB, photochemical hole burning; PSB, phonon sideband; PSBH, phonon sideband hole; *S*, Huang–Rhys factor; $S_0$ and $S_1$, electronic ground and first excited singlet state, respectively, TLC, thin-layer chromatography; ZPL, zero-phonon line(s); $\Gamma$, phonon bandwidth; $\Gamma_{hom}$, homogeneous line width; $\Gamma_{inh}$, site-inhomogeneous line broadening; $\lambda_B$, burn wavelength; $\omega'_\alpha$ and $\omega'_\beta$, excited-state vibronic frequencies; $\omega_{(0,0)}$, the center frequency of the (0,0) absorption band; $\omega_L$, frequency of laser; $\omega_m$, phonon frequency.

2428–5/92/0182$06.00/0

of PAH in crystalline and amorphous hosts has been extensively studied. In crystalline hosts $\Gamma_{inh}$ is about 2 orders of magnitude smaller than in glasses. Thus, one might say that, in the application of FLNS to molecules imbedded in amorphous hosts, a narrow-line laser is used to "trick" an amorphous host into behaving like a well-ordered crystalline host, albeit at a loss of about 2 orders of magnitude in sensitivity.

Fluorescence line narrowing emerged from the work of Szabo (8) and Personov et al. (9) on inorganic and organic doped solids, respectively. Since then it has been applied to a variety of spectroscopic and photophysical problems (10–12) very often associated with the lowest excited singlet $(S_1)$ state of planar aromatic molecules. From these studies it became evident that the (0,0) or origin absorption band of a chromophore imbedded in an amorphous host is generally site inhomogeneously broadened with $\Gamma_{inh} \sim$ 100–300 cm$^{-1}$ and a contribution to the profile that is Gaussian. One could expect, therefore, that the inhomogeneous line width of a specific DNA–PAH adduct imbedded in a glass should lie in this range. More recently, FLNS has been applied to numerous biomolecules and biological systems: chlorophylls *a* and *b* (13–15) pheophytin *a* (16, 17), protochlorophyll and protopheophytin (18), heme proteins (19), iron-free cytochrome *c* (20), etiolated leaves (21), and bacteriochlorophyll *a* and bacteriopheophytin *a* (22). The closely related line narrowing technique of spectral hole burning (10, 23–25) has now been applied to a wider variety of photosynthetic protein–pigment complexes (14, 22, 26–30). For such complexes, $\Gamma_{inh}$ also lies in the aforementioned range and it is clear that there is a significant contribution to $\Gamma_{inh}$ from statistical fluctuations in protein–pigment structure.

The first analytical applications of FLNS with glasses emerged from this laboratory and were concerned with the distinction between and quantitation of PAH in complex mixtures (31). That FLN is operative for a macromolecular DNA–PAH adduct was first established in 1984. The adduct studied was the dominant adduct from the (+)-anti stereoisomer of the benzo[a]pyrene diol epoxide (32). Its structure is given later in this review. Shortly thereafter, a detection limit for this adduct (which has pyrene as the fluorescent chromophore) of about 1 damaged base pair in $10^8$ bases for 100 μg of DNA was demonstrated at a spectral resolution of 10 cm$^{-1}$ (33). It was shown that structurally similar DNA adducts could be distinguished by FLNS (34). Comparable selectivity was demonstrated for polar PAH metabolites (35). It was also shown that FLNS can be applied to nucleoside adducts sorbed on TLC plates and that absolute quantitation is possible (36).

In this review we will discuss only our most recent work that pertains to the versatility and selectivity of FLNS for adduct analysis. Attention will be focused on benzo[a]-pyrene (BP), the most extensively studied PAH in chemical carcinogenesis. The physicochemical properties of DNA adducts derived from BP and its biological effects have been recently reviewed by Geacintov (37, 38) and Gräslund and Jernström (39). The remainder of this review is organized as follows: In section II a nonmathematical discussion of the principles of line narrowing spectroscopies is given that covers both FLN and "competitive" spectral hole burning. In section III the FLNS apparatus currently employed in our laboratory is described. In section IV the mechanistic aspects of the initial phase of chemical carcinogenesis from PAH are briefly reviewed. Section V covers five recent applications of FLNS to the analysis of carcinogen–macromolecular (DNA, globin),

–polynucleotide, and –nucleoside adducts; included here is a description of a powerful methodology that combines FLNS with laser-excited $S_2 \leftarrow S_0$ fluorescence spectroscopy at 77 K and fluorescence quenching with acrylamide. Section VI is devoted to some final remarks and a discussion of future prospects. It is in this section that the question of the relevance of DNA adduct structure types determined in a glass at 4.2 K to structure at biological temperatures is considered.

## II.  Principles of Solid-State Line Narrowing Spectroscopies

In condensed-phase molecular systems that are several mechanisms by which optical transitions and spectra are broadened. However, all broadening mechanisms can be categorized as homogeneous or inhomogeneous. Often homogeneous broadening and inhomogeneous broadening are defined and contrasted in terms of a single well-defined (i.e., all relevant quantum numbers for the initial and final states specified) transition. Homogeneous broadening is the broadening that is the same for each and every chemically identical molecule in the ensemble. In the gas phase one could consider, for example, absorption from the rotationless and vibrationless (zero-point) level of the ground electronic state to a particular rotational–vibrational level of an electronically excited state. At sufficiently low pressure this rovibronic transition would be homogeneously broadened due to the finite lifetime $(\tau_1)$ of the excited state with a full width half-maximum of

$$\text{FWHM (cm}^{-1}) = (2\pi c \tau_1)^{-1} \qquad (1)$$

where $c$ is the speed of light. For $\tau_1 = 50$ ns, which is typical for the lowest excited electronic state $(S_1)$ of polycyclic aromatic hydrocarbons, FWHM $\approx 0.0001$ cm$^{-1}$. At room temperature this lifetime broadening is negligible compared to the Doppler broadening, which is a manifestation of the distribution of molecular velocities which leads to a Guassian distribution of transition frequencies. This "heterogeneity" in frequencies is an example of inhomogeneous line broadening. However, gas-phase spectra of large molecules exhibit vibronic line widths that are most often determined by overlapping rotational transitions. This structure may also be viewed as a homogeneous broadening mechanism since the structure is common to each and every molecule.

The solid-state analogue of rotational structure is phonon (lattice–vibrational) sideband structure; it is a manifestation of the change in the lattice equilibrium structure which accompanies the electronic excitation. The analogue of Doppler broadening is site-inhomogeneous broadening, which is the result of an analyte or probe molecule in a solid host adopting a very large number of energetically inequivalent sites. This leads to a Gaussian distribution of frequencies for any given vibronic transition. For amorphous hosts, such as glasses and polymers, the inhomogeneous broadening, $\Gamma_{inh}$, is $\approx 100$–300 cm$^{-1}$. For crystalline hosts, $\Gamma_{inh}$ is reduced by about 2 orders of magnitude.

Fluorescence line narrowing spectroscopy (FLNS) is a member of a class of laser-based spectroscopies that can eliminate or significantly reduce the contribution of $\Gamma_{inh}$ to the vibronic bandwidths. Spectral hole burning spectroscopy is another member of this class. It should be noted that spectral hole burning experiments have shown that for chlorophylls of light harvesting and reaction center complexes $\Gamma_{inh} \approx 100$–200 cm$^{-1}$ (40–44), essentially the same as for chlorophyll embedded in organic glass hosts (14).

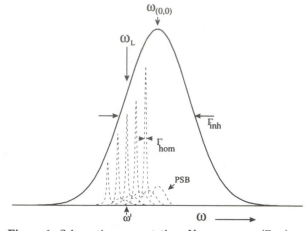

**Figure 1**. Schematic representation of homogeneous ($\Gamma_{hom}$) and inhomogeneous ($\Gamma_{inh}$) broadening. Profiles of the zero-phonon lines (ZPL) and their associated phonon sidebands (PSB) for specific sites at different frequencies have been enlarged compared to the inhomogeneous line to provide more detail. $\omega_L$ is the laser frequency that selectively excites a narrow isochromat of an inhomogeneously broadened absorption band.

This is also the case for PAH metabolites covalently bound to DNA.

**A. Principles of Fluorescence Line Narrowing.** Fluorescence line narrowing spectroscopy has been the subject of several reviews (10–12). Below, a nonmathematical discussion is presented (for a theoretical discussion see ref 10). Figure 1 is a schematic representation of an inhomogeneously broadened electronic absorption origin band, (0,0), at low temperature. The relatively sharp dashed bands depict the (0,0) transitions of the "guest" molecule occupying inequivalent sites. It is customary to refer to the dashed band as the zero-phonon line (ZPL). A zero-phonon transition is one for which no net change in the number of phonons accompanies the electronic transition. Building to higher energy on each ZPL in Figure 1 is a broader phonon wing. This is the phonon sideband (PSB). We return to a discussion of the PSB later. As mentioned earlier, $\Gamma_{inh} \approx 100$–300 cm$^{-1}$ for glassy hosts. Even for a picosecond excited-state lifetime ($\tau_1$), the pure lifetime broadening is well over an order of magnitude less than $\Gamma_{inh}$. Each single site ZPL carries a homogeneous line width, $\Gamma_{hom}$, which is determined by the total dephasing time $\tau_2$ of the optical transition:

$$\frac{1}{\tau_2} = \frac{1}{2\tau_1} + \frac{1}{\tau_2'} \qquad (2)$$

$\tau_1$ is defined above, and $\tau_2'$ is the pure dephasing time. This equation would be familiar to practitioners of magnetic resonance spectroscopies. The time $\tau_2'$ is a fundamentally important and interesting quantity, best understood by means of the density matrix formulation of spectroscopic transitions (45, 46). For our purposes it suffices to say that $\tau_2'$ is due to a modulation of the single-state transition frequency that results from the interaction of the excited state with the bath phonons [and other low-energy excitations in glasses (23)]. This interaction does not lead to electronic relaxation of the excited state but rather to a decay of the phase coherence of the superposition state initially created by the photon. Loosely speaking, one can say that $\tau_2'$ leads to an uncertainty broadening associated with the time required for the

system to "forget" how it was excited. In units of cm$^{-1}$

$$\Gamma_{hom} = (\pi \tau_2 c)^{-1} \qquad (3)$$

where $c$ is the speed of light in cm s$^{-1}$.

Obviously $\Gamma_{hom}$ determines the ultimate spectral resolution (selectivity) attainable by line narrowing techniques and since, for moderate to good fluorescers, the broadening from $\tau_1$ is very small, it is $\tau_2'$ that is of concern. The key point for analytical fluorescence spectroscopy is that $\tau_2'$ is strongly temperature dependent. Pure dephasing theories are now well developed (23, 47, 48), and photon echo (49–51, 53) and spectral hole burning (23–25, 52–55) have been used to study the temperature dependence of $\tau_2'$ in a wide variety of molecular systems. At room temperature, $\Gamma_{hom}$ from $\tau_2'$ (determined by the interaction with both phonons) is $\approx kT$, i.e., $\approx 200$ cm$^{-1}$, which is comparable to $\Gamma_{inh}$ for glasses. Line narrowing spectroscopies cannot eliminate $\Gamma_{hom}$, which means that low temperatures are required to minimize the number of thermally populated low-frequency phonons responsible for $\Gamma_{hom}$. For glass hosts it is now firmly established that $\Gamma_{hom}$ from pure dephasing is $\lesssim 0.1$ cm$^{-1}$ at 4.2 K (23, 24, 53–55).

The basic principles of FLN can be understood from Figure 1. First, we note that if a broad-band classical excitation source is used to excite the (0,0) band, all sites (ZPL) will be excited and all sites will fluoresce, resulting in a broad fluorescence spectrum characterized by vibronic bandwidths equal to $\Gamma_{inh}$. When a laser of frequency $\omega_L$ and line width $\Delta\omega_L \ll \Gamma_{inh}$ is used, only ZPL (sites) whose transition frequency overlaps the laser profile will be excited. In the absence of intermolecular energy transfer only this "isochromat" will fluoresce, resulting in a "line-narrowed" fluorescence spectrum. The spectrum consists of an origin ZPL coincident with $\omega_L$ and, to lower energy, numerous ZPL corresponding to transitions to intramolecular vibrational sublevels of the ground electronic states. Note that as $\omega_L$ is tuned across the inhomogeneously broadened absorption, the origin ZPL and, therefore, the entire fluorescence spectrum will "track" $\omega_L$. For (0,0) or origin band excitation the origin ZPL is not a useful analytical line due to interference from scattered laser light. For this reason and others related to selectivity, it proves advantageous to excite into vibronic bands (vibronically excited FLN).

To demonstrate this, we turn to Figure 2A in which the electronic ground and fluorescent states are labeled as $S_0$ and $S_1$. Without loss of generality and for simplicity, the zero-point vibrational level of $S_0$ (solid horizontal line) is shown as isoenergetic for all sites. The existence of site-inhomogeneous broadening for the (0,0) absorption transition is indicated by the "slanted" solid zero-point line for $S_1$. The magnitude of $\Gamma_{inh}$ is indicated to the left of the $S_1$ state. The slanted dashed lines for the $S_1$ state denote vibrational sublevels $\alpha$, $\beta$, $\gamma$. The vibronic absorption transitions to these levels are labeled as $(1_\alpha,0)$, etc., where the zero indicates that the absorption originates from zero-point level of $S_0$. Since $\Gamma_{inh}$ for these transitions is essentially the same as for the (0,0) transition, they will not be resolved or well resolved in absorption, even at very low temperatures. The thick solid vertical arrow is the laser excitation frequency ($\omega_L$) chosen to excite isochromats A and B belonging to the $(1_\beta,0)$ and $(1_\alpha,0)$ transitions, respectively. The intersections of the horizontal dash–dot line with the dashed lines are the locations of the isochromats in the site excitation energy distributions for the two vibronic transitions. Because intramolecular vibrational relaxation occurs on a picosecond time scale, the two initially excited isochromats rapidly relax to their re-

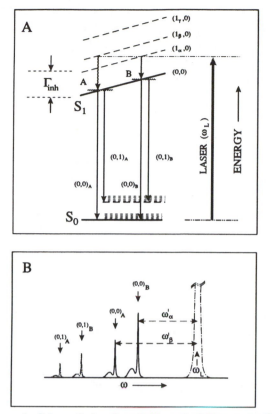

**Figure 2.** Schematic for vibronically excited fluorescence line narrowing. (A) The slope of excited-state levels represents the variation of their energies as a function of the site. $\Gamma_{inh}$ denotes the inhomogeneous broadening of the (0,0) transition. Using the laser excitation, $\omega_L$, two subsets of molecules within $\Gamma_{inh}$ are selectively excited. (B) Schematic of the resulting fluorescence spectrum. Fluorescence from two isochromats results in a doubling of the origin or (0,0) transitions as well as the (0,1) vibronic transitions.

spective and correlated positions (A and B) in the zero-point distribution (squiggly downward arrows). The fluorescence transitions from these two positions to the zero point and a vibrational level of $S_0$ are shown as solid arrows. The resulting line-narrowed fluorescence spectrum is given in Figure 2B. Note that the (0,0) and (0,1) transitions are doublets. More importantly, the spectrum yields *both* the ground and excited-state vibrational frequencies. For example, the displacements of $(0,0)_A$ and $(0,0)_B$ from $\omega_L$ yield the excited-state frequencies $\omega_\alpha'$ and $\omega_\beta'$.

The $(0,0)_A$ and $(0,0)_B$ lines comprise what will be referred to as *multiplet origin structure*. Several examples of multiplet structure with more than two lines will be shown later. It is because $\Gamma_{inh} \gg \Gamma_{hom}$ at low $T$ that one can observe such structure and, therefore, determine excited-state vibrational frequencies. For reasons not completely understood (*56*), excited-state vibrational structure is often far more sensitive to minor structural perturbations than ground-state vibrational structure.

With Figure 2A one can see that, for (0,0) band excitation, the FLN spectrum provides information on ground-state vibrations only and, as mentioned earlier, the (0,0) fluorescence band is not generally useful as an analytical line due to laser light scatter.

It is usually possible to determine the (0,0) absorption band. Let the center frequency of the band be $\omega_{(0,0)}$. How

does one generate a multiplet origin structure in fluorescence corresponding to excited vibrations in the range $\Omega \pm \Gamma_{inh}/2$? The answer is apparent from Figure 2A. The excitation frequency should be chosen so that $\omega_L \approx \omega_{(0,0)} + \Omega$. In our laboratory we routinely employ vibronically excited FLN and focus on the origin multiplet structure. This structure is associated with intramolecular vibrational sublevels of $S_1$ initially excited. Such levels have a lifetime of $\approx 1$ ps due to vibrational relaxation. Therefore, the multiplet origin bands will generally be no sharper than $\approx 5$ cm$^{-1}$ (eq 1).

We consider now the relatively broad PSB that builds on the ZPL in absorption and fluorescence (Figures 1 and 2B). The PSB is contributed to by 1-, 2-, ... phonon transitions; one speaks of the 1-, 2-, ... phonon profiles. The 1-phonon profile is given by $g(\omega) D(\omega)$, where $\omega$ is measured relative to the ZPL. Here $g(\omega)$ is the phonon density of states and $D(\omega)$ is the electron–phonon coupling constant. For PAH in glasses, the 1-phonon profile has a maximum at $\omega_m \approx 25$ cm$^{-1}$ and a width of $\Gamma \approx 40$ cm$^{-1}$. The $r$-phonon profile has a maximum at $r\omega_m$ and a width of $\approx r^{1/2}\Gamma$ (*57*). Of particular importance to FLN is that the Franck–Condon factor for the ZPL is $\exp(-S) \equiv P_0$ while that for the $r$-phonon process is $[\exp(-S)]S^r/r! \equiv P_r$. Noting that $\sum P_r = 1$, it is easy to show that for $S < 1$ (weak coupling) the ZPL is dominant while, for $S > 1$ (strong coupling), the PSB is dominant. For large $S$, the ZPL becomes Franck–Condon forbidden and FLN will not be possible. Except for intercalated DNA–PAH adducts, we have found that adduct spectra are characterized by weak coupling.

An example of weak coupling and the selectivity achievable with FLN is shown in Figure 3. Adducts A, B, and C are derived from *trans*-1,2-dihydroxy-*anti*-3,4-epoxy-1,2,3,4-tetrahydrochrysene, *trans*-3,4-dihydroxy-*anti*-1,2-epoxy-1,2,3,4-tetrahydrobenz[a]anthracene, and *trans*-8,9-dihydroxy-*anti*-10,11-epoxy-8,9,10,11-tetrahydrobenz[a]anthracene, respectively. Binding is at the exocyclic N2 of guanine for all adducts, and the adducts have phenanthrene as the parent fluorophore (*1, 3, 34, 37, 58*). The standard adduct spectra shown were obtained with $\lambda_{ex} = 343.1$ nm, which provides vibronic excitation for all three adducts. Their multiplet origin structures are dominated by ZPL which are labeled with $S_1$ vibrational frequencies. From the earlier discussion of FLN it is apparent, from Figure 3, that the $S_1$-state energies increase in the order B, C, A. These adducts plus two others in a mixture were shown to be readily distinguishable by FLN (*34*). The mixture spectrum for $\lambda_{ex} = 343.1$ nm shown in Figure 3 is clearly the superposition of the three standard adduct spectra. It would not be possible to distinguish between these adducts on the basis of room temperature fluorescence.

It is important to realize that different multiplet origin structure will be observed for different $\lambda_{ex}$; by appropriate variations of $\lambda_{ex}$ it is possible to determine all active excited-state fundamental intramolecular vibrations.

**B. Spectral Hole Burning.** Solid-state spectral hole burning (*9, 10, 23, 24, 51, 53*) is another example of a line narrowing technique which can be applied to biomolecular systems. For the cofactors of photosynthetic antenna and reaction center complexes hole burning has improved the resolution of optical spectra by 2–4 orders of magnitude (*40–44*). As will be seen, hole burning can be a complication for FLN.

There are two types of hole burning (HB): nonphotochemical (NPHB) and photochemical (PHB). With an understanding of site-inhomogeneous broadening and

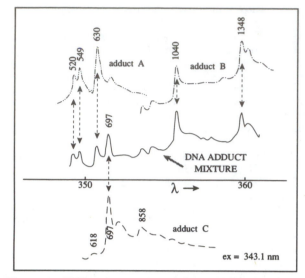

**Figure 3.** FLN spectra of the three individual adducts in a mixture of five. (A) CDE–DNA, (B) B[a]ADE-1–DNA, and (C) B[a]ADE-8–DNA adducts. Solid line represents the fluorescence spectrum of the DNA adduct mixture. Spectra were obtained at $T = 4.2$ K with $\lambda_{ex} = 343.1$ nm. Adducts were generated in vitro at comparable modification levels. See ref 34 for details.

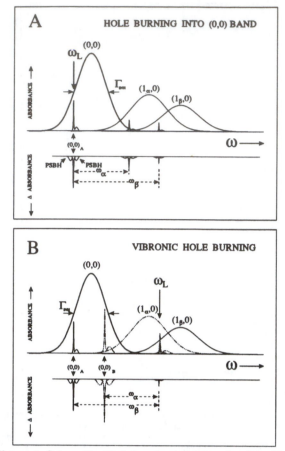

**Figure 4.** Schematic of hole burning (selective photobleaching) into origin band (A) and into vibronic region (B) (see explanation in text).

isochromat selection by a narrow-line laser, PHB follows in a natural way from FLN. What is required is that the absorbing chromophore be photoreactive. Only the ZPL of sites contributing to the isochromat at $\omega_L$ will be excited. As a consequence, isochromat-selective photobleaching occurs and a zero-phonon hole (ZPH) will appear at $\omega_L$ [see the ZPH at $\omega_L$ in the $\Delta$ (absorbance (difference) spectrum of Figure 4A]. Provided the photochemistry is irreversible, photochemical holes are persistent provided the sample is held at a low temperature. The literature on PHB is very extensive (24, 54), and PHB can be observed for amorphous and crystalline hosts. Photoreactivity is not required in NPHB. However, NPHB is generally observed only for amorphous hosts such as glasses, polymers, and proteins. A mechanism for persistent NPHB was first proposed in 1978 (59). It is based on the inherent structural disorder of a glass and phonon-assisted tunneling between different impurity–glass configurations. In simplest terms: upon completion of the ground state → excited state → ground state cycle, the host configuration around the impurity is altered, more or less permanently; the change in environment produces a shift of the ZPL frequency at $\omega_L$ to other regions of the inhomogeneously broadened absorption profile. However, the absorber (impurity) does not undergo photodecomposition. Ionic dye molecules in alcoholic glasses and polymers (60, 61) and chlorophylls in antenna complexes (40) are two classes that exhibit facile NPHB.

In the absorbance spectrum of Figure 4A the burn frequency excites an isochromat in the (0,0) band. However, the sites that contribute to the origin isochromat also contribute to the $(1_\alpha,0)$ and $(1_\beta,0)$ vibronic bands. Thus, a ZPH burnt at $\omega_L$ will be accompanied by higher energy vibronic satellite holes, as indicated in the $\Delta$(absorbance) spectrum. Because the ZPL in absorption is accompanied by a PSB, the ZPH is accompanied by phonon sideband holes (PSBH). The PSBH to higher energy of the ZPH is readily understood and is referred to as the real PSBH. That a pseudo-PSBH to lower energy of the ZPH should appear is not so obvious. The pseudo-PSBH is due to sites whose ZPL frequencies lie to lower energy of $\omega_L$ and which

absorb the laser light by virtue of their PSB. The phonons excited rapidly relax to the zero-point level after which NPHB causes. A recent example of ZPH and PSBH structure is shown in Figure 5B. The dashed spectrum is the absorption of the D1–D2 reaction center complex of photosystem II at 4.2 K (44). The hole structure shown is actually due to the $Q_y$-state of the pheophytin $a$ molecule active in primary charge separation (44, 62). The hole structure persists indefinitely at 4.2 K but can be eliminated by annealing at much higher temperatures.

In the same manner that pseudo-PSBH can be produced, pseudo-vibronic hole structure can be generated. The basic idea is very similar to that involved in vibronically excited FLN. In Figure 4B the burn frequency ($\omega_L$) excites isochromats belonging to vibrations $\alpha$ and $\beta$. Since the inverse of the average rate constant for NPHB is no larger than $\approx 1$ $\mu$s, the vibrational isochromats relax to their respective zero-point positions in the (0,0) band prior to hole burning. Two ZPH [at $(0,0)_A$ and $(0,0)_B$] are produced, which, in turn, lead to a hole at $\omega_L$. The relative intensities of the former two and latter holes depend, in part, on the Franck–Condon factors for vibrations $\alpha$ and $\beta$. A recent example of this type of hole burning is shown for the photosystem II reaction center (in the presence of Triton X-100 detergent) in Figure 5A. Because the burn wavelength, $\lambda_B$, lies $\approx 1000$ cm$^{-1}$ to higher energy of the absorption origin region, the pseudovibronic hole structure is associated with vibrations possessing frequencies in the vicinity of 1000 cm$^{-1}$. The displacements of the vibronic "satellite" holes from the burn frequency yield the excited-state vibrational frequencies. The structure shown is

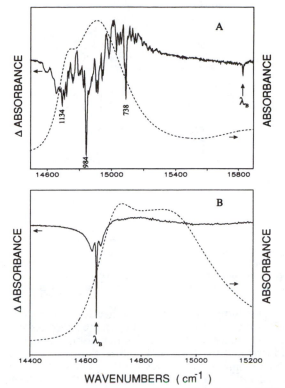

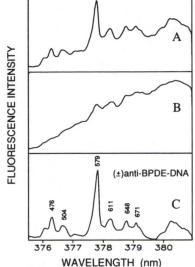

**Figure 5.** (A) Vibronic satellite hole burned spectrum (solid line) of PS II reaction center. Several of the satellite holes (e.g., 738, 984, and 1134) are labeled with their excited-state vibrational frequencies in cm$^{-1}$; $T_B$ = 4.2 K, $\lambda_B$ = 632 nm (unpublished data). Dashed line represents the 4.2 K absorption spectrum of PS II RC with ~0.05% Triton X-100 detergent. (B) Dashed spectrum corresponds to the absorption spectrum of PS II RC without Triton X-100. Solid line shows an example of nonphotochemical hole burned spectrum of PS II PC burned into (0,0) band at $\lambda_B$ = 682.5 nm. The % $\Delta A$ change for the sharp zero-phonon hole is ~20%. Hole burning conditions: $I_B$ = 200 mW/cm$^2$; $\tau_B$ = 20 min; $T_B$ = 4.2 K.

due mainly to the accessory chlorophyll *a* of the photosystem II reaction center.

Although NPHB is a powerful spectroscopic technique for photosynthetic protein–pigment complexes (*40–44*), it, as well as PHB, can lead to a severe degradation in the quality (resolution) of FLN spectra. If hole burning is too efficient, the ZPL at $\omega_L$ responsible for FLN will be removed prior to recording of the FLN spectrum, resulting in broad emission from PSB excitation. In our studies of macromolecular DNA–PAH adducts we have found that PHB from photooxidation is often facile. Such hole burning, however, has often been largely eliminated by degassing of the sample. Nonphotochemical hole burning can actually be used to improve the resolution of FLN spectra (*33, 63*), vide infra, and a synchronous scanning method has been developed to circumvent hole burning during the recording of the fluorescence spectrum (*33*).

Our approach for utilizing hole burning to improve the resolution of FLN is an extension of a procedure used by Bogner and Schwarz (*69*) and Fünfschilling et al. (*65*). It is one that minimizes the broad fluorescence from PSB excitation (Figure 1). The method involves recording the FLN spectrum at various stages of NPHB for a fixed $\lambda_{ex}$. As NPHB proceeds, the FLN spectrum degrades because of the diminution of fluorescence originating from sites excited by their zero-phonon transitions (see Figure 1). An example for (±)-*anti*-BPDE–DNA is shown in Figure 6.

**Figure 6.** FLN spectra of (±)-*anti*-BPDE–DNA, $\lambda_{ex}$ = 369.6 nm, $T$ = 4.2 K. (A) Spectrum obtained with a laser excitation power density of $I$ = 85 mW/cm$^2$ and an exposure time of ~100 s. (B) Spectrum obtained under experimental conditions as in (a), following a 100-s exposure to 400 mW/cm$^2$ excitation power density. (C) The "difference" spectrum (A – B) obtained as discussed in detail in the text. The numbers correspond to excited-state vibrations in cm$^{-1}$.

Spectrum A was obtained at an earlier irradiation time than spectrum B. The difference spectrum A – B represents the FLN spectrum of the molecules that have been burned out by irradiation during the time interval between the recording of the two spectra. Therefore, spectrum C is considerably sharper (better resolved) since it is not contributed to by fluorescence from sites excited by phonon sideband transitions. Careful comparison of spectrum C with recently generated FLN spectra for different stereoisomers of BPDE (unpublished data) indicates that spectrum C is the FLN spectrum of the trans isomer of (+)-*anti*-BPDE–DNA, although the contribution from (−)-*anti*-BPDE isomer cannot be excluded (compare also with Figure 11A,B). This method is most effective for weak electron–phonon coupling. The difference spectra methodology can also be applied to short-lived fluorescing species for which gated fluorescence detection is not possible and scattered laser light poses a problem (*33*).

To conclude this section, we note that hole burning can be used to study the photochemical and/or photophysical properties of drug molecules bound to DNA. Results of experiments on the antitumor drug daunomycin bound to two different oligonucleotides [d(AT)$_5$ and d(CG)$_5$] have recently been reported (*66*). A factor of 60 difference in the hole burning quantum yields of the two daunomycin–oligonucleotide complexes was observed and ascribed to the significant differences in the environment of daunomycin.

## III. Instrumentation

The FLNS system employed in our laboratory was designed to provide a spectral resolution of ~3 cm$^{-1}$ at 400 nm with photon-counting sensitivity. The equipment needed to perform FLN can be divided into four major components: a spectrally narrow excitation source (tunable laser), an optically accessible low-temperature sample chamber, a dispersion device, and a detection system. A block diagram of the FLNS instrumentation is shown in

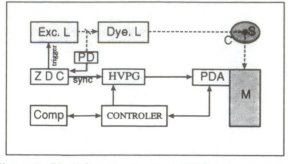

**Figure 7.** Block diagram of present FLNS instrumentation: excimer laser (Exc. L), dye laser (Dye. L), photodiode (PD), zero-drift control (ZDC), cryostat (C), sample (S), photodiode array (PDA), monochromator (M), high-voltage pulse generator (HVPG), controller, and computer (Comp).

Figure 7. Because most PAH metabolites covalently bound to globin and DNA absorb in the 340–450-nm region, a Lambda Physik EMG 102 MSO excimer (XeCl gas) pumped-dye laser (Lambda Physik FL-2002) with a 10-ns pulse width provides a convenient excitation source. Average power densities used typically ranged from 1 to 100 mW/cm$^2$ at a pulse repetition rate of 30 Hz. A 1-m focal length McPherson 2061 monochromator (f/7.0) with a reciprocal linear dispersion of 0.42 nm/mm (2400 grooves/mm grating) is used to disperse the fluorescence. Luminescence is detected with a Princeton Instruments IRY-1024/G/R/B intensified blue-enhanced gateable photodiode array (PDA), which determines the optimum resolution, interfaced with a Princeton Instruments ST-120 controller. The PDA and the monochromator provide an ~9-nm segment of the fluorescence spectrum for a given monochromator setting. Gated detection in fluorescence studies is accomplished by using a Lambda Physik EMG-97 zero-drift controller (ZDC) to trigger a high-voltage pulse generator (HVPG), Princeton Instruments Model FG-100, which provides adjustable delay and width of the detector's temporal observation window. The ZDC compares the timing of the laser firing as measured by the photodiode, PD, with the synchronization pulse from the trigger circuit. Long-term temporal drift of the laser firing is avoided by maintaining a constant time difference between these pulses. With the system employed it is possible to obtain gating times as short as 5 ns. Fluorescence spectra at 77 K were detected by the same experimental system except that photodiode array was replaced by a photomultiplier tube (PMT). The signal from PMT is sent to an SRS Model SR 250 boxcar averager. For the fluorescence data presented in this paper the delay time for fluorescence measurements was ~20 ns and the width of the observation window was 60 ns.

Absorption and hole burned spectra were obtained with a Bruker IFS 120 HR Fourier transform spectrometer operating at a resolution of 2 cm$^{-1}$ for extended scans shown in Figure 5A,B. A Coherent 699-21 CW ring dye laser (DCM dye) pumped by an Innova 90-5 argon ion laser with a line width of <0.002 cm$^{-1}$ was used for hole burning. Burn intensities and times are given in the figure captions.

For FLNS a double-nested glass liquid helium Dewar (with fused-quartz optical windows) was designed to eliminate liquid nitrogen from the optical pathways. For hole burning studies, a Janis Research Model 8-DT Super Vari-Temp liquid helium cryostat equipped with an optical access tail section was used. Sample temperatures were measured with a silicon diode thermometer (Lake Shore Cryogenic Model DT-500K).

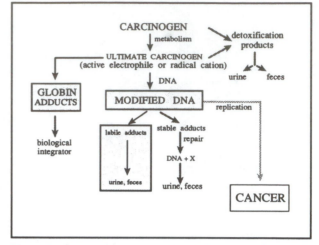

**Figure 8.** Schematic representation of the principal events in chemical carcinogenesis. Note that in fact FLNS can be directly applied not only for studies of intact modified DNA but also for globin, depurinated DNA adducts, and detoxification products.

A glass-forming solvent of 50% glycerol, 40% water, and 10% ethanol by volume is most often employed for studies on macromolecular DNA and globin adducts, nucleoside adducts, and the PAH metabolites themselves. Quartz tubing (3 mm o.d. × 2 mm i.d. × 1 cm) is used to contain the solvent (~30 μL total volume). For the hole burning studies of photosystem II reaction center, samples with appropriate optical density were prepared by dilution with buffered glycerol/H$_2$O (60:40) in polystyrene centrifuge tubes (path length ~ 1 cm) and cooled rapidly to liquid helium temperature (for details, see refs 44 and 62). The samples for FLNS studies are also rapidly cooled from 300 to 4.2 K in about 2 min by direct immersion in boiling helium. To avoid photodecomposition, the sample was subjected to several freeze–pump–thaw cycles and sealed under vacuum. Macromolecular adducts (DNA, globin) and reaction center samples were stored at −10 and −80 °C until use, respectively. All samples were prepared in dim light and at ambient temperature, except for reaction center samples, which were prepared at 0 °C.

## IV. Mechanisms of PAH Chemical Carcinogenesis

A schematic representation of the principal events in chemical carcinogenesis is given in Figure 8. Polycyclic aromatic hydrocarbons are not reactive toward DNA and/or globin and must be metabolized to a reactive electrophiles which can bind to DNA. Covalent binding of PAH metabolites to DNA is generally believed to be the first critical step in the multistage process that leads to tumor formation (1–3). Although globin adducts are not considered to be pathobiological lesions, they provide useful complementary data to DNA adduct levels, serving as dosimeters for carcinogen exposure (67, 68) (hemoglobin modified by PAH can be obtained from blood samples more easily and in greater amounts than DNA). The characterization and determination of labile adducts which undergo elimination (see section V-C) is also very important (69–73). As shown in Figure 8, there are nontumorigenic pathways. For example, carcinogenic metabolites can be detoxified and removed by excretion into urine and/or feces. Damaged DNA can be enzymatically repaired (74). All pathways in Figure 7 are strongly influ-

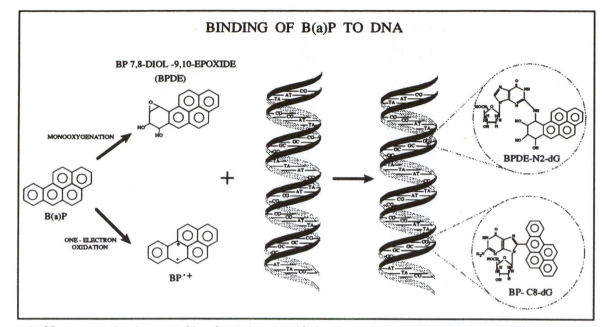

**Figure 9.** Monooxygenation (upper path) and one-electron oxidation (lower) in the metabolic activation. Two types of adducts, BPDE-N2-dG and BP-C8-dG, are formed with the DNA base guanine acting as the nucleophile.

enced and often determined by host-dependent factors which may vary according to cell type, tissue, strain, species, and individual (*74*).

The current view for PAH is that metabolic activation to electrophilic intermediates occurs by two pathways: *monooxygenation* to yield diol epoxides (*4, 5*) and *one-electron oxidation* to produce radical cations (*6, 7*). Benzo[*a*]pyrene (BP) is a good example of a PAH metabolized according to this dual pathway, as illustrated in Figure 9. For BP these pathways of activation yield six adducts, the stable one (BPDE-N2-dG as shown in Figure 9) and several labile, e.g., N7-Gua or N7-Ade type adducts (not shown) and BP-C8-dG shown in Figure 9, which rapidly depurinate from DNA, leaving an apurinic site (refs 72 and 73, and unpublished data). The major identified BP-diol epoxide adduct (due to monooxygenation) arises from a covalent bonding between C-10 of the BP-diol epoxide and the 2-amino group of dG (BPDE-N2-dG) (*4, 5, 38, 39*) (see upper part of Figure 9). When BP is activated by one-electron oxidation to its radical cation, the C-6 of BP can, for example, bind to the C-8 position of dG, forming BP-C8-dG (lower part of Figure 9). In addition to cellular peroxidases and prostaglandin H synthase, cytochrome P-450 catalyzes the one-electron oxidation (*69, 75*). The microsomal enzyme systems present in cells (including cytochrome P-450 and aryl hydrocarbon hydroxylase) catalyze the monooxygenation of PAH to more water-soluble oxygenated derivatives. Arene oxides (epoxides) have received primary attention in this regard and are the likely precursors of numerous PAH metabolites: phenols, dihydrodiols, diol epoxides, tetrols, and conjugation products (*38, 39*). Considerable support for the monooxygenation pathway has come from the determination in numerous in vitro and in vivo experiments (*5, 39*) that the structure of a major adduct from BPDE is BPDE-N2-dG. This adduct is extremely stable in DNA and exists in two different DNA conformations (*76*) (see section V-B). The BP-C8-dG adduct, which is quite stable in DNA, has been identified in vitro (*63*). Labile N7-Gua adducts from both activation pathways have also been observed (*72–74*).

## V. Recent Applications of FLNS to the Analysis of Macromolecular and Nucleoside Adducts

This section is divided into five parts: the first two underscore the superior selectivity of FLNS, and the last three establish the potential of FLNS for in vivo studies.

**A. Stereoisomeric DNA Adducts from BPDE.** Although selectivity to DNA adducts of the type shown in Figure 3 is notable, more challenging is the distinction between different stereoisomers of a DNA adduct. The stereoisomers of BPDE are particularly interesting since they exhibit very different tumorigenic and mutagenic activities. The structures are shown in Figure 10 along with the structures of the tetraols formed by hydrolysis. Monoxygenation type adducts from BP and DNA appear to occur primarily at guanine, and the relative yields from the *anti*-BPDE and *syn*-BPDE isomers are dependent on the species under study (*78–81*). In the anti diastereomer (*anti*-BPDE) the benzylic hydroxyl group and the epoxide oxygen atom are on opposite faces of the molecule, whereas in the syn isomer (*syn*-BPDE) these groups are on the same face. Because each diastereomer may exist as a pair of enantiomers, four stereoisomers of BPDE are possible.

FLN spectra for native DNA adducts derived from (+)- and (−)-*anti*-BPDE and *syn*-BPDE (*12*) are shown in Figure 11. These spectra were obtained with an excitation wavelength of 369.6 nm (vibronic excitation); the bands in each spectrum represent the multiplet origin structure. The adducts can be distinguished primarily by vibronic intensity distributions (*12, 33, 82*). We have shown (unpublished results) that the differences in vibronic intensity distribution between spectra A and B of Figure 11 are due to a difference in the distribution of DNA adduct sites. For example, the concentration of site I type adducts [designated as (−)-**3** or (+)-**3**] for (−)-*anti*-BPDE–DNA is about 30–40%, which in the case of (+)-*anti*-BPDE–DNA is only ~10% (*37*). Higher relative concentrations of (−)-2–DNA [from (−)-*anti*-BPDE] as compared to (+)-2–DNA [from (+)-*anti*-BPDE], though not quantitated, were also observed. Further details can be found in refs 77, 82, and 85 (see also section V-B). The spectra in Figure 11 are

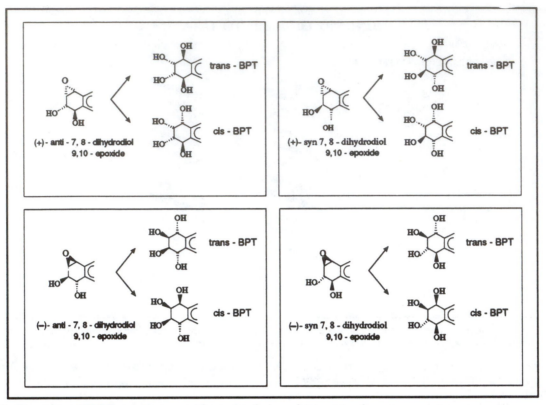

**Figure 10.** Structures of BPDE stereoisomers and associated *cis*- and *trans*-tetraols.

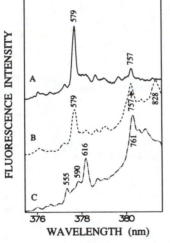

**Figure 11.** Comparison of the FLN spectra of three different DNA adducts from different BPDE stereoisomers. All spectra were obtained in the standard glass at $T = 4.2$ K, $\lambda_{ex} = 369.6$ nm. (A) (+)-*anti*-BPDE–DNA; 0.5% bases modified. (B) (−)-*anti*-BPDE–DNA; 1.5% bases modified. (C) (±)-*syn*-BPDE–DNA; ∼1 adduct in $10^7$ bases. The FLN peaks are labeled with their corresponding excited-state vibrational frequencies (cm⁻¹).

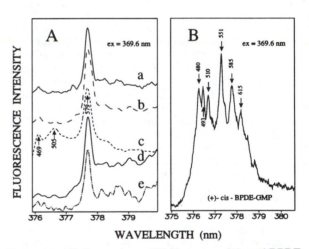

**Figure 12.** (A) Comparison of FLN spectra: (+)-*anti*-BPDE–DNA, (+)-1 adduct (a); (+)-*anti*-BPDE–(dG-dC)₂ (b); trans isomer of (+)-*anti*-BPDE-N2-GMP (c); (+)-*anti*-BPDE–(dG)·(dC) (d); (+)-*anti*-BPDE–poly(dG) (e). $\lambda_{ex} = 369.6$ nm, $T = 4.2$K. (B) FLN spectrum of (+)-*anti*-*cis*-BPDE–GMP obtained at $T = 4.2$ K for $\lambda_{ex} = 369.6$ nm. FNL peaks are labeled with their excited-state vibrational frequencies (cm⁻¹).

distinctly different from the FLN spectra of tetraols. It has also been shown that trans-isomer adducts can be distinguished from cis-isomer adducts (*77*).

The vibronically excited FLN spectra of Figure 11 prove that the $S_1$-state energy increases in the following order: *syn*-BPDE < (−)-*anti*-BPDE < (+)-*anti*-BPDE < tetraol (spectra given in ref 12). The ability of FLNS to resolve different stereoisomeric adducts results from DNA–metabolite intermolecular interactions. We refer to this as "DNA host-engineered selectivity."

Very recently it has been shown that FLNS can distinguish between the cis and trans isomers of (+)-*anti*-BPDE–DNA (*77*). Figure 12A shows that the FLN spectrum of the trans isomer of (+)-*anti*-BPDE-N2-GMP (guanosine monophosphate) is very similar to those of the exterior type II adducts of (+)-*anti*-BPDE–DNA, (+)-*anti*-BPDE–poly(dG-dC)·poly(dG-dC), (+)-*anti*-BPDE–poly(dG)·poly(dC), and (+)-*anti*-BPDE–poly(dG) (*77*). However, the spectrum of the cis isomer of (+)-*anti*-BPDE-N2-GMP shown in Figure 12B is distinctly different from that of the trans isomer. Therefore, by *direct analysis of macromolecular DNA*, it is possible to conclude that the major DNA (and above polynucleotides) adduct from (+)-*anti*-BPDE is of the *trans*-N2-dG type, a conclusion consistent with those of earlier studies (*83, 84*).

The above studies and those to be discussed in the following subsection were performed in collaboration with Geacintov and co-workers at New York University.

**B. Different DNA Binding Configurations for the Same Stereoisomeric Adduct.** FLNS in combination with 77 K $S_2 \leftarrow S_0$ laser-excited fluorescence and fluorescence quenching with acrylamide (ACR) has been used to show that the DNA adduct distribution from (–)-*anti*-BPDE is more heterogeneous than the distribution from (+)-*anti*-BPDE (*82, 87*). Five and three adducts were identified from (–)-*anti*-BPDE and (+)-*anti*-BPDE, respectively. Here we only consider the finding that the trans isomer of N2-dG from (+)-*anti*-BPDE exists in two different site configurations, (+)-1 and (+)-2. The former is the major adduct. Fluorescence quenching studies showed that the (+)-1 adduct is solvent accessible and, therefore, exterior with a relatively weak interaction between the pyrene chromophore and the DNA bases. The (+)-2 adduct has a significant BPDE–base stacking interaction.

The same conclusions were reached for the (+)-*anti*-BPDE adducts of poly(dG-dC)·poly(dG-dC) [abbreviation (dG-dC)$_2$], and moreover, the (+)-1 and (+)-2 adducts of this polynucleotide exhibit FLN and 77 K non-line-narrowed fluorescence spectra and quenching properties that are essentially identical with those of the (+)-1 and (+)-2–DNA adducts (*82*). We consider a few of the results for (dG-dC)$_2$. On the basis of $S_2 \leftarrow S_0$ laser-excited fluorescence it was determined that the $S_1$-state origins of (+)-1 and (+)-2 lie at 377.6 and 379.4 nm, respectively. The 77 K non-line-narrowed (+)-1 fluorescence spectrum is shown as spectrum c in Figure 13A. For $\lambda_{ex}$ = 355 nm and in the presence of ACR, the fluorescence is due almost entirely to (+)-2 (spectrum b of Figure 13) since (+)-1 is significantly quenched. In the absence of ACR and for $\lambda_{ex}$ = 346 nm, fluorescence is dominated by (+)-1 (spectrum c, scaled here to show that the sum of spectra b and c yields spectrum a), but, in the presence of ACR, spectrum a of Figure 13A is obtained, for which (+)-1 is not completely quenched and is responsible for the shoulder at ~377.8 nm (solid arrow). The (+)-2 adduct makes a significant contribution to spectrum a because it is not appreciably quenched by ACR. Results of this type and many others (*77*) establish that selective laser-excited $S_1 \leftarrow S_0$ fluorescence spectra obtained in the absence and presence of ACR are a useful complement to FLN. In fact, it is advisable to obtain such spectra prior to detailed FLN studies.

FLN spectra obtained in the absence and presence of ACR are also informative. An example is given in Figure 13B, which shows spectra for $\lambda_{ex}$ = 369.6 nm (vibronic excitation). Because the dominant N2-dG adduct for poly(dG-dC)·poly(dG-dC) is (+)-1 [as is the case for DNA adducts from (+)-*anti*-BPDE (*82*)], spectrum a is dominated by (+)-1, except for the 763- and 830-cm$^{-1}$ bands which are due to quasi-intercalated tetraol produced by photodegradation (*77*). [Spectral contamination by "free" and quasi-intercalated tetraols must always be considered as a possibility, and standard fluorescence spectra for such tetraols must be obtained (*12, 76, 82, 85*).] Spectrum b, obtained in the presence of ACR, is due primarily to the (+)-2 adduct. The (+)-1 adduct probably contributes to the 581-cm$^{-1}$ band. Since the $S_1$ origin band of (+)-2 is at

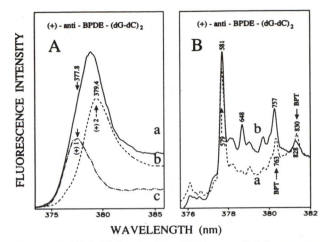

**Figure 13.** (A) Origin bands of the 77 K laser-excited fluorescence spectra of (+)-*anti*-BPDE–(dG-dC)$_2$ in the presence of 1 M acrylamide, obtained with $\lambda_{ex}$ = 346 nm (a) and $\lambda_{ex}$ = 355 nm (b), respectively. Spectrum c represents the incompletely quenched contribution from (+)-1–(dG-dC)$_2$ which is dominant in the absence of quencher for $\lambda_{ex}$ = 346 nm. (B) Comparison of FLN spectra of (+)-*anti*-BPDE–(dG-dC)$_2$ for experimental conditions which provide as the major adduct (+)-1 (a) and (+)-2 (b), respectively. $\lambda_{ex}$ = 369.6 nm, $T$ = 4.2 K. ZPL are labeled with excited-state vibrational frequencies. ZPL at 763 and 830 cm$^{-1}$ correspond to tetraols.

379.4 nm [versus 377.6 nm for (+)-1, vide supra], one predicts considerable FLN activity near 380 nm as is observed.

**C. Conformation of the BP Radical Cation Pathway in Vivo.** As mentioned earlier, in vitro studies with BP have convincingly shown that the one-electron oxidation or radical cation mechanism is catalyzed by cytochrome P-450 (*69*). However, until very recently (*73*) this mechanism had not been demonstrated to be operative in vivo. In this subsection we discuss the FLN data, obtained in collaboration with Rogan, Cavalieri, and co-workers at the Eppley Institute for Research in Cancer, that demonstrated the pathway in vivo.

Specifically, we consider the BP-N7-Gua adduct which had earlier been proven (*75*) to readily depurinate due, presumably, to the instability of the guanine glycosidic bond induced by binding at N7. Similar instability has been reported for the BPDE-N7-Gua adduct (*86*) as well as other adducts (*38, 82*).

The standard FLN spectrum of the electrochemically synthesized BP-N7-Gua adduct from ref 63 is shown as the upper spectrum in Figure 14. The excitation wavelength of 386.5 nm provides vibronic excitation of the $S_1$ state of the BP chromophore, and four prominent multiple origin bands due to the 1164-, 1218-, 1250-, and 1316-cm$^{-1}$ vibrations are observed. The middle and lowest spectra are for a chromatographically isolated adduct from the urine and feces of rats treated with [$^{14}$C]BP (intraperitoneal injection) (*73*). These spectra are identical with that of the synthesized BP-N7-Gua. It is apparent that BP-N7-Gua is depurinated into the urine and feces of rats treated with BP. Additional support was provided by FLN spectra obtained with other excitation wavelengths. The structure was also confirmed by cochromatography with synthesized BP-N7-Gua (*73*). It was found that 12% of the administered $^{14}$C was excreted in the urine and 12% in the feces in the 5 days after treatment. About 0.1% of these amounts were recovered as BP-N7-Gua, indicating that BP-N7-Gua is not an insignificant depurination adduct. In future work, an attempt will be made to char-

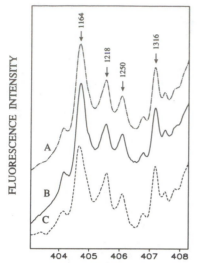

**Figure 14.** Comparison of the vibronically excited FLN spectra ($\lambda_{ex}$ = 386.5 nm, $T$ = 4.2 K) of the synthesized standard BP-N7-Gua adduct (A) with adducts isolated from urine (B) and feces (C) of rats. FLN peaks are labeled with their excited-state vibrational frequencies (cm$^{-1}$).

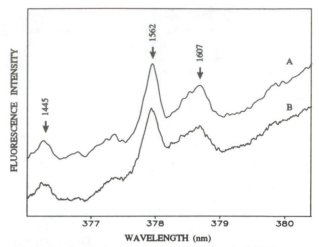

**Figure 15.** Comparison of the vibrationally excited FLN spectra ($\lambda_{ex}$ = 356.9 nm, $T$ = 4.2 K) of intact human globin (A) with the synthesized ester ALA-BP (B) (oxygen type adduct).

acterize and determine the BPDE-N7-Gua adduct so that its importance relative to BP-N7-Gua may be assessed.

**D. Human Hemoglobin Adducts from BP.** Characterization and determination of hemoglobin (Hb)–PAH adducts have potential for body burden assessment. In vivo studies with BP have focused on mouse Hb adducts, but purification of adducts by acidic and enzymatic digests of the protein was unsuccessful, leading only to isolation and characterization of isomeric BP tetraols (*86, 87*). Recent $^{18}$O isotope incorporation experiments have suggested that (*89*) carboxylic adducts are the major adducts formed by BPDE in vitro and in vivo. This has now been confirmed by direct analysis with FLNS of macromolecular Hb damaged by BP (*68*). We briefly discuss some of the FLN results obtained in collaboration with Tannenbaum and co-workers at the Massachusetts Institute of Technology.

Globin was obtained from nonsmoking donors, who were exposed to ambient levels of BP. Standard FLN spectra were obtained for appropriate C-10 amino, thioether, and carboxylic ester synthetic model adducts. The spectrum for the latter, 7,8,9-trihydroxy-*r*-7,*t*-8,*t*-9,*c*-10-tetrahydrobenzo[a]pyren-10-yl *N*-*t*-BOC-L-alaninate (ALA-BP), as obtained with $\lambda_{ex}$ = 356.9 nm is shown as the lower spectrum in Figure 15. The upper spectrum is that of ~30 fmol of an adduct from the human Hb which was believed to possess pyrene as a chromophore. This spectrum is very similar to that of the standard, and both are distinct for the FLN spectra of the nitrogen and sulfur synthetic adducts and their tetraols. FLN spectra obtained with $\lambda_{ex}$ = 363.4 nm further confirm that the major globin adduct from BP is an ester, indicating that the adduct formation occurs at the C termini of the $\alpha$ or $\beta$ chains or the side chain of aspartate or glutamate in hemoglobin. Finally, fluorescence quenching studies with ACR established that the carboxylic ester adduct is interior and would not be recognized by the monoclonal antibody methodology.

**E. Additional in Vivo Experiments.** We now illustrate another attribute of FLNS: namely, that it can provide identification of unanticipated or nontargeted adducts from in vivo studies.

In collaboration with Varanasi and co-workers at Northwest and Alaska Fisheries Center we have initiated FLNS studies of liver DNA from English sole exposed to various dosages of BP (*90*). There is considerable interest in DNA adducts formed in aquatic species since it is believed that bottom-feeding fish species which exhibit frequencies of neoplasm may serve as early warning indicators of carcinogenic hazards to man (*81*). Some of the results are shown in Figure 16A. Spectrum a is that of liver DNA from fish exposed to a high dosage, 100 mg of BP/kg body weight (bw). For comparison, the standard spectra (for the same $\lambda_{ex}$ = 369.6 nm) for *syn*-BPDE–DNA and (+)-*anti*-BPDE–DNA are given, spectra b and d. Contrary to our expectations, the standard spectra reveal that the major adduct from the monooxygenation pathway is *syn*-BPDE–DNA, not (+)-*anti*-BPDE–DNA. However, it was also observed that at lower dosage levels it is the latter that is dominant, see spectrum c (50 mg of BP/kg bw dosage). This interesting dosage level dependence was observed earlier by Varanasi et al. (*81*), who utilized $^{32}$P-postlabeling for nucleotide adduct analysis. They showed that while the liver DNA of fish exposed to 2–15 mg of BP/kg bw contained mainly the *anti*-BPDE–DNA adduct, the DNA of fish exposed to 100 mg of BP/kg bw contained a high proportion of *syn*-BPDE–DNA adducts. Experiments which may provide an understanding of the dosage level dependence are planned.

In collaboration with Maccubbin and co-workers at Roswell Park Memorial Institute, we have performed preliminary experiments on liver DNA of carp from the Buffalo River. The results are shown in Figure 16B for $\lambda_{ex}$ = 369.6 nm [targeted for (+)-*anti*-BPDE–DNA]. The upper spectrum is that of liver DNA of carp, and for easy comparison, the standard spectra for (+)-*anti*-BPDE–DNA (b) and *syn*-BPDE–DNA (c) are given again. Although small contributions from the two adducts to spectrum a cannot be excluded, it is clear that the dominant contributor is a different adduct. Our analysis indicates that this uncharacterized adduct processes pyrene as the parent fluorescing chromophore. Spectrum a is not similar to the FLN spectra of "free" or quasi-intercalated BP tetraols. To add to the story, we have studied, in collaboration with Santella and co-workers at Columbia University, placental DNA from smokers (*90*). ELISA had failed to identify the *anti*-BPDE–DNA adducts in the placental samples. The

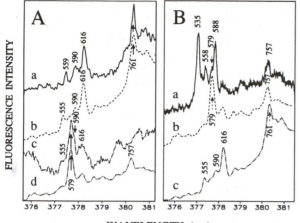

**Figure 16.** (A) FLN spectra of liver DNA from English sole exposed to 100 mg (spectrum a) and 50 mg (spectrum c) of BP/kg body weight of fish. The modification level is approximately 1 adduct in $10^7$–$10^8$ nucleotides. (b) and (d) show FLN spectra of standard *syn*-BPDE–DNA adduct with a modification level of approximately 1 adduct in $10^7$ bases (determined radiometrically) and of (+)-*anti*-BPDE–DNA with a modification level of 1 adduct in ~200 bases, respectively. Conditions: $\lambda_{ex}$ = 369.6 nm, $T$ = 4.2 K. (B) Comparison of FLN spectrum of the liver DNA of carp (from the Buffalo River) (a) with the standard spectra for (+)-*anti*-BPDE–DNA (b) (modification ~1 adduct in 8 × $10^7$ bases) and (±)-*syn*-BPDE–DNA (c) (modification ~1 adduct in $10^7$–$10^8$ bases). All spectra were obtained in the standard glass at $T$ = 4.2 K for $\lambda_{ex}$ = 369.6 nm.

FLN studies revealed that the spectra of the placental DNA are essentially identical with those from carp DNA (for $\lambda_{ex}$ = 369.6 nm and several other $\lambda_{ex}$ values). The results of recent experiments (*91*) on the binding of [³H]BP metabolites to DNA of hamster cheek pouch and liver may also be relevant. Melikian et al. (*91*) identified two major deoxyribonucleoside adducts, with one being the well-known N2-dG adduct from (+)-*anti*-BPDE. The second adduct has not been characterized, although it was suggested that it might be derived from the activation of (+)-*anti*-7,8-diol to a metabolite other than BPDE. It will be interesting to determine by FLNS whether or not this adduct is the same as the adduct from carp.

Finally, we remark that stable one-electron oxidation macromolecular adducts were not observed in liver DNA from fish or in placental DNA. This is not surprising since recent studies have shown that (*92*) all major adducts from the one-electron oxidation pathway are lost from DNA by depurination.

## VI. Final Remarks and Future Prospects

Although it has been known for several years that FLNS possesses the sensitivity and high selectivity required for detailed studies of DNA and other macromolecular PAH adducts, it has only been recently that the technique has been shown to be applicable to DNA and protein from in vivo exposure to PAH. In this review we have demonstrated that FLNS can provide definite answers to important and long-standing problems associated with in vivo exposure (sections V-C and -D). At the same time, the high selectivity of FLNS can be expected to raise new questions (section VI-E).

However, a question that often arises is whether or not the identification of different DNA configurations (sites) for a base–metabolite adduct of specified chemical struc-

ture at low temperatures in glasses is relevant to structure at biological temperatures. We have in mind, for example, our results that show that there are two configurations for the *trans*-N2-dG isomer from (+)-*anti*-BPDE (*77*) (section V-B). HPLC analysis of the nucleotide adducts showed the existence of only one *trans*-N2-dG adducts (*74, 93*). The answer to the question emerges when one recognizes that a DNA adduct configuration can only be trapped at low temperature provided it is thermally accessible at high temperatures. The only relevant question is whether what one observes in a solid at low temperatures reflects *all* of the configurations accessible at room temperature. Most if not all such configurations are likely to be in dynamic equilibrium at room temperature. This is not the case at liquid helium temperatures. There is no reason why all configurations that are thermally accessible at room temperature should be trapped at low temperatures. However, it is interesting to note that our findings concerning the heterogeneity of DNA adducts from (+)-*anti*-BPDE and (–)-*anti*-BPDE are in qualitative accord with those based on HPLC analysis of nucleotide adducts (*74, 93*). We hasten to add that low-temperature linear dichroism and kinetic studies of photosynthetic reaction centers have proven to be extremely relevant to their structures and dynamics at room temperature (*41–44*).

It was also our intent to demonstrate that $S_2 \leftarrow S_0$ laser-excited fluorescence spectroscopy at 77 K can provide valuable insights into the structures of DNA, globin, and nucleoside adducts. This spectroscopy, FLNS, and the fluorescence quenching methodology of Geacintov and co-workers constitute a powerful combination for the study of the initial phases of chemical carcinogenesis. It is a combination that should be useful for future studies of DNA, globin, nucleotide, and nucleoside adducts from PAH and other classes of fluorescent carcinogens. Application of the following problems can also be anticipated: base sequence specificity (through the study of polynucleotide adducts), DNA repair mechanisms, and body burden assessment.

**Acknowledgment.** The Ames Laboratory is operated for the U.S. Department of Energy by Iowa State University under Contract W-7405-Eng-82. This work was supported by the Office of Health and Environmental Research, Office of Energy Research, and partly by Grant P01-CA49210 from the National Cancer Institute. We would like to acknowledge P. Lu, D. Tang, and H. Jeong for their valuable experimental contribution to the research discussed herein.

## References

(1) Miller, E. C., and Miller, J. A. (1981) Searches for ultimate chemical carcinogens and their reactions with cellular macromolecules. *Cancer* **47**, 2327–2345.

(2) *World Health Organization Monograph on the Evaluation of the Carcinogenic Risks of the Chemical to Man: Polynuclear Aromatic Compounds* (1983) Vol. 32, International Agency for Research in Cancer, World Health Organization, Lyon, France.

(3) Miller, J. A. (1970) Carcinogenesis by chemicals: an overview—G. M. A. Clowes memorial lecture. *Cancer Res.* **30**, 558–576.

(4) Sims, P., and Grover, P. L. (1981) Involvement of dihydrodiols and diol epoxides in the metabolic activation of polycyclic hydrocarbons other than benzo[a]pyrene. In *Polycyclic Hydrocarbons and Cancer* (Gelboin, N. V., and Ts'o, P. O. P., Eds.) Vol. 3, pp 117–181, Academic Press, New York.

(5) Weinstein, I. B., Jeffrey, A. M., Jennette, K. W., Blobstein, S. H., Harvey, R. G., Harris, C., Autrup, H., Kasai, H., and Nakanishi, K. (1976) Benzo[a]pyrene diol epoxides as intermediates in nucleic acid binding in vitro and in vivo. *Science* **193**, 592–595.

(6) Cavalieri, E., and Rogan, E. (1985) Role of radical cations in aromatic hydrocarbon carcinogenesis. *Environ. Health Perspect.* **64**, 69–84.

(7) Cavalieri, E., and Rogan, E. (1985) One-electron oxidation in aromatic hydrocarbon carcinogenesis. In *Polycyclic Hydrocarbons and Carcinogeneis* (Harvey, R. G., Ed.) pp 289–305, ACS Symposium Series 283, American Chemical Society, Washington, DC.

(8) Szabo, A. (1970) Laser-induced fluorescence-line narrowing in Ruby. *Phys. Rev. Lett.* **25**, 924–928.

(9) Personov, R. I., Al'Shits, E. I., and Bykovskaya, L. A. (1972) The effect of fine structure appearance in laser-excited spectra of organic compounds in solid solutions. *Opt. Commun.* **6**, 169–174.

(10) Personov, R. I. (1983) Site selection spectroscopy of complex molecules in solution and its applications. In *Spectroscopy and Excitation Dynamics of Condensed Molecular Systems* (Agranovich, V. M., and Hochstrasser, R. M., Eds.) Vol. 4, pp 555–619, North-Holland, Amsterdam.

(11) Hofstraat, J. W., Gooijer, C., and Velthorst, N. H. (1988) in *Molecular Luminescence Spectroscopy, Methods and Applications. Part 2* (Schulman, S. G., Ed.) pp 283–459 John Wiley & Sons, New York.

(12) Jankowiak, R., and Small, G. J. (1989) Fluorescence line-narrowing spectroscopy in the study of chemical carcinogenesis. *Anal. Chem.* **61**, 1023A–1029A.

(13) Fünfschilling, J., and Williams, D. F. (1977) Fluorescence spectra of chlorophyll-a and chlorophyll-b from site selection spectroscopy. *Photochem. Photobiol.* **26**, 109–113.

(14) Avarmaa, R. A., and Rebane, K. K. (1985) High-resolution optical spectra of chlorophyll molecules. *Spectrochim. Acta* **41A**, 1365–1369.

(15) Hala, J., Pelant, I., Ambroz, M., Pancoska, P., and Vacek, K. (1985) Site-selection and Shpolskii spectroscopy of model photosynthetic systems. *Photochem. Photobiol.* **41**, 643–648.

(16) Plattenkamp, R. J., den Blanken, H. F., and Hoff, A. J. (1980) Single-site absorption spectroscopy of pheophytin-a and chlorophyll-a in a n-octane matrix. *Chem. Phys. Lett.* **76**, 35–41.

(17) Rebane, K. K., and Avarmaa, R. A. (1982) Sharp line vibronic spectra of chlorophyll and its derivatives in solid solutions. *Chem. Phys.* **68**, 191–200.

(18) Renge, I., Mauring, K., Sarv, P., and Avarmaa, J. (1986) Vibrationally resolved optical spectra of chlorophyll derivatives in different solid media. *Chem. Phys.* **90**, 6611–6616.

(19) Koloczek, H., Fidy, J., and Vanderkooi, J. M. (1987) Fluorescence line-narrowing spectra of Zn-cytochrome c. Temperature dependence. *J. Chem. Phys.* **87**, 4388–4394.

(20) Angiolillo, P. J., Leigh, J. S., and Vanderkooi, J. M. (1982) Quasi-line emission spectra of porphyrin in iron-free heme cytochrome c. *Photochem. Photobiol.* **63**, 133–137.

(21) Avarmaa, R., Renge, I., and Mauring, K. (1984) Sharp-line structure in the fluorescence and excitation spectra of greening ethiolated leaves. *FEBS Lett.* **167**, 186–190.

(22) Renge, I., Mauring, K., and Avarmaa, R. (1987) Site-selection optical spectra of bacteriochlorophyll and bacteriopheophytin in frozen solution. *J. Lumin.* **37**, 207–214.

(23) Hayes, J. M., Jankowiak, R., and Small, G. J. (1987) Two-level system relaxation in amorphous solids as probed by nonphotochemical hole-burning in electronic transitions. In *Topics in Current Physics, Persistent Spectral Hole Burning: Science and Applications* (Moerner, W. E., Ed.) Vol. 44, Chapter 5, pp 153–202, Springer-Verlag, New York; see also references therein.

(24) Jankowiak, R., and Small, G. J. (1987) Hole-burning spectroscopy and relaxation dynamics of amorphous solids at low temperatures. *Science* **237**, 618–625.

(25) Haarer, D. (1987) Photochemical hole-burning in electronic transition. In *Topics in Current Physics, Persistent Spectral Hole Burning: Science and Applications* (Moerner, W. E., Ed.) Chapter 3, pp 79–123, Springer-Verlag, New York.

(26) Avarmaa, R., Mauring, K., and Suisalu, A. (1981) Reversible resonant hole burning in the fluorescence spectra of protochlorophyll. *Chem. Phys. Lett.* **77**, 88–92.

(27) Rebane, K. K., and Avarmaa, R. A. (1982) Sharp line vibronic spectra of chlorophyll and its derivatives in solid solutions. *Chem. Phys.* **68**, 191–200.

(28) Renge, I., Mauring, K., and Vladkova, R. (1988) Zero-phonon transitions of chlorophyll *a* in mature plant leaves revealed by spectral hole-burning method at 5 K. *Biochim. Biophys. Acta* **935**, 333–336.

(29) Renge, I., Mauring, K., and Avarmaa, R. (1984) High resolution optical spectra in vivo photoactive protochlorophyllide in etiolated leaves at 5 K. *Biochim. Biophys. Acta* **766**, 501–504.

(30) Mauring, K., Renge, I., and Avarmaa, R. (1987) A spectral hole-burning study of long-wavelength chlorophyll *a* forms in greening leaves at 5 K. *FEBS Lett.* **223**, 165–168.

(31) Brown, J. C., Duncanson, J. A., Jr., and Small, G. J. (1980) Fluorescence line narrowing spectrometry in glasses for direct determination of polycyclic aromatic hydrocarbons in solvent-refined coal. *Anal. Chem.* **52**, 1711–1715.

(32) Heisig, V., Jeffrey, A. M., McGlade, M. J., and Small, G. J. (1984) Fluorescence line-narrowing spectra of polycyclic aromatic carcinogen–DNA adducts. *Science* **223**, 288–291.

(33) Jankowiak, R., Cooper, R. S., Zamzow, D., Small, G. J., Doskocil, G., and Jeffrey, A. M. (1988) Fluorescence line narrowing-nonphotochemical hole burning spectrometry: femtomole detection and high selectivity for intact DNA–PAH adducts. *Chem. Res. Toxicol.* **1**, 60–68.

(34) Sanders, M. J., Cooper, R. S., Jankowiak, R., Small, G. J., Heisig, V., and Jeffrey, A. M. (1986) Identification of polycyclic aromatic hydrocarbon metabolites and DNA adducts in mixtures using fluorescence line narrowing spectrometry. *Anal. Chem.* **58**, 816–820.

(35) Sanders, M. J., Cooper, R. S., Small, G. J., Heisig, V., and Jeffrey, A. M. (1985) Identification of polycyclic aromatic hydrocarbon metabolites in mixtures using fluorescence line narrowing spectrometry. *Anal. Chem.* **57**, 1148–1152.

(36) Cooper, R. S., Jankowiak, R., Hayes, J. M., Lu, P., and Small, G. J. (1988) Fluorescence line narrowing spectrometry of nucleoside–polycyclic aromatic hydrocarbon adducts on thin-layer chromatographic plates. *Anal. Chem.* **60**, 2692–2694.

(37) Geacintov, N. E. (1988) Mechanisms of reaction of polycyclic aromatic epoxide derivatives with nucleic acids. In *Polycyclic Aromatic Hydrocarbon Carcinogenesis: Structure–Activity Relationships* (Yang, S. K., and Silverman, B. D., Eds.) Vol. 2, pp 181–206. CRC Press, Boca Raton, FL.

(38) Geacintov, N. E. (1985) Mechanisms of interaction of polycyclic aromatic diol epoxides with DNA and structures of the adducts. In *Polycyclic Aromatic Hydrocarbon Carcinogenesis* (Harvey, R. G., Ed.) pp 107–124, American Chemical Society, Washington, DC.

(39) Gräslund, A., and Jernström, B. (1989) DNA–carcinogen interaction: covalent DNA adducts of benzo(a)pyrene 7,8-dihydrodiol 9,10-epoxides studied by biochemical and biophysical techniques. *Q. Rev. Biophys.* **22**, 1–37.

(40) Gillie, J. K., Small, G. J., and Golbeck, J. H. (1989) Nonphotochemical hole burning of the native antenna complex of photosystem I (PSI-200). *J. Phys. Chem.* **93**, 1620–1627.

(41) Johnson, S. G., Tang, D., Jankowiak, R., Hayes, J. M., and Small, G. J. (1989) Spectral hole burning: a window on energy transfer in photosynthetic units. In *Proceedings of VIth International Conference on Energy and Electron Transfer* (Skala, L., Ed.) Charles University Press, Prague, Czechoslovakia.

(42) Johnson, S. G., Tang, D., Jankowiak, R., Hayes, J. M., Small, G. J., and Tiede, D. M. (1989) Structure and marker mode of the primary electron donor state absorption of photosynthetic bacteria: hole-burned spectra. *J. Phys. Chem.* **93**, 5953–5957.

(43) Johnson, S. G., Tang, D., Jankowiak, R., Hayes, J. M., Small, G. J., and Tiede, D. M. (1989) Primary donor state mode structure and energy transfer in bacterial reaction centers. *J. Phys. Chem.* **94**, 5849–5855.

(44) Tang, D., Jankowiak, R., Seibert, M., Yocum, C. F., and Small, G. J. (1990) Excited-state structure and energy-transfer dynamics of two different preparations of the reaction center of photosystem II: a hole burning study. *J. Phys. Chem.* **94**, 6519–6522.

(45) Sargent, M., III, Scully, M. O., and Lamb, W. E., Jr. (1974) *Laser Physics*, Addison-Wesley, Reading, MA.

(46) Blum, K. (1981) Density matrix theory and applications. In *Physics of Atoms and Molecules* (Burke, P. G., and Keimpoppen, M., Series Eds.) Plenum Press, New York and London.

(47) Lyo, S. K. (1986) Dynamical theory of optical linewidths in glasses. In *Optical Spectroscopy of Glasses* (Zschokke, I., Ed.) pp 1–21, D. Reidel Publishing Co., see also references therein.

(48) Huber, D. L., Broer, M. M., and Golding (1984) Low-temperature optical dephasing of rare-earth ions in glass. *Phys. Rev. B* **52**, 2281–2284.

(49) Hesselink, W. H., and Wiersma, D. A. (1983) in *Modern Problems in Condensed Matter* (Agranovich, V. M., and Maradudin, A. A., Eds.) Vol. 4, pp 249–263, North-Holland, Amsterdam.

(50) Skinner, J. L., Andersen, H. C., and Fayer, M. D. (1981) Theory of photon echoes from a pair of coupled two level systems: impurity dimers and energy transfer in molecular crystals. *J. Chem. Phys.* **75**, 3195–3202.

(51) Walsh, C. A., Berg, M., Narasimhan, L. R., and Fayer, M. D. (1987) Probing intermolecular interactions with picosecond photon echo experiments. *Acc. Chem. Res.* **20**, 120–126.

(52) Rebane, K. K. (1970) *Impurity Spectra of Solids*, Plenum Press, New York.

(53) Berg, M., Walsh, C. A., Narasimhan, L. R., Littau, K. A., and Fayer, M. D. (1988) Dynamics in low temperature glasses: theory and experiments on optical dephasing, spectral diffusion, and hydrogen tunneling. *J. Chem. Phys.* **88**, 1564–1587.

(54) Völker, S. (1989) Structured hole-burning in crystalline and amorphous organic solids. Optical relaxation processes at low temperature. In *Relaxation Processes in Molecular Excited States* (Fünfschilling, J., Ed.) 1st ed., pp 113–242, Kluwer Academic Publishers, Dordrecht.

(55) Walsh, C. A., Berg, M., Narasimhan, L. R., and Fayer, M. D. (1986) Optical dephasing of chromophores in an organic glass: picosecond photon echo and hole burning experiments. *Chem. Phys. Lett.* **130**, 6–11; (1987) Dynamics in low temperature glasses: theory and experiments on optical dephasing, spectral diffusion and hydrogen tunneling. *J. Chem. Phys.* **86**, 77–87.

(56) Warren, J. A., Hayes, J. M., and Small, G. J. (1986) Vibronic mode mixing in the $S_1$ state of $\beta$-methylnaphthalene. *Chem. Phys.* **102**, 313–323.

(57) Hayes, J. M., Gillie, J. K., Tang, D., and Small, G. J. (1987) Theory for spectral hole burning of the primary electron donor state of photosynthetic reaction centers. *Biochim. Biophys. Acta* **932**, 287–305.

(58) Nicolini, C., Ed. (1982) *Chemical Carcinogenesis*, Plenum Press, New York.

(59) Hayes, J. M., and Small, G. J. (1978) Non-photochemical hole burning and impurity site relaxation processes in organic glasses. *Chem. Phys.* **27**, 151–157.

(60) Fearey, B. L., Carter, T. P., and Small, G. J. (1983) Efficient nonphotochemical hole burning of dye molecules in polymers. *J. Phys. Chem.* **87**, 3590–3592.

(61) Fearey, B. L., Carter, T. P., and Small, G. J. (1983) Nonphotochemical hole burning of organic dyes and rare-earth ions in polymers and glasses: a probe of the amorphous state, Ph.D. Dissertation, ISU, 1986.

(62) Tang, D., Jankowiak, R., Seibert, M., and Small, G. J. (1991) Effects of detergent on the excited state structure and relaxation dynamics of the photosystem II reaction center: a high resolution hole burning study. *Photosynth. Res.* **27**, 19–29.

(63) Zamzow, D., Jankowiak, R., Cooper, R. S., Small, G. J., Tibels, S. R., Cremonosi, P., Devanesan, P., Rogan, E. G., and Cavalieri, E. L. (1989) Fluorescence line narrowing spectrometric analysis of benzo[a]pyrene–DNA adducts formed by one-electron oxidation. *Chem. Res. Toxicol.* **2**, 29–34.

(64) Bogner, U., and Schwarz, R. (1981) Laser-induced changes in the sideband shape of selectively excited dyes in noncrystalline organic solids at 1.3 K. *Phys. Rev. B: Condens. Matter* **B24**, 2846–2849.

(65) Fünfschilling, J., Glatz, D., and Zschokke-Gränacher, I. (1986) Hole-burning spectroscopy as a tool to eliminate inhomogeneous broadening. *J. Lumin.* **36**, 85–92.

(66) Flöser, G., and Haarer, D. (1988) The photochemistry of daunomycin in solution and intercalated into DNA studies by photochemical hole burning. *Chem. Phys. Lett.* **147**, 288–292.

(67) Calleman, C. J. (1984) Hemoglobin as a dose monitor and its application to the risk estimation of ethylene oxide, Ph.D. Thesis, University of Stockholm, Sweden; see also references therein.

(68) Jankowiak, R., Day, B. W., Lu, P., Doxtader, M. M., Skipper, P. L., Tannenbaum, S. R., and Small, G. J. (1990) Fluorescence line narrowing spectral analysis of in vivo human hemoglobin-benzo[a]pyrene adducts: comparison to synthetic analogues. *J. Am. Chem. Soc.* **112**, 5866–5869 and references therein.

(69) Cavalieri, E. L., Devanesan, P. D., Cremonesi, P., Cerny, R. L., Gross, M. L., and Bodell, W. J. (1990) Binding of benzo[a]pyrene to DNA by cytochroem P-450 catalyzed one-electron oxidation in rat liver microsomes and nuclei. *Biochemistry* **29**, 4820–4827.

(70) Hamminki, K. (1982) Dimethylnitrosamine adducts excreted in rat urine. *Chem.-Biol. Interact.* **39**, 139–148.

(71) Bennett, R. A., Essignmann, J. M., and Wogan, G. N. (1981) Excretion of an aflatoxin–guanine adduct in the urine of aflatoxin B1 treated rats. *Cancer Res.* **41**, 650–654.

(72) Tierney, B., Martin, C. N., and Garner, R. C. (1987) Topical treatment of mice with benzo[a]pyrene or parenteral administration of benzo[a]pyrene diol epoxide–DNA to rats results in fecal excretion of a putative benzo[a]pyrene diol epoxide–deoxyguanosine adduct. *Carcinogenesis* **8**, 1189–1192.

(73) Rogan, E. G., Ramakrishna, N. V. S., Higginbotham, S., Cavalieri, E. L., Jeong, H., Jankowiak, R., and Small, G. J. (1990) Identification and quantitation of 7-(benzo[a]pyren-6-yl) guanine in the urine and feces of rats treated with benzo[a]pyrene. *Chem. Res. Toxicol.* **3**, 441–444.

(74) Osborne, M. E., Jacobs, S., Harvey, R. G., and Brookes, P. (1981) Minor products from the reaction of (+)- and (−)-benzo[a]pyrene-*anti*-diol-epoxide with DNA. *Carcinogenesis* **2**, 553–558.

(75) Bartsch, H., Hemminki, and O'Neill, I. K., Eds. (1988) *Methods for Detecting DNA Damaging Agents in Humans: Applications in Cancer Epidemiology and Prevention* IARC Scientific Publications 89, IARC Lyon.

(76) Rogan, E. G., Cavalieri, E. L., Tibbels, S. R., Cremonesi, P., Warner, C. D., Nagel, D. L., Tomer, K. B., Cerny, R. L., and Gross, M. L. (1988) Synthesis and identification of benzo[a]pyrene–guanine nucleoside adducts formed by electrochemical oxidation and horseradish peroxidase-catalyzed reaction of benzo[a]pyrene with DNA. *J. Am. Chem. Soc.* **110**, 4023–4029.

(77) Lu, P., Jeong, H., Jankowiak, R., Small, G. J., Kim, S. K., Cosman, M., and Geacintov, N. E. (1991) Comparative laser spectroscopic study of DNA and polynucleotide adducts from the (+)-*anti*-diol epoxide of benzo[a]pyrene. *Chem. Res. Toxicol.* **4**, 58–69.

(78) Newbold, R. F., and Brookes, P. (1976) Exceptional mutagenicity of a benzo[a]pyrene diol epoxide in cultured mammalian cells. *Nature (London)* **261**, 53–55.

(79) Slaga, T. J., Bracken, W. J., Gleason, G., Levin, W., Yagi, H., Jerina, D. M., and Conney, A. H. (1979) Marked differences in the skin tumor-initiating activities of the optical enantiomers of the diastereomeric benzo[a]pyrene 7,8-diol-9,10-epoxides. *Cancer Res.* **39**, 67–71.

(80) Alexandrov, K., Sala, M., and Rojas, M. (1988) Differences in the DNA adducts formed in cultured rabbit and rat dermal fibroblasts by benzo[a]pyrene and (−)-benzo[a]pyrene-7,8-diol. *Cancer Res.* **48**, 7132–7139.

(81) Varanasi, V., Reichert, W. L., Eberhart, B. T. L., and Stein, J. E. (1989) Formation and persistence of benzo[a]pyrene–diol-epoxide–DNA adducts in liver of English sole (Parophrys vetulus). *Chem.-Biol. Interact.* **69**, 203–216.

(82) Jankowiak, R., Lu, P., Small, G. J., and Geacintov, N. E. (1990) Laser spectroscopic studies of DNA adduct structure types from enantiomeric diol epoxides of benzo[a]pyrene. *Chem. Res. Toxicol.* **3**, 39–46.

(83) Cheng, S. C., Hilton, B. D., Roman, J. M., and Dipple, A. (1988) DNA adducts from carcinogenic and noncarcinogenic enantiomers of benzo[a]pyrene dihydrodiol epoxide. *Chem. Res. Toxicol.* **2**, 334–338.

(84) Meehan, T., and Straub, K. (1979) The covalent binding of enantiomeric BaPDE's to double stranded DNA is stereoselective. *Nature* **261**, 410–412.

(85) R. Jankowiak, P. Lu, N. E. Geacintov, and G. J. Small, in preparation.

(86) King, H. W. S., Osborne, M. R., and Brookes, P. (1979) The in vitro and in vivo reaction at the $N^7$-position of guanine of the ultimate carcinogen derived from benzo[a]pyrene. *Chem.-Biol. Interact.* **24**, 345–353.

(87) Shugart, L. (1985) Quantitating exposure to chemical carcinogens: in vivo alkylation of hemoglobin by benzo[a]pyrene. *Toxicology* **34**, 211–220.

(88) Wallin, H., Jeffrey, A. M., and Santella, R. M. (1987) Investi-

gation of benzo[a]pyrene–globin adducts. *Cancer Lett.* **35**, 139–146.

(89) Skipper, P. L., Naylor, S., Gau, L.-S., Day, B. W., Pastorelli, R., and Tannenbaum, S. R. (1989) Origin of the tetrahydrotetrals derived from human hemoglobin adducts of benzo[a]pyrene. *Chem. Res. Toxicol.* **2**, 280–281.

(90) R. Jankowiak, P. Lu, M. Nishimoto, V. Varansi, A. E. Maccubbin, R. M. Santella, and G. J. Small. Unpublished results.

(91) Melikian, A. A., Fudern Goldin, B., Prahalad, A. K., and Hecht, S. S. (1990) Modulation of benzo[a]pyrene–DNA adducts in hamster cheek pouch by chronic ethanol consumption. *Chem. Res. Toxicol.* **3**, 138–143.

(92) P. Devaneson, N. Ramakrishna, E. Cavalieri, E. Rogan, H. Jeong, R. Jankowiak, and G. J. Small. In preparation.

(93) Brookes, P., and Osborne, M. R. (1982) Mutation in mammalian cells by stereoisomers of *anti*-benzo[a]pyrene-diolepoxide in relation to the extent and nature of the DNA reaction products. *Carcinogensis* **3**, 1223–1226.

**Macromolecular
Modification**

# Chapter 16

# Site-Specifically Modified Oligodeoxynucleotides as Probes for the Structural and Biological Effects of DNA-Damaging Agents

Ashis K. Basu and John M. Essigmann*

*Department of Applied Biological Sciences, Massachusetts Institute of Technology, Cambridge, Massachusetts 02139*

Reprinted from *Chemical Research in Toxicology,* Vol. 1, No. 1, January/February, 1988

## Introduction

Early studies mainly from the laboratory of Miller and Miller showed that the carcinogenic activity of most chemicals is dependent upon their ability to act as electrophiles (*1*). They found that some carcinogens are inherently electrophilic, whereas others spontaneously hydrolyze to active species in solution or are activated by cellular enzymes. The resultant reactive intermediates modify the constituents of DNA, RNA, and protein, forming covalent adducts in which the carcinogen residue is joined to nucleophilic atoms of the constituent nucleotides or amino acids. Although all damaged macromolecules can potentially compromise cellular welfare, adducts within DNA have special significance in view of their potential to force replication or repair errors and thus be the chemical progenitors of heritable genetic alterations. An attractive but as yet incompletely proved hypothesis postulates that some of the resultant mutations may constitute the initial step leading normal cells along the pathway to neoplastic transformation.

Genotoxic agents such as chemical carcinogens and radiation commonly generate more than one and usually many activated species (*2*). Moreover, DNA, the target of these species, is a chemical composite made up of many different nucleophilic atoms. Accordingly the range of the resultant adduct structures can be vast, making it a challenging task to (i) assess the impact of any individual adduct on local DNA structure and chemistry, (ii) determine the contribution of each adduct to the spectrum of mutations induced by chemical or radiation treatment, (iii) assess the relative abilities of DNA adducts to act as cytotoxic lesions, and (iv) ascertain the identities of the DNA

repair proteins responsible for protecting cells from specific forms of DNA damage. Advances of the past decade in molecular biology and nucleic acid chemistry have made it possible to explore the contributory roles of individual DNA adducts to these biological endpoints. Of central importance to the development of this field has been the synthesis of oligodeoxynucleotides containing DNA adducts at defined sites. This review will critically analyze the state of knowledge on the preparation, characterization, and uses of site-specifically modified DNA segments.[1] Although these substrates have begun to have an impact upon several fields, some of which were enumerated above and are depicted in Figure 1, our review will focus upon their applications in site-directed mutagenesis studies and for defining the effect of chemical damage upon DNA structure.

How does one determine which DNA adducts are candidates for study using the tools described in this review? Candidate lesions can be identified by using information from three areas. The first is an examination of the structures of the adducts formed by in vivo treatment with a DNA-damaging agent. Occasionally certain changes in normal base structure, such as fixation of the enol tautomer of a base, will herald a likely change in Watson-Crick base-pairing potential. Second, examination of the

---

[1] Numerous studies of carcinogen–nucleic acid adducts have been carried out at the level of dinucleotide phosphates, and the results are often valid for longer sequences. It is beyond the scope of this article, however, to include these studies, and we shall concentrate primarily on longer oligonucleotides. We also shall largely ignore studies on randomly modified polynucleotides used as templates for DNA or RNA synthesis; again, this field is too large to review here.

2428–5/92/0198$06.00/0

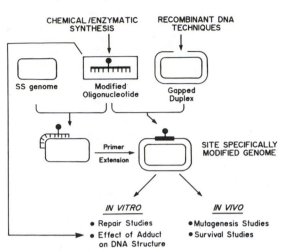

**Figure 1.** Applications of site-specifically modified oligonucleotides.

spectrum of mutations within a target gene treated with a DNA-damaging agent can implicate the base pairs and DNA sequence contexts most prone to mutation (3). These data in turn provide clues as to the identities of the premutagenic lesions. Third, facile methods are available for building random DNA and RNA polymers in which adducts can be inserted biochemically by using various polymerases (4–6). With these polymers as templates for nucleic acid synthesis, the type and amount of base misincorporation, or termination of chain elongation, can be assessed. Such in vitro methodology has been very useful for defining the range of mutagenic possibilities expected from single lesions in vivo, although it lacks the fine control over lesion placement of approaches described below and cannot provide insight into the modulating influence of DNA repair on adduct toxicology. Data from all three approaches described above—adduct analysis coupled with chemical model building, mutational spectrum analysis, and biochemical assessment of mutagenic potential by copying random polymers in vitro—are of value in formulating hypotheses as to which DNA adducts should be studied individually.

The first step toward evaluating the genetic and structural effects of an adduct involves its synthesis as part of an oligodeoxynucleotide. Generally three synthetic routes have been used. The first is *total synthesis*. This method involves preparation of a protected monomer (or dimer) of the adduct deoxynucleoside followed by solution- or solid-phase synthesis of the oligomer by a DNA synthesis protocol (usually the "phosphotriester" or "phosphoramidite" methods; see Gait (7) for a review of conventional synthetic procedures). The key advantages of this approach include high yields (milligram quantities, as would be needed for NMR or X-ray diffraction studies), precise adduct placement in the oligonucleotide chain, and, often, better control over final purity than other methods. Unfortunately not all DNA adducts are stable to the harsh conditions of synthesis and deprotection, so other methods are necessary. The second method is *modification of a preformed oligonucleotide*. This involves treatment of an unmodified oligonucleotide with a chemical carcinogen or its activated derivative, or with radiation, followed by purification of the desired adducted oligomer. Although the conditions of formation are milder than the previous approach, yields can be low, sometimes too low to enable complete oligonucleotide characterization. Moreover, DNA sequences must be chosen carefully to limit the number of potential adduction sites. The third method is *enzy-*

*matic synthesis.* DNA polymerases and bacteriophage T4 RNA ligase have been used to build site-specifically modified oligonucleotides biochemically (8, 9). The enzymatic approaches have the advantage of mild reaction conditions and, with some constraints, enable precise adduct placement. They are limited, however, both in overall yield and, likely, in the range of adduct nucleotide structures capable of being joined in phosphodiester bonds to abutting normal nucleotides. Clearly, no single approach stands out as the method of choice for preparation of all modified DNA segments. However, total synthesis offers the best opportunities for marked advances in the coming years as new deprotection schemes are devised that preserve the structural integrity of the adduct.

All synthetic oligonucleotides are subject to some level of contamination depending upon the method of preparation. If they are to be used in NMR studies to evaluate the effect of the adduct on DNA conformation, the purity requirement is not strict since levels of contamination of 1–2% would be unlikely to affect interpretation of the results. It was pointed out by Chambers (10) that if the oligonucleotides are to be used for genetic studies, the levels of tolerable impurities are one or two orders of magnitude lower, because it is in this lower range that most DNA adducts are likely to be mutagenic in DNA repair-proficient cells. Consider, for instance, a mutagenesis experiment in which an adduct shows a 1% mutation frequency. If the oligonucleotide used to produce the site-specifically modified genome had a purity of 99%, it would be impossible to distinguish between the following two possibilities: the adduct could have caused the mutations at a 1% efficiency, or the 1% impurity could have been mutagenic with 100% efficiency.

The serious problem of adduct-containing oligonucleotide purity for genetic studies is not an intractable one. Oligonucleotide preparation by several synthetic routes should generate different levels or different types of impurities relative to the desired product. Obviously, the conclusion that an adduct is mutagenic would be buttressed if the biological results on it prove to be independent of the route of modified oligonucleotide preparation. As a second point, we and some of our colleagues (11–15) have emphasized the use of *short* oligonucleotides (4- to 7-mers), which are much easier to purify than the 14- to 20-mers used by others (16, 17). It is also easier to assess the levels of contamination of short oligonucleotides since minor differences in composition usually cause substantial differences in mobility characteristics by conventional separation tools; these differences decrease in magnitude as the length of the oligonucleotide increases.

Oligonucleotides have been synthesized containing alkylated DNA bases, aromatic amine adducts, base oxidation products, cyclic nucleic acid adducts, model apurinic/apyrimidinic sites, UV and psoralen photoadducts, and several antitumor drug–DNA covalent complexes. Below, we shall describe the progress to date on synthesis of site-specifically modified DNA segments and how these oligonucleotides have been used to further our understanding of the roles of individual DNA adducts in toxicology. The structures of the DNA adducts and adduct-derived products discussed in this review are presented in Figure 2.

## Alkylated Bases

$O^6$-**Alkylguanines.** The $O^6$-alkylguanine derivatives have received much attention, dating from the original proposal of Loveless (18) that reaction of guanine (Gua)[2]

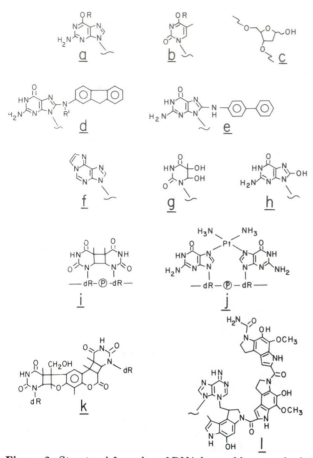

**Figure 2.** Structural formulae of DNA-base adducts, and adduct-derived lesions, described in this review: a, $O^6$-alkylguanines, R = alkyl group; b, $O^4$-alkylthymines, R = alkyl group; c, a formal apurinic/apyrimidinic (AP) site; d, N-guanin-8-yl-2-aminofluorene (R = H) or N-guanin-8-yl-2-(acetylamino)fluorene (R = acetyl); e, N-guanin-8-yl-4-aminobiphenyl; f, $1,N^6$-ethenoadenine; g, cis-thymine glycol; h, 8-hydroxyguanine; i, cyclobutyl–thymine photodimer; j, cis-[Pt(NH$_3$)$_2${d(GpG)}] cross-link; k, cross-link of 4'-(hydroxymethyl)-4,5',8-trimethylpsoralen between two thymine bases; l, CC-1065 adduct at the N3 of adenine.

at the $O^6$-atom would fix the enol tautomer of the base and facilitate base pairing with thymine (Thy) (Figure 3, A). In recent years the prediction of this model, a G→A transition, has undergone rigorous evaluation through both chemical experimentation and biological studies. In early work, the mutagenic potentials of $O^6$-methylguanine ($O^6$MeGua) and $O^6$-ethylguanine ($O^6$EtGua) were shown

2 The abbreviations used are as follows: $O^6$MeGua, $O^6$-methylguanine; $O^4$MeThy, $O^4$-methylthymine; $O^6$EtGua, $O^6$-ethylguanine; MT, $O^6$MeGua–DNA methyltransferase; DBU, 1,8-diazabicyclo[5.4.0]undec-7-ene; THF, tetrahydrofuran; MNNG, N-methyl-N'-nitro-N-nitrosoguanidine; $O^6$BuGua, $O^6$-butylguanine; $T_m$, melting temperature; COSY, two-dimensional J-correlated spectroscopy; MSNT, 1-(mesitylenesulfonyl)-3-nitro-1,2,4-triazole; C$^5$Me, 5-methyldeoxycytidine; CD, circular dichroism; IR, infrared; NOE, nuclear Overhauser enhancement; NOESY, two-dimensional $^1$H–$^1$H NOE spectroscopy; AMV, avian myoblastosis virus; bp, base pair; ss, single stranded; ds, double stranded; Cyt, cytosine; Ade, adenine; Gua, guanine; Thy, thymine; AF, 2-aminofluorene; AAF, 2-(acetylamino)fluorene; ABP, 4-aminobiphenyl; AP site, apurinic/apyrimidinic site; C8-OH-Gua, 8-hydroxyguanine; εAde, $1,N^6$-ethenoadenine; t' (or cis-thymine glycol), cis-5,6-dihydroxy-5,6-dihydrothymine; UV, ultraviolet light; HMT, 4'-(hydroxymethyl)-4,5',8-trimethylpsoralen; cis-DDP, cis-diamminedichloroplatinum(II); 2D, two dimensional. A consistent nomenclature is used to indicate adducts in oligonucleotides. For example, the abbreviation G$^{O^6Me}$ in the oligonucleotide d-(NNNG$^{O^6Me}$NNN) (N = any deoxynucleotide) denotes an oligomer containing $O^6$-methyldeoxyguanosine. The prefix "d" indicates an oligodeoxyribonucleotide, whereas the lack of designation indicates an oligoribonucleotide; all sequences proceed 5' to 3' unless otherwise indicated.

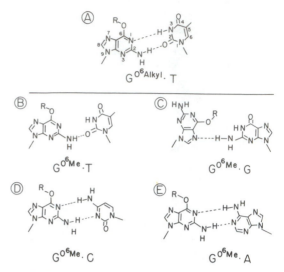

**Figure 3.** Postulated base-pairing modes of $O^6$-methylguanine: A, scheme proposed on the basis of the original model of Loveless (18); B–E, the base pairing of $O^6$-methylguanine with thymine, guanine, cytosine, and adenine, respectively, predicted from NMR studies.

in vitro in templates copied by RNA and DNA polymerases (19, 20). In accordance with the Loveless model, significant misincorporation of Thy opposite the alkylated bases occurs, although some competition with the normal base-pairing partner, cytosine (Cyt), is observed (20). More recently, it has been proposed that the DNA synthesis precursor pool is a target for alkylation (21). The $O^6$-alkyl dGTP formed within cells in this way could be incorporated opposite Thy residues in DNA, yielding, as the anticipated mutagenic event, a T→C transition. Both in mammalian and bacterial cells, however, genetic experiments indicate that alkylating agents predominantly give rise to G·C→A·T substitutions and rarely to T·A→C·G transitions (22, 23). For this and other reasons (24), the relevance of DNA precursor pool alkylation to mutagenesis by alkylating agents is doubtful.

Support for the role of $O^6$-alkylguanines in carcinogenesis came initially from experiments by the laboratory of Rajewsky. A strong correlation was observed between the incidence of brain tumors in N-ethyl-N-nitrosourea treated rats and the persistence of $O^6$EtGua in brain DNA; in a nontarget organ such as liver, this lesion rapidly disappeared (25). The work of Margison and co-workers (26) and later others showed similar correlations in other mammalian systems. Further support for the role of $O^6$MeGua in carcinogenesis has come more recently from experiments in which it was shown that the methylating agent N-methyl-N-nitrosourea induces mammary carcinomas that suffer ras gene mutations of exactly the type caused by $O^6$MeGua (27). Taken together, the data from most in vitro and in vivo studies have suggested a causative role for $O^6$-alkylguanine residues in mutagenesis and carcinogenesis. As a result, $O^6$MeGua was chosen by several laboratories as the prototype for synthesis of site-specifically modified oligonucleotides (28–30, 10) and later for adduct-directed mutagenesis (31, 16, 30, 32, 33).

Traditionally, alkylation of the $O^6$-position of Gua has been carried out by reaction with diazoalkanes (34) or via the 6-chloro derivative of 2'-deoxyguanosine (19). The former reaction is cumbersome in that it generates a mixture of alkylated products and hence rigorous chromatographic purification is necessary to obtain the desired product. Although preferable to the diazoalkane approach, the chlorination reaction of the 2'-deoxynucleoside can result in extensive depuration. A third synthetic approach involves conversion of the modified base to the corresponding nucleoside (*28*); however, several tedious synthetic steps are necessary to convert the modified base to the β-anomer of the nucleoside. These disadvantages led Gaffney and Jones (*35*) to introduce a more convenient route of synthesis involving the facile sulfonylation of the 6-oxygen of Gua to the 6-*O*-(triisopropylphenyl)sulfonyl derivative. The sulfonate function is displaced by a trimethylamino group which, in turn, can be easily replaced by an alkyl functionality. This protocol, now the method of choice for synthesis of $O^6$-substituted Gua adducts, also provides a means for protecting the $O^6$-position of unmodified Gua residues during oligonucleotide synthesis by introducing base-labile (4-nitrophenyl)ethyl or 2-cyanoethyl groups. Despite its advantages, it is noteworthy that this synthetic method may generate certain side products such as 2-amino-6-(dimethylamino)purines, especially when longer (or more sterically hindered) alkyl groups are introduced (*36*, *10*).

The first total synthesis of an oligodeoxynucleotide containing $O^6$MeGua was performed by a solution-phase phosphotriester approach (*28*). The modified base prepared from 2-amino-6-chloropurine was condensed with 2-deoxy-3,5-di-*O*-*p*-toluoyl-D-*erythro*-pentosyl chloride. Isolation of the β-anomer of the nucleoside derivative followed by protection of the hydroxyl and amino functionalities generated a completely protected $O^6$-methyldeoxyguanosine monomer. The protected dimers of d-(TG$^{O^6Me}$) and d(CA) were made by condensation of the respective monomers in the presence of mesitylenesulfonyl tetrazolide. These in turn were coupled to prepare a tetranucleotide, which was deblocked sequentially by phenylsulfonic acid, oximate, and ammonium hydroxide. After purification by HPLC, the tetramer was characterized by $^1$H NMR, by analysis of enzymatic digestion products, and by two-dimensional homochromatography. The presence of $O^6$MeGua was shown to inhibit the enzyme activities of snake venom phosphodiesterase and nuclease P$_1$.

A disadvantage of ammonia deprotection of oligonucleotides was discovered by Jones and co-workers, who used both phosphotriester and phosphoramidite chemistries to prepare several $O^6$MeGua-containing oligonucleotides (*29*, *37*). It was noted that $O^6$-alkylation significantly stabilizes the $N^2$-isobutyryl group and that the $O^6$-alkyl groups are prone to displacement by thiophenol and ammonia. In the work described above (*28*) this possibly led to the production of an oligomer containing 2,6-diaminopurine; this putative product (observed but not identified) elutes by HPLC later than the tetranucleotide containing $O^6$MeGua, affording facile purification of the latter. However, since purification of longer oligonucleotides is comparatively difficult, Jones and his colleagues employed the nonaqueous deblocking conditions of oximate and 1,8-diazabicyclo[5.4.0]undec-7-ene (DBU) in methanol/THF to avoid these side reactions. The $N^2$-isobutyryl group can be removed by this mixture if a prolonged reaction time is used; alternatively, a more base-labile acetyl functionality can be employed to protect the exocyclic amino group. A recent study on these side products details the chromatographic purification steps necessary to separate the desired oligomer from unwanted contaminants (*10*).

Five laboratories have built synthetic oligonucleotides containing $O^6$-alkylguanines into viral or plasmid genomes and investigated the genetic fate of the lesions. In all reported studies of site-specific mutagenesis by $O^6$MeGua,

the type of mutation induced was exclusively the G→A transition. The first such study was from our own laboratory and investigated the mutagenicity and repair of $O^6$MeGua situated at a unique site in the genome of virus M13mp8 (*31*). Mutation was investigated initially in a repair competent strain of *Escherichia coli* [*ada*$^+$; the *ada* gene encodes the $O^6$MeGua DNA methyltransferase (MT), which repairs $O^6$MeGua, several other *O*-alkyl bases, and one methylphosphotriester diastereomer (*38*, *39*)]. The adduct in single-stranded (ss) DNA has a mutation efficiency (i.e., the fraction of progeny phage with a mutation) of approximately 0.4%. The mutation efficiency of the lesion in double-stranded (ds) DNA is 10- to 20-fold lower (*40*), probably because ds genomes are better substrates for $O^6$MeGua repair than ss DNA (*41*). To investigate the modulating influence of DNA repair more directly, the cells were challenged with *N*-methyl-*N*-nitrosoguanidine (MNNG) before uptake of the modified viral genome into the cell; this treatment alkylates the *E. coli* chromosome and thereby diminishes $O^6$MeGua repair capacity by suicidally inactivating the MT repair protein. Diminution of $O^6$MeGua repair capability was paralleled by an increase in the mutagenic efficiency of the lesion to approximately 20%. A mutational strand-bias exists, in that $O^6$MeGua in the (+) strand of ss M13 DNA is approximately three times more likely to yield mutants than it would if it is in the complementary (−) strand. Exactly the opposite bias is observed when the adduct is in ds DNA. Although seemingly contradictory, these data are possibly explained by the fact that, for ss DNA, the (+) strand of the virus is more infectious than the (−) strand. Introduced into cells in duplex form, however, the genotype of the (−) strand predominates among progeny because the virus employs an asymmetric replication strategy favoring the (−) strand (*42*). An alternative explanation of these data has been noted by Topal (*33*), who, using a bacteriophage f1/pBR322 chimera site-specifically modified with $O^6$MeGua, has shown that the repair efficiency of $O^6$MeGua is not uniform and in fact varies three- to fourfold from site to site.

Identical qualitative and similar quantitative features of mutagenesis were obtained by Hill-Perkins et al. (*30*) and Bhanot and Ray (*32*), who investigated $O^6$MeGua mutagenesis in the (−) strand of M13mp9 and φX174 duplex templates, respectively. Neither group was able to discern measurable mutation when adducted DNA was introduced into repair-competent cells. However, substantial mutagenesis resulted when replication of $O^6$MeGua containing genomes was carried out in MNNG-treated cells (*30*, *32*) or in cells possessing a kinetically slow $O^6$MeGua DNA methyltransferase (*32*). As expected, induction of the *ada* gene by continuous exposure of cells to low, adapting doses of MNNG reduces the $O^6$MeGua mutation frequency by at least 17-fold (*30*). The results of these studies, together with those of Loechler et al. (*31*), show that the principal fate of a single $O^6$MeGua lesion in *E. coli* is *repair* and that a high level of mutagenesis, as is observed when $O^6$MeGua-containing oligonucleotides are copied by polymerases in vitro (*43*), occurs only when cellular repair competence is low. At odds with this conclusion are the findings of Chambers and colleagues, who built a φX174 duplex containing $O^6$MeGua in the third codon of gene *G* (*16*). They observed a strikingly high mutation frequency of at least 15% for $O^6$MeGua in repair competent cells. These data are especially surprising in view of the fact that both Bhanot (*32*) and Chambers (*16*) built $O^6$MeGua into the same codon of the φX174 genome, seemingly ruling out local genomic structure and trivial differences between the experimental systems as being responsible for the disparity. Only further experimentation will resolve this apparent conflict.

$O^6$-Butylguanine ($O^6$BuGua) has been built into the $\phi$X174 genome, and it induces transition mutations at a modest frequency (~0.3%; ref 16). Interestingly, its mutation efficiency is elevated eightfold in an excision repair-deficient (*uvrA*) background, showing that UvrA protein plays a role in modulating $O^6$BuGua mutagenesis in vivo. When parallel studies were conducted on $O^6$MeGua, the unexpected result was obtained that *uvrA* and, to a lesser extent, *recA* defects cause a large (up to 40-fold) reduction in mutagenesis. These data suggest a hitherto unknown enhancement of methylating agent mutagenesis by the Uvr system.

Can the miscoding properties of the $O^6$-alkylguanines be rationalized by their effects on DNA structure? While the above studies utilized $O^6$-alkyldeoxyguanosine containing oligonucleotides for genetic evaluation of the lesion, Jones and co-workers used them for physical studies. In one study, the methylated base was built into the central dinucleotide unit of a d(CG)$_3$ hexamer which, when unmodified, undergoes a salt-induced B→Z transition in DNA conformation. A 40 °C decrease in melting temperature ($T_m$) of d(CGCG$^{O^6Me}$CG) compared to the unmodified sequence indicates that the methylated base greatly destabilizes the hexanucleotide (29). Interestingly, the $T_m$ of the mispairing sequence d(CGTG$^{O^6Me}$CG), in which $O^6$MeGua can only pair with Thy in duplex form, is even lower. This suggests that neither G$^{O^6Me}$·C nor G$^{O^6Me}$·T pairing is very efficient. An examination of the circular dichroism (CD) spectra of these oligonucleotides indicates that the salt-induced B→Z transition is partially inhibited by the presence of the alkylated base. Further studies involving a set of self-complementary dodecanucleotides d(CGNGAATTCG$^{O^6Me}$CG) where N = A, C, G, or T showed that the $T_m$s of these adduct-containing sequences are 19–26 °C lower than the corresponding unmodified oligomer, suggesting once again the inefficient pairing of $O^6$MeGua regardless of the opposing base (37). In fact the mispairing sequence d(GCTGAATTCGCG), which contains two G·T mismatches in the duplex form, is more stable than all four sequences containing $O^6$MeGua, although each of the 12-mers maintains the B conformation. These data collectively suggest a local destabilization of the helix induced by $O^6$MeGua. Furthermore, the order of $T_m$ is N = C > A > G > T, which constituted the first in a series of experimental data to argue against the original hypothesis by Loveless of facile and stable G$^{O^6Me}$·T base pairing (Figure 3, A).

Proton NMR spectra of d(CGNGAATTCG$^{O^6Me}$CG) duplexes, where N = C, T, A, or G, provided further information on the effect of $O^6$MeGua on DNA structure (44–47). The directionality of the NOE between base protons and the sugar H1' proton indicates that all four duplexes remain as right-handed helices in solution. The glycosyl bond torsion angles at the site of modification show small alterations, but retain the anti conformation except in the G$^{O^6Me}$·G duplex, where the orientation of the alkylated base is syn. The syn orientation of $O^6$MeGua also results in significantly better stacking as shown by a 0.36–0.45 ppm upfield chemical shift of the respective H8 proton compared to other duplexes. In all of the duplexes both $O^6$MeGua and its complementary base N are well stacked in the helices. This is evident from the absence of NOEs between the two neighboring Gua imino protons and from the normal NOEs between the protons of G$^{O^6Me}$·N and the two flanking G·C base pairs. Also, the

strong NOE exhibited by the methoxy group to the H5 of the 5'-Cyt residue in all four cases further demonstrates stacking of the alkylated base and establishes that the methoxy group is oriented into the major groove. The upfield shifts of the methoxy protons indicate favorable stacking of this functionality into the helices when N = C, T, or A. In contrast, in the G$^{O^6Me}$·G 12-mer, the methoxy group does not appear to be as well stacked with its neighboring bases. The flanking base pairs to the site of modification are stable at –5 °C and do not experience significant distortions.

Probably the most significant observation is the unusually large upfield shift for the imino protons of Thy and Gua in G$^{O^6Me}$·T and G$^{O^6Me}$·G duplexes, respectively (8.5–9.0 ppm compared to 12.5–14.5 ppm for Watson–Crick base pairs). The data suggest that the imino protons are not engaged in hydrogen bonds with the complementary $O^6$MeGua residue and, as a result, these pairs probably are stabilized by a solitary hydrogen bond. Patel et al. (45) propose a G$^{O^6Me}$·T interaction consistent with the NMR data involving one amino–carbonyl hydrogen bond in the minor groove (Figure 3B). To relieve an unfavorable methoxy–carbonyl interaction, the bases open out in the major groove. Similarly, they suggest that pairing of G$^{O^6Me}$·G primarily involves the amino functionality of Gua and the N7 ring nitrogen of $O^6$MeGua (i.e., the hydrogen-bonding scheme in Figure 3, C; see ref 46). A weak hydrogen bond between the imino proton of Gua and the methoxy group oxygen might also be possible in this configuration. In the G$^{O^6Me}$·C duplex, small alterations in base-pair overlaps with the flanking G·C base pairs are noted by the respective upfield and downfield shifts of the base protons. A possible mode of pairing involves both bases in anti conformation with connecting hydrogen bonds between each amino group and a ring nitrogen (Figure 3, D; see ref 47). A similar pairing interaction was also suggested for the G$^{O^6Me}$·A duplex (Figure 3, E; see ref 46).

The proton NMR data, viewed together with the $T_m$ experiments, suggest that the hydrogen bonding in G$^{O^6Me}$·T is weaker than originally believed. Interestingly, on the basis of hydrogen bonding alone, one would predict that $O^6$MeGua would induce a G→T transversion, because G$^{O^6Me}$·A base pairing does *not* induce major distortions in the helix configuration. Furthermore, this base pair (bp) appears to be stronger than the G$^{O^6Me}$·T pair. In biological studies, however, the G→A transition is the exclusive mutagenic event observed in vivo and the predominant one in vitro; only with RNA polymerase in vitro is a small proportion of misincorporation of Ade observed (43).

Examination of the $^{31}$P NMR spectra of these duplexes reveals another interesting conformational feature that may be of central importance in rationalizing genetic effects of $O^6$MeGua (44). Only in the G$^{O^6Me}$·T 12-mer duplex is there no distortion of the phosphodiester backbone. The phosphorus dispersion of this duplex in the range of 3.9–4.5 ppm is reminiscent of Watson–Crick base pairing in normal B-DNA helices. In marked contrast, all other G$^{O^6Me}$·N duplexes exhibit signals outside this region providing evidence of alteration in the phosphodiester backbone presumably at the site of modification.

It has been suggested that repair proteins may be able to recognize an altered sugar–phosphate backbone architecture while the G$^{O^6Me}$·T pairing may remain cryptic, thus enabling the observed mutagenic specificity of the methylated base (44). However, this is unlikely because in vitro studies using purified polymerase and *no* DNA repair proteins show that $O^6$MeGua prefers to base pair with Thy

*(48).* A more likely possibility may be that DNA polymerase is, by some unknown mechanism, sensitive to distortion of the phosphodiester backbone. The ability of the polymerase to incorporate complementary bases on the basis of preferred hydrogen-bonding interactions is, as yet, the only feature that has been well established. However, physicochemical factors such as base stacking and conformation may also be important criteria polymerases employ to assess accurate formation of base pairs during replication. In this regard, the predominant factor in the mutational specificity of $O^6$MeGua may *not* be its hydrogen bonding characteristics, but rather its property of forming a consistently uniform phosphodiester backbone.

In contrast to the conclusions of the above studies, a recent investigation on the base-pairing capabilities of $O^6$MeGua in hydrophobic media showed quite different results *(49).* To mimic the base-pairing properties of $O^6$MeGua in the hydrophobic core of duplex DNA, nucleosides were derivatized at the sugar hydroxyls with triisopropyl groups so that $^1$H NMR studies could be performed in $CDCl_3$. The downfield shifts of the amino proton of $O^6$MeGua and the imino proton of Thy suggest formation of two hydrogen bonds within the base pair (Figure 3, A). Interestingly, no evidence of hydrogen-bond formation between $O^6$MeGua and Cyt was found, although Cyt, when protonated, could base pair efficiently. Since protonation promotes the deamination of Cyt, a new mechanism of mutation was suggested involving $O^6$-alkylguanine-induced conversion of the opposite-strand Cyt to uracil. By this scheme, cross-strand protonation followed by deamination and replication, would lead to the genetically observed G→A transitions. It should be pointed out that this novel model is not mutually exclusive of the more conventional one involving simple misreplication of the adduct.

**$O^4$-Alkylthymines.** As with alkylation of the $O^6$-atom of guanine, the formation of $O^4$-alkylthymines is believed to play a role in alkylating agent-induced mutagenesis and, possibly, carcinogenesis (for reviews, see ref 50–52). In vitro studies establish that DNA polymerases preferentially incorporate Gua opposite this lesion, in accordance with the mispairing scheme suggested in Figure 3A *(53).* Modeling studies support the feasibility of $T^{O^4alkyl} \cdot G$ base pairs *(54).* Moreover, in the *lacI* gene of *E. coli*, alkylating agents induce A·T→G·C transitions at a low, but significant frequency in addition to the predominant G·C→A·T events *(22).* The contributing role of the $O^4$-alkylthymines in carcinogenesis was suggested by the work of Swenberg, who showed that certain tissues susceptible to alkylating agent carcinogenesis can repair $O^6$-alkylguanines rapidly, whereas $O^4$-alkylthymines are relatively refractory to removal *(55).*

To assess the mutagenic potential of the $O^4$-alkylthymines unambiguously, Preston et al. *(56, 57)* recently performed site-specific mutagenesis studies on single adducts enzymatically incorporated into the *amber* codon of the bacteriophage $\phi$X174 *am3* duplex genome. The modified deoxynucleoside triphosphate was introduced at the 3′-terminus of a 15-mer by using *E. coli* DNA polymerase I. The comparatively mild conditions of enzymatic synthesis (30–37 °C, pH 8) allowed facile synthesis of several oligonucleotides containing chemically labile adducts, including $O^4$-methylthymine ($O^4$MeThy), $O^4$-ethylthymine, $O^2$-methylthymine, $O^4$-isopropylthymine, $O^6$MeGua, 1-methyladenine, and 3-methylcytosine *(56, 57).* The $O^4$MeThy-containing primer was annealed to the $\phi$X174 ss genome (see Figure 1), extended by incorporating unmodified nucleotides, and the resulting partial duplex

was transfected into *E. coli* spheroplasts. In *ada⁻* cells (i.e., cells defective in the repair of $O^6$MeGua and $O^4$MeThy), a tenfold higher mutation frequency over the control was noted. The mutagenic efficiency of $O^4$-ethylthymine is approximately half that of $O^4$MeThy, and, for each lesion, no mutation is detected in repair proficient *(ada⁺) E. coli*. The mutant populations derived from both adducts exclusively contain the T→C transition at the site of modification. $O^2$-Methylthymine and $O^4$-isopropylthymine are not detectably mutagenic. Not yet investigated but of interest would be a comparison in the same genetic system of the mutagenic potential of $O^6$MeGua and $O^4$MeThy, since both are mutagenic and both are simultaneously present in the genomes of cells treated with certain carcinogenic and mutagenic alkylating agents. Such a comparison would allow direct determination of the relative importance of the two lesions in a specific biological system.

Recent physical studies on the base-pairing characteristics of $O^4$MeThy utilize oligodeoxynucleotides synthesized by the solution-phase phosphotriester method. A tetramer d($T^{O^4Me}$GCG) and two self-complementary dodecamers d(CGCAAGCTT$^{O^4Me}$GCG) and d-(CGCGAGCTT$^{O^4Me}$GCG) have been prepared by Swann and colleagues *(58).* The nucleoside $O^4$-methylthymidine, prepared from the 3-nitro-1,2,4-triazolo derivative, was protected with a 9-phenylxanthen-9-yl functionality at the 5′-hydroxy, and a 2-chlorophenyl phosphate was introduced at the 3′-terminus. Dimers, tetramers, octamers, and dodecamers were synthesized in a stepwise fashion by the condensation reaction of a free 5′-hydroxyl group with a 3′-phosphodiester in the presence of 1-(mesitylenesulfonyl)-3-nitro-1,2,4-triazole (MSNT) followed by selective removal of either the 5′- or the 3′-protecting groups. Because of the protic acid-labile nature of $O^4$-alkyl groups, deblocking of the 5′-O-(9-phenylxanthen-9-yl) functionality was performed with zinc bromide. Cyanoethyl groups protecting the 3′-terminus were easily removed by a triethylamine–pyridine mixture. After the synthesis was complete, subsequent deprotection was by oxime treatment following which the N-acyl groups were removed with DBU/methanol. The latter reagent, as in the synthesis of $O^6$MeGua containing oligomers, was used to preserve the structural integrity of $O^4$MeThy which, under ammoniacal conditions, is prone to conversion to 5-methylcytosine.

Hypochromicity measurements on the dodecamers d-(CGCAAGCTT$^{O^4Me}$GCC) and d(CGCGAGCTT$^{O^4Me}$GCG) containing $O^4$MeThy opposite adenine (Ade) and Gua, respectively, reveal not only lower $T_m$s than their unmodified counterparts, but their melting curves display a biphasic shape. These data indicate that the methylated Thy has a destabilizing effect on the helix, as was also shown for $O^6$MeGua *(58).* The exact reason for the biphasic character of the melting curve was not established, although it is possible that preferential melting of the base pairs around the modified base is responsible.

High-resolution two-dimensional NMR studies were performed on the self-complementary dodecamer duplexes containing $O^4$MeThy *(59, 60).* The dodecanucleotide d-(CGCAAGCTT$^{O^4Me}$GCG) containing a $T^{O^4Me}$·A base pair assumes a duplex structure similar to an A·C mismatched duplex d(CGCAAGCTCGCG) at pH 5.5 (Figure 4, B and C). The glycosyl torsion angles of both A and $T^{O^4Me}$ are anti, and the duplex remains as a right-handed helix. The base pairing in A·$T^{O^4Me}$ appears to involve a lateral displacement concurrent with the formation of a hydrogen bond between the 6-amino group of Ade and the N3 ring

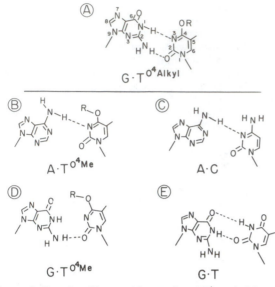

**Figure 4.** Postulated base-pairing modes of $O^4$-methylthymine: A, conventional scheme for mispairing of $O^4$-methylthymine with guanine (53); B and C compare the base pairing of adenine with $O^4$-methylthymine and cytosine, respectively, as predicted from NMR studies; D and E compare the base pairing of guanine with $O^4$-methylthymine and thymine, respectively, as predicted by NMR.

nitrogen of $O^4$MeThy similar to the one predicted (61) for the wobble base-pair structure of the A·C mismatch (Figure 4, C). No significant distortion in spatial orientation of the neighboring base pairs to the modification site is evident. The methoxy group of $O^4$MeThy assumes a syn orientation relative to N3 and is directed toward the complementary Ade residue, as indicated by the magnitude of the observed NOEs.

For the dodecamer d(CGCGAGCTT$^{O^4Me}$GCG), the connectivities between the nonexchangeable protons indicate a stacked right-handed helical structure (60). Also, both G and T$^{O^4Me}$ in the duplex are in anti orientation. The expanded NOESY plots suggest only small alterations in the structure of the helix due to the alkylated base. Some interesting differences were noted when this 12-mer was compared to the corresponding G·T mismatched 12-mer d(CGCGAGCTTGCG). In contrast to the Gua and Thy imino proton resonances at 10.57 and 11.98 ppm, respectively, in the wobble base-paired G·T mismatch, the Gua imino proton of the G·T$^{O^4Me}$ duplex resonates at 8.67 ppm. The large upfield shift of the latter argues against the previously suggested G·T$^{O^4Me}$ pairing scheme (Figure 4, A) involving the imino proton and N3 of $O^4$MeThy. Lack of hydrogen bonding by this imino proton also rules out a pairing similar to the G·T wobble base pair (Figure 4, E). The magnitude of NOEs between the imino proton of Gua and the methoxy group of $O^4$MeThy, when compared with other neighboring protons, indicates a syn orientation of the methoxy group with respect to N3. On the basis of the NMR data, Patel and co-workers (60) suggested that the most likely pairing between Gua and $O^4$MeThy is that involving a short hydrogen bond between an amino proton of Gua and the carbonyl oxygen of $O^4$MeThy (Figure 4, D). Interestingly, as shown for the G$^{O^6Me}$·T duplex, phosphorus NMR establishes an unperturbed phosphodiester backbone demonstrated by the localized signals between 3.9–4.4 ppm. It is also noteworthy that the dodecamer containing the G·T wobble base pair shows significantly more backbone distortion than the oligonucleotide containing G·T$^{O^4Me}$.

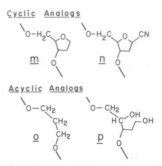

**Figure 5.** Cyclic and acyclic analogues of apurinic/apyrimidinic sites.

In summary, the physical data from these studies presents a situation similar to that seen with the O⁶MeGua containing oligonucleotides, in that the physicochemical studies argue against the commonly accepted mispairing schemes. In an attempt to rationalize the mutational specificity observed for O⁴MeThy, Patel has suggested that its base pair with Gua is very lipophilic and that hydrophobic forces, in addition to, or rather than, hydrogen bonds hold the base pair together within the DNA polymerase molecule (60). A great opportunity for research in the future will be to design experiments that test this intriguing model.

### Apurinic/Apyrimidinic Sites

Cleavage of the $N$-glycosyl linkage between a base and its deoxyribose in DNA generates apurinic/apyrimidinic (AP) sites (Figure 2, C). Spontaneous loss of a purine (depurination) occurs more rapidly than loss of a pyrimidine (reviewed in ref 62), because protonation of the N7 atom of Ade and Gua, and of N3 of Ade, renders the N-glycosylic bond particularly sensitive to hydrolysis. Many chemical–DNA adducts are known to labilze the sugar–base linkage and give rise to abasic sites (63). In addition, damaged DNA is often subject to repair in vivo by DNA glycosylases, which remove selected lesions generating AP sites (64). Formation of AP sites spontaneously or indirectly by chemical modification is of concern because such lesions are known to be mutagenic in bacterial and mammalian cells (65, 66). The roles of AP sites in genetic toxicology have been reviewed recently by Loeb and Preston (67).

AP sites give rise to a highly unstable cyclic carboxonium ion, which hydrolyzes to the $\alpha$- and $\beta$-anomers of 2-deoxyribose. These anomers exist in ring–chain tautomeric equilibrium with the respective ring-opened aldehyde form. Each of these species is highly base-labile and easily generates a scission in the phosphodiester backbone by a $\beta$-elimination reaction. AP sites have been introduced at random into DNA by acid/heat treatment or by uracil glycosylase treatment of uracil-containing DNA templates, but, because of their unstable nature, as yet no reports have appeared on chemical synthesis of a formal abasic site as part of an oligonucleotide. However, synthesis of oligodeoxynucleotides containing chemically stable analogues of AP sites have been reported by a number of workers. Figure 5 displays the structures of a series of cyclic and acyclic AP site analogues that have been introduced into oligonucleotides by total synthesis. These analogues are stable enough to undergo standard protection and deprotection by contemporary DNA synthesis protocols.

Preparation of analogue $\underline{m}$ was reported by two laboratories using different synthetic schemes (68, 69). Following conventional protection of the hydroxyl groups, several oligodeoxynucleotides were synthesized by either phosphotriester or phosphite-triester approaches. The purified oligonucleotides display a faster mobility in a polyacrylamide electrophoresis gel compared to a marker of one unit smaller size; this high mobility presumably is due to an increase in negative charge without an offsetting increase in molecular weight (69). When model AP sites are situated in a template and used for DNA synthesis, Ade and to a lesser extent Gua are inserted opposite the lesion (68, 69).

Hypochromicity measurements on 15-mer duplexes containing a single 1,2-dideoxyribofuranose (structure m, Figure 5) positioned opposite the four common bases have been made (70). A reduction of $T_m$ of ~40% is observed in each case. It is likely that this large change in $T_m$ not only indicates a diminution of base-pairing energy pertaining to hydrogen bonding but also reflects a loss of base-stacking interactions. A comparison of different bases opposite the model AP site analogue displays the following order in $T_m$: G > T > A $\simeq$ C. The introduction of a nonpolar aromatic ring in the form of a phenyl group is also highly destabilizing. This suggests that the hydrophobic nature of an ring is insufficient to account alone for the effective stacking interactions and presumably indicates a role of dipole-induced dipole interactions in stacking phenomena.

Other studies have suggested the possibility that AP sites might provide a flexible joint separating different conformational domains in DNA (71). A 13-mer d-(CGC5MeGCGXACATGT) was synthesized where X represents the AP site analogue, 1-cyano-2-deoxy-$\beta$-D-ribose (structure n, Figure 5), and C5Me is 5-methyldeoxycytidine. Since the two hexamers flanking X are self-complementary, they form duplex concatamers. Right- and left-handed helical conformations, separated by the AP site analogue, coexist within this DNA molecule. CD measurements reveal the left-handed Z form only in high salt, whereas at an order of magnitude higher concentration, 1H NMR studies indicate approximately 30% Z form at room temperature in 0.1 M NaCl. Infrared (IR) studies in film show that the d(CGC5MeGCG) segment is always in Z conformation at low humidity. A single AP site, therefore, appears to be able to release the torsional stress resulting from multiple conformational domains in the same duplex.

Proton and phosphorus NMR studies have been performed on a nonadeoxynucleotide duplex d-(CATGAGTAC)·d(GTACmCATG) containing the AP site analogue m (72). NOE connectivities in the 1H NMR establish a stacked right-handed helical structure at and adjacent to the AP site with all glycosyl torsion angles in the anti range. Watson–Crick base pairing adjacent to the AP site remains intact, and the imino protons of the two guanines flanking the Ade residue opposite the AP site exhibit NOEs of similar magnitude to the H2 protons of the adenines in the adjacent base pairs. These observations indicate that the Ade opposite the abasic site is inserted into the helix, is stacked with the adjacent base pairs, and remains in an anti orientation. Further evidence of stacking has been obtained from the chemical shifts and NOE connectivities. The 31P NMR spectrum displays two conspicuous downfield-shifted peaks, indicating structural perturbation of two phosphates near the AP site. Another 1H NMR study on a different nonadeoxynucleotide duplex, d(CGTGmGTGC)·d(GCACACACG), shows similar results (73). In this sequence the non-hydrogen-bonded Ade and the A·T base pair have comparable $T_m$ values, and the conformation of the Ade opposite the model AP site is similar to that in a normal A·T pair.

A number of biochemical studies has been performed (68) on AP site analogues (structures m, o, p, Figure 5). Both exonuclease III and endonuclease IV can cleave on the 5'-side of these analogues. Apparently the absence of a base is detected by these enzymes, rather than the presence of a formal deoxyribose moiety, and it has been suggested that local changes in DNA conformation may be the recognition signal for the endonucleases. In other studies, the acyclic analogue 1,3-propanediol (structure o, Figure 5) has been introduced in a set of four dodecamers d(CGCGAATTCGCG) where each of the Ade and Thy residues was replaced by this acyclic moiety (74). The phosphodiester bond between 1,3-propanediol and its 5'-nucleoside is refractory to snake venom phosphodiesterase, while the other phosphodiester bonds are hydrolyzed completely. Helix-melting studies with these oligonucleotides indicate an enhancement of hairpin formation due to the presence of the acyclic structure within the loop region.

Extension of primed templates containing AP site analogues by E. coli DNA polymerase I (Klenow fragment), calf-thymus DNA polymerase-$\alpha$, and avian myoblastosis virus (AMV) reverse transcriptase has been studied (68). The polymerases are initially blocked immediately 3' to the abasic analogue, as evidenced by the accumulation of an intermediate. Subsequently, a dNMP is incorporated opposite the lesion and another intermediate starts to accumulate. Eventually some of the nascent chains are extended to completion. Efficiency of bypass or "read through" is lower with DNA polymerase I as compared to $\alpha$ and AMV polymerases. The process of polymerase "read through" is more efficient when a ring structure is used as an AP site analogue (structure m, Figure 5) rather than an acyclic form. As in the in vivo and in vitro studies on AP sites (75, 76), dAMP is the major nucleotide incorporated opposite the AP site analogues. In general, all of these in vitro studies on the genetic effects of AP site analogues are in accord with previous studies in which the behavior of AP sites formed at random in DNA sequences has been investigated (67).

## Aromatic Amines and Amides

Aromatic amines and amides have been widely used in industry and are common environmental pollutants (77). Many of these compounds have been shown to be carcinogenic in experimental animals and in humans, and most require metabolic activation in order to interact with DNA and form adducts (1, 78). Purines in DNA are the principal targets of aromatic amines, with the reaction usually occurring at the C8 atom of Gua and somewhat less readily at exocyclic amino functionalities (79).

Oligodeoxyribonucleotides containing C8 Gua adducts of 2-aminofluorene (AF), 2-(acetylamino)fluorene (AAF), and 4-aminobiphenyl (ABP) have been synthesized (Figure 2; ref 80, 81, 14, 12). Stohrer and co-workers first reported the synthesis of a tetradecamer containing a C8-Gua-AF adduct by total synthesis using a solid-phase phosphoramidite approach (80). A protected monomer of the 2'-deoxyguanosin-8-yl-AAF was prepared and used in an automated DNA synthesizer to form the desired protected oligomer with an overall yield of 72%. Deprotection by standard procedures, however, results in total loss of the fluorene moiety. Apparently, this is the result of an oxidative degradation, because treatment with ammonium hydroxide in the presence of mercaptoethanol in inert atmosphere resulted only in the loss of the acetyl function and generated an AF-adducted 14-mer. Following chromatographic purification, the AF-containing deoxyoligonucleotide preparation consisted of a single homogeneous band on an electrophoresis sizing gel. The UV spectrum exhibits an additional maximum at ~300 nm as expected for the fluorene moiety. DNA sequencing by the Maxam–Gilbert method of the presumed AF-containing oligonucleotide and its oxidized counterpart provided evidence of altered bases at the expected sites of modification. Although it is likely that this synthetic route generated the desired AF-containing 14-mer, no definitive evidence of the integrity of the adducted base was shown. This protocol was used to synthesize several oligodeoxynucleotides for biological studies, but to our knowledge no further characterization was reported.

In a second study by Romano and co-workers, a heptamer d(ATCCGTC) containing a single Gua was treated with the activated derivative N-acetoxy-N-(trifluoroacetyl)-2-aminofluorene which, under the conditions of the experiment, generates N-guanin-8-yl-2-AF (C8-Gua-AF; Figure 2, d) as the only detectable adduct (14). The adducted heptamer was purified from the unreacted oligonucleotide by HPLC. Piperidine induced cleavage of the AF-adducted Gua indicated that the site of modification was indeed the expected one, while trifluoroacetic acid digestion of the modified heptamer produced a product chromatographically identical with the C8-Gua-AF. It was demonstrated that the adduct was not detectably altered by exposure to several buffer environments typical of those encountered during insertion of oligonucleotides into genomes (82). The AF-adducted oligodeoxynucleotide was introduced into an M13 genome (14) for in vitro replication blockage studies and to investigate site-directed mutagenesis in vivo (vide infra).

A similar approach utilizing selective reaction of the activated carcinogen with an oligonucleotide was used for the synthesis of an AAF-containing oligonucleotide. Sharma and Box (81) modified d(TACGTA) with N-acetoxy-2-(acetylamino)fluorene under mild conditions (pH 7.4) at room temperature to produce an N-guanin-8-yl-2-AAF (C8-Gua-AAF) containing hexamer as the major product. The absorption spectrum of the purified AAF-adducted hexamer displays the 304-nm absorption maximum characteristic of the carcinogen chromophore. Identification of C8-Gua-AAF was by acid hydrolysis of the oligomer followed by TLC analysis. Proton NMR of the modified hexamer demonstrated the conspicuous absence of the C8 proton of Gua as expected for this C8-Gua adducted molecule. An upfield shift of the Thy methyl protons of ~0.65 ppm after modification suggests a Z-like conformation of this hexamer in duplex form; these data are in accordance with others on randomly modified DNA polymers (83). In addition the temperature-dependent chemical shifts of the Thy methyl protons, which would signal the helix-coil transition of the unmodified hexamer, are absent in the AAF-adducted oligonucleotide. Downfield chemical shifts noted in the $^{31}$P NMR spectrum of the AAF-adducted hexamer suggest a change from a normal gauche,gauche conformation to a gauche,trans orientation. These data are all typical characteristics of a Z-DNA conformation. CD spectra provided additional evidence of a left-handed helical geometry even in low-salt solution. This result correlates well with previous studies on randomly AAF-adducted poly(dG·dC) and modeling studies with other adducted oligomers, which indicate that C8-Gua-AAF changes the normal anti glycosyl bond conformation to syn (reviewed in ref 62). In addition, the adducted Gua is almost coplanar with the 5′-Cyt, with the AAF residue twisted perpendicular to the Gua base. The

biological significance of the Z-like conformation of covalently bound AAF is unknown.

Another established conformation of DNA containing C8-Gua-AAF involves base displacement with accompanying denaturation and helix bending (84, 85). It is noteworthy that despite structural similarities, AF and AAF induce entirely different conformational effects in DNA. The AF adduct usually induces very little distortion in the B-DNA helix and prefers to reside in the major groove. As indicated above, the effects of the AAF lesion are far more severe. Genetic studies on AAF in a pBR322 forward mutation assay indicate that >90% of the mutations induced by AAF adducts are frameshifts occurring mainly in strings of G·C base pairs; in contrast, ~85% of the AF-induced mutations are base-pair substitutions, predominantly G·C→T·A transversions (86–88). Most of the mutations of both adducts are dependent on induced SOS functions. The different biological effects of these two adducts may be related to their distinct conformational effects.

A tetranucleotide containing a single N-guanin-8-yl-ABP (C8-Gua-ABP; Figure 2, e), the major DNA adduct of 4-aminobiphenyl, was synthesized as part of a collaboration between our laboratory and that of F. Kadlubar by reaction of N-acetoxy-N-(trifluoroacetyl)-4-ABP with the unmodified tetranucleotide d(TGCA) (12). Fast atom bombardment mass spectrometry confirmed the addition of a single ABP molecule by a prominent molecular ion corresponding to the disodium salt of the monoadducted tetranucleotide. High-field $^{1}$H NMR established the absence of the C8 proton of Gua, thereby indicating the C8 carbon as the site of adduction. Upfield shifts of the ABP ring protons indicated significant stacking interactions of the biphenyl residue with the DNA bases. Interestingly, the Ade and Thy ring proton resonances shift upfield, whereas a downfield shift of the Cyt protons occurs. This suggests that the biphenyl residue is stacked primarily with the Ade and Thy bases. The interaction of ABP with Cyt is more complex and will require further investigation. As with other DNA adducts, the presence of the ABP moiety inhibits cleavage of the tetramer by nuclease $P_1$. The tetranucleotide was introduced into the (−) strand of a duplex M13mp10 virus, where it also was observed to inhibit partially the activity of a restriction endonuclease.

Most of the aromatic amine-modified oligonucleotides described above have been built into viral or plasmid genomes and used in site-directed mutagenesis studies. In vitro studies by Stohrer using his C8-Gua-AF-containing oligonucleotide demonstrated that DNA polymerase I (Klenow fragment) can bypass this lesion and that purines (Gua > Ade) are preferentially incorporated opposite the adduct (89). After incorporating the oligomer into pBR322, mutagenesis was studied in vivo (90). A startlingly high mutation efficiency of 25–40% is observed when the adduct is in one strand of duplex pBR322, and it increases to nearly 100% when both strands of the duplex are modified. The explanation offered for the high mutation frequency in the doubly modified genomes was that the AAF residues are close to one another in this system and that their physical proximity may have inhibited excision repair. No subsequent studies on singly modified genomes containing aromatic amine adducts show quantitative results as striking as these. The mutants induced in this system by C8-Gua-AF are exclusively deletions (predominantly of one base), either at or immediately preceding the site of modification. These mutations were obtained without prior induction of SOS functions, which is surprising in view of other studies showing that

AAF induced frameshifts are largely SOS dependent (87).

Similar polymerase-bypass and site-specific mutagenesis studies on AF- and AAF-C8-Gua lesions in duplex M13 bacteriophage DNA have been conducted by Romano and King and their colleagues (91–93). The C8-Gua-AF adduct was confirmed not to inhibit DNA synthesis by several polymerases (91). Mutagenesis studies revealed that each adduct induces mutations at more than tenfold above the background level when the phage are replicated in *E. coli* (92). Further enhancement of mutagenicity occurs upon induction of SOS functions. In contrast to the results of Mitchell and Stohrer (90), most of the mutations are nontargeted frameshifts of +1 Gua in a string of four Gua's beginning four nucleotides 5′ to the adduct site. From M13 genomes containing the AAF lesion replicated in *uvrA* cells, two mutants were isolated that had a −1 Gua frameshift in the same region. With both AAF and AF adducts, base substitutions, mostly transversions, also were found. They occur at a lower frequency than the frameshifts, however. In contrast to the studies on AAF, the AF-induced population includes base changes one nucleotide 3′ and four nucleotides 5′ to the adduct site. Again nontargeted mutagenesis is an unexpected finding in view of the mutational spectra of the respective parent carcinogens (87, 88).

A third study, which involved incorporation of an AAF-modified decamer into the tetracycline resistance gene of a shuttle vector (pAM86), shows an overall mutagenic efficiency for the AAF adduct of ~6% in *E. coli*. Sixty to seventy percent of the mutants are transversions or one-base deletions at the site of previous AAF-modification (94).

The three studies reviewed above disagree with one another and, perhaps most importantly, with the mutational spectra of these compounds generated in vivo (86–88). The obvious differences between the qualitative and quantitative experimental results of the different laboratories may reflect differences in the viral, plasmid, or cell systems, local base-sequence effects, or technical factors such as the presence of undetected oligonucleotide contaminant(s) that affect the nature and/or degree of mutagenesis.

## Cyclic Nucleic Acid Adducts

Bifunctional compounds can form covalent linkages at two atoms on the same base generating cyclic adducts (for a recent review on the formation and biological effects of this class of DNA adducts, see ref 95). Included in this group are certain dicarbonyl compounds such as methyl glyoxal and malonaldehyde, $\alpha,\beta$-unsaturated carbonyls such as acrolein, crotonaldehyde and related compounds, vinyl halides and their metabolites, carbamate esters, and (haloalkyl)nitrosoureas. A large number of these compounds are mutagenic and a few have been shown to be carcinogenic (95). Most of the cyclic adducts are highly base labile, and, therefore, their synthesis as part of an oligonucleotide poses experimental challenges.

Syntheses of oligoribonucleotides containing the vinyl chloride induced etheno adducts and $1,N^2$-(2-methyl-allylidene)guanosine (the $\alpha$-methylmalonaldehyde–guanosine adduct) have been reported (8, 96–98). Oligoribonucleotides containing $1,N^6$-ethenoadenine ($\epsilon$Ade; Figure 2, f) and $3,N^4$-ethenocytosine were first enzymatically prepared using *E. coli* polynucleotide phosphorylase (8). Later, $T_4$ RNA ligase was shown to be a superior reagent for this synthesis (98). Efficient incorporation of 3′,5′-bisphosphates onto the 3′ end of an oligonucleotide by RNA ligase was demonstrated by synthesis of several ol-

igoribonucleotides containing these and other modified bases (98). Subsequently the reaction conditions were adapted for the preparative synthesis of oligodeoxyribonucleotides (99, 100).

A solution-phase phosphotriester synthesis of a ribotrinucleotide diphosphate has been performed by condensing 5′-O-(monomethoxytrityl)-2′-O-benzoyl-1,$N^6$-ethenoadenosine 3′-phosphate with the triester, 2′-O-benzoyluridylyl-(3′-5′)-N,N,2′-O,3′-O-tetrabenzoyladenosine phosphate (96). After deprotection with acid followed by methanolic ammonia treatment, the trimer ($\epsilon$A)UA was purified on a DEAE-cellulose column with an overall yield of 24%. The trinucleotide is resistant to digestion by RNase M, and $\epsilon$Ade fluorescence is quenched, as shown by comparison of fluorescence emission before and after enzymatic digestion (96).

A deoxyhexanucleotide d{GCT($\epsilon$A)GC} containing a single $\epsilon$Ade was synthesized recently by a solid-phase phosphotriester approach (13). Nonaqueous deblocking conditions were employed for deprotection because of the unstable nature of $\epsilon$Ade in aqueous basic solutions. Efficient stacking of the adduct is indicated, as above, by quenching of fluorescence, by an increment of the minimum in the CD spectrum, and by upfield shifts of the $\epsilon$Ade protons in the $^1$H NMR spectrum. Increased stacking is especially evident with the 5′-Thy. Helix-denaturation studies and CD profiles indicate that almost complete loss of base-pairing capabilities accompany the formation of the etheno ring. The hexamer, despite its central unpaired bases, forms duplexes at high concentrations and maintains a B-DNA conformation. The modified hexamer was incorporated into an M13 bacteriophage genome, and site-directed mutagenesis studies in *E. coli* established $\epsilon$Ade to be mutagenic in single-stranded DNA (mutagenic efficiency, ~1%; Basu and Essigmann, unpublished results). Induction of SOS functions had no influence on the mutation frequency, and the adduct is not detectably mutagenic when present in a duplex genome.

## DNA Lesions Induced by Radiation

**Ionizing Radiation.** Ionizing radiation causes numerous DNA lesions, including strand breaks, base modifications, protein–DNA cross-links, and AP sites (101, 102). The hydroxyl radical is the major reactive species responsible for many DNA damages in irradiated, aerated aqueous solutions (103). This radical is highly reactive toward Thy residues, generating *cis*-5,6-dihydroxy-5,6-dihydrothymine or *cis*-thymine glycol (t′; Figure 2, g) as one of the major stable products (104). Oxidation, either through hydroxyl radical attack or by other mechanisms, also gives rise to this Thy modification (105). Several in vitro studies have established that t′, in most sequence contexts, inhibits DNA synthesis (106–109). Interestingly, certain local sequences such as 5′-Ct′A-3′,5′-Ct′G-3′ and, to a lesser extent, 5′-Ct′C-3′ allow bypass by polymerases in vitro (108).

To assess further the biological effects of this lesion in vivo and in vitro, t′ has been introduced into DNA at preselected sites by several workers. Total synthesis of t′-containing oligonucleotides would be difficult because of the noted susceptibility of the base to alkaline conditions; at pH 12, t′ rapidly degrades to urea, acetol, carbon dioxide, and other fragments (107). As a result, the most convenient approach is selective oxidation of an oligonucleotide containing a single Thy. With mild conditions, $KMnO_4$ oxidizes Thy to 5-hydroxy-5-methylbarbituric acid and t′ without modifying other DNA bases, whereas $OsO_4$ generates t′ as the only major product. Clark and

Beardsley (110) oxidized two 18-mer sequences containing a solitary Thy, d(GATGGGCCGACGAAAAGG) (18A) and d(GCTGGGCCGACGAAAAGG) (18C), with osmium tetraoxide in the presence of pyridine. After purification through a size-exclusion column, the lesion frequency in the DNA templates was determined by piperidine cleavage of the 5′-end-labeled (with [³²P]phosphate) oxidized 18-mers followed by electrophoretic analysis in polyacrylamide gels. Because t′ is base labile, piperidine induces a strand break at the site of modification. The extent of modification of Thy by this method was judged to be >95%, although a low level of alteration of some other sites also was noted. An end-labeled 14-mer d-(CCTTTTCGTCGGCC) was annealed to the oxidized templates and extended by various polymerases. DNA synthesis by E. coli DNA polymerase I (Klenow fragment), T₄ DNA polymerase, and DNA polymerase α₂ was inhibited opposite the site of the lesion. AMV reverse transcriptase, in contrast, could bypass the lesion measurably. The change in sequence context from 5′-At′G-3′ to 5′-Ct′G-3′ did not increase extent of bypass by any of the polymerases, contradicting the earlier prediction of Hayes and LeClerc (108).

In all experiments, >90% of the time the correct nucleotide, dAMP, is inserted opposite t′ (110). It is of interest that the 3′→5′ exonuclease activity of E. coli DNA polymerase I is more efficient in removing the 3′-terminal dAMP from an A·t′ base pair than from a normal A·T base pair. These results suggest that t′ predominantly pairs with Ade and that increased nucleotide turnover opposite the site of modification, as shown in the case of polymerase I, may be responsible for the inhibitory effect of t′ on DNA synthesis. A modeling and molecular mechanical study involving one of the above sequences (18A) shows that hydrogen bonding of A·t′ is not significantly different from A·T, although a local perturbation in DNA is introduced (111). A major destabilizing effect, especially an impaired stacking interaction, on the neighboring base pair on the 5′ side of t′ is evident. This, in part, is attributed to an unfavorable steric overlap of the 5′-base and the 5-methyl group of t′. The 3′-base pairs, in contrast, experience a stabilizing effect. This suggests that the formation of the next base pair beyond the lesion may be energetically unfavorable especially in this 5′-At′G-3′ sequence, as experimentally reflected in the in vitro studies showing inhibition of DNA synthesis.

In another recent study, a deoxyhexanucleotide d-(GCTAGC) was oxidized by either KMnO₄ or OsO₄ to generate d(GCt′AGC) (Basu and Essigmann, unpublished results). This modified hexanucleotide was introduced into an M13 bacteriophage genome, which was replicated in E. coli. Progeny phage from this experiment were predominantly wild type, although mutants carrying the T→C transition were detected at a frequency of ~0.5%. Mutants were detected only when ss DNA carrying t′ is used to infect cells.

8-Hydroxyguanine (C8-OH-Gua; Figure 2, h) is formed in DNA both by ionizing radiation and by agents that produce oxygen radicals (112). In a recent study from Nishimura's laboratory, oligodeoxynucleotides containing C8-OH-Gua were chemically synthesized by a solid-phase phosphotriester method (113). The protected monomer, N²-acetyl-5′-(monomethoxytrityl)-8-methoxy-2′-deoxyguanosine-3′-(o-chlorophenyl) phosphate, was prepared from 8-methoxy-2′-deoxyguanosine by conventional synthetic procedures. This monomer was used to synthesize a 13-mer d(GCCAGCTAG^{C8-OH}TCAT) and a 16-mer d-(CAGCCAATCAGTG^{C8-OH}CAC) containing a single C8-OH-Gua residue. Deprotection of the oligomers was carried out in concentrated ammonia at 55 °C. The 8-methoxy functionality was subsequently displaced by treating the aqueous concentrate with an aqueous mixture of N,N-dimethylformamide and triethylamine at 35 °C in an inert atmosphere. Both oligodeoxynucleotides were ligated to the 5′-end of a 30-mer to yield partially duplex structures for in vitro mutagenesis studies. Extension of an oligonucleotide primer by E. coli DNA polymerase I shows no inhibition of DNA synthesis due to C8-OH-Gua. When the lesion is located between Thy and Cyt, it directs misincorporation of all four bases at almost equal frequency, indicating a localized region of high infidelity during DNA synthesis. Interestingly, the adjacent bases such as the 3′-Cyt and, to a lesser extent, the 5′-Thy also direct misincorporation. In the second template, which carried C8-OH-Gua between Ade and Thy, the only misinsertion at the site of modification is Thy instead of the normal Cyt, whereas the 3′-Thy induces misincorporation of all bases. In the latter template, however, the 5′ neighboring Ade is correctly read. The results of these in vitro studies indicate that C8-OH-Gua creates a region of ambiguity that partly depends on the local base sequence. Its mutagenic effect may be related to the fact that C8-OH-Gua predominantly exists as a 6,8-diketo form and that it adopts a syn conformation. Further studies are needed to reveal the relationship of the structural and configurational effects of this adduct with its biological consequences.

**Photoproducts Induced in DNA by Ultraviolet Light.** The relative roles of the DNA lesions induced by ultraviolet (UV) light in mutagenesis and carcinogenesis have been a matter of debate (114). The two major UV-induced photoproducts are the cyclobutane-type dipyrimidine and pyrimidine–pyrimidone (6–4) dimers, both of which form between adjacent pyrimidine bases in DNA. Although four isomeric forms of Thy photodimers are possible, only the cis-syn isomers are formed in reasonable yield in biological systems. Most mutations (60–65%) induced by UV light in E. coli are targeted, are SOS dependent, and occur mainly at the 3′-linked nucleotide of TT, TC, and CC sites, supporting the role of pyrimidine dimers at such sites in mutagenesis (115–117). The mutations in bacterial systems are G·C→A·T and A·T→G·C transitions; similar mutagenic specificity is seen in mammalian cells (118–122). In E. coli, a fraction (~30%) of the mutations induced by UV are targeted frameshifts, again occurring mainly at sites of potential pyrimidine dimer formation (117).

Comparison of the distributions of cyclobutyl pyrimidine dimers and (6–4) lesions in the lacI gene of E. coli reveals that the sites of cyclobutane dimer formation (123) do not correlate well with the sites destined to mutate upon UV irradiation (117). Since the frequency distribution of (6–4) lesions better matches the distribution of mutations, it appears that (6–4) lesions are likely to be the principal premutagenic UV lesions (123, 124). The current view is that the more abundant cyclobutyl pyrimidine dimers, although also mutagenic (125), are the chief killing lesions induced by UV and are possibly also the main inducers of the SOS functions required for realizing the mutagenic potential of the complete set of premutagenic lesions in irradiated DNA (reviewed in ref 114).

Recently, an octamer d(GCGTTGCG) containing a Thy–Thy cyclobutyl photodimer (Figure 2, i) has been synthesized (126). The octadeoxynucleotide d-(GCGTTGCG) was irradiated in aqueous solution with a mercury lamp equipped with two filters, one to limit short-wavelength light and the other to prevent heating. Approximately 20% of the starting material was converted to and purified as an oligonucleotide containing a cyclobutyl Thy–Thy dimer. Following 1:1 titration with the complementary strand, d(CGCAACGC), a duplex was generated and studied in comparison to its unmodified counterpart by two-dimensional (2D) $^1$H NMR. A three-bond $J$ coupling and strong cross-peaks between the H6 protons of the two thymines establishes the syn orientation of the Thy moieties at the cyclobutane ring junction, whereas the cis stereochemistry is ascertained from the magnitude of NOEs between the 5-CH$_3$ groups and the H6 protons of the thymines. A comparison of the chemical shift data and NOE spectra indicate a normal B-DNA structure with distortions confined only to the Thy dimer region. Absence of an NOE between the H6 of the 5′ Thy and the H1′ of the adjacent deoxyguanosine suggests an increase in the glycosyl torsion angle of the Thy, concurrent with conformational alteration in its vicinity. NOE measurements show a change in the 3′-Thy moiety of the dimer from the preferred C2′-endo to the C1′-exo sugar pucker, while the 5′-Thy maintains the C2′-endo conformation. These observations are further supported by the $^1$H NMR spectrum of the exchangeable protons (*127*). NOE connectivities and chemical shift values provide evidence that all six G·C base pairs remain essentially unaltered after photodimer formation. The A·T imino protons, however, exhibit large upfield shifts, suggesting a likely weakening of hydrogen bonding between the Thy and Ade, especially for the 5′ A·T pair. The $T_m$ of the octamer is reduced 13 °C as a consequence of dimer formation. Taken together, the NMR data strongly suggest that Thy dimerization, at least in this local sequence, does not induce severe distortions in DNA with the exception of at the dimer site. In general, the results of this study correlate well with previous modeling studies on oligonucleotides containing a cis-syn cyclobutyl Thy dimer (*128*).

Although the arguments promoting (6–4) photodimers as the major premutagenic lesions induced by UV light are quite compelling (*114*), there also is a body of evidence indicating significant and even important roles for other lesions (*117, 125*). This class includes the cyclobutyl dimers. Even though, as indicated above, chemical methods exist to address the unresolved issue of which lesion is the progenitor of most mutations, comparative mutagenic studies on the (6–4) lesions and cyclobutyl pyrimidine dimers have not been reported by workers in the site-directed mutagenesis field. Such studies would be of much value.

**Psoralen Photochemical Addition Products.** Psoralens are a group of clinically important tricyclic drugs consisting of a furan ring fused linearly to a coumarin moiety. These bifunctional photoreagents have been used in the treatment of skin diseases such as psoriasis and vitiligo (*129*). Psoralen derivatives have also been used extensively to probe the secondary and tertiary structure of nucleic acids, and as models for studying how cross-links in DNA are repaired (for a review, see ref. *129*). Thymine bases in DNA, particularly those in 5′-TA-3′ sequences, are preferred targets for psoralen photobinding. Interaction of psoralens with DNA usually involves three steps: first, psoralen intercalates in DNA; then, the intercalated psoralen moiety absorbs a photon of 320–400-nm light and forms a photoaddition product with the 5,6-double bond of Thy (or, at a much lower rate, with Cyt) through the pyrone or furan ring of psoralen; finally, the Thy furan-side adduct undergoes a second addition with a Thy in the complementary strand—if positioned properly—generating

an interstrand cross-link. The structure of a Thy-psoralen-Thy interstrand cross-link is shown in Figure 2, k.

The configuration of the monoadduct and cross-links is "cis-syn" at both ends of the psoralen moiety, indicating stringent restrictions on its mode of interaction with DNA (*130, 129*). Molecular modeling predicts that psoralen-DNA cross-links induce large conformational changes, as reflected by substantial kinking and unwinding of the helix (*130*). This kinking should be more severe than that induced by a Thy-Thy cyclobutane photodimer.

In vitro experiments have shown that the psoralen cross-links are inhibitory to replication by blocking DNA polymerase I (*131, 132*). Psoralen monoadducts, as well, block replication of ss templates by several polymerases, whereas they act only as attenuators of DNA synthesis when located in the template strand of duplex DNA. The monoadducts have little if any effect on the rate of DNA synthesis when situated opposite the template strand. In other studies, using a DNA sequence with a site-specific interstrand cross-link, it was shown that transcription by RNA polymerase is blocked one nucleotide before the adducted Thy in the template strand (*133*). Inhibition of transcription is also noted with the furan-side monoadduct.

Cantor and co-workers developed a technique for introducing psoralen cross-links into genomes through a sulfhydryl linkage; although their procedure is not technically "site-specific", they were able to localize the DNA adducts to a very small area of the genome, thus facilitating evaluation of the genetic effects of the adduct (*134*). Using the tetracycline resistance gene of pBR322 as the target, it was established that the ability of the plasmid to survive the psoralen modification is increased if the *E. coli* host is excision repair-proficient (*uvrA*$^+$) and *recA*$^+$ and that survival occurs only if SOS functions are induced. Repair is accompanied by substantial mutagenesis (frequency ∼ 4%), creating predominantly untargeted G·C→A·T transitions (*135, 136*).

In a second study, a single Thy-4,5′,8-trimethylpsoralen-Thy interstrand cross-link was site-specifically introduced into plasmid pUC19 (*137*). Again, induction of SOS functions results in increased survival, although the overall transformation efficiency is still only ∼6% compared to the control. Two mutants containing T→A transversions targeted at the previously cross-linked site were determined by DNA sequencing. These genetic data do not necessarily contradict those of the more complete studies of Saffran and Cantor (described above), because one cannot generalize a mutational mechanism on the basis of two mutants and the experiment was not done in such a way that the nontargeted genetic changes described by Saffran and Cantor (*135, 136*) would have been detected.

Recently, oligonucleotides containing each of the psoralen monoadducts and interstrand cross-links were synthesized by Hearst and co-workers (*138, 139*). An octanucleotide d(TCGTAGCT) was hybridized to a dodecamer d(GAAGCTACGAGC), and the duplex (with two base overhangs on each end) was irradiated with 320–380-nm light at 4 °C in the presence of 4′-(hydroxymethyl)-4,5′,8-trimethylpsoralen (HMT) to provide almost 70% cross-linked molecules (see Figure 2, k, for the structure of the Thy-HMT-Thy cross-link). Both orientations of the cross-links were obtained in the 5′-TA-3′ sequence. Following gel purification, the cross-linked molecules were partially photoreversed with 254-nm light to generate the monoadducts. The monoadducts and diadducts were characterized on the basis of their mobility pattern on a polyacrylamide gel and by their differential behavior to near-UV (360 nm) and far-UV (254 nm) light. The adducted oligomers were ligated to unmodified flanking oligonucleotides to form a 40-bp duplex containing a centrally located HMT monoadduct or a cross-link. These DNA fragments were used to study the reaction mechanism of the *E. coli* UvrABC excision nuclease (the products of the *uvrA*, *uvrB*, and *uvrC* genes). Previous studies with randomly modified carcinogen-damaged DNA fragments showed that this ATP-dependent enzyme complex cleaves the eighth phosphodiester bond 5′ to, and the fourth or fifth phosphodiester bond 3′ to, a diadduct; monoadducts are cut at the eighth 5′ and fifth 3′ phosphodiester bonds (*140, 141*). The study using the site-specifically modified DNA confirmed the mode of incision for monoadducts. However, a different mechanism exists for the cross-links, in that the UvrABC complex cuts only one of the cross-linked strands. The cleavage occurs at the ninth phosphodiester bond 5′ to, and at the third phosphodiester bond 3′ to, the furan-side thymine of the cross-link.

Thermodynamic studies on two non-self-complementary duplex dodecanucleotides containing either a furan-side or a pyrone-side HMT monoadduct in only one strand indicate that the base-stacking interactions are not perturbed and that the psoralen moiety remains stacked with the neighboring bases (*142*). This results in a net stabilization of the helix, albeit only slightly. In contrast, when an HMT monoadduct is introduced into each strand of a self-complementary duplex d(GGGTACCC), significant destabilization occurs (1.8 kcal/mol at 25 °C, or a 10 °C reduction in $T_m$ at 100 $\mu$M concentration). It was speculated that this may be a result of one HMT-adducted Thy stacked in the helix while the adducted Thy in the complementary strand is forced out. Also, as expected, HMT cross-linked oligomers stabilize the helix.

Hearst and co-workers also have used the oligonucleotides containing the furan-side HMT monoadducted thymidine as a DNA hybridization probe (*143, 144*). The ability to introduce cross-links into the target DNA provides improvements over the standard Southern hybridization protocol.

## Antitumor Agents

Many cancer chemotherapeutic agents act by binding to DNA, inhibiting replication or transcription and thereby blocking the division of tumor cells. As is the case with the mutagenic and carcinogenic agents discussed above, these DNA-damaging agents often form many DNA adducts. Hence, it has been of interest to build DNA segments with defined antitumor agent adducts in order to assess the impact of each adduct on DNA structure and to assign the relative roles of individual adducts in the cytotoxic events they presumably induce. We shall describe studies on two drug–DNA adducts: those of *cis*-diamminedichloroplatinum(II) (*cis*-DDP) and CC-1065.[3]

*cis*-**Diamminedichloroplatinum(II).** *cis*-DDP is one of the most clinically effective antitumor drugs commercially available and is used to combat testicular, ovarian, bladder, and head and neck tumors (*145*). It is a bifunctional reagent that reacts with DNA to form a variety of intra- and interstrand cross-links (for a review, see Sherman and Lippard, ref 146). The N7 atoms of the purine bases are the principal targets for *cis*-DDP adduction. Its major adducts include *cis*-[Pt(NH$_3$)$_2${d(GpG)}] (Figure 2,

j) and *cis*-[Pt(NH$_3$)$_2${d(ApG)}] intrastrand cross-links, representing almost two-thirds and one-fourth, respectively, of the binding to DNA in vitro (*147, 148*). Several indirect lines of evidence have identified the *cis*-[Pt(NH$_3$)$_2${d(GpG)}] intrastrand cross-link as the adduct likely to be responsible for the chemotherapeutic effectiveness of *cis*-DDP. This evidence includes the observation that DNA synthesis in vitro is blocked primarily at $G_n$ sequences ($n \geq 2$) on platinated templates (*149*) and the fact that the amount of the *cis*-[Pt(NH$_3$)$_2${d(GpG)}] adduct in the DNA of *cis*-DDP treated cancer patients correlates with the clinically observed response to drug treatment (*150, 151*).

Early studies on the mutagenicity of *cis*-DDP in the *lacI* gene of *E. coli* indicated that GpNpG sequences are hot spots for point mutations. Approximately 70% of the mutations in this system are G·C→T·A transversions and G·C→A·T transitions (*152*). The *lacI* system is biased against detecting certain mutations, however, and more recent studies using pBR322 in a forward mutation assay showed that the principal sites of base-pair substitutions are at GpG and ApG sequences (*153*). A fivefold higher mutagenic efficiency is detected at the ApG sites. The mutations of *cis*-DDP are SOS dependent, and the predominant base change in this region is the A·T→T·A transversion at ApG sequences.

The effects of the *cis*-DDP adducts on DNA structure have been reviewed recently (*146*), and only the pertinent highlights will be described here. These studies were carried out mainly in the laboratories of Altona, Chottard, Lippard, and Reedijk.

Adduction of *cis*-DDP at the N7 position of Gua prevents protonation at this site and lowers the p$K_a$ for deprotonation at N1 by almost two units (*154*). This enables one to determine the Pt-binding sites by monitoring the pH dependence of the base proton signals. N7 coordination of Pt to two Gua residues also results in local destacking and a destabilization of the double helix, although a normal B-like architecture is retained (*146*). Both guanines of the cross-link remain in head-to-head, anti conformation with the directed angle between the bases approaching perpendicularity. The deoxyribose sugar of the 5′-deoxyguanosine adopts a C3′-endo conformation whereas the neighboring 3′-sugar of the vicinal platinated Gua retains predominantly the C2′-endo sugar pucker; B-DNA in aqueous solution adopts the C2′-endo sugar pucker. In most cases of Pt adduction, the 3′-sugar displays more conformational flexibility than the 5′-sugar moiety.

Molecular mechanics studies indicate that both kinked and unkinked modes of the *cis*-DDP–GpG diadduct are possible with energetically comparable values (*155–157*). In the unkinked mode, however, more disruption of Watson–Crick hydrogen bonding occurs, and this is accompanied by tilting of the 5′-Gua. Two-dimensional NMR data suggest that a 50–60° kinked structure without major disruption of hydrogen bonding is the more likely of the two models. The extent of disruption of hydrogen bonding is still a matter of controversy, although there is no doubt that the therapeutically ineffective *trans*-DDP disrupts hydrogen bonding to a much greater extent than the *cis*-DDP isomer (*146*).

The severe effects of *cis*-[Pt(NH$_3$)$_2${d(GpG)}] on local DNA structure presented a formidable challenge to the construction of a site-specifically modified genome. It was anticipated that the structural perturbations induced by adduct formation would preclude the formation of a stable duplex between short, adduct-containing oligonucleotides

---

[3] Many antibiotics and chemotherapeutic agents form noncovalent complexes with DNA, and it is believed that the nature of these interactions is of importance for their biological activity. A large literature exists on NMR and X-ray studies on oligonucleotides complexed with these agents. The noncovalent interactions of such agents with oligonucleotides will not be reviewed here.

and their unmodified complementary strand (see Figure 1). Our laboratory, in collaboration with that of Stephen Lippard, recently developed a strategy by which oligonucleotides of any sequence and reasonable length can be introduced into the genome of bacteriophage M13 (*158*). With this strategy, a *cis*-[Pt(NH$_3$)$_2$[d(GpG)]] intrastrand cross-link was introduced into the genome at a unique recognition site for restriction endonuclease *Stu* I. It was found that the effects of *cis*-[Pt(NH$_3$)$_2$[d(GpG)]] on DNA structure noted above were reflected in part in the inability of *Stu* I to cleave at its adduct-containing recognition sequence. Sensitivity to *Stu* I is fully restored following incubation of the platinated genome with cyanide to remove platinum as [Pt(CN)$_4$]$^{2-}$. Gradient denaturing gel electrophoresis of a 298 base-pair fragment encompassing the site of adduction reveals that the presence of the *cis*-[Pt(NH$_3$)$_2$[d(GpG)]] cross-link induces localized weakening of the DNA double helix (*159*). This site-specifically modified genome, together with others constructed to contain the other *cis*-DDP adducts, will be a valuable tool used to probe the mechanisms by which cells repair cytotoxic DNA damage and the mechanisms by which cells acquire resistance to this antitumor drug.

**CC-1065.** CC-1065 is a potent antitumor agent isolated from *Streptomyces zelensis* cultures (reviewed recently by Reynolds et al., ref 160). The antitumor activity of CC-1065 is likely to result from its direct interaction with DNA (*160*). The drug consists of three benzopyrrole systems connected by amide bonds, and it reacts with DNA directly and with high sequence selectivity to form an adduct at the N3 atom of Ade (N3-Ade-CC-1065; Figure 2, l) (*161*). The CC-1065 component of the resultant adduct resides in the minor groove of DNA, without causing a major disruption of helical architecture.

The drug has a right-handed twist and is likely to stabilize the B conformation of DNA by tracking along the minor groove. The non-helix-disrupting effects of the adduct may render it poorly recognized by DNA repair proteins in vivo. Furthermore, the long-range interaction of the drug in the minor groove (vide infra) may function as a clamp to inhibit unwinding of the helix during replication. The helix-stabilizing property of the CC-1065 adduct is suggested by the striking increase of 51 °C in the $T_m$ of poly(dA·dT) saturated with CC-1065, by the inability of nuclease S1 to hydrolyze DNA in the vicinity of an adduct, and by reduced ability of ethidium bromide to unwind CC-1065 treated duplexes (*162*).

The binding of CC-1065 at the N3 of Ade in DNA labilizes the deoxyglycosidic bond, leaving a base-cleavable AP site. The ability to cleave DNA at the former site of adduction makes it possible to map the binding sites of the drug within defined-sequence segments of DNA. Adduct mapping studies indicate that the drug binds primarily at 5'-PuNTTA-3' and 5'-AAAAA-3' sequences in vitro (*163*). The sequence selectivity of CC-1065 binding facilitated preparation of a site-specifically modified oligonucleotide containing N3-Ade-CC-1065 by reaction of the drug with a 14-mer d(CGGAG<u>TTA</u>GGGGCG) containing the consensus binding site, 5'-TTA-3' (*164*). The product was characterized by its similar CD spectrum to CC-1065-treated calf thymus DNA and by piperidine cleavage of its thermal hydrolysis product followed by electrophoretic analysis on a sizing gel. By similar techniques a 117 bp duplex DNA fragment containing N3-Ade-CC-1065 has been prepared, and the base pairs protected by the drug have been mapped by methidium-propyl-EDTA–iron(II) cleavage (*165*). This analysis reveals that the DNA-bound drug is indeed located in the minor

groove and oriented in the 5' direction away from the Ade reaction site. Approximately four bp are protected by the drug, and the helix on both sides of the drug binding site displays enhanced sensitivity to the cleavage reagent. DNase I footprinting and restriction endonuclease sensitivity analysis shows that the effect of CC-1065 on local DNA structure is asymmetric, with the drug most severely affecting the modified strand (*166*).

## Concluding Comments

Our goal in writing this review was not only to organize and evaluate the literature on the preparation and applications of site-specifically modified oligonucleotides but also to examine the extent to which chemical studies probing the effects of an adduct on DNA structure have been able to anticipate successfully the biological effects of the lesion in vivo, as measured by lethality or mutation. An accurate assessment of the success or failure of these studies is difficult at this time, but a few generalizations are nonetheless possible. First, many impressive results have emerged from NMR and other studies on modified oligonucleotides, but it is clear from the paucity of in vivo data on adducts that the biological studies have lagged behind. In part this results from the high degree of difficulty involved in building a site-specifically modified oligonucleotide into a genome. It was pointed out earlier that successful genetic studies require starting materials, i.e., modified oligonucleotides, of the utmost purity. Moreover, the genetic engineering methodology used to situate the adduct site-specifically within a larger genetic context must be surgically precise. Few laboratories have interfaced the chemical and molecular biological technologies with the level of success necessary to generate valid biological data.

A second generalization is that this field is much in need of novel synthetic approaches that will enable preparation of adduct-containing DNA segments that do not survive conventional oligonucleotide assembly protocols. It is unfortunate that many adducts of suspected biological importance have not been studied, and may not be studied, because they are unstable to the conditions of synthesis and deprotection. The principal problem that has hindered the development of this aspect of the field is that many adducts are unstable under the alkaline conditions commonly used to deprotect exocyclic amino groups. Rapoport and colleagues have developed reagents that enable deprotection under neutral conditions and hence avoid this problem (*167, 168*). Unfortunately the application of these tools thus far has been limited. The design of novel protecting groups that maintain the structural integrity of adducts during oligonucleotide assembly is an area of great opportunity.

Finally, a fundamental assumption toxicologists often are forced to make is that dose–response relationships are linear from the high doses at which it is convenient to make biological measurements down to the much lower doses at which normal human exposure to chemicals or radiation occurs. High doses are used in conventional genotoxicity studies because the genetic target in which a measurement is to be made is usually small, and a high overall level of modification of the cellular genome ensures that enough damage will fall within the target to generate a measurable response. Unfortunately, high doses often result in much higher levels of genome modification than would be encountered normally, and DNA repair and other genoprotective systems can saturate, giving rise to a non-linear dose–response curve. One of the attractive but often overlooked features of making genetic measurements using site-specifically modified genomes is that they enable the effects of DNA damage to be measured in cells that have a damage dose as little as one adduct per cell. Hence, genetic measurements made by using the probes described in this review are likely to reflect accurately the ways that a given organism might respond in its natural environment, at the doses of chemicals and radiation routinely encountered.

**Acknowledgment.** We are very grateful to our colleagues who made available to us their unpublished results, preprints, and recently published material. We also thank the members of our laboratory at MIT for critical reading of the manuscript and for their continued support and enthusiasm. Financial support for our laboratory comes from the National Institutes of Health (Grants 5P01ES00597, 5P01ES03926, CA40817, CA33821, and CA43066).

## References

(1)  Miller, E. C. (1978) "Some current perspectives on chemical carcinogenesis in humans and experimental animals: presidential address". *Cancer Res.* **38**, 1479–1496.

(2)  Singer, B., and Kusmierek, J. T. (1982) "Chemical mutagenesis". *Annu. Rev. Biochem.* **52**, 655–693.

(3)  Miller, J. H. (1983) "Mutational specificity in bacteria". *Annu. Rev. Genet.* **17**, 215–238.

(4)  Ludlum, D. B., and Wilhelm, R. C. (1968) "Ribonucleic acid polymerase reactions with methylated polycytidylic acid templates". *J. Biol. Chem.* **253**, 2750–2753.

(5)  Ludlum, D. B. (1970) "Alkylated polycytidylic acid templates for RNA polymerase". *Biochim. Biophys. Acta* **213**, 142–148.

(6)  Singer, B., and Fraenkel-Conrat, H. (1970) "Messenger and template activities of chemically modified polynucleotides". *Biochemistry* **9**, 3694–3701.

(7)  Gait, M. J., Ed. (1984) *Oligonucleotide Synthesis, A Practical Approach*, IRL Press, Washington, DC.

(8)  Walker, G. C., and Uhlenbeck, O. C. (1975) "Stepwise enzymatic oligoribonucleotide synthesis including modified nucleotides". *Biochemistry* **14**, 817–824.

(9)  Brennan, C. A., and Gumport, R. I. (1985) "T4 RNA ligase catalyzed synthesis of base analogue-containing oligodeoxyribonucleotides and a characterization of their thermal stabilities". *Nucleic Acids Res.* **13**, 8665–8684.

(10)  Borowy-Borowski, H., and Chambers, R. W. (1987) "A study of side reactions occurring during synthesis of oligodeoxynucleotides containing $O^6$-alkyldeoxyguanosine residues at preselected sites". *Biochemistry* **25**, 2465–2471.

(11)  Green, C. L., Loechler, E. L., Fowler, K. W., and Essigmann, J. M. (1984) "Construction and characterization of extrachromosomal probes for mutagenesis by carcinogens: site-specific incorporation of $O^6$-methylguanine into viral and plasmid genomes". *Proc. Natl. Acad. Sci. U.S.A.* **81**, 13–17.

(12)  Lasko, D. D., Basu, A. K., Kadlubar, F. F., Evans, F. E., Lay, J. O., and Essigmann, J. M. (1987) "A probe for the mutagenic activity of the carcinogen 4-aminobiphenyl: synthesis and characterization of an M13mp10 genome containing the major carcinogen–DNA adduct at a unique site". *Biochemistry* **26**, 3072–3081.

(13)  Basu, A. K., Niedernhofer, L. J., and Essigmann, J. M. (1987) "Deoxyhexanucleotide containing a vinyl chloride induced DNA lesion, 1,$N^6$-ethenoadenine: synthesis, physical characterization, and incorporation into a duplex bacteriophage M13 genome as part of an *amber* codon". *Biochemistry* **26**, 5626–5635.

(14)  Johnson, D. L., Reid, T. M., Lee, M.-S., King, C. M., and Romano, L. J. (1986) "Preparation and characterization of a viral DNA molecule containing a site-specific 2-aminofluorene adduct: a new probe for mutagenesis by carcinogens". *Biochemistry* **25**, 449–456.

(15)  Benasutti, M., Ezzedine, D., and Loechler, E. L. (1987) "A viral genome containing the major adduct of benzo[a]pyrene to study *in vivo* mutagenesis". *Proc. Am. Assoc. Cancer Res.* **28**, 107.

(16)  Chambers, R. W., Sledziewska-Gojska, E., Hirani-Hojatti, S., and Borowy-Borowski, H. (1985) "*uvr*A and *rec*A mutations inhibit a site-specific transition produced by a single $O^6$-methylguanine in gene G, of bacteriophage $\phi$X174". *Proc. Natl. Acad. Sci. U.S.A.* **82**, 7173–7177.

(17)  O'Connor, D., and Stohrer, G. (1985) "Site-specifically modified oligodeoxyribonucleotides as templates for *Escherichia coli* DNA polymerase I". *Proc. Natl. Acad. Sci. U.S.A.* **82**, 2325–2329.

(18)  Loveless, A. (1969) "Possible relevance of O-6 alkylation of deoxyguanosine to the mutagenicity and carcinogenicity of nitrosamines and nitrosamides". *Nature (London)* **223**, 206–207.

(19)  Mehta, J. R., and Ludlum, D. B. (1978) "Synthesis and properties of $O^6$-methyldeoxyguanylic acid and its copolymers with deoxycytidylic acid". *Biochim. Biophys. Acta* **521**, 770–778.

(20)  Abbott, P. J., and Saffhill, R. (1979) "DNA synthesis with methylated poly(dC-dG) templates: Evidence for a competitive nature to miscoding by $O^6$-methylguanine". *Biochim. Biophys. Acta* **562**, 51–61.

(21)  Topal, M. D. (1985) "Mutagenesis by incorporation of alkylated nucleotides". In *Genetic Consequences of Nucleotide Pool Imbalance* (deSerres, F. J., Ed.) pp 339–351, Plenum, New York.

(22)  Coulondre, C., and Miller, J. H. (1977) "Genetic studies of the *lac* repressor. IV. Mutagenic specificity in the *lacI* gene of *Escherichia coli*". *J. Mol. Biol.* **177**, 577–606.

(23)  DuBridge, R. B., Tang, P., Hsia, H. C., Leong, P. M., Miller, J. H., and Calos, M. P. (1987) "Analysis of mutation in human cells by using an Epstein–Barr virus shuttle system". *Mol. Cell. Biol.* **7**, 379–387.

(24)  Snow, E. T., and Mitra, S. (1987) "Do carcinogen-modified deoxynucleotide precursors contribute to cellular mutagenesis?" *Cancer Invest.* **5**, 119–125.

(25)  Goth, R., and Rajewsky, M. F. (1974) "Molecular and cellular mechanisms associated with pulse-carcinogenesis in the rat nervous system by ethylnitrosourea: ethylation of nucleic acids and elimination rates of ethylated bases from the DNA of different tissues". *Z. Krebsforsch.* **82**, 37–64.

(26)  Kleihues, P., and Margison, G. P. (1974) "Carcinogenicity of N-methyl-N-nitrosourea: possible role of excision repair of $O^6$-methylguanine from DNA". *J. Natl. Cancer Inst.* **53**, 1839–1841.

(27)  Zarbl, H., Sukumar, S., Arthur, A. V., Martin-Zanca, D., and Barbacid, M. (1985) "Direct mutagenesis of H-*ras*-1 oncogenes by nitrosomethylurea during initiation of mammary carcinogenesis in rats". *Nature (London)* **318**, 382–385.

(28)  Fowler, K. W., Buchi, G., and Essigmann, J. M. (1982) "Synthesis and characterization of an oligonucleotide containing a carcinogen-modified base: $O^6$-methylguanine". *J. Am. Chem. Soc.* **104**, 1050–1054.

(29)  Kuzmich, S., Marky, L. A., and Jones, R. A. (1983) "Specifically alkylated DNA fragments. Synthesis and physical characterization of d[CGC($O^6$Me)GCG] and d[CGT($O^6$Me)GCG]". *Nucleic Acids Res.* **11**, 3393–3404.

(30)  Hill-Perkins, M., Jones, M. D., and Karran, P. (1986) "Site-specific mutagenesis in vivo by single methylated or deaminated purine bases". *Mutat. Res.* **162**, 153–163.

(31)  Loechler, E. L., Green, C. L., and Essigmann, J. M. (1984) "*In vivo* mutagenesis by $O^6$-methylguanine built into a unique site in a viral genome". *Proc. Natl. Acad. Sci. U.S.A.* **81**, 6271–6275.

(32)  Bhanot, O. S., and Ray, A. (1986) "The in vivo mutagenic frequency and specificity of $O^6$-methylguanine in $\phi$X174 replicative form DNA". *Proc. Natl. Acad. Sci. U.S.A.* **83**, 7348–7352.

(33)  Topal, M. D., Eadie, J. S., and Conrad, M. (1986) "$O^6$-Methylguanine mutation and repair is nonuniform; selection for DNA most interactive with $O^6$-methylguanine". *J. Biol. Chem.* **261**, 9879–9885.

(34)  Friedman, O. M., Mahapatra, G. N., Dash, B., and Stevenson, R. (1965) "Studies of the action of diazomethane on deoxyribonucleic acid. The action of diazomethane on deoxyribonucleosides". *Biochim. Biophys. Acta* **103**, 286–297.

(35)  Gaffney, B. L., and Jones, R. A. (1982) "Synthesis of $O^6$-alkylated deoxyguanosine nucleosides". *Tetrahedron Lett.* **23**, 2253–2256.

(36)  Jones, R. (1984) "Preparation of protected deoxyribonucleosides". In *Oligonucleotide Synthesis, a Practical Approach* (Gait, M. J., Ed.) pp 22–34, IRL Press, Washington, DC.

(37)  Gaffney, B. L., Marky, L. A., and Jones, R. A. (1984) "Synthesis and characterization of a set of four dodecadeoxyribonucleoside undecaphosphates containing $O^6$-methylguanine opposite adenine, cytosine, guanine, and thymine". *Biochemistry* **23**, 5686–5691.

(38)  McCarthy, T., and Lindahl, T. (1985) "Methyl phosphotriesters in alkylated DNA are repaired by the Ada regulatory protein of *E. coli*". *Nucleic Acids Res.* **13**, 2683–2698.

(39)  Weinfeld, M., Drake, A. F., Saunders, J. K., and Paterson, M. C. (1985) "Stereospecific removal of methyl phosphotriesters from DNA by an *Escherichia coli* ada$^+$ extract". *Nucleic Acids Res.* **13**, 7067–7077.

(40) Essigmann, J. M., Loechler, E. L., and Green, C. L. (1986) "Genetic toxicology of O⁶-methylguanine". In *Genetic Toxicology of Environmental Chemicals, Part A: Basic Principles and Mechanisms of Action* (Ramel, C., Lambert, B., and Magnusson, J., Eds.) pp 433–440, Alan, R. Liss, Inc., New York.

(41) Lindahl, T., Demple, B., and Robins, P. (1982) "Suicide inactivation of the *E. coli* O⁶-methylguanine–DNA methyltransferase". *EMBO J.* 1, 1359–1363.

(42) Dressler, D., Hourcade, D., Koths, K., and Sims, J. (1978) "The DNA replication cycle of the isometric phages". In *The Single Stranded DNA Phages* (Denhardt, D. T., Dressler, D., and Ray, D. S., Eds.) pp 187–214, Cold Spring Harbor Laboratory, Cold Spring Harbor, NY.

(43) Gerchman, L. L., and Ludlum, D. B. (1973) "The properties of O⁶-methylguanine in templates for RNA polymerase". *Biochim. Biophys. Acta* 308, 310–316.

(44) Patel, D. J., Shapiro, L., Kozlowski, S. A., Gaffney, B. L., Kuzmich, S., and Jones, R. A. (1985) "Covalent carcinogenic lesions in DNA: NMR studies of the O⁶-methylguanosine containing oligonucleotide duplexes". *Biochimie* 67, 861–886.

(45) Patel, D. J., Shapiro, L., Kozlowski, S. A., Gaffney, B. L., and Jones, R. A. (1986) "Structural studies of the O⁶meG·T interaction in the d(C-G-T-G-A-A-T-T-C-O⁶meG-C-G) duplex". *Biochemistry* 25, 1036–1042.

(46) Patel, D. J., Shapiro, L., Kozlowski, S. A., Gaffney, B. L., and Jones, R. A. (1986) "Covalent carcinogenic O⁶-methylguanosine lesions in DNA: structural studies of the O⁶meG·A and O⁶meG·G interactions in dodecanucleotide duplexes". *J. Mol. Biol.* 188, 677–692.

(47) Patel, D. J., Shapiro, L., Kozlowski, S. A., Gaffney, B. L., and Jones, R. A. (1986) "Structural studies of the O⁶meG·C interaction in the d(C-G-C-G-A-A-T-T-C-O⁶meG-C-G) duplex". *Biochemistry* 25, 1027–1036.

(48) Toorchen, D., and Topal, M. D. (1983) "Mechanisms of chemical mutagenesis and carcinogenesis: effects on DNA replication of methylation at the O⁶-guanine position of dGTP". *Carcinogenesis* 4, 1591–1597.

(49) Williams, L. D., and Shaw, B. R. (1987) "Protonated base pairs explain the ambiguous pairing properties of O⁶-methylguanine". *Proc. Natl. Acad. Sci. U.S.A.* 84, 1779–1783.

(50) Pegg, A. (1984) "Is O⁶-alkylguanine necessary for initiation of carcinogenesis by alkylating agents?" *Cancer Invest.* 2, 221–231.

(51) Singer, B. (1984) "Alkylation of the O⁶ of guanine is only one of many chemical events that may initiate carcinogenesis". *Cancer Invest.* 2, 233–238.

(52) Singer, B. (1986) "O-Alkyl pyrimidines in mutagenesis and carcinogenesis: occurrence and significance". *Cancer Res.* 46, 4879–4885.

(53) Lawley, P. D. (1984) "Carcinogenesis by alkylating agents". *Chemical Carcinogens* (Searle, C. E., Ed.) ACS Monograph 182, Vol. 1, pp 325–484, American Chemical Society, Washington, DC.

(54) Brennan, R. G., Pyzalska, D., Blonski, W. J., Hruska, F. E., and Sundaralingam, M. (1986) "Crystal structure of the promutagen O⁴-methylthymidine: importance of the anti conformation of the O⁴ methoxy group and possible mispairing of O⁴-methylthymidine and guanine". *Biochemistry* 25, 1181–1185.

(55) Swenberg, J. A., Dyroff, M. C., Bedell, M. A., Popp, J. A., Huh, N., Kirstein, U., and Rajewsky, M. F. (1984) "O⁴-Ethyldeoxythymidine, but not O⁶-ethyldeoxyguanosine, accumulates in hepatocyte DNA of rats exposed continuously to diethylnitrosamine". *Proc. Natl. Acad. Sci. U.S.A.* 81, 1692–1695.

(56) Preston, B. D., Singer, B., and Loeb, L. A. (1986) "Mutagenic potential of O⁴-methylthymine *in vivo* determined by an enzymatic approach to site-specific mutagenesis". *Proc. Natl. Acad. Sci. U.S.A.* 83, 8501–8505.

(57) Preston, B. D., Singer, B., and Loeb, L. A. (1987) "Comparison of the relative mutagenicities of O-alkylthymidines site-specifically incorporated into φX174 DNA". *J. Biol. Chem.* 262, 13821–13827.

(58) Li, B. F. L., Reese, C. B., and Swann, P. F. (1987) "Synthesis and characterization of oligodeoxynucleotides containing 4-O-methylthymine". *Biochemistry* 26, 1086–1093.

(59) Kalnik, M., Kouchakdjian, M., Li, B. F. L., Swann, P. F., and Patel, D. J. (1988) "Base pair mismatches and carcinogen-modified bases in DNA: an NMR study of A·C and A·O⁴meT pairing in dodecanucleotide duplexes". *Biochemistry* 27, 100–108.

(60) Kalnik, M. W., Kouchakdjian, M., Li, B. F. L., Swann, P. F., and Patel, D. J. (1988) "Base mismatches and carcinogen-modified bases in DNA: an NMR study of G·T and G·O⁴meT pairing in dodecanucleotide duplexes". *Biochemistry* 27, 108–115.

(61) Patel, D. J., Kozlowski, S. A., Ikuta, S., and Itakura, K. (1984) "Deoxyadenosine–deoxycytidine pairing in the d(C-G-G-A-A-T-T-C-A-C-G) duplex: conformation and dynamics at and adjacent to the dA·dC mismatch site". *Biochemistry* 23, 3218–3226.

(62) Singer, B., and Grunberger, D. (1983) *Molecular Biology of Mutagens and Carcinogens*, Plenum, New York.

(63) Drinkwater, N. R., Miller, E. C. and Miller, J. A. (1980) "Estimation of apurinic/apyrimidinic sites and phosphotriesters in deoxyribonucleic acid treated with electrophilic carcinogens and mutagens". *Biochemistry* 19, 5087–5092.

(64) Foster, P. L., and Davis, E. F. (1987) "Loss of an apurinic/apyrimidinic site endonuclease increases the mutagenicity of *N*-methyl-*N'*-nitro-*N*-nitrosoguanidine to *Escherichia coli*". *Proc. Natl. Acad. Sci. U.S.A.* 84, 2891–2895.

(65) Schaaper, R. M., and Loeb, L. A. (1981) "Depurination causes mutations in SOS-induced cells". *Proc. Natl. Acad. Sci. U.S.A.* 78, 1773–1777.

(66) Gentil, A., Margot, A., and Sarasin, A. (1984) "Apurinic sites cause mutations in simian virus 40". *Mutat. Res.* 129, 141–147.

(67) Loeb, L. A., and Preston, B. D. (1986) "Mutagenesis by apurinic/apyrimidinic sites". *Annu. Rev. Genet.* 20, 201–230.

(68) Takeshita, M., Chang, C.-N., Johnson, F., Will, S., and Grollman, A. P. (1987) "Oligodeoxynucleotides containing synthetic abasic sites: model substrates for DNA polymerases and AP endonucleases". *J. Biol. Chem.* 262, 10171–10179.

(69) Eritja, R., Walker, P. A., Randall, S. K., Goodman, M. F., and Kaplan, B. E. (1987) "Synthesis of oligonucleotides containing the abasic site model compound 1,4-anhydro-2-deoxy-d-ribitol". *Nucleosides Nucleotides* 6, 803–814.

(70) Millican, T. A., Mock, G. A., Chauncey, M. A., Patel, T. P., Eaton, M. A. W., Gunning, J., Cutbush, S. D., Neidle, S., and Mann, J. (1984) "Synthesis and biophysical studies of short oligodeoxynucleotides with novel modifications: a possible approach to the problem of mixed base oligodeoxynucleotide synthesis". *Nucleic Acids Res.* 12, 7435–7453.

(71) Pochet, S., Huynh-Dinh, T., Neumann, J. M., Tran-Dinh, S., Adam, S., Taboury, J., Taillandier, E., and Igolen, J. (1986) "NMR, CD, and IR spectroscopies of a tridecanucleotide containing a no-base residue; coexistence of B and Z conformations". *Nucleic Acids Res.* 14, 1107–1126.

(72) Kalnik, M. W., Chang, C. N., Grollman, A. P., and Patel, D. J. "NMR studies of abasic sites in DNA duplexes: adenosine stacks into helix opposite cyclic analog of 2-deoxyribose". *Biochemistry* (in press).

(73) Cuniasse, P., Sowers, L. C., Eritja, R., Kaplan, B., Goodman, M. F., Cognet, J. A. H., Le Bret, M., Guschlbauer, W., and Fazakerley, G. V. "An abasic site in DNA. Solution conformation determined by proton NMR and molecular mechanics calculations". *Nucleic Acids Res.* (in press).

(74) Seela, F., and Kaiser, K. (1987) "Oligodeoxyribonucleotides containing 1,3-propanediol as nucleoside substitute". *Nucleic Acids Res.* 15, 3113–3129.

(75) Sagher, D., and Strauss, B. (1983) "Insertion of nucleotides opposite apurinic/apyrimidinic sites in deoxyribonucleic acid during in vitro synthesis: uniqueness of adenine nucleotides". *Biochemistry* 22, 4518–4526.

(76) Kunkel, T. A. (1984) "Mutational specificity of depurination". *Proc. Natl. Acad. Sci. U.S.A.* 81, 1494–1498.

(77) Parkes, H. G., and Evans, A. E. J. (1984) "Epidemiology of aromatic amine cancers". In *Chemical Carcinogens* (Searle, C. S., Ed.) ACS Monograph 182, Vol. 1, pp 277–301, American Chemical Society, Washington, DC.

(78) Garner, R. C., Martin, C. N., and Clayson, D. B. (1984) "Carcinogenic aromatic amines and related compounds". In *Chemical Carcinogens* (Searle, C. E., Ed.) ACS Monograph 182 Vol. 1, pp 175–276, American Chemical Society, Washington, DC.

(79) Beland, F. A., and Kadlubar, F. F. (1985) "Formation and persistence of arylamine DNA adducts *in vivo*". *EHP, Environ. Health Perspect.* 62, 19–30.

(80) Stohrer, G., Osband, J. A., and Alvarado-Urbina, G. (1983) "Site-specific modification of the lactose operator with acetylaminofluorene". *Nucleic Acids Res.* 11, 5093–5102.

(81) Sharma, M., and Box, H. C. (1985) "Synthesis, modification with N-acetoxy-2-acetylaminofluorene and physicochemical studies of DNA model compound d(TACGTA)". *Chem.-Biol. Interact.* 56, 73–88.

(82) Johnson, D. L., Reid, T. M., Lee, M.-S., King, C. M., and Romano, L. J. (1987) "Chemical stability of oligonucleotides containing the acetylated and deacetylated adducts of the carcinogen N-2-acetylaminofluorene". *Carcinogenesis* 8, 619–623.

(83) Santella, R. M., Grunberger, D., Broyde, S., and Hingerty, B. E. (1981) "Z-DNA conformation of N-2-acetylaminofluorene modified poly(dG-dC)·poly(dG-dC) determined by reactivity with anti cytidine antibodies and minimized potential energy calculations". *Nucleic Acids Res.* 9, 5459–5467.

(84) Grunberger, D., Nelson, J. H., Cantor, C. R., and Weinstein, I. B. (1970) "Coding and conformational properties of oligonucleotides modified with the carcinogen N-2-acetylaminofluorene". *Proc. Natl. Acad. Sci. U.S.A.* 66, 488–494.

(85) Fuchs, R. P. P., and Daune, M. (1972) "Physical studies on deoxyribonucleic acid after covalent binding of a carcinogen". *Biochemistry* 11, 2659–2666.

(86) Fuchs, R. P. P., Schwartz, N., and Daune, M. P. (1981) "Hot spots of frameshift mutations induced by the ultimate carcinogen N-acetoxy-N-2-acetylaminofluorene". *Nature (London)* 294, 657–659.

(87) Fuchs, R. P. P., Schwartz, N., and Daune, M. P. (1983) "Analysis at the sequence level of mutations induced by the ultimate carcinogen N-acetoxy-N-2-acetylaminofluorene". *EHP, Environ. Health Perspect.* 49, 135–140.

(88) Bichara, M., and Fuchs, R. P. P. (1985) "DNA binding and mutation spectra of the carcinogen N-2-aminofluorene in *Escherichia coli*". *J. Mol. Biol.* 183, 341–351.

(89) O'Connor, D., and Stohrer, G. (1985) "Site-specifically modified oligodeoxyribonucleotides as templates for *Escherichia coli* DNA polymerase I". *Proc. Natl. Acad. Sci. U.S.A.* 82, 2325–2329.

(90) Mitchell, N., and Stohrer, G. (1986) "Mutagenesis originating in site-specific DNA damage". *J. Mol. Biol.* 191, 177–180.

(91) Michaels, M. L., Johnson, D. L., Reid, T. M., King, C. M., and Romano, L. J. (1987) "Evidence for *in vitro* translesion DNA synthesis past a site-specific aminofluorene adduct". *J. Biol. Chem.* 262, 14648–14654.

(92) Romano, L. J., Johnson, D. L., Gupta, P., Reid, T. M., Lee, M. S., and King, C. M. (1987) "Mutagenesis by site-specific aminofluorene (AF) and acetylaminofluorene (AAF) adducts in M13 DNA". *Proc. Am. Assoc. Cancer Res.* 28, 107.

(93) Gupta, P. K., Johnson, D. L., Reid, T. M., Lee, M. S., Romano, L. J., and King, C. M. (1987) "Identification of base-substitution (BS) mutations induced by site-specific DNA adducts using *in situ* hybridization". *Proc. Am. Assoc. Cancer Res.* 28, 104.

(94) Takeshita, M., Johnson, F., Peden, K., Will, S., and Grollman, A. P. "Targeted mutations induced by a single acetylaminofluorene–DNA adduct in mammalian cells and bacteria". *Proc. Natl. Acad. Sci. U.S.A.* (in press).

(95) Singer, B., and Bartsch, H., Eds. (1986) *The Role of Cyclic Nucleic Acid Adducts in Carcinogenesis and Mutagenesis*, IARC Scientific Publication 70, IARC, Lyon, France.

(96) Ohtsuka, E., Tanaka, T., and Ikehara, M. (1976) "Studies on transfer ribonucleic acid and related compounds. XIII. Synthesis of trinucleoside diphosphates, A-U-A and εA-U-A *via* a triester intermediate". *Chem. Pharm. Bull.* 24, 2143–2148.

(97) Zhenodarova, S. M., and Klyagina, V. P. (1977) "Stepwise synthesis of oligonucleotides XXV. Synthesis of trinucleoside diphosphates containing a fluorescent label". *Bioorg. Khim.* 3, 1192–1194.

(98) Barrio, J. R., Barrio, M. C. G., Leonard, N. J., England, T. E., and Uhlenbeck, O. C. (1978) "Synthesis of modified nucleoside 3′,5′-biphosphates and their incorporation into oligoribonucleotides with T4 RNA ligase". *Biochemistry* 17, 2077–2081.

(99) Hinton, D. M., Brennan, C. A., and Gumport, R. I. (1982) "The preparative synthesis of oligodeoxyribonucleotides using RNA ligase". *Nucleic Acids Res.* 10, 1877–1894.

(100) Brennan, C. A., Manthey, A. E., and Gumport, R. I. (1983) "Using T4 RNA ligase with DNA substrates". *Methods Enzymol.* 100, 38–52.

(101) Huttermann, J., Kohnlein, W., Teoule, R., and Bertinchamps, A. J., Eds. (1978) *Effects of Ionizing Radiation on DNA*, Springer-Verlag, New York.

(102) Hutchinson, F. (1985) "Chemical changes induced in DNA by ionizing radiation". *Prog. Nucleic Acid Res. Mol. Biol.* 32, 115–154.

(103) Teoule, R. (1987) "Radiation-induced DNA damage and its repair". *Int. J. Radiat. Biol.* 51, 573–589.

(104) Teoule, R., Bonicel, A., Bert, C., Cadet, J., and Polverelli, M. (1974) "Identification of radioproducts resulting from the breakage of thymine moiety by gamma irradiation of *E. coli* DNA in an aerated aqueous solution". *Radiat. Res.* 57, 46–58.

(105) Cathcart, R., Schwiers, E., Saul, R. L., and Ames, B. N. (1984) "Thymine glycol and thymidine glycol in human and rat urine: a possible assay for oxidative DNA damage". *Proc. Natl. Acad. Sci. U.S.A.* 81, 5633–5637.

(106) Rouet, P., and Essigmann, J. M. (1985) "Possible role of thymine glycol in the selective inhibition of DNA synthesis on oxidized DNA templates". *Cancer Res.* 45, 6113–6118.

(107) Ide, H., Kow, Y. W., and Wallace, S. S. (1985) "Thymine glycols and urea residues in M13 DNA constitute replicative blocks *in vitro*". *Nucleic Acids Res.* 13, 8035–8052.

(108) Hayes, R. C., and LeClerc, J. E. (1986) "Sequence dependence for bypass of thymine glycols in DNA by DNA polymerase I". *Nucleic Acids Res.* 14, 1045–1061.

(109) Clark, J. M., and Beardsley, G. P. (1986) "Thymine glycol lesions terminate chain elongation by DNA polymerase I *in vitro*". *Nucleic Acids Res.* 14, 737–749.

(110) Clark, J. M., and Beardsley, G. P. (1987) "Functional effects of *cis*-thymine glycol lesions on DNA synthesis in vitro". *Biochemistry* 26, 5398–5403.

(111) Clark, J. M., Pattabiraman, N., Jarvis, W., and Beardsley, G. P. (1987) "Modeling and molecular mechanical studies of the *cis*-thymine glycol radiation damage lesion in DNA". *Biochemistry* 26, 5404–5409.

(112) Dizdaroglu, M. (1985) "Formation of an 8-hydroxyguanine moiety in deoxyribonucleic acid on γ-irradiation in aqueous solution". *Biochemistry* 24, 4476–4481.

(113) Kuchino, Y., Mori, F., Kasai, H., Inoue, H., Iwai, S., Miura, K., Ohtsuka, E., and Nishimura, S. (1987) "Misreading of DNA templates containing 8-hydroxydeoxyguanosine at the modified base and at adjacent residues". *Nature (London)* 327, 77–79.

(114) Franklin, W. A., and Haseltine, W. A. (1986) "The role of the (6-4) photoproduct in ultraviolet light-induced transition mutations in *E. coli*". *Mutat. Res.* 165, 1–7.

(115) LeClerc, J. E., Istock, N. L., Saran, B. R., and Allen, R., Jr. (1984) "Sequence analysis of ultraviolet-induced mutations in M13 *lacZ* hybrid phage DNA". *J. Mol. Biol.* 180, 217–237.

(116) Wood, R. D., Skopek, T. R., and Hutchinson, F. (1984) "Changes in DNA base sequence induced by targeted mutagenesis of lambda phage by ultraviolet light". *J. Mol. Biol.* 173, 273–291.

(117) Miller, J. H. (1985) "Mutagenic specificity of ultraviolet light". *J. Mol. Biol.* 182, 45–68.

(118) Lebkowski, J. S., Clancy, S., Miller, J. H., and Calos, M. P. (1985) "The *lacI* shuttle: rapid analysis of the mutagenic specificity of ultraviolet light in human cells". *Proc. Natl. Acad. Sci. U.S.A.* 82, 8606–8610.

(119) Hauser, J., Seidman, M. M., Sidur, K., and Dixon, K. (1986) "Sequence specificity of point mutations induced during passage of a UV-irradiated shuttle vector plasmid in monkey cells". *Mol. Cell. Biol.* 6, 277–285.

(120) Protic-Sabljic, M., Tuteja, N., Munson, P. J., Hauser, J., Kraemer, K. H., and Dixon, K. (1986) "UV light-induced cyclobutane pyrimidine dimers are mutagenic in mammalian cells". *Mol. Cell. Biol.* 6, 3349–3356.

(121) Bredberg, A., Kraemer, K. H., and Seidman, M. M. (1986) "Restricted ultraviolet mutational spectrum in a shuttle vector propagated in xeroderma pigmentosum cells". *Proc. Natl. Acad. Sci. U.S.A.* 83, 8273–8277.

(122) Brash, D. E., Seetharam, S., Kraemer, K. H., Seidman, M. M., and Bredberg, A. (1987) "Photoproduct frequency is not the major determinant of UV base substitution hot spots or cold spots in human cells". *Proc. Natl. Acad. Sci. U.S.A.* 84, 3782–3786.

(123) Brash, D. E., and Haseltine, W. (1982) "UV-induced mutation hotspots occur at DNA damage hotspots". *Nature (London)* 298, 189–192.

(124) Glickman, B. W., Schaaper, R. M., Haseltine, W. A., Dunn, R. L., and Brash, D. E. (1986) "The C-C (6-4) UV Photoproduct is mutagenic in *Escherichia coli*". *Proc. Natl. Acad. Sci. U.S.A.* 83, 6945–6949.

(125) Bridges, B. A., and Woodgate, R. (1985) "Mutagenic repair in *Escherichia coli*. RecA and umuC,D gene products act at different steps in UV mutagenesis". *Proc. Natl. Acad. Sci. U.S.A.* 82, 4193–4197.

(126) Kemmink, J., Boelens, R., Koning, T. M. G., Kaptein, R., van der Marel, G. A., and van Boom, J. H. (1987) "Conformational changes in the oligonucleotide duplex d(GCG*TT*GCG)·d-(CGCAACGC) induced by formation of a *cis–syn* thymine dimer. A two-dimensional NMR study". *Eur. J. Biochem.* 162, 37–43.

(127) Kemmink, J., Boelens, R., Koning, T., van der Marel, G. A., van Boom, J. H., and Kaptein, R. (1987) "¹H NMR study of the exchangeable protons of the duplex d(GCG*TT*GCG)·d-(CGCAACGC) containing a thymine photodimer". *Nucleic Acids Res.* 15, 4645–4653.

(128) Rao, S. N., Keepers, J. W., and Kollman, P. (1984) "The structure of d(CGCGAAT[|]TCGCG)·d(CGCGAATTCGCG): the

incorporation of a thymine photodimer into a B-DNA helix". *Nucleic Acids Res.* **12**, 4789–4807.

(129) Cimino, G. D., Gamper, H. B., Isaacs, S. T., and Hearst, J. E. (1985) "Psoralens as photoactive probes of nucleic acid structure and function: organic chemistry, photochemistry, and biochemistry". *Annu. Rev. Biochem.* **54**, 1151–1193.

(130) Pearlman, D. A., Holbrook, S. R., Pirkle, D. H., and Kim, S. H. (1985) "Molecular models for DNA damaged by photoreaction". *Science (Washington, DC)* **227**, 1304–1308.

(131) Piette, J. G., and Hearst, J. E. (1983) "Termination sites of the *in vitro* nick-translation reaction on DNA that had photoreacted with psoralen". *Proc. Natl. Acad. Sci. U.S.A.* **80**, 5540–5544.

(132) Piette, J., and Hearst, J. (1985) "Sites of termination of *in vitro* DNA synthesis on psoralen phototreated single-stranded templates". *Int. J. Radiat. Biol.* **48**, 381–388.

(133) Shi, Y. B., Gamper, H., and Hearst, J. E. (1987) "The effects of covalent additions of a psoralen on transcription by *E. coli* RNA polymerase". *Nucleic Acids Res.* **15**, 6843–6854.

(134) Saffran, W. A., Goldenberg, M., and Cantor, C. R. (1982) "Site-directed psoralen crosslinking of DNA". *Proc. Natl. Acad. Sci. U.S.A.* **79**, 4594–4598.

(135) Saffran, W. A., and Cantor, C. R. (1984) "The complete pattern of mutagenesis arising from the repair of site-specific psoralen crosslinks: analysis by oligonucleotide hybridization". *Nucleic Acids Res.* **24**, 9237–9248.

(136) Saffran, W. A., and Cantor, C. R. (1984) "Mutagenic SOS repair of site-specific psoralen damage in plasmid pBR322". *J. Mol. Biol.* **178**, 595–609.

(137) Zhen, W.-P., Jeppesen, C., and Nielsen, P. E. (1986) "Repair in *Escherichia coli* of a psoralen–DNA interstrand crosslink site specifically introduced into $T_{410}A_{411}$ of the plasmid pUC19". *Photochem. Photobiol.* **44**, 47–51.

(138) Van Houten, B., Gamper, H., Holbrook, S. R., Hearst, J. E., and Sancar, A. (1986) "Action mechanism of ABC excision nuclease on a DNA substrate containing a psoralen crosslink at a defined position". *Proc. Natl. Acad. Sci. U.S.A.* **83**, 8077–8081.

(139) Van Houten, B., Gamper, H., Hearst, J. E., and Sancar, A. (1986) "Construction of DNA substrates modified with psoralen at a unique site and study of the action mechanism of ABC excinuclease on these uniformly modified substrates". *J. Biol. Chem.* **261**, 14135–14141.

(140) Sancar, A., Franklin, K. A., Sancar, G., and Tang, M.-S. (1985) "Repair of psoralen and acetylaminofluorene DNA adducts by ABC excinuclease". *J. Mol. Biol.* **184**, 725–734.

(141) Beck, D., Popoff, S., Sancar, A., and Rupp, W. D. (1985) "Reactions of the UVRABC excision nuclease with DNA damaged by diamminedichloroplatinum(II)". *Nucleic Acids Res.* **13**, 7395–7412.

(142) Shi, Y. B., and Hearst, J. E. (1986) "Thermostability of double-stranded deoxyribonucleic acids: effects of covalent additions of a psoralen". *Biochemistry* **25**, 5895–5902.

(143) Gamper, H. B., Cimino, G. D., Isaacs, S. R., Ferguson, M., and Hearst, J. E. (1986) "Reverse southern hybridization". *Nucleic Acids Res.* **14**, 9943–9954.

(144) Gamper, H. B., Cimino, G. D., and Hearst, J. E. (1987) "Solution hybridization of crosslinkable DNA oligonucleotides to M13 DNA: effect of secondary structure on hybridization kinetics and equilibria". *J. Mol. Biol.* **197**, 349–362.

(145) Prestayko, A. W., Crooke, S. T., and Carter, S. K. (1980) *Cisplatin: Current Status and New Developments*, Academic, New York.

(146) Sherman, S. E., and Lippard, S. J. (1987) "Structural aspects of platinum anticancer drug interactions with DNA". *Chem. Rev.* **87**, 1153–1181.

(147) Fichtinger-Schepman, A.-M. J., van der Veer, J. L., Lohman, P. H. M., and Reedijk, J. (1985) "Adducts of the antitumor drug *cis*-diamminedichloroplatinum(II) with DNA: formation, identification, and quantitation". *Biochemistry* **24**, 707–713.

(148) Eastman, A. (1986) "Reevaluation of interaction of *cis*-dichloro(ethylenediamine)platinum(II) with DNA". *Biochemistry* **25**, 3912–3915.

(149) Pinto, A. L., and Lippard, S. J. (1985) "Binding of the antitumor drug *cis*-diamminedichloroplatinum(II) (cisplatin) to DNA". *Biochim. Biophys. Acta* **780**, 167–180.

(150) Reed, E., Ozols, R. F., Tarone, R., Yuspa, S. H., and Poirier, M. C. (1987) "Platinum–DNA adducts in leukocyte DNA correlate with disease response in ovarian cancer patients receiving platinum-based chemotherapy". *Proc. Natl. Acad. Sci. U.S.A.* **84**, 5024–5028.

(151) Fichtinger-Schepman, A. M. J., Van Oosterom, A. T., Lohman, P. H. M. Berends, F. (1987) "*cis*-Diamminedichloroplatinum-(II)-induced DNA adducts in peripheral leukocytes from seven cancer patients: quantitative immunochemical detection of the adduct induction and removal after a single dose of *cis*-diamminedichloroplatinum(II)". *Cancer Res.* **47**, 3000–3004.

(152) Brouwer, J., van de Putte, P., Fichtinger-Schepman, A. M. J., and Reedijk, J. (1981) "Base-pair substitution hotspots in GAG and GCG nucleotide sequences in *Escherichia coli* K-12 induced by *cis*-diamminedichloroplatinum(II)". *Proc. Natl. Acad. Sci. U.S.A.* **78**, 7010–7014.

(153) Burnouf, D., Daune, M., and Fuchs, R. P. P. (1987) "Spectrum of cisplatin-induced mutations in *Escherichia coli*". *Proc. Natl. Acad. Sci. U.S.A.* **84**, 3758–3762.

(154) Altona, C. (1982) "Conformational analysis of nucleic acids. Determination of backbone geometry of single-helical RNA and DNA in aqueous solutions". *Recl. Trav. Chim. Pays-Bas* **101**, 413.

(155) Kozelka, J., Petsko, G., Lippard, S. J., and Quigley, G. J. (1985) "Molecular mechanics calculations on *cis*-[Pt(NH$_3$)$_2${d-(GpG)}] adducts in two oligonucleotide duplexes". *J. Am. Chem. Soc.* **107**, 4079–4081.

(156) Kozelka, J., Petsko, G. A., Quigley, G. J., and Lippard, S. J. (1986) "High-salt and low-salt models for kinked adducts of *cis*-diamminedichloroplatinum(II) with oligonucleotide duplexes". *Inorg. Chem.* **25**, 1075–1077.

(157) Kozelka, J., Archer, S., Petsko, G. A., Lippard, S. J., and Quigley, G. J. (1987) "Molecular mechanics modeling of oligonucleotide adducts of the antitumor drug *cis*-diamminedichloroplatinum(II)". *Biopolymers* **26**, 1245–1271.

(158) Pinto, A. L., Naser, L. J., Essigmann, J. M., and Lippard, S. J. (1986) "Site-specifically platinated DNA, a new probe of the biological activity of platinum anticancer drugs". *J. Am. Chem. Soc.* **108**, 7405–7407.

(159) Naser, L. J., Pinto, A. L., Lippard, S. J., and Essigmann, J. M. "Structural, physical, and biological properties of *cis*-[Pt-(NH$_3$)$_2${d(G$_p$G)}] built site-specifically into a viral genome". *Biochemistry* (in press).

(160) Reynolds, V. L., McGovern, J. P., and Hurley, L. H. (1986) "The chemistry, mechanism of action and biological properties of CC-1065, a potent antitumor antibiotic". *J. Antibiot.* **39**, 319–334.

(161) Hurley, L. H., Reynolds, V. L., Swenson, D. H., Petzold, G. L., and Scahill, T. A. (1984) "Reaction of the antitumor antibiotic CC-1065 with DNA: structure of a DNA adduct with DNA sequence specificity". *Science (Washington, DC)* **226**, 843–844.

(162) Swenson, D. H., Li, L. H., Hurley, L. H., Rokem, J. S., Petzold, G. L., Dayton, B. D., Wallace, T. L., Lin, A. H., and Krueger, W. C. (1982) "Mechanism of interaction of CC-1065 (NSC 298223) with DNA". *Cancer Res.* **42**, 2821–2828.

(163) Reynolds, V. L., Molineux, I. J., Kaplan, D. J., Swenson, D. H., and Hurley, L. H. (1985) "Reaction of the antitumor antibiotic CC-1065 with DNA. Location of the site of thermally induced strand breakage and analysis of DNA sequence specificity". *Biochemistry* **24**, 6228–6237.

(164) Needham-VanDevanter, D. R., Hurley, L. H., Reynolds, V. L., Theriault, N. Y., Krueger, W. C., and Wierenga, W. (1984) "Characterization of an adduct between CC-1065 and a defined oligodeoxynucleotide duplex". *Nucleic Acids Res.* **12**, 6159–6168.

(165) Needham-VanDevanter, D. R., and Hurley, L. H. (1986) "Construction and characterization of a site-directed CC-1065-N3-adenine adduct within a 117 base-pair DNA restriction fragment". *Biochemistry* **25**, 8430–8436.

(166) Hurley, L. H., Needham-VanDevanter, D. R., and Chong-Soon, L. (1987) "Demonstration of the asymmetric effect of CC-1065 on local DNA structure using a site-directed adduct in a 117-base-pair fragment from M13mpl". *Proc. Natl. Acad. Sci. U.S.A.* **84**, 6412–6416.

(167) Watkins, B. E., and Rapoport, H. (1982) "Synthesis of benzyl and benzyloxycarbonyl base-blocked 2′-deoxyribonucleosides". *J. Org. Chem.* **47**, 4471–4477.

(168) Watkins, B. E., Kiely, J. S., and Rapoport, H. (1982) "Synthesis of oligodeoxyribonucleotides using *N*-benzyloxycarbonyl-blocked nucleosides". *J. Am. Chem. Soc.* **104**, 5702–5708.

## Chapter 17

# Sequence Selectivity of DNA Covalent Modification

Martha A. Warpehoski*,† and Laurence H. Hurley*,‡

*Cancer and Infectious Diseases Research, The Upjohn Company, Kalamazoo, Michigan 49001,
and Drug Dynamics Institute, College of Pharmacy, University of Texas at Austin, Austin,
Texas 78712*

Reprinted from *Chemical Research in Toxicology,* Vol. 1, No. 6, November/December, 1988

### Introduction

DNA sequence selectivity is achieved by both proteins (1, 2) and small molecular weight ligands (3, 4). DNA-binding proteins have evolved to recognize sequence-dependent features of DNA in order to participate in a variety of genetic events such as control of gene expression and replication of DNA (5). Structural biologists are currently unmasking the recognition motifs that allow the sequence-specific recognition of DNA by binding proteins such as endonucleases (6) and repressors (7–12). The central function of any sequence-specific protein is its ability to discriminate its cognate sequence from the vast excess of noncognate DNA sequences in which it is embedded. This is achieved by a number of different molecular mechanisms including specific hydrogen bonding (2) and hydrophobic interactions (13) which occur primarily between amino acid residues in the protein and major or minor groove substituents on DNA. Electrostatic interactions (6), and in some cases DNA flexibility or elasticity (14), may also contribute to the overall sequence selectivity. In the case of *Eco*RI endonuclease covalent modification of DNA permits a higher degree of sequence specificity than simple protein binding (4, 6).

While their overall sequence selectivity is generally lower, small molecular weight ligands such as DNA-reactive drugs and carcinogens may also exhibit DNA sequence recognition (3, 4). These drugs and carcinogens may share common DNA recognition motifs with proteins, and because of their relatively small molecular weight and corresponding reduced structural complexity may serve as useful models for the more complex protein–DNA recognition mechanisms. Like DNA-binding proteins,

---

†The Upjohn Co.
‡The University of Texas.

drugs and carcinogens may possess one or more different molecular mechanisms for DNA sequence recognition (4). A practical way of classifying these molecular mechanisms is to divide them into noncovalent and covalent recognition processes. In the drug–DNA interaction literature, the term "DNA binding" frequently has loosely encompassed both covalent and noncovalent interactions. However, in recent years experiments studying those interactions have become increasingly sophisticated and generally can distinguish quite clearly the manifestations of covalency vs those of physical interactions. Since we wish to understand the relative importance of covalent and noncovalent interactions in the drug–DNA interaction area, we will use the terms "binding" and "bonding" to differentiate between noncovalent and covalent intermolecular interactions, respectively. (These are the commonly accepted usages of these terms in chemistry, notwithstanding the somewhat misleading term "hydrogen bond". The energy of this noncovalent "bond" is typically 3–7 kcal/mol, in contrast to covalent bond energies which are on the order of 100 kcal/mol.)

Considerable effort has been invested into both elucidating the sequence recognition mechanisms for noncovalent binding (15–20) and designing noncovalent sequence-specific probes for duplex DNA (21–24). While the information content or "readout" in the minor groove of DNA is poorer than in the major groove (25), to date the most successful nonprotein DNA sequence specific probes for duplex DNA are minor groove binders (4). This situation may change as the more complex protein recognition motifs for major groove specificity are elucidated and can then be mimicked by small molecular weight ligands. In contrast, covalent bonding as a molecular mechanism for sequence recognition has so far received much less attention, even though many of the most potent DNA interactive agents like CC-1065 and the anthramycins, or metabolically activated carcinogens, and many clinically useful antitumor agents, react covalently with DNA.

In principle, ligands such as drugs or carcinogens which become covalently bonded to DNA in a sequence-selective fashion may recognize the sequence by either noncovalent (binding) or covalent (bonding) mechanisms, or perhaps more likely by a combination of these mechanisms. In terms of a generalized kinetic scheme (eq 1) the binding

$$\text{drug} + \text{DNA} \underset{\longleftarrow}{\overset{K_b}{\rightleftharpoons}} (\text{drug}\cdot\text{DNA})_{noncov} \xrightarrow{k_r} (\text{drug}-\text{DNA})_{cov} \tag{1}$$

step is sequence selective if different sequences have different binding affinities for the drug (a situation with ample precedent among noncovalently binding drugs). The bonding step is sequence selective if the rate constant ($k_r$) for covalent bond formation is different for different sequences, i.e., if the free energy of activation to the transition state leading to the covalent adduct depends upon the nucleotide sequence around it. While the common perception is that covalent or bonding mechanisms can only give rise to low sequence discrimination, we illustrate here that with certain drugs, such as CC-1065 analogues, relatively marked sequence selectivity can be achieved predominantly by a bonding mechanism.

This perspective weighs the relative importance of the noncovalent and covalent mechanisms in DNA sequence selection by certain small molecular weight molecules which become covalently bonded to DNA. We begin by consideration of the experimental approaches that have been used to reveal sequence selectivity of such agents, since experimental conditions may have a significant effect on the "selectivity" described. Next we summarize the pertinent results reported for a number of important drugs and carcinogens, along with the interpretations of the sources of observed selectivities as presented by the investigators. The extensive recent work on DNA sequence selectivity of CC-1065 and its analogues is described in detail, since this example has given us considerable insight into the molecular basis of its sequence-selective reaction with DNA. From this survey, we attempt to extract some general principles regarding the interplay of noncovalent and covalent mechanisms in determining the sequence discrimination observed for drugs and carcinogens that react covalently with DNA.

To the extent that this selectivity impinges upon biological action, we propose that this analysis has an obvious bearing on drug design. We also comment on the possible relevance of the bonding reaction between small molecular weight ligands and DNA, to protein–DNA interactions. Finally, we propose that covalent mechanisms which result in selection of a base (e.g., adenine) in a particular sequence may have biological relevance to the mechanism of action of certain DNA-reactive drugs.

## Experimental Measurements of Sequence Selectivity

Over the last decade the availability of defined DNA restriction fragments and of methods to determine their nucleotide sequences has enabled investigators to examine the sequence selectivity of a large number of DNA-interactive compounds. The experiments that provide sequence selectivity information are quite intricate, and statements about "selectivity" must be qualified by describing some experimental parameters. It is instructive to begin the discussion of these parameters by describing the widely used Maxam and Gilbert sequencing method, which pro-

vides the "sequence ladder" of DNA for virtually all of these sequence selectivity studies (*26*). In this method a single- or double-stranded DNA restriction fragment with a single end label ($^{32}$P) is subjected to four different sets of chemical reactions, each set resulting in cleavage of a fraction of the DNA molecules at a particular base (G or C) or pairs of bases (G+A or C+T). The cleaved fragments are separated according to size by high-resolution polyacrylamide gel electrophoresis and visualized by autoradiography. The nucleotide sequence is read by ordering each successive cleavage fragment according to the base-specific reaction set that produced it.

For exemplary purposes we consider in detail the guanine-specific set of reactions. First, the single end-labeled DNA restriction fragment is reacted to a small extent (only a few percent of the DNA molecules are modified) with the electrophilic reagent dimethyl sulfate, which alkylates the nucleophilic N-7 position of guanine bases. The partially modified DNA is then treated with hot aqueous piperidine, during which the series of transformations depicted in Scheme I takes place (*26*, *27*), which ultimately leads to strand cleavage. That segment of the original DNA restriction fragment extending from the radioactively labeled end (e.g., 5′ in Scheme I) up to the modified guanine will appear as a discrete band in the guanine-specific lane of the sequence ladder (as well as in the guanine + adenine lane resulting from another set of reactions) after electrophoresis and autoradiography. Several points concerning the procedure warrant additional comments for the purposes of this article.

(1) It is the consequence of the entire set of chemical reactions leading to strand cleavage, under carefully specified reaction conditions, that results in base specificity. For instance, in addition to N-7 of guanine, dimethyl sulfate also alkylates N-3 of adenine to some extent, but these modified adenines are not converted to backbone strand breaks during the piperidine reaction and thus do not give rise to fragments in the guanine-specific lane. In other words, dimethyl sulfate itself is not "specific" for N-7 of guanine in a purely chemical sense.

(2) It has been established for the Maxam–Gilbert procedures that the intensity of the observed band (i.e., the amount of radioactivity in each fragment) is proportional to the frequency of modification (rate of covalent reaction) at the base involved. That is, once the N-7 atom of the guanine is alkylated, the chemical transformations (Scheme I) leading to strand breakage are quantitative; further, those transformations cannot occur when the N-7 of guanine is not alkylated (*26*).

(3) In determining the nucleotide sequence of a given region of DNA, the goal is to visualize each possible subfragment resulting from specific cleavage at each base. If a given labeled DNA molecule were to be methylated by dimethyl sulfate at two of its guanine nucleotides, only the fragment corresponding to cleavage at the guanine nearest the labeled end would be detected. Thus, the more extensively the DNA is alkylated, the less likely it is that the more distant bases from the labeled end will be visualized; experimentally a "fade-out" toward the high molecular weight end of the autoradiogram would appear, and the intensity of a band would no longer reflect the true frequency of alkylation at the site relative to other bases. When very limited reaction is allowed, with most of the DNA remaining unreacted there is greater probability that each DNA molecule that reacts does so at only one of its potential sites.

The highly reactive dimethyl sulfate ensures that there is sufficient probability that each guanine in the DNA will

Scheme I. Chemical Transformations of the Guanine-Specific Maxam and Gilbert Sequencing Reaction (*26, 27*)

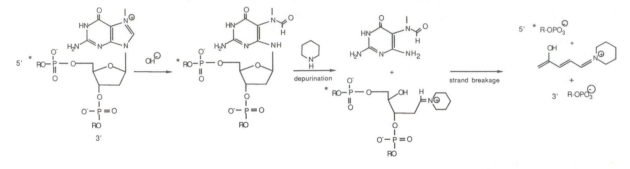

react in enough of the molecules to allow detection. Although conditions are chosen such that all of the guanines can be detected, in fact not all of the bands observed in a dimethyl sulfate guanine sequencing lane may be equally intense, indicating that alkylation and consequent cleavage occur more frequently at some guanines than at others. Hartley et al. have observed slight enhancement of cleavage at runs of guanines in double-stranded DNA, suggesting some degree of sequence selectivity for the alkylation reaction (*28*). In an alternative sequencing method, guanines are photooxidized in the presence of methylene blue or rose bengal to products that lead to strand scission upon treatment with piperidine (*29*). This guanine-specific set of reactions is reported to give somewhat more uniform bands on sequencing gels than does dimethyl sulfate, indicating that singlet oxygen is an even less DNA sequence selective reagent than dimethyl sulfate.

Most of the studies on sequence selectivity of agents that covalently modify DNA have been more or less modeled after Maxam-Gilbert type methodology. DNA alkylation leads to strand cleavage (or inhibition of cleavage, as in footprinting; vide infra) which is then located and quantified. The three same points discussed above must be considered when evaluating and integrating these studies.

For example, a number of drugs and carcinogens react at more than one heteroatom or more than one base. The sequence selectivity for one type of reaction may be different than that for another (this appears to be the case for benzpyrenediol epoxide, as will be discussed below). Postalkylation reaction steps leading to strand breakage may visualize only one type of alkylation chemistry, although in some cases, such as methods using exonuclease cleavage or DNA footprinting, this adduct selectivity cannot be assumed. Thus the chemical and enzymatic procedures used in determining a given "sequence selectivity" are an important part of its description and of the evaluation of its possible biological significance.

Likewise, it is important to establish that the sequence selectivity observed is due to the initial covalent modification event, and not to sequence effects on subsequent transformations which lead to the observed strand breakage. For a given level of covalent reaction it should be shown experimentally that the follow-up reactions are quantitative for all sequences under the conditions chosen (*27*).

Finally, "sequence selectivity" as presently understood refers not only to which bases in which sequence contexts *can be made* to react with a given drug or carcinogen at *high ligand concentration*, but also to how readily a base in a particular sequence reacts relative to the same base in another sequence, i.e., to what extent is one sequence preferred to other sequences? Ideally this means that, in addition to quantitative cleavage as discussed in (2),

"single-hit statistics" are followed, i.e., that most fragments represent DNA molecules experiencing only one covalent reaction. Thus, conditions should be chosen such that only a small percentage of the DNA has reacted at all.

Only a few years ago a review on DNA sequence selective drugs dismissed covalently modifying agents and in particular nitrogen mustards and aflatoxin as being poorly sequence selective (*17*). These agents primarily alkylate the guanine N-7 position (*30, 31*) and thus can be made to give strand breaks that can reveal the alkylation site, in the same way that 7-methylguanine is detected in Maxam–Gilbert sequencing. However, the early sequencing studies, to which the reviewer referred, were primarily concerned with establishing the *base* specificity of these agents and used relatively high drug concentrations which would permit detection of all reasonably attainable covalent products (*32, 33*). Consequently, it was little wonder that all guanines were susceptible to alkylation under such conditions and that this aspect of the experimental results was emphasized. Later studies focused in on the markedly different intensities of some of the bands observed with these agents, and experiments were designed to maximize sequence selectivity information.

## Sequence Selectivity of DNA-Reactive Agents

**Aflatoxin.** Aflatoxin $B_1$ is a highly mutagenic and carcinogenic mycotoxin which creates a significant problem in many parts of the world as a food chain contaminant (*34*). The possibility that inherent selectivity for reaction at certain sequences in DNA might contribute to the extraordinary genotoxic potency of this natural product prompted detailed investigations of its sequence selectivity. Humayun explored the "sequence contexts" of aflatoxin $B_1$ which was oxidized in situ to its epoxide (Figure 1a), the putative reactive species. Its reaction at guanine N-7 with several single- and double-stranded DNA restriction fragments, under conditions of minimal DNA cleavage, was determined (*35, 36*). He found up to 10-fold increases in the rate of alkylation of certain guanines relative to others in double-stranded DNA, but not in single-stranded DNA. The sequence patterns were remarkably consistent, allowing the formulation of a set of rules for the sequences immediately flanking the reacted guanine. Single-stranded DNA reacted sluggishly with activated aflatoxin, with all guanines equally susceptible to the slow reaction. Humayun presented two dichotomous interpretations: (1) aflatoxin oxide diffuses freely, and the observed selectivity is due to sequence-dependent conformational features of DNA favoring reaction with the drug [in essence, rate constant ($k_r$ in eq 1) differences among guanines], or (2) all guanine residues are equally reactive, irrespective of

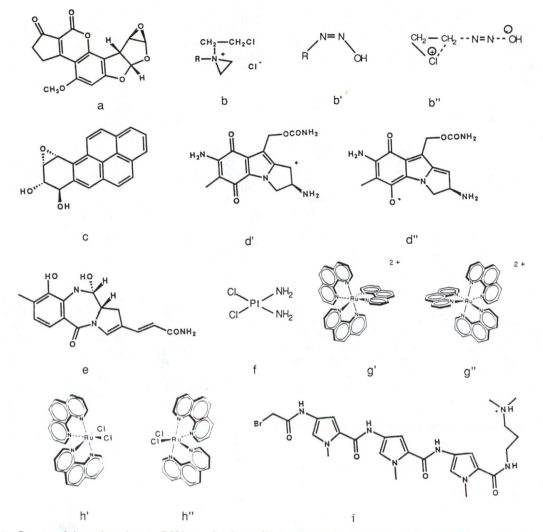

**Figure 1.** Structural formulas of some DNA covalently modifying agents (*a–i*) or reactive intermediates described in the text.

sequence, but the relative frequency of alkylation is due to variations in local drug concentration (mass action effect). This variation would be a result of noncovalent binding of aflatoxin oxide, which would thus be the sequence-dependent factor (i.e., $K_b$ in eq 1). Humayun favored the latter interpretation, both because of the greater activity of double-stranded DNA and because of partial inhibition of the alkylation by nonreactive structural analogues of aflatoxin $B_1$ (*36*).

Recently, Benasutti et al. described extensive and carefully executed experiments revealing the sequence selectivity of activated aflatoxin $B_1$ reaction with DNA, examining 190 guanine residues (*37*). Statistical treatment of the data gave trinucleotide priority to 5'GG*G, followed closely by 5'GG*T and 5'CG*G. The most reactive sequence was about 20-fold more frequently alkylated than was the least reactive sequence. Like Humayun, these investigators invoked a sequence-dependent noncovalent (precovalent) binding site for aflatoxin oxide in the DNA major groove. The importance of binding interactions was also supported by other lines of evidence which these authors listed (*37*). However, the role of binding was not merely postulated to favor mass action, but also to "orient (the drug) for covalent reaction". Thus, "significant enhancement of reaction would be expected with nucleophilic centers in DNA compared to reaction with simple nu-

cleophiles of similar basicity" (*37*). The precovalent binding is postulated to enhance not only the effective concentration of the carcinogen but also the rate constant for covalent reaction, in keeping with current understanding of the sources of reactivity of bound complexes (*38*).

**Alkylating Agents.** Nitrogen mustards and alkylnitrosoureas are highly efficacious antitumor agents in animal tumor systems, and some have become important clinically used drugs (*39*). Nitrogen mustards are bifunctional agents that are postulated to react as aziridinium cation intermediates (Figure 1b) with, principally, the nucleophilic N-7 atoms of guanines (*30*). Nitrosoureas are known to decompose in water, at neutral pH, via electrophilic alkyldiazohydroxide species (Figure 1b'), and these have been proposed as intermediates in DNA alkylation (*28*), although this is still an area of debate (*40, 41*).

Kohn has made a careful study of the sequence selectivity of guanine N-7 alkylation of a wide range of nitrogen mustards, using experimental conditions that allow an accurate assessment of relative reaction rates of guanines in different sequence contexts (*42, 43*). The recurring finding was an increased reaction frequency at guanines within guanine clusters. Typically, these reaction intensities varied by a factor of 2, but in runs of three guanines each was, on average, three to four times as reactive as an isolated guanine. The fact that many agents differing significantly in their nonalkylating structure show the same sequence preference argues against a role for specific binding interactions of the nonalkylating moieties in influencing this selectivity. Intrastrand cross-linking does not appear to be a causal factor in the enhanced reaction at adjacent guanines. A cross-link would be visualized as a break nearest the labeled end. Analysis of the same fragment labeled at either the 3′ or the 5′ end showed no such pattern of weak and strong bands (42).

Theoretical calculations by Pullman and Pullman on the electrostatic potential of DNA indicate that this potential is sequence dependent (44). In particular, guanines surrounded by other guanines are associated with a considerably more negative potential at N-7 than are isolated guanines. Using the molecular electrostatic potential (MEP) calculated for various XG*Y sequences, Kohn observed a remarkable correlation of negative MEP with the reaction intensity of most nitrogen mustards at these sequences (43, 44). The molecular rationale advanced for this phenomenon was that the "positively charged aziridinium group ... would be drawn selectively towards the more electronegative regions" (43). (Note that in Kohn's paper the term "electronegative" was used to mean "having a more negative MEP", rather than Pauling's definition of "the power of an atom in a molecule to attract electrons to itself". The implication is that the electrostatic attraction between negative regions on DNA and positively charged reactive intermediates serves a "collection" function, i.e., increases the rate by increasing the effective local concentration of cationic intermediates. A similar sequence selectivity for clusters of guanines was found for (chloroethyl)nitrosourea and led to the suggestion that a "partial chloronium" ion intermediate (Figure 1b″), with a partial positive charge, was involved (28). These investigators had reported a lack of similar guanine N-7 sequence selectivity for ethylnitrosourea and (chloroethyl)methanesulfonate and felt this to be consistent with the inability of these agents to form an analogous partially charged intermediate.

Very recently, the guanine N-7 sequence selectivity of methylnitrosourea (MNU) has been reported (40) and found to resemble the same selectivity for oligo(dG) sequences shown by the nitrogen mustards and (chloroethyl)nitrosourea. The inhibition of guanine alkylation by salt or by the cationic DNA binders spermine, distamycin, and ethidium bromide, which weaken the negative MEP (45), was interpreted as consistent with "a strong electrostatic attraction (at low salt) between the polyanionic DNA backbone and a positively charged alkylating intermediate" (from hydrolysis of MNU).

These data appear to fan a controversy since a number of kinetic, product structure, and isotope exchange experiments argue against carbocationic hydrolysis products of nitrosoureas as intermediates in DNA alkylation by these agents (41). Indeed, an intriguing mechanism for DNA alkylation was postulated by Buckley, whereby a 5′ guanine, in lieu of the hydroxide ion (involved in the hydrolysis mechanism), attacked an imido form of the alkylnitrosourea, allowing subsequent regioselective attack on the electrophilic carbon by the 3′ guanine, without the intermediacy of a carbocation. In effect, this GG sequence would catalyze its own alkylation. Either N-7 or O-6 of the 3′ guanine was proposed as likely nucleophiles. However, as pointed out by Wurdeman and Gold (40), the observed sequence preference for guanine N-7 alkylation

is not consistent with this mechanism, since the guanine nearest the 3′ end of a run of guanines is least likely to be alkylated. The sequence selectivity of O-6 guanine alkylation, which is probably the more critical lesion for mutagenesis and carcinogenesis (46), cannot be determined by this Maxam–Gilbert type experiment and thus is still a matter of speculation. Experiments quantifying N-7- and O-6-methylguanine, and other modified bases, from reaction of MNU with various synthetic polydeoxynucleotides show clear sequence dependence of alkylation product ratios, consistent with unique sequence preferences for each type of adduct (47). In support of his mechanism, Buckley stated that certain mutational studies show "sequence-specific alkylation of O6-dG$_2$ in a 5′-dGdGdN-3′ DNA codon" (41). However, it should be clarified that these data refer to sequence-specific mutational events and not sequence-specific alkylation per se. Mutational selectivity could be introduced in sequence-dependent repair and mispairing capabilities as well as in the alkylation step (48).

While the above examples imply a great deal of similarity in the sequence preferences of a number of alkylating agents, in reactions with N-7 of guanine, several exceptions merit a more detailed description. Of the rather broad structural range of nitrogen mustards studied, two presented unique sequence specificities which must be attributed to their nonalkylating moieties (42, 43). In an impressive piece of detective work, Kohn showed that uracil mustard had enhanced reactivity with 5′PyG*C sequences. For most mustards, however, a 3′C greatly reduced guanine N-7 reaction, presumably because the cytosine amino group nearly overlies the guanine N-7, exerting, with its dipole, a positive electrostatic field. Modeling the uracil mustard interaction with the perferred nucleotide sequence indicated that the uracil O-4 atom could interact with and counter the suppressive effects of the 3′ cytosine amino group and, further, that such interaction would be facilitated by "a displacement of the reacting guanine towards the sugar–phosphate backbone"—a process in turn favored by the flanking pyrimidines (43). Thus this special sequence selectivity is attributed to a specific hydrogen bond (uracil O-4 to cytosine NH2) which enhances the negative electrostatic potential of the target guanine, and to sequence-dependent DNA flexibility in a "sliding" mode.

The other anomalous nitrogen mustard, quinacrine mustard, showed an even greater degree of sequence discrimination (43). Here the reacting guanine must be flanked on the 3′ side by either guanine or thymine, and then by a purine, in order to show maximum reactivity. Kohn's model invokes rapid, noncovalent intercalation of the quinacrine ring between the second and third bases to the 3′ side of the guanine target. The covalently reactive portion of the drug must then successfully reach over the edge of the middle base in the major groove to approach the guanine N-7. Evidently the amino groups of adenine or cytosine which would occupy this edge in 5′G*APu or 5′G*CPu sequences interfere with this required conformation of the linking hydrocarbon chain, while the oxygens of guanine or thymine occupying this edge do not. Again, although precovalent binding is indicated, the sequence selectivity is not simply dictated by the preferred sites of intercalation, since intercalation of the quinacrine ring may be occurring at many other sites with equal or greater affinity. Rather, in this model, the selectivity (for a central guanine or thymine) is dictated by the conformational needs of the intercalated drug in approaching the correct trajectory for the covalent reaction.

**Benzpyrenediol Epoxide.** Benzo[*a*]pyrene is one of a number of environmentally prevalent polycyclic aromatic hydrocarbons implicated in human cancer (*49*). It is metabolized to the highly tumorigenic (+)-diol epoxide (BPDE, Figure 1c), which reacts covalently with DNA, predominantly at the exocyclic 2-$NH_2$ group of guanine (*49*).

Sequence selectivity studies with this potent carcinogen present an instance where different sequence preferences are shown for the covalent reaction of the same electrophile at two different atoms, N-7 or N-2, of guanine. Boles and Hogan (*50*) exploited the long-wavelength UV photolability of the covalent BPDE–DNA adduct to produce single-strand cuts (independent of diffusible oxygen species) locating the reaction site. Unlike the usual sequencing studies, in this case the carcinogen was reacted with the entire supercoiled plasmid, which was then cleaved with restriction enzymes, end labeled, laser photolyzed, and analyzed on a denaturing electrophoresis gel. Since the N-2 guanine adduct is the major one (80–90%) formed in DNA by BPDE (*51*), it is presumed to be the major origin of the observed photolytic cuts, although low levels of cutting at adenine and cytosine suggest that the photolysis can operate on a variety of adduct structures. The most pronounced sequence preference was for adjacent guanines, such that strings of three to five guanines, and one sequence of 16 guanines, were cleaved three to four times as readily as isolated guanines. These investigators proposed that this sequence preference reflected favored intercalation binding sites for BPDE. Alternatively, they suggested that the sequence selectivity could reflect a tendency for A-like helix geometry in tracts of poly-(dG)·poly(dC) (*52*), which would expose guanine N-2, making it more accessible to attack (*50*).

A very different preferred sequence context for BPDE reaction at guanine N-7 with DNA restriction fragments was determined by Lobenenkov et al. (*53*). These workers analyzed the strand breaks produced by piperidine and heat treatment of the covalently modified DNA, and hence they presumably detected only guanine N-7 adducts, which are expected to be formed in much lesser amounts than guanine N-2 adducts (*51*). Preferred cleavages occurred at the central guanines of 5'PyGN sequences. While densitometric scans were not run to show relative intensity of the various sites, the absence of cleavage at the 3' guanine of 5'GG sequences was marked, and longer strings of guanines were likewise devoid of evidence of cleavage-inducing alkylation, in sharp contrast to the outcome of Boles and Hogan's results for guanine N-2 alkylation (*50*). Interestingly, these investigators also attributed this very different sequence selectivity to preferred intercalation sites for BPDE (*53*)!

Physical studies offer some insights into the reaction of BPDE with DNA, which may have an important bearing on the sequence determinants of covalent reaction. In spite of apparently reasonable, if conflicting, speculations about preferred intercalation of BPDE described above, Harvey and Geacintov have shown that BPDE intercalation is actually favored at dA-dT sequences (*49*). The location of the intercalated complex does not seem to be critical except to protect the very reactive electrophile from non-productive chemical reactions with other cellular nucleophiles. The BPDE may in fact "diffuse" from one binding site to another along the helix (*49*). Only when it reaches an appropriately reactive guanine will covalent reaction occur, i.e., when the "requirements for proper orientation and sufficiently close approach of the reacting electrophile

(i.e., a triol carbonium intermediate) to an appropriate base site on DNA" (*49*) are met.

Some insight into what might be the "requirements" for covalent reaction between DNA and BPDE might be gained from studying another reaction between DNA and BPDE, namely, DNA-catalyzed hydrolysis of the diol epoxide (*54, 55*). Spectroscopic evidence shows initial rapid noncovalent binding to DNA, and while a small fraction of the carcinogen then reacts covalently with DNA, most of it undergoes hydrolytic ring opening of the epoxide at a rate much faster than hydrolysis in the absence of DNA. This DNA-catalyzed hydrolysis is hydronium ion dependent, either through protonation of the epoxide or the kinetically equivalent general-acid catalysis. It is also extremely sensitive to salt concentration. Thus even at physiological ranges of ionic strength, the ability of DNA to catalyze BPDE hydrolysis is significantly impeded. This has been interpreted to reflect electrostatic stabilization by the polyelectrolytic DNA of the *transition state* in the rate-determining step of BPDE hydrolysis and "suggests that such an interaction may provide a large part of the driving force for the observed catalysis by DNA" (*54*). Michaud et al. compared the rate constants for acid-catalyzed hydrolysis by calf thymus DNA and by poly(dA), concluding that the reaction was not highly dependent on DNA secondary structure or sequence. They proposed general-acid catalysis by a protonated phosphodiester group of DNA as one possible mechanism (*54*). MacLeod et al. reported an increasing ability of various natural and synthetic DNAs to catalyze BPDE hydrolysis with increasing GC content, which did not correlate with noncovalent association constants. They postulated that general-acid catalysis of BPDE hydrolysis might be provided by small concentrations of protonated exocyclic $NH_2$ groups of guanines (*55*). They reported a similar dependence of the extent of covalent adduct formation (with titrated BPDE) on the base composition of the DNA (*55*). Thus, in addition to the suggested sequence-dependent binding of the drug, there may be sequence-dependent electrostatic stabilization of the transition state for adduct formation and sequence-dependent general-acid catalysis of the covalent reaction.

**Mitomycin C.** Mitomycin C is a potent antitumor antibiotic which reacts covalently with DNA following activation by chemical or enzymatic reduction, or by protonation. Recently Tomasz has investigated the regiochemistry of the guanine adduct with mitomycin C, formed under either reductive or acidic activation (*56*). The heteroatom selectivity is quite high for the covalent reaction with the dinucleotide dGpC: *acid activation* gives 95:5 guanine N-7:N-2 alkylation, while *reductive activation* leads to 15:85 guanine N-7:N-2 derived products. This has been explained according to the electronic character of the reactive mitomycin species believed to be formed under each set of conditions (*56*). Acid activation produces a species (Figure 1d') with relatively localized carbocationic character and is thus a "hard" alkylating agent, which is expected to react at guanine N-7, the site of greatest electron density. Reductive activation forms a quinone methide (Figure 1d") with highly delocalized electropositive character, i.e., a "soft" alkylator. Delocalized electrophiles are known to react preferentially at guanine N-2 (*57*).

However, in double-stranded DNA, guanine N-2 alkylation is enhanced relative to guanine N-7 alkylation, regardless of the activation mode. Thus, in the acidic reaction, only half of the guanine adducts from calf thymus DNA derive from N-7 alkylation, while reductive activation leads exclusively to guanine N-2 adducts (56, 58). The rationale suggested for this striking influence of double-helical DNA structure on the chemical nature of the covalent reaction was the possible steric hindrance of attack on the guanine N-7 atom in the major groove (56).

Differences in sequence selectivity for formation of guanine N-2 (50) and guanine N-7 (53) adducts of BPDE suggest that such differences may also be found for mitomycin and might aid in evaluating the significance of acid activation of mitomycin C in vivo.

The sequence selectivity studies published on mitomycin C (59, 60) leave a number of unanswered questions. In these studies strand breaks were created, mostly at guanines, but also to a significant extent at adenines, by incubation of DNA fragments with mitomycin C and sodium borohydride, the chemical reducing agent, under *aerobic* conditions, and then briefly heating (90 °C, 5 min) the isolated, modified DNA. When the heating was extended to 30 min and done in the presence of piperidine, strand cleavage was much more extensive, although the pattern of fragments was judged to be the same. Ueda et al. (59) noted that under reductive conditions the known products of mitomycin–DNA covalent reaction (N-2 guanine adducts) are not heat labile. Guanine N-7 alkylated products, which might give such cleavage, are presumed not to be formed under reductive conditions (58). However, they also pointed out that the heat-labile sites which they observed are evidently not generated under anaerobic reducing conditions (59). Oxygen radical and singlet oxygen scavengers, and metal chelating agents, inhibited this heat-labile site induction. A specific chemical rationale uniting all of these observations was not formulated, although the suggestion was made that intercalation of the mitomycin semiquinone was the basis of the observed sequence selectivity (60). These selectivities were very modest. Almost all guanines and many adenines reacted to some extent, with central guanines in some (but not all) 5′GGT sequences appearing to react three to four times as readily as others (59). In view of Tomasz' more recent characterization of chemical adduct differences depending on mitomycin activation conditions (56), the need for further work in evaluating the sequence selectivity of mitomycin C is clear.

**Anthramycin.** Anthramycin (Figure 1e) is a carbinol-amine-containing antitumor antibiotic that bonds covalently to guanine N-2, with the resulting adduct fitting snugly in the minor groove (61, 62). The sequence selectivity of this and related pyrrolo[1,4]benzodiazepines [P(1,4)B's] has been examined by methidium-EDTA–Fe (MPE·Fe) footprinting (63). In this technique, in principle applicable to either covalent or noncovalent complexes, the DNA fragment is incubated with the drug of interest and then exposed to the non-sequence-selective intercalator, methidium, which in turn is linked to EDTA–Fe. Under the appropriate redox conditions the MPE·Fe reagent generates diffusible reactive oxygen species which cause strand cleavage (ideally, one cleavage per DNA molecule) at every possible nucleotide except those protected by the presence of the drug complex (64). This technique is attractive in that it can reveal sequence selectivity for the formation of adducts (such as those of N-2 guanine) which are not readily susceptible to direct chemical cleavage. More recently, an exonuclease III stop assay has been used to examine the sequence selectivity of a variety of natural and synthetic P(1,4)B's (65). Similar results to that pre-

viously revealed by MPE·Fe footprinting were found. The P(1,4)B drugs preferentially protected 5′PuGPu sequences (bonding to the middle guanine) and enhanced MPE·Fe cleavage in regions between footprints (63). The close van der Waals contacts between the twisted drug adducts and the floor and sides of the minor groove indicated by CPK models were considered as possibly of key importance in sequence recognition (63). Theoretical calculations, however, indicated that the actual residual binding energies of the covalent adducts in various sequences were quite similar but that there were pronounced differences in the requisite distortion energies of the different sequences, with 5′PuGPu sequences showing the lowest deformation energy costs on binding of the already bonded anthramycin (66). The nature of this proposed requisite deformation resembled the parameters of A-DNA, in which the minor groove is wider and shallower than in B-DNA (67). It is interesting that Boles and Hogan had also remarked on the possibility that the contiguous guanine preference or BPDE (with respect to guanine N-2 adducts) might be promoted by the A-like helical tendencies of poly(dG)·poly(dC) tracts (50). The hypothesis implied here, in any event, is that the inherent dynamic conformational properties of a given sequence of nucleotides, i.e., flexibility of deformability along certain coordinates, may be important in affecting the rate of a covalent reaction at that sequence (66). Whether it does so by creating a favorable noncovalent binding site for the drug, increasing its effective local concentration, or whether the deformability is necessary to bring the reactive atoms of the drug and the target nucleotide into bonding distance (i.e., binding of the transition state) is a question potentially answerable by analogue studies. The distortion indirectly observed or calculated for a final covalent adduct must be interpreted with some caution, until direct structural data are available.

**Transition Metal Complexes.** Transition metal complexes can readily coordinate with "soft" endocyclic nitrogen lone pair electrons on DNA bases, forming organometallic covalent bonds (68). The clinically important antitumor agent *cis*-diamminedichloroplatinum(II) (cisDDP) (Figure 1f) has been shown to form a coordination complex with two adjacent guanines through their N-7 atoms, in effect creating a single atom (Pt) cross-link (69). Early sequencing studies using exonuclease III inhibition to locate cisDDP on DNA fragments reported inhibition by both cisDDP and transDDP at every guanine, which was attributed to monoguanine–platinum coordination (70). In addition, the exonuclease was inhibited one (70) or three (71) nucleotides preceding runs of two or more guanines, and this phenomenon, which was shown only by the biologically active cisDDP, and not by the inactive trans isomer, was attributed to the locally distorting bifunctionally coordinated complex. Again, rather large drug loading was used (8–16 Pt atoms per 165 base pair fragment in reference 71), with the object of detecting all possible sites rather than discerning the most reactive sites. Hence, other possible sequence effects on relative reaction rates at different guanines have not yet been explored.

Although not yet subjected to specific sequence analyses, the chiral transition metal complexes studied by Barton provide some interesting insight into covalent DNA modification as a recognition mechanism (68). The left- and right-handed tris(phenanthroline) complexes of ruthenium(II) (structures g′ and g″ of Figure 1), which bind noncovalently to DNA by intercalation of one of the phenanthroline rings, with the other ligands lying along the DNA groove, show a modest degree of difference in their affinity for B-DNA. Dialysis of calf thymus DNA against the racemic mixture produced enrichment of the left-handed isomer in the dialysate, consistent with preferential intercalative binding of the *right*-handed *isomer* (*72*). The left-handed isomer appears to interact with B-DNA in a more loosely held hydrophobic groove binding mode (*73*). Replacement of a phenanthroline ligand (Figure 1h', h'') with two chlorine ligands, on the other hand, gives complexes that can now covalently bond to guanine N-7 atoms, forming cross-links in much the same manner as cisDDP, except that they are octahedral and chiral (*68*, *74*). When racemic bis(phenanthroline)dichlororuthenium(II) was incubated with calf thymus DNA, and the DNA-bonded ruthenium complexes were separated out, the supernatant solution was enriched in the right-handed isomer, indicating that DNA covalent bonding to the *left-handed isomer* was more efficient. Not only was the enrichment of recovered metal complex in solution opposite to that found with the complexes which were merely intercalated, but the magnitude of the enrichment was 5-fold greater (*74*). Thus the covalent interaction is a much more enantiospecific one. Furthermore the degree of stereoselectivity with synthetic polymers depended on DNA sequence, being highest for poly-(dG)·poly(dC) (*68*). Modeling of this left-handed isomer showed that selective covalent reaction with adjacent guanines in the major groove is reminiscent of the groove binding interactions of the left-handed tris(phenanthroline)ruthenium(II) noncovalent complex (*68*). The broad picture emerging from this and related work, as Barton insightfully pointed out (*68*), is that this class of simple coordination complexes recognizes specific regions of DNA by "matching of shapes and symmetries". Such recognition can be "site selective", i.e., for Z-DNA, but is not necessarily sequence selective. However, metal coordination (covalent bonding) provides a mechanism for sequence selectivity, since it involves specific reactions with bases while still maintaining, in the optimum situation, the shape and symmetry complementarity.

***N*-(Bromoacetyl)distamycin.** In order to separate the contributions of noncovalent "physical" binding and those of covalent bonding to the sequence selectivity of agents that are capable of both interactions (as are most of the compounds discussed above), one would ideally want to compare the sequence selectivity of the noncovalent binding forces independently of the covalent bonding reaction. Dervan has achieved this by a kinetic separation of binding and bonding. The tripeptide tris-(*N*-methylpyrrolecarboxamide) unit of the natural product distamycin was shown to bind (noncovalently) four A-T sites, each covering five base pairs, on a 167 base pair DNA restriction fragment, by footprinting and by affinity cleavage methods (*75*). Incorporating a bromoacetyl moiety on the amino end of this tripeptide gave a compound (Figure 1i) which, after short (0.5-h) interaction times with the same DNA fragment, gave identical footprints. But after longer (10-h) incubation times, *N*-(bromoacetyl)distamycin (BD) produced a single cleavage following heat and piperidine treatment at one adenine of one of the four binding sites (*76*). Presumably BD alkylated the adenine N-3, as does CC-1065 (see below). An α-bromoacetyl amide is not an extremely reactive electrophile, unlike the unisolatable reactive intermediates described in the foregoing discussion. The covalent reaction of BD with DNA is slow relative to those agents, but this serves to amplify the rate differences that underlie the phenomenon of sequence selectivity. Only one nucleotide in 334, and only one of a minimum of seven ad-

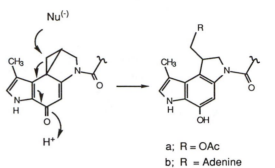

**Figure 2.** Reaction of CC-1065 with acetic acid or DNA to give rise to the cyclopropane ring opened products (a) and (b), respectively (*81*, *82*). For production of the CC-1065-(N3-adenine) product (b) subsequent thermal treatment is required (Figure 3).

enines near the noncovalently (and also sequence-selectively) bound drug, underwent detectable covalent reaction in 10 h. At longer times, a second covalent site began to appear at another footprinted binding site (*76*). Clearly, noncovalent interactions of this molecule are quite sequence selective and most likely resemble those described in the crystal structure of netropsin–oligonucleotide complex (*16*). They localize the drug at the four binding sites. But even further selectivity is imparted at the covalent reaction stage, where the sequence-dependent differences of local DNA structure must meet the "stereoelectronic requirement in the transition state for the backside nucleophilic displacement reaction" (*76*). Dervan suggested that BD may be a mechanistic mimic of the antitumor antibiotic CC-1065, which alkylates adenine N-3 in AT-rich regions of DNA. In the remaining discussion we shall focus on this extraordinary DNA-interactive natural product and some of its synthetic analogues.

**CC-1065.** CC-1065, one of the most potent antitumor antibiotics known (*77*), interacts with double-helical DNA both covalently and through unusually strong noncovalent binding in the minor groove at AT-rich regions (*78–80*). Among DNA-interactive natural products, CC-1065 appears to be exceptionally selective, both with respect to the covalent adduct that it forms and with respect to the sequences in which that adduct occurs.

**(A) Structure of the CC-1065–DNA Adduct.** Definitive evidence for a covalent adduct between CC-1065 and DNA was obtained from a [1]H and [13]C NMR analysis of the major product (comprising 85% of the total drug chromophore) obtained after thermal treatment of calf thymus DNA previously incubated with CC-1065 (*81*). The structure of the CC-1065–base adduct was analogous to the acetic acid addition product (Figure 2a) (*82*). In place of an acetoxy group, adenine was now covalently attached at C-4 via its N-3 position, i.e., structure b in Figure 2; and for reaction of CC-1065 to form the CC-1065-(N3-adenine)–DNA adduct see Figure 3. Radiolabeling experiments showed adenine to be the only base detectably alkylated (*81*). Since N-3 of deoxyadenosine is located in the minor groove of DNA, this implied that on DNA the drug molecule is cradled within this groove following the right-handed twist of B-type DNA (*81*) (Figure 4) and corroborated earlier evidence of minor groove interaction (*79*). More recently, two-dimensional [1]H NMR studies on CC-1065–oligomer duplex adducts have confirmed minor groove occupancy through drug–nucleotide interproton nuclear Overhauser effect connectivities (*83*). In early 1985 the absolute stereochemistry of the cyclopropyl ring of CC-1065 was still unknown; however, this was tentatively assigned on the basis of the directionality of the DNA sequence specificity in relation

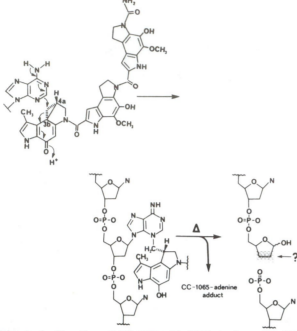

**Figure 3.** Reaction of CC-1065 with DNA to form the CC-1065–(N3-adenine)–DNA adduct and products of thermally induced strand breakage (*81, 84*). The structure of the modified deoxyribose on the 3′ side of the strand break is not definitely known.

to the covalent binding site (*84*). We assumed on the basis of available evidence that the DNA sequence specificity was entirely a consequence of the binding interactions (close van der Waals contacts between the inside edge of the drug molecule and the floor of the minor groove of DNA), and therefore directionality of sequence specificity would be akin to directional orientation in the minor groove. As luck would have it, our stereochemical prediction was later determined to be correct (*85*), but our argument was partially flawed, since more recent data have shown that the molecular basis for DNA sequence selectivity of CC-1065 is primarily due to the *bonding reaction* (see later) and is consequently independent of the drug's directional orientation in the minor groove of DNA.

**(B) DNA Sequence Selectivity of CC-1065.** Thermal treatment (e.g., 100 °C for 30 min) of CC-1065–DNA adducts produces cleavage of the N-glycosidic bond that connects the CC-1065-modified adenine to its deoxyribose (*84*). The resulting apurinic site then presumably undergoes a β-elimination to give rise to a DNA backbone strand breakage, leaving a 5′-phosphate on the 3′ side of the strand break and a modified deoxyribose on the 5′ side (Figure 3). This thermal strand breakage assay has been used to determine the consensus bonding sequences for CC-1065 (*84*). In all, over 1000 base pairs from SV40 and T7 DNA restriction fragments were used in the analysis. A wide range (7 orders of magnitude) of drug concentrations were examined to give some indication of the relative reactivities of different sequences. At the lowest drug levels at which any alkylation-mediated strand cleavage could be detected (0.14 pmol/mL), only the 3′ adenine of the 5′AGTTA sequences in the 21 base pair repeats of the SV40 early promoter region was alkylated. At 10–100-fold higher drug concentrations (1.4–14 pmol/mL), 32 additional alkylation sites in the various fragments could be detected under the same incubation conditions. At these concentrations, most of the DNA remained as high molecular weight material on the gels, indicating a high

likelihood that each strand break observed represented a single "hit" on the DNA molecule. Remarkably, in 18 of these sites the adenine was flanked on the 5′ side by thymine, and 14 of these had a second 5′ thymine. The other alkylated adenines were flanked by two more adenines (in 13 out of 14 instances). At a 10-fold higher concentration the 11 additional sites still maintained a two base pair specificity (AA* or TA*). Still higher concentrations of CC-1065 led to more extensive DNA reaction, presumably resulting in multiple hits and with less regard for their intrinsic reactivity. From the sequence data obtained at the lower drug concentrations (140 pmol/mL or less), two subsets of sequences were identified, 5′AAAAA* and 5′PuNTTA*, where A* represents the covalently modified adenine, and the most stringent sequence specificity lies to the 3′ end of the sequences. (i.e., AAA* and TTA*).

It is important to recognize that these are the consensus sequences for the *covalent bonding* reaction, since this assay depends upon thermal labilization of the N-glycosidic bond as a consequence of N-3 alkylation. At the time we determined these consensus sequences for the covalent reaction, we assumed, because of the drug overlap with these sequences, that the noncovalent interactions were selecting the adenines that underwent alkylation, i.e., that the binding interactions directed the bonding selectivity. Specifically, we proposed a model in which close van der Waals contacts between the inside edge of the drug molecule and the floor of the minor groove of DNA precisely aligned the drug molecule in the minor groove, and as a consequence of this preferred physical alignment a subset of CC-1065 noncovalent binding sites were selected for covalent bonding by a proximity to an adenine N-3 atom. Indeed, we referred to the most reactive sites as "highest affinity binding sites" on the basis of this assumption.

Subsequent attempts to detect noncovalent binding sites of CC-1065 on a DNA restriction enzyme fragment by "chemical footprinting" with MPE·Fe (*64*) revealed no new sites other than the covalent reaction sites (Figure 5) (*86*). Attempts to detect footprints with the nonalkylating species des-ABC (Figure 6) proved unsuccessful presumably because the stability of its binding to DNA was insufficient to allow detection by this technique (*86*). However, spectroscopic methods can be used to distinguish noncovalent binding from covalent bonding modes. It has been shown that synthetic polydeoxynucleotides that bind CC-1065 reversibly (noncovalently) exhibit a longer wavelength maximum induced circular dichroism (CD) than those polymers that undergo irreversible (covalent) reaction with CC-1065 (390 vs 370 nm), due to the different chromophores of intact and ring-opened drug (*87–89*). The covalent reaction could be followed by slow conversion of the 390-nm CD band to a 370-nm band, for some polymers.

The observation of CC-1065 sequence selectivity with DNA restriction fragments (*84*) prompted Theriault and Krueger to exploit these spectroscopic tags of the bonded and unbonded (but physically bound) drug in an elegant and powerfully precise manner, with a series of defined oligonucleotides (*90*). Unlike the modified Maxam–Gilbert sequencing technique, this study allowed examination of sequence effects both on covalent bonding and on noncovalent binding. [While chemical footprinting had shown no additional CC-1065 sites other than the adenine N-3 alkylation sites, it is possible that noncovalently bound CC-1065 would not be detected by this technique (*86*).] Consistent with the significantly more facile alkylation-mediated strand cleavage at 5′TTA* sequences, duplex dodecamers containing these sequences showed rapid (less than 24-h) conversion from the 392-nm absorbing species

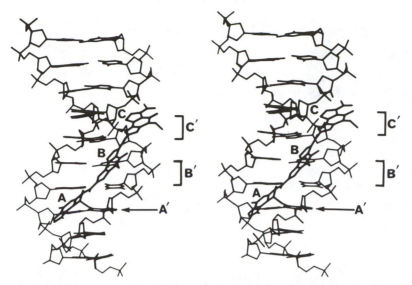

**Figure 4.** Stereo diagram of a CC-1065–DNA adduct. The computer graphic representation of a CC-1065–DNA adduct using the X-ray crystal coordinates for ring-opened CC-1065 (*78*) and B-form DNA generated by the MIDAS graphics program. The sequence used is that of the center of a 14 base pair oligodeoxy duplex shown to covalently bond a single CC-1065 molecule. $\sigma$ bonds around amide linkages between the drug subunits (see Figure 5) were rotated to reduce close drug–DNA contacts upon docking. CC-1065 A, B, and C subunits are labeled.

to the 370-nm species, i.e., rapid alkylation. In contrast, closely related dodecamers with 5'ATAT or 5'TATA sequences required 1–2 weeks to convert to the alkylated form, although the high magnitude of the induced CD indicated substantial noncovalent binding interaction for all of these oligomers. Dodecamers with 5'AAGAA and particularly 5'GAATT showed little or no tendency to form covalent species, although intense induced CD at 392 nm indicated very stable, noncovalently bound complexes. In addition, the longer the 5'TTA*-containing oligomer, particularly in the 5' direction from the target adenine, the more intense was the induced CD (reflecting tighter noncovalent binding) and the faster was the 392 nm to 370 nm transition (i.e., the rate of the covalent reaction). These experiments suggest that it is the rate constant for the covalent reaction itself which is so different for, say, 5'TTAA and 5'AATT. The stereoelectronic requirements for covalent reaction with CC-1065 reflected in that rate constant are simply not energetically accessible for 5'AATT, even though a very strong and sequence-selective (*90*) noncovalent complex forms.

It is interesting to note that the dodecamer 5'CGCGAATTCGCG, which binds CC-1065 so well, but only noncovalently, is the same one used in the crystal structure of a netropsin–DNA binding site (*16*). From that crystal structure, primary sequence recognition for this cationic, noncovalent minor groove binding molecule was attributed to favorable van der Waals contacts between protons on adenine and the inside edge of the drug. The A-T binding preference of netropsin and similarly shaped minor groove binding drugs, including CC-1065, has for some time been considered to reflect the steric inhibition to such close binding presented by exocyclic $NH_2$ groups of guanines protruding into the minor groove (*84, 91*). The strong noncovalent binding of CC-1065 to this dodecamer could support the commonality of these factors in noncovalent groove interaction. On the other hand, a comparison of distamycin MPE footprints with those of CC-1065 (which also correspond to covalent bonding sites) on a 117 base pair restriction fragment for M13mp1 (Figure 5) shows significant differences in both the positions and relative protection of these sites; e.g., the *best* CC-1065

bonding site in Figure 5 is located in proximity to the *poorest* distamycin binding site (*86*). Thus the introduction of covalent bond formation in the case of CC-1065 strongly modifies the effect of what might possibly be similar noncovalent recognition processes for these agents.

A complementary approach to that of Theriault and Krueger, who analyzed the complex reaction components by simplifying the DNA, is to simplify the CC-1065 molecule, in order to separately evaluate its binding and bonding contributions to sequence selectivity. The availability of synthetic analogues of (+)-CC-1065 (Figure 6) allowed us to dissect the structural requirements for both DNA sequence selectivity and biological potency. The analogues (+)-A, (+)-AB, and (+)-ABC (Figure 6) exemplify three main groups of a large number of synthetic cyclopropylpyrroloindole (CPI) analogues (*92–95*). Compounds like (+)-A do not show evidence of significant noncovalent DNA binding by CD or thermal melting measurements (*92, 96*). These compounds, although efficacious as antitumor agents in vivo, have relatively low potency in vivo or in vitro (Table I). At the other extreme are compounds resembling (+)-ABC, which show intense induced CD bands with DNA and cause large melting temperature increases, indicative of strong physical binding interactions. These compounds rival or exceed CC-1065 itself in biological potency. Finally, compounds of intermediate length, like (+)-AB, tend to show intermediate binding parameters and potencies (Table I).

In view of our early assumptions that strong, specific noncovalent binding interactions dictated the location of the ultimate alkylation sites, we expected a progressive loss of site discrimination with analogues of progressively less DNA binding ability [i.e., proceeding from (+)-ABC to (+)-AB to (+)-A]. Surely the surrounding sequence context could matter very little to a molecule as small as (+)-A. To our great surprise, (+)-A, (+)-AB, and (+)-ABC, so different in the degree of their physical binding interactions with DNA, showed virtually the same sequence specificity (Figure 7)!! While the absolute rates of alkylation were quite different, judged by the $10^2$ and $10^4$ higher molar concentrations of (+)-AB and (+)-A required to achieve the equivalent intensity of strand breakage

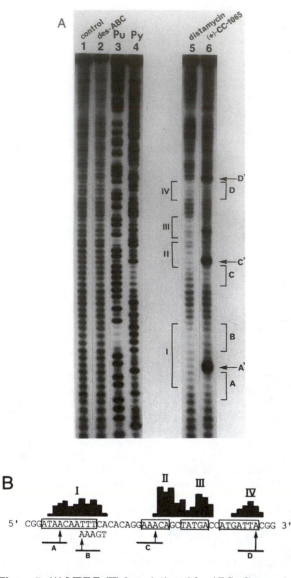

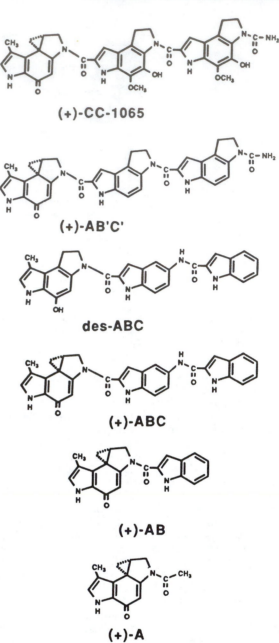

**Figure 6.** Structural formulas for (+)-CC-1065 and synthetic analogues (92–95) described in the text.

**Figure 5.** (A) MPE·Fe(II) footprinting of des-ABC-, distamycin-, and (+)-CC-1065-modified 117-bp *Msp*I–*Bst*NI fragments from M13mp1 DNA (*86*). Single 5′-³²P-labeled aliquots of M13mp1 117-bp DNA fragments were modified with 2.8 μM solutions of des-ABC (lane 2), distamycin (lane 5), and (+)-CC-1065 (lane 6). Reactions were electrophoresed adjacent to Maxam–Gilbert purine and pyrimidine lanes (lanes 3 and 4). Brackets correspond to the MPE·Fe(II) footprints for distamycin (I–IV) and (+)-CC-1065 (A–D), and arrows (A′, C′, and D′) correspond to the covalent bonding sites for (+)-CC-1065. (B) Diagrammatic representation of the distamycin and (+)-CC-1065 binding sites and relative affinities in the readable portion of the sequencing gel shown in (A). The sequences enclosed by boxes and the histogram bars represent the position and relative inhibition of MPE·Fe(II) cutting for the distamycin binding sites on the 117-bp fragment of DNA. Data are taken from lanes 1 and 5 in (A), which were scanned by an integrating laser densitometer. Following normalization of total lane areas, the relative inhibition of cutting was calculated and plotted on the histogram such that maximum inhibition is given the highest histogram rating. The three-base offset of binding sites (boxes) relative to the inhibition zones (histogram bars) is in accord with the model for asymmetric footprint patterns. Distamycin footprints I–IV correspond to zones of inhibition I–IV in (A). The bars A–D below the DNA sequence correspond to the (+)-CC-1065 alkylation sites shown in (A). The length of the arrows corresponds to the relative sensitivities of sites A–D for (+)-CC-1065, and the arrowheads point to the covalently modified adenines. The longest arrows represent the most sensitive sites. Since the (+)-CC-1065 covalent bonding site corresponding to B is on the strand opposite to that which is 5′-³²P-labeled, the complementary sequence is shown for this site.

shown by (+)-ABC (Figure 7), the relative rates among the various potential alkylation sites remained basically the same, i.e., (+)-A "preferred" the same sequences that (+)-ABC did. This implies that the DNA sequence selectivity of (+)-ABC, and presumably of (+)-CC-1065, is primarily mediated not by its strong, noncovalent binding interactions, but by the bonding reaction, which it shares in common with (+)-A and (+)-AB. Gratifyingly, the corresponding higher levels of (+)-AB and (+)-A that were required to produce equivalent DNA bonding also produced similar levels of cytotoxic potency (Table I). This suggests that the thermally induced DNA strand breakage assay, a gauge of the covalent bonding reaction, is a good indicator of biological potency.

While the (+)-A subunit of (+)-ABC is structurally sufficient to mediate the primary basis for the sequence selectivity of the entire molecule, we noted that different "B" and "C" subunits modulate or fine-tune the precise

**Table I. Biological Activities of CC-1065 and Its Analogues (86)**

| | in vitro L1210 cell growth inhibition, $ID_{50}$[a] | | in vivo P388 leukemia[b] | |
|---|---|---|---|---|
| | nM | rel to CC-1065 | %ILS | O.D., μg/(kg·day) |
| (+)-CC-1065 | 0.03 | 1 | 67 | 100 |
| (+)-AB'C' | 0.05 | 2 | 77 | 100 |
| des-ABC | 160 | 5300 | 0 | 6000[d] |
| (+)-ABC | 0.004 | 1/7 | (4/6)[e] | 25 |
| (+)-AB | 0.1 | 3 | 100 | 250 |
| (+)-A | 12 | 400 | 45[c] | 3000[c] |

[a] Drug concentration causing 50% inhibition of cell growth with 3-day drug incubation. [b] I.p.-implanted P388 leukemia with i.p.-drug treatment on days 1,5,9. %ILS = percent increase in life span of treated compared to control mice. O.D. = optimum dose. [c] Tested as racemic mixture of (+)-A and (−)-A. [d] Highest dose tested. [e] Ratio of the number of day 30 survivors/total number mice in test group.

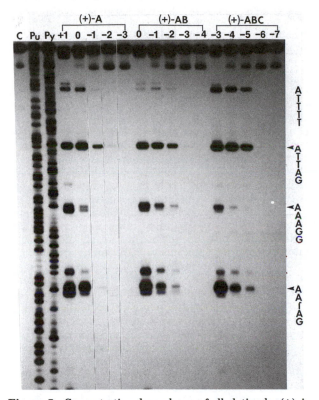

**Figure 7.** Concentration dependency of alkylation by (+)-A, (+)-AB, and (+)-ABC of the 5'-$^{32}$P-labeled (+) strand in the 117-bp *Msp*I–*Bst*NI fragment of M13mp1 DNA (86). Aliquots of single 5'-$^{32}$P-labeled 117-bp DNA fragments were modified with $1 \times 10^n$ (n = +1 through −7) dilutions (labeled +1 through −7) of 280 μM stock of (+)-A, (+)-AB, or (+)-ABC solution. Modified DNA was heated at 90 °C for 30 min in DSC buffer and electrophoresed adjacent to Maxam–Gilbert purine- and pyrimidine-specific DNA cleavage reactions. The drug alkylation sequences are shown, and arrows indicate adenines modified by (+)-A, (+)-AB, and (+)-ABC.

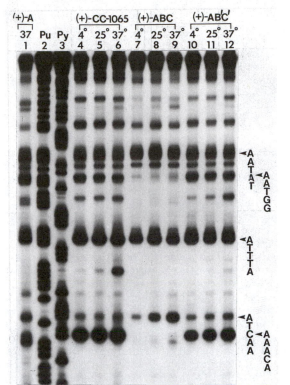

**Figure 8.** Temperature dependency of alkylation by (+)-CC-1065, (+)-ABC, and (+)-AB'C' of the 118-bp *Mbo*I–*Hin*fI fragment of SV40 DNA (86). Aliquots of single 5'-$^{32}$P-labeled 118-bp fragments were incubated with 2.8 mM (+)-A or 28 μM (+)-CC-1065, (+)-ABC, and (+)-AB'C' and incubated at 4, 25, or 37 °C for 2 h. Drug-modified DNA was heated at 90 °C for 30 min in DSC buffer and electrophoresed adjacent to Maxam–Gilbert purine- and pyrimidine-specific DNA cleavage reactions. The drug bonding sequences are shown, and arrows indicate adenines modified by either (+)-A, (+)-CC-1065, (+)-ABC, or (+)-AB'C'.

sequences to which CC-1065 analogues covalently bond (Figure 8). Thus there are a few sequences which (+)-ABC and (+)-CC-1065, for example, do not share in common [although (+)-A at sufficiently high concentrations alkylates both sets of potential bonding sites]. Presumably steric inhibition to binding of CC-1065 can effectively exclude some sites available to (+)-ABC; alternatively, specific favorable binding interactions of CC-1065 could promote reaction at some sties that might alkylate (+)-

ABC too slowly. An analogue of intermediate structural complexity to (+)-ABC and (+)-CC-1065, namely, (+)-AB'C', which shows physical binding parameters (induced CD, $\Delta T_m$) closely resembling those of (+)-ABC (95, 97), nevertheless closely matches CC-1065 in sequence specificity. Thus, changing the inside edge substituents of the B and C subunits [i.e., (+)-CC-1065 or (+)-AB'C' to (+)-ABC] is sufficient to modulate the DNA sequence selectivity of the parent drug molecule. The outside-edge hydrophilic substituents of (+)-CC-1065 apparently have little effect in this regard. Thus we conclude that while the *bonding reaction* is the main determinant of DNA sequence specificity of (+)-CC-1065, certain *binding interactions* (close van der Waals contacts between the inside edge of the drug molecule and the floor of the minor groove of DNA) can modulate or fine-tune this sequence selectivity. The "fine tuning" in this instance assumes added significance from the observation that (+)-AB'C' also shares with (+)-CC-1065 the unusual and potent biological action of causing delayed death in rodents (95).

## Molecular Basis for the DNA Sequence Selectivity of DNA Covalent Modification

The recent literature on the sequence selectivity of drugs and carcinogens that bond covalently to DNA establishes a strong case for the existence of such selectivity. That is, the rate of covalent reaction of a given drug with a nucleotide base in one sequence may be much faster than that with the same base in a different sequence. While some drugs (e.g., CC-1065) are much more sequence discriminating than others (e.g., BPDE or nitrogen mustards), very few appear to be non sequence selective.

Examination of most of the published literature would seem to suggest, however, that these observed selectivities for reaction at certain sequences are the consequence of sequence–specific binding, preceding the bonding reaction (35, 50, 53, 59, 63, 84). In terms of eq 1, the sequence

$$\text{drug} + \text{DNA} \underset{}{\overset{K_b}{\rightleftharpoons}} (\text{drug·DNA})_{noncov} \xrightarrow{k_r} (\text{drug–DNA})_{cov}$$
(1)

information is generally presumed to be important only in the requisite initial binding step. A priori, this is a reasonable notion in light of the extensive literature establishing the ability of noncovalently binding molecules to discriminate among DNA sequences (15–20). Thus the faster rates of covalent reaction at certain sequences are presumed to reflect higher concentrations of the noncovalently held complex at those sequences.

In principle, however, the bonding step might also be sequence selective; that is, certain sequences may be better than others at stabilizing the transition state for covalent reaction. Some of the ways that this bonding sequence selectivity might be brought about have already been mentioned in one context or another: electrostatic stabilization of charge development in the transition state [as in DNA-catalyzed BPDE hydrolysis (55)], DNA conformational flexibility [proposed for anthramycin (66) and uracil mustard (43)], orientation-dependent multifunctional catalysis [proposed for MNU (41) and modeled by nucleotide-catalyzed BPDE hydrolysis as discussed below (98)], and steric accessibility [proposed for quinacrine mustard (43)]. In most cases the experimental work required to distinguish among these possibilities has not yet been undertaken. Nevertheless, we believe it is possible to form some useful, if tentative, generalizations about the origins of selectivity in some of these instances.

It might be supposed that sources of bonding selectivity could be most easily discerned when binding selectivity is likely to be weak or nonspecific. Small molecular weight alkylating agents in general have few noncovalent DNA binding capabilities except for charge attraction. The observed MEP-dependent sequence selectivity of most nitrogen mustards was attributed to selective attraction of aziridinium cations to negative potential regions of DNA (43). However, the detailed study of salt effects on DNA alkylation by the mustards (43) suggested that electrostatic potential might be performing more than this collecting function. Negatively charged mustards (with therefore net neutral aziridinium intermediates) still showed selectivity for guanine clusters, even though they were not inhibited by salts as were other mustards. In general, salts reduced the magnitude but not the order of selectivity (i.e., rate differences) for different sequences. Thus bonding selectivity remained even under ionic strength conditions which made ionic binding interactions extremely weak (43). Similarly, inhibition of MNU guanine N-7 alkylation by salt of DNA binding cations still retained relative selectivity for certain guanine sequences (40). Kohn had suggested (42) that the observed selectivity for runs of guanines shown by aflatoxin (35–37) might be due to the more negative electrostatic potential at these sequences, even though a discrete cation of sufficient lifetime to allow equilibrium binding seems unlikely, since the aflatoxin epoxide itself cannot be isolated.

These observations suggest that a sequence-dependent molecular electrostatic potential may operate on the bonding step, perhaps by stabilizing an incipient positive charge on the electrophilic substrate in the transition state.

This has precedent in the DNA-catalyzed hydrolysis reaction of BPDE (55) discussed above. It implies some degree of catalysis of the alkylation by the electronegative DNA sequence. It has been noted that guanine N-7 in double-stranded DNA reacts with aflatoxin oxide much more rapidly than does guanosine (37). Similarly, $N^1$-methyl-$N$-nitro-$N^1$-nitrosoguanidine (MNNG) does not alkylate guanosine but readily alkylates DNA (99). The rate constant for hydrolysis of $n$-propylnitrosourea (PNU) is 3-fold lower than that for its alkylation of calf thymus DNA (41). Numerous other instances of alkylation rate acceleration by DNA have been enumerated by the Pullmans and tentatively ascribed to MEP effects (44). These instances suggest that in some way DNA is acting to *catalyze* its own alkylation by these reagents. While the interesting sequence-dependent catalytic mechanism proposed by Buckley (41) for MNU alkylation of DNA is inconsistent with the observed sequence selectivity, at least for guanine N-7 alkylation (40), an alternative electrostatic stabilization of a cationic transition state could perhaps reconcile the sequence selectivity (40), salt effects (40), and kinetic and product studies (41) on alkylnitrosourea reactions with DNA.

The concept of electrostatic stabilization of a cationic transition state by a sequence of negative electrostatic potential, as opposed to that of rate acceleration simply by virtue of the local concentration of discrete cations near such a sequence, helps to explain the interesting behavior of uracil mustard. It was proposed that a specific hydrogen bond (uracil O-4 and 3′ cytosine $NH_2$) facilitated reaction at the 5′ guanine by in essence unmasking its potentially electron-rich N-7 (43). The drug must already be present at this site in order for this N-7 activation to occur, so the guanine electrostatic potential is not serving to collect the drug. Rather, this model appears to present an intriguing picture of sequence-dependent bifunctional catalysis of DNA alkylation with one part of the substrate (uracil) interacting with a "suppressor" (the amino group of the 3′ cytosine) to allow "activation" of the adjacent guanine N-7, which can then stabilize positive charge formation in the alkylating portion of the substrate.

A clear case for DNA catalysis of DNA alkylation by these agents is difficult to demonstrate, because they are such highly reactive molecules to begin with. It would seem that such high reactivity (high $k_r$ in eq 1) is necessary to offset the lack of strong, specific DNA binding capabilities (low $K_b$) in these small molecules, and still retain biological impact. Quinacrine mustard provides a fascinating example of what can happen to sequence discrimination when a significant DNA binding (in this case, by intercalation) component is present in the alkylating drug (43). First, this drug efficiently alkylates DNA at a much lower concentration than that required for the other mustards (42), reflecting the collection of drug from bulk solution by its binding to DNA. Second, the degree of discrimination among sequences is larger, and finally, the preferred sequences are quite different from those of other mustards and do not show a straightforward correlation with sequence MEP (43). However, the model presented to rationalize the sequence selectivity observed is not a simple case of preferred binding at those sequences (magnitude of $K_b$) but rather represents a compromise between binding and bonding interactions. The strong requirement for a G or T 3′ to the reacting guanine is postulated to be due to unfavorable stereoelectronic interactions between the base and the alkyl chain of the drug when A or C occurs in that position, i.e., the transition state for alkylation of the intercalated drug would be of considerably higher energy in these sequences, and hence alkylation is less favored. Intercalation at those sequences may be as favorable or even more favorable, but the final discriminant is the relative free energy of activation of the bonding reaction of the (noncovalently) bound drug. Thus, binding ($K_b$) in this case does not dictate sequence specificity, but it does constrain the selectivity of the purely bonding reaction shown by other nitrogen mustards and introduces additional requirements.

Quinacrine mustard may be a good model of the determinants of sequence selectivity for many natural products and metabolites such as aflatoxin oxide and BPDE, which are highly reactive alkylators that are capable of significant noncovalent binding to DNA. Thus the AT intercalation preference reported for BPDE (49), for example, is not inconsistent with the observed preference for guanine N-2 alkylation on G-rich regions if it is allowed that less favored binding sites (lower $K_b$) might nevertheless provide a much better stereoelectronic environment for covalent reaction (higher $k_r$). The interaction of ruthenium complexes with DNA (68) may represent a case in which bonding ultimately selects sites with lower affinity binding. Thus while DNA preferentially binds (noncovalently) the right-handed isomer tris(phenanthroline)ruthenium complexes, with a lower affinity binding mode for the left-handed isomer (72, 73), it is the left-handed isomer of the bis(phenanthroline)dichlororuthenium complex that undergoes faster alkylation. While DNA binding affinity is likely to increase potency in general by mass action, its effect on sequence selectivity might be quite indirect and dependent on the covalent bonding ability of the noncovalent complex.

One might expect binding affinity to exert a predominant effect on sequence selectivity for those agents inherently capable of strong, sequence-selective noncovalent interactions with DNA, such as the minor groove binders. Dervan has engineered molecules in which a highly sequence selective binder, distamycin, is tethered to EDTA. In the presence of Fe(II), oxygen, and a reducing agent, this produces a sequence-selective DNA cleaving molecule, whose sequence selectivity is determined totally by noncovalent binding. The "covalent" component, in this case a diffusible oxygen radical induced cleavage rather than base alkylation, is totally sequence nonspecific (64). (Bromoacetyl)distamycin (BD), in contrast, contains the same selective binding moiety, but in this case tethered to a relatively unreactive $\alpha$-bromoacetyl alkylating moiety. Covalent reaction with N-3 of adenine occurs at only one of the four binding sites examined, and that only after a 5–10-h incubation. The covalent reaction, by as yet undescribed mechanisms, adds another set of requirements to further discriminate among the sets of sequences permitted by binding.

The alkylation of DNA by BD is extremely slow (very low $k_r$), and thus it becomes experimentally feasible to distinguish between sequences having relatively small differences in $k_r$. The second alkylation site of BD, which becomes detectable after 10 h, may not be significantly less reactive than the first site, even though each alkylation event may be experimentally easy to isolate. Thus the sequence selectivity attributable to the alkylation step may not be very high. To the extent that such bonding selectivity is present, it would seem most likely to reflect steric constraints imposed by binding. Indeed, it would not be surprising if a similar electrophile devoid of the strong DNA binding component, for instance a simple $\alpha$-bromoacetyl amide, failed to alkylate adenine N-3 of DNA under conditions resembling those used for BD. In this regard,

the BD model differs fundamentally from the quinacrine mustard model, in which the alkylating moiety has sufficient intrinsic reactivity to alkylate DNA. While in the quinacrine mustard model, binding facilitates potency and constrains the covalent reaction to certain sequences, in the example of BD, strong specific binding may be essential for covalent reaction to occur at all. If this is the case, then the concept championed by Boger, originally to describe CC-1065, namely, "accessible hydrophobic binding driven binding" (100), would seem particularly suitable to describe DNA alkylation by BD. According to this concept, as we understand it, an energetically unfavorable covalent process can be "promoted and rendered irreversible (bonding) by the strong, stabilizing noncovalent association" (101) of the drug with DNA.

CC-1065 also exhibits strong noncovalent binding selective for AT-rich regions of the minor groove of DNA. Like BD, it contains a potential alkylating moiety. For quite some time we, like Boger, regarded CC-1065 as a molecule possessing a reactive alkylating moiety "superimposed" (101) on the sequence-selective DNA binding structure and, as discussed earlier, interpreted the initial sequence selectivity studies as revealing "high-affinity binding sites" (84).

However, further experimentation has forced us to reinterpret the sequence selectivity observed for CC-1065. The oligomer studies have shown that very high affinity binding is not a sufficient condition for bonding. Specifically, both 5′AGTTA- and 5′GAATT-containing oligomers exhibit strong binding to CC-1065 (high $K_b$), yet the former is alkylated readily while the latter is not alkylated at all (90). Clearly, $k_r$ is very different for those two noncovalent complexes. The nearly identical sequence selectivity of small CPI analogues such as (+)-A, which can contribute very little to the hydrophobic and van der Waals noncovalent interactions shown by (+)-ABC or by CC-1065, clearly demonstrates that such A-T-selective noncovalent interaction is *not* essential for the observed sequence selectivity of the covalent reaction of DNA with the alkylating segment. While this binding equilibrium does appear to have an important effect on the overall rate of the covalent reaction, in that strongly binding analogues detectably react at much lower concentrations than nonbinding analogues (86), as might be expected from a mass action effect of binding, it is *not* the important determinant of the sequence selectivity of the CPI agents. Therefore, the phrase "accessible hydrophobic binding driven bonding" (100) is inappropriate to generally describe the reaction of (+)-CC-1065 and analogues with DNA, since it implies that the hydrophobic binding provided by the concave surface of these agents is both necessary and sufficient for the bonding reaction to take place, which is contrary to the experimental results (86, 90). While noncovalent binding in the minor groove of DNA must take place as a prelude to reaction, this binding might, as in the case of (+)-A, be very weak, and need not be sequence selective. When noncovalent binding is stronger and can exert sequence selectivity [e.g., (+)-CC-1065 or (+)-AB′C′], it may modulate reaction at those sequences where $k_r$ is not prohibitively small.

If sequence-dependent binding interactions do not solely determine the preferred sequences alkylated by CC-1065 and its CPI analogues, this selectivity must originate in the bonding reaction. Several models can be proposed that, together or separately, can rationalize sequence discrimination in the covalent bonding reaction of CC-1065 with DNA.

One mechanism by which the bonding reaction can exert

sequence selectivity is through a required alteration in conformation of the local region of DNA in the transition state for alkylation. The energetics of DNA helix distortions are highly sequence dependent. Reactive sequences would be those capable, at the least cost of distortion energy, of adopting an "active conformation" required for covalent reaction. In the most obvious situation the conformational change would bring the nucleophilic atom within bonding distance of the electrophilic center. Studies of a site-directed (+)-CC-1065–DNA adduct (102, 103) provided evidence for an asymmetric effect on DNA structure, based on cleavage patterns of DNase I and AluI restriction enzymes. In the case of AluI, strand scission was selectively inhibited on the noncovalently modified strand 14 bases away from the (+)-CC-1065 covalent bonding site on the 5′ side of the adduct. Likewise, DNase I cleavage was selectively inhibited on the noncovalently modified strand and also on the 5′ side of the adduct. On the basis of the X-ray structure of DNase I (104) we interpreted this to mean that the deoxyribose–phosphate backbone conformation of DNA was changed to the 5′ side of the covalent binding site (86). A more recent structure of a DNase I–oligomer complex (105) offers a second possibility for interpretation of our results. In this case, reduced DNA flexibility induced by (+)-CC-1065 bonding in the vicinity of the CC-1065–DNA adduct could inhibit DNase I cleavage. Experimental data which differentiate between these possibilities are as yet unavailable. If DNA flexibility is important for (+)-CC-1065 and protein reaction with DNA, one might speculate that the reason (+)-CC-1065 and (+)-A alkylate only certain adenines is that these are embedded in flexible regions of DNA, and once alkylation by (+)-CC-1065 occurs DNA flexibility is lost and enzymatic cleavage by DNase I and AluI is selectively inhibited in the vicinity of the adduct. An important prediction from this speculation is that the (+)-A subunit should show the same pattern of DNase I inhibition as (+)-CC-1065. We have shown this to be true (86). We suggested that this altered DNA structure might represent a "trapped" reactive conformation of DNA, a conformation that certain sequences could presumably adopt more readily than others and that could thus be a basis for sequence selectivity (86).

DNA conformational mobility has also been suggested for the sequence selectivity of anthramycin (66), and as a component in the selectivity of uracil mustard (43). Examining the features, experimentally or theoretically, of a postalkylation adduct does not necessarily give information on the transition state for the alkylation step. However, the finding of similar alterations in DNA by the truncated (+)-A segment and by (+)-CC-1065 argues against a role for the noncovalent binding interactions in causing these conformational changes in DNA since these would be expected to be quite different for (+)-A and (+)-CC-1065.

The sequence-dependent catalysis of the alkylation is a second mechanism through which sequence selectivity can be expressed by a covalent reaction. This type of mechanism was recently proposed, but without direct experimental support, for MNU alkylation (41). CC-1065 and its analogues are, because of their chemical properties, also attractive candidates for sequence-dependent catalysis. Unlike the mustards, nitrosoureas, and reactive epoxides already discussed, CC-1065 and its CPI analogues are quite stable, when appropriately solubilized in neutral aqueous solutions (92, 106). They do not undergo cyclopropyl ring opening reactions under these conditions, even with fairly good nucleophiles. Yet, in the presence of DNA, covalent

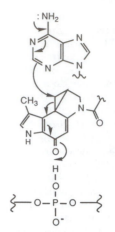

**Figure 9.** Proposed mechanism for general-acid-catalyzed suicide inactivation of DNA by CC-1065.

reaction occurs rapidly (80, 107). In this regard they also differ from the poorly reactive alkylating moiety of BD. CC-1065 and the CPI analogues do undergo ready opening of the cyclopropane ring by nucleophiles (even such weak nucleophiles as acetic acid or water) in the presence of acids (82), or in aqueous solutions at low pH (92). It is possible that certain sequences of DNA may catalyze the irreversible ring-opening reaction of CPI, by providing the prerequisite activation simultaneously with nucleophilic attack.

The recently described nucleotide-catalyzed hydrolysis of BPDE (98) provides an elegant model of how multifunctional catalysis can express an electronic and spatial "recognition" event in a very simple system. While the acidic phosphate group of the nucleotide provides general-acid catalysis for BPDE hydrolysis, large additional rate enhancements occur when simultaneous stacking interactions with the nucleotide base can occur. The magnitude of the rate enhancement depends on the base. The electron-donating ability of the base appears to stabilize the transition state for hydrolysis of BPDE, while the phosphate, when located in certain positions on the sugar, provides acidic activation.

For CPI analogues and CC-1065, we might picture an "active site", consisting of the minor groove along 5′AAA or 5′TTA duplex nucleotide sequences, which, when occupied by CPI, would position the N-3 of adenine within bonding distance of the cyclopropyl methylene carbon and, at the same time (or slightly before), would position a proton, perhaps from a phosphate group, or an electron-attracting ion such as $Mg^{2+}$ close to the carbonyl oxygen atom which becomes a phenolic group in the final adduct (Figure 9). While only minute amounts of protonated phosphates are present on DNA at neutral pH, it is at least conceivable that their strong acidity might allow them to be kinetically relevant. Thus catalysis for the hydrolysis of certain carcinogenic epoxides by phosphoric acid, at pH's that are 4 orders of magnitude above its $pK_a$, has been reported (108). Such bifunctional catalysis might account for the observed rapid nucleophilic ring opening of CPI under neutral conditions. This process also resembles suicide enzyme inactivation, in which the inhibitor is a relatively unreactive molecule until it becomes activated at the enzyme's catalytic site (109).

It would be necessary to postulate that this orientation of functionalities was sequence dependent. In this regard, it is interesting to note that, in molecular diagrams based on theoretical calculations of CC-1065, covalent adducts with several oligonucleotides showed a "stabilizing" hydrogen bond between the new phenolic group and an anionic oxygen of the phosphate on the unmodified strand, distant by two base pairs from the covalent site, "in all the optimal complexes (5'AAA, 5'TTA) obtained" (*110*). In such a model for sequence selectivity, binding would be necessary only to permit approach and desolvation of the "active site" and might be very weak, while still permitting the sequence selection of the bonding step, accounting for the behavior of (+)-A.

Thus, the main determinant of which sequences of all the possible binding regions will become covalently modified by (+)-CC-1065 is not the affinity of the various binding sites ($K_b$), but rather the energetic requirements of the bonding reaction ($k_r$). Existing evidence on CC-1065–DNA adduct local structure points to a possible role for *sequence–dependent conformational flexibility* of DNA, and the observed acceleration of CPI ring opening by DNA also suggests a possible role for *sequence-dependent catalytic activation* of the drug.

These examples illustrate the many possible mechanisms for sequence selectivity available to agents that covalently modify DNA. Table II is a tentative "classification" of some of these examples according to the degree of influence exerted by binding interactions ($K_b$) vs bonding interactions ($k_r$) on the overall sequence selectivity of the covalent reaction.

The only clear instance of exclusive selectivity control by noncovalent binding interactions is distamycin–EDTA, which is not an alkylating agent but rather mediates non-sequence-specific DNA cleavage by diffusible oxygen radical species. Such a process may generally operate for agents, like bleomycin, that cleave DNA through this type of mechanism. It appears that any alkylation reaction [recall even the reaction of guanine N-7 with dimethyl sulfate (*27*)] exerts some degree of sequence discrimination. While there may be agents for which the relative rate of alkylation at various sites may parallel the binding affinity at these sites (BD may be an example), and hence which fit the term "binding driven bonding" (*100*), most agents, including many natural products, fall into the more complex category in which both binding forces and bonding requirements play an essential role in sequence discrimination. Unraveling the molecular details of these processes promises to be a fascinating exploration into chemical evolution [e.g., CC-1065 (*86*), aflatoxin (*37*)] and into drug design [e.g., quinacrine mustard (*43*)].

## What Is the Structural, Biochemical, and Biological Significance of Sequence-Specific Covalent Modification of DNA?

The major impetus behind investigations of the sequence specificity of DNA-reactive drugs and carcinogens is the search for clues to their biological mechanisms of action. Certainly, the potential for nucleotide sequence selective recognition exists at various stages of the biochemical processes linking the chemical reaction to the biological expression. Topoisomerase II inhibitors of the 9-amino-acridine class, for example, cleave DNA at sequences that appear to be selected by the topoisomerase enzyme, rather than by the drugs themselves (*111*). The lack of correlation of mutational "hot spots" on DNA with sites of guanine C-8 reaction with *N*-acetoxy-*N*-acetyl-2-aminofluorene has been interpreted as indicating that mutation specificity depends on protein recognition of covalently modified mutation-prone sequences (*112*). In addition to the many levels of protein intervention in drug action, the correlation

**Table II. Influence of Binding ($K_b$) on Sequence Selectivity of Covalent Reactions**

| influence of binding ($K_b$) | examples |
|---|---|
| exclusive (nonselective $k_r$) | distamycin–EDTA |
| predominant (selectivity parallels $K_b$) | *N*-(bromoacetyl)distamycin (BD) |
| constraining (binding imposes limitations on bonding sites) | CC-1065 |
| | DNA-binding CPI's |
| | anthramycin |
| | quinacrine mustard |
| | aflatoxin |
| | BPDE |
| | bis(phenanthroline)-dichloro-Ru(II) |
| minimal (selectivity parallels $k_r$) | most nitrogen mustards |
| | methylnitrosourea |
| | (+)-A (nonbinding CPI's) |

of DNA–drug complexes or adducts with biological consequences is further complicated, in most cases, by the multiplicity of complexes and adducts formed.

An observation that may be of general significance is that drugs and carcinogens which bond covalently to DNA are typically far more biologically potent than related structures devoid of alkylating ability; e.g., des-ABC is 4 $\times 10^4$ less potent than (+)-ABC (Table I). The synthetic elaboration of doxorubicin to the electrophilic (cyanomorpholinyl)doxorubicin changes a moderately potent cytotoxic anthracycline into a potentially covalently bonding drug which is one thousand times more potent than the noncovalently binding drug (*113*). Potentially covalently bonding bromo derivatives of the distamycins have been synthesized that are one to two hundred times more potent than the nonbonding parent molecule (*114*). Metabolic activation is already a well-established concept in chemical carcinogenesis since the Millers in their pioneering work (*115*) demonstrated that metabolic activation of aromatic compounds to electrophilic species can occur, the products of which can then react covalently with DNA. BPDE discussed above is one example of oxidative activation of a precarcinogen benzpyrene to an alkylating species (*49*, *54*). The antitumor drug mitomycin C is, on the other hand, the prototype of "bioreductive alkylating agents" (*116*), which are metabolically reduced to give alkylating species. These examples serve to illustrate the importance of metabolic activation in bioconversion of DNA binding ligands to DNA bonding agents. It is interesting to speculate that some of the biological effects normally attributed to DNA binding mechanisms may be due to very minor amounts of DNA bonding species produced by metabolic activation.

However, even when the discussion is restricted to covalent DNA adducts, the variety of alkylation products usually formed complicates any attempted correlation. The most abundant adducts may not be the most biologically significant. From a purely experimental viewpoint, the problem is enormous.

If Nature has given us any assistance in this matter, surely the extraordinarily potent, highly sequence selective CC-1065 molecule is a part of it. Designed by evolution, it is among the most highly efficient toxins known. The selectivity of its chemical reaction with DNA allows us to ask several questions that have important implications for mechanistic understanding as well as for drug design. First, is this sequence-dependent reactivity of select adenines for bonding to CC-1065, which apparently depends at least partially upon DNA flexibility, also a general aspect of sequence recognition by DNA binding proteins? In this regard Hogan and Austin have proposed (*14*) that certain DNA sequences do have increased flexibility, and this may be important in sequence recognition by DNA binding proteins. It will be important to determine whether CC-1065 bonding sequences are also protein recognition sequences; i.e., does (+)-CC-1065 have a "fine nose" for regions of increased DNA flexibility and is it consequently a sensitive probe (exogenous substrate?) for DNA sequences recognized by certain DNA binding proteins? Intriguingly, we have previously noted that the hierarchy of Sp1 binding sites in the 21 base pair repeats of SV40 DNA is predicted by the sequence selectivity of CC-1065 in this region of SV40 DNA (83, 85). Second, if a conformational variant of DNA is entrapped by the covalent bonding of CC-1065, how does this affect DNA recognition by DNA binding proteins such as those involved in DNA repair or transcriptional control? In this regard, the CC-1065–DNA adduct is poorly recognized by the UVRABC nuclease (117), although more recent work suggests that the particular sequence alkylated by CC-1065 may be an additional factor (118). For the particular sequence examined in this study (117) nuclease scission occurs predominately on the 5' side of the CC-1065–DNA adduct rather than on both sides as is common with other DNA adducts. A possible reason for the predominantly unilateral scission is that the proposed entrapped conformational change is also preferentially on this side of the covalently bonded site (103). Last, can the sequence-specific modification of DNA by CC-1065 and its analogues be related to biological end points such as antitumor efficacy and the delayed death effects of select CC-1065 analogues (95)? In this regard, the biological properties of CC-1065 analogues vary quite widely. Is it possible that the precise tuning of DNA sequence specificity which is mediated by variation in the structure of the nonalkylating subunits (86) can account for these variations in biological effects? While such questions are not easily answered, they are of considerable current interest to us.

The recent advances in high-field NMR, computational chemistry, and molecular biology provide excellent opportunities to study both the structural and biochemical consequences of ligand binding and bonding to DNA. For example, sufficient amounts of defined DNA sequence molecules can be prepared by automated synthesis methods and then characterized by X-ray crystallography and high-field NMR. The application of high-field NMR to elucidation of the three-dimensional structure of carcinogen–DNA adducts has been recently reviewed by Harris et al (119). Recombinant DNA technology provides both the methods and materials to test both old and new concepts on how DNA-reactive ligands may exert their potent biological effects. The construction of site-directed adducts in DNA to examine structural, biochemical, and biological effects such as mutagenicity is a very important aspect of molecular toxicology and pharmacology. This has recently been insightfully reviewed by Essigmann (120). This latter technology permits the study of ternary complexes (drug–DNA–protein), a procedure that we believe will be a major contributor in the future to the elucidation of the molecular basis of carcinogen or drug action.

**Acknowledgment.** The research carried out at the University of Texas was supported by grants from the Public Health Service (CA-30349, CA-35318) and the Welch Foundation. We would be remiss if we were not to gratefully acknowledge the stimulating discussions with our past and present colleagues in Kalamazoo and Austin, from which many of the ideas and concepts presented in this review emerged. We thank Kurt Kohn for his critical reading of the manuscript and valuable suggestions for its improvement. We also thank Donna Jacobsen and Melissa Winstead for their patience in the careful preparation of the manuscript.

### References

(1) Ollis, D. L., and White, S. W. (1982) Structural basis of protein-DNA interactions. *Chem. Rev.* **87**, 981–995.

(2) von Hippel, P. H., and Berg, O. G. (1986) On the specificity of DNA–protein interactions. *Proc. Natl. Acad. Sci. U.S.A.* **83**, 1608–1612.

(3) Dabrowiak, J. C. (1983) Sequence specificity of drug–DNA interactions. *Life Sci.* **32**, 2915–2931.

(4) Hurley, L. H., and Boyd, F. L. (1987) Approaches towards the design of sequence specific drugs for DNA. *Annu. Rep. Med. Chem.* **22**, 259–268.

(5) Watson, J. D., Hopkins, N. H., Roberts, J. W., Steitz, J. A., and Weiner, A. M. (1987) *Molecular Biology of the Gene*, 4th ed., Benjamin/Cummings, Menlo Park, CA.

(6) McClarin, J. A., Frederick, C. A., Wang, B.-C., Geene, P., Boyes, H. W., Grable, J., and Rosenberg, J. M., (1986) Structure of the DNA-EcoRI endonuclease recognition complex at 3Å resolution. *Science* **234**, 1526–1541.

(7) Anderson, W. F., Ohlendorf, D. H., Takeda, Y., and Matthews, B. W. (1981) Structure of the cro repressor from bacteriophage λ and its interaction with DNA. *Nature*, **290**, 754–758.

(8) Anderson, W. F., Takeda, Y., Ohlendorf, D. H., and Matthews, B. W. (1982) Proposed α-helical super-secondary structure associated with protein–DNA recognition. *J. Mol. Biol.* **159**, 745–751.

(9) Pabo, C. O., and Lewis, M. (1982) The operator-binding domain of λ repressor: Structure and DNA recognition. *Nature* **298**, 443–447.

(10) McKay, D. B., and Steitz, T. A. (1981) Structure of catabolite gene activator protein at 2.9 Å resolution suggests binding to left-handed B-DNA. *Nature* **290**, 744–749.

(11) Steitz, T. A., Ohlendorf, D. H., McKay, D. B., Anderson, W. F., and Matthews, B. W. (1982) Structural similarity in the DNA-binding domains of catabolite gene activator and cro repressor proteins. *Proc. Natl. Acad. Sci. U.S.A.* **79**, 3097–3100.

(12) Schevitz, R. W., Orwinowski, A., Joachimiak, A., Lawson, C. L., and Sigler, P. B. (1985) The three-dimensional structure of trp repressor. *Nature* **317**, 782–786.

(13) Zhang, R., Joachimiak, A., Lawson, C. L., Schevitz, R. W., Otwinoski, Z., and Sigler, P. B. (1987) The crystal structure of *trp* aporepressor at 1.8Å shows how binding tryptophan enhances DNA affinity. *Nature* **327**, 591–597.

(14) Hogan, M. E., and Austin, R. H. (1987) Importance of DNA stiffness in protein–DNA binding specificity. *Nature* **329**, 263–266.

(15) Robbie, M., and Wilkins, R. J. (1984) Identification of the Specific sites of interaction between intercalating drugs and DNA. *Chem.-Biol. Interact.* **49**, 189–207.

(16) Kopka, M. L., Yoon, C., Goodsell, D., Pjura, P., and Dickerson, R. E., The molecular origin of DNA-drug specificity in netropsin and distamycin. *Proc. Natl. Acad. Sci. U.S.A.* **82**, 1376–1380.

(17) Wilkins, R. J. (1984) Sequence specificities in the interactions of chemicals and radiation with DNA. *Mol. Cell. Biochem.* **64**, 111–126.

(18) Pullman, B. (1984) Electrostatics and specificity in nucleic acid reactions. In *Specificity in Biological Interactions. International Symposium at the Pontifical Academy of Sciences* (Chagas, C., and Pullman, B., Eds.) pp 1–20, Vatican Press and Adenine Press, New York.

(19) Zimmer, C., and Wahnert, U. (1986) Nonintercalating DNA-binding ligands: specificity of the interaction and their use as tools in biophysical, biochemical and biological investigations of the genetic material. *Prog. Biophys. Mol. Biol.* **47**, 31–112.

(20) Wakelin, L. P. G. (1986) Polyfunctional DNA Intercalating Agents. Med. Res. Rev. **6**, 275–340.

(21) Dervan, P. B. (1987) Design of synthetic sequence specific DNA binding molecules. In *Molecular Mechanisms of Carcinogenic and Antitumor Activity. International Symposium at the Pontifical Academy of Sciences* (Chagas, C., and Pullman, B., Eds.) pp 365–384, Vatican Press and Adenine Press, New York.

(22) Lee, M., Krowicki, K., Hartley, J. A., Pon, R. T., and Lown, J. W. (1988) Molecular recognition between oligopeptides and nucleic acids: influence of van der Waals contacts in determining the 3'-terminus of DNA sequences read by monocationic lexitropsins. *J. Am. Chem. Soc.* **110**, 3641–3649.

(23) Lee, M., Chang, D.-K., Hartley, J. A., Pon, R. T., Krowicki, K.,

and Lown, J. W. (1988) Structural and dynamic aspects of binding of a prototype lexitropsin to the decadeoxyribonucleotide d-(CGCAATTGCG)$_2$ deduced from high-resolution $^1$H NMR studies. *Biochemistry* **17**, 445–455.

(24) Moser, H. E., and Dervan, P. B. (1987) Sequence-specific cleavage of double helical DNA by triple helix formation. *Science* **238**, 645–650.

(25) Saenger, W. (1983) in *Principles of Nucleic Acid Structure* (Cantor, R., Ed.) p 391, Springer-Verlag, New York.

(26) Maxam, A. M., and Gilbert W. (1980) Sequencing end-labeled DNA with base-specific chemical cleavages. *Methods Enzymol.* **65**, 499-560.

(27) Mattes, W. B., Hartley, J. A., and Kohn, K. W. (1986) Mechanism of DNA strand breakage by piperdine at sites of N7-alkylguanines. *Biochim. Biophys. Acta* **868**, 1-76.

(28) Hartley, J. A., Gibson, N. W., Kohn, K. W., and Mattes, W. B. (1986) DNA sequence selectivity of guanine-N7 alkylation by three antitumor chloroethylating agents. *Cancer Res.* **46**, 1943–1947.

(29) Friedman, T., and Brown, D. M. (1978) Base-specific reactions useful for DNA sequencing: methylene blue-sensitized photo-oxidation of guanine and osmium tetroxide modification of thymine. *Nucleic Acids Res.* **5**, 615–622.

(30) Singer, B. (1975) The chemical effects of nucleic acid alkylation and their relation to mutagenesis and carcinogenesis. *Prog. Nucleic Acids Res. Mol. Biol.* **15**, 219–284.

(31) Essigmann, J. M., Croy, R. G., Nadzan, A. M., Busby, W. F., Jr., Reinhold, V. N., Buchi, G., and Wogan, G. N. (1977) Structural Identification of the major DNA adduct formed by aflatoxin B$_1$ *in vitro*. *Proc. Natl. Acad. Sci. U.S.A.* **74**, 1870–1874.

(32) D'Andrea, A. D., and Haseltine, W. A. (1978) Modification of DNA by aflatoxin B$_1$ creates alkali-labile lesions in DNA at positions of guanine and adenine. *Proc. Natl. Acad. Sci. U.S.A.* **75**, 4120–4124.

(33) Grunberg, S. M., and Haseltine, W. A. (1980) Use of an indicator sequence of human DNA to study DNA damage by methylbis(2-chloroethyl)amine. *Proc. Natl. Acad. Sci. U.S.A.* **77**, 6546–6550.

(34) Busby, W. F., and Wogan, G. N., Jr., (1984) In *Chemical Carcinogenesis* (Searle, C. E., Ed.) Vol. 2, pp 945–1136, American Chemical Society, Washington, DC.

(35) Muench, K. F., Misra, R. P., and Humayun, M. Z. (1983) Sequence specificity of aflatoxin B$_1$–DNA interactions. *Proc. Natl. Acad. Sci. U.S.A.* **80**, 6–10.

(36) Misra, R. P., Muench, K. F., and Humayun, M. Z. (1983) Covalent and noncovalent interactions of aflatoxin with defined deoxyribonucleic acid sequences. *Biochemistry* **22**, 3351–3359.

(37) Benasutti, M., Ejadi, S., Whitlow, M. D., and Loechler, E. L. (1988) Mapping the binding site of aflatoxin B$_1$ in DNA: Systematic analysis of the reactivity of aflatoxin B$_1$ with guanines in different DNA sequences. *Biochemistry* **27**, 472–481.

(38) Menger, F. M. (1985) On the source of intramolecular and enzymatic reactivity. *Acc. Chem. Res.* **18**, 128–134.

(39) Pratt, W. B., and Ruddon, R. W. (1979) in *The Anticancer Drugs*, pp 64–97, Oxford University Press, New York.

(40) Wurdeman, R. L., and Gold, B. (1988) The effect of DNA sequence, ionic strength, and cationic DNA affinity binders on the methylation of DNA by *N*-methyl-*N*-nitrosourea. *Chem. Res. Toxicol.* **1**, 146–147.

(41) Buckley, N. (1987) A regioselective mechanism for mutagenesis and oncogenesis caused by alkylnitrosourea sequence-specific DNA alkylation. *J. Am. Chem. Soc.* **109**, 7918–7920.

(42) Mattes, W. B., Hartley, J. A., and Kohn, K. W. (1986) DNA sequence selectivity of guanine N7 alkylation by nitrogen mustards. *Nucleic Acids Res.* **14**, 2971–2987.

(43) Kohn, K. W., Hartley, J. A., and Mattes, W. B. (1987) Mechanisms of DNA sequence selective alkylation of guanine-N7 positions by nitrogen mustards. *Nucleic Acids Res.* **15**, 10531–10544.

(44) Pullman, A., and Pullman, B., (1981) Molecular electrostatic potential of the nucleic acids. *Q. Rev. Biophys.* **14**, 289–380.

(45) Zakrzewska, K., and Pullman, B. (1985) The effects of spermine binding on the reactivity of DNA towards carcinogenic alkylating agents. *J. Biomol. Struct. Dyn.* **3**, 437–444.

(46) Dipple, A., Moschel, R. C., and Hadgins, W. R. (1982) Selectivity of alkylation and aralkylation of nucleic acid components. *Drug Metab. Rev.* **13**, 249–268.

(47) Briscoe, W. T., and Cotter, L. E. (1985) DNA sequence has an effect on the extent and kinds of alkylation of DNA by a potent carcinogen. *Chem.-Biol. Interact.* **56**, 321–331.

(48) Richardson, K. K., Richardson, F. C., Crosby, R. M., Swenberg, J. A., and Skopek, T. R. (1987) DNA base changes and alkylation following *in vivo* exposure of *Escherichia coli* to N-methyl-N-nitrosourea or N-ethyl-N-nitrosourea. *Proc. Natl. Acad. Sci. U.S.A.* **84**, 344–348.

(49) Harvey, R. G., and Geacintov, N. E. (1988) Intercalation and binding of carcinogenic hydrocarbon metabolites to nucleic acids. *Acc. Chem. Res.* **21**, 66–73.

(50) Boles, T. C., and Hogan, M. E. (1986) High-resolution mapping of carcinogen binding sites on DNA. *Biochemistry* **25**, 3039–3043.

(51) Osborne, M., and Merrifield, K. (1985) Depurination of benzo[a]pyrene-diol epoxide treated DNA. *Chem.-Biol. Interact.* **53**, 183–195; and references cited therein.

(52) Wang, A. H.-J., Fugii, S., van Bloom, J. H., and Rich, A. (1982) Molecular structure of the octamer d(G-G-C-C-G-G-C-C): Modified A-DNA. *Proc. Natl. Acad. Sci. U.S.A.* **79**, 3968–3972.

(53) Lobanenkov, V. V., Plumb, M., Goodwin, G. H., and Grover, P. L. (1986) The effect of neighboring bases on G-specific DNA cleavage mediated by treatment with the anti-diol epoxide of benzo(a)pyrene *in vitro*. *Carcinogenesis* **7**, 1689–1695.

(54) Michaud, D. P., Gupta, S. C., Whalen, D. L., Sayer, J. M., and Jerina, D. M. (1983) Effects of pH and salt concentration on the hydrolysis of a benzo(a)pyrene 7,8-diol-9,10-epoxide catalyzed by DNA and polyadenylic acid. *Chem.-Biol. Interact.* **44**, 41–52.

(55) MacLeod, M. C., and Zachary, K. L. (1985) Involvement of the exocyclic amino group of deoxyguanosine in DNA-catalyzed carcinogen detoxification. *Carcinogenesis* **6**, 147–149.

(56) Tomasz, M., Lipman, R., Lee, M., Verdine, G., and Nakanishi, K. (1987) Reaction of acid-activated mitomycin C with calf thymus DNA and model guanines: Elucidation of the base-catalyzed degradation of N-7 alkylguanine nucleosides. *Biochemistry* **26**, 2010–2027.

(57) Singer, B., and Grunberger, D. (1983) *Molecular Biology of Mutagens and Carcinogens*, Plenum Press, New York.

(58) Tomasz, M., Chowdary, D., Lipman, R., Shimotakahara, S., Verio, D., Walker, V., and Verdine G. (1986) Reaction of DNA with chemically or enzymatically activated mitomycin C: Isolation and structure of the major covalent adduct. *Proc. Natl. Acad. Sci. U.S.A.* **83**, 6702–6706.

(59) Ueda, K., Morita, J., and Komano, T. (1984) Sequence specificity of heat-labile sites in DNA induced by mitomycin C. *Biochemistry* **23**, 1634–1640.

(60) Ueda, K., and Komano, T. (1984) Sequence-specific DNA damage induced by reduced mitomycin C and 7-N-(p-hydroxyphenyl) mitomycin C. *Nucleic Acids Res.* **12**, 6673–6683.

(61) Graves, D. E., Pattaroni, C., Krishnan, B. S., Ostrander, J. M., Hurley, L. H., and Krugh, T. R. (1984) The reaction of anthramycin with DNA. Proton and carbon nuclear magnetic resonance studies on the structure of the anthramycin–DNA adduct. *J. Biol. Chem.* **259**, 8202–8209.

(62) Thurston, D. T., and Hurley, L. H. (1983) A rational basis for development of antitumor agents in the pyrrolo[1,4]benzodiazepine group. *Drugs Future CIPS* **8**, 957–971; and references cited therein.

(63) Hertzberg, R. P., Hecht, S. M., Reynolds, V. L., Molineux, I. J., and Hurley, L. H. (1986) DNA sequence specificity of the pyrrolo[1,4]benzodiazepine antitumor antibiotics. Methidium-propyl-EDTA–iron(II) footprinting analysis of DNA binding sites for anthramycin and related drugs. *Biochemistry* **25**, 1249–1258.

(64) Dervan, P. B. (1986) Design of sequence specific DNA binding molecules. *Science* **232**, 464–471.

(65) Hurley, L. H., Reck, T., Thurston, D. E., Langley, D. R., Holden, K. G., Hertzberg, R. P., Hoover, J. R. E., Gallagher, G., Jr., Faucette, L. F., Mong, S.-M., and Johnson, R. K. (1988) Pyrrolo[1,4]benzodiazepine antitumor antibiotics: Relationship of DNA alkylation and sequence specificity to the biological activity of natural and synthetic compounds. *Chem. Res. Toxicol.* **1**, 258–268.

(66) Zakrzewska, K., and Pullman, B. (1986) A theoretical investigation of the sequence specificity in the binding of the antitumor drug anthramycin to DNA. *J. Biomol. Struct. Dyn.* **4**, 127–136.

(67) Dickerson, R. E., Drew, H. R., Conner, B. N., Wing, R. M., Fratini, A. Y., and Kopka, M. L. (1982) The Anatomy of A-, B-, and Z-DNA. *Science* **216**, 475–485.

(68) Barton, J. K. (1986) Metals and DNA: Molecular left-handed complements. *Science* **233**, 727–734.

(69) Sherman, S. E., Gibson, D., Wang, A. H. J., and Lippard, S. J. (1985) X-ray structure of the major adduct of the anticancer drug

Cisplatin with DNA: cis[P$^+$(NH$_3$)$_2${d(pGpG)}]. *Science* **230**, 412–417.

(70) Royer-Pokora, M., Gordon, L. K., and Haseltine, W. A. (1981) Use of exonuclease III to determine the site of stable lesions in defined sequences of DNA: the cyclobutane pyrimidine dimer and *cis* and *trans* dichlorodiammine platinum II examples: *Nucleic Acids Res.* **9**, 4595–4609.

(71) Tullius, T. D., and Lippard, S. J. (1981) cis-Diaminedichloro-platinum(II) binds in a unique manner to oligo(dG)·oligo(dC) sequences in DNA—A new assay using exonuclease III. *J. Am. Chem. Soc.* **103**, 4620–4622.

(72) Barton, J. K., Danishefsky, A. T., and Goldberg, J. M. (1984) Tris(phenanthroline)ruthenium(II): Stereoselectivity in binding to DNA. *J. Am. Chem. Soc.* **106**, 2172–2176.

(73) Barton, J. K., Goldberg, J. M., Kumar, C. V., and Turro, N. J. (1986) Binding modes and base specificity of tris(phenanthroline)ruthenium(II) enantiomers with nucleic acids: Tuning the stereoselectivity. *J. Am. Chem. Soc.* **108**, 2081–2088.

(74) Barton, J. K., and Lolis, E. (1985) Chiral discrimination in the covalent binding of bis(phenanthroline)dichlororuthenium(II) to B-DNA. *J. Am. Chem. Soc.* **107**, 708–709.

(75) Schultz, P. G., and Dervan, P. B. (1984) Distamycin and penta-N-methylpyrrolecarboxamide binding sites on native DNA. A comparison of methidiumpropyl-EDTA-Fe(II) footprinting and DNA affinity cleaving. *J. Biomol. Struct. Dyn.* **1**, 1133–1147.

(76) Baker, B. F., and Dervan, P. B. (1985) Sequence specific cleavage of double-helical DNA N-bromoacetyldistamycin. *J. Am. Chem. Soc.* **107**, 8266–8268.

(77) Hanka, L. J., Dietz, A., Gerpheide, S. A., Kuentzel, S. L., and Martin, D. G. (1978) CC-1065 (NSC-298223), a new antitumor antibiotic. Production *in vitro* biological activity, microbiological assays and taxonomy of the producing microorganism. *J. Antibiot.* **31**, 1211–1217.

(78) Chidester, C. G., Krueger, W. C., Mizsak, S. A., Duchamp, D. J., and Martin, D. G. (1981) The structure of CC-1065, a potent antitumor agent, and its binding to DNA. *J. Am. Chem. Soc.* **103**, 7629–7635.

(79) Swenson, D. H., Li, L. H., Hurley, L. H., Rokem, J. S., Petzold, G. L., Dayton, B. D., Wallace, T. L., Lin, A. H., and Krueger, W. C. (1982) Mechanism of interaction of CC-1065 (NSC 298223) with DNA. *Cancer Res.* **42**, 2821–2828.

(80) Li, L. H., Swenson, D. H., Schpok, S. L. F., Kuentzel, S. L., Dayton, B. D., and Krueger, W. C. (1982) CC-1065 (NSC-298223), a novel antitumor agent that interacts strongly with double-stranded DNA. *Cancer Res.* **42**, 999–1004.

(81) Hurley, L. H., Reynolds, V. L., Swenson, D. H., and Scahill, T. (1984) Reaction of the antitumor antibiotic CC-1065 with DNA: Structure of a DNA adduct with DNA sequence specificity. *Science* **226**, 843–844.

(82) Martin, D. G., Mizsak, S. A., and Krueger, W. C. (1985) CC-1065 transformations. *J.Antibiot.* **38**, 746–752.

(83) T. Scahill, unpublished results.

(84) Reynolds, V. L., Molineux, L. J., Kaplan, D., Swenson, D. H., and Hurley, L. H. (1985) Reaction of the antitumor antibiotic CC-1065 with DNA, location of the site of thermally induced strand breakage and analysis of DNA sequence specificity. *Biochemistry* **24**, 6228–6237.

(85) Martin, D. G., Kelly, R. C., Watt, W., Wicnienski, N., Mizsak, J. A., Nielsen, J. W., and Prairie, M. D., (1988) The absolute configuration of CC-1065 by X-ray crystallography on a derivatized chiral fragment (CPI) from the natural antibiotic. *J. Org. Chem.* **53**, 4610–4613.

(86) Hurley, L. H., Lee, C.-S., McGovren, J. P., Mitchell, M., Warpehoski, M. A., Kelly, R. C., and Aristoff, P. A. (1988) Molecular basis for the DNA sequence specificity of CC-1065. *Biochemistry* **27**, 3886–3892.

(87) Krueger, W. C., Li, L. H., Moscowitz, A. Prairie, M. D., Petzold, G., and Swenson, D. H. (1985) Binding of CC-1065 to poly- and oligonucleotides. *Biopolymers* **24**, 1549–1572.

(88) Krueger, W. C., Duchamp, D. J., Li, L. H., Moscowitz, A., Petzold, G. L., Prairie, M. D., and Swenson, D. H. (1986) The binding of CC-1065 to thymidine and deoxyadenosine oligonucleotides and to poly(dA)·poly(dT). *Chem.-Biol. Interact.* **59**, 55–72.

(89) Krueger, W. C., and Prairie, M. D. (1987) A circular dichroism study of the binding of CC-1065 to B and Z form poly(dI-5BrdC)·poly(dI-5BrdC). *Chem.-Biol. Interact.* **62**, 281–295.

(90) Theriault, N. Y., Krueger, W. C., and Prairie, M. D. (1988) Studies on the base pair binding specificity of CC-1065 to oligomer duplexes. *Chem.-Biol. Interact.* **65**, 187–201.

(91) Braithwaite, A. W., and Baguley, B. C. (1980) Existence of an extended series of antitumor compounds which bind to deoxyribonucleic acid by nonintercalative means. *Biochemistry* **19**, 1101–1106.

(92) Warpehoski, M. A., Gebbard, I., Kelly, R. C., Krueger, W. C., Li, L. H., McGovren, J. P., Prairie, M. D., Wicnienski, N., and Wierenga, W. (1988) Stereoelectronic factors influencing the biological activity and DNA interaction of synthetic antitumor agents modeled on CC-1065. *J. Med. Chem.* **31**, 590–603.

(93) Warpehoski, M. A. (1986) Total synthesis of U-71184, a potent new antitumor agent modeled on CC-1065. *Tetrahedron Lett.* **27**, 4103–4106.

(94) Kelly, R. C., Gebhard, I., Wicnienski, N., Aristoff, P. A., Johnson, P. D., and Martin, D. G. (1987) Coupling of cyclopropylpyrroloindole (CPI) derivatives. The preparation of CC-1065, ent-CC-1065, and analogs. *J. Am. Chem. Soc.* **109**, 6837–6838.

(95) Warpehoski, M. A., and Bradford, V. S. (1988) Bis-des-hydroxy, bis-des-methoxy CC-1065. Synthesis, DNA binding, and biological activity. *Tetrahedron Lett.* **29**, 131–144.

(96) Wierenga, W., Bhuyan, B. K., Kelly, R. C., Krueger, W. C., Li, L. H., McGovren, J. P., Swenson, D. H., and Warpehoski, M. A. (1986) "Antitumor activity and biochemistry of novel analogs of the antibiotic CC-1065. *Adv. Enzyme Regul.* **25**, 141–155.

(97) M. D. Prairie, unpublished results.

(98) Gupta, S. C., Iskim, N. B., Whalen, D. L., Yagi, H., and Jerina, D. M. (1987) Bifunctional catalysis in the nucleotide-catalyzed hydrolysis of (±)-7β,8α-Dihydroxy-9α,10α-epoxy-7,8,9,10-tetrahydrobenzo(a)pyrene. *J. Org. Chem.* **52**, 3812–3815.

(99) Pullman, B., and Pullman, A. (1980) in *Carcinogenesis: Fundamental Mechanisms and Environmental Effects* (Pullman, B., Ts'O., P. O. P., and Gelboin, H., Eds.) pp 55–66, D. Reidel, New York.

(100) Bogar, D. L., and Coleman, R. S. (1988) Total synthesis of (±)-N$^2$-(phenylsulfonyl)-CP1,(±)-CC-1065,(±)-CC-1065, ent-(-)-CC-1065 and the precise functional agents (±)-CPI-CDPI$_2$, (+)-CPI-CDPI$_2$ and (-)-CPI-CDPI$_2$ [(±)-(3bR*,4aS*)-, (+)-(3bR,4aS)- and (-)-(3bS,4aR)-deoxy-CC-1065]. *J. Am. Chem. Soc.* **110**, 4796–4807.

(101) Boger, D. L., and Coleman, R. S. (1988) Total Synthesis of (+) and (-) CPI-CDPI$_2$: (+)-(3bR,4aS)- and (-)-(3bS,4aR)-deoxy-CC-1065. *J. Org. Chem.* **53**, 695–698.

(102) Needham-VanDevanter, D. R., and Hurley, L. H. (1986) Construction and characterization of a site directed CC-1065 (N3-adenine) DNA adduct within a 117 bp DNA restriction fragment. *Biochemistry* **25**, 8430–8436.

(103) Hurley, L. H., Needham-VanDevanter, D. R., and Lee, C.-S. (1987) Demonstration of the asymmetric effect of CC-1065 on local DNA structure using a site-directed adduct in a 117 base pair fragment from M13mpl. *Proc. Natl. Acad. Sci. U.S.A.* **84**, 6412–6416.

(104) Suck, D., and Oefner, C. (1986) Structure of DNase I at 2.0 Å resolution suggests a mechanism for binding to and cutting DNA. *Nature* **321**, 620–624.

(105) Suck, D., Lahm, A., and Oefner, C. (1988) Structure refined to 2Å of a nicked DNA octanucleotide complex with DNase I. *Nature* **332**, 464–468.

(106) McGovren, J. P., Clarke, G. L., Pratt, E. A., and DeKoning, T. F. (1983) Preliminary toxicity studies with the DNA-binding antibiotic, CC-1065. *J. Antibiot.* **37**, 63–70.

(107) M. A. Warpehoski, unpublished results.

(108) Sayer, J. M., Yagi, H., Croisy-Delcey, M., and Jerina, D. M. (1981) Novel bay-region diol epoxides from benzo(c)-phenanthrene. *J. Am. Chem. Soc.* **103**, 4970–4972.

(109) Abels, R. H., and Maycock, A. L. (1976) Suicide enzyme inactivators. *Acc. Chem. Res.* **9**, 313–319.

(110) Zakrzewska, K., Randrianarivelo, M., and Pullman, B. (1987) Theoretical study of the sequence specificity in the covalent binding of the antitumor drug CC-1065 to DNA. *Nucleic Acids Res.* **15**, 5775–5785.

(111) Pommier, Y., Covey, J. Kerrigan, D., Mattes, W., Markovits, J., and Kohn, K. W. (1987) Role of DNA intercalation in the inhibition of purified mouse leukemia (L1210) DNA topoisomerase II by 9-aminoacridines. *Biochem. Pharmacol.* **36**, 3477–3486.

(112) Fuchs, R. P. P. (1984) DNA binding spectrum of the carcinogen N-acetoxy-N-2-acetylaminofluorene significantly differs from the mutation spectrum. *J. Mol. Biol.* **177**, 173–180.

(113) Sikic, B. I., Ehsan, M. N., Harker, W. G., Friend, N. F., Brown, B. W., Newman, R. A., Hacker, M. P., and Acton, E. M. (1985) Dissociation of antitumor potency from anthracycline cardiotoxicity in a doxorubicin analog. *Science* **228**, 1544–1546.

(114) Giuliani, F. C., Barberi, B., Biasoli, L., Geroni, C., Menozzi, M., and Mongelli, N. (1988) Distamycin A derivatives: *In vitro* and *in vivo* activity of a new class of antitumor agents. *Proc. AACR* **29**, 330.

(115) Miller, E. C., and Miller, J. A. (1981) Mechanisms of chemical carcinogenesis. *Cancer* **47**, 1055–1065.

(116) Moore, H. W. (1977) Bioactivation as a model for drug design bioreductive alkylation. Science **197**, 527–532.

(117) Tang, M.-S., Lee, C.-S., Doisy, R., Ross, L., Needham-Van-Devanter, D. R., and Hurley, L. H. (1988) Recognition and repair of the CC-1065-(N3-adenine)-DNA adduct by the UVRABC nucleases. *Biochemistry* **27**, 893–901.

(118) Selby, C. P., and Sancar, A. (1988) ABC excinonuclease incises both 5′ and 3′ to the CC-1065–DNA adduct and its incision activity is stimulated by DNA helicase II and DNA polymerase 1. *Biochemistry* **27**, 7184–7188.

(119) Harris, T. M., Stone, M. P., and Harris, C. M. (1988) Applications of NMR spectroscopy to studies of reactive intermediates and their interactions with nucleic acids. *Chem. Res. Toxicol.* **1**, 79–96.

(120) Basu, A. K., and Essigmann, J. M. (1988) Site-specifically modified oligodeoxynucleotides as probes for the structural and biological effects of DNA damaging agents. *Chem. Res. Toxicol.* **1**, 1–18.

# Chapter 18

# Covalent Bonding of the Prosthetic Heme to Protein: A Potential Mechanism for the Suicide Inactivation or Activation of Hemoproteins

Yoichi Osawa* and Lance R. Pohl

*Laboratory of Chemical Pharmacology, National Heart, Lung, and Blood Institute, National Institutes of Health, Building 10, Room 8N110, Bethesda, Maryland 20892*

Reprinted from *Chemical Research in Toxicology*, Vol. 2, No. 3, May/June, 1989

## Introduction

Since the first description of the phenomenon almost two decades ago, a vast array of enzymes have been documented to undergo suicide or mechanism-based inactivation (1–3). Among these are the P-450 cytochromes (P-450) (4), a family of hemoprotein monooxygenases that play a vital role in the metabolism of a variety of xenobiotics, including drugs and environmental pollutants, as well as endogenous compounds, such as steroids, prostaglandins, and fatty acids. The metabolites of P-450 are usually stable hydroxylated products; in certain cases, however, depending on the structure of the substrates, highly reactive intermediates are formed, which are deleterious to the enzyme.

During the past decade two pathways for the mechanism-based destruction of liver microsomal P-450 cytochromes have been extensively characterized (Scheme I). One pathway involves the modification of the heme moiety to products that can dissociate from the protein; this is best exemplified by the formation of the well-documented green pigments that result from N-alkylation of the iron porphyrin apparently by reactive radical or cation radical metabolites of the substrate (for review see ref 4). In the case of 3,5-dicarbethoxy-2,6-dimethyl-4-ethyl-1,4-dihydropyridine (DDEP), it is thought that this process occurs by a one-electron oxidation of the nitrogen of DDEP by cytochrome P-450 to yield a radical cation intermediate that subsequently aromatizes and releases an ethyl radical (4–6), which ethylates the pyrrole nitrogen of the prosthetic heme moiety (5).

The second pathway (Scheme I) involves the covalent modification of the protein moiety by a reactive metabolite.

**Scheme I. Mechanisms for Suicide Inactivation**[a]

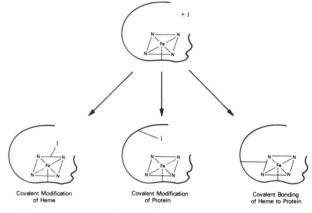

Covalent Modification of Heme     Covalent Modification of Protein     Covalent Bonding of Heme to Protein

[a] I represents a reactive intermediate.

The best example is the suicide inactivation caused by chloramphenicol, an antibiotic that is hydroxylated, in part, to form an oxamyl chloride intermediate that acylates lysine residue(s) (7, 8). This modification seems to interfere with the electron transport from NADPH-cytochrome P-450 reductase to the cytochrome (9). Other examples include the suicide inactivation of cytochrome P-450b, the major phenobarbital-inducible form, by N-methylcarbazole (10) and lauric acid hydroxylase by an acetylenic fatty acid (11).

More recently, a third mechanism (Scheme I) was discovered, which involves the irreversible binding of the prosthetic heme to the protein presumably involving a covalent bond (12). This pathway occurs with a variety

**Table I. Destruction of Cytochrome P-450 and Irreversible Binding to Protein of [³H]Heme Radiolabel in Rat Liver Microsomes 1 h after CCl₄ Treatment[a]**

| treatment | % loss of cytochromes P-450 | irreversible binding of $^3H$ label to protein, % of total microsomal $^3H$ label |
|---|---|---|
| control | 0 | 4 |
| CCl₄ | 64 | 28 |

[a] Rats were pretreated with phenobarbital (80 mg/kg) for 4 days and then administered $NaH^{14}CO_3$ and [3,5-³H]ALA to radiolabel microsomal protein and heme, respectively (21). Two rats received CCl₄ (26 mmol/kg, ip in 50% sesame oil) and two rats sesame oil as control. The radioactivity (dpm/mg of protein; average of two values) of the microsomal suspensions prepared from these rats was as follows: control group, ³H label 280 000, ¹⁴C label 35 000; CCl₄ group, ³H label 213 000, ¹⁴C label 32 000. Microsomes were precipitated with 5 volumes of acetone containing 0.5 M HCl and washed with the same solution. The numbers represent the average of single determinations (22).

of structurally diverse compounds, including those that cause N-alkylation of the prosthetic heme.

In this perspective, we will discuss the following: (1) describe how this pathway was first discovered with the hepatotoxic agent CCl₄; (2) illustrate that the formation of heme–protein adducts can be mediated by a variety of other xenobiotics, as well as endogenous compounds such as hydrogen peroxide and linoleic hydroperoxide; (3) describe how this pathway can occur with other hemoproteins; (4) propose possible mechanisms for the formation of heme–protein adducts; (5) speculate on the physiological and pathological consequences of such an alteration of the heme prosthetic group of hemoproteins. It will be illustrated that heme–protein adduct formation does not necessarily produce an inactive protein, but instead may result in the formation of an activated protein, a process we define as "suicide activation".

## Covalent Bonding of the Prosthetic Heme to P-450 Cytochromes

**Carbon Tetrachloride.** This hepatotoxic agent is one of the most extensively studied xenobiotics and is a classic example of a compound that causes free-radical-mediated tissue injury (13, 14). The initial event involved in its hepatotoxicity is the metabolism to the trichloromethyl radical by P-450 cytochromes (14–16). One consequence of this reaction is the inactivation of the P-450 and formation of uncharacterized heme breakdown products (14, 17–20). The mechanism for the CCl₄-induced metabolism of heme is independent of the heme oxygenase pathway of heme catabolism, since CO is not formed from the heme during this process (18).

Our interest in the metabolism and toxicity of various halogenated hydrocarbons led to the observation that, after incubation of liver microsomal cytochrome P-450 preparations with CCl₄, a reddish brown chromophore persisted in the protein fraction after precipitation of the reaction mixture with acidic acetone. This finding indicated that heme products were irreversibly bound to protein. The definitive experiments to demonstrate that heme was metabolized to protein-bound adducts were carried out with cytochrome P-450 containing radiolabeled heme. This was accomplished by administering radiolabeled δ-aminolevulinic acid (ALA), a precursor to heme, to phenobarbital-pretreated rats (21). The P-450 apoprotein was also labeled in vivo by $NaH^{14}CO_3$ so that simultaneous measurements of radiolabeled protein and heme could be made (21). Administration of CCl₄ to these prelabeled rats caused an extensive loss of P-450 as measured by its fer-

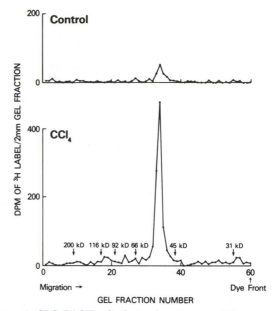

**Figure 1.** SDS–PAGE radioelectrophoretograms of 40-µg samples of liver microsomes from control and CCl₄-treated rats (22). Rats were administered [³H]ALA to radiolabel microsomal heme and given CCl₄ as described in Table I.

rous–carbonyl complex (Table I) (22). Precipitation and extensive washing of the liver microsomal protein from these rats with acetone–HCl, which removes noncovalently bound heme from hemoproteins, revealed that 28% of the total heme was irreversibly bound to the protein fraction (Table I) (22). This indicated that approximately half of the destroyed P-450 could be accounted for by covalent modification of heme to protein. Analysis of the labeled microsomal proteins by sodium dodecyl sulfate polyacrylamide gel electrophoresis (SDS–PAGE) and radioisotope counting of gel slices revealed that the protein-bound heme-derived material was confined to a region corresponding to molecular masses of 47–56 kDa, the range where P-450 cytochromes dominate (Figure 1) (22). Interestingly, a small amount of heme-derived material was found to be irreversibly bound even in untreated animals (Figure 1, control), suggesting an endogenous level of these adducts.

The target of the irreversible binding of heme was confirmed to be cytochrome P-450 in immunoprecipitation studies. Immunoprecipitation of P-450b with specific antibody followed by SDS–PAGE and radioisotope counting revealed that heme was bound to the P-450b protein in liver microsomes of both control (Figure 2, panels A and B) and CCl₄-treated rats (Figure 2, panels C and D) (12). The amount of irreversibly bound [³H]heme label, however, was at a much higher level in the livers of the CCl₄-treated rats; in this case 39% of the total heme irreversibly bound to microsomal protein was bound to this isozyme (12). Moreover, it appears that the activation of heme and its subsequent irreversible binding occurs before the heme dissociates from the apoprotein. This was shown by using a reconstituted system containing purified cytochrome P-450b, which contained a [³H]heme prosthetic group, NADPH-cytochrome P-450 reductase, NADPH, and CCl₄. After incubation for 5 min, the reaction mixture was separated by SDS–PAGE and gel slices were analyzed for radioactivity (Figure 3). Under these conditions, approximately 90% of the cytochrome P-450 was lost, as measured by the absorbance of the ferrous–CO complex, and 44% of the total heme was irreversibly bound to P-450;

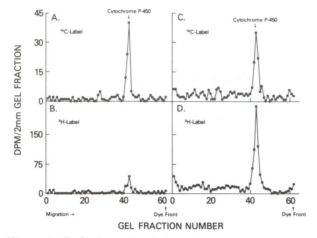

**Figure 2.** Radioelectrophoretograms of anti-P-450b IgG immunoprecipitates of liver microsomes from control (panels A and B) and CCl$_4$-treated (panels C and D) rats (*12*). Rats were treated as described in Table I.

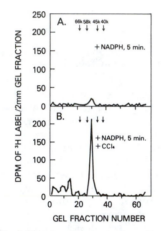

**Figure 3.** Radioelectrophoretograms of reaction mixtures containing a reconstituted system of purified [$^3$H]heme-labeled P-450b (6060 dpm/nmol of P-450) and NADPH-cytochrome P-450 reductase. Panel A, control; panel B, CCl$_4$-treated. The reaction mixture contained 2.0 nmol of P-450b, 1.0 nmol of NADPH-cytochrome P-450 reductase, 60 μg of dilauroylglycero-3-phosphocholine, 200 μg of sodium cholate, and an NADPH-regenerating system consisting of 0.8 mM NADP$^+$, 8.0 mM glucose 6-phosphate, 6.0 mM MgCl$_2$, and 0.8 unit of glucose-6-phosphate dehydrogenase, in a total volume of 0.8 mL of 50 mM Hepes, pH 7.4. The incubations were at 37 °C for 5 min in sealed vials, and CCl$_4$ was 5.0 mM. A portion (80 μL) was taken and the protein precipitated and washed with HCl–acetone as before (Table I). SDS–PAGE was run on 200 μL of sample, and 2-mm gel slices were taken and counted. A portion was also taken for determination of cytochrome P-450. Rats were administered [$^3$H]ALA to radiolabel microsomal heme as described in Table I, and cytochrome P-450 and NADPH-cytochrome P-450 reductase was purified by published methods (*23, 24*).

very little if any of heme-derived products were irreversibly bound to the flavoprotein, which has a molecular mass of approximately 76 000.

In order to learn more about the mechanism of inactivation and formation of heme–protein adducts, liver microsomal suspensions prepared from rats pretreated with [$^3$H]ALA were treated with CCl$_4$ under various conditions (Table II) (*25*). The irreversible binding of heme to protein required metabolic activation of CCl$_4$ since the reaction was dependent on NADPH, a cofactor of the NADPH-cytochrome P-450 reductase. Since it can occur in the absence of molecular oxygen, at least a part of the adduct formation is independent of lipid peroxidation, as was

**Table II. Destruction of Cytochrome P-450 Heme, Irreversible Binding to Protein of [$^{14}$C]Heme Radiolabel, and Malondialdehyde Formation in Incubations of Rat Liver Microsomes**[a]

| incubation conditions | % loss of cytochrome P-450 heme[b] | irreversible binding of $^{14}$C label to protein, % of initial $^{14}$C dpm added to incubations[c] | MDA,[b] μM |
|---|---|---|---|
| aerobic | | | |
| −NADPH | 0 | 2 | 0.03 |
| +NADPH | 10 | 6 | 0.40 |
| +CCl$_4$ | 3 | 2 | <0.03 |
| +CCl$_4$ + NADPH | 69 | 31 | 5.80 |
| anaerobic | | | |
| −NADPH | 0 | 3 | 0.08 |
| +NADPH | 4 | 3 | 0.28 |
| +CCl$_4$ | 4 | 3 | 0.19 |
| +CCl$_4$ + NADPH | 71 | 37 | 0.55 |

[a] The incubation mixtures contained 12 mg of microsomal protein, 1.5 mM EDTA, and 2 mM NADPH in a total volume of 3.0 mL in 100 mM potassium phosphate, pH 7.4 (*25*). CCl$_4$ was 5 mM. Incubations were for 20 min. Loss of P-450 heme was measured as its pyridine hemochrome complex. [b] Average of four determinations. [c] Average of two determinations.

clearly confirmed by the measurements of malondialdehyde formation. This finding implicated the trichloromethyl radical as the initiator of covalent heme binding, since it had been established that CCl$_4$ is reductively dechlorinated by P-450 cytochromes to trichloromethyl radical, in a reaction that does not require molecular oxygen (*15, 16*). It also suggested that other agents that could be metabolized to radical intermediates might also mediate the irreversible binding of the heme to protein.

**Other Xenobiotics.** To date, 18 compounds that inactivate P-450 cytochromes have been documented by various investigators to produce heme-derived protein adducts (Chart I) (*12, 22, 25–34*). Only in the cases of CCl$_4$, HOOH, and 2-isopropyl-4-pentenamide (AIA), however, has the heme adduct been shown conclusively to be bound to the protein moiety of the P-450 cytochrome. Many of the same compounds also cause the formation of green pigments. When AIA was incubated with liver microsomes prepared from phenobarbital-treated rats, the amount of heme found in the green pigment fraction containing N-alkylated porphyrins was about equal to that found in the heme-derived protein adducts; together, these products accounted for virtually all of the lost heme (*28*). SDS–PAGE and immunoprecipitation experiments on microsomal suspensions, similar to those carried out with CCl$_4$, have further indicated that cytochrome P-450 is also the target for heme adduct formation in the reaction with AIA (*28*). Indeed, immunoprecipitation of P-450b revealed that 64% of the total [$^3$H]heme label bound to protein was associated with this isozyme (*28*). Similar results were obtained when AIA was incubated for 5 min with the reconstituted enzyme system containing [$^3$H]heme-labeled P-450b, NADPH-cytochrome P-450 reductase, and NADPH (Figure 4). Under these conditions approximately 19% of the P-450 that was destroyed could be accounted for as heme irreversibly bound to the protein moiety. As seen above with CCl$_4$, the majority of the heme was bound to the P-450 cytochrome while the flavoprotein reductase was not appreciably altered, if at all. When AIA was administered to rats, approximately 6% of the total liver microsomal heme was detected as being irreversibly bound to protein (*31*). In addition to the two pathways of cytochrome P-450 heme destruction, the covalent modification

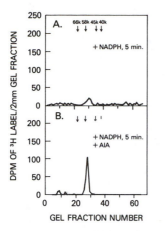

**Figure 4.** Radioelectrophoretograms of AIA-treated reaction mixtures containing purified [³H]heme-labeled P-450b, NADPH-cytochrome P-450 reductase, and NADPH. Panel A, control; panel B, AIA-treated (10 mM). The methods are the same as in Figure 3.

**Chart I. Structures of Compounds That Cause the Irreversible Binding of the Prosthetic Heme to Protein**

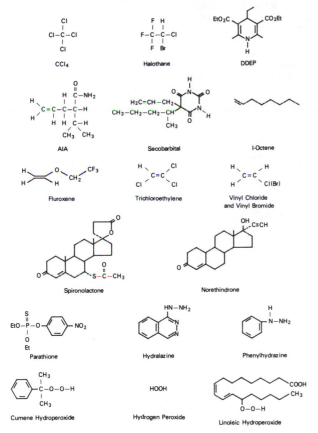

tered to rats pretreated with dexamethasone, approximately 8% of the liver microsomal heme was irreversibly bound to proteins (*32*). Whether the P-450 cytochromes were the actual targets of heme alkylation in these studies has not been definitively established, but on the basis of the studies with CCl₄ and AIA this hypothesis seems reasonable.

The antimineralocorticoid spironolactone has been shown to mediate the inactivation of P-450 cytochromes in the liver, adrenal, and testis (*29, 36*). When liver microsomal suspensions from dexamethasone-pretreated rats were incubated with spironolactone, the amount of the heme irreversibly bound to microsomal protein could account for about 54% of the P-450 destroyed (*29*). Another 10% of the P-450 destroyed could be accounted for in the fraction containing polar heme metabolites, presumably mono- and dipyrrole derivatives (*29*). It appears that significant amounts of the heme products formed in this reaction remain to be identified.

Secobarbital, a sedative–hypnotic agent, has been shown to inactivate cytochrome P-450b, the major phenobarbital-inducible form in rat liver microsomes (*34*). When secobarbital was administered to rats, the formation of green pigments accounted for 24% of the P-450 destroyed, whereas heme–protein adducts accounted for another 7% of the destroyed cytochrome P-450. The fate of the remaining 69% of the altered prosthetic heme was not determined. It was additionally found that approximately 0.45 nmol of drug was bound to microsomal protein for each nmol of P-450 destroyed. This barbiturate is another example of a compound that may inactivate P-450 cytochromes by all three pathways described in Scheme I.

Similarly, incubation of microsomal suspensions from phenobarbital-treated rats with norethindrone, hydralazine, or phenylhydrazine was found to cause a decrease in the amount of spectral cytochrome P-450 of 16, 36, and 76%, respectively; concomitantly, 56, 36, and 30%, respectively, of the destroyed P-450 heme became irreversibly bound to microsomal protein (*28*). Fluroxene, 1-octene, vinyl bromide, vinyl chloride, trichloroethylene, and parathion, under similar conditions, caused the irreversible binding of heme to microsomal protein, as well as the formation of propentdyopent and maleimide degradation products, albeit at a lower level than that bound to protein (*30*). Only small amounts of bilirubin or biliverdin were detected. This finding, as well as that with DDEP (*32*) and spironolactone (*29*), indicates that degradation products resulting from the cleavage of the heme ring represent only a minor pathway for the xenobiotic-mediated inactivation of cytochrome P-450.

**Peroxides.** The marked autoinactivation of P-450 cytochromes is well recognized (*26, 27, 37, 38*). For example, when [³H]heme-labeled P-450b was incubated with NADPH-cytochrome P-450 reductase and NADPH, 83% of the P-450 was destroyed. Under these conditions 36% of the total heme was found to be irreversibly bound to the P-450 (Figure 5, panels A and B), confirming the results of Guengerich (*26*), who was the first to demonstrate the formation of heme–protein adducts. The prosthetic heme of P-450b was also shown to form water-soluble mono- and dipyrrole products, but at most these products accounted for only 15% of the heme destroyed (*26, 27*). In experiments with microsomal suspensions, heme destruction was only observed in the presence of sodium azide, which inhibits catalase (*26, 27*). This finding suggests that hydrogen peroxide, resulting from the oxidase activity of P-450 cytochromes, can mediate these transformations.

of heme (*35*) and the heme bonding to the protein, AIA has also been reported to irreversibly bind to the protein moiety of cytochrome P-450 (*31*). Thus AIA may inactivate P-450 cytochromes by all three of the pathways described in Scheme I.

DDEP, which destroys P-450 cytochromes in part by forming *N*-ethylporphyrins (*4–6*), also produces irreversibly bound heme adducts. In fact, the heme adducts formed in vitro accounted for 61% of the destroyed P-450 in microsomes from phenobarbital-treated rats (*28*) and about 73% of the P-450 lost in liver microsomes from dexamethasone-treated rats (*32*). When DDEP was adminis-

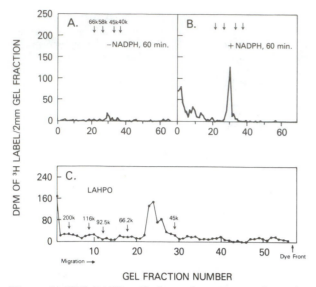

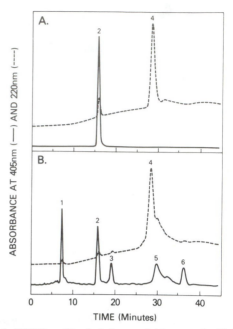

**Figure 5.** SDS–PAGE radioelectrophoretograms of reaction mixtures. Panel A, control; panel B, treated with NADPH (2 mM); panel C, treated with linoleic hydroperoxide (200 μM). The conditions used for experiments reported in panels A and B with the reconstituted system containing purified P-450b were as described in Figure 3, except that the reactions were allowed to proceed for 60 min. Experiments reported in panel C were performed with liver microsomes (8.0 nmol of cytochrome P-450) suspended in 100 mM potassium phosphate, pH 7.4, in a total volume of 2.0 mL. Liver microsomes were prepared from rats pretrerated with [³H]ALA as previously described in Table I. The reaction mixture was incubated for 40 min at 37 °C. A sample of 40 μg (1000 dpm) of microsomal protein was analyzed by SDS–PAGE.

In the study of the mechanism of the peroxidase activity associated with P-450 cytochromes with cumene hydroperoxides, it was noted that the enzyme was unstable and that bleaching of the chromophore of heme occurred (*39*, *40*). It was subsequently shown that during this process a major product was protein-bound heme (*30*, *32*). This was also demonstrated by incubating linoleic hydroperoxide with liver microsomes from phenobarbital-treated rats (Figure 5, panel C). Under these conditions, 58% of the P-450 that was lost could be accounted for by heme products irreversibly bound to protein. Although most of the heme products appeared to be associated with the P-450 region of the gel, a significant amount of the label was found in the high molecular weight region of the gel, which likely corresponds to polymeric products. Perhaps the interaction of endogenous peroxides, such as hydrogen peroxide and lipid hydroperoxides, with P-450 cytochromes in vivo may explain, at least in part, the low levels of heme-derived protein adducts found in control microsomes (see controls in Figures 1–4). The irreversible binding of heme to protein may also play a role in the destruction of P-450 caused by lipid peroxidation (*41–44*).

## Other Hemoproteins

**BrCCl₃.** Because of the complexity of the reactions that occur with cytochrome P-450 enzyme systems, a model system for the formation of heme–protein adducts has been developed. Recently, it has been found that the reductive metabolism of BrCCl₃ by myoglobin, under anaerobic conditions, leads to formation of heme products bound to the protein as well as those that could be dissociated from the protein (Figure 6) (*45*). Panel A shows the HPLC profile of untreated myoglobin; the major fraction with absorption at 220 nm (fraction 4) is native apomyoglobin, and that with absorption at 405 nm (frac-

**Figure 6.** HPLC profile of whale myoglobin treated with BrCCl₃ (*45*). Panel A, untreated myoglobin: panel B, myoglobin treated with BrCCl₃. A reaction mixture containing 140 μM myoglobin, 2.5 μM FMN (flavin mononucleotide), 10 mM EDTA, and 75 mM potassium phosphate (pH 7.4) in a total volume of 2.0 mL was placed in a cuvette stoppered with a rubber septum. The mixture was made anaerobic, and myoglobin was photoreduced to the ferrous state. BrCCl₃ was added in large excess (3 mM). After 60 min, the reaction mixture was applied directly to a C4 column (Bio-Rad HiPore 0.46 × 25 cm) at a flow rate of 1.0 mL/min. The mobile phase was water (A) vs acetonitrile/2-propanol, 1:1 (B), with 0.1% TFA throughout. A linear gradient from 35% B to 50% B (0.6%/min) and then to 55% B (0.1%/min) was used. The eluting products were detected by their absorption at 405 and 220 nm.

tion 2) is the prosthetic heme group that has dissociated from the protein under the acidic conditions of the chromatography. After the reaction of myoglobin with BrCCl₃, four new fractions are evident (panel B). Fractions 1, 3, and 6 have been identified as β-carboxyvinyl, α-hydroxy-β-(trichloromethyl)ethyl, and α,β-bis(trichloromethyl)ethyl heme derivatives of the I-vinyl group, respectively. Fraction 5 represents a heme product irreversibly bound to the protein, presumably by a covalent bond, because it absorbed at both 405 and 220 nm and extraction of the reaction mixture with acidic 2-butanone (*46*) left only fraction 5 and fraction 4 (apomyoglobin) in the aqueous phase. Moreover, treatment of the reaction mixture with trypsin resulted in the loss of only fraction 5 and apomyoglobin (fraction 4) (results not shown).

**Peroxides.** In a recent preliminary study, the reaction of myoglobin and HOOH has been shown to lead to the formation of a heme–protein adduct (*47*). A similar reaction occurred with hemoglobin and HOOH (Figure 7). Panel B shows the HPLC profile of untreated hemoglobin with the major fraction absorbing at 405 nm representing the dissociated prosthetic group and the two major fractions absorbing at 220 nm representing the α- and β-subunits of the hemoglobin tetramer (*48*). A small amount of heme chromophore was associated with the polypeptide fractions, perhaps representing an endogenous level of heme–protein adducts. After a 10-fold excess of HOOH was added to hemoglobin, an increased amount of heme product became associated with the protein fraction (panel A). Alteration of the protein was also evident by the profile of the peptide fractions absorbing at 220 nm, likely due

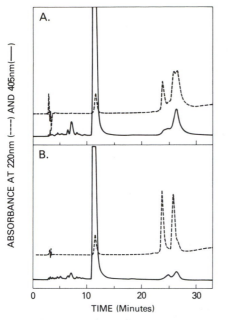

**Figure 7.** HPLC profile of hemoglobin (bovine) treated with hydrogen peroxide. Panel A, HOOH treated; panel B, control. Hemoglobin (15 μM) was treated with HOOH (150 μM) for 30 min at room temperature. A portion of the reaction mixture (11.3 nmol of hemoglobin) was applied to a C4 column (Bio-Rad HiPore 0.46 × 25 cm) at a flow rate of 1.0 mL/min. The mobile phase was water (A) vs acetonitrile (B) with 0.1% TFA throughout. The separation procedure was similar to that described previously (48). A nonlinear gradient (no. 8 of Waters Systems) from 36 to 51% B over 20 min and then a linear gradient to 75% B over the next 5 min was run. The eluting products were detected by their absorption at 405 and 220 nm.

to peroxide-mediated cleavage of the protein (49). Only small amounts of nonbound heme-derived products appeared to be formed in this reaction, in contrast to the reaction of myoglobin with BrCCl₃ (Figure 6).

Various other hemoproteins, such as prostaglandin H synthase (50, 51), lactoperoxidase (52, 53), and myeloperoxidase (54, 55), have been reported to undergo inactivation by oxidative damage due to HOOH. It remains to be determined whether these reactions also result in the alteration of the prosthetic group to form protein-bound adducts.

## Mechanism

Because of the diverse structures of the compounds in Chart I that produce heme–protein adducts a unifying mechanism is not readily apparent. One possible common property appears to be their metabolism to form radical intermediates (Table III) (4–6, 15–16, 39, 40, 56–64). Radical metabolites could conceivably promote the formation of heme–protein adducts by two general pathways (Scheme II). If the radical reacted with the heme prosthetic group, an activated radical or, upon delocalization of the electron to the heme iron, a cation species could be formed (path a). The activated cation heme derivative could then covalently bind to an amino acid side chain of histidine, tyrosine, lysine, cysteine, serine, or threonine. This schema would be consistent with the products isolated from the reaction of myoglobin with BrCCl₃ (Scheme III) (45). In this reaction, the first step involves the reductive dehalogenation of BrCCl₃ to the trichloromethyl radical, similar to that found for the activation of CCl₄ or BrCCl₃ by P-450 cytochromes (15, 16, 57). The trichloromethyl radical next appears to add to

**Table III. Radical Intermediates That Have Been Proposed To Form in the Metabolism of Various Compounds**

| compound | radical intermediate | ref |
|---|---|---|
| CCl₄ and BrCCl₃ | ·CCl₃ | 15, 16 |
| halothane | CF₃C·HCl | 56, 57 |
| olefins and acetylenes | PPIXFeO—C—C· | 4, 60, 63 |
| | PPIXFeO—C=C· | |
| DDEP | CH₃CH₂· | 4–6 |
| hydrazines | RNHNH₂ → R· | 4, 59, 61, 62, 64 |
| peroxides | RO· | 39, 40, 58 |

**Scheme II. Proposed Mechanisms for the Covalent Bonding of Heme to Protein**

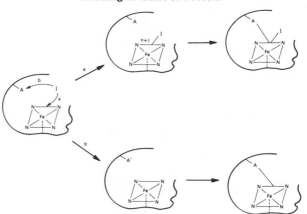

the β-carbon of the I-vinyl group followed by delocalization of the electron to the ferric iron to form a cationic species (panel A). Addition of water yields the trichloromethyl alcohol adduct, which has been characterized (compound **3**). The cationic species or perhaps even compound **3** might undergo desaturation or dehydration, respectively, to form the proposed vinyl intermediate A. Another equivalent of BrCCl₃ appears to be reductively activated by this ferrous intermediate to yield the bis(trichloromethyl) adduct compound **6**, which has also been characterized. The vinyl intermediate A might also undergo a peripheral reductive dechlorination to form a carbon-centered radical, which after delocalization of the electron to the iron would form a dichloroheme cationic species (panel B). Addition of water to this intermedite would yield an unstable dichloro alcohol species. Dehydrochlorination would give an acyl chloride, and hydrolysis of this species would yield the acrylic acid adduct compound **1**, which has also been characterized. The acyl chloride would be a good candidate for the intermediate that covalently binds to protein. However, it has been found by peptide mapping and mass spectrometry that an intact heme moiety containing a carbon and two chlorine atoms from BrCCl₃ is covalently bound to the proximal histidine (unpublished results). This finding is consistent with the activated heme derivative being the dichloroheme cationic species.

In light of the above finding of an intact heme moiety bound to the protein, it should be noted that, in all studies with P-450 cytochromes discussed earlier, spectral measurements were used to quantify the heme loss. Clearly, this may result in an underestimation of the amounts of P-450 heme destroyed if intact heme products with differing absorptivities are formed.

The second possible pathway for the formation of protein-bound heme adducts involves the initial attack of the radical metabolite on the protein (Scheme II, path b), to produce an amino acid radical, which can then react covalently with the heme moiety. The reaction of hydrogen peroxide with the ferric form of myoglobin or hemoglobin

**Scheme III. Proposed Mechanism for the Formation of Altered Heme Products of Myoglobin**

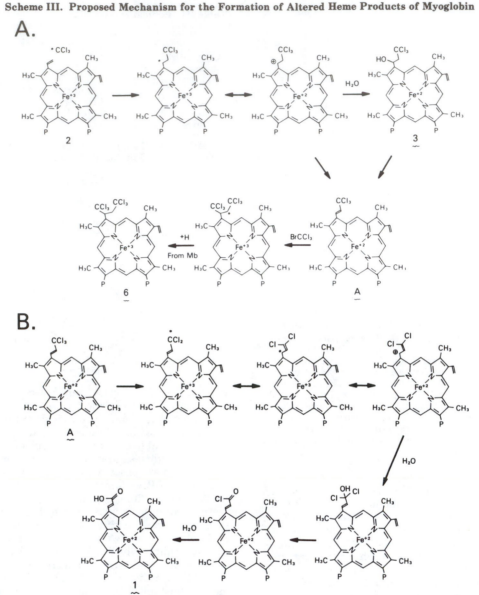

has been postulated to result in the formation of a $Fe^{4+}=O$ species and a protein radical (65–67). The formation of a protein radical is substantiated by EPR studies (68, 69), and a tyrosine residue has been implicated as the site of the radical (67). Moreover, a tyrosine residue has been tentatively assigned as the site of heme attachment in the reaction of myoglobin with HOOH (47).

The site of attack of either the protein or metabolite radical on the heme moiety leading to the covalent binding will likely be dependent upon the reactivity of the radical and the geometry of the protein around the prosthetic group. In addition to the vinyl group, radicals generated from the metabolism of various xenobiotics by hemoproteins are also known to covalently interact at the pyrrole nitrogen or the meso carbon, or may abstract a hydrogen atom from a methyl group of the prosthetic heme (5, 61, 70).

### Suicide Activation: A Potential Biological Effect of Heme-Derived Protein Adduct Formation

The formation of protein-bound heme adducts is routinely discussed in light of the inactivation process. These products, however, may play an important role in the activation of a protein for proteolytic attack or in the formation of a more reactive catalyst, both of which would eventually lead to the demise of the hemoprotein.

**Possible Marker for Proteolysis and Turnover.** It is known that the administration of $CCl_4$, AIA, or DDEP to rats results not only in the destruction of the heme moiety of P-450 cytochromes but also in the loss of the protein moiety of specific P-450 cytochromes from the endoplasmic reticulum, presumably due to the enhanced proteolysis of the protein (32, 71–74). For example, a major phenobarbital-inducible P-450 isozyme, which is a form known to metabolize $CCl_4$ to the trichloromethyl radical (16), has been shown to be selectively lost as an early and specific consequence of $CCl_4$ administration to rats (71). It has also been found that damage to other hemoproteins, such as hemoglobin and prostaglandin H synthase, by oxidants, such as hydrogen peroxide and phenylhydrazine, results in the enhanced proteolysis of the protein (51, 75, 76). Although it has been proposed that the increased rate of degradation of these proteins is due to either oxidative damage (51, 75–82) or the formation of apoprotein after alteration of the heme moiety (72, 74, 83), another possible

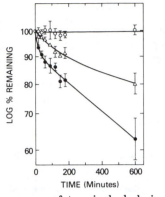

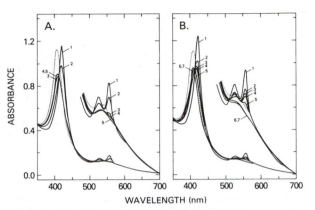

**Figure 8.** Time course of trypsin hydrolysis of native vs BrCCl₃-treated myoglobin. The reaction of BrCCl₃ with photoreduced myoglobin was conducted as described in Figure 6. After 60 min of reaction, 450 nmol of K₃Fe(CN)₆, 0.3 mL of 1.0 M potassium phosphate, pH 8.1, and 1.7 mL of water were added. This mixture was preincubated for 10 min at 30 °C, at which time trypsin was added (1:40 w/w ratio of trypsin to myoglobin). Aliquots of 200 μL were taken over time and added to a test tube on ice containing 10-fold excess of trypsin inhibitor to stop the hydrolysis. These samples were analyzed by HPLC as described in Figure 6. Control experiments were performed exactly as above, except that, after photoreduction, K₃Fe(CN)₆ was added prior to BrCCl₃ to prevent formation of altered heme products. (O) Native myoglobin control [determined from areas corresponding to apomyoglobin (fraction 4) on panel A of Figure 6]; (△) myoglobin that has been altered in a process other than covalent bonding of heme to protein [determined from areas corresponding to apomyoglobin (fraction 4) on panel B of Figure 6]; (●) heme covalently bound to protein (determined from areas corresponding to fraction 5 of panel B of Figure 6). Each point represents the mean of three independent values, with standard deviations as indicated.

**Figure 9.** Absorption spectra of the myoglobin heme–protein adduct in the reaction with molecular oxygen (panel A) or carbon tetrachloride (panel B). (···) Oxidized form; (—) reduced and treated. The heme–protein adduct was isolated from other heme products by extraction of the reaction mixture, described in Figure 6, with 2-butanone under acidic conditions (46). This extract (1.36 mg of protein/mL) was photoreduced with 1.0 μM FMN and 10 mM EDTA in 50 mM potassium phosphate, pH 7.5, in a total volume of 2.0 mL. In Panel A: 1, photoreduced deoxo form; 2–5, 2.0-nmol increments of oxygen were added every 10 s and spectra determined. After the last addition of oxygen, 110 nmol of K₃Fe(CN)₆ was added. No changes in the spectrum were observed. In panel B, carbon tetrachloride (250 μM) was added and spectra taken at (2) time of mixing, or after (3) 10 s, (4) 20 s, (5) 60 s, or (6) 60 min. Spectrum 7 was taken after the addition of K₃Fe(CN)₆.

consideration is the role of the irreversible binding of heme to protein as a "tag" for protein degradation (22).

In light of this possibility, the susceptibility of the heme–protein adduct of myoglobin to trypsin hydrolysis has been investigated (Figure 8). Under the conditions of the reaction, native myoglobin was found to be highly resistant to proteolysis. After reaction with BrCCl₃, however, the protein became more susceptible to hydrolysis. The most rapidly degraded protein component of the reaction mixture corresponded to protein that was covalently altered by the modified heme moiety. Approximately 35% of this compound was hydrolyzed in 600 min. Interestingly, an intermediate rate of hydrolysis was noted for the apoprotein fraction with approximately 20% of this component hydrolyzed in 600 min. The hydrolysis of this fraction could be due to apomyoglobin formed in the reaction or myoglobin containing the dissociable altered heme products (compounds 1, 3, and 6). Although it is known that apomyoglobin is hydrolyzed more rapidly than the holoprotein, it probably does not account for the increased rate of hydrolysis, because the amount of apoprotein present in the reaction mixture was less than 2% of the myoglobin as determined by isoelectric focusing of the reaction mixture and quantification by laser densitometry of the Coomassie Blue stained protein vs. a standard curve of apoprotein (data not shown). This finding indicates that the noncovalently bound heme products, compounds 1, 3, and 6, remain bound to the protein, in contrast to the current dogma that altered heme products are released from hemoproteins to form apoproteins, which are then degraded (72, 74, 83). Therefore, the increased rate of degradation of myoglobin may be due to both covalently bound and noncovalently bound heme products that alter the three-dimensional structure of the protein.

In this regard, it is known that reconstitution of I-vinyl-substituted hemes can cause large movements of certain amino acids in myoglobin (84). The covalent modification of the protein moiety by the trichloromethyl radical may also play a role in the increased hydrolysis of the protein but was not investigated.

The enhanced susceptibility of the covalently bound heme product to hydrolysis may explain why only low levels of the protein-bound material were detected after the administration of AIA (31) or DDEP (32) to rats, whereas higher levels were found under in vitro conditions (28, 31, 32).

**Alteration of Catalytic Activity.** Since an altered but intact heme moiety was found to be covalently bound to the protein moiety of myoglobin in the reaction with BrCCl₃, it seemed possible that such an altered heme–protein might still retain the ability to carry out redox reactions. To test this possibility, the redox properties of the altered heme–protein was investigated. Panel A (Figure 9) shows a spectrum of the oxidized state of the covalently altered myoglobin. The prominent Soret absorbance at 408 nm, similar to native myoglobin, clearly shows that the heme group is intact. Photoreduction of this fraction gave a ferrous deoxy complex with a red shift of the Soret to 420 nm and distinct α and β bands at 556 and 526 nm, respectively, clearly indicating that the altered heme still may be reduced. Addition of oxygen in 2-nmol increments in buffer to this ferrous form caused reversion to the ferric form, suggesting that the reduced altered myoglobin was readily oxidized by molecular oxygen. This is clearly unlike native myoglobin, which forms a stable, reversible ferrous–O₂ complex. The reduced altered hemoprotein was also appreciably oxidized by CCl₄ over a period of 60 s (panel B); under these conditions native myoglobin remained predominantly in the reduced state even after 60 min, as previously found (85). These experiments indicate that covalently altered heme–proteins can have enhanced reducing activity.

The potential for enhanced reactivity of the covalently

bound heme–proteins may have significant toxicological and pathological consequences. In oxygen reperfusion injury, especially in the myocardium, heme–protein adducts may be formed by the interaction of myoglobin with hydrogen peroxide or lipid hydroperoxides. These activated heme–proteins may enhance the formation of oxygen metabolites, such as superoxide, hydrogen peroxide, and hydroxyl radicals, which would exacerbate the tissue injury. Indeed, it has already been proposed that myoglobin may react with HOOH to initiate peroxidative reactions that may play a role in the myocardial injury due to ischemia and reperfusion (86–88). A similar process may also be responsible in part for the hepatotoxicity associated with $CCl_4$ and $BrCCl_3$.

The enhanced redox capabilities of the altered heme may also lead to accelerated destruction of the heme. The destruction of heme to mono- and dipyrrole products would most likely require multiple catalytic cycles. Studies on the P-450 cytochromes all involve multiple reducing equivalents, and one can readily envision that if a more reactive entity were to form initially, it might participate in the fission of the heme ring. In contrast, in the reaction of myoglobin with $BrCCl_3$, only stoichiometric amounts of reducing equivalents were used, and thus this destructive pathway was limited. Moreover, upon a second redox cycle by the altered heme–protein, there was a loss of the absorbance in the oxidized state, at the Soret, indicating further alterations of the heme group (Figure 9). One can envision that, under conditions where multiple catalytic cycles could take place, a bleaching of the chromophore would result, similar to that observed for other hemoproteins. In turn, the destruction of the heme moiety by such processes could lead to the release of free iron, which may also play a role in tissue injury. For example, hemoglobin-dependant peroxidation, presumably mediated by free iron, has been implicated in the oxidative tissue damage to the central nervous system (89).

## Summary

In this perspective we have described a newly characterized pathway for the metabolism of the prosthetic heme of cytochrome P-450, which results in the formation of protein-bound adducts. This reaction occurs when the cytochrome P-450 metabolizes a variety of xenobiotics as well as endogenous compounds such as hydrogen peroxide and lipid hydroperoxides. It also takes place during the reactions catalyzed by other hemoproteins, such as myoglobin and hemoglobin. In the case of the reaction of ferrous myoglobin with $BrCCl_3$, under single-turnover conditions, an intact heme moiety becomes covalently bound to an active-site amino acid. This covalently altered protein has significantly enhanced reductive activity compared to that of native myoglobin, as demonstrated by its rapid reduction of molecular oxygen and $CCl_4$. It also is more rapidly proteolyzed than myoglobin. These findings may have relevance to the P-450 cytochromes in which suicide inactivation, destruction of the heme prosthetic group, and loss of the protein is observed. The activation of hemoproteins to heme–protein adducts may also have toxicological significance, perhaps in oxygen reperfusion injury in the myocardium as well as other tissues by enhancing the production of oxygen-derived radicals from molecular oxygen and lipid hydroperoxides. Clearly, further research in the characterization of heme–protein adducts is necessary before their importance in protein turnover and oxygen-induced injury can be determined.

**Acknowledgment.** We thank Dr. Kaori Maeda for the purification of P-450b and NADPH-cytochrome P-450 reductase and for experiments performed with the use of these components in the reconstituted system. We also thank Dr. Helen W. Davies for the experiments with linoleic hydroperoxide. We are grateful to Dr. James R. Gillette for critically reviewing the manuscript.

## References

(1) Walsh, C. (1977) Recent developments in suicide substrates and other active site-directed inactivating agents of specific enzymes. *Horiz. Biochem. Biophys.* **3**, 36–81.

(2) Alston, T. A. (1981) Suicide substrates for mitochondrial enzymes. *Pharmacol. Ther.* **12**, 1–41.

(3) Abeles, R. H. (1983) Suicide enzyme inactivators. *Chem. Eng. New* **61**, 48–56.

(4) Ortiz de Montellano, P. R., and Correia, M. A. (1983) Suicidal destruction of cytochrome P-450 during oxidative drug metabolism. *Annu. Rev. Pharmacol. Toxicol.* **23**, 481–503.

(5) Ortiz de Montellano, P. R., Beilan, H. S., and Kunze, K. L. (1981) N-Alkylprotoporphyrin IX formation in 3,5-dicarbethoxy-1,4-dihydrocollidine-treated rats. Transfer of the alkyl group from the substrate to the porphyrin. *J. Biol. Chem.* **256**, 6708–6713.

(6) Augusto, O., Beilan, H. S., and Ortiz de Montellano, P. R. (1982) The catalytic mechanism of cytochrome P-450. Spin-trapping evidence for one-electron substrate oxidation. *J. Biol. Chem.* **257**, 11288–11295.

(7) Halpert, J., and Neal, R. A. (1980) Inactivation of purified rat liver cytochrome P-450 by chloramphenicol. *Mol. Pharmacol.* **17**, 427–431.

(8) Halpert, J. (1982) Further studies on the suicide inactivation of purified rat liver cytochrome P-450 by chloramphenicol. *Mol. Pharmacol.* **21**, 166–172.

(9) Halpert, J., Miller, N. E., and Gorsky, L. D. (1985) On the mechanism of the inactivation of the major phenobarbital-inducible isozyme of rat liver cytochrome P-450 by chloramphenicol. *J. Biol. Chem.* **260**, 8397–8403.

(10) Hollenberg, P., Kuemmerle, S., and Gurka, D. (1988) Suicide inactivation of cytochrome P-450 b by N-methylcarbazole results in the covalent modification of P-450 apoprotein. *FASEB J.* **2**, a563.

(11) CaJacob, C. A., Chan, W. K., Shephard, E., and Ortiz de Montellano, P. R. (1988) The catalytic site of rat hepatic lauric acid ω-hydroxylase. Protein versus prosthetic heme alkylation in the ω-hydroxylation of acetylenic fatty acids. *J. Biol. Chem.* **263**, 18640–18649.

(12) Davies, H. W., Satoh, H., Schulick, R. D., and Pohl, L. R. (1985) Immunochemical identification of an irreversibly bound heme-derived adduct to cytochrome P-450 following $CCl_4$ treatment of rats. *Biochem. Pharmacol.* **34**, 3203–3206.

(13) Fantone, J. C., and Ward, P. A. (1985) Oxygen-derived radicals and their metabolites: relationship to tissue injury. Scope Publication, Kalamazoo, MI.

(14) Recknagel, R. O., and Glende, E. A. (1989) The carbon tetrachloride hepatotoxicity model: free radicals and calcium homeostasis. In *CRC Handbook of Free Radicals and Antioxidants in Biomedicine* (Miquel, J., Quintanilha, A. T., and Weber, H., Eds.) pp 3–16, CRC press, Boca Raton, FL.

(15) Poyer, J. L., Floyd, R. A., McCay, P. B., Janzen, E. G., and Davis, E. R. (1978) Spin-trapping of the trichloromethyl radical produced during enzymatic NADPH oxidation in the presence of carbon tetrachloride or bromotrichloromethane. *Biochim. Biophys. Acta* **539**, 402–409.

(16) Noguchi, T., Fong, K., Lai, E. K., Alexander, S. S., King, M. M., Olson, L., Poyer, J. L., and McCay, P. B. (1982) Specificity of a phenobarbital-induced cytochrome P-450 for metabolism of carbon tetrachloride to the trichloromethyl radical. *Biochem. Pharmacol.* **31**, 615–624.

(17) Levin, W., Jacobson, M., and Kuntzman, R. (1972) Incorporation of radioactive δ-aminolevulinic acid into microsomal cytochrome P-450: Selective breakdown of the hemoprotein by allylisopropylacetamide and carbon tetrachloride. *Arch. Biochem. Biophys.* **148**, 262–269.

(18) Guzelian, P. S., and Swisher, R. W. (1979) Degradation of cytochrome P-450 haem by carbon tetrachloride and 2-allyl-2-isopropylacetamide in rat liver in vivo and in vitro. *Biochem. J.* **184**, 481–489.

(19) De Groot, H., and Haas, W. (1981) Self-catalyzed, $O_2$-independent inactivation of NADPH- or dithionite-reduced microsomal cytochrome P-450 by carbon tetrachloride. *Biochem. Pharmacol.* **30**, 2343–2347.

(20) Manno, M., De Matteis, F., and King, L. J. (1988) The mechanism of the suicidal, reductive inactivation of microsomal cytochrome P-450 by carbon tetrachloride. *Biochem. Pharmacol.* **37**, 1981–1990.

(21) Parkinson, A., Thomas, P. E., Ryan, D. E., and Levin, W. (1983) The in vivo turnover of rat liver microsomal epoxide hydrolase and both the apoprotein and heme moieties of specific cytochrome P-450 isozymes. *Arch. Biochem. Biophys.* **225**, 216–236.

(22) Davies, H. W., Thomas, P. E., and Pohl, L. R. (1986) Activation of cytochrome P-450 heme in vivo. In *Biological Reactive Intermediates III* (Kocsis, J. J., Jollow, D. J., Witmer, C. M., Nelson, J. O., and Snyder, R., Eds.) pp 253–261, Plenum Publishing Corp., New York.

(23) Yasukochi, Y., and Masters, B. S. S. (1976) Some properties of a detergent-solubilized NADPH-cytochrome c (cytochrome P-450) reductase purified by biospecific affinity chromatography. *J. Biol. Chem.* **251**, 5337–5344.

(24) Kawano, S., Kamataki, T., Maeda, K., Kato, R., Nakao, T., and Mizoguchi, I. (1985) Activation and inactivation of a variety of mutagenic compounds by the reconstituted system containing highly purified preparations of cytochrome P-450 from rat liver. *Fundam. Appl. Toxicol.* **5**, 487–498.

(25) Davies, H. W., Britt, S. G., and Pohl, L. R. (1986) Carbon tetrachloride and 2-isopropyl-4-pentenamide-induced inactivation of cytochrome P-450 leads to heme-derived protein adducts. *Arch. Biochem. Biophys.* **244**, 387–392.

(26) Guengerich, F. P. (1978) Destruction of heme and hemoproteins mediated by liver microsomal reduced nicotinamide adenine dinucleotide phosphate cytochrome P-450 reductase. *Biochemistry* **17**, 3633–3639.

(27) Schaefer, W. H., Harris, T. M., and Guengerich, F. P. (1985) Characterization of the enzymatic and nonenzymatic peroxidative degradation of iron porphyrins and cytochrome P-450 heme. *Biochemistry* **24**, 3254–3263.

(28) Davies, H. W., Britt, S. G., and Pohl, L. R. (1986) *Inactivation of cytochrome P-450 by 2-isopropyl-4-pentenamide and other xenobiotics leads to heme-derived protein adducts. Chem.-Biol. Interact.* **58**, 345–352.

(29) Decker, C., Sugiyama, K., Underwood, M., and Correia, M. A. (1986) Inactivation of rat hepatic cytochrome P-450 by spironolactone. *Biochem. Biophys. Res. Commun.* **136**, 1162–1169.

(30) Guengerich, F. P. (1986) Covalent binding to apoprotein is a major fate of heme in a variety of reactions in which cytochrome P-450 is destroyed. *Biochem. Biophys. Res. Commun.* **138**, 193–198.

(31) Bornheim, L. M., Underwood, M. C., Caldera, P., Rettie, A. E., Trager, W. F., Wrighton, S. A., and Correia, M. A. (1987) Inactivation of multiple hepatic cytochrome P-450 isozymes in rats by allylisopropylacetamide: mechanistic implications. *Mol. Pharmacol.* **32**, 299–308.

(32) Correia, M. A., Decker, C., Sugiyama, K., Caldera, P., Bornheim, L., Wrighton, S. A., Rettie, A. E., and Trager, W. F. (1987) Degradation of rat hepatic cytochrome P-450 heme by 3,5-dicarbethoxy-2,6-dimethyl-4-ethyl-1,4-dihydropyridine to irreversibly bound protein adducts. *Arch. Biochem. Biophys.* **258**, 436–451.

(33) Satoh, H., Davies, H. W., Takemura, T., Gillette, J. R., Maeda, K., and Pohl, L. R. (1987) An immunological approach to investigating the mechanism of halothane-induced hepatotoxicity. In *Progress in Drug Metabolism* (Bridges, J. W., Chasseaud, L. F., and Gibson, G. G., Eds.) pp 187–206, Taylor and Francis, Ltd., New York.

(34) Lunetta, J. M., Sugiyama, K., and Correia, M. A. (1989) Secobarbital-mediated inactivation of rat liver cytochrome P-450b: a mechanistic reappraisal. *Mol. Pharmacol.* **35**, 10–17.

(35) Ortiz de Montellano, P. R., Stearns, R. A., and Langry, K. C. (1984) The allylisopropylacetamide and novonal prosthetic heme adducts. *Mol. Pharmacol.* **25**, 310–317.

(36) Menard, R. H., Guenthner, T. M., Kon, H., and Gillette, J. R. (1979) Studies on the destruction of adrenal and testicular cytochrome P-450 by spironolactone. *J. Biol. Chem.* **254**, 1726–1733.

(37) Loosemore, M., Light, D. R., and Walsh, C. (1980) Studies on the autoinactivation behavior of pure, reconstituted phenobarbital-induced cytochrome P-450 isozyme from rat liver. *J. Biol. Chem.* **255**, 9017–9020.

(38) Osawa, Y., Yarborough, C., and Osawa, Y. (1982) Norethisterone, a major ingredient of contraceptive pills, is a suicide inhibitor of estrogen biosynthesis. *Science* **215**, 1249–1251.

(39) Blake, R. C., and Coon, M. J. (1980) On the mechanism of action of cytochrome P-450. Spectral intermediates in the reaction of P-450 LM$_2$ with peroxy compounds. *J. Biol. Chem.* **255**, 4100–4111.

(40) Coon, M. J., White, R. E., and Blake, R. C. (1982) Mechanistic studies with purified liver microsomal cytochrome P-450: Comparison of O$_2$- and peroxide-supported hydroxylation reactions. In *Oxidases and Related Redox Systems* (King, T. E., Mason, H. S., and Morrison, M., Eds.) pp 857–885, Pergamon Press, New York.

(41) Levin, W., Lu, A. Y. H., Jacobson, M., Kuntzman, R., Poyer, J. L., and McCay, P. B. (1973) Lipid peroxidation and the degradation of cytochrome P-450 heme. *Arch. Biochem. Biophys.* **158**, 842–852.

(42) De Matteis, F., Gibbs, A. H., and Unseld, A. (1977) Loss of haem from cytochrome P-450 caused by lipid peroxidation and 2-allyl-2-isopropylacetamide. *Biochem. J.* **168**, 417–422.

(43) Klimek, J., Schaap, P., and Kimura, T. (1983) The relationship between NADPH-dependant lipid peroxidation and degradation of cytochrome P-450 in adrenal cortex mitochondria. *Biochem. Biophys. Res. Commun.* **110**, 559–566.

(44) Iba, M. M., and Mannering, G. J. (1987) NADPH- and linoleic acid hydroperoxide-induced lipid peroxidation and destruction of cytochrome P-450 in hepatic microsomes. *Biochem. Pharmacol.* **36**, 1447–1455.

(45) Osawa, Y., Highet, R. J., Murphy, C. M., Cotter, R. J., and Pohl, L. R. (1989) Formation of heme-derived products by the reaction of ferrous deoxymyoglobin with BrCCl$_3$. *J. Am. Chem. Soc.* **111**, 4462–4467.

(46) Teale, F. W. J. (1959) Cleavage of the haem–protein link by acid methylethylketone. *Biochim. Biophys. Acta* **35**, 543.

(47) Choe, Y. S., Catalano, C. E., and Ortiz de Montellano, P. R. (1988) Crosslinking of heme to protein in peroxide-treated horse myoglobin. *FASEB J.* **2**, a585.

(48) Schafer, M. P. (1988) Reversed-phase high-performance liquid chromatographic separation and quantitation of reticulocyte α- and β-globin chains from normal and β-thalassemic mice. *J. Chromatogr.* **431**, 177–183.

(49) Hunt, J. V., Simpson, J. A., and Dean, R. T. (1988) Hydroperoxide-mediated fragmentation of proteins. *Biochem. J.* **250**, 87–93.

(50) Kulmacz, R. J. (1986) Prostaglandin H synthase and hydroperoxides: peroxidase reaction and inactivation kinetics. *Arch. Biochem. Biophys.* **249**, 273–285.

(51) Chen, Y.-N. P., Bienkowski, M. J., and Marnett, L. J. (1987) Controlled tryptic digestion of prostaglandin H synthase: characterization of protein fragments and enhanced rate of proteolysis of oxidatively inactivated enzyme. *J. Biol. Chem.* **262**, 16892–16899.

(52) Jenzer, H., and Kohler, H. (1986) The role of superoxide radicals in lactoperoxide-catalysed H$_2$O$_2$-metabolism and in irreversible enzyme inactivation. *Biochem. Biophys. Res. Commun.* **139**, 327–332.

(53) Jenzer, H., Kohler, H., and Broger, C. (1987) The role of hydroxyl radicals in irreversible inactivation of lactoperoxidase by excess H$_2$O$_2$. *Arch. Biochem. Biophys.* **258**, 381–390.

(54) Vissers, M. C. M., and Winterbourn, C. C. (1987) Myeloperoxidase-dependent oxidative inactivation of neutrophil neutral proteinases and microbicidal enzymes. *Biochem. J.* **245**, 277–280.

(55) Edwards, S. W., Nurcombe, H. L., and Hart, C. A. (1987) Oxidative inactivation of myeloperoxidase released from human neutrophils. *Biochem. J.* **245**, 925–928.

(56) Sipes, I. G., Gandolfi, A. J., Pohl, L. R., Krishna, G., and Brown, B. R. (1980) Comparison of the biotransformation and hepatotoxicity of halothane and deuterated halothane. *J. Pharmacol. Exp. Ther.* **214**, 716–720.

(57) Kubic, V. L., and Anders, M. W. (1981) Mechanism of the microsomal reduction of carbon tetrachloride and halothane. *Chem.-Biol. Interact.* **34**, 201–207.

(58) Blake, R. C., and Coon, M. J. (1981) On the mechanism of action of cytochrome P-450. Evaluation of homolytic and heterolytic mechanisms of oxygen–oxygen bond cleavage during substrate hydroxylation by peroxides. *J. Biol. Chem.* **256**, 12127–12133.

(59) Jonen, H. G., Werringloer, J., Prough, R. A., and Estabrook, R. W. (1982) The reaction of phenylhydrazine with microsomal cytochrome P-450. Catalysis of heme modification. *J. Biol. Chem.* **257**, 4404–4411.

(60) Ortiz de Montellano, P. R., Kunze, K. L., Beilan, H. S., and Wheeler, C. (1982) Destruction of cytochrome P-450 by vinyl fluoride, fluroxene, and acetylene. Evidence for a radical intermediate in olefin oxidation. *Biochemistry*. 21, 1331–1339.

(61) Augusto, O., Kunze, K. L., and Ortiz de Montellano, P. R. (1982) N-Phenylprotoporphyrin IX formation in the hemoglobin–phenylhydrazine reaction. Evidence for a protein-stabilized iron–phenyl intermediate. *J. Biol. Chem.* 257, 6231–6241.

(62) Ortiz de Montellano, P. R., Augusto, O., Viola, F., and Kunze, K. L. (1983) Carbon radicals in the metabolism of alkyl hydrazines. *J. Biol. Chem.* 258, 8623–8629.

(63) Ortiz de Montellano, P. R. (1985) Alkenes and alkynes. In *Bioactivation of Foreign Compounds* (Anders, M. W., Ed.) pp 121–155, Academic Press, New York.

(64) Ortiz de Montellano, P. R., and Watanabe, M. D. (1987) Free radical pathways in the in vitro hepatic metabolism of phenelzine. *Mol. Pharmacol.* 31, 213–219.

(65) George, P., and Irvine, D. H. (1952) The reaction between metmyoglobin and hydrogen peroxide. *Biochem. J.* 52, 511–517.

(66) Ortiz de Montellano, P. R., and Catalano, C. E. (1985) Epoxidation of styrene by hemoglobin and myoglobin. *J. Biol. Chem.* 260, 9265–9271.

(67) Tew, D., and Ortiz de Montellano, P. R. (1988) The myoglobin protein radical. Coupling of Tyr-103 to Tyr-151 in the $H_2O_2$-mediated cross-linking of sperm whale myoglobin. *J. Biol. Chem.* 263, 17880–17886.

(68) King, K., and Winfield, M. E. (1963) The mechanism of metmyoglobin oxidation. *J. Biol. Chem.* 238, 1520–1528.

(69) Yonetani, T., and Schleyer, H. (1967) Studies on cytochrome c peroxidase. IX. The reaction of ferrimyoglobin with hydroperoxides and a comparison of peroxide-induced compounds of ferrimyoglobin and cytochrome c peroxidase. *J. Biol. Chem.* 242, 1974–1979.

(70) Ator, M. A., and Ortiz de Montellano, P. R. (1987) Protein control of prosthetic heme reactivity. Reaction of substrates with the heme edge of horseradish peroxidase. *J. Biol. Chem.* 262, 1542–1551.

(71) Noguchi, T., Fong, K., Lai, E. K., Olson, L., and McCay, P. B. (1982) Selective early loss of polypeptides in liver microsomes of CCl$_4$-treated rats. *Biochem. Pharmacol.* 31, 609–614.

(72) Liem, H. H., Johnson, E. F., and Muller-Eberhard, U. (1983) The effect in vivo and in vitro of allylisopropylacetamide on the content of hepatic microsomal cytochrome P-450 2 of phenobarbital treated rabbits. *Biochem. Biophys. Res. Commun.* 111, 926–932.

(73) Moody, D. E., Taylor, L. A., and Smuckler, E. A. (1986) Immunohistochemical evidence for alterations in specific forms of rat hepatic microsomal cytochrome P-450 during acute carbon tetrachloride intoxication. *Drug Metab. Dispos.* 14, 709–713.

(74) Tephly, T. R., Black, K. A., Green, M. D., Coffman, B. L., Dannan, G. A., and Guengerich, F. P. (1986) Effect of the suicide substrate 3,5-diethoxycarbonyl-2,6-dimethyl-4-ethyl-1,4-dihydropyridine on the metabolism of xenobiotics and on cytochrome P-450 apoproteins. *Mol. Pharmacol.* 29, 81–87.

(75) Fligiel, S. E. G., Lee, E. C., McCoy, J. P., Johnson, K. J., and Varani, J. (1984) Protein degradation following treatment with hydrogen peroxide. *Am. J. Pathol.* 115, 418–425.

(76) Fagan, J. M., Waxman, L., and Goldberg, A. L. (1986) Red blood cells contain a pathway for the degradation of oxidant-damaged hemoglobin that does not require ATP or ubiquitin. *J. Biol. Chem.* 261, 5705–5713.

(77) Rivett, A. J., Roseman, J. E., Oliver, C. N., Levine, R. L., and Stadtman, E. R. (1984) Covalent modification of proteins by mixed-function oxidation: recognition by intracellular proteases. *Prog. Clin. Biol. Res.* 180, 317–328.

(78) Stadtman, E. R. (1986) Oxidation of proteins by mixed-function oxidation systems: implications in protein turnover, ageing and neutrophil function. *Trends Biochem. Sci.* 11, 11–12.

(79) Davies, K. J. A., and Golberg, A. L. (1987) Oxygen radicals stimulate intracellular proteolysis and lipid peroxidation by independent mechanisms in erythrocytes. *J. Biol. Chem.* 262, 8220–8226.

(80) Davies, K. J. A. (1987) Protein damage and degradation by oxygen radicals: I. General aspects. *J. Biol. Chem.* 262, 9895–9901.

(81) Davies, K. J. A., and Delsignore, M. E. (1987) Protein damage and degradation by oxygen radicals: III. Modification of secondary and tertiary structure. *J. Biol. Chem.* 262, 9908–9913.

(82) Davies, K. J. A., Lin, S. W., and Pacifici, R. E. (1987) Protein damage and degradation by oxygen radicals: IV. Degradation of denatured protein. *J. Biol. Chem.* 262, 9914–9920.

(83) Shiraki, H., and Guengerich, F. P. (1984) Turnover of membrane proteins: kinetics of induction and degradation of seven forms of rat liver microsomal cytochrome P-450, NADPH-cytochrome P-450 reductase, and epoxide hydrolase. *Arch. Biochem. Biophys.* 235, 86–96.

(84) Miki, K., Harada, S., Hato, Y., Iba, S., Kai, Y., Kasai, N., Katsube, Y., Kawabe, K., Yoshida, Z., and Ogoshi, H. (1986) Crystal structures of modified myoglobins. II. Relation between oxygen affinity properties and structural changes around heme in myoglobins reconstituted with 2,4-diisopropyldeuteroheme,2-isopropyl-4-vinyldeuteroheme and 2-vinyl-4-isopropyldeuteroheme. *J. Biochem.* 100, 277–284.

(85) Bartnicki, E. W., Belser, N. O., and Castro, C. E. (1978) Oxidation of heme proteins by alkyl halides: a probe for axial inner sphere redox capacity in solution and in whole cells. *Biochemistry* 17, 5582–5586.

(86) Grisham, M. B. (1985) Myoglobin-catalyzed hydrogen peroxide dependent arachidonic acid peroxidation. *J. Free Radicals Biol. Med.* 1, 227–232.

(87) Puppo, A., and Halliwell, B. (1988) Formation of hydroxyl radicals in biological systems. Does myoglobin stimulate hydroxyl radical formation from hydrogen peroxide? *Free Radical Res. Commun.* 4, 415–422.

(88) Mitsos, S. E., Kim, D., Lucchesi, B. R., and Fantone, J. C. (1988) Modulation of myoglobin–$H_2O_2$-mediated peroxidation reactions by sulfhydryl compounds. *Lab. Invest.* 59, 824–830.

(89) Sadrzadeh, S. M. H., and Eaton, J. W. (1988) Hemoglobin-mediated oxidant damage to the central nervous system requires endogenous ascorbate. *J. Clin. Invest.* 82, 1510–1515.

# Chapter 19

# Biochemical, Structural, and Functional Properties of Oxidized Low-Density Lipoprotein

Hermann Esterbauer,* Martina Dieber-Rotheneder, Georg Waeg, Georg Striegl, and Günther Jürgens

*Institute of Biochemistry, University of Graz, Schubertstrasse 1, A-8010 Graz, Austria*

Reprinted from *Chemical Research in Toxicology*, Vol. 3, No. 2, March/April, 1990

## Introduction

Human low-density lipoprotein (LDL)[1] is a main carrier for cholesterol in the blood stream, and it is well established that cholesterol deposits in the arteries stem primarily from LDL and that increased levels of plasma LDL correlate with an increased risk of atherosclerosis. Various lines of research provide strong but not conclusive evidence that LDL may become oxidized in vivo and that oxidized LDL (oLDL) is the species involved in the formation of early atherosclerotic lesions (for review see refs 1–3). The most crucial findings in this context are the following: (1) oLDL has chemotactic properties and if present in the intimal space of the arteries would recruit blood monocytes which then can develop into tissue macrophages; (2) macrophages take up oLDL unregulated to form lipid laden foam cells; and (3) oLDL is highly cytotoxic and could be responsible for damage of the endothelial layer and for the destruction of smooth muscle cells.

Early atherosclerotic lesions are characterized by massive accumulation of cells filled with lipid droplets consisting of cholesterol and cholesteryl esters. Because of their foamy appearance such cells are called foam cells (4–8). In the progression of atherosclerosis these lesions can develop into fatty streaks and plaques. The foam cell precursors are mostly monocyte-macrophages (4–8) which have entered into the intima, although some may also develop from smooth muscle cells (SMC) (9). In culture, macrophages take up native LDL only slowly, and even if incubated over long periods with high LDL concentrations, they do not accumulate cholesteryl esters and transform to lipid-laden cells (10, 11). The uptake of native

LDL by monocyte-macrophages occurs through the LDL receptor (apo B/E receptor) mediated pathway which is efficiently downregulated if the level of intracellular cholesterol increases (10, 11). As discovered by Brown and Goldstein (10), macrophages express also a scavenger receptor which mediates the endocytosis of several forms of modified LDL. This receptor is not under the control of intracellular cholesterol. In culture, the uptake of modified LDL by the scavenger receptor can therefore lead to the accumulation of cholesterol which is then stored in the form of lipid droplets, i.e., the macrophages develop into cells with an appearance typical of foam cells. This scavenger receptor pathway was discovered (10) in experiments with LDL which had been pretreated in vitro with acetic acid anhydride, i.e., acetylated LDL. Such a treatment leads among other things to the acetylation of ε-amino groups of lysine residues at the apo B, the protein part of LDL, and as a consequence to a loss of positive charges and a net increase of negative surface charges in LDL.

The classical LDL receptor recognizes a specific domain of positive charges from lysine, arginine, and histidine residues at the apo B (for review see ref 11). If this domain is altered as, for example, by acetylation of free amino groups, the recognition is decreased or completely lost. On the other hand, the binding of altered LDL to the sca-

---

* Correspondence should be addressed to this author.

[1] Abbreviations: LDL, low-density lipoprotein; oLDL, oxidized low-density lipoprotein; apo B, apolipoprotein B; MDA, malonaldehyde; HNE, 4-hydroxynonenal; EC, endothelial cells; BHT, butylated hydroxytoluene; HPLC, high-performance liquid chromatography; PUFA, polyunsaturated fatty acid; cDNA, complementary DNA; TBARS, thiobarbituric acid reactive substances expressed as MDA equivalents; BHA, butylated hydroxyanisole; EDTA, ethylenediaminetetraacetate; REM, relative electrophoretic mobility; PBS, phosphate-buffered saline; kDa, kilodalton(s).

2428–5/92/0247$06.00/0

venger receptor appears to afford certain clusters of negative charges at the apo B (*12*), and an artificial increase of the negative surface charges by chemical modifications usually, though not always, results in an increased recognition by the scavenger receptor and an unlimited uptake of cholesterol. LDL recognizable by this receptor can be made by chemical modifications such as acetylation, maleylation, acetoacetylation, carbamylation, and succinylation and by treatment with glutaraldehyde (for review see ref 13). All these procedures have in common that they mainly attack free amino groups, e.g., $\epsilon$-amino groups from lysine residues. However, none of these modifications occur in vivo, and thus it remained obscure what the biological modifier might be. Fogelman et al. proposed in 1980 (*14*) that malonaldehyde (MDA) generated by inflammatory cells at the site of arterial injury is the biological agent that converts LDL in vivo into a form recognizable by the scavenger receptor. LDL modified directly by incubation with MDA leads in fact to cholesterol loading of macrophages (*13–16*). The necessary MDA concentrations, however, are extremely high (10–100 mM), and it is unlikely that inflammatory cells can locally produce such concentrations. With the discovery that cells of the vascular system can induce lipid peroxidation in the LDL particle, it appears now more reasonable that reactive aldehydes including MDA are generated within the LDL particle itself and interact at or near the site of their origin with neighboring free amino groups and probably also with other functional groups (SH, OH) of the apo B polypeptide chain.

Many other changes are associated with the oxidative degradation of the LDL lipids as, for example, an extensive fragmentation of the apo B to smaller peptides (*17, 18, 55*), and it seems very likely that these together with the covalent binding of aldehydes lead to a complete structural rearrangement of the protein creating new epitopes which do not bind to the B/E receptor but to the scavenger receptor. Oxidized LDL is toxic toward endothelial cells, smooth muscle cells, and fibroblasts; proliferating cells are more susceptible than quiescent cells, and it appears that the susceptibility to oLDL is higher in the S-phase of the cell cycle, i.e., during DNA synthesis (*19–21*). The toxic principle is not yet identified, but it is proven that the protein moiety is not required. The toxic component(s) can be extracted with organic solvents such as chloroform and therefore resides (reside) in the lipid phase of oLDL (*19*). Some of the toxic characteristics of oLDL can be mimicked by direct chemical modification of LDL with 4-hydroxynonenal (HNE) (*21*). HNE-treated LDL shows a comparable growth-inhibitory activity toward fibroblasts as oLDL, and the inhibitory properties solely reside in the lipid phase. Kaneko et al. (*22, 23*) have recently tested in cultured fibroblasts and endothelial cells (EC) the cytotoxicity and inhibition of growth of several propagation products of lipid peroxidation. The most toxic compounds were 2,4-alkadienals (nonadienal, decadienal) and 4-hydroxynonenal which produced 50% inhibition of endothelial cell proliferation at 10–25 $\mu$M; linoleic acid hydroperoxides had a similar toxicity (25 $\mu$M), whereas 2-alkenals (hexenal to nonenal) and alkanals are less toxic. In a homologous series of aldehydes, the toxicity increases with the chain length, i.e., lipophilicity, of the aldehydes. Lysophosphatides which are also present in oLDL (*24*) have been shown to produce a number of deleterious effects in cells as, for example, inhibition of the K/Na pump. At present, it is not known whether the toxicity of oLDL relies on a single compound or is produced by the combined action of several compounds generated from the

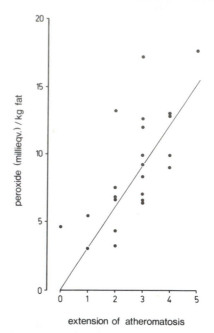

**Figure 1.** Relationship between extension of atherosclerotic lesions and peroxide content of the extractable lipids. Washed pieces of human aortic walls (post-mortem material) were extracted with chloroform, and the peroxide content of the extracted lipids was determined (redrawn from a figure in ref 30).

LDL lipids during the lipid peroxidation process. Since the cytotoxicity of LDL is an important aspect for the hypothesis that oLDL is atherogenic, the identification of the toxin(s) and the clarification of the detailed mechanisms of how products of oLDL would kill the cells would deserve more attention. The same is true for the chemotactic principle of oLDL (*25*). It has been suggested (*24*) that oxidation of LDL results in an activation of a previously masked phospholipase $A_2$, which results in the release of lysophosphatides and oxidized fatty acids. In in vitro experiments (Boyden-Chamber) lysophosphatides act as chemoattractants for blood monocytes (*25*). A great variety of other potentially chemotactic substances may additionally be present in oLDL. For example, 4-hydroxyalkenals are chemotactic toward polymorphonuclear leukocytes in extremely low concentrations of $10^{-6}$–$10^{-10}$ M (*26–28*). 12-Oxododeca-5,8,10-trienoic acid, another aldehyde with the CH=HCHO functional group, can be produced by lipoxygenase from arachidonic acid and is a powerful chemoattractant for leukocytes (*29*). Thus it appears that the chemotactic properties of oLDL probably also rely on more than one compound.

The hypothesis that oxidized LDL plays a crucial role in the pathogenesis of arteriosclerosis is further suggested by a number of other indirect evidences. For example, Glavind and co-workers reported already in 1952 (*30*) that atherosclerotic lesions of the human aorta (post-mortem material) contained lipid peroxides and that the extent of peroxidation is connected with the extension of the atheromatas (Figure 1). Increased levels of serum lipid peroxides were found in patients with heart disease (*31*) and diabetes and in chronic smokers (*31*). The plasma peroxide value appeared to increase with age (*31, 32*). The antiatherogenic effect of probucol is now ascribed, at least in part, to its ability to be incorporated into the LDL and to protect it against oxidation (*33, 34*). Probucol has a structure very similar to the chain-breaking antioxidant butylated hydroxytoluene (BHT). Epidemiological studies on plasma antioxidants and risk of cardiovascular diseases

**Table I. Composition of Native Human LDL[a]**

| | n | μg/mg of LDL | approximate molecules/ LDL particle |
|---|---|---|---|
| phospholipids | 6 | 217 ± 19 | 700 |
| free cholesterol | 6 | 94 ± 6 | 600 |
| cholesteryl esters | 6 | 414 ± 14 | 1600 |
| triglycerides | 6 | 37 ± 10 | 100 |
| protein | 6 | 237 ± 20 | 1 |
| myristic acid | 4 | 6.6 ± 2.0 | 70 |
| palmitic acid | 19 | 70.9 ± 20.9 | 700 |
| palmitoleic acid | 19 | 4.3 ± 2.5 | 50 |
| stearic acid | 19 | 16.3 ± 7.4 | 150 |
| oleic acid | 19 | 57.2 ± 18.3 | 450 |
| linoleic acid | 19 | 120.8 ± 32.4 | 1100 |
| arachidonic acid | 19 | 16.1 ± 8.8 | 150 |
| docosahexaenoic acid | 4 | 2.1 ± 2.0 | 20 |
| total fatty acids | 19 | 290 | 2700 |
| total PUFAs | 19 | 138 | 1300 |
| α-tocopherol | 25 | 1.03 ± 0.27 | 6 |
| γ-tocopherol | 25 | 0.087 ± 0.033 | 0.5 |
| β-carotene | 20 | 0.069 ± 0.048 | 0.3 |
| lycopene | 20 | 0.037 ± 0.017 | 0.2 |

[a] The numbers of molecules/LDL were calculated on the basis of a molecular size of 2500 kDa. n is the number of LDL samples from the different donors analyzed.

in the European population revealed a very strong inverse relationship between plasma vitamin E levels and mortality from ischemic heart diseases (*35–37*).

## Chemical Composition of LDL

Human low-density lipoprotein is a spherical particle with a diameter of about 22 nm and an average molecular weight of 2.5 million (for review see refs 38 and 39). The LDL isolated by us from human plasma has a hydrated density of 1.020–1.050 g/mL. According to our analysis (Table I) the LDL is composed of 20–24% phospholipids, 9–10% free cholesterol, 40–44% cholesteryl esters, 3–5% triglycerides, and 21–26% protein. The neutral lipids (cholesteryl ester, triglycerides) form a hydrophobic core which is surrounded by a surface monolayer consisting of phospholipids and cholesterol. Embedded in the monolayer is a large protein, i.e., apolipoprotein B, with a molecular weight of about 500 000 (*38, 39*). With our analysis (Table I) and a molecular weight of 2.5 million the outer monolayer on average consists of 700 molecules of phospholipid and 600 molecules of cholesterol, whereas the interior core is composed of 1600 molecules of cholesteryl ester and 100 molecules of triglyceride. The fatty acid composition as determined by us by capillary gas chromatography of the methyl esters prepared from a Folch extract of LDL (*3, 40–42*) is also shown in Table I. The total number of fatty acid molecules bound in the different lipid classes is 2700 on average, of which about half are polyunsaturated fatty acids (PUFAs), mainly linoleic acid (18:2) and arachidonic acid (20:4). Some of the LDL preparations also contained traces of other PUFAs such as 18:3, 20:3, and 22:6. The analysis of the distribution of the PUFAs revealed an unequal distribution in the different lipid classes (*3*). Linoleic acid is always predominantly (65%) bound in the cholesteryl esters, whereas arachidonic acid is mostly (68%) bound to the phospholipids. Docosahexaenoic acid was only found in phospholipids. The number of fatty acid molecules which can indirectly be calculated from the lipid composition (phospholipids, cholesteryl ester, triglycerides) amounts to about 3300. This is more than we have determined by direct fatty acid analysis, which on average gave 2700 fatty acid molecules per LDL particle. This difference probably

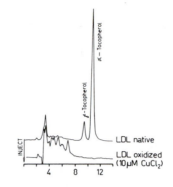

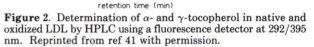

**Figure 2.** Determination of α- and γ-tocopherol in native and oxidized LDL by HPLC using a fluorescence detector at 292/395 nm. Reprinted from ref 41 with permission.

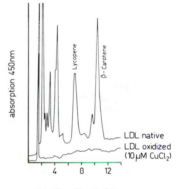

**Figure 3.** Determination of carotenoids in native and oxidized LDL by HPLC using a UV/vis detector at 450 nm. Reprinted from ref 41 with permission.

relies, at least in part, on the large donor-dependent variations in the fatty acid composition (*3, 45*) and in the different reference systems used for calculating the total amount of LDL in a particular preparation. In the case of chemical composition (first part of Table I) the sum of all determined components is taken as total LDL. It is not considered here, for example, that apo B is highly glycosylated and contains about 8–10% carbohydrates (*46*). Fatty acid analyses are based on total dry weight or on the total cholesterol content with a conversion factor of 3.16 (LDL in mg/mL = 3.16 × total cholesterol in mg/mL). The conversion factor 3.16 was computed by us from about 20 different analyses found in the literature. (Total cholesterol determined by the Monotest of Boehringer Mannheim.)

On a molar base the major antioxidant in LDL is α-tocopherol (Figure 2), which has a mean value of 1.03 μg/mg of LDL (*3*). This corresponds to about 6 molecules of α-tocopherol/LDL particle. The other antioxidants that have so far been identified (*3, 42, 45*) are present in much smaller quantities and include γ-tocopherol, β-carotene, and lycopene. Since more than 20 different carotenoids have been reported to be present in plasma (*43*), it is conceivable that LDL contains besides β-carotene and lycopene a number of other carotenoids which may also have antioxidant properties. The HPLC chromatogram (Figure 3), for example, always showed a number of other peaks which are not yet identified. From their chromatographic behavior these peaks are likely α-carotene and oxycarotenoids such as zeaxanthin and cryptoxanthin. Native LDL also contains a fluorescent compound with an excitation maximum at 360 nm, an emission maximum

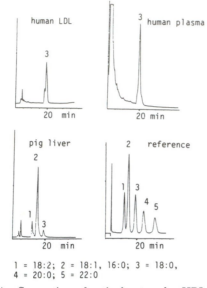

**Figure 4.** Separation of retinyl esters by HPLC using a fluorescence detector 350/510 nm. Injected into the HPLC were a chloroform/methanol extract from isolated LDL, human plasma, pig liver, and a mixture of synthetic retinyl esters with 18:2 (1), 18:1 (2), 18:0 (3), 20:0 (4), 22:0 (5). Peak 3 in LDL and human plasma which has the same retention time as retinyl stearate was later (45) identified as phytofluene. Conditions for separation: ODS column, methanol/2-propanol, 3:1.

at 470 nm, and a shoulder at 510 nm (41). On the basis of the retention time in reversed-phase HPLC and the fluorescence behavior, we assumed that the compound is retinyl stearate (3, 42, 44) (Figure 4). Recent more detailed studies including mass spectroscopy performed in collaboration with F. Gey from F. Hoffmann-La Roche, however, indicate that the compound is phytofluene (42, 45). This compound also seems to act as an antioxidant. Values below 1.0 antioxidant/LDL (Table I) mean of course that only a certain fraction (in the case of $\beta$-carotene about 30%) of the LDL molecules contain these antioxidants, whereas the others do not. It is not known whether a rapid intermolecular exchange of the antioxidants can occur between LDL molecules in solution. The ratio of antioxidants to PUFAs in native LDL is about 1:200.

Both the total PUFA and the vitamin E content of LDL showed a rather large donor-dependent variation (3, 45). The lowest and highest PUFA content, for example, were 700 and 1470 molecules/LDL molecule. $\alpha$-Tocopherol ranged from 3.15 to 9.9 molecules/LDL particle. The vitamin E content of LDL appears to be positively associated with the PUFA content of LDL as shown in Figure 5.

The apo B amino acid sequence has been directly determined on the protein and also deduced from the cDNA (for review see ref 46). The number of amino acid residues per apo B (512.937 Da) are as follows: Asp + Asn 478, Thr 298, Ser 393, Glu + Gln 529, Pro 169, Gly 207, Ala 266, Cys 25, Val 251, Met 78, Ile 288, Leu 523, Tyr 152, Phe 223, His 115, Lys 356, Arg 148, and Trp 37.

The carbohydrate components in apo B are mannose, galactose, glucosamine, and sialic acid; the carbohydrate can amount to 8–10% of the total dry mass of LDL.

## Oxidation of LDL by Cultured Vascular Cells

In the experiments that led in 1981 to the discovery that cells can modify LDL to a form that is taken up by the macrophage scavenger receptor, an established endothelial

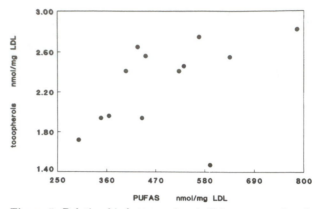

**Figure 5.** Relationship between vitamin E ($\alpha$- + $\gamma$-tocopherol) in LDL and its content of polyunsaturated fatty acids 18:2 + 20:4.

cell line derived from rabbit thoracic aorta was used (47). LDL (1 mg/mL) preincubated for 24 h with these endothelial cells was taken up by mouse peritoneal macrophages 3–4 times more rapidly than control LDL incubated under identical conditions in the absence of cells. Since the uptake could be inhibited by excess acetylated LDL, it was concluded that the endothelial cells modified LDL was endocytosed by the acetyl LDL receptor, i.e., the scavenger receptor of the macrophages.

The possible mechanism by which endothelial cells had modified LDL became clear to some extent when two groups (18, 48) independently discovered several years later that the LDL incubation medium contained thiobarbituric acid reactive substances (TBARS). From these findings it was concluded that the cells had initiated a lipid peroxidation process in LDL and that as a consequence of this oxidation process the apo B became altered into a form recognizable by the scavenger receptor. In view of the previous observations made by Fogelman et al. (14) already in 1980 that native LDL treated with MDA is also taken up by the scavenger mechanism, the general assumption was that MDA generated in the lipid peroxidation process reacted with amino acid side chains (e.g., lysine residues) of the apo B. Subsequent work by a number of different laboratories (Table II) showed that all three cell types of the vascular system, i.e., endothelial cells, smooth muscle cells, and macrophages, can oxidize LDL in culture and transform it into a form leading to an uncontrolled uptake by the macrophage scavenger receptor pathway.

A brief chronological summary on LDL oxidation studies published between 1981 and 1988 is given in Table II. With the exception of the classical work by Henriksen et al. (47) only those studies are listed that also have reported the TBARs or MDA values for the cell-modified LDL. Since the different laboratories use widely different indices, such as, for example, TBARS/mL, TBARS/mg of protein, TBARS/mg of cholesterol, or TBARS/mg of LDL, we have converted all values into nmol of MDA/mg of total LDL to facilitate comparison of the various oxidation studies. In the case of the endothelial cells, species-related differences appear to exist since human umbilical vein and rabbit aorta endothelial cells modified LDL, whereas bovine aorta endothelial cells did not (48). Cultured fibroblasts also did not oxidize LDL (48, 53). The assumption that free-radical reactions and lipid peroxidation were causally involved in the cell-mediated modification of LDL was supplemented by inhibitory studies with antioxidants. Inclusion of 100 $\mu$M vitamin E into the culture medium largely prevented the formation of TBARS over 24 h, and the LDL reisolated from the vitamin E supplemented culture could not be degraded more rapidly by

**Table II. List of Studies on Cell-Mediated Oxidation of LDL**

| ref | cell line | medium | TBARS[a] |
|---|---|---|---|
| Henriksen et al., 1981 (47) | rabbit aortic EC | F10 | |
| Morel et al., 1984 (48) | human umbilical vein EC | M199 | 1.86 |
| | bovine aortic EC | M199 | 0.42 |
| | bovine aortic SMC | M199 | 3.49 |
| | human skin fibroblast | M199 | 0.42 |
| Steinbrecher et al., 1984 (18) | rabbit aortic EC | F10 | 10.0 |
| | no cell control | F10 | 2.0 |
| | rabbit aortic EC | DME + 5 $\mu$M Cu | 8.0 |
| | no cell control | DME + 5 $\mu$M Cu | 1.0 |
| Heinecke et al., 1984 (49) | human aortic SMC | DMEM | 0.9 |
| | human aortic SMC | DMEM + 1 $\mu$M Cu | 6.0 |
| | human aortic SMC | DMEM + 2 $\mu$M Fe | 4.6 |
| | human aortic SMC | M199 | 0.0 |
| Parthasaraty et al., 1985 (24) | rabbit aortic EC | F10 | 14.6 |
| | cell-free control | F10 | 4.0 |
| | cell-free control | F10 + 5 $\mu$M Cu | 9.2 |
| Cathcart et al., 1985 (50) | human monocytes | M199 + latex | 0.96 |
| | human neutrophils | M199 | 2.5 |
| Heinecke et al., 1986 (51) | human aortic SMC | MEM | 0.48 |
| | human aortic SMC | MEM + 10 $\mu$M Cu | 11.6 |
| | human aortic SMC | MEM + 10 $\mu$M Fe | 9.8 |
| | no cell control | MEM | 0.2 |
| | no cell control | MEM + 10 $\mu$M Cu | 2.6 |
| Montgomery et al., 1986 (52) | rabbit aortic EC | F10 | 2.0 |
| | no cell control | F10 | 5.0 |
| | rabbit aortic EC | F10 + 10 $\mu$M Cu | 0.6 |
| | no cell control | F10 + 10 $\mu$M Cu | 6.4 |
| van Hinsbergh et al., 1986 (53) | human umbilical artery EC | M199 + 1% BSA | 6.2 |
| | human carotid artery EC | M199 + 1% BSA | 4.6 |
| | human vena carva EC | M199 + 1% BSA | 4.8 |
| | human foreskin fibroblasts | M199 + 1% BSA | 0.6 |
| | no cell control | M199 + 1% BSA | 3.0 |
| Masana et al., 1987 (54) | human umbilical vein EC | F10 | 8.2 |
| | no cell control | F10 | 5.8 |
| Fong et al., 1987 (55) | rabbit aortic EC | F10 | 12.2 |
| | no cell control | F10 + 5 $\mu$M Cu | 16.4 |
| Hiramatsu et al., 1987 (56) | human monocytes | RPMI | 1.6 |
| | human monocytes | RPMI + PMA | 4.4 |
| Heinecke et al., 1987 (57) | monkey arterial SMC | MEM + cystine/EDTA | 0 |
| | monkey arterial SMC | MEM + cystine/EDTA, 5 $\mu$M Cu | 13.8 |
| Steinbrecher, 1988 (58) | rabbit aortic EC | F10 | 9.8 |

[a] The TBAR values are given as nmol of MDA/mg of total LDL; this is a generally accepted unit. To convert to nmol/nmol of LDL, multiply by 2.5.

macrophages than the LDL from the cell-free controls (18, 48). A very effective protection was also observed by 20 $\mu$M butylated hydroxytoluene (18, 50, 51), butylated hydroxyanisole (51), and glutathione (50). A general consensus exists that cell-mediated modification of LDL has a strict requirement for traces of copper and/or iron ions. Other transition-metal ions such as Zn are ineffective (49). A blockage of oxidation of LDL over 24 h could be achieved when 50–100 $\mu$M EDTA (18, 51) or 10–100 $\mu$M desferrioxamine (51, 56) was added to the culture medium whereas supplementation with copper or iron ions above the level normally present in the medium significantly enhanced the potential of the cells to oxidatively modify LDL. Interestingly, EDTA in equimolar concentration to $Fe^{2+}$ (10 $\mu$M) stimulated oxidation of LDL in cultures of smooth muscle cells (51). Heinecke et al. (49) showed that the amount of TBARS formed by incubation of LDL with human arterial smooth muscle cells strongly depended on the copper and iron concentrations. At concentrations below 0.8 $\mu$M no TBARS were formed; the critical concentration leading to a significant oxidation appeared to be between 1 and 2 $\mu$M copper or iron. Concentrations of 5 or 10 $\mu$M were equally effective, indicating some kind of saturation. The requirement for traces of copper and/or iron ions together with the inhibitory effects of antioxidants also explains that the composition of the culture medium has a significant effect on the extent to which LDL is modified by incubation with cells (Tables II and

**Table III. Concentration of Iron and Copper Ions and Antioxidants in Culture Media[a]**

| | M199 | F10 | DMEM | MEM | PRMI |
|---|---|---|---|---|---|
| iron ions | 1.80 | 3.0 | 0.25 | 0 | 0 |
| copper ions | 0 | 0.01 | 0 | 0 | 0 |
| cysteine | 0 | 157.0 | 0 | 0 | 0 |
| glutathione | 0.16 | 0 | 0 | 0 | 3.25 |
| vitamin E | 0.20 | 0 | 0 | 0 | 0 |
| ascorbic acid | 0.28 | 0 | 0 | 0 | 0 |
| vitamin A acetate | 3.39 | 0 | 0 | 0 | 0 |

[a] Concentrations are given in $\mu$M.

III). Steinbrecher et al. reported in 1984 (18), for example, that modification of LDL occurred in F10 medium which contains micromolar concentrations of iron and copper (Table III) and in DMEM supplemented with 5 $\mu$M copper, yet not in plain DMEM, medium 199, $\alpha$ME medium, PRMI 1640, or NCTC 109. On the other hand, Morel et al. (48) and van Hinsberg et al. (53) reported that endothelial cells and also smooth muscle cells could modify LDL in medium 199. The medium DME contains only 0.25 $\mu$M Fe and no copper (according to the supplier); and Heinecke et al. (49) in fact claimed that Fe and Cu were undetectable (<0.2 $\mu$M) in this medium. The medium 199 contains various antioxidants which should prevent LDL oxidation (Table III).

The somewhat inconsistent results regarding the importance of the culture medium probably result in part

Table IV. Time Dependence of the Formation of MDA in Some Selected Cell Cultures and Cell-Free Controls[a]

| | nmol of MDA/mg of total LDL after an incubation time, h, of | | | | | | | |
|---|---|---|---|---|---|---|---|---|
| | 0 | 2 | 4 | 6 | 12 | 16/18 | 24 | 48 |
| rabbit aorta EC, F10 (55) | 0.4 | 5.9 | 11.3 | 11.9 | 12.2 | 12.2 | 12.1 | |
| human umbilical vein EC, M199 (48) | 0 | | | 1.1 | 2.3 | 3.4 | 4.2 | 5.1 |
| human aortic SMC, DMEM + 1 µM Cu (49) | 0.3 | | | 0.3 | 3.4 | | 5.6 | 7.3 |
| human aortic SMC, DMEM + 2 µM Fe (49) | 0.0 | | | 0.3 | 1.4 | | 4.5 | 4.9 |
| bovine aortic SMC, M199 (48) | | 0.0 | | | 0.0 | 0.3 | 1.2 | 1.4 |
| human monocytes, PRMI + 1 µg PMA/mL (56) | | | 1.9 | 2.7 | | | 4.2 | |
| no cell control, DMEM + 1 µM Cu (49) | | | | | | | 1.0 | |
| no cell control, DMEM + 10 µM Cu (49) | | | | | | | 3.0 | |
| no cell control,[b] PBS + 6.6 µM Cu | 0 | 7.6 | 12.5 | 20 | | 20 | | |

[a] Most of these values were not given explicitly but were read off by us from figures in the cited papers and converted into nmol/mg of total LDL. [b] Unpublished results from our laboratory.

from contamination of the culture medium (i.e., DMEM, M199) with traces of copper and/or iron. To overcome the problem with purely defined concentrations of transition-metal ions, most researchers now seem to use media supplemented with additional copper in concentrations of 3, 5, or 10 µM (see Table II). The strict requirement for copper and or iron, however, also leads at least in our opinion to a significant uncertainty regarding the true role of cells in the oxidation process with copper/iron-supplemented media. It appears that the conditions with respect to the metal ion concentration and incubation time must be exactly balanced to see a clear difference between cells and no cell controls. For example, Heinecke et al. (49) tested the effect of $Cu^{2+}$ concentrations (0.2–10 µM) on LDL modification (24-h incubation in DMEM) by human smooth muscle cells and in cell-free dishes. Copper concentrations below 1 µM were ineffective. At 1 µM the cell incubation clearly gave more TBARS than the cell-free dishes, but at 10 µM the difference was much less pronounced, i.e., 34 nmol of MDA/mg of protein vs 16 nmol. The uptake of macrophages was 4.2- (cell modified in 10 µM $Cu^{2+}$) and 2.6-fold (cell free in 10 µM $Cu^{2+}$) higher compared with the native nonincubated LDL. Montgomery et al. (52) used F10 supplemented with 10 µM $Cu^{2+}$ and found that a cell-free control contained even about 10 times more TBARS than the incubation with endothelial cells; also, in plain F10 medium more TBARS were formed in the cell-free incubation medium. This is in clear contradiction to the results obtained by Steinbrecher (18) or Parthasarathy (24) and would deserve further investigation. The list of somewhat inconsistent results could be continued, and it seems likely that some laboratories had difficulty in obtaining clear differences between cells and cell-free controls and compare therefore cell-modified LDL with LDL incubated in the presence or absence of cells in media supplemented with antioxidants such as EDTA, butylated hydroxytoluene, or high concentrations of vitamin E. It is clear that in such cases cell-modified LDL shows a much higher degree of oxidation as measured by TBARS than the respective control. It should also be noted here that most researchers make only a single determination of TBARS in the culture medium, usually after an incubation time of 24 or 48 h.

The time course of the formation of TBARS during incubation of LDL with cells has only occasionally been measured. Examples that can be found in the literature are human aorta smooth muscle cells (49), human umbilical vein endothelial cells (48), rabbit thoracic aorta endothelial cells (55), and stimulated human monocytes (56) (Table IV). Table IV shows that among these cell lines the rabbit aorta endothelial cells are most efficient, yielding the highest TBARS in the shortest time. It is regrettable that the kinetics of the formation of TBARS in cell-free controls was not given for these experiments

and the contribution of cells to the oxidation process is therefore not clear. From our results (unpublished) with human umbilical vein endothelial cells in plain F10 medium (i.e., without $Cu^{2+}$), we conclude that no general difference exists between oxidation in the presence of cells and oxidation in the same medium in the absence of cells. As in the case of oxidation of LDL in phosphate-buffered saline supplemented with $Cu^{2+}$ (see below), the onset of TBARS formation was preceded by a lag phase where no TBARS were formed. In the incubation with cells the lag phase lasted for about 3 h whereas in the absence of cells the lag phase was longer and lasted for about 6 h. The rate of TBARS formation after the lag phase was more or less the same in both cases. The final TBARS values reached after more than 12 h were only slightly lower in the absence of cells. A clear difference in TBARS values between cells and cell-free controls can therefore only be seen within a rather narrow time window of an incubation time of about 3–6 h. It is reasonable to assume that the situation is similar with the other cell lines and other media. Nevertheless, it appears to us that the question of the true role of cells in the oxidation process deserves more attention.

It should also be mentioned that the TBARS found by different laboratories with the same cell line in the same medium differ in a wide range (Table II). With endothelial cells in M199 values of 1.85 and 6.2 nmol of MDA/mg of LDL were found; in F10 medium endothelial cells gave values of 2.0–14.6 nmol/mg of LDL. The reason for this strong variation may be severalfold as, for example, differences in the determination of TBARS, interferences by background absorption in the thiobarbituric acid assay caused by phenol red present in the culture medium, metabolism of MDA by cells, generation of MDA by damaged and dead cells, and differences in the fatty acid content of LDL. It is known (59) that MDA can only be formed from fatty acids with more than two double bonds, that is, in the case of LDL mainly from arachidonic acid. The amount of this fatty acid in LDL can vary considerably depending on the donor. Additionally, differences in various cell preparations and/or cell lines may also influence TBARS formation.

### Relationship between TBARS, Relative Electrophoretic Mobility, and Macrophage Uptake

During oxidation the negative surface charge of the LDL particle increases (for review see ref 2). The relative increase of the negative charge can be easily determined by electrophoretic separation of LDL on agarose gels and by measuring the migration distance to the anode. The relative electrophoretic mobility (REM) is defined as the ratio of migration distance of oxidized LDL to native LDL. The reason for the increase of the negative charge is not fully

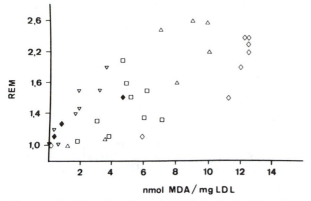

**Figure 6.** Relationship between relative electrophoretic mobility (REM) of LDL and extent of oxidation. The figure was constructed from data reported by Morell et al. (1984) (*48*) (▽), Ball et al. (1986) (*61*) (◆), van Hindsberg et al. (1986) (*53*) (□), Fong et al. (1987) (*55*) (◇), and Steinbrecher (1988) (*58*) (△).

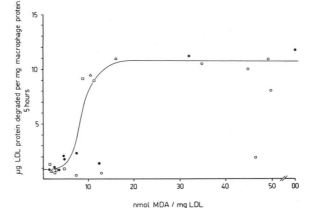

**Figure 7.** Relationship between the rate of degradation of LDL by macrophages and extent of oxidation. The figure was constructed from data reported by Steinbrecher et al. (1984) (*18*) (□), Parthasaraty et al. (1985) (*24*) (△), Fong et al. (1987) (*55*) (●), and Steinbrecher (1988) (*58*) (O). Note that MDA values are in nmol/mg of LDL protein; to convert into nmol/mg of total LDL, divide by 5.

clear. One possible explanation is that the binding of aldehydic lipid peroxidation products to the ε-amino groups of lysine residues in apo B neutralizes positive charges and thus leads to a net increase of negative charges. Alternatively, one should also consider (*60*) that reactive oxygen species generated during the oxidation process attack histidine or proline residues and convert these residues into negatively charged aspartic acid (from His) and glutamic acid (from Pro) residues (*60*). Several laboratories have reported TBARS for oLDL and the associated REM. From these reports (*48, 53, 55, 58, 61*) we have constructed Figure 6. The relationship is obviously not so clear as it might appear from a single report. The data from Fong et al. (*55*) suggested that there is an abrupt increase of REM if TBARS increase above a threshold level of about 10 nmol of MDA/mg of LDL. The values reported by Morel et al. (*48*) on the other hand rather fit into a hyperbolic curve with a more or less linear increase at a low degree of LDL oxidation and a plateau at higher MDA values. Nevertheless, the data from all five laboratories which were used in Figure 6 show that the REM increased with the extent of oxidation as measured by TBARS.

The rate by which oxidized LDL is taken up by macrophages apparently also strongly depends on the extent of oxidation. We have again compiled data from four laboratories (*18, 24, 55, 58*) into one figure which shows the uptake by macrophages as a function of MDA/mg of LDL protein (Figure 7). The interesting conclusion that can be drawn from this plot is that LDL is not taken up by macrophages unless it is oxidized to a certain extent. The threshold level is about 8 nmol of MDA/mg of LDL protein, corresponding to 1.6 nmol of MDA/mg of total LDL. Above this value there is an abrupt 10-fold increase of the uptake rate which then remains at this level independent of the extent of oxidation.

The relationship between REM and uptake rate by macrophages is shown in Figure 8. This graph constructed from two reports (*18, 58*) shows that the uptake rate increases almost linearly with increasing negative surface charge of the LDL particle as determined by its REM. This good correlation also explains that many researchers determine the extent of LDL modification due to oxidation or direct treatment with chemicals by its REM. It should be considered, however, that the REM does not always allow the prediction of scavenger receptor uptake. A large number of negatively charged residues at the apo B by itself is not necessarily associated with increased scavenger

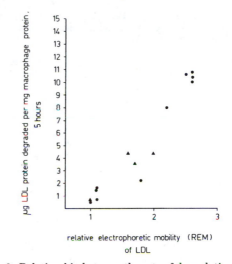

**Figure 8.** Relationship between the rate of degradation of LDL by macrophages and its relative electrophoretic mobility. The figure was constructed from data reported by Steinbrecher et al. (1984) (*18*) (▲) and Steinbrecher (1988) (*58*) (●).

uptake. It appears that rather the charge density at specific regions of the apo B is critical (for review see ref 2).

## Oxidation of LDL in Phosphate-Buffered Saline

Already in 1978, Schuh et al. (*17*) reported about an oxygen-mediated heterogeneity of low-density apolipoprotein. In these experiments, a LDL solution contained in a dialysis bag was exposed for 24 h to air-saturated phosphate-buffered saline, pH 7.4. It was observed by polyacrylamide–sodium dodecyl sulfate electrophoresis that this treatment converted the apo B from a homogeneous component of 500 kDa into a mixture of lower molecular weight peptides. The degradation of apo B correlated temporally with the increase in TBARS and in 450-nm fluorescence of apo B. Inclusion of EDTA (1 mM), propyl gallate (5 mM), or butylated hydroxytoluene (2 mM) largely prevented the degradation of apo B and formation of TBARS, and it was therefore concluded that the conversion of apo B to smaller peptides was due to

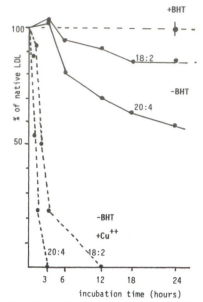

**Figure 9**. Kinetics of the degradation of the polyunsaturated fatty acids 18:2 and 20:4 during oxidation of LDL. LDL (1.5 mg/mL) was incubated with PBS in the presence of 10 $\mu$M $Cu^{2+}$ (--- ● ---), in plain PBS (—●—), and in PBS supplemented with 20 $\mu$M BHT (see also ref 41).

metal ion catalyzed free-radical reactions and to the reaction of the protein moiety with autoxidizing lipids. With the same incubation model, i.e., LDL solution in a dialysis bag, we showed (3, 40–42, 62–66) a temporal relationship between loss of the endogenous antioxidants in LDL (vitamin E, carotenoids), degradation of polyunsaturated fatty acids (18:2, 20:4), and increase in MDA and other aldehydic lipid peroxidation products. All these alterations could completely be prevented, if antioxidants such as butylated hydroxytoluene or EDTA were included in the phosphate-buffered saline. On the other hand, when PBS was supplemented with $Cu^{2+}$ ions, the polyunsaturated fatty acids were very rapidly oxidized, arachidonic acid was completely degraded within 3 h, and linoleic acid, within 12 h (Figure 9). The inhibitory effect of EDTA and the prooxidative effect of copper again indicated the importance of traces of transition-metal ions for the initiation of oxidative modification of LDL. With plain PBS the situation was therefore very similar to cell-mediated oxidation of LDL, where also a strict requirement existed for iron and/or copper ions.

In the continuation of our work we had examined the prooxidative effects of a great variety of transition-metal ions and other systems (Table V). It was rather surprising to find that among all transition-metal ions only copper ions had the ability to induce LDL oxidation in PBS. An effective prooxidant was also ferritin/cysteine and a mixture of hemoglobin and methemoglobin. Ineffective were all other systems with iron ions, including the iron/ascorbic acid mixture, a system that is a strong prooxidant, inducing lipid peroxidation in most other biological samples. When the ratio of copper ions to LDL was varied over 4 orders of magnitude (Figure 10), it was found that about 2 copper ions/LDL particle were sufficient to obtain the maximum rate of oxidation. To be on the safe side, we now routinely use a somewhat higher ratio of 16 $Cu^{2+}$/LDL. This corresponds to 6.4 $\mu$M $Cu^{2+}$ in incubation systems with 1 mg of total LDL/mL, which is very close to the copper concentration used by others in cell-mediated oxidation of LDL. The curve in Figure 10 has similarities with a titration curve and suggests, but of course does not

**Table V. Effect of Various "Prooxidants" on Oxidation of LDL[a]**

| | |
|---|---|
| $Cu^{2+}$, 1.66 $\mu$M | + |
| $Fe^{2+}$, 1.66 $\mu$M | 0 |
| $Fe^{3+}$, 1.66 $\mu$M | 0 |
| $Co^{2+}$, 1.66 $\mu$M | 0 |
| $Mn^{2+}$, 1.66 $\mu$M | 0 |
| $Zn^{2+}$, 1.66 $\mu$M | 0 |
| $Sn^{2+}$, 1.66 $\mu$M | 0 |
| $Pb^{2+}$, 1.66 $\mu$M | 0 |
| $Fe^{2+}$/EDTA, 20 $\mu$M/22 $\mu$M | 0 |
| $Fe^{2+}$/$Fe^{3+}$/EDTA, 20 $\mu$M/2 $\mu$M/24 $\mu$M | 0 |
| $Fe^{2+}$/EDTA/ascorbate, 20 $\mu$M/22 $\mu$M/100 $\mu$M | 0 |
| $Fe^{2+}$/EDTA/cysteine, 20 $\mu$M/22 $\mu$M/100 $\mu$M | 0 |
| $Fe^{2+}$/EDTA/$H_2O_2$, 20 $\mu$M/22 $\mu$M/1 mM | 0 |
| $H_2O_2$, 200 $\mu$M, 1 mM, 30 mM | 0 |
| $Fe^{2+}$/$H_2O_2$, 20 $\mu$M/1 mM | 0 |
| cumene hydroperoxide, 1 mM | 0 |
| azobis(amidinopropane)·2HCl (AAPH), 1 mM | + |
| xanthine/xanthine oxidase, 0.1–3 mM/66–660 units | 0 |
| ferritin, 500 $\mu$M Fe | 0 |
| ferritin/1 mM cysteine | + |
| ferritin/1 mM glutathione | 0 |
| hemoglobin + methemoglobin, 500 $\mu$M Fe | + |

[a] (0) Lag phase lasts longer than 3 h; (+) lag phase is over after 3 h.

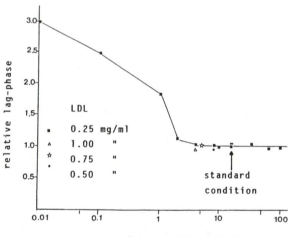

**Figure 10**. Relationship between duration of lag phase and ratio of $Cu^{2+}$ ions to LDL.

prove, that LDL has two specific copper binding sites which are crucial for the initiation of lipid peroxidation in LDL. Cell culture media supplemented with 5 or 10 $\mu$M copper appear to be much less effective in oxidizing LDL than PBS supplemented with the same concentration of copper (see Table IV). The most reasonable explanation is that in such culture media $Cu^{2+}$ is complexed to components of the medium such as, for example, histidine, which would compete for the binding of $Cu^{2+}$ to LDL. In the culture media DMEM or MEM the histidine concentration is 220 $\mu$M.

## Kinetics of Oxidation of LDL in PBS Supplemented with $Cu^{2+}$

Whereas the time course of the changes of the chemical properties of LDL during incubation with vascular cells is only poorly characterized, a rather good knowledge exists of the sequence of events, if LDL is oxidized in plain PBS in the presence or absence of pro- and antioxidants. The oxidation of LDL is a rather complex process during which all chemical and functional properties of the lipid and protein moiety of LDL are altered. The temporal rela-

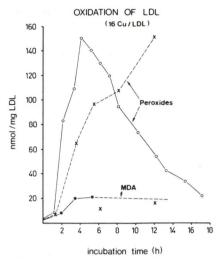

**Figure 11.** Kinetics of $Cu^{2+}$-stimulated oxidation of LDL. LDL (0.25 mg/mL) in PBS supplemented with 1.66 $\mu$M $Cu^{2+}$ was incubated for 18 h. (O) LDL from donor A; (×) LDL from donor B. Peroxides were measured as described in ref 76. MDA was measured by the TBA assay. Reprinted from ref 3 with permission.

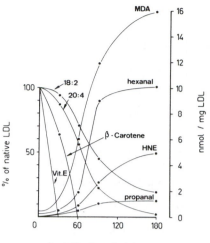

**Figure 12.** Temporal relationship in decrease of antioxidants (vitamin E and $\beta$-carotene), polyunsaturated fatty acids (18:2, 20:4), and formation of aldehydes during $Cu^{2+}$-stimulated oxidation of LDL. Experimental conditions as in Figure 11. Reprinted from ref 3 with permission.

tionship between these alterations should allow us to make conclusions regarding the importance of the individual events in the context of the whole process. The onset and progression of lipid peroxidation in LDL solutions can be followed by measuring the increase of MDA, lipid peroxides, conjugated dienes, aldehydes, and fluorescent protein or lipids. Other possibilities are the measurement of the disappearance of antioxidants and polyunsaturated fatty acids, the fragmentation of apo B, and the increase of the relative electrophoretic mobility. None of these methods gives by its own a full and satisfactory picture of the stage of oxidation. Figure 11, for example shows a copper-stimulated LDL oxidation where the lipid hydroperoxides and TBARS were measured over a period of 18 h. The time curve of the lipid peroxides clearly shows that the oxidation is a dynamic process, whereby the state of LDL continuously changes. TBARS on the other hand approaches a constant maximum value, which could be misinterpreted that oxidation has reached a final and defined stage of oxidation, which is clearly not the case. It is also important that each LDL preparation differs somewhat in the rate of the formation of peroxides as indicated by the second trace in Figure 11. A single measurement of lipid peroxides does not allow us to conclude whether LDL oxidation is in its early phase, i.e., before the peroxide maximum, or late phase, i.e., after the peroxide maximum, and the same is true in the case of a single determination of MDA. This simple example in Figure 11 already shows that an "oxidized LDL" with defined chemical properties does not exist. The properties will always depend on the time point of oxidation at which the LDL is analyzed.

Figure 12 shows another example of $Cu^{2+}$-stimulated oxidation of LDL. In this case the loss of the endogenous antioxidants vitamin E and $\beta$-carotene and the polyunsaturated fatty acids 18:2 and 20:4 and the increase of aldehydic lipid peroxidation products were measured at different time points over a period of 3 h. The temporal relationship suggested that the degradation of polyunsaturated fatty acids by lipid peroxidation is preceded by the loss of the endogenous antioxidants and that the formation of aldehydes is temporally associated with the degradation of PUFAs.

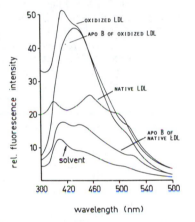

**Figure 13.** Fluorescence emission spectra at 360-nm excitation of native and oxidized LDL and its apo B (from ref 40 with permission). LDL (4.5 mg) was incubated in 3 mL of plain PBS (without $Cu^{2+}$) for 24 h. Thereafter, the fluorescence spectra were measured. After delipidation the protein was dissolved in 3 mL of 3% aqueous SDS, and the fluorescence spectra of the apo B was measured again.

At this point we intended to determine the sequence of the changes more precisely than is shown in Figures 11 and 12. But this brought us to a discouraging dilemma. As mentioned above, each LDL preparation had its own characteristic kinetics. Some were fully oxidized in about 60 min. Others needed about 180 min. So it occurred in some experiments that we missed for the analysis the phase where the antioxidants were consumed and in other experiments that the LDL sample was completely consumed by the various analyses before the onset of lipid peroxidation. One must also keep in mind that the analyses shown in Figures 11 and 12 are rather time consuming and only in retrospect give an idea of how the oxidation process proceeded in a particular experiment. To overcome these difficulties, procedures for the continuous monitoring of the LDL oxidation were developed. One possibility is the measurement of the increase of the 430-nm fluorescence (excitation 360 nm). It has already been observed by Schuh et al. (*17*), and this was later confirmed by Steinbrecher et al. (*67*) and by our group (*2, 40, 64*), that oxidized LDL has a strong fluorescence at

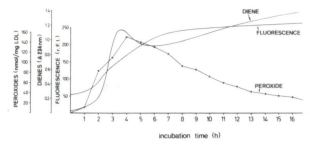

**Figure 14.** Kinetics of Cu²⁺-stimulated oxidation of LDL measured by the change of the 430-nm fluorescence (ex 360 nm), dienes, and lipid peroxides. LDL (0.25 mg/mL) was incubated in PBS + 1.66 μM Cu²⁺. Reprinted from ref 3 with permission.

around 430–450 nm. This fluorescence is most entirely associated with the protein moiety apo B (Figure 13). The chemical structure of the fluorophore has so far not been determined. From analogous studies with other peroxidized systems, it is reasonable to assume that the fluorophore is generated by binding of aldehydic lipid peroxidation products to amino groups of the protein. In fact, if native, not oxidized, LDL is treated with 4-hydroxynonenal, the resulting fluorescent chromophore is in its spectral characteristics nearly identical with that of apo B from oxidized LDL (2, 63, 64). This was clearly proven by three-dimensional fluorescence spectroscopy, a technique that allows one to determine the coordinates of a fluorophore in a complex fluorescent matrix. According to these analyses (63), the fluorophore of apo B in oxidized LDL and in LDL treated with 4-hydroxynonenal had an excitation maximum at 360 nm and an emission maximum at 430 nm (2, 63, 64). A similar fluorophore was also formed when LDL was treated with 2,4-alkadienals. On the other hand, MDA-treated LDL revealed a fluorescence maximum at 470 nm with an excitation maximum at 400 nm. At these wavelengths oxidized LDL showed only a weak shoulder. The studies on the chemical nature of these fluorophores are still at their early stages, but provisional data suggest that aldehydic lipid degradation products such as 4-hydroxynonenal and similar aldehydes formed during oxidation of LDL are able to react with apo B and thereby form fluorescent products. The kinetics of the generation of these fluorescent products can easily be followed in a fluorimeter as shown in Figure 14. In comparison with the formation of lipid peroxides measured in a parallel LDL sample, one can see that the formation of the 430-nm fluorophore is temporally associated with the increase of the peroxides during the first 4 h of the oxidation. Thereafter, the peroxides decrease, whereas the fluorescence increases, approaching a maximum after about 8 h. This was expected since the aldehydes producing these fluorophores are generated by decomposition reactions (mainly by so-called β-cleavage reactions) at the expense of lipid hydroperoxides.

An even better method for continuously monitoring the oxidation of LDL is the measurement of the increase of the 234-nm absorption (44). This absorption develops in LDL during oxidation by the conversion of the polyunsaturated fatty acids with isolated double bonds (18:2, 20:4) into fatty acid hydroperoxides with conjugated double bonds. LDL is fully soluble in PBS and remains clear in solution during the oxidation. The experimental conditions for monitoring the increase of the 234-nm absorption are very simple and, briefly, are as follows: 2 mL of an LDL solution (0.25 mg of total LDL/mL) in PBS is transferred into a 1-cm quartz cuvette and put in a UV spectrometer set to 234 nm and connected with a recorder. The basal absorption is adjusted to 0.0 or 0.1, the oxidation

is initiated by addition of 10 μL of a 0.33 mM solution of CuCl₂ in water (final concentration of Cu²⁺ is 1.66 μM), and the change of the absorption is recorded (see diene curve in Figure 14). The 234-nm absorption vs time clearly shows three phases: a lag phase or induction period, where the absorption does not increase or only slightly increases, indicating that lipid peroxidation is low or absent; a rapid increase (propagation phase) of the absorption which indicates that the lipid peroxidation has entered into a propagating chain reaction; and a decomposition or terminal phase which begins immediately after the conjugated dienes have reached the maximum value. At this period most of the fatty acids have been oxidized, and therefore, the rate of the decomposition of conjugated lipid hydroperoxides exceeds the rate of their formation, and as a consequence the 234-nm absorption decreases. The maximum peroxide content coincides with the maximum diene content, and also the first period of peroxide decomposition is temporally correlated with the decrease of the 234-nm absorption. At later stages of the oxidation process this correlation is lost, since the peroxide content steadily decreases, whereas the 234-nm absorption starts to increase again, reaching a plateau. This late increase of the 234-nm absorption is not due to the formation of new conjugated lipid hydroperoxides, but to the accumulation of lipid degradation products absorbing in the 230–210-nm region. It has also been proven (44) that during the lag phase and propagation phase conjugated dienes correlate with TBARS and lipid hydroperoxides with r = 0.99. In the decomposition phase, i.e., shortly after the diene maximum, this correlation is lost. The continuous monitoring of the LDL oxidation process by the 234-nm absorption has proven to be a very useful method for the LDL oxidation studies. Of particular importance is that only a small sample size (about 0.25–0.5 mg of LDL) of the valuable LDL is needed and that the onset of lipid peroxidation can exactly be determined by the duration of the lag phase. This enabled us to very exactly determine the temporal relationship between the consumption of endogenous antioxidants and the onset of lipid peroxidation.

## The Role of Antioxidants in Retarding Oxidation of LDL

As mentioned previously, inclusion of high concentrations of vitamin E (100 μM) in the culture medium largely prevented cell-mediated oxidation of LDL over 24 h. A concentration of 100 μM vitamin E is very high and corresponds to about 200 nmol of vitamin E/mg of LDL, which is about 100-fold higher than the endogenous vitamin E content of LDL. We wished to determine the antioxidant capacity of the endogenous antioxidants contained in LDL. For that a batch of LDL (0.25 mg/mL) was divided into two portions prior to the initiation of oxidation by the addition of Cu²⁺ (1.66 μM), one portion was used to measure the kinetics of lipid peroxidation by the increase of the 234-nm absorption, and the other portion was used for antioxidant analysis. Figure 15 shows the resulting time curves. It is evident that the first protective barrier is vitamin E, i.e., α- and γ-tocopherol. If vitamin E is consumed, the carotenoids (lycopene, β-carotene, and phytofluene) become effective, and only when this second defense line is destroyed does the lipid peroxidation process enter into a propagating chain reaction as indicated by the rapid increase of the 234-nm absorption. This sequence of destruction of endogenous antioxidants was observed in all other oxidation experiments performed so far. The sequence remained also the

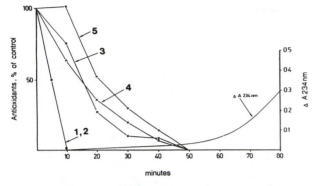

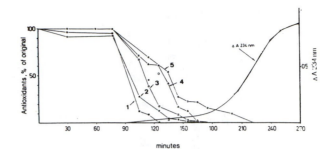

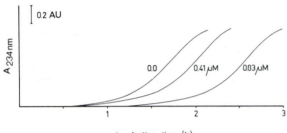

**Figure 15.** Temporal sequence of the consumption of endogenous antioxidants in LDL and onset of lipid peroxidation as measured by the diene absorption (from ref 44 with permission). (1 and 2) α- and γ-tocopherol; (3) lycopine; (4) phytofluene; (5) β-carotene. Experimental conditions as in Figure 14. Reprinted from ref 42 with permission.

**Figure 16.** Effect of urate on Cu²⁺-stimulated oxidation of LDL as measured by the increase of the diene absorption. Experimental conditions as in Figure 14 except that the LDL solution was supplemented with 0, 0.41, and 0.83 μM urate. Reprinted from ref 42 with permission.

**Figure 18.** Temporal sequence of the consumption of endogenous antioxidants in LDL and increase of diene absorption in Cu²⁺-stimulated oxidation of an LDL solution supplemented with 10 μM ascorbic acid. (1) α-Tocopherol; (2) γ-tocopherol; (3) lycopine; (4) phytofluene; (5) β-carotene. Reprinted from ref 42 with permission.

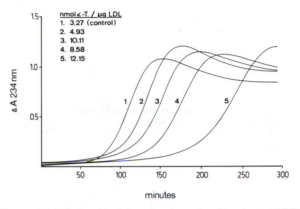

**Figure 19.** Relationship between α-tocopherol content of LDL and rate of Cu²⁺-stimulated oxidation measured by the increase of the diene absorption. Plasma was supplemented with α-tocopherol (125–1000 μM) prior LDL isolation, and the concentration of α-tocopherol in the isolated LDL is shown in the insert. Experimental conditions as in Figure 14.

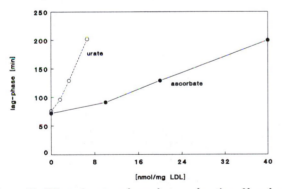

**Figure 17.** Effect of urate and ascorbate on duration of lag phase in Cu²⁺-stimulated oxidation of LDL. Experimental conditions as in Figure 14 except that the LDL solution was supplemented with 0, 0.41, 0.83, and 1.66 μM urate or 0.0, 2.5, 5.0, and 10.0 μM ascorbate.

same when water-soluble antioxidants were included in the PBS. For example, urate and also ascorbate prolong the lag period in a concentration-dependent manner (Figures 16 and 17), and even more importantly, these water-soluble antioxidants can retard the destruction of the endogenous antioxidants in LDL in a concentration-dependent manner. A typical experiment with 10 μM ascorbate is shown in Figure 18. Here the endogenous antioxidants remained virtually unchanged for 90 min; during this time the ascorbate decreased to zero. Thereafter, vitamin E and carotenoids decreased in the same sequence as in the absence of ascorbate, and lipid peroxidation entered into a propagating chain reaction when the LDL was depleted from its endogenous lipophilic antioxidants.

Vitamin E is a chain-breaking antioxidant which prevents the propagation of lipid peroxidation by scavenging lipid peroxyl radicals according to LOO• + vit E → LOOH + vit E•. Vitamin E is most likely located in the LDL in the outer phospholipid layer with the chromanol ring facing the acqueous phase. It is reasonable that the protective effect of ascorbic acid relies on its capacity of reactivating vitamin E according to vit E• + ascorbic acid → vit E + ascorbyl radical. The temporal relationship of the disappearance of the endogenous antioxidants also suggests that vitamin E has a protective effect on the carotenoids. The mechanism for vitamin E reactivation of carotenoid radicals or oxidized carotenoids is however unclear. It is also not clear by which mechanism urate retards copper-stimulated LDL oxidation.

To further prove that vitamin E is a true antioxidant in LDL, we have supplemented plasma with increasing concentrations of D,L-α-tocopherol (125–1000 μM) prior to the isolation of LDL (45). By this procedure the α-tocopherol content of LDL could be increased by nearly 4-fold, from 3.27 to 12.15 nmol/mg of LDL. When the kinetics of the copper-stimulated oxidation of these LDLs were compared (Figure 19), it became clearly evident that the lag phase increased proportionally with the α-tocopherol content of the LDL. This proves that the oxidation resistance of LDL is, at least in part, determined by its content of vitamin E. One must however consider that the other antioxidants in LDL are of equal or of possibly greater importance for the protection of LDL. Additionally, other factors, such as, for example, the content and

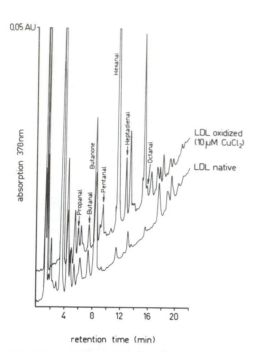

**Figure 20.** HPLC separation of the 2,4-dinitrophenylhydrazones of alkanals and alkadienals in native and 3-h Cu$^{2+}$-oxidized LDL. Reprinted from ref 41 with permission. ODS, 77.5–100% MeOH, 378 nm.

distribution of polyunsaturated fatty acids and the amount of preformed lipid peroxides, probably contribute to the susceptibility of LDL toward prooxidants. In agreement with others (68), we also find that each LDL shows its own characteristic behavior with respect to oxidation resistance. In LDL samples from different donors we have observed lag phases from 19 to 105 min (45). This wide variation cannot fully be explained by the antioxidants and polyunsaturated fatty acid content; therefore, LDL must possess additional donor-specific factors which increase or decrease the resistance to oxidation.

## Aldehydes in Oxidized LDL

It is well established from many studies on lipid peroxidation in biological samples that a great variety of aldehydes are formed through the decomposition of lipid hydroperoxides. In the first report (40) on aldehydic products in LDL, an LDL sample exposed for 24 h to an oxygen-saturated buffer was analyzed. The total amount of aldehydes on average was 5.76 nmol/mg of LDL. The molar distribution of the individual aldehydes was as follows: 36.6% malonaldehyde, 25% hexanal, 8.9% propanal, 8.2% 4-hydroxynonenal, 7.6% butanal, 4.1% 2,4-heptadienal, 3.4% pentanal, 3.4% 4-hydroxyhexenal, and 3.5% 4-hydroxyoctenal. Considerably higher concentrations of aldehydes are present in LDL exposed for 3 h to 1.66 µM Cu$^{2+}$ ions in PBS (41, 65). In this case, the total amount of aldehydes was about 40 nmol/mg of LDL; 41% of that was malonaldehyde and 8.8% 4-hydroxynonenal. It has also been shown (40) that about 90% of the TBARS is free malonaldehyde. Malonaldehyde is a hydrophilic substance and is nearly entirely released from the lipophilic LDL particle into the surrounding aqueous phase; in contrast, most of the other aldehydes are highly lipophilic and remain associated with the LDL particle where they can reach millimolar concentrations (40). Such concentrations are certainly high enough to explain chemical reactions with amino acid side chains of the apo B, e.g., binding to free amino groups.

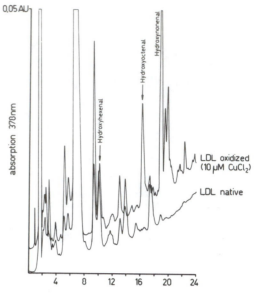

**Figure 21.** HPLC separation of the 2,4-dinitrophenylhydrazones of 4-hydroxyalkenals in native and 3-h Cu$^{2+}$-oxidized LDL. Reprinted from ref 41 with permission. ODS, 77.5–100% MeOH, 378 nm.

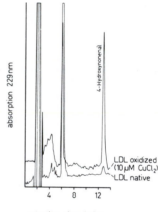

**Figure 22.** Determination by HPLC of 4-hydroxynonenal in its free form in native and 3-h Cu$^{2+}$-oxidized LDL. Reprinted from ref 41 with permission. ODS, 23% acetonitrile, 223 nm.

The method for determining the aldehydes in oLDL is in principle the same as used for other biological samples (for review see ref 69). Briefly, the LDL is treated with 2,4-dihydrophenylhydrazine in dilute hydrochloric acid. The aldehydes are converted into dinitrophenylhydrazones which are extracted, preseparated by thin-layer chromatography, and then separated by HPLC (Figures 20 and 21). 4-Hydroxynonenal can also be determined by HPLC as a free aldehyde, i.e., without prior conversion to the hydrazone (Figure 22). The direct determination usually gives somewhat lower values which suggest that in the hydrazone method also some bound or masked 4-hydroxynonenal is determined. Table VI lists the aldehydes thus far detected in copper-oxodized LDL together with their approximate concentrations, as well as the fatty acids from which they are formed and some of their properties relevant for properties of oxidized LDL. It should be mentioned here that the aldehydes identified so far are fragmentation products of the fatty acid hydroperoxides derived from the methyl terminus of the fatty acids. The

**Table VI. Aldehydes in LDL Oxidized with $Cu^{2+}$ for 3 h[a]**

| aldehyde | nmol/mg of LDL | comments |
|---|---|---|
| 4-hydroxynonenal (HNE) | 5.0–6.5 | formed from ω-6 PUFAs (18:2, 20:4); highly reactive toward SH; binds to apo B (Lys, Ser, Tyr, His); gives with apo B, proteins, and phosphatidylethanolamine a fluorophore with 430-nm emission; chemotactic for PMN at $10^{-7}$ M; cytotoxic to most cells at micromolar concentrations; potentiates aggregation and thromboxane formation in platelets; increases REM of LDL; HNE-treated LDL has lowered affinity for LDL receptor and leads to cholesterol accumulation in macrophages (*80*) |
| 4-hydroxoctenal | 0.5–1.5 | formed from unknown PUFAs; properties similar to those of HNE; chemotactic for PMN at $10^{-12}$ M |
| 4-hydroxyhexenal | 1.5–2.5 | formed from ω-3 PUFAs (22:6); less cytotoxic than HNE; also formed by hepatic metabolism from naturally occurring pyrrolizidine alkaloids (seneca alkaloids) |
| octanal | 0.5 | likely formed from 18:1; properties similar to those of hexanal |
| hexanal | 10–12 | formed from ω-6 PUFAs (18:2, 20:4); forms Schiff base with $NH_2$; gives with polylysine or erythrocyte ghost proteins a fluorophore with 430-nm emission and a similar fluorophore with phenylalanine (*81*); slightly increases REM of LDL; low cytotoxicity |
| pentanal | 1.0–2.0 | source unknown; properties similar to those of hexanal |
| butanal | 0.5–1.0 | source unknown; properties similar to those of hexanal; can condense to 2-ethyl-2-hexenal |
| propanal | 1.5–2.0 | formed from ω-3 PUFAs (22:6); properties similar to those of hexanal |
| 2,4-heptadienal | 1.0–2.0 | likely formed from ω-3 PUFAs; highly cytotoxic; likely forms cross-links in erythrocyte ghost proteins as do hexadienal and decadienal (*81*); increases REM of LDL similarly to HNE |
| malonaldehyde | 16–25 | formed from PUFAs with 3 or more double bonds (20:4, 20:5, 22:6); forms cross-links in proteins; gives with LDL, proteins, and phosphatidylethanolamine fluorophores with 460-nm emission; uptake of oxidized LDL by macrophages and its cytotoxicity correlate with MDA content; increases REM of LDL; LDL treated with 0.05–0.1 M MDA is taken up by scavenger receptor; low cytotoxicity; low mutagenicity (cytotoxicity and mutagenicity of MDA solutions can result from contaminants of methoxyacrolein) |

[a] For general review on the origin and properties of these aldehydes, see refs 2, 59, and 77–79.

counterparts where the aldehyde function is at the acyl chain linked to the parent lipid molecule must also be present in oxidized LDL, but so far no method for their analysis exists.

The oxidation hypothesis presumes that aldehydes generated by the degradation of the lipids bind covalently to amino acid side chains, in particular, to free amino groups of the apo B. The most likely reaction is the formation of Schiff bases through reaction of the aldehyde function with $NH_2$ groups. Aldehydes with the α,β-unsaturated CH=CHCHO function are bifunctional and can react with nucleophiles ($NH_2$, SH, OH) to Michael adducts (*70–73*) where the nucleophile XH is bound to carbon 3 as $CHXCH_2CHO$. The CHO group may then further condense with $NH_2$ groups to a Schiff base. In the case of 4-hydroxyalkenals, the Michael conjugates can also be stabilized through the formation of cyclic hemiacetals. The reaction of malonaldehyde has been studied intensively. Most prominent is the formation of aminoiminopropene structures RN=CHCH=CHNHR. Many consecutive reactions likely occur, once the aldehydes have been bound to the apo B. This can be concluded, for example, from the development of the 430-nm fluorescence (Figure 14). A simple Schiff base or Michael adduct between 4-hydroxynonenal and an ε-amino group would not give a fluorescent chromophore. Since a very intensive 430-nm fluorescence develops if LDL is directly treated with HNE, it is clear that additional reactions occur. The nature of these reactions is not clear. The occurrence of aldehyde or carbonyl functions in the protein apo B was demonstrated by reacting oLDL with 2,4-dihydrophenylhydrazine followed by separation of the apo B from the lipid moiety. The isolated apo B is yellow colored due to covalently bound hydrazones and gives an absorption maximum at 365 nm (Figure 23). From the absorption the amount of aldehyde function produced by the oxidation in the apo B can be calculated. The value found is 68 nmol/mg of LDL protein, in the case of LDL exposed for 3 h to 1.66 μM $Cu^{2+}$ in PBS and 0.25 mg of total LDL/mL. The carbonyl function in oxidized LDL could result from Michael adducts, i.e., 2-alkenals, which reacted with free amino or thiol groups according to RCH=CHCHO + XH

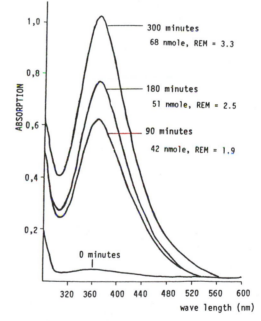

**Figure 23.** Carbonyl functions in apo B of native and oxidized LDL. LDL (1.5 mg/mL) was incubated in PBS containing 10 μM $Cu^{2+}$ for 0, 90, 180, and 300 min. The samples were treated with 2,4-dinitrophenylhydrazine reagent, the protein was recovered and dissolved in 5% SDS (600 μg of protein/mL), and the spectra were measured. The amount of aldehyde functions was calculated with the molar extinction coefficient of 25 000. The values found were 42, 52, and 68 nmol of aldehyde functions/mg of apo B after 90-, 180-, and 300-min oxidation. Given also are the REM values of the LDL samples analyzed.

→ $RCHXCH_2CHO$. Alternatively, it should also be considered that carbonyl functions are produced in proteins by the direct attack of oxygen radicals or lipid alkoxy radicals.

Antisera and monoclonal antibodies raised against malonaldehyde- (*1, 74, 75*) or 4-hydroxynonenal-modified LDL (*1, 75, 82*) cross-react with oxidized LDL, which further proves that during the oxidation of LDL these aldehydes are formed and are covalently bound to the protein moiety of LDL. The characterization of these antibodies is still at a very early stage, and the chemical structure of the antigen-determining group introduced by the reaction of the aldehydes with the apo B is not clear. Nevertheless, we think that the use of antibodies in studies of the oxidation of LDL can lead to a major breakthrough and reveal new insights regarding the role of oxidized LDL in atherosclerosis. In the arteries of hyperlipidemic Watanabe rabbits, the presence of malonaldehyde- and 4-hydroxynonenal-modified proteins was demonstrated by immunostaining (1, 74, 75), and it was also reported that autoantibodies against oxidized and aldehyde-modified LDL occur in human plasma. All these findings are strong evidence that oxidation of LDL and modification by aldehydes does in fact take place in vivo.

**Acknowledgment**. Our work has been supported by the Association for International Cancer Research (AICR) and by the Austrian Science Foundation.

## References

(1) Steinberg, D., Parthasarathy, S., Carew, T. E., Khoo, J. C., and Witztum, J. L. (1989) Beyond cholesterol. Modifications of low-density lipoprotein that increase its atherogenicity. *N. Engl. J. Med.* **320**, 915–924.

(2) Jürgens, G., Hoff, H. F., Chisolm, G. M., and Esterbauer, H. (1987) Modification of human serum low density lipoprotein by oxidation-characterization and pathophysiological implications. *Chem. Phys. Lipids* **45**, 315–336.

(3) Esterbauer, H., Rotheneder, M., Striegl, G., Waeg, G., Ashy, A., Sattler, W., and Jürgens, G. (1989) Vitamin E and other lipophilic antioxidants protect LDL against oxidation. *FAT Sci. Technol.* **91**, 316–324.

(4) Ross, R., and Glomset, J. A. (1976) The pathogenesis of atherosclerosis (Part I). *N. Engl. J. Med.* **295**, 369–377.

(5) Fowler, S., Shio, H., and Haley, N. J. (1979) Characterization of lipid-laden aortic cells from cholesterol-fed rabbits. IV. Investigation of macrophage-like properties of aortic cell populations. *Lab. Invest.* **41**, 372–378.

(6) Gerrity, R. G. (1981) The role of the monocyte in atherogenesis. I. Transition of blood-borne monocytes into foam cells in fatty lesions. *Am. J. Pathol.* **103**, 181–190.

(7) Aqel, N. M., Ball, R. Y., Waldman, H., and Mitchinson, M. J. (1984) Monocyte origin of foam cells in human atherosclerotic plaques. *Atherosclerosis* **53**, 265–271.

(8) Rosenfeld, M. E., Tsukada, T., Gown, A. M., and Ross, R. (1987) Fatty streak initiation in Watanabe heritable hyperlipidemic and comparably hypercholesterolemic fat-fed rabbits. *Arteriosclerosis* **1**, 9–23.

(9) Wissler, R. W. (1978) Progression and regression of atherosclerotic lesions. *Adv. Exp. Med. Biol.* **104**, 77–109.

(10) Brown, M. S., and Goldstein, J. L. (1983) Lipoprotein metabolism in the macrophage: Implications for cholesterol deposition in atherosclerosis. *Annu. Rev. Biochem.* **52**, 223–261.

(11) Gianturco, S. H., and Bradley, W. A. (1987) Lipoprotein receptors. In *Plasma Lipoproteins* (Gotto, A. M., Ed.) pp 183–220, Elsevier Science Publishers B.V., Amsterdam.

(12) Brown, M. S., Basu, S. P., Falck, J. R., Ho, Y. R., and Goldstein, J. L. (1980) The scavenger pathway for lipoprotein degradation: specificity of the binding site that mediates the uptake of negatively charged LDL by macrophages. *J. Supramol. Struct.* **13**, 67–81.

(13) Haberland, M. E., Olch, C. L., and Folgelman, A. M. (1984) Role of lysines in mediating interaction of modified low density lipoproteins with the scavenger receptor of human monocyte macrophages. *J. Biol. Chem.* **259**, 11305–11311.

(14) Fogelman, A. M., Shechter, I., Saeger, J., Hokom, M., Child, J. S., and Edwards, P. A. (1980) Malondialdehyde alteration of low density lipoproteins leads to cholesteryl ester accumulation in human monocyte-macrophages. *Proc. Natl. Acad. Sci. U.S.A.* **77**, 2214–2218.

(15) Haberland, M. E., Fogelman, A. M., and Edwards, P. A. (1982) Specificity of receptor-mediated recognition of malondialdehyde-modified low density lipoprotein. *Proc. Natl. Acad. Sci. U.S.A.* **79**, 1712–1716.

(16) Haberland, M. E., and Fogelman, A. M. (1987) The role of altered lipoproteins in the pathogenesis of atherosclerosis. *Am. Heart J.* **113**, 573–577.

(17) Schuh, J., Fairclough, G. F., and Haschemeyer, R. H. (1978) Oxygen-mediated heterogenity of apo-low-density lipoprotein. *Proc. Natl. Acad. Sci. U.S.A.* **75**, 3173–3177.

(18) Steinbrecher, U. P., Parthasarathy, S., Leake, D. S., Witztum, J. L., and Steinberg, D. (1984) Modification of low density lipoprotein by endothelial cells involves lipid peroxidation and degradation of low density lipoprotein phospholipids. *Proc. Natl. Acad. Sci. U.S.A.* **81**, 3883–3887.

(19) Hessler, J. R., Morel, D. W., Lewis, L. J., and Chisolm, G. M. (1983) Lipoprotein oxidation and lipoprotein-induced cytotoxicity. *Arteriosclerosis* **3**, 215–222.

(20) Morel, D. W., Hessler, J. R., and Chisolm, G. M. (1983) Low density lipoprotein cytotoxisity induced by free radical peroxidation of lipid. *J. Lipid Res.* **24**, 1070–1076.

(21) Hoff, H. F., Chisolm, G. M., III, Morel, D. W., Jürgens, G., and Esterbauer, H. (1988) Chemical and functional changes in LDL following modification by 4-hydroxynonenal. In *Oxy-Radicals in Molecular Biology and Pathology* (Cerutti, P. A., McCord, J. M., and Fridovich, T., Eds.) pp 459–472, Alan R. Liss, New York.

(22) Kaneko, T., Honda, S., Nakano, S. I., and Matsuo, M. (1987) Lethal effects of a linoleic acid hydroperoxide and its autoxidation products, unsaturated aliphatic aldehydes, on human diploid fibroblasts. *Chem.-Biol. Interact.* **63**, 127–137.

(23) Kaneko, T., Kaji, K., and Matsuo, M. (1988) Cytotoxicities of a linoleic acid hydroperoxide and its related aliphatic aldehydes toward cultured human umbilical vein endothelial cells. *Chem.-Biol. Interact.* **67**, 295–304.

(24) Parthasarathy, S., Steinbrecher, U. P., Barnett, J., Witztum, J. L., and Steinberg, D. (1985) Essential role of phospholipase $A_2$ activity in endothelial cell-induced modification of low density lipoprotein. *Proc. Natl. Acad. Sci. U.S.A.* **82**, 3000–3004.

(25) Quinn, M. T., Parthasarathy, S., Fong, L. G., and Steinberg, D. (1987) Oxidatively modified low density lipoproteins: A potential role in recruitment and retention of monocyte/macrophages during atherogenesis. *Proc. Natl. Acad. Sci. U.S.A.* **84**, 2995–2998.

(26) Curzio, M., Esterbauer, H., and Diazani, M. U. (1985) Chemotactic activity of hydroxyalkenals on rat neutrophils. *Int. J. Tissue React.* **7**, 137–142.

(27) Curzio, M., Esterbauer, H., Di Mauro, C., Cecchini, G., and Dianzani, M. U. (1986) Chemotactic activity of the lipid peroxidation product 4-hydroxynonenal and homologous hydroxyalkenals. *Biol. Chem. Hoppe-Seyler* **367**, 321–329.

(28) Curzio, M. (1988) Interaction between neutrophils and 4-hydroxyalkenals and consequences on neutrophil motility. *Free Radical Res. Commun.* **5**, 55–66.

(29) Glasgow, W. C., Harris, T. M., and Brash, A. R. (1986) A short-chain aldehyde is a major lipoxygenase product in arachidonic acid stimulated porcine leukocytes. *J. Biol. Chem.* **261**, 200–204.

(30) Glavind, J., Hartmann, S., Clemmensen, J., Jessen, K. E., and Dam, H. (1952) Studies on the role of lipoperoxides in human pathology. *Acta Pathol. Microbiol. Scand.* **30**, 1–6.

(31) Goto, Y. (1982) Lipid peroxides as a cause of vascular disease. In *Lipid Peroxides in Biology and Medicine* (Yagi, K., Ed.) pp 295–303, Academic Press, Orlando, San Diego, and San Francisco.

(32) Yagi, K. (1986) A biochemical approach to atherogenesis. *Trends Biochem. Sci. (Pers. Ed.)* **11**, 18–19. Yagi, K. (1986) Increased serum lipid peroxides initiate atherogenesis. *BioEssays* **1**, 58–60.

(33) Carew, T. E., Schwenke, D. C., and Steinberg, D. (1987) Antiatherogenic effect of probucol unrelated to its hypocholesterolemic effect: Evidence that antioxidants in vivo can selectively inhibit low density lipoprotein degradation in macrophage-rich fatty streaks and slow the progression of atherosclerosis in the Watanabe heritable hyperlipidemic rabbit. *Proc. Natl. Acad. Sci. U.S.A.* **84**, 7725–7729.

(34) Kita, T., Nagano, Y., Yokode, M., Ishi, K., Kume, N., Ooshima, A., Yoshida, H., and Kawai, C. (1987) Probucol prevents the progression of atherosclerosis in Watanabe heritable hyperlipidemic rabbit, an animal model for familial hypercholesterolemia. *Proc. Natl. Acad. Sci. U.S.A.* **84**, 5928–5931.

(35) Gey, K. F. (1986) On the antioxidant hypothesis with regard to arteriosclerosis. *Bibl. "Nutr. Dieta"* **37**, 53–91.

(36) Gey, K. F. (1989) Inverse Correlation of vitamin E and ischemic heart disease. In *Elevated Dosages of Vitamins. Benefits and Hazards* (Waler, P., Brubacher, G., and Stähelin, H., Eds.) pp 224–231, Huber, Toronto, Lewiston, NY, Bern, and Stuttgart.

(37) Gey, K. F., and Puska, P. (1989) Plasma vitamins E and A inversely related to mortality from ischemic heart disease in cross-cultural epidemiology. *Ann. N.Y. Acad. Sci.* **570**, 268–282.

(38) Goldstein, J. L., and Brown, M. S. (1978) Low-density lipoprotein pathways and its relation to atherosclerosis. *Annu. Rev. Biochem.* **46**, 897–930.

(39) Gotto, A. M., Ed. (1987) *Plasma Lipoproteins*, Elsevier, Amsterdam, New York, and Oxford.

(40) Esterbauer, H., Jürgens, G., Quehenberger, O., and Koller, E. (1987) Autoxidation of human low density lipoprotein: Loss of polyunsaturated fatty acids and vitamin E and generation of aldehydes. *J. Lipid Res.* **28**, 495–509.

(41) Esterbauer, H., Quehenberger, O., and Jürgens, G. (1988) Oxidation of human low density lipoprotein with special attention to aldehydic lipid peroxidation products. In *Free Radicals: Methodology and Concepts* (Rice-Evans, C., and Halliwell, B., Eds.) pp 243–268, The Richelieu Press, London.

(42) Esterbauer, H., Striegl, G., Puhl, H., Oberreither, S., Rotheneder, M., El-Saadani, M., and Jürgens, G. (1989) The Role of vitamin E and carotenoids in preventing oxidation of low density lipoproteins. In *Vitamin E. Biochemistry and Health Implications. Ann. N.Y. Acad. Sci.* **570**, 254–267.

(43) Di Mascio, P., Kaiser, S., and Sies, H. (1989) Lycopene as the most efficient biological carotenoid singlet oxygen quencher. *Arch. Biochem. Biophys.* **274**, 532–538.

(44) Esterbauer, H., Striegl, G., Puhl, H., and Rotheneder, M. (1989) Continuous monitoring of in vitro oxidation of human low density lipoprotein. *Free Radical Res. Commical* **6**, 67–75.

(45) Esterbauer, H., Dieber-Rotheneder, M., Striegl, G., and Waeg, G. (1990) Role of vitamin E in preventing the oxidation of low density lipoprotein. *Am. J. Clin. Nutr.* (in press).

(46) Yang, C. Y., Chan, L., and Gotto, A. M. (1987) The complete structure of human apolipoprotein B-100 and its messenger RNA. In (Gotto, A. M., Ed.) *Plasma Lipoproteins*, pp 77–93, Elsevier Science Publishers B.V., Amsterdam.

(47) Henriksen, T., Mahoney, E. M., and Steinberg, D. (1981) Enhanced macrophage degradation of low density lipoprotein previously incubated with cultured endothelial cells: recognition by receptors for acetylated low density lipoproteins. *Proc. Natl. Acad. Sci. U.S.A.* **78**, 6499–6503.

(48) Morel, D. W., DiCorleto, P. E., and Chisolm, G. M. (1984) Endothelial and smooth muscle cells alter low density lipoprotein in vitro by free radical oxidation. *Arteriosclerosis* **4**, 357–364.

(49) Heinecke, J. W., Rosen, H., and Chait, A. (1984) Iron and copper promote modification of low density lipoprotein by human arterial smooth muscle cells in culture. *J. Clin. Invest.* **74**, 1890–1894.

(50) Cathcart, M. K., Morel, D. W. and Chisolm, G. M. (1985) Monocytes and neutrophils oxidize low density lipoprotein making it cytotoxic. *J. Leukocyte Biol.* **38**, 341–350.

(51) Heinecke, J. W., Baker, L., Rosen, H., and Chait, A. (1986) Superoxide-mediated modification of low density lipoprotein by arterial smooth muscle cells. *J. Clin. Invest.* **77**, 757–761.

(52) Montgomery, R. R., Nathan, C. F., and Cohn, Z. A. (1986) Effects of reagent and cell-generated hydrogen peroxide on the properties of low density lipoprotein. *Proc. Natl. Acad. Sci. U.S.A.* **83**, 6631–6635.

(53) van Hinsbergh, V. W. M., Scheffer, M., Havekes, L., and Kempen, H. J. M. (1986) Role of endothelial cells and their products in the modification of low-density lipoproteins. *Biochim. Biophys. Acta* **878**, 49–64.

(54) Masana, L. L., Shaikh, M., La Ville, A., and Lewis, B. (1986) Lipid peroxidation of low density lipoprotein by human endothelial cells modifies its metabolism in vitro. *Rev. Esp. Fisiol.* **42**, 99–104.

(55) Fong, L. G., Parthasarathy, S., Witztum, J. L., and Steinberg, D. (1987) Nonenzymatic oxidative cleavage of peptide bonds in apoprotein B-100. *J. Lipid Res.* **28**, 1466–1477.

(56) Hiramatsu, K., Rosen, H., Heinecke, J. W., Wolfbaur, G., and Chait, A. (1987) Superoxide initiates oxidation of low density lipoprotein by human monocytes. *Arteriosclerosis* **7**, 55–60.

(57) Heinecke, J. W., Rosen, H., Suzuki, L. A., and Chait, A. (1987) The role of sulfur-containing amino acids in superoxide production and modification of low density lipoprotein by arterial smooth muscle cells. *J. Biol. Chem.* **262**, 10098–10103.

(58) Steinbrecher, U. P. (1988) Role of superoxide in endothelial-cell modification of low-density lipoproteins. *Biochim. Biophys. Acta* **959**, 20–30.

(59) Esterbauer, H., Zollner, H., and Schaur, R. J. (1989) Aldehydes formed by lipid peroxidation: Mechanism of formation, occurrence and determination. *Membrane Lipid Oxidation* (Vigo-Pelfrey, C., Ed.) Vol. I, pp 239–268, CRC Press, Boca Raton, FL.

(60) Stadtman, E. R. (1989) Oxygen radical mediated damage of enzymes: Biological implications. In *Medical, Biochemical and Chemical Aspects of Free Radicals* (Hayaishi, O., Niki, E., Kondo, M., and Yoshikawa, T., Eds.) pp 11–19, Elsevier Science Publishers B.V., Amsterdam.

(61) Ball, R. Y., Bindmann, J. P., Carpenter, K. L. H., and Mitchinson, M. J. (1986) Oxidized low density lipoprotein induces ceroid accumulation in murine peritoneal macrophages in vitro. *Atherosclerosis* **60**, 173.

(62) Esterbauer, H., Jürgens, G., Puhl, H., and Quehenberger, O. (1989) Role of oxidatively modified LDL in atherogenesis. In *Medical, Biochemical and Chemical Aspects of Free Radicals* O. (Hayaishi, E. E., Niki, M., Kondo, and T., Yoshikawa, Eds.) Elsevier, Amsterdam, pp. 1203-1209.

(63) Quehenberger, O., Koller, E., Jürgens, G., and Esterbauer, H. (1987) Investigation of lipid peroxidation in human low density lipoprotein. *Free Radical Res. Commun.* **3**, 233–242.

(64) Esterbauer, H., Quehenberger, O., and Jürgens, G. (1988) Effect of peroxidative conditions on human plasma low-density lipoproteins. In *Eicosanoids, Lipid Peroxidation and Cancer* (Nigam, S. K., McBrien, D. C. H., and Slater, T. F., Eds.) pp 203–213, Springer-Verlag, Berlin and Heidelberg.

(65) Esterbauer, H., Jürgens, G., and Quehenberger, O. (1988) Modification of human low density lipoprotein by lipid peroxidation. In *Oxygen Radicals in Biology and Medicine* (Simic, M., Taylor, K., Ward, J., and Sonntag, C. v., Eds.) pp 369–373, Plenum Press, New York and London.

(66) Quehenberger, O., Jürgens, G., Zadravec, S., and Esterbauer, H. (1988) Oxidation of human low density lipoprotein. Initiated by copper(II) chloride. In *Oxygen Radicals in Biology and Medicine* (Simic, M., Taylor, K., Ward, J., and von Sonntag, C., Eds.) pp 387–390, Plenum Press, New York and London.

(67) Steinbrecher, U. P. (1987) Oxidation of human low density lipoprotein results in derivatisation of lysine residues of apolipoprotein B by lipid peroxide decomposition products. *J. Biol. Chem.* **262**, 3603-3608.

(68) Jessup, W., Rankin, S. M., de Whalley, C. V., Hoult, J. R. S., Scott, J., Harding, A., and Leake, D. S. (1989) Alpha-tocopherol consumption during low density lipoprotein oxidation. *Biochem. J.* **265**, 399–405.

(69) Esterbauer, H., and Zollner, H. (1989) Methods for determination of aldehydic lipid peroxidation products. *Free Radical Biol. Med.* **7**, 197–203.

(70) Esterbauer, H. (1985) Lipid peroxidation products: formation, chemical properties and biological activities. In *Free Radicals in Liver Injury* (Poly, G., Cheeseman, K. H., Dianzani, M. U., Slater, T. F. Eds.) pp 29–47, IRL Press, Oxford, England.

(71) Esterbauer, H., Ertl, A., and Scholz, N. (1976) The reaction of cysteine with $\alpha,\beta$-unsaturated aldehydes. *Tetrahedron* **32**, 285–289.

(72) Schauenstein, E., Esterbauer, H., and Zollner, H. (1977) *Aldehydes in biological systems. Their natural occurrence and biological activities*, Pion Limited, London.

(73) Jürgens, G., Lang, J., and Esterbauer, H. (1986) Modification of human low-density lipoprotein by the lipid peroxidation product 4-hydroxy nonenal. *Biochim. Biophys. Acta* **875**, 103–114.

(74) Haberland, M. E., Fong, D., and Cheng, L. (1988) Malondialdehyde-altered protein occurs in atheroma of watanabe heritable hyperlipidemic rabbits. *Science* **241**, 215–218.

(75) Palinski, W., Rosenfeld, M. E., Ylä-Herttuala, S., Gurtner, G. C., Socher, S. S., Butler, S. W., Parthasarathy, S., Carew, T. E., Steinberg, D., and Witztum, J. L. (1989) Low density lipoprotein undergoes oxidative modification in vivo. *Proc. Natl. Acad. Sci. U.S.A.* **86**, 1372–1376.

(76) El-Saadani, M., Esterbauer, H., El-Sayed, M., Goher, M., Nasser, A. Y., and Jürgens, G. (1989) A spectrophotometric assay for lipid peroxides in serum lipoproteins using a commercially available reagent. *J. Lipid Res.* **30**, 627–630.

(77) Esterbauer, H., Zollner, H., and Schaur, R. J. (1989) Hydroxyalkenals: Cytotoxic products of lipid peroxidation. ISI Atlas of Science. *Biochemistry* **1**, 311–317.

(78) Esterbauer, H. (1982) Aldehydic products of lipid peroxidation. In *Free Radicals, Lipid Peroxidation and Cancer* (McBrien, D. C., and Slater, T. F., Eds.) pp 101–128, Academic Press, London and New York.

(79) Witz, G. (1989) Biological interactions of $\alpha,\beta$-unsaturated aldehydes. *Free Radical Biol. Med.* **7**, 333–349.

(80) Hoff, H. F., O'Neil, J., Chisolm, G. M., III, Cole, T. B., Quehenberger, O., Esterbauer, H., and Jürgens, G. (1989) Modification of low density lipoprotein with 4-hydroxynonenal induces uptake by macrophages. *Arteriosclerosis* **9**, 538–549.

(81) Kikugawa, K., and Beppu, M. (1987) Involvment of lipid oxidation products in the formation of fluorescent and cross-linked proteins. *Chem. Phys. Lipids* **44**, 277–296.

(82) Jürgens, G., Ashy, A., and Esterbauer, H. (1990) Detection of new epitopes formed upon oxidation of low-density lipoprotein, lipoprotein (a) and very-low-density lipoprotein. Use of an antiserum against 4-hydroxynonenal-modified low-density lipoprotein. *Biochem. J.* **265**, 605–608.

# Chapter 20

# Metallothionein and Other Cadmium-Binding Proteins: Recent Developments

Michael P. Waalkes*,[†] and Peter L. Goering[‡]

*Inorganic Carcinogenesis Section, Laboratory of Comparative Carcinogenesis, Division of Cancer Etiology, National Cancer Institute, Frederick Cancer Research Facility, Frederick, Maryland 21701, and Division of Life Sciences, Office of Science and Technology, Center for Devices and Radiological Health, Food and Drug Administration, 12709 Twinbrook Parkway, Rockville, Maryland 20857*

Reprinted from *Chemical Research in Toxicology,* Vol. 3, No. 4, July/August, 1990

## I. Introduction

Cadmium (Cd) is an extremely toxic element of continuing concern because environmental levels have risen steadily with continued worldwide industrialization. This heavy metal can cause a remarkable diversity of toxic effects including carcinogenicity, teratogenicity, mutagenicity, and endocrine and reproductive damage. Like most metals, Cd is not biotransformed, and hence biologic tolerance must therefore be of a different nature than breakdown into less toxic or more readily excreted products. In fact, most unbound metals are potentially toxic. However, the essentially of many metals, which are frequently available only in trace amounts, created the need for homeostatic mechanisms within biological systems to transport and store these metals in a nontoxic form. Such transport and storage moieties are frequently proteins. Although Cd is not essential for growth and development, at least in mammalian systems, it does, when presented to a biological system, follow the pathways of the essential elements zinc (Zn) and to a lesser extent copper (Cu). Hence Cd will be found in association with proteins whose functions are probably involved with essential trace element metabolism. Metallothionein (MT) is certainly the most well-known of such proteins and has essentially been used as the standard for which to compare other Cd-binding proteins.

[†] Frederick Cancer Research Facility.
[‡] Food and Drug Administration.

## II. Metallothionein—Background and Physiological Roles

The protein most frequently associated with cadmium exposure is MT. Since it was first detected by Margoshes and Vallee in the late 1950s as a Cd-binding protein from equine renal cortex (*1*), MT has proven to be of interest to a variety of scientists including toxicologists, biochemists, physical chemists, nutritionists, physiologists, and more recently, molecular biologists. What follows is in part a distillation of an enormous amount of work on this remarkable protein, and the interested reader is referred to several valuable reviews for a more extensive treatment of this background material (*2–6*).

MTs are a class of low molecular weight, metal-binding proteins found in a variety of prokaryotes and eukaryotes (*2–4*). As their name implies, MTs contain numerous cysteinyl thiol groups, a characteristic that allows for their high-affinity binding of several metals including Cd and Zn. Typically, some of the highest concentrations of MT are found in the liver of the mature animal after metal exposure (*7, 8*) or of perinatal animals (*2, 5, 9, 10*). The pancreas and the kidney are also tissues that contain a high amount of MT following Zn or Cd exposure (*7, 8*). MT synthesis is highly inducible by metals, especially Cd and Zn (*2–5, 7, 8, 11*). MTs are also induced by other agents, such as hormones, pharmaceuticals, alcohols, cytokines, and other diverse chemical and physical treatments (*12*; see Table I). These include ethanol (*13*), acetaminophen (*14*), urethane (*15*), formaldehyde (*16*), estradiol (*17*),

**Table I. Treatments That Induce Metallothionein Synthesis[a]**

| | |
|---|---|
| metals: Cd, Zn, Mg, Cu, etc. | formaldehyde |
| acetaminophen | 2-propanol |
| alkylating agents | glucocorticoids |
| azacytidine | catecholamines |
| butyric acid | interleukin I |
| carbon tetrachloride | interferon |
| carrageenan | food deprivation |
| chloroform | hypothermia |
| dextran | infection |
| bis(ethylhexyl)phthalate | inflammation |
| endotoxin | X irradiation |
| estradiol | UV irradiation |
| ethanol | elevated oxygen tension |
| ethionine | |

[a] Modified from ref 12.

**Table II. Criteria for Classification as a Metallothionein[a]**

1. molecular weight 6000–7000
2. metal clusters with bridging thiolate ligands
3. high cysteine content (approximately 30%)
4. high metal content
5. high metal affinity
6. deficient in aromatic amino acids and histidine
7. optical properties characteristic of metal–thiolate bonds
8. highly ordered and conserved cysteine arrangement
9. inducible by various chemical and physical treatments including metals
10. typically polymorphic

[a] See refs 4, 20, and 102 for review.

ethionine (18), and D-penicillamine (19), several of which act as cytotoxic hepatoxins. The biological half-lives of MTs generally range from 1 to 4 days depending on the type of metal bound (2, 3, 5). MTs have been characterized from a wide variety of mammalian and nonmammalian sources (2–5, 20, 21), including striped dolphin kidney (22), dogfish (23), rhesus monkey (24), and guinea pig (25) liver MT.

The physiological function of MTs may be related to Zn and Cu homeostasis, and the highest tissue concentrations of these proteins occur during periods of high Zn demand such as the perinatal period (26) or during tissue regeneration such as after partial hepatectomy (27). It is thought that MT supplies Zn within rapidly growing tissues (2, 26). Zn is an element essential for various aspects of cell division including DNA, RNA, and protein synthesis (26). MT levels are very high at birth in various tissues (2, 9, 10, 26) and then decay rapidly during postpartum growth and development. The MT gene appears to be expressed very early in mammalian development as MT mRNA has been detected in mouse endodermal yolk sacs (28). Clough et al. (29) recently observed that MT levels in human fetal liver correlate well with Zn levels and gestational age, reinforcing the concept that MT functions physiologically as a Zn storage depot particularly during periods of rapid growth. Such a function may not apply, however, during rapid growth associated with malignancy, as it has been shown that relatively low MT levels occur in the malignant portions versus the surrounding areas of surgically resected human livers (30), although levels in normal liver were not determined. This may well depend on the tumor type, however, as elevated MT expression has been detected in several tumors of nonhepatic origin such as thyroid tumors (31), testicular embryonal carcinomas (32), and skin papillomas (33). MT is also elevated during regenerative proliferation induced after chronic exposure to the hepatotoxins and tumor promoters acetaminophen and bis(2-ethylhexyl) phthalate (DEHP) (34). Tumor promoters that induce cellular enlargement (barbiturates), and not cellular hyperplasia, do not elevate hepatic MT, indicating a relationship to the enhancement of cell turnover (34). Likewise, the tumor-promoting phorbol ester 12-O-tetradecanoylphorbol 13-acetate (TPA) enhances MT mRNA expression (33). The role of MT in tumor cell pathobiology will require further definition.

The potential role of MT as an active donor of Zn and Cu to other sites within the cell has recently been reviewed (20). In support of this hypothesis, various apoenzymes which require Zn and/or Cu as cofactors, such as carbonic anhydrase, alkaline phosphatase, and superoxide dismutase, have been reactivated in vitro after incubation with Zn– and Cu–thioneins (35–37). Recently, it has also been shown that Zn–thionein in kidney activates aminolevulinic acid dehydratase, a Zn-requiring enzyme involved in heme biosynthesis, by donating Zn to the enzyme (38). It should be noted that such an effect has yet to be demonstrated utilizing an in vivo model. Further evidence suggesting that MT is not essential for direct metal donation to enzymes with subsequent activation includes the demonstration that cultured cells lacking MT function normally (39, 40). Further work on animal models (e.g., transgenic mice containing aberrant MT genes) may help to define the physiological function of this protein (41).

Another biological role for MT may involve a "detoxication" function for reactive intermediates. Thus, MT may constitute a cellular defense mechanism in the acute response to electrophilic agents such as free radicals and reactive metabolites. This capacity is most likely related to the high sulfhydryl content of MT which would provide a high degree of "neutralizing" nucleophilic equivalents (12). MT has been shown to be an effective scavenger of hydroxyl radicals in vitro (42). Toxicity resulting from exposure to carbon tetrachloride or X irradiation, which is mediated by reactive metabolites and free radicals, respectively, is markedly reduced by preinduction of MT (43–45). It is also possible that MT may provide a source and storage depot for cysteine, especially during development (46). None of these functions directly relate to Cd detoxication, and the reader should refer to recent studies and reviews for more details of these possible functions of MT (46–48).

It should be kept in mind that, despite the immense amount of research on MT, the exact function or functions of MT are not known. Further research is therefore required to determine the function of this remarkable protein. The following sections review what is known about the biochemistry, molecular biology, physiology, and pathophysiology of MT as it applies to Cd metabolism.

## III. Biochemistry of Metallothionein

The characteristics of MT (Table II) have been extensively examined. The primary structure of MT is characterized by a high content of cysteine (one-third of the total residues), serine, and glycine and is lacking in aromatic amino acids and histidine. The cysteines are arranged in a highly ordered manner; the 61-residue chain is interspersed with a series of C-X-C, C-X-X-C, and C-X-X-X-C units (C = cysteine; X = other amino acid). This confers the capacity to bind a high number of metal atoms (up to 7 in the case of Cd and Zn, 12 in the case of Cu) in two oligonuclear clusters (vide infra). There are no disulfide bridges, and all cysteinyl sulfurs directly participate in metal complexation. Generally, there are two major isoforms, referred to as MT 1 and 2, which can be resolved chromatographically and have closely related but distinct amino acid sequences (49).

**Cluster A**

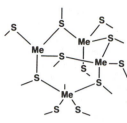

**Cluster B**

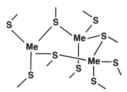

**Figure 1.** Proposed structures of the four-metal and three-metal clusters of mammalian MTs based on $^{113}$Cd NMR data. Me = metal binding site; S = cysteine sulfhydryl group. (Adapted from ref 104.)

MT is primarily an elongated, single-chain molecule with a high degree of random structure. The main contribution to the secondary structure of MT arises almost exclusively from β-turns, with some α-helical and extended β-sheet segments as determined by X-ray crystallography (50), 2D NMR (51), and Raman and infrared spectroscopy (52). Metal complexation imparts a conformational transition from predominantly random apometallothionein to MT, consisting primarily of β-turns and random segments (52).

All vertebrate MTs possess two highly organized domains, the amino-terminal domain, designated as β, consisting of residues 1–30, and the carboxy-terminal domain, designated as α, consisting of residues 31–61 (53, 54). A total of seven metal atoms are chelated into the α- and β-domains to form cluster A and cluster B, respectively, each coordinated tetrahedrally to bridging and terminal cysteines (Figure 1). Cluster A consists of four divalent metal atoms coordinated by six terminal and five bridging cysteine thiolate ligands. Cluster B consists of three divalent metal atoms coordinated by six terminal and three bridging cysteine thiolate ligands (54, 55). Binding of various metals to each cluster is specific and is a function of known differences in binding affinity and selectivity of each domain. For example, a Cd,Zn–MT is invariably induced in tissues after Cd administration with a distribution of Cd ions in cluster A and Zn ions in cluster B (54). The basis for this model rests on data obtained by several methods, including $^{113}$Cd homonuclear NMR (54, 56), proteolysis followed by sequencing (53), and X-ray crystallography (50). Recently, the "rigid" cluster model for MT has taken on an exciting new perspective. Data from 2D $^{113}$Cd NMR and $^{1}$H NMR studies, from extended X-ray absorption fine structure (EXAFS) studies, and from perturbed angular correlation of γ-ray measurements indicate that the metal–thiolate clusters of MT possess a high degree of dynamic freedom (56–58). Thus, this model projects MT as interconverting between a number of isomers of similar thermodynamic stability, involving the continual breaking and re-forming of coordination bonds while maintaining the overall spatial organization of the molecule (56). These dynamic processes allow for intramolecular metal exchange primarily within cluster B (57, 58) and possibly intermolecular metal exchange between

clusters of different MT molecules (58). While the overall structural organization of MT clusters in solution based on the above studies is in agreement with X-ray crystallography studies (50), differences revealed by the two methodologies with regard to the sequence-specific cysteine-metal binding have yet to be resolved (57). At present, it is highly likely that the dynamic nature of MT is related to the biological function of these proteins. This unique structure may facilitate intermolecular metal ion transfer between MTs and other metal-requiring proteins, including enzymes.

The nature and order of metal binding to MT have been elucidated. Earlier data suggested that Cd binding to apothionein was a sequential two-step process with the four-metal cluster A formed prior to the three-metal cluster B and that the process was strictly cooperative, with binding of one metal enhancing the binding of others (59). However, while some cooperative metal binding is apparent, Cd binding to MT has also been shown to be a multistep process involving random and ordered events with no definitive data for cooperativity in metal binding to cluster B (60). On the basis of this evidence, cluster formation proceeds only after binding of more than 3 equiv of Cd. Other evidence that metal binding events occurring within the two domains are related stems from data in which it appears that the β-domain (cluster B) is required to facilitate formation of the four-metal cluster A (61). A more detailed review of MT cluster formation can be found in ref 57.

## IV. Molecular Biology of Metallothionein

The molecular biology of the MT gene has received a great deal of attention (6, 41, 62). One of the most remarkable properties of MTs is the marked enhancement of their synthesis by exposure to metals, hormones, and other treatments (see Table I). Inducibility is regulated at the transcriptional level by various cis-acting DNA sequences located in the promoter region of MT genes (41, 62). Such sequences are responsible for MT induction by metals [metal regulatory elements (MREs)] and glucocorticoids [glucocorticoid regulatory elements (GREs)]. In addition, trans-acting DNA-binding proteins, which bind MT-inducing metals, are believed to facilitate metal interaction with the metal-regulatory elements to promote MT gene transcription (41, 62).

MT gene expression appears to be correlated with the methylation status of the DNA. Hypomethylating agents, such as 5-azacytidine, deoxyazacytidine, butyric acid, and dimethyl sulfoxide, cause enhanced synthesis of MT mRNA and protein (Figure 2) and confer tolerance to Cd (39, 41, 63–66). In contrast, hypermethylation of the MT gene correlates well with susceptibility to Cd, both in vitro and in vivo. For example, the MT gene in mouse testes is relatively highly methylated (67), an observation that correlates well with the apparent deficiency of MT protein in testes (68) and with testicular susceptibility to the acute and chronic carcinogenic effects of Cd.

Since expression of the MT gene is so readily induced by exposure to metals, the use of MT fusion genes has found very promising applications in the field of genetic engineering. In these gene constructs, the MT promoter region is linked to a specific structural gene of interest which will allow for subsequent regulation of the gene product by exposing the system to metals. In particular, the development of transgenic mice that express a MT–growth hormone fusion gene and show dramatic growth (2-fold compared to normal littermates) when fed Zn-en-

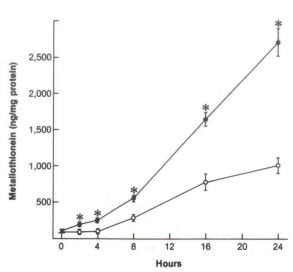

**Figure 2**. Time course of MT induced by Cd in cultured TRL 1215 cells with (●) or without (○) previous exposure to 5-aza-2′-deoxycytidine (AZA-CdR). Cells in log-phase growth were exposed to 4 μM AZA-CdR, and Cd (10 μM) was added 48 h later. MT was measured at the times indicated after Cd addition. Data represent the mean ± SE of three separate determinations, and an asterisk indicates a significant difference from cells not treated with AZA-CdR. (Adapted from ref 64.)

riched diets (69) has potential utility in biomedical and agricultural areas.

## V.  Quantitation of Metallothionein

Of the available assays for MT quantitation several take advantage of the protein's ability to avidly bind metals. Levels are measured indirectly by metal content after saturation with a metal of known high affinity such as Cd (7, 8), mercury (11), or more recently, silver (70). Problems of over- or underestimation by metal saturation assays have occurred, although the Cd-binding assay appears to be similar in precision and recovery of MT to assays that directly measure the protein moiety, such as radioimmunoassay (71). Metal saturation assays also have a potential problem of nonspecificity, although the use of silver, with its very high affinity for sulfhydryls, in part alleviates this difficulty (70). A recently developed system of discontinuous gradient polyacrylamide gel for quantitation of MT appears promising in its specificity although it would be, perhaps, cumbersome in routine use (72). Molecular biology techniques have been developed to determine MT mRNA (39, 41) and are now widely used. The clinical use of MT determination by RIA in urine of Cd-exposed individuals has been recently addressed (73). Elevated levels of urinary MT appear to have potential as a biomarker indicative of Cd-induced renal damage (73) and may prove of value in diagnosis of chronic Cd toxicity in humans. Its value in prognostication of Cd-induced nephropathy has not been thoroughly defined.

## VI.  Metallothionein and Cadmium Detoxification

MT is thought to play a very important role in the biologic detoxification of Cd in many species, and tolerance occurs through high-affinity sequestration of the metal. Piscator (74) demonstrated that following Cd exposure the metal was predominantly associated with MT. Later it was shown that pretreatment with metals known to stimulate MT synthesis will prevent the toxicity of subsequent Cd exposure (75, 76). This includes pretreatment with Zn and Cd (75, 76). It is believed that after initial exposure

to Cd or Zn the synthetic mechanisms for MT are induced, and thus newly synthesized protein provides the sequestrating capacity necessary to prevent, or substantially reduce, interactions of Cd with target molecules. That MT's primary function is for detoxication of Cd would seem, however, quite dubious. For a protein to evolve specifically for protection from a metal that has only recently been concentrated within the biosphere through industrialization is improbable. Rather, it would seem a fortuitous occurrence that MT is capable of reducing Cd toxicity and is probably related to the chemical/metabolic similarities of Zn and Cd.

## VII.  Toxicity of Cadmium–Metallothionein

Although the Cd–MT complex is relatively biologically inert when localized intracellularly, it becomes a potent nephrotoxin when it reaches the systemic circulation. Extracellular Cd–MT is available systemically either through release from liver after toxic insult or after gastrointestinal absorption, since the Cd–MT complex is absorbed at least partially intact (77). The Cd–MT complex is highly toxic to the kidney after tubular reabsorption (78, 79). Human Cd nephrotoxicity may be related to Cd–MT since this may be a major form of Cd in the diet. Smoking, another major source of human exposure to Cd, results in elevated renal Cd–MT (80) although its role in Cd-induced nephropathy in highly exposed individuals is not clearly defined.

Cd absorbed from the gastrointestinal tract or lungs is initially transported to the liver, where it induces the synthesis of MT. Continual exposure to Cd results in liver injury with leakage of Cd–thionein from damaged hepatocytes into the circulation (81). The metal–protein complex is transported to the kidney, filtered by the glomerulus, and subsequently reabsorbed by the proximal tubule cell (PTC) in a manner similar to other low molecular weight proteins (79, 81–84). The reabsorbed complex is rapidly degraded by the renal PTC lysosome system with release of Cd ion (82–84), which induces the synthesis of renal MT. This process continues until the capacity of the PTC to synthesize MT is exceeded and Cd can readily react with sensitive biological target systems.

The most critical ligands in the PTC may include those that appear to play a role in the fusion of primary lysosomes with pinocytotic vesicles to yield mature secondary lysosomes. Cadmium selectively disrupts this fusion process, leading to loss of lysosomal proteolytic enzyme activity, development of tubular proteinuria, and calcuria (85, 86) similar to what is observed in workers chronically exposed to cadmium (87).

The molecular mechanism responsible for the perturbation of the renal PTC lysosome system may involve the activation of calmodulin by non-MT-bound Cd ion (88). Activation of calmodulin, the major calcium-binding receptor in eukaryotic cells, results in a biochemical cascade leading to disruption of cell cytoskeletal components. Fusion of primary lysosomes with pinocytotic vesicles is believed to be mediated by cytoskeletal components. Disruption of the fusion process would inhibit reabsorption of proteins by the PTC, resulting in tubular proteinuria.

An alternative mechanism for Cd–MT nephrotoxicity may involve damage of the renal brush border membrane during pinocytosis of the metal–protein complex with secondary lysosomal disruption (78). This hypothesis is based on patterns of urinary enzyme excretion after Cd–MT injection which are indicative of early damage to the renal tubule brush border (89).

**Table III. Amino Acid Composition (Residues/Molecule) of Rat Hepatic Metallothionein and Testicular Metal-Binding Proteins**

| amino acid | liver | | testes | |
|---|---|---|---|---|
| | $L_1$ | $L_2$ | $T_1$ | $T_2$ |
| Asp | 5.0 (5) | 4.8 (5) | 5.6 (6) | 5.1 (5) |
| Thr | 3.1 (3) | 3.3 (3) | 3.7 (4) | 5.3 (5) |
| Ser | 10.0 (10) | 9.6 (10) | 4.9 (5) | 3.4 (3) |
| Glu | 3.3 (3) | 4.8 (5) | 12.1 (12) | 11.3 (11) |
| Pro | 2.2 (2) | 1.8 (2) | 2.9 (3) | 2.5 (3) |
| Gly | 6.2 (6) | 3.8 (4) | 1.7 (2) | 1.7 (2) |
| Ala | 5.7 (6) | 4.1 (4) | 2.9 (3) | 3.7 (4) |
| Val | 1.4 (1) | 0.8 (1) | | |
| $^1/_2$-Cys | 18.4 (18) | 17.6 (18) | 0.8 (1) | 0.6 (1) |
| Met | 0.8 (1) | 1.4 (1) | 0.9 (1) | 0.8 (1) |
| Ile | 0.8 (1) | 1.3 (1) | 2.4 (2) | 3.0 (3) |
| Leu | | | 2.5 (3) | 2.4 (2) |
| Tyr | | | | |
| Phe | | | 1.3 (1) | 1.2 (1) |
| His | | | | |
| Lys | 8.0 (8) | 8.0 (8) | 10.0 (10) | 9.0 (9) |
| total residues | 64 | 62 | 53 | 47 |
| minimum $M_r$ | 6670 | 6510 | 5850 | 5600 |

$^a$ Results are the analysis of proteins hydrolyzed with 6 N HCl containing 0.1% phenol at 110 °C for 20 h. Half-cystine was determined as cysteic acid after performic acid oxidation and hydrolysis. Values in parentheses are the nearest whole numbers. Compositions are given relative to lysine content (*93*).

## VIII.  Metallothionein Deficiency and Cadmium Toxicity

A deficiency of MT appears to occur in several mammalian tissues that are highly susceptible to the toxic effects of Cd. In particular, rat, mouse, and monkey testes, rat ventral prostate, and hamster ovary are all known to be susceptible to either the acute and/or chronic carcinogenic effects of Cd (*90, 91*) and appear to be deficient in MT as assessed by biochemical analysis of the Cd-binding protein (*68, 92–96*). The Cd-binding protein available in these tissues is uninducible (*68, 92, 95*) and thus may not afford the detoxication capacity associated with the MT. These proteins also appear to have a much lower cysteine content (Table III), the amino acid thought to be responsible for the high binding affinity of Cd for MT. Since the MT system is thought to detoxicate Cd through inducible high-affinity binding, the lack of inducibility and the low levels of cysteine in these proteins may reduce affinity as well as capacity for Cd sequestration in target tissues. This may allow Cd greater access to specific molecular targets. In the case of Cd carcinogenesis, it is presumed that DNA is the molecular target, and in this respect, treatments that prevent Cd carcinogenesis in testes, such as Zn (*97*), reduce Cd interactions with testicular DNA (*98*).

## IX.  Nonmammalian Cadmium-Binding Proteins

It is now evident that the family of Cd-binding proteins is large and diverse, particularly when nonmammalian proteins are considered. While mammalian MTs serve as a basis for comparison, it is essential that each of the nonmammalian systems be evaluated individually without tacitly assuming that these metal-binding proteins share the physical, chemical, and functional properties of MT. Cd-binding proteins have been isolated in a number of nonmammalian and plant species as diverse as *Pseudomonas*, yeast, mushroom, scallop, whelk, oyster, algae, crab, rice, and rainbow trout (*20, 21*). Many of these proteins possess properties that exclude their classification as MTs, such as differences in molecular weight, isoelectric points,

sulfhydryl and/or amino acid content, and metal-binding properties. Other Cd-binding macromolecules exist, such as the cysteine-rich polypeptides that accumulate high amounts of Cd in various plant cells and fungal organisms (*99–101*). The poly($\gamma$-glutamylcysteinyl)glycines function as the major metal ion detoxication mechanism in some plants, fungi, and unicellular organisms (*99–101*).

Evolutionary relationships between MT and other Cd-binding proteins are potentially important in elucidating the functional roles of these proteins. Sequence data for a number of proteins reveal a high degree of homology with conservation of cysteine sequences. The evolution of these proteins may have involved addition of exons to the MT gene and/or substitution of amino acids (*102*). The comparative biological approach utilized in these studies (*20, 21*) provides a tool for evaluating the potential role of these molecules in permitting such organisms to bioaccumulate toxic metals, such as Cd, resulting in an eventual contamination of the food chain. Analysis of high-affinity metal-binding proteins may also be of potential value as biological indicators of metal pollution, particularly in the marine environment (*103*). Fundamental to studying Cd-binding proteins from various phyla is the understanding that basic physiology may influence metal-binding patterns of proteins to a much higher degree in nonmammalian organisms than in mammals. It has been shown that normal seasonal processes such as alterations in water pH, temperature, and salinity as well as reproductive and molting cycles strongly influence the molecular binding patterns of metals in nonmammalian organisms (*20*).

## X.  Future Directions

Although MT must be generally considered as one of the most prominent of the Cd-binding proteins, it is now evident that other important Cd-binding proteins do exist. Further research is required to determine the extent and variation of such proteins in both mammalian and nonmammalian systems. The apparent correlation between tissue-specific MT and deficiencies and susceptibility to various aspects of Cd toxicity, especially carcinogenicity deserves further study, and the role of the available non-MT proteins in such tissues should be assessed. Specific cell types within a tissue may also be found to be deficient in MT, which may help explain certain observations concerning cell specificity in Cd toxicity, such as its apparent propensity for endothelial damage in certain organs. The study of the chronic toxicity of Cd–protein complexes, especially those found in human diets, also merits research effort. This would include the toxicity of Cd–MT as well as non-MT Cd-binding protein complexes such as those found in edible invertebrates, vegetables, and grain products. While the physiological function of MT remains enigmatic, research must continue to focus on elucidating the biological roles of MT and other Cd-binding proteins.

## References

(1) Margoshes, M., and Vallee, B. L. (1957) A cadmium protein from equine kidney cortex. *J. Am. Chem. Soc.* **79**, 4813–4814.

(2) Webb, M. (1979) *The Chemistry, Biochemistry and Biology of Cadmium*, Elsevier/North-Holland, Amsterdam.

(3) Kagi, J. H. R., and Nordberg, M. (1979) Metallothionein. *Experientia*, Suppl. 34, Birkhäuser Verlag, Basel.

(4) Kagi, J. H. R., and Kojima, Y. (1987) Metallothionein II. *Experientia*, Suppl. 52, Birkhäuser Verlag, Basel.

(5) Foulkes, E. C. (1981) *Biological roles of metallothionein*, Elsevier/North-Holland, Amsterdam.

(6) Hamer, D. H., and Winge, D. R. (1989) Metal Ion Homeostasis. *Molecular Biology and Chemistry*, Alan R. Liss, New York.

(7) Onosaka, S., and Cherian, M. G. (1981) The induced synthesis of metallothionein in various tissues of rat in response to metals. I. Effects of repeated injection of cadmium salts. *Toxicology* **22**, 91–101.

(8) Onosaka, S., and Cherian, M. G. (1981) The induced synthesis of metallothionein in various tissues of rats in response to metals. II. Influence of zinc status and specific effect on pancreatic metallothionein. *Toxicology* **23**, 11–20.

(9) Wong, K.-L., and Klaassen, C. D. (1979) Isolation and characterization of metallothionein which is highly concentrated in newborn rat liver. *J. Biol. Chem.* **254**, 12399–12403.

(10) Waalkes, M. P., and Klaassen, C. D. (1984) Postnatal ontogeny of metallothionein in various organs of the rat. *Toxicol. Appl. Pharmacol.* **74**, 314–320.

(11) Piotrowski, J. K., Bolanowska, W., and Sapota, A. (1973) Evaluation of metallothionein content in animal tissues. *Acta Biochim. Pol.* **20**, 207–215.

(12) Kagi, J. H. R., and Schaffer, A. (1988) Biochemistry of metallothionein. *Biochemistry* **27**, 8509–8515.

(13) Waalkes, M. P., Hjelle, J. J., and Klaassen, C. D. (1984) Transient induction of hepatic metallothionein following oral ethanol administration. *Toxicol. Appl. Pharmacol.* **74**, 230–236.

(14) Wormser, U., and Calp, D. (1988) Increased levels of hepatic metallothionein in rat and mouse after injection of acetaminophen. *Toxicology* **53**, 323–329.

(15) Brzeznicka, E. A., Lehman, L. D., and Klaassen, C. D. (1987) Induction of hepatic metallothionein following administration of urethane. *Toxicol. Appl. Pharmacol.* **87**, 457–463.

(16) Goering, P. L. (1989) Acute exposure to formaldehyde induces hepatic metallothionein synthesis in mice. *Toxicol. Appl. Pharmacol.* **98**, 325–337.

(17) Nishiyama, S., Taguchi, T., and Onosaka, S. (1987) Induction of zinc–thionein by estradiol and protective effects on inorganic mercury-induced renal toxicity. *Biochem. Pharmacol.* **36**, 3387–3391.

(18) Waalkes, M. P. (1985) Elevation of hepatic metallothionein in rats chronically exposed to dietary ethionine. *Toxicol. Lett.* **26**, 133–138.

(19) Heilmaier, H. E., Jiang, J. L., Greim, H., Schramel, P., and Summer, K. H. (1986) D-Penicillamine induces rat hepatic metallothionein. *Toxicology* **42**, 23–31.

(20) Petering, D. H., and Fowler, B. A. (1986) Roles of metallothionein and related proteins in metal metabolism and toxicity, problems and perspectives. *Environ. Health Perspect.* **65**, 217–224.

(21) Stone, H. C., and Overnell, J. (1985) Non-metallothionein cadmium binding proteins. *Comp. Biochem. Physiol.* **80C**, 9–14.

(22) Kwohn, Y. T., Yamazaki, S., Okubo, A., Yoshimura, E., Tatsukawa, R., and Toda, S. (1986) Isolation and characterization of metallothionein from kidney of striped dolphin, *stenella-coeruleoalba*. *Agric. Biol. Chem.* **50**, 2881–2885.

(23) Hidalgo, J., and Flos, R. (1986) Dogfish metallothionein. I. Purification and characterization and comparison with rat metallothionein. *Comp. Biochem. Physiol.* **83**, 99–103.

(24) Paliwal, V. K., Kohli, K. K., Sharma, M., and Nath, R. (1986) Purification and characterization of metallothionein from liver of cadmium exposed Rhesus-monkeys. *Mol. Cell Biochem.* **71**, 139–147.

(25) Stillman, M. J., Law, A. Y. C., Lui, E. M. K., and Cherian, M. G. (1986) Isolation and characterization of metallothionein from guinea-pig liver. *Inorg. Chim. Acta* **124**, 29–35.

(26) Brady, F. O. (1982) The physiological function of metallothionein. *Trends Biochem. Sci.* (*Pers. Ed.*) **7**, 143–145.

(27) Ohtake, H., and Koga, M. (1979) Purification and characterization of zinc-binding protein from the liver of the partially hepatectomized rat. *Biochem. J.* **183**, 683–690.

(28) Andrews, G. K., Adamson, E. D., and Gedamu, L. (1984) The ontogeny of expression of murine metallothionein: Comparison with α-fetoprotein gene. *Dev. Biol.* **103**, 294–303.

(29) Clough, S. R., Mitra, R. S., and Kulkarni, A. P. (1986) Qualitative and quantitative aspects of human-fetal liver metallothioneins. *Biol. Neonate* **49**, 241–254.

(30) Onosaka, S., Min, K.-S., Fukuhara, C., Tanaka, K., Tashiro, S.-I., Shimizu, I., Furuta, M., Yasutomi, T., Kobashi, K., and Yamamoto, K.-I. (1986) Concentrations of metallothionein and metals in malignant and non-malignant tissues in human liver. *Toxicology* **38**, 261–268.

(31) Kontozoglou, T. E., Banerjee, D., and Cherian, M. G. (1989) Immunohistochemical localization of metallothionein in human testicular embryonal carcinoma cells. *Virchows Arch. A: Pathol.* *Anat. Histopathol.* **415**, 545–549.

(32) Nartley, N., Cherian, M. G., and Banerjee, D. (1987) Immunohistochemical localization of metallothionein in human thyroid tumors. *Am. J. Pathol.* **129**, 177–182.

(33) Hashiba, H., Hosoi, J., Karasawa, M., Yamada, S., Nose, K., and Kuroki, T. (1989) Induction of metallothionein mRNA by tumor promoters in mouse skin and its constitutive expression in papillomas. *Mol. Carcinog.* **2**, 95–100.

(34) Waalkes, M. P., and Ward, J. M. (1989) Induction of hepatic metallothionein in male B6C3F1 mice exposed to hepatic tumor promoters: Effects of phenobarbital, acetaminophen, sodium barbital and di(2-ethylhexyl) phthalate. *Toxicol. Appl. Pharmacol.* **100**, 217–226.

(35) Udom, A. O., and Brady, F. O. (1980) Reactivation of in vitro of zinc-requiring apoenzymes by rat liver zinc–thionein. *Biochem. J.* **187**, 329–335.

(36) Li, T. Y., Kraker, A. J., Shaw, C. F., and Petering, D. A. (1980) Ligand substitution reactions of metallothioneins with EDTA and apocarbonic anhydrase. *Proc. Natl. Acad. Sci. U.S.A.* **77**, 6334–6338.

(37) Geller, B. L., and Winge, D. R. (1982) Metal binding sites of rat liver Cu–thionein. *Arch. Biochem. Biophys.* **219**, 109–117.

(38) Goering, P. L., and Fowler, B. A. (1987) Kidney zinc–thionein regulation of aminolevulinic acid dehydratase inhibition by lead. *Arch. Biochem. Biophys.* **253**, 48–55.

(39) Compere S. J., and Palmiter, R. D. (1981) DNA methylation controls the inducibility of the mouse metallothionein-I gene in lymphoid cells. *Cell* **25**, 233–240.

(40) Crawford, B. D., Enger, M. D., Griffith, B. B., Griffith, J. K., Hanners, J. L., Longmire, J. L., Munk, A. C., Stallings, R. L., Tesmer, J. G., Walters, R. A., and Hildebrand, C. E. (1985) Coordinate amplification of metallothionein I and II genes in cadmium-resistant Chinese hamster cells: Implications for mechanisms regulating metallothionein gene expressions. *Mol. Cell Biol.* **5**, 320–329.

(41) Hamer, D. H. (1986) Metallothionein. *Annu. Rev. Biochem.* **55**, 913–951.

(42) Thornalley, P. J., and Vasak, M. (1985) Possible role for metallothionein in protection against radiation-induced oxidative stress. Kinetics and mechanism of its reaction with superoxide and hydroxyl radicals. *Biochim. Biophys. Acta* **827**, 36–44.

(43) Cagen, S. Z., and Klaassen, C. D. (1989) Protection of carbon tetrachloride-induced hepatotoxicity by zinc: Role of metallothionein. *Toxicol. Appl. Pharmacol.* **51**, 107–116.

(44) Clarke, I. S., and Lui, E. M. K. (1986) Interaction of metallothionein and carbon-tetrachloride on the protective effect of zinc on hepatotoxicity. *Can. J. Physiol. Pharmacol.* **64**, 1104–1110.

(45) Matsubara, J., Tajima, Y., and Karasawa, M. (1987) Promotion of radioresistance by metallothionein induction prior to irradiation. *Environ. Res.* **43**, 66–74.

(46) Zlotkin, S. H., and Cherian, M. G. (1988) Hepatic metallothionein as a source of zinc and cysteine during the first year of life. *Pediatr. Res.* **24**, 326–329.

(47) Bremner, I. (1987) Interactions between metallothionein and trace elements. *Prog. Food Nutr. Sci.* **11**, 1–37.

(48) Cousins, R. J. (1985) Absorption, transport, and hepatic metabolism of copper and zinc: Special reference to metallothionein and ceruloplasmin. *Physiol. Rev.* **65**, 238–309.

(49) Kagi, J. H. R., Vasak, M., Lerch, K., Gilg, D. E. O., Hunziker, P., Bernhard, W. R., and Good, M. (1984) Structure of metallothionein. *Environ. Health Perspect.* **54**, 93–104.

(50) Furey, W. F., Robbins, A. H., Clancy, L. L., Winge, D. R., Wang, B. C., and Stout, C. D. (1986) Crystal structure of Cd, Zn metallothionein. *Science* **231**, 704–710.

(51) Braun, W., Wagner, G., Worgotter, E., Vasak, M., Kagi, J. H. R., and Wuthrich, K. (1986) Polypeptide fold in the two metal clusters of metallothionein-2 by nuclear magnetic resonance in solution. *J. Mol. Biol.* **187**, 125–129.

(52) Pande, J., Pande, C., Gilg, D., Vasak, M., Callender, R., and Kagi, J. H. R. (1986) Raman, infrared, and circular dichroism spectroscopic studies on metallothionein: a predominantly "turn"-containing protein. *Biochemistry* **25**, 5526–5532.

(53) Winge, D. R., and Miklossy, K. A. (1982) Domain nature of metallothionein. *J. Biol. Chem.* **257**, 3471–3476.

(54) Otvos, J. D., and Armitage, I. M. (1980) Structure of the metal clusters in rabbit liver metallothionein. *Proc. Natl. Acad. Sci. U.S.A.* **77**, 7094–7098.

(55) Vasak, M., and Kagi, J. H. R. (1983) Spectroscopic properties of metallothionein. In *Metal Ions in Biological Systems* (Sigel, H., Ed.) Vol. 15, pp 213–273, Marcel Dekker, New York.

(56) Vasak, M. (1986) Dynamic metal–thiolate cluster structure of metallothioneins. *Environ. Health Perspect.* **65**, 193–197.

(57) Vasak, M. (1989) Solution structure determination and pathways of cluster formation in metallothionein. In *Metal Ion Homeostasis, Molecular Biology and Chemistry* (Winge, D., and Hamer, D., Eds.) pp 207–225, Alan R. Liss, New York.

(58) Otvos, J., Chen, S., and Liu, X. (1989) NMR insights into the dynamics of metal interaction with metallothionein. In *Metal Ion Homeostasis, Molecular Biology and Chemistry* (Winge, D., and Hamer, D., Eds.) pp 197–206, Alan R. Liss, New York.

(59) Nielson, K. B., and Winge, D. R. (1983) Order of metal binding in metallothionein. *J. Biol. Chem.* **258**, 13063–13069.

(60) Bernhard, W. R., Vasak, M., and Kagi, J. H. R. (1986) Cadmium binding and metal cluster formation in metallothionein: a differential modification study. *Biochemistry* **25**, 1975–1980.

(61) Good, M., and Vasak, M. (1986) Spectroscopic properties of the cobalt (II)-substituted α-fragment of rabbit liver metallothionein. *Biochemistry* **25**, 3328–3334.

(62) Palmiter, R. (1987) Molecular biology of metallothionein gene expression. *Experientia*, Suppl. 52, 63–80.

(63) Waalkes, M. P., Wilson, M. J., and Poirier, L. A. (1985) Reduced cadmium-induced cytotoxicity in cultured liver cells by 5-azacytidine pretreatment. *Toxicol. Appl. Pharmacol.* **81**, 250–257.

(64) Waalkes, M. P., Miller, M., Wilson, M. J., Bare, R. M., and McDowell, A. E. (1988) Increased metallothionein gene expression in 5-aza-2′-deoxycytidine-induced resistance to cadmium cytotoxicity. *Chem.-Biol. Interact.* **66**, 189–204.

(65) Waalkes, M. P., and Wilson, M. J. (1986) Enhancement of cadmium-induced metallothionein synthesis in cultured liver cells by butyric acid pretreatment. *Toxicol. Lett.* **32**, E121–E126.

(66) Waalkes, M. P., and Wilson, M. J. (1987) Increased synthetic capacity for metallothionein in cultured liver cells following dimethyl sulfoxide pretreatment. *Exp. Cell Res.* **169**, 25–31.

(67) Bhave, M. R., Wilson, M. J., and Waalkes, M. P. (1988) Methylation status of the metallothionein-I gene in liver and testes of strains of mice resistant and susceptible to cadmium. *Toxicology* **50**, 231–245.

(68) Waalkes, M. P., Perantoni, A., Bhave, M. R., and Rehm, S. (1988) Strain dependence in mice of resistance and susceptibility to the testicular effects of cadmium: Assessment of the role of testicular cadmium-binding proteins. *Toxicol. Appl. Pharmacol.* **93**, 47–61.

(69) Palmiter, R. D., Brinster, R. L., Hammer, R. E., Trumbauer, M. E., Rosenfeld, M. G. Birnberg, N. C., and Evans, R. M. (1982) Dramatic growth of mice that develop from eggs microinjected with metallothionein-growth hormone fusion genes. *Nature* **300**, 611–615.

(70) Scheuhammer, A. M., and Cherian, M. G. (1986) Quantification of metallothioneins by a silver-saturation method. *Toxicol. Appl. Pharmacol.* **82**, 417–425.

(71) Dieter, H. H., Muller, L., Abel, J., and Summer, K.-H. (1986) Determination of Cd-thioneine in biological-materials: comparative standard recovery by 5 current methods using protein nitrogen for standard calibration. *Toxicol. Appl. Pharmacol.* **85**, 380–388.

(72) Lin, L.-Y., and McCormick, C. C. (1986) Quantitation of chick tissue zinc-metallothionein by gel-electrophoresis and silver stain enhancement. *Comp. Biochem. Physiol.* **85**, 75–84.

(73) Tohyama, C., Shaikh, Z. A., Nogawa, K., Kobayashi, E., and Honda, R. (1981) Elevated urinary excretion of metallothionein due to environmental cadmium exposure. *Toxicology* **20**, 289–297.

(74) Piscator, M. (1964) On cadmium in normal human kidney together with a report on the isolation of metallothionein from livers of cadmium exposed rabbits. *Nord. Hyg. Tidskr.* **45**, 76–82.

(75) Leber, A. P., and Miya, T. S. (1976) A mechanism for cadmium and zinc-induced tolerance to cadmium toxicity: Involvement of metallothionein. *Toxicol. Appl. Pharmacol.* **37**, 403–414.

(76) Yoshikawa, H. (1973) Preventive effects of pretreatment with cadmium on acute cadmium poisoning in rats. *Ind. Health* **11**, 113–119.

(77) Klein, D., Greim, H., and Summer, K. H. (1986) Stability of metallothionein in gastric-juice. *Toxicology* **41**, 121–129.

(78) Cherian, M. G., Goyer, R. A., and Delaquerriere-Richardson, L. (1976) Cadmium–metallothionein-induced nephrotoxicity. *Toxicol. Appl. Pharmacol.* **38**, 399–408.

(79) Foulkes, E. C. (1978) Renal tubular transport of cadmium–metallothionein. *Toxicol. Appl. Pharmacol.* **45**, 505–512.

(80) Summer, K. H., Drasch, G. A., and Heilmaier, H. E. (1986) Metallothionein and cadmium in human-kidney cortex: Influence of smoking. *Hum. Toxicol.* **5**, 27–33.

(81) Dudley, R. E., Gammal, L. M., and Klaassen, C. D. (1985) Cadmium-induced hepatic and renal injury in chronically exposed rats: Likely role of hepatic cadmium–metallothionein in nephrotoxicity. *Toxicol. Appl. Pharmacol.* **77**, 414–426.

(82) Squibb, K. S., Ridlington, J. W., Carmichael, N. G., and Fowler, B. A. (1979) Early cellular effects of circulating cadmium–thionein on kidney proximal tubules. *Environ. Health Perspect.* **28**, 287–296.

(83) Squibb, K. S., and Fowler, B. A. (1984) Intracellular metabolism and effects of circulating cadmium–metallothionein in the kidney. *Environ. Health Perspect.* **54**, 31–35.

(84) Squibb, K. S., Pritchard, J. B., and Fowler, B. A. (1984) Cadmium metallothionein nephropathy: Ultrastructural/biochemical alterations and intracellular cadmium binding. *J. Pharmacol. Exp. Ther.* **229**, 311–321.

(85) Goering, P. L., Squibb, K. S., and Fowler, B. A. (1986) Calcuria and proteinuria during cadmium–metallothionein-induced proximal tubule cell injury. In *Trace Substances in Environmental Health* (Hemphill, D. D., Ed.) Vol. 19, pp 22–35, University of Missouri Press, Columbia.

(86) Fowler, B. A., Goering, P. L., and Squibb, K. S. (1987) Mechanism of cadmium–metallothionein-induced nephrotoxicity: Relationship to altered renal calcium metabolism. *Experientia*, Suppl. 52, 661–668.

(87) Kazantzis, G. (1979) Renal tubular dysfunction and abnormalities of cadmium metabolism in cadmium workers. *Environ. Health Perspect.* **28**, 155–159.

(88) Chao, S. H., Suzuki, Y., Zysk, J. R., and Cheung, W. Y. (1984) Activation of calmodulin by various metal cations as a function of ionic radius. *Mol. Pharmacol.* **26**, 75–82.

(89) Suzuki, C. A. M., and Cherian, M. G. (1987) Renal toxicity of cadmium–metallothionein and enzymuria in rats. *J. Pharmacol. Exp. Ther.* **240**, 314–319.

(90) Rehm, S., and Waalkes, M. P. (1988) Species and estrous cycle dependent ovarian toxicity induced by cadmium chloride in hamsters, rats, and mice. *Fundam. Appl. Toxicol.* **10**, 635–647.

(91) Waalkes, M. P., Rehm, S., Riggs, C., Bare, R. M., Devor, D. E., Poirier, L. A., Wenk, M. L., Henneman, J. R., and Balaschak, M. S. (1988) Cadmium carcinogenesis in the male Wistar [Crl:(WI)-BR] rats; Dose–response analysis of tumor induction in the prostate, the testes and at the injection site. *Cancer Res.* **48**, 4656–4663.

(92) Waalkes, M. P., Rehm, S., and Perantoni, A. (1988) Deficiency of metallothionein in the ovaries of Syrian hamsters. *Biol. Reprod.* **39**, 953–961.

(93) Waalkes, M. P., and Perantoni, A. (1986) Isolation of a novel metal-binding protein from rat testes: Characterization and distinction from metallothionein. *J. Biol. Chem.* **261**, 13097–13103.

(94) Waalkes, M. P., Perantoni, A., and Palmer, A. E. (1988) Isolation of a novel low molecular weight metal-binding protein from the testes of the patas monkey (*Erythrocebus patas*); Characterization and distinction from metallothionein. *Biochem. J.* **256**, 131–137.

(95) Waalkes, M. P., and Perantoni, A. (1989) Apparent deficiency of metallothionein in the Wistar rat prostate. *Toxicol. Appl. Pharmacol.* **101**, 83–94.

(96) Deagen, J. T., and Whanger, P. D. (1985) Properties of cadmium-binding proteins in rat testes: Characteristics unlike metallothionein. *Biochem. J.* **231**, 279–283.

(97) Waalkes, M. P., Rehm, S., Riggs, C. W., Bare, R. M., Devor, D. E., Poirier, L. A., Wenk, M. L., and Henneman, J. R. (1989) Cadmium carcinogenesis in male Wistar [Crl:(WI)BR] rats: Dose–response analysis of effects of zinc on tumor induction in the prostate, in the testes, and at the injection site. *Cancer Res.* **49**, 4282–4288.

(98) Koizumi, T., and Waalkes, M. P. (1989) Interactions of cadmium with rat testicular interstitial cell nuclei: Alterations induced by zinc pretreatment and cadmium binding proteins. *Toxicol. in Vitro* **3**, 215–220.

(99) Steffens, J. C., Hunt, D. F., and Williams, B. G. (1986) Accumulation of non-protein metal-binding polypeptides ($\gamma$-glutamyl-cysteinyl)$_n$glycine in selected cadmium-resistant tomato cells. *J. Biol. Chem.* **261**, 13879–13882.

(100) Weber, D. N., Shaw, C. F., and Petering, D. H. (1987) *Euglena gracilis* cadmium-binding protein-II contains sulfide ions. *J. Biol. Chem.* **262**, 6962–6969.

(101)  Winge, D. R., Reese, R. N., Mehra, R. K., Torbet, E. B., Hughes, A. K., and Dameron, C. T. (1989) Structural aspects of metal-γ-glutamyl peptides. In *Metal Ion Homeostasis, Molecular Biology and Chemistry* (Winge, D., and Hamer, D., Eds.) pp 301–311, Alan R. Liss, New York.

(102)  Vasak, M., and Armitage, I. (1986) Nomenclature and possible evolutionary pathways of metallothionein and related proteins. *Environ. Health Perspect.* **65**, 215–216.

(103)  Hennig, H. F.-K. O. (1986) Metal-binding proteins as metal pollution indicators. *Environ. Health Perspect.* **65**, 175–188.

(104)  Hunt, C. T., Boulanger, Y., Fesik, S. W., and Armitage, I. M. (1984) NMR analysis of the structure and metal sequestering properties of metallothioneins. *Environ. Health Perspect.* **54**, 135–145.

**Chapter 21**

# Loss of Cancer Suppressors, a Driving Force in Carcinogenesis

Noel P. Bouck* and Benjamin K. Benton

*Department of Microbiology–Immunology and Cancer Center, Northwestern University Medical and Dental Schools, Chicago, Illinois 60611*

Reprinted from *Chemical Research in Toxicology,* Vol. 2, No. 1, January/February, 1989

Cancer has been a toxicology problem since the identification of the first carcinogens in the 1700s. Today one out of three individuals faces a diagnosis of cancer, and the majority of these lesions are thought to be environmentally induced (1). A vast number of structurally diverse compounds are able to induce and promote the development of tumors. Understanding the roles and relative importance of this array of carcinogens and promoters not only requires an understanding of their chemistry but also demands both a knowledge of exactly what types of damage actually contribute to the conversion of a cell from normal to tumorigenic and an estimate of the relative frequency with which each type of damage occurs in the development of a single tumor. Some of the most useful information on these points is currently coming from studies in cancer genetics.

Cancer in humans and other organisms is fundamentally a genetic disease resulting from stable, heritable changes in the cellular genome. It is a genetic disease in the classic sense in that almost all the different types of tumors can occur not only sporadically but also in a hereditary form whose passage from generation to generation can be followed in classic family studies (2–4). It is a genetic disease at the cellular level as tumors themselves are clonal outgrowths of cells that differ genetically from adjacent normal tissue (5). In addition, a number of diseases whose primary defect results in increased sensitivity to DNA damage have cancer as their major outcome (4), and most carcinogens and promoters are capable of damaging DNA or changing the chromosome complements of cells (6–11).

The classification of cancer as a genetic disease does not mean that tumor development is immune to epigenetic influences that alter gene expression without altering the nucleotide sequence of the gene. Regulatory toggle switches sensitive to extracellular controls operate in higher organisms (12) and are thought to orchestrate the normal process of development (13) Genes that can contribute to tumorigenicity are active during development (14). The sensitivity of such cancer genes to epigenetic changes can be demonstrated experimentally by the ability of a variety of agents, including tumor promoters, to induce whole populations of normal cells to temporarily display the phenotype of transformed cells or vice versa (15–19). Phenotypic shifts like this taking place in vivo can provide opportunities for setting up stable regulatory changes, for selection, and for further genetic changes, any or all of which might be crucial to the development of a tumor. Genes involved in neoplasia are also sensitive to changes in expression that are due to modifications of DNA rather than to changes in primary sequence. The potent demethylating agent 5-aza-2-deoxycytidine can induce tumors (20). It transforms cultured BHK hamster cells in the absence of detectable mutations by reversibly altering a gene(s) in the same functional complementation group as a suppressor gene that is mutated when transformation is induced by mutagenic carcinogens like nitrosomethylurea (NMU) and 4-nitroquinoline N-oxide (4-NQO) (21). Fascinating arguments can be made that methylation changes play a role in carcinogenesis (22), but as there is little hard data giving an indication of how frequent or how effective these or other epigenetic influences are, this discussion will deal primarily with tumor-specific changes in gene structure.

Genes that have been identified so far that are found to be altered when tumor cells are compared to normal cells fall into two general classes: oncogenes and cancer suppressor genes. Oncogenes are derived from normal cellular genes usually involved in mitogenic stimulation (14). In tumor cells these cellular genes are activated by mutations, amplifications, or rearrangements of DNA that

*Address correspondence to this author at the Department of Microbiology–Immunology, 303 East Chicago Ave., Chicago, IL 60611.

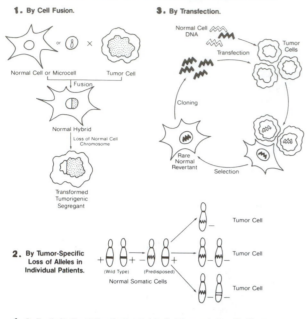

**Figure 1. Operational detection of cancer suppressor genes.** In part 4, □ = male; ○ = female; and ■ and ● = individuals with tumors, all of whom have inherited the same specific chromosome, presumably one carrying a defective suppressor gene.

result in an increase in the amount or the activity of their protein product (*14, 23, 24*). Weinberg has effectively classified oncogenes by the location of their protein products and by their ability to perform complementary functions into nuclear oncogenes and cytoplasmic oncogenes (*25*). While some oncogenes can function in both categories when overexpressed or when present in cells isolated from their neighbors, most fit in well and it is generally found that primary rodent cells can be transformed to tumorigenicity by transfection with a combination of one activated oncogene from each category. Cooperating oncogenes also occur together in some transforming viruses and a number of human tumors (*14, 23–25*).

Cancer suppressor genes, also called tumor suppressor genes, antioncogenes, or emergenes, are genes that suppress or block the expression of malignancy or of transformation (parameters of malignancy measured in cultured cells; *26*). Unlike oncogenes whose activity increases in tumor cells, suppressor genes are most active in normal cells; their function is lost in tumor cells. Their existence can be experimentally demonstrated in several ways, illustrated in Figure 1. Cell fusion studies pioneered by Harris and Klein have shown that, with few exceptions, when a normal cell is fused to a tumor cell the hybrid cell containing all of the chromosomes from each parent has a normal, nonmalignant phenotype (*27–30*). As such hybrids grow and lose chromosomes from the normal parent, malignancy is reexpressed (Figure 1, part 1). The imposition of the normal phenotype in hybrids is due to trans suppression of the transformed, malignant phenotype for (i) the type of transformation that reemerges in hybrid segregants (for example, temperature sensitivity for anchorage independence) matches that present in the original transformed parent (*31, 32*) and (ii) the malignancy of a human tumor line can be sequentially blocked, restored,

and blocked again by the addition, loss, and readdition of a single human chromosome (*33*).

Perceptive interpretation of epidemiological studies of sporadic and inherited childhood tumors by Alfred Knudson led to a second independent line of evidence supporting the existence of suppressor genes. He concluded in 1971 (*34*) that children at high risk for such tumors inherited one defective allele from their affected parent and that tumors arose when the functional, wild-type allele at this same locus, inherited from the unaffected parent, was also inactivated. Rare spontaneous tumors on the other hand required inactivation of both wild-type alleles of this gene in a single clone of somatic cells. The suppressor gene whose loss can be responsible for one of these tumors, retinoblastoma, has now been cloned (*35–37*) and Knudson's hypothesis validated by the demonstration that a defective allele, RB-1, is indeed inherited with disease susceptibility (*37, 38*). In both spontaneous and inherited retinoblastoma tumors the wild-type RB-1 alleles are often lost, either physically or functionally, leaving a hemizygous (top tumor cell, Figure 1, part 2) or homozygous (middle tumor cell, Figure 1, part 2) defective gene at the DNA level that results in the absence of the protein product of the suppressor gene in the tumor tissue (*35–37, 39*). Replacement of the lost RB-1 gene in tumor cell lines via a retroviral vector slows growth and delays tumorigenicity (*39a*).

Transfection of normal DNA into transformed cells is a third way to identify suppressor genes (Figure 1, part 3). Dominant tumor suppressor genes have recently been isolated both by Noda (*40a,b*) and by Schaefer (*41*) who transfected normal human cDNA, or DNA, into rodent cells transformed by multiple copies of an active *ras* oncogene, selected for those that reverted to normal upon uptake of human DNA, and rescued the active human sequences. Fourth, genes that dominantly predispose to a particular tumor type or subset of types (see, for example, reference 42) that have been mapped to specific chromosomes by their segregation in family pedigrees may also be suppressor genes (Figure 1, part 4). Chromosomes inherited along with tumor susceptibility are thought to carry an inactivated suppressor gene that predisposes the carrier to cancer by virtue of the fact that only one allele, present on the homologous chromosome inherited from the unaffected parent, must be inactivated for loss of suppression gene function. Such genes may be quite common in some tumor types. A recent kindred analysis suggests that over half of common colorectal cancers are associated with a dominantly inherited susceptibility (*43*) that may result from the inheritance of a defective suppressor gene.

Since the intellectual framework for the concept of suppressor genes was laid by Ohno (*44*) and Comings (*45*) in the 1970s and the field defined in its present form by reviews in 1985–6 (*26, 46–48*), summaries from a variety of points of view have been published (*4, 30, 49–55*). Data contained in these reviews indicate that multiple suppressor genes exist in the genomes of humans and rodents as they do in *Drosophila* (*56*) and in the fish *Xiphophorus* (*57*). Fifteen putative suppressor genes scattered over twelve different chromosomes have been identified so far in the human genome by detection of tumor-specific hemi- or homozygosities similar to those that accompany the loss of the RB-1 suppressor gene during the development of retinoblastomas. Verification that an actual suppressor function is associated with these chromosomes has been obtained for a total of six different chromosomes and seven genes: four genes by cell fusion or chromosome transfer [on chromosomes 1 (*58–60*), 11, which appears to carry two suppressor genes (*33, 58, 61, 62*), and 13 (*58*)] and six genes by their linkage to an inherited predisposition for the same tumor type [on chromosomes 1 (*4*), 3 (*63–65*), 5 (*66, 67*), 11 (*4, 42*), 13 (*4, 38*), and 22 (*68*)].

The consistent recessiveness of the malignant phenotype in cell fusion indicates that most if not all tumors lose both alleles of at least one suppressor gene during their development. Even cells transformed in vitro by *myc* plus *ras* or expressing both oncogenes in vivo need additional changes to become fully tumorigenic (*25, 71, 70*), changes that in hamster cells are satisfied by loss of what is presumably a suppressor on hamster chromosome 15 (*71*).

Loss of both alleles of a wild-type suppressor gene is required for the development of retinoblastoma. This is likely true for many other suppressor genes, although a failure to detect the expected allelic deletions on chromosome 5 in colonic adenomas of patients with familial polyposis coli, a predisposing condition inherited on chromosome 5, suggests this may not always be the case (*72–75*). Loss of the activity of both alleles of a single suppressor gene can promote development of tumors in a wide but not unlimited range of different tissue types (*76, 77*).

A number of functions have been envisioned as suitable for suppressor genes on the basis of the fact that their presence could block and/or their loss could contribute to a neoplastic phenotype (*26*). There is some evidence available to support the possibilities that different suppressor genes may act to do the following:

(1) **Block Oncogene Action.** In HeLa cells carrying human papilloma virus 18, a suppressor on chromosome 11 seems to mediate the sensitivity of viral oncogene transcription to methylation control (*78*) as well as to block tumorigenicity (*33*). Fibroblasts cultured from individuals with a deletion of the region of chromosome 11 associated with suppression of Wilms' tumor are sensitive to partial transformation by oncogenic proteins from the human papova virus BK (*79*) and from HPV-16 (*80*) whereas normal human fibroblasts lacking this deletion are not. Sensitivity correlates with high levels of HPV-16 RNA expression. In mouse cells a dominant function has been identified that is able to block transformation by *ras* and a subset of other oncogenes by actively interfering with either a necessary modification of the oncogene product or with a proximal or distal *ras* target (*81, 82*; see also below), and DNAs and cDNAs able to block *ras* transformation have been derived from normal human cells (*40, 41*). It seems doubtful that the normal, nonactivated allele of oncogenes like H-*ras* routinely serves as a suppressor of their activated allele as high expression of the normal allele does not limit the efficiency of transformation in vitro by the activated allele (*83, 83a*), and in several tumors where a mutated *ras* allele occurs with a very high frequency and hence probably plays a causative role, it coexists with the unaltered wild-type allele (*84–87*).

(2) **Allow normal growth control**, for example, by mediating the receipt and processing of negative growth-regulating signals. There are a number of inhibitory growth factors (see reference 14) and several retinoblastoma lines have been shown to be deficient in receptors for one of them, TGF-β (*88*). Other growth-controlling signals are able to block expression of transformation in vitro or tumorigenicity in vivo (*89–94*). Some of these seem to be passed directly from cell to cell, suggesting genes required for the synthesis of such compounds or for the integrity of the junctions through which they pass would be able to suppress malignancy. One cell variant deficient in such junctions is unusually sensitive to in vitro transformation (*91*). Maintenance of the proper

intracellular pH may also be required to maintain normal growth control (*95*).

(3) **Mediate Normal Differentiation.** Leukemias lacking the ability to respond to normal differentiation signals have been extensively characterized by Sachs and others (*96–98*), and tumor suppressor genes in flies and fish have clear defects in differentiation (*56, 57, 99*). The failure of some suppressed hybrids to form tumors has been attributed to their restored ability to differentiate in vivo (*93, 100*).

(4) **Enable Immune Rejection.** Transplantation antigens have been suggested by Sager to be suppressor genes (*26*), and their functional loss can be shown to enable cells that would otherwise be rejected to grow into tumors in an immune-competent animal (*101–103*). Whether such loss is helpful or harmful to tumor growth may vary with the experimental setting (*104*).

(5) **Promote Senescence.** Senescence, the limited capacity for cell division exhibited by mammalian cells, is a dominant cellular trait mediated in some cells by sensitivity to differentiation factors (*98*). It can be shown to block tumorigenicity (*105*) as cells stop dividing.

(6) **Prevent Angiogenesis.** All solid tumors require neovascularization to grow progressively (*106*), and in BHK hamster cells a suppressor gene controls the extracellular activity of an inhibitor able to block angiogenesis (*107*).

With the broad outlines of the concept of suppressor genes firmly established, it is now possible to look beyond verification of their existence and to begin to ask how much influence these genes have on the overall process of carcinogenesis. Evidence rapidly accumulating on two different fronts suggests that the influence of suppressor genes will be startlingly wider than originally suspected. This evidence and some of its implications are summarized below.

### Loss or Inactivation of Multiple Suppressor Genes Contributes to the Development of a Single Tumor

Recent studies suggest that individual tumors from various tissues lose suppressor gene alleles from several different specific chromosomal locations during their progression to malignancy. Although such secondary losses may not be frequent in tumors of childhood and adolescence (*2*), they are being seen with increasing frequency in a number of common adult tumors. In the case of the retinoblastoma suppressor it is possible to define the actual locus that is lost, but most other suggestions of suppressor gene loss in human material are based on detecting alleles lost in tumor tissue but present in normal cells of the same individual using anonymous DNA probes defined only by their general location on a specific chromosome. Except for chromosomes 1, 3, 5, 11, 13, and 22 (see above), arguments that tumor-specific homozygosities detected in this way really represent loss of tumor suppressor genes linked to these probes rest on observations that these tumor-specific changes occur with high frequency and considerable tumor-type specificity and consistently represent a loss of genetic material.

Vogelstein has cataloged losses from multiple chromosomes occurring at high frequency in individual colon tumors (*72*). Allelic deletions present in the DNA of colon tumor tissue but absent in DNA prepared from adjacent normal mucosa were detected on chromosome 17p (p indicates the petite, short arm of the chromosome, q the long arm) in 75% of sporadic carcinomas and 83% of these also had deletions on 18q. About 50% of these had an activated *ras* and 36% had lost material on chromosome 5q where the autosomal locus for familial adenomatous polyposis resides. Multiple loss in colon tumors was also seen by Monpezat et al. (73). These data on colon tumors are particularly compelling because almost 200 individual tumors were analyzed and because it is one of the few studies to minimize the contribution to tumor DNA preparations of stromal cells, which can obscure tumor-specific losses. It raises the possibility that modest frequencies of allele loss seen in some other studies and other tumor types might in reality be much higher. In addition, correlating the various changes in individual tumors with histopathological staging suggested that sequential *ras* activation/5q loss, 18q loss, and 17p loss often occurred as colon tumors progressed clinically from benign class I adenomas to malignant carcinomas (72).

In a smaller study of uncultured tumor tissue from small cell carcinoma of the lung, Yokota (108) found typical 3p deletions in tumors from 7/7 patients and in addition saw deletions of 13q in 10/11 individuals and 17p in 5/5. These changes also appeared sequential as 3p loss was frequent in adenocarcinomas and both 3p and 13q loss occurred prior to the amplification of the N-*myc* oncogene that is characteristic of the majority of these tumors (109). The loss of chromosome 13q may reflect loss of the retinoblastoma suppressor as in one study 80% of cell lines derived from small cell lung cancers expressed little or no mRNA specific for the Rb gene and apparently homozygous structural rearrangements of this specific locus could be detected in DNA from 1/8 primary tumors examined (110).

Provocative if not yet thoroughly convincing data suggest that multiple suppressor genes may be lost from many other tumors. Predisposition to multiple endocrine neoplasia type 2, presumably due to inheritance of a suppressor gene, is linked to chromosome 10 (111, 112), yet in 7/14 tumor samples genes are deleted from the short arm of chromosome 1 (113). Twenty (114) to twenty-seven (115) percent of primary breast tumors have tumor-specific homozygosities on chromosome 11, and 4/6 lines and 3/41 tumors have homozygous rearrangements detectable by a retinoblastoma probe from chromosome 13 (116). Only one of these tumors seems to derive from the subset of ductal breast tumors associated with a loss of DNA from chromosome 13 rather than from chromosome 11 (117). Heterozygous deletions have been seen in an additional three tumors, suggesting loss of a gene near to but not identical with the Rb gene (77). Karyotypes of eight primary anal canal carcinomas obtained prior to treatment showed seven had lost chromosomal material from chromosome 11, and five from 3p (118). Loss of chromosome 10 in 28/29 glioblastomas was accompanied by nonrandom loss of sequences from one of three other chromosomes (119). In one family where Wilms' tumor susceptibility was not linked to chromosome 11, a tumor-specific loss of alleles on chromosome 11p was seen (120).

Seventy-seven percent of osteosarcomas, a tumor associated by a number of criteria with loss of the retinoblastoma gene, also show loss of alleles on chromosome 17 (121), and 12/19 meningiomas with losses on chromosome 22 showed additional loss of genetic material on one to three other chromosomes (122). So far all of the osteosarcomas and meningiomas with secondary losses fall into the subset of such tumors that have already lost some other (presumed) suppressor allele. This raises the possibility that secondary losses in these cases could represent random segregation from a tetraploid stem line, but at least in the case of osteosarcoma, the high frequency with which stem lines with a loss of chromosome 17 are maintained in tumors suggests its loss confers a genuine growth advantage.

Interpreting those multiple tumor-specific allele losses that are seen with high frequency in human tumors as indicative of multiple suppressor gene loss seems reasonable in light of data on the in vitro transformation of cultured cells, which clearly lose multiple suppressors as they progress to malignancy. Cultured cells transform in vitro (123), as do developing tumors in vivo (5), by stepwise changes. Acquisition of immortality is frequently a first step. Immortality itself is a recessive trait (124, 125) achieved by the loss of alleles with a dominant phenotype that assure senescence and that can be classified as cancer suppressor genes (26, 105, 126). Immortal lines that have lost these genes are not yet tumorigenic but must undergo further changes, which in the case of Syrian hamster BHK (31) and SHE cells (127), as well as various mouse lines (27), involve loss of another suppressor gene detectable by cell fusion. Even *abl*-transformed lymphoid cells require additional changes, possibly suppressor loss, to become tumorigenic (128). Tumorigenic Chinese hamster cells, CHEF-16, have acquired three independent recessive traits, immortality, anchorage independence, and the ability to grow in low serum, suggesting loss of three suppressor genes (129).

Even the immortal mouse line NIH/3T3, that is so exquisitely sensitive to spontaneous transformation that it is thought to have progressed to a stage very close to tumorigenicity, loses a suppressor function when transformed by *ras* oncogenes. Transformed clones can be reverted to a normal phenotype when they are fused to parental NIH/3T3 (see Table I), although mutated *ras* genes are retained in the transformants and hybrids (Cohen, Tolsma, and Bouck, unpublished data). A DMBA-transformed mouse line and NIH/3T3 transformed by polyoma virus middle t, both known to behave dominantly in cell fusions (130), serve as controls, showing the ability of this protocol to detect lack of suppression. Additional evidence that *ras* transformants of NIH/3T3 cells have lost suppressor activity comes from (i) Harris' demonstration that NIH/3T3 but not *ras*-transformed NIH/3T3 can suppress the high malignancy of mouse melanoma PG19 (100); (ii) his finding that the tumorigenicity of NIH/3T3 *ras* transformants evolves in vivo to become independent of the *ras* gene product (131); and (iii) the ability of Noda to clone from normal cells genes that dominantly block the transformation of Ki-*ras*-transformed NIH/3T3 (40). It can also explain the observations that differentiated functions of thyroid cells transformed by a temperature-sensitive Ki-*ras* remain blocked at transcription when cells are shifted to the temperature that inactivates *ras* and that such a block can be alleviated by fusion of the transformed cell to a nontransformed thyroid cell (132, 133).

One can speculate that the other immortal cells that are not converted to malignancy solely by an active *ras* oncogene, for example, the rat line REF 32 (134) and hamster CHEF-18 (135), may also require suppressor gene inactivation. The observation that SV-40 large T antigen, a protein that complexes with the retinoblastoma suppressor gene protein, will convert ras-transformed REF 32 cells to tumorigenicity supports this interpretation (136). *Ras*-transformed CHEF-18 cells, on the other hand, seem to become tumorigenic upon duplication of DNA of chromosome 3q, not suggestive of suppressor loss (135).

## Oncogenes Can Interact with Suppressor Genes and Mimic Phenotypes of Suppressor Loss

The oncogenic proteins of several DNA viruses have

**Table I. Suppression of Oncogene Transformed NIH/3T3[a]**

| fusions | agar plating eff. of parents, % | hybrids plating in agar, % | agar plating eff. of cloned hybrids, % | |
|---|---|---|---|---|
| **NIH/3T3** | <0.07 | | | |
| × NIH/3T3 | <0.05 | <0.3 | <0.03 | (2)[b] |
| × A31 DMBA | 99 | 6.8 | 43 | (3) |
| × polyoma mT transformed NIH | 81 | 13 | 7.4 | (3) |
| **NIH/3T3 transfected with pEJ** | | | | |
| clone S2 × NIH/3T3 | 7.7 | <0.07 | <0.05 | (4) |
| clone S1b × NIH/3T3 | 49 | <0.01 | | |
| × polyoma mT transformed NIH | | 42 | 4.0 | (2) |
| clone B × NIH/3T3 | 8.6 | <0.09 | <0.4 | (3) |
| **NIH/3T3 transfected with HT1080 DNA** | 7.1 | | | |
| × NIH/3T3 | | <2.0 | 0.06 | (2) |
| × polyoma mT transformed NIH | | 113 | | |
| **NIH/3T3 transfected with T24 DNA** | | | | |
| clone 12 × NIH/3T3 | 21 | <0.05 | <0.05 | (3) |
| × polyoma mT transformed NIH | | 48 | 7.5 | (3) |
| clone 21 × NIH/3T3 | 38 | <2.5 | 0.03 | (3) |
| × polyoma mT transformed NIH | | 129 | 1.6 | (2) |

[a] NIH/3T3 cells were transformed by transfection with plasmid containing activated H-*ras* (pEJ), high molecular weight DNA from human tumors containing active H-*ras* (T24) or N-*ras* (HT1080), or polyoma middle t antigen (mT) and transformants recloned in agar. Selectable markers on plasmids pSV2neo and pSV2gpt were fixed in parental lines and cells fused pairwise as indicated by using polyethylene glycol. Twenty-four hours after fusion, cultures were assayed, under conditions that only hybrids could survive, for total hybrids by cloning on plastic and for transformed hybrids by cloning in soft agar. Hybrid clones arising on plastic were also picked, recloned, and assayed for expression of transformation by ability to clone in soft agar. [b] The number of clones tested is given in parentheses.

recently been found to interact physically with the one identifiable suppressor gene product, the ~105-kDa retinoblastoma (Rb) protein. Stable protein–protein complexes between the Rb protein and SV-40 large T antigen and between the Rb protein and the adenovirus E1A proteins have been detected by their coimmunoprecipitation from extracts of virus-transformed cells (*137–139*). This associated with Rb may be necessary for viral transformation for (i) the Rb protein is one of three whose coprecipitation with E1A is lost when the ability of E1A to cooperate with *ras* in transformation is lost; (ii) transformation-defective SV-40 large T antigen mutants with alterations at residues 105–114 failed to coprecipitate Rb; and (iii) sequences in this region of large T are homologous to those in E1A region 2, which is necessary for E1A transformation, and their substitution for E1A region 2 results in only a modest loss of its transforming activity (*139*). Other nuclear oncogenes may also interact with Rb, for E7 of human papilloma virus 16 and the large T antigens of other papova viruses also have strong homologies to regions of SV-40 large T and E1A that interact with Rb (*139*). These viral oncogenes and the Rb cellular suppressor gene are apparently constituents of intersecting regulatory pathways within the cell.

This raises the possibility that these viral oncogenes may perform a function in transformation equivalent to that of the loss of the Rb suppressor and that nuclear oncogenes in general may function as antisuppressors, imposing on the cell the same phenotype it would have if it lost a suppressor gene function. Phenotypes induced by activation of nuclear oncogenes do resemble phenotypes predicted for suppressor gene loss. Transplantation antigens that enable immune rejection can be down regulated by cellular oncogenes c-*myc* and N-*myc* as well as by the viral E1A (*140–142*), differentiation can be blocked by *myc* (*143, 144*), and nuclear oncogenes in general can block senescence (*25*).

There are a number of ways that activated oncogenes could be imposing phenotypes parallel to those resulting from suppressor gene inactivation. (i) Some oncogenes could be acting via paths that are independent of those used by suppressor genes. There are apparently several different pathways by which cells can differentiate (*96, 97,*

*145*) and therefore presumably independent ways to block differentiation. (ii) Other oncogenes may interfere with suppressor gene activity either at the protein level by interacting directly with their products as SV-40 large T and E1A may do with Rb or at transcription/translation as *myc*, *fos*, and E1A with the major histocompatibility complex and other cellular genes. (iii) Active oncogenes could induce suppressor gene loss at the DNA level. Chromosomal changes are seen in rat lines expressing the nuclear oncogenes polyoma large T or *myc*, but not in those expressing cytoplasmic polyoma middle T or *ras* oncogenes (*146*). (iv) Some activated oncogenes may actually be wayward suppressor genes whose products, when mutated or overexpressed, are able to dominantly interfere with the function of their own wild-type gene product. Models for such effects exist. The Prm promoter of λ is inactive when the CI repressor is either absent or present in excessive concentrations (*147*). Perhaps more pertinent, evidence has recently been summarized by Ptashne that transcription can be squelched by excessive concentrations of transcriptional activators, although it is promoted by lower concentrations of the same activator and does not proceed at all in its absence (*12*). Most nuclear oncogenes appear to be components of transcriptional complexes and can be activated by overexpression. One of them, the p53 oncogene, seems to be able to promote malignancy either when it is overexpressed or when its function is lost (*148, 149*). Both alleles of p53 were recently found to be inactivated by double, independent viral insertions in a Friend virus-induced erythroleukemia (*149*). In an intriguing parallel to the Rb suppressor protein, p53 is bound by transforming proteins of adenovirus and SV-40 (*150*), and some clones of p53 capable of immortalizing cells in culture have activating mutations enabling them to bind a cellular protein, hsc-70 (*151*). Some members of the *myc* and *jun* families could also be suppressor genes that can be activated to oncogenes. One *jun* locus maps to the region of human chromosome 1 that is deleted in the majority of neuroblastomas (*152*) and hence thought to contain a suppressor locus.

Cytoplasmic oncogenes may also interact with suppressor genes. Noda has isolated revertants of NIH/3T3 cells doubly transformed by Ki-*ras* that block the ex-

**Table II. Cellular Blocks to Transformation by Cell-Derived Oncogenes[a]**

| Species | Oncogenes — Sensitive to block [ - - - - - - - - - - - - - - - - - - - - - - - ] Resistance to block | Reference |
|---|---|---|
| Mouse | src   ras     fes [-] fms     sis     mos | (81,82) |
|  | src [-] ras     mos | (156) |
|  | ras   abl   fes [ - - - - - - - - - - - ] mos | (157) |
|  | ras [- - - - - - - - - - - - - - - - - - - - - - - - - - - - - - - | (158,159) |
| Mink | ras     fes   fms [ - - - - - - - ] mos | (160) |
| Hamster | neu [-] ras     mos | (b) |
|  | ras [- - - - - - - - - - - - - - - - - - - - - - - - - - - - - - | (126) |
| Rat | ras [- - - - - - - - - - - - - - - - - - - - - - - - - - - - - - - | (161,41) |
|  | ras   abl     mos   fos [-] trk | (162) |
|  | mos [-] ras | (163,164) |
|  | ras     mos [ - - - - - - - - - - | (163,164) |

[a] Multifunctional oncogenes of DNA viruses and blocks that appear to be confined to LTR-mediated transcription of oncogenes of RNA viruses have been omitted. [b] Benton and Bouck, unpublished data.

pression of transformation when fused to cells transformed by *ras*, *fes*, or *src*, but not by *mos*, *sis*, *fms*, polyoma, or SV-40, yet remain normal upon fusion to the NIH/3T3 from which they were derived (*81*, *82*). These revertants behave as if they contain a suppressor gene able to actively block either some necessary activation of *ras*, *src*, and *fes* proteins or, more likely, impair a proximal or distal target of *ras*, *fes*, and *src* not used by the other oncogenes to which the line is sensitive.

A number of variant cell lines have been isolated that have lesions that confer resistance to transformation by some but not all oncogenes. Such lesions indicate points at which suppressor genes may interfere with oncogene action. Table II summarizes the variant lines reported to date that have cellular blocks preventing oncogene transformation and the activated oncogenes known to be sensitive or resistant to each block. The set of oncogenes sensitive to a particular cellular block are listed to the left in the table. Such sets of oncogenes may be sensitive to the same cellular lesion because they transform cells via pathways that converge at some point. Alternatively, groups of oncogenes sensitive to the same block may have a common requirement for some cellular activity, such as acylation. Although the function of a *ras* homologue in yeast can be impaired by loss of the yeast cell's ability to acylate proteins (*155*), there do not seem to be any consistent modifications that we know of so far that tie together sets of oncogenes sensitive to the same cellular block. It thus seems more likely that common sensitivity is the result of common pathways, a concept supported by the ability of injected anti-*ras* antibodies to inhibit several aspects of the transformed phenotype in cells transformed by *src*, *fms*, and *fes* but not in those transformed by *mos*, *raf*, or SV-40 (*153*) and also by common alterations in the inositol phospholipid cycle seen in cells transformed by *ras*, *src*, *met*, *trk*, *mos*, or *raf*, but not by *fos* or *myc* (*154*). The set of oncogenes to the right of the blocks indicated in Table II may act beyond the block or by some unrelated pathway.

Considering all of the available data, it is possible to arrange many of the oncogenes in a sequential pathway dictated by their sensitivity to known blocks and to define

**Scheme I. One Simple Sequence of Oncogene Action Compatible with Known Oncogene Blocks**

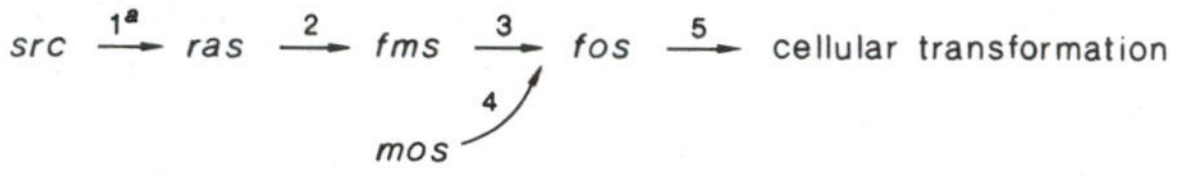

src —1[a]→ ras —2→ fms —3→ fos —5→ cellular transformation
           4↗ mos

*neu* acts in this pathway prior to *ras* or via another pathway
*fes* and *abl* act in this pathway prior to *fms*
*sis* acts in this pathway distal to the block at 2 or via another pathway
*trk* acts in this pathway distal to the block at 5 or via another pathway

[a] Numbers indicate possible positions of blocks that obstruct the ability of an activated oncogene to transform cells as predicted by data in Table II. The number of intervening steps between oncogenes is unknown and probably variable.

limits as to where in this pathway other oncogenes might interact. A sequential arrangement that is hypothetical, but compatible with available data, is illustrated in Scheme I. This arrangement suggests that a minimum of five independent cellular functions exist that when altered obstruct oncogene action without loss of cell viability (numbered in Scheme I). The cellular lesion defining block 2 is dominant (*81*), and that at block 3 in mink cells may also be dominant as some hybrids between blocked cells and transformed cells appear normal (*160*). Dominant blocks suggest (re-) activation of a suppressor. Variants defining blocks 1 and 5 seem recessive, suggesting they arose upon loss of modifying genes or loss of proximal or distal oncogene targets (*156*, *162*). Such targets could themselves either be products of suppressor genes or be influenced by suppressor genes. Suppressor involvement in oncogene function is further supported by a hamster line in which loss of suppressor genes as defined by cell fusion is accompanied by the loss of a block to transformation by activated H-*ras* (*127*).

Thus two lines of evidence indicate that, of the known genetic changes that take place as a normal cell progresses to malignancy, oncogene activation and suppressor gene loss, suppressor loss may be the more influential. First, genetic considerations lead to the conclusion that, during the development of any given tumor, genetic events that inactivate suppressor genes outnumber events that activate

**Table III. Genetic Changes Expected To Contribute to Suppressor Inactivation**

**First and/or Second Allele**
frame-shift or point mutation
deletion in coding or noncoding region
inactivating chromosomal translocation
disruption by insertion of viral or cellular element
silencing by parental imprinting

**Second Allele**
mitotic recombination
chromosome loss with or without reduplication
gene conversion

oncogenes. Suppressor gene loss seems to occur in virtually every tumor and usually requires inactivation of two alleles. Often individual tumors lose multiple suppressor genes. Genetic changes thought at the moment to be involved in the development of a malignant colon tumor consist of the activation of one *ras* allele and the inactivation of five alleles of suppressor genes on three different chromosomes. Second, studies aimed at analyzing functions of oncogenes and suppressors indicate that some oncogene products can interact with a suppressor gene product, many can impose on the cell a phenotype that mimics suppressor loss, and a number of cellular blocks to oncogene action exist that define further points at which suppressor genes may determine oncogene effects.

The kind of genetic damage expected to be effective at inactivating suppressor genes is summarized in Table III. It differs in several respects from that required to activate oncogenes. As there is no need to maintain function, any lethal hit to the gene will be effective. The target is thus larger than for oncogene activation and likely to be hit more frequently by any given agent or event and to be sensitive to a broader range of more disruptive damage. Carcinogens that activate oncogenes by mutation can also inactivate suppressor genes as seen by the ability of NMU and 4NQO to knock out one suppressor gene in the already immortal BHK hamster cells (*31*) and by the induction of immortality in primary hamster cultures by a variety of carcinogens (*165*) and in human cells by 4NQO or benzo-[a]pyrene (*166, 167*). In contrast to the expectations for oncogene activation, suppressor genes are as effectively inactivated by frame-shift carcinogens as by point mutagens in vitro (*31*) and frequently suffer deletions and gross structural rearrangements in tumors (*35, 38, 50, 168–170*). Chromosomal translocations do inactivate suppressor genes with some frequency in renal cell carcinoma (*63, 64, 171*), but no targeted patterns like those seen in translocations that activate oncogenes (*172*) have been discerned.

Inactivation by parental imprinting could be involved in loss of function of the first allele of some suppressor genes. In mice the expression of both endogenous and experimentally introduced genes can be determined by whether the gene is inherited from the male or female parent (*173, 174*). Such inactivation by parental imprinting is associated with increased methylation (*174–176*). A series of recent observations on the inheritance of susceptibility to Wilms' tumor raises the possibility that imprinting could be influencing the suppressor gene that is lost with high frequency from chromosome 11p in these tumors. First, in a study of five random families and in several other isolated reports, tumor-specific loss always involves loss of the maternal chromosome 11p (*177, 120*). Second, the region of 11p to which the Wilms' tumor suppressor maps is frequently highly methylated in tumor lines (*178*). Third, genes in this region of chromosome 11 are homologous to those on a part of mouse chromosome 7 that is subject to inactivation by parental imprinting

(*173*). Fourth, three families have recently been reported in which inherited predisposition to Wilms' tumor fails to map to chromosome 11p, suggesting alterations elsewhere in the genome can also contribute to this tumor (*120, 179*). These embryonic tumors occur so early in life that it is uncomfortable to propose that a large number of somatic changes must take place before tumors can develop, yet a suppressor from 11p likely also must be lost for tumorigenicity. If the activity of one suppressor allele on chromosome 11 was lost by genomic imprinting, then only a single additional inactivation would be required after fertilization (*120*). Instead of gene inactivation by genomic imprinting, loss of maternal chromosome 11p could unveil an inactivating mutation on the paternal chromosome 11p that occurred early in development as a result of the deamination of 5-methylcytosine to thymine in the DNA (*179a*). It is well-known that 5-methylcytosine occurs with a very high frequency in sperm DNA compared to the DNA of the egg (*179b*).

The usual need to inactivate both alleles of a suppressor gene in the same cell lineage in order to abrogate its function means that chromosomal changes assume an important role in carcinogenesis. In about 60% of retinoblastoma tumors one allele of the RB-1 suppressor gene is lost as a result of loss of the whole chromosome 13 carrying the wild-type allele. In about one-tenth of retinoblastomas, one defective allele has become homozygous due to mitotic recombination between the defective and wild-type chromosome 13s (*50, 2*). In about half of the Wilms' tumors (*180*) and rhabdomyosarcomas (*181*) suppressor gene homozygosity is achieved by mitotic recombination. The fact that tumor-specific homozygosities have been seen for 13 different chromosomes suggests that such mechanisms operate for other suppressor genes on other chromosomes. A sharp increase in aneuploidy also accompanies development of carcinomas in skin tumor models (*182, 183*). All these findings make agents that destabilize chromosomes, the clastogens and aneuploidogens (*184, 185*), seem increasingly important in carcinogenesis. In addition, carcinogen assays like the micronucleus test that measure malsegregation of chromosomes and that have been suggested as supplements to the combined chemical structure plus *Salmonella* mutagenicity assays used to detect carcinogens take on added meaning (*6*).

Although tumors appear to succeed due to the acquisition of new abilities that enable them to grow in adverse circumstances, a major driving force behind their evolution at the genetic level appears to be not gain, but loss. What are lost in developing tumors are functional alleles of cancer suppressor genes that in normal cells play vital roles in maintaining the orderly growth, differentiation, and repair that are the hallmarks of normal tissue. The crucial questions of what agents induce such gene loss, how they act, and how one might protect against their effects are already being studied, often in other contexts. The increasing influence of suppressor gene loss in tumorigenesis brings a new focus to the process of carcinogenesis and to the range of chemicals that induce and promote malignancy.

**Acknowledgment.** We are grateful for the support of NIH Grants CA 27306 (to N.P.B.) and CA 09560, which contributed to the support of B.K.B.

## References

(1) Doll, R., and Peto, R. (1981) The causes of cancer: quantitative estimates of avoidable risks of cancer in the United States today. *J. Natl. Cancer Inst.* **66**, 1190–1380.

(2) Hansen, M. F., and Cavenee, W. K. (1987) Genetics of cancer predisposition. *Cancer Res.* **47**, 5518–5527.

(3) Mulvihill, J. J. (1977) Genetic repertoire of human neoplasia. In *Genetics of Human Cancer* (Mulvihill, J. J., Miller, R. W., and Fraumeni, J. F., Eds.) pp 137–143, Raven Press, New York.

(4) Knudson, A. G. (1986) Genetics of human cancer. *Annu. Rev. Genet.* **20**, 231–251.

(5) Nowell, P. C. (1986) Mechanisms of tumor progression. *Cancer Res.* **46**, 2203–2207.

(6) Ashby, J., and Tennant, R. W. (1988) Chemical structure, Salmonella mutagenicity and extent of carcinogenicity as indicators of genotoxic carcinogenesis among 222 chemicals tested in rodents by the US NCI/NTP. *Mutat. Res.* **204**, 17–115.

(7) Yuspa, S. H., and Poirier, M. C. (1988) Chemical carciniogenesis: from animal models to molecular models in one decade. *Adv. Cancer Res.* **50**, 25–70.

(8) Parry, J. M., Parry, E. M., and Barrett, J. C. (1981) Tumor promoters induce mitotic aneuploidy in yeast. *Nature* **294**, 263–265.

(9) Cerutti, P. A. (1987) Genotoxic oxidant tumor promoters. In *Nongenotoxic Mechanisms in Carcinogenesis* (Butterworth, B. E., and Slaga, T. J., Eds.) pp 325–331, Cold Spring Harbor Laboratory, Cold Spring Harbor, NY.

(10) Dzarlieva, R. T., and Fusenig, N. E. (1982) Tumor promoter 12-O-tetradecanoyl-phorbol-13-acetate enhances sister chromatid exchanges and numerical and structural chromosome aberrations in primary mouse epidermal cell cultures. *Cancer Lett.* **16**, 7–17.

(11) Aldaz, C. M., Conti, C. J., Klein-Szanto, A. J. P., and Slaga, T. J. (1987) Progressive dysplasia and aneuploidy are hallmarks of mouse skin papillomas: relevance to malignancy. *Proc. Natl. Acad. Sci. U.S.A.* **84**, 2029–2032.

(12) Ptashne, M. (1988) How eukaryotic transcriptional activators work. *Nature* **335**, 683–689.

(13) Blau, H. M. (1988) Hierarchies of regulatory genes may specify mammalian development. *Cell* **53**, 673–674.

(14) Heldin, C.-H., Betsholtz, C., Claesson-Welsh, L., and Westermark, B. (1987) Subversion of growth regulatory pathways in malignant transformation. *Biochim. Biophys. Acta* **907**, 219–244.

(15) Weinstein, I. B. (1987) Growth factors, oncogenes, and multistage carcinogenesis. *J. Cell. Biochem.* **33**, 213–224.

(16) Flatow, U., Willingham, M. C., and Rabson, A. S. (1984) Butyrate prevents harvey sarcoma virus focus formation but permits oncogene expression. *Cancer Lett.* **22**, 203–210.

(17) Samid, D., Chang, E. H., and Friedman, R. M. (1985) Development of transformed phenotype induced by a human *ras* oncogene is inhibited by interferon. *Biochem. Biophys. Res. Commun.* **126**, 509–516.

(18) Hosoi, J., Kato, K., and Kuroki, T. (1987) Induction of anchorage-independent growth of mouse JB6 cells by cholera toxin. *Carcinogenesis* **8**, 377–380.

(19) Leavitt, J., Barrett, J. C., Crawford, B. D., and Ts'o, P. O. P. (1978) Butyric acid suppression of the in vitro neoplastic state of Syrian hamster cells. *Nature* **271**, 262–265.

(20) Cavaliere, A., Bufalari, A., and Vitali, R. (1987) 5-Azacytidine carcinogenesis in Balb/c mice. *Cancer Lett.* **37**, 51–58.

(21) Bouck, N., Kokkinakis, D., and Ostrowsky, J. (1984) Induction of a step in carcinogenesis that is normally associated with mutagenesis by nonmutagenic concentrations of 5-azacytidine. *Mol. Cell. Biol.* **4**, 1231–1237.

(22) Holliday, R. (1987) The inheritance of epigenetic defects. *Science* **238**, 163–170.

(23) Varmus, H. E. (1984) The molecular genetics of cellular oncogenes. *Annu. Rev. Genet.* **18**, 553–612.

(24) Bishop, J. M. (1987) The molecular genetics of cancer. *Science* **235**, 305–311.

(25) Weinberg, R. A. (1985) The action of oncogenes in the cytoplasm and nucleus. *Science* **230**, 770–776.

(26) Sager, R. (1986) Genetic suppression of tumor formation: a new frontier in cancer research. *Cancer Res.* **46**, 1573–1580.

(27) Ozer, H. L., and Jha, K. K. (1977) Malignancy and transformation: expression in somatic cell hybrids and variants. *Adv. Cancer Res.* **25**, 53–93.

(28) Sager, R. (1985) Genetic suppression of tumor formation. *Adv. Cancer Res.* **44**, 43–68.

(29) Stanbridge, E. J. (1987) Genetic regulation of tumorigenic expression in somatic cell hybrids. *Adv. Viral Oncol.* **6**, 83–101.

(30) Harris, H. (1988) The analysis of malignancy by cell fusion: the position in 1988. *Cancer Res.* **48**, 3302–3306.

(31) Bouck, N., and di Mayorca, G. (1982) Chemical carcinogens transform BHK cells by inducing a recessive mutation. *Mol. Cell. Biol.* **2**, 97–105.

(32) Marshall, C. J. (1980) Suppression of the transformed phenotype with retention of the viral "src" gene in cell hybrids between Rous sarcoma virus-transformed rat cells and untransformed mouse cells. *Exp. Cell Res.* **127**, 373–384.

(33) Saxon, P. J., Srivatsan, E. S., and Stanbridge, E. J. (1986) Introduction of human chromosome 11 via microcell transfer controls tumorigenic expression of HeLa cells. *EMBO J.* **5**, 3461–3466.

(34) Knudson, A. G. (1971) Mutation and cancer: statistical study of retinoblastoma. *Proc. Natl. Acad. Sci. U.S.A.* **68**, 820–823.

(35) Friend, S. H., Bernards, R., Rogelj, S., Weinberg, R. A., Rapaport, J. M., Albert, D. M., and Dryja, T. P. (1986) A human DNA segment with properties of the gene that predisposes to retinoblastoma and osteosarcoma. *Nature* **323**, 643–646.

(36) Lee, W.-H., Bookstein, R., Hong, F., Young, L.-J., Shew, J.-Y., and Lee, E. Y.-H. P. (1987) Human retinoblastoma susceptibility gene: cloning, identification and sequence. *Science* **235**, 1394–1399.

(37) Fung, Y.-K. T., Murphree, A. L., T'Ang, A., Qian, J., Hinrichs, S. H., and Benedict, W. F. (1987) Structural evidence for the authenticity of the human retinoblastoma gene. *Science* **236**, 1657–1661.

(38) Strong, L. C., Riccardi, V. M., Ferrell, R. E., and Sparkes, R. S. (1981) Familial retinoblastoma and chromosome 13 deletion transmitted via an insertional translocation. *Science* **213**, 1501–1503.

(39) Lee, W.-H., Shew, J.-Y., Hong, F. D., Sery, T. W., Donoso, L. A., Young, L.-J., Bookstein, R., and Lee, E. Y.-H. P. (1987) The retinoblastoma susceptibility gene encodes a nuclear phosphoprotein associated with DNA binding activity. *Nature* **329**, 642–645. (a) Huang, H.-J. S., Yee, J.-K., Shew, J.-Y., Chen, P.-L., Bookstein, R., Friedmann, T., Lee, E. Y.-H., P., and Lee, W.-H. (1988) Suppression of the neoplastic phenotype by replacement of the RB gene in human cancer cells. *Science* **242**, 1563–1566.

(40) (a) Noda, M., Kitayama, H., Matsuzaki, T., Sugimoto, Y., Okayama, H., Bassin, R. H., and Ikawa, Y. (1988) Detection of genes with a potential for suppressing the transformed phenotype associated with activated ras genes. *Proc. Natl. Acad. Sci. U.S.A.* (in press). (b) Kitayama, H., Sugimoto, Y., Matsuzaki, T., Ikawa, Y., and Noda, M. (1989) A ras-related gene with transformation suppressor activity. *Cell* (in press).

(41) Schaefer, R., Iyer, J., Iten, E., and Nirkko, A. C. (1988) Partial reversion of transformed phenotype in HRAS-transfected tumorigenic cells by transfer of a human gene. *Proc. Natl. Acad. Sci. U.S.A.* **85**, 1590–1594.

(42) Larsson, C., Skogseid, B., Oberg, K., Nakamura, Y., and Nordenskjold, M. (1988) Multiple endocrine neoplasia type 1 gene maps to chromosome 11 and is lost in insulinoma. *Nature* **332**, 85–87.

(43) Cannon-Albright, L. A., Skolnick, M. H., Bishop, D. T., Lee, R. G., and Burt, R. W. (1988) Common inheritance of susceptibility to colonic adenomatous polyps and associated colorectal cancers. *N. Engl. J. Med.* **319**, 533–537.

(44) Ohno, S. (1971) Genetic implication of karyological instability of malignant somatic cells. *Physiol. Rev.* **51**, 496–526.

(45) Comings, D. E. (1973) A general theory of carcinogenesis. *Proc. Natl. Acad. Sci. U.S.A.* **70**, 3324–3328.

(46) Knudson, A. G. (1985) Hereditary cancer, oncogenes, and antioncogenes. *Cancer Res.* **45**, 1437–1443.

(47) Stanbridge, E. J. (1986) A case for human tumor-suppressor genes. *Bioessays* **3**, 252–255.

(48) Cavenee, W. K., Koufos, A., and Hansen, M. F. (1986) Recessive mutant genes predisposing to human cancer. *Mutat. Res.* **168**, 3–14.

(49) Schafer, R. (1987) Suppression of the neoplastic phenotype and "anti-oncogenes". *Blut* **54**, 257–265.

(50) Brodeur, G. M. (1987) The involvement of oncogenes and suppressor genes in human neoplasia. *Adv. Pediatr.* **34**, 1–44.

(51) Willman, C. L., and Fenoglio-Preiser, C. M. (1987) Oncogenes, suppressor genes, and carcinogenesis. *Hum. Pathol.* **18**, 895–902.

(52) Friend, S. H., Dryja, T. P., and Weinberg, R. A. (1988) Oncogenes and tumor-suppressing genes. *N. Engl. J. Med.* **318**, 618–622.

(53) Green, A. R. (1988) Recessive mechanisms of malignancy. *Br. J. Cancer* **58**, 115–121.

(54) Hansen, M. F., and Cavenee, W. K. (1988) Tumor suppressors: recessive mutations that lead to cancer. *Cell* **53**, 172–173.

(55) Klein, G. (1987) The approaching era of the tumor suppressor genes. *Science* 238, 1539–1545.

(56) Gateff, E. (1978) Malignant neoplasms of genetic origin in Drosophila melanogaster. *Science* 200, 1448–1459.

(57) Schwab, M. (1987) Oncogenes and tumor suppressor genes in Xiphophorus. *Trends Genet.* 3, 38–42.

(58) Pasquale, S. R., Jones, G. R., Doersen, C.-J., and Weissman, B. E. (1988) Tumorigenicity and oncogene expression in pediatric cancers. *Cancer Res.* 48, 2715–2719.

(59) Benedict, W. F., Weissman, B. E., Mark, C., and Stanbridge, E. J. (1984) Tumorigenicity of human HT1080 fibrosarcoma X normal fibroblast hybrids: chromosome dosage dependency. *Cancer Res.* 44, 3471–3479.

(60) Stoler, A., and Bouck, N. (1985) Identification of a single chromosome in the normal human genome essential for suppression of hamster cell transformation. *Proc. Natl. Acad. Sci. U.S.A.* 82, 570–574.

(61) Weissman, B. E., and Stanbridge, E. J. (1983) Complementation of the tumorigenic phenotype in human cell hybrids. *J. Natl. Cancer Inst.* 70, 667–672.

(62) Weissman, B. E., Saxon, P. J., Pasquale, S. R., Jones, G. R., Geiser, A. G., and Stanbridge, E. J. (1987) Introduction of a normal human chromosome 11 into a Wilms' tumor cell line controls its tumorigenic expression. *Science* 236, 175–180.

(63) Cohen, A. J., Li, F. P. Berg, S., Marchetto, D. J., Tsai, S., Jacobs, S. C., and Brown, R. S. (1979) Hereditary renal-cell carcinoma associated with a chromosomal translocation. *N. Engl. J. Med.* 301, 592–595.

(64) Pathak, S., Strong, L. C., Ferrell, R. E., and Trindade, A. (1982) Familial renal cell carcinoma with a 3;11 chromosome translocation limited to tumor cells. *Science* 217, 939–941.

(65) Seizinger, B. R., Rouleau, G. A., Ozelius, K. J., et al.[1] (1988) Von Hippel–Lindau disease maps to the region of chromosome 3 associated with renal cell carcinoma. *Nature* 332, 268–269.

(66) Bodmer, W. F., Bailey, C. J., Bodmer, J., et al. (1987) Localization of the gene for familial adenomatous polyposis on chromosome 5. *Nature* 328, 614–616.

(67) Leppert, M., Dobbs, M., O'Connell, P., et al. (1987) The gene for familial polyposis coli maps to the long arm of chromosome 5. *Science* 238, 1411–1414.

(68) Rouleau, G. A., Wertelecki, W., Haines, J. L., et al. (1987) Genetic linkage of bilateral acoustic neurofibromatosis to a DNA marker on chromosome 22. *Nature* 329, 246–248.

(69) Ono, M., Yakushinji, M., Segawa, K., and Kuwano, M. (1988) Transformation by viral and cellular oncogenes of a mouse BALB/3T3 cell mutant resistant to transformation by chemical carcinogens. *Mol. Cell. Biol.* 8, 4190–4196.

(70) Sinn, E., Muller, W., Pattengale, P., Tepler, I., Wallace, R., and Leder, P. (1987) Coexpression of MMTV/v-Ha-ras and MMTV/c-myc genes in transgenic mice: synergistic action of oncogenes in vivo. *Cell* 49, 465–475.

(71) Oshimura, M., Koi, M., Ozawa, N., Sugawara, O., Lamb, P. W., and Barrett, J. C. (1988) Role of chromosome loss in ras/myc-induced Syrian hamster tumors. *Cancer Res.* 48, 1623–1632.

(72) Vogelstein, B., Fearon, E. R., Hamilton, S. R., et al. (1988) Genetic alterations during colorectal-tumor development. *N. Engl. J. Med.* 319, 525–532.

(73) Monpezat, J. P., Delattre, O., Bernard, A., et al. (1988) Loss of alleles on chromosome 18 and on the short arm of chromosome 17 in polyploid colorectal carcinomas. *Int. J. Cancer* 41, 404–408.

(74) Solomon, E., Voss, R., Hall, et al. (1987) Chromosome 5 allele loss in human colorectal carcinomas. *Nature* 328, 616–619.

(75) Okamoto, M., Sasaki, M., Sugio, K., et al. (1988) Loss of constitutional heterozygosity in colon carcinoma from patients with familial polyposis coli. *Nature* 331, 273–276.

(76) Meadows, A. T., Baum, E., Fossati-Bellani, F., et al. (1985) Second malignant neoplasms in children: an update from the late effects study group. *J. Clin. Oncol.* 3, 532–538.

(77) Friend, S. H., Horowitz, J. M., Gerber, M. R., Wang, X.-F., Bogenmann, E., Li, F. P., and Weinberg, R. A. (1987) Deletions of a DNA sequence in retinoblastomas and mesenchymal tumors: organization of the sequence and its encoded protein. *Proc. Natl. Acad. Sci. U.S.A.* 84, 9059–9063.

(78) Rosl, F., Durst, M., and zur Hausen, H. (1988) Selective suppression of human papilloma virus transcription in non-tumorigenic cells by 5-azacytidine. *EMBO J.* 7, 1321–1328.

(79) de Ronde, A., Mannens, M., Slater, R. M., et al. (1988) Morphological transformation by early region human polyomavirus BK DNA of human fibroblasts with deletions in the short arm of one chromosome 11. *J. Gen. Virol.* 69, 467–471.

(80) Smits, H. L., Raadsheer, E., Rood, I., Mehendale, S., Slater, R. M., van der Noordaa, J., and ter Schegget, J. (1988) Induction of anchorage-independent growth of human embryonic fibroblasts with a deletion in the short arm of chromosome 11 by human papillomavirus type 16 DNA. *J. Virol.* 62, 4538–4543.

(81) Bassin, R. H., and Noda, M. (1987) Oncogene inhibition by cellular genes. *Adv. Viral Oncol.* 6, 103–127.

(82) Noda, M., Selinger, Z., Scolnick, E. M., and Bassin, R. H. (1983) Flat revertants isolated from Kirsten sarcoma virus-transformed cells are resistant to the action of specific oncogenes. *Proc. Natl. Acad. Sci. U.S.A.* 80, 5602–5606.

(83) Ricketts, M. H., and Levinson, A. D. (1988) High level expressin of c-H-ras-1 fails to fully transform rat-1 cells. *Mol. Cell. Biol.* 8, 1460–1468. (a) Paterson, H., Reeves, B., Brown, R., Hall, A., Furth, M., Bos, J., Jones, P., and Marshall, C. (1987) Activated N-ras controls the transformed phenotype of HT1080 human fibrosarcoma cell. *Cell* 51, 803–812.

(84) Barbacid, M. (1987) Ras genes. *Annu. Rev. Biochem.* 56, 779–827.

(85) Bos, J. L., Fearon, E. R., Hamilton, S. R., Verlaan-deVries, M., van Boom, J. H., van der Eb, A. J., and Vogelstein, B. (1987) Prevalence of ras gene mutations in human colorectal cancers. *Nature* 327, 293–297.

(86) Forrester, K., Almoguera, C., Han, K., Grizzle, W. E., and Perucho, M. (1987) Detection of high incidence of K-ras oncogenes during human colon tumorigenesis. *Nature* 327, 298–303.

(87) Bos, J. L., Verlaan-de Vries, M., Jansen, A. M., Veeneman, G. H., van Boom, J. H., and van der Eb, A. J. (1984) Three different mutations in codon 61 of the human N-ras gene detected by synthetic oligonucleotide hybridization. *Nucleic Acids Res.* 12, 9155–9163.

(88) Kimchi, A., Wang, X. F., Weinberg, R. A., Cheifetz, S., and Massague, J. (1988) Absence of TGF-B receptors and growth inhibitory responses in retinoblastoma cells. *Science* 240, 196–199.

(89) Stoker, M. G. P. (1967) Transfer of growth inhibition between normal and virus-transformed cells: autoradiographic studies using marked cells. *J. Cell Sci.* 2, 293–304.

(90) Mehta, P. P., Bertram, J. S., and Loewenstein, W. R. (1986) Growth inhibition of transformed cells correlates with their junctional communication with normal cells. *Cell* 44, 187–196.

(91) Yamasaki, H., Enomoto, T., Shiba, Y., Kanno, Y., and Kakunaga, T. (1985) Intercellular communication capacity as a possible determinant of transformation sensitivity of BALB/c3T3 clonal cells. *Cancer Res.* 45, 637–641.

(92) Herschman, H. R., and Brankow, D. W. (1986) Ultraviolet irradiation transforms C3H10T1/2 cells to a unique, suppressible phenotype. *Science* 234, 1385–1388.

(93) Stanbridge, E. J., and Ceredig, R. (1981) Growth-regulatory control of human cell hybrids in nude mice. *Cancer Res.* 41, 573–580.

(94) Bignami, M., Rosa, S., Falcone, G., Tato, F., Katoh, F., and Yamasaki, H. (1988) Specific viral oncogenes cause differential effects on cell-to-cell communication, relevant to the suppression of the transformed phenotype by normal cells. *Mol. Carcinog.* 1, 67–75.

(95) Perona, R., and Serrano, R. (1988) Increased pH and tumorigenicity of fibroblasts expressing a yeast proton pump. *Nature* 334, 438–440.

(96) Sachs, L. (1987) Development and suppression of malignancy. *Adv. Viral Oncol.* 6, 129–142.

(97) Sachs, L. (1987) Cell differentiation and bypassing of genetic defects in the suppression of malignancy. *Cancer Res.* 47, 1981–1986.

(98) Guan, X.-P., and Block, A. (1988) Differentiation-induced ML-1 cells as targets for transformation by a chemical carcinogen. *Cancer Res.* 48, 4389–4394.

(99) Jacob, L., Opper, M., Metzroth, B., Phannavong, B., and Mechler, B. M. (1987) Structure of the l(2)gl gene of Drosophila and delimitation of its tumor suppressor domain. *Cell* 50, 215–225.

(100) Harris, H. (1985) Suppression of malignancy in hybrid cells: the mechanism. *J. Cell Sci.* 79, 83–94.

(101) Wallich, R., Bulbuc, N., Hammerling, G. J., Katzav, S., Segal, S., and Feldman, M. (1985) Abrogation of metastatic properties of tumor cells by de novo expression of H-2K antigens following H-2 gene transfection. *Nature* 315, 301–305.

---

[1] In the reference list, when there were nine or more authors on a single publication, only the first three were listed, followed by et al.

(102) Tanaka, K., Isselbacher, K. J., Khoury, G., and Jay, G. (1985) Reversal of oncogenesis by the expression of a major histocompatibility complex class 1 gene. *Science* **228**, 26–30.

(103) Hui, K., Grosveld, F., and Festenstein, H. (1984) Rejection of transplantable AKR leukemia cells following MHC DNA-mediated cell transformation. *Nature* **311**, 750–752.

(104) Tanaka, K., Yoshioka, T., Bieberich, C., and Jay, G. (1988) Role of the major histocompatibility complex class I antigens in tumor growth and metastasis. *Annu. Rev. Immunol.* **6**, 359–380.

(105) O'Brien, W., Stenman, G., and Sager, R. (1986) Suppression of tumor growth by senescence in virally transformed human fibroblasts. *Proc. Natl. Acad. Sci. U.S.A.* **83**, 8659–8663.

(106) Folkman, J., and Cotran, R. (1976) Relation of vascular proliferation to tumor growth. *Int. Rev. Exp. Pathol.* **16**, 207–248.

(107) Rastinejad, F., Polverini, P. J., and Bouck, N. P. (1989) Regulation of the activity of a new inhibitor of angiogenesis by a cancer suppressor gene. *Cell* (in press).

(108) Yokota, J., Wada, M., Shimosato, Y., Terada, M., and Sugimura, T. (1987) Loss of heterozygosity on chromosomes 3, 13, and 17 in small-cell carcinoma and on chromosome 3 in adenocarcinoma of the lung. *Proc. Natl. Acad. Sci. U.S.A.* **84**, 9252–9256.

(109) Wong, A. J., Ruppert, J. M., Eggleston, J., Hamilton, S. R., Baylin, S. B., and Vogelstein, B. (1986) Gene amplification of c-myc and N-myc in small cell carcinoma of the lung. *Science* **233**, 461–464.

(110) Harbour, J. W., Lai, S.-L., Whang-Peng, J., Gazdar, A. F., Minna, J. D., and Kaye, F. J. (1988) Abnormalities in structure and expression of the human retinoblastoma gene in SCLC. *Science* **241**, 353–356.

(111) Simpson, N. E., Kidd, K. K., Goodfellow, P. J., et al. (1987) Assignment of multiple endocrine neoplasia type 2A to chromosome 10 by linkage. *Nature* **328**, 528–530.

(112) Mathew, C. G. P., Chin, K. S., Easton, D. F., et al. (1987) A linked genetic marker for multiple endocrine neoplasia type 2A on chromosome 10. *Nature* **328**, 527–528.

(113) Mathew, C. G. P., Smith, B. A., Thorpe, K., Wong, Z., Royle, N. J., Jefferys, A. J., and Ponder, B. A. J. (1987) Deletion of genes on chromosome 1 in endocrine neoplasia. *Nature* **328**, 524–526.

(114) Ali, I. U., Lidereau, R., Theillet, C., and Callahan, R. (1987) Reduction to homozygosity of genes on chromosome 11 in human breast neoplasia. *Science* **238**, 185–188.

(115) Theillet, C., Lidereau, R., Escot, C., Hutzell, P., Brunet, M., Gest, J., Schlom, J., and Callahan, R. (1986) Loss of a c-H-ras-1 allele and aggressive human primary breast carcinomas. *Cancer Res.* **46**, 4776–4781.

(116) T'Ang, A., Varley, J. M., Chakraborty, S., Murphree, A. L., and Fung, Y.-K. T. (1988) Structural rearrangement of the retinoblastoma gene in human breast carcinoma. *Science* **242**, 263–266.

(117) Lundberg, C., Skoog, L., Cavenee, W. K., and Nordenskjold, M. (1987) Loss of heterozygosity in human ductal breast tumors indicates a recessive mutation on chromosome 13. *Proc. Natl. Acad. Sci. U.S.A.* **84**, 2372–2376.

(118) Muleris, M., Salmon, R.-J., Girodet, J., Zafrani, B., and Dutrillaux, B. (1987) Recurrent deletions of chromosomes 11q and 3p in anal canal carcinoma. *Int. J. Cancer* **39**, 595–598.

(119) James, C. D., Carlbom, E., Dumanski, J. P., Hansen, M., Nordenskjold, M., Collins, V. P., and Cavenee, W. K. (1988) Clonal genomic alterations in glioma malignancy stages. *Cancer Res.* **48**, 5546–5551.

(120) Grundy, P., Koufos, A., Morgan, K., Li, F. P., Meadows, A. T., and Cavenee, W. K. (1988) Familial predisposition to Wilms' tumour does not map to the short arm of chromosome 11. *Nature* **336**, 374–378.

(121) Toguchida, J., Ishizake, K., Sasake, M. S., Idenaga, M., Sugimoto, M., Kotoura, Y., and Yamamuro, T. (1988) Chromosomal reorganization for the expression of recessive mutation of retinoblastoma susceptibility gene in the development of osteosarcoma. *Cancer Res.* **48**, 3939–3943.

(122) Dumanski, J. P., Carlbom, E., Collins, V. P., and Nordenskjold, M. (1987) Deletion mapping of a locus on human chromosome 22 involved in the oncogenesis of meningioma. *Proc. Natl. Acad. Sci. U.S.A.* **84**, 9275–9279.

(123) Barrett, J. C., Oshimura, M., and Koi, M. (1987) Role of oncogenes and tumor suppressor genes in a multistep model of carcinogenesis. *Annu. Symp. Fundam. Cancer Res.*, [*Proc.*] **39**, 45–55.

(124) Bunn, C. L., and Tarrant, G. M. (1980) Limited lifespan in somatic cell hybrids and cybrids. *Exp. Cell Res.* **127**, 385–396.

(125) Pereira-Smith, O. M., and Smith, J. R. (1988) Genetic analysis of indefinite division in human cells: identification of four complementation groups. *Proc. Natl. Acad. Sci. U.S.A.* **85**, 6042–6046.

(126) Shkolnik, T., and Sachs, L. (1978) Suppression of the in vivo malignancy and in vitro cell multiplication of myeloid leukemic cells by hybridization with normal macrophages. *Exp. Cell Res.* **113**, 197–204.

(127) Koi, M., and Barrett, J. C. (1986) Loss of tumor-suppressive function during chemically induced neoplastic progression of Syrian hamster embryo cells. *Proc. Natl. Acad. Sci. U.S.A.* **83**, 5992–5996.

(128) McLaughlin, J., Chianese, E., and Witte, O. N. (1987) In vitro transformation of immature hematopoietic cells by the p210 BCR/ABL oncogene product of the Philadelphia chromosome. *Proc. Natl. Acad. Sci. U.S.A.* **84**, 6558–6562.

(129) Smith, B. L., and Sager, R. (1985) Genetic analysis of tumorigenesis: XXI. Suppressor genes in CHEF cells. *Somatic Cell Mol. Genet.* **11**, 25–34.

(130) Small, M. B., Simmons, E., Jha, K. K., Oxwee, H. L., Feldmen, L. A., Pyati, J., Hann, S., and Dina, D. (1984) Genetic analysis of the transformed phenotype in mouse cells. In *Gene Transfer and Cancer* (Pearson, M. L., and Sternberg, N. L., Eds.) pp 227–235, Raven Press, New York.

(131) Gilbert, P. X., and Harris, H. (1988) The role of the ras oncogene in the formation of tumors. *J. Cell Sci.* **90**, 433–446.

(132) Avvedimento, V. E., Musti, A., Fusco, A., Bonapace, M. J., and Di Lauro, R. (1988) Neoplastic transformation inactivates specific trans-acting factor(s) required for the expression of the thyroglobulin gene. *Proc. Natl. Acad. Sci. U.S.A.* **85**, 1744–1748.

(133) Colletta, G., Pinto, A., di Fiore, P. P., Fusco, A., Ferrentino, M., Avvedimento, V. E., Tsuchida, N., and Vecchio, G. (1983) Dissociation between transformed and differentiated phenotype in rat thyroid epithelial cells after transformation with a temperature-sensitive mutant of the Kirsten murine sarcoma virus. *Mol. Cell. Biol.* **3**, 2009–2109.

(134) Franza, B. R., Maruyama, K., Garrels, J. I., and Ruley, H. E. (1986) In vitro establishment is not a sufficient prerequisite for transformation by activated ras oncogenes. *Cell* **44**, 409–418.

(135) Stenman, G., Delorme, E. O., Lau, C. C., and Sager, R. (1987) Transfection with plasmid pSV2gptEJ induces chromosome rearrangements in CHEF cells. *Proc. Natl. Acad. Sci. U.S.A.* **84**, 184–188.

(136) Hirakawa, T., and Ruley, H. E. (1988) Rescue of cells from ras oncogene-induced growth arrest by a second, complementing, oncogene. *Proc. Natl. Acad. Sci. U.S.A.* **85**, 1519–1523.

(137) Whyte, P., Buchkovich, K. J., Horowitz, J. M., Friend, S. H., Raybuck, M., Weinberg, R. A., and Harlow, E. (1988) Association between an oncogene and an anti-oncogene: the adenovirus E1A proteins bind to the retinoblastoma gene product. *Nature* **334**, 124–129.

(138) DeCaprio, J. A., Ludlow, J. W., Figge, J., Shew, J.-Y., Huang, C.-M., Lee, W.-H., Marsilio, E., Paucha, E., and Livingston, D. M. (1988) SV40 large tumor antigen forms a specific complex with the product of the retinoblastoma susceptibility gene. *Cell* **54**, 275–283.

(139) Moran, E. (1988) A region of SV40 large T antigen can substitute for a transforming domain of the adenovirus E1A products. *Nature* **334**, 168–170.

(140) Bernards, R., Dessain, S. K., and Weinberg, R. A. (1986) N-myc amplification causes down-modulation of MHC class 1 antigen expression in neuroblastoma. *Cell* **47**, 667–674.

(141) Versteeg, R., Noordermeer, I. A., Kruse-Wolters, M., Ruiter, D. J., and Schrier, P. I. (1988) c-myc down regulates class 1 HLA expression in human melanomas. *EMBO J.* **7**, 1023–1029.

(142) Vaessen, R. T. M. J., Houweiling, A., and van der Eb, A. J. (1987) Post-transcriptional control of class I MHC mRNA expression in adenovirus 12-transformed cells. *Science* **235**, 1486–1488.

(143) Cole, M. D. (1986) The myc oncogene: its role in transformation and differentiation. *Annu. Rev. Genet.* **20**, 361–384.

(144) Prochownik, E. V., and Kukowska, J. (1986) Deregulated expression of c-myc by murine erythroleukaemia cells prevents differentiation. *Nature* **322**, 848–850.

(145) Symonds, G., Klempnauer, K.-H., Evan, G. I., and Bishop, J. M. (1984) Induced differentiation of avian myeloblastosis virus-transformed myeloblasts: phenotypic alteration without altered expression of the viral oncogene. *Mol. Cell. Biol.* **4**, 2587–2593.

(146) Cerni, C., Mougneau, E., and Cuzin, F. (1987) Transfer of immortalizing oncogenes into rat fibroblasts induces both high rates of sister chromatid exchange and appearance of abnormal

karyotypes. *Exp. Cell Res.* **168**, 439–446.

(147) Ptashne, M. (1986) *A Genetic Switch*, Cell Press, Cambridge, MA, and Blackwell Scientific Publications, Palo Alto.

(148) Eliyahu, D., Michalovitz, D., and Oren, M. (1985) Overproduction of p53 antigen makes established cells highly tumorigenic. *Nature* **316**, 158–160.

(149) Hicks, G. G., and Mowat, M. (1988) Integration of Friend murine leukemia virus into both alleles of the p53 oncogene in an erythroleukemic cell line. *J. Viol.* **62**, 4752–4755.

(150) Oren, M. (1985) The p53 cellular tumor antigen: gene structure, expression and protein properties. *Biochim. Biophys. Acta* **823**, 67–78.

(151) Finlay, C. A., Hinds, P. W., Tan, T.-H., Eliyahu, D., Oren, M., and Levine, A. J. (1988) Activating mutations for transformation produce a gene product that forms a hsc70–p53 complex with an altered half-life. *Mol. Cell. Biol.* **8**, 531–539.

(152) Haluska, F. G., Huebner, K., Isobe, M., Nishimura, T., Croce, C. M., and Vogt, P. K. (1988) Localization of the human JUN protooncogene to chromosome region 1p31-32. *Proc. Natl. Acad. Sci. U.S.A.* **85**, 2215–2218.

(153) Smith, M. R., DeGudicibus, S. J., and Stacey, D. W. (1986) Requirement for c-ras proteins during viral oncogene transformation. *Nature* **320**, 540–543.

(154) Alonso, T., Morgan, R. O., Marvizon, J. C., Zarbl, H., and Santos, E. (1988) Malignant transformation by ras and other oncogenes produces common alterations in inositol phospholipid signaling pathways. *Proc. Natl. Acad. Sci. U.S.A.* **85**, 4271–4275.

(155) Powers, S., Michaelis, S., Broek, D., Santa Anna-A., S., Field, J., Herskowitz, I., and Wigler, M. (1986) RAM, a gene of yeast required for a functional modification of ras proteins and for production of mating pheromone a-factor. *Cell* **47**, 413–422.

(156) Ono, M., Yakushinji, M., Segawa, K., and Kuwano, M. (1988) Transformation by viral and cellular oncogenes of a mouse BALB/3T3 cell mutant resistant to transformation by chemical carcinogens. *Mol. Cell. Biol.* **8**, 4190–4196.

(157) Samid, D., Flessate, D. M., and Friedman, R. M. (1987) Interferon-induced revertants of ras-transformed cells: resistance to transformation by specific oncogenes and retransformation by 5-azacytidine. *Mol. Cell. Biol.* **7**, 2196–2200.

(158) Morris, A., Clegg, C., Jones, J., Rogers, B., and Avery, R. J. (1980) Isolation and characterization of a clonally related series of murine retrovirus-infected mouse cells. *J. Gen. Virol.* **49**, 105–113.

(159) Norton, J. D., Cook, F., Roberts, P. C., Clewley, J. P., and Avery, R. J. (1984) Expression of Kirsten murine sarcoma virus in transformed nonproducer and revertant NIH/3T3 cells: evidence for cell-mediated resistance to a viral oncogene in phenotypic reversion. *J. Virol.* **50**, 439–444.

(160) Haynes, J. R., and Downing, J. R. (1988) A recessive cellular mutation in v-fes-transformed mink cells restores contact inhibition and anchorage-dependent growth. *Mol. Cell. Biol.* **8**, 2419–2427.

(161) Stephenson, J. R., Reynolds, R. K., and Aaronson, S. A. (1973) Characterization of morphologic revertants of murine and avian sarcoma virus-transformed cells. *J. Virol.* **11**, 218–222.

(162) Zarbl, H., Latreille, J., and Jolicoeur, P. (1987) Revertants of v-fos-transformed fibroblasts have mutations in cellular genes essential for transformation by other oncogenes. *Cell* **51**, 357–369.

(163) Inoue, H., Yutsudo, M., and Hakura, A. (1983) Rat mutant cells showing temperature sensitivity for transformation by wild type Moloney murine sarcoma virus. *Virology* **125**, 242–245.

(164) Inoue, H., Kizaka, S., Yutsudo, M., and Hakura, A. (1988) Temperature-sensitive cellular mutant for expression of mRNA from murine retrovirus. *J. Virol.* **62**, 106–113.

(165) Newbold, R. F., Overell, R. W., and Connell, J. R. (1982) Induction of immortality is an early event in malignant transformation of mammalian cells by carcinogens. *Nature* **299**, 633–635.

(166) Kakunaga, T. (1978) Neoplastic transformation of human diploid fibroblast cells by chemical carcinogens. *Proc. Natl. Acad. Sci. U.S.A.* **75**, 1334–1338.

(167) Stampfer, M. R., and Bartley, J. C. (1985) Induction of transformation and continuous cell lines from normal human mammary epithelial cells after exposure to benzo[a]pyrene. *Proc. Natl. Acad. Sci. U.S.A.* **82**, 2394–2398.

(168) Dunn, J. M., Phillips, R. A., Becker, A. J., and Gallie, B. J. (1988) Identification of germline and somatic mutations affecting the retinoblastoma gene. *Science* **241**, 1797–1800.

(169) Lee, E. Y.-H., Bookstein, R., Young, L.-J., Lin, C.-J., Rosenfeld, M. G., and Lee, W.-H. (1988) Molecular mechanism of retinoblastoma gene inactivation in retinoblastoma cell line Y79. *Proc. Natl. Acad. Sci. U.S.A.* **85**, 6017–6021.

(170) Goddard, A. D., Balakier, H., Canton, M., Dunn, J., Squire, J., Reyes, E., Becker, A., Phillips, R. A., and Gallie, B. L. (1988) Infrequent genomic rearrangement and normal expressiono of the putative RB1 gene in retinoblastoma tumors. *Mol. Cell. Biol.* **8**, 2082–2088.

(171) Yoshida, M. A., Ohyashiki, K., Ochi, H., Gibas, Z., Pontes, J. E., Prout, G. R., Huben, R., and Sandberg, A. A. (1986) Cytogenetic studies of tumor tissue from patients with nonfamilial renal cell carcinoma. *Cancer Res.* **46**, 2139–2147.

(172) Haluska, F. G., Tsujimoto, Y., and Croce, C. M. (1987) Oncogene activation by chromosome translocation in human malignancy. *Annu. Rev. Genet.* **21**, 321–345.

(173) Searle, A. G., and Beechey, C. V. (1985) Noncomplementation phenomena and their bearing on nondisjunctional effects. In *Aneuploidy, Etiology and Mechanisms* (Dellarco, V. L., Voytek, P. E., and Hollaender, A., Eds.) pp 363–376, Plenum Press, New York and London.

(174) Swain, J. L., Stewart, T. A., and Leder, P. (1987) Parental legacy determines methylation and expression of an autosomal transgene: a molecular mechanism for parental imprinting. *Cell* **50**, 719–727.

(175) Sapienza, C., Peterson, A. C., Rossant, J., and Balling, R. (1987) Degree of methylation of transgenes is dependent on gamete of origin. *Nature* **328**, 251–254.

(176) Reik, W., Collick, A., Norris, M. L., Barton, S. C., and Surani, M. A. (1987) Genomic imprinting determines methylation of parental alleles in transgenic mice. *Nature* **328**, 248–251.

(177) Schroeder, W. T., Chao, L.-Y., Dao, D. D., Strong, L. C., Pathak, S., Riccardi, V., Lewis, W. H., and Saunders, G. F. (1987) Nonrandom loss of maternal chromosome 11 alleles in Wilms tumors. *Am. J. Hum. Genet.* **40**, 413–420.

(178) de Bustros, Nelkin, B. D., Silverman, A., Erlich, G., Poiesz, B., and Baylin, S. B. (1988) The short arm of chromosome 11 is a "hot spot" for hypermethylation in human neoplasia. *Proc. Natl. Acad. Sci. U.S.A.* **85**, 5693–5697.

(179) Huff, V., Compton, D. A., Chao, L.-Y., Strong, L. C., Geiser, C. F., and Saunders, G. F. (1988) Lack of linkage of familial Wilms' tumour to chromosomal band 11p13. *Nature* **336**, 377–378. (a) Wang, R. Y.-H., Kuo, K. C., Gehrke, C. W., Huang, L.-H., and Ehrlich, M. (1982) Heat- and alkali-induced deamination of 5-methylcytosine and cytosine residues in DNA. *Biochim. Biophys. Acta* **697**, 371–377. (b) Solter, D. (1988) Differential Imprinting and Expression of Maternal and Paternal Genomes. *Annu. Rev. Genet.* **22**, 127–146.

(180) Dao, D. D., Schroeder, W. T., Chao, L.-Y., et al. (1987) Genetic mechanisms of tumor-specific loss of 11p DNA sequences in Wilms' tumor. *Am. J. Hum. Genet.* **41**, 202–217.

(181) Scrable, H. J., Witte, D. P., Lampkin, B. C., and Cavenee, W. K. (1987) Chromosomal localization of the human rhabdomyosarcoma locus by mitotic recombination mapping. *Nature* **329**, 645–647.

(182) Conti, C. J., Aldaz, C. M., O'Connell, J., Klein-Szanto, A. J. P., and Slaga, T. J. (1986) Aneuploidy, an early event in mouse skin tumor development. *Carcinogenesis* **7**, 1845–1848.

(183) Aldaz, C. M., Conti, C. J., Klein-Szanto, A. J. P., and Slaga, T. J. (1987) Progressive dysplasia and aneuploidy are hallmarks of mouse skin papillomas: relevance to malignancy. *Proc. Natl. Acad. Sci. U.S.A.* **84**, 2029–2032.

(184) Galloway, S. M., and Ivett, J. L. (1986) Chemically induced aneuploidy in mammalian cells in culture. *Mutat. Res.* **167**, 89–105.

(185) Oshimura, M., and Barrett, J. C. (1986) Chemically induced aneuploidy in mammalian cells: mechanisms and biological significance in cancer. *Environ. Mutagen.* **8**, 129–159.

# Author Index

# Affiliation Index

# Subject Index

Vulnerable intracellular targets, effect on selective
  localization of nephrotoxicity, 90

**X**

Xenobiotic(s)
  effect on kidney function, 74
  history of interest in metabolic fate, 125
Xenobiotic chemicals, definition, 98

**Y**

Yeast, nucleotide excision repair of DNA damage, 143
Yeast AP endonucleases, base excision repair of DNA
  damage, 140
Yeast redoxyendonuclease, base excision repair of DNA
  damage, 139

**Z**

Zero-phonon hole, example, 186,187*f*
Zero-phonon transition, definition, 184

Production: *Betsy Kulamer and Donna Lucas*
Indexing: *Deborah H. Steiner*
Acquisition: *Barbara C. Tansill*
Cover design: *Amy Meyer Phifer*

Printed and bound by BookCrafters, Fredericksburg, VA